# Principles of Neutron Scattering
# from Condensed Matter

# Principles of Neutron Scattering from Condensed Matter

## Andrew T. Boothroyd

UNIVERSITY PRESS

Great Clarendon Street, Oxford, OX2 6DP,
United Kingdom

Oxford University Press is a department of the University of Oxford.
It furthers the University's objective of excellence in research, scholarship,
and education by publishing worldwide. Oxford is a registered trade mark of
Oxford University Press in the UK and in certain other countries

Published in the United States of America by Oxford University Press
198 Madison Avenue, New York, NY 10016, United States of America

British Library Cataloguing in Publication Data
Data available

Library of Congress Control Number: 2020939592
ISBN 978–0–19–886231–4

Printed and bound by
CPI Group (UK) Ltd, Croydon, CR0 4YY

*To Margaret, Ben, Mary, and in memory of Brian*

# Preface

Neutron scattering as an experimental probe of condensed matter was born over eighty years ago. The formative work was done by a small band of pioneers working within nuclear programmes at national laboratories. Starting in the 1960s and 70s, large-scale neutron scattering facilities with user access programmes began to appear. The first generation of users, largely solid-state physicists, tended to employ neutron scattering as their principal research activity, and as well as performing experiments they would often be closely involved in the development of instrumentation. Through immersion in the technique they acquired great expertise.

Today, the neutron community is very different. Of course there are still experts, especially among the beamline scientists who operate and develop the instrumentation at the facilities, but the majority of neutron scatterers are external users who make short-term visits to the facilities for specific experiments selected via a proposal system. The latter could be working on problems in physics, chemistry, materials science, biology, engineering, and even archeology, not to mention a host of interdisciplinary areas of science and technology, and will generally deploy neutron scattering alongside an array of other complementary techniques.

As neutron scattering has evolved, from the first pioneering experiments to the superlative probe of atomic-scale structure and dynamics that it has now become, so the number and variety of instruments available at today's leading facilities has burgeoned. Neutron scattering is no longer a single technique; it is a suite of techniques, each with its own dedicated instrumentation and methodology. This diversification in both the technology and its user base has created an increasing need for learning opportunities, including training schools, pedagogical literature, and online resources, to equip neutron users with the requisite skills to design and perform their experiments to the best degree. It is hoped that this book will contribute something towards this important educational effort.

The first two chapters give an introduction to the properties of the neutron and how it interacts with condensed matter, emphasising the physical basis of neutron scattering from crystalline, disordered, and molecular materials. Later chapters develop the formal theory of neutron scattering, and cover the main areas of application of nuclear and magnetic scattering at a level suitable for advanced undergraduate and research students. Chapter 10 describes a wide range of practical matters that might be encountered when performing experiments or analysing data. For the later chapters, it will be an advantage for the reader to be acquainted with some specific mathematical results, and with the basic principles of quantum mechanics and linear response theory. The relevant material is summarized in appendices for convenience.

At the end of each chapter there is a set of exercises. These range from simple drills intended to cement understanding of basic concepts, to more advanced extensions of the theory. The exercises are included primarily to encourage the reader to have a go for themselves, since an active engagement with the subject matter is generally the best way to learn and leads to a deeper understanding. The exercises also form a resource that could be drawn upon for a course on neutron scattering. Indeed, some of the material is based on the content of tutorial sessions given at the long-running Oxford School on Neutron Scattering. Solutions to the exercises are available to course tutors via the book's web page.

As far as possible, the material in the book is presented within a common framework and with a consistent notation, in order to make it easier to appreciate the connections between the different strands of neutron scattering. It is my sincere hope that the book will make the theoretical aspects of the technique more accessible, and make it as easy as possible for readers to lay their hands on the expressions they need to interpret their data.

Formulae and experimental variables are expressed in SI units whenever practical. The main difference from most traditional accounts is in the use of inverse nanometres ($nm^{-1}$) for wavevectors. This may upset readers who, like me, spent their formative years with $\text{Å}^{-1}$, but my sense is that this is the direction of travel. In any case, the conversion is very straightforward: $1\,\text{Å}^{-1} = 10\,nm^{-1}$. I have not gone as far as to express energies in joules (or some sub-division thereof). The electron-volt (eV) remains the natural unit of energy at the atomic scale, and there is currently no move within the neutron-scattering community to replace it with an equivalent SI unit.

More importantly, all relevant formulae for scattering cross-sections are given with the numerical prefactors needed to calculate scattering intensities in absolute units. This is to make it easy to take advantage of a very powerful feature of neutron scattering, namely, the ability to calibrate scattering intensities and compare them with theoretical predictions without the need for arbitrary scaling factors.

I wish that this book were error-free. Alas, this is extremely unlikely, despite my best efforts and those of many friends and colleagues who kindly read sections and provided feedback. Any corrections which are found after going to press will be posted on the book's web page.

Andrew T. Boothroyd
University of Oxford
July 2020

Book web page: `https://groups.physics.ox.ac.uk/Boothroyd/PNS/`

# Acknowledgements

I have learned about neutron scattering from many people, and I am indebted to each and every one. I wish especially to acknowledge the influence of the late Gordon Squires, my PhD supervisor, whose clarity of thinking and simplicity of presentation, as found throughout his own book on neutron scattering, should be admired by any writer of non-fiction and has been a constant source of inspiration to me. Many individuals kindly sent me data and permitted me to re-plot it in the figures. Their names are given in the figure captions, and I am extremely grateful for their help. I wish to thank the following for providing constructive comments on the text: Stephen Blundell, Steven Bramwell, Russell Ewings, Björn Fåk, Mike Glazer, Henrik Jacobsen, David Keen, Martin Rotter, Ross Stewart, and Andrew Wildes. I also wish to thank Sönke Adlung at OUP, for his patience and encouragement over the many years of this project. Finally, I am grateful to the Institut Laue–Langevin, the Paul Scherrer Institut, and the European Spallation Source for hospitality and support during extended visits that made possible the drafting or re-drafting of large portions of the text.

# Contents

# Basic Concepts

**1**

In this first chapter we introduce the basic characteristics of the neutron. We also define some of the terminology and formal tools needed to describe the interaction of neutrons with matter. We shall discover that a simple picture based on the geometrical properties of waves provides an intuitive understanding of how neutrons probe the structure of matter on an atomic scale.

## 1.1 Particles and waves

Table 1.1 lists some of the basic properties of the neutron.[1] Mass, charge, spin, etc, are properties we readily associate with particles. Quantum mechanics, on the other hand, informs us that particles can also be represented as waves. This split personality is known as *wave–particle duality*. It requires that neutrons possess not only the attributes normally associated with moving objects, such as momentum and kinetic energy, but also the characteristics of a wave, namely wavelength and frequency.

The representation of a particle as a wave, or *matter wave* as first introduced by de Broglie,[2] is illustrated in Figure 1.1. The curve is oscillatory, with a wavelength roughly equal to the distance between adjacent wave crests and a frequency that could be obtained from the rate at which the crests pass a given point on the horizontal axis. At the same time, the outline of the oscillations can easily be imagined to represent a localized object.

According to quantum theory, the momentum $p$ and kinetic energy $E$ of a particle are linked to the wavelength $\lambda$ and frequency $\nu$ of the

[1] A neutron walks into a bar and asks 'How much for a pint of beer?' The bartender looks at the neutron, and replies 'For you, no charge!'

[2] Louis de Broglie (1892–1987) won the Nobel Prize for Physics in 1929 for his hypothesis that all matter has wave properties.

**Table 1.1** Basic properties of the neutron.

| | |
|---|---|
| mass, $m_n$ | $1.675 \times 10^{-27}\,\mathrm{kg}$ |
| charge | 0 |
| spin, $s_n$ | $\frac{1}{2}$ |
| magnetic moment, $\mu_n$ | $-1.913\,\mu_N = -0.0010\,\mu_B$ |
| mean lifetime, $\tau_n$ | $880\,\mathrm{s} = (\text{half-life})/\ln 2$ |
| quark composition | udd |

**Fig. 1.1** Graphical representation of a matter wave.

corresponding matter wave via the de Broglie equation,

$$p = \frac{h}{\lambda} = \hbar k, \tag{1.1}$$

and the Planck–Einstein equation,

$$E = h\nu, \tag{1.2}$$

where $h = 6.626 \times 10^{-34}$ J s is Planck's constant, $\hbar = h/2\pi$ and $k = 2\pi/\lambda$. Neutrons used for condensed matter studies are non-relativistic, so

$$E = \frac{p^2}{2m_\mathrm{n}} = \frac{\hbar^2 k^2}{2m_\mathrm{n}}. \tag{1.3}$$

Equations (1.1)–(1.3) together with the value of the mass $m_\mathrm{n}$ given in Table 1.1 enable us to estimate the typical wavelength and frequency of neutrons employed in condensed matter studies. Suppose the neutrons have reached thermal equilibrium by multiple collisions in a substance at room temperature, $T \approx 300\,\mathrm{K}$. The average kinetic energy of the neutrons is given by $E \sim k_\mathrm{B}T$, where $k_\mathrm{B} = 1.381 \times 10^{-23}\,\mathrm{J\,K^{-1}}$, so that $E \sim 4.1 \times 10^{-21}\,\mathrm{J} \sim 26\,\mathrm{meV}$. This energy translates via eqn (1.2) to a frequency of $6 \times 10^{12}\,\mathrm{Hz}$. The typical velocity of these neutrons, calculated from (1.3) with $p = m_\mathrm{n}v$, works out to be $v \sim 2200\,\mathrm{m\,s^{-1}}$, and from eqn (1.1) the typical de Broglie wavelength is $\lambda \sim 1.8 \times 10^{-10}\,\mathrm{m}$.

These *ball park* figures reveal that neutrons thermalized at room temperature, whose kinetic energies are therefore comparable with the typical thermal excitations of atoms in condensed matter, have a wavelength which is comparable with the separation of atoms in condensed matter. This happy coincidence, which is thanks to the particular value of the neutron mass, means that neutrons can probe both the atomic-scale structure and the atomic-scale dynamics in substances at normal operating temperatures.[3]

The other properties listed in Table 1.1 give neutron scattering additional strengths as a tool for studying the microscopic processes responsible for the macroscopic properties of materials. Charge neutrality means that neutrons interact weakly with condensed matter and penetrate deeply (typically $\sim$ cm) into most materials. This has three important benefits. First, the technique is non-destructive. This is a particular advantage for studies of soft matter. Second, neutrons are a bulk probe of matter and not sensitive to effects localized at the surface.[4] Third, neutrons can be used to study samples in complex sample environments such as cryostats, magnets, and pressure cells, or to perform *in situ* measurements, e.g. for engineering applications. The spin and magnetic moment of the neutron enable neutron scattering to probe magnetism, and the lifetime of the neutron is relevant insofar as it is much longer than the time for a scattering measurement to take place so the effect of neutron decay can be neglected. These points are elaborated upon in Section 1.4.

[3]The neutron is so well suited to the study of condensed matter that Nobel laureate B. N. Brockhouse quipped: '*It might well be said that, if the neutron did not exist, it would need to be invented!*' (Brockhouse, 1983.)

[4]An exception is the technique of neutron reflectometry which is designed specifically for the investigation of surfaces and interfaces — see Chapter 9.

## 1.1.1   Uncertainty in position and wavelength

The notion of wavelength as the distance between adjacent wave crests seems reasonable until we look more carefully at a wave such as that illustrated in Fig. 1.2(a). On close inspection we would find is that the distance between adjacent maxima is not exactly the same everywhere on the wave. This apparent variation in wavelength arises because the oscillations grow and decay so rapidly that the shape of the wave changes significantly over one period of the oscillation. Moreover, as the length $\Delta x$ of the matter wave decreases it becomes ever more difficult to assign a value to the wavelength, and so the spread of wavelengths $\Delta\lambda$ becomes larger. Conversely, as $\Delta x$ increases $\lambda$ becomes better defined and $\Delta\lambda$ decreases, see Fig. 1.2(b).   This anticorrelation between the spread in wavelength and the length of the matter wave is a manifestation of Heisenberg's famous *Uncertainty Principle* which asserts that there exists a fundamental limit to the precision with which certain pairs of observables called *conjugate variables* can be known simultaneously. Position and momentum are two such variables, and applied to them the Uncertainty Principle can be expressed through the inequality

$$\Delta p \Delta x \gtrsim \hbar. \tag{1.4}$$

Using (1.1) one can write this as $\Delta(1/\lambda)\Delta x \gtrsim 1/2\pi$.

Matter waves such as those depicted in Figs. 1.2(a) and (b) describe particles whose positions and wavelengths are known to within certain limits.  A matter wave can be represented mathematically as a *wave packet* or *wave group*, formed from a superposition of pure sinusoidal waves of different wavelengths, amplitudes, and phases. The theory of neutron scattering from condensed matter is most easily developed for a *monochromatic* neutron beam, i.e. a beam described by single sinusoidal wave that is infinite in extent and of constant wavelength and amplitude, rather than a wave packet. Our approach will be to develop the theory for such a monochromatic beam, and to average the scattering over the particular wavelength distribution in an experiment when required. This is a valid procedure because, owing to a certain randomness in the neutron production process, there is no special phase relationship between the different wavelength components of the neutron beam when averaged over time.

Neutron beams are in reality quasi-monochromatic, and a useful picture is that of a series of finite wave trains resembling Fig. 1.2(b). Such a disturbance is nearly sinusoidal on length scales below a certain threshold called the *longitudinal coherence length*, $l_{\mathrm{coh}}$. In the present case $l_{\mathrm{coh}} \sim \Delta x$. For a nearly monochromatic beam $l_{\mathrm{coh}} \gg \lambda$, Fig. 1.2(b), whereas a beam containing a wider range of wavelengths has a shorter $l_{\mathrm{coh}}$, Fig. 1.2(a). As neutron beams extend in three dimensions we also have a *lateral coherence length*, which is the distance perpendicular to the direction of propagation over which the wavefronts remain nearly continuous. In broad terms, the coherence length(s) set the range over which structural information can be obtained in a scattering experiment.

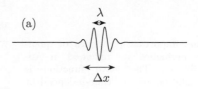
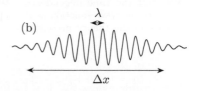

**Fig. 1.2** (a) A well-localised particle with an ill-defined wavelength. (b) A particle whose position is rather indeterminate but whose wavelength is well defined.

5Some important principles of quantum mechanics are reviewed in Appendix C. The wave function is the solution of the time-independent Schrödinger equation for a free particle. We omit the time dependence $\exp(-iEt/\hbar)$ of the wave function as it plays no part in the theory presented here.

### 1.1.2 Free particle wave function

We shall represent a monochromatic neutron beam by a *wave function* written in complex exponential form,[5]

$$\psi(\mathbf{r}) = A \exp(i\mathbf{k} \cdot \mathbf{r}), \tag{1.5}$$

where $\mathbf{k}$ is the neutron wavevector (a vector of magnitude $k = 2\pi/\lambda$ and direction parallel to the neutron's velocity), and $A$ is the amplitude. As the wave function is complex,

$$\begin{aligned}
|\psi(\mathbf{r})|^2 &= \psi^*(\mathbf{r})\,\psi(\mathbf{r}) \\
&= A^* \exp(-i\mathbf{k} \cdot \mathbf{r}) A \exp(i\mathbf{k} \cdot \mathbf{r}) \\
&= |A|^2.
\end{aligned} \tag{1.6}$$

6The interpretation of $|\psi(\mathbf{r})|^2$ as a probability is due to Max Born (1882–1970), who won the 1954 Nobel Prize in Physics. It is a fundamental assumption of quantum mechanics and cannot be proved, but it is intuitively plausible since each neutron represents a packet of energy and the energy density of any wave motion, such as a vibrating string, is proportional to the square of the amplitude of vibration.

The squared modulus of the wave function has a special significance in quantum theory.[6] If $dV$ is a very small volume element then $|\psi(\mathbf{r})|^2 dV$ is interpreted to be the probability of finding the particle in the volume $dV$ centred at position $\mathbf{r}$. It follows that $|\psi(\mathbf{r})|^2$ corresponds to the neutron density in the beam at the point $\mathbf{r}$.

### 1.1.3 Normalization of the wave function

According to eqn (1.6) there is an equal chance of finding a neutron at any point in space. This is clearly an idealization. There will always be physical constraints on the position of the neutrons, such as the size of the beam tubes and the length of the apparatus, and these confine the neutron to a finite volume $V$. Within this region eqn (1.5) is a very good approximation to the wave function of the neutron, but outside of it the wave function must vanish.[7] Because $|\psi(\mathbf{r})|^2$ is the neutron density it follows that the integral of $|\psi(\mathbf{r})|^2$ over the confinement volume is unity, and therefore

7If the reader feels uneasy about the slightly vague notion of a confinement volume, then it will be a comfort to learn that the normalization of the wave function eventually drops out of all formulae of practical interest.

$$|A| = (1/V)^{1/2}. \tag{1.7}$$

If there are on average $N$ neutrons in the beam at any one time then we can use eqn (1.5) for the wave function of the beam as a whole, but we change the normalization so that $|A|$ becomes $(N/V)^{1/2}$.

### 1.1.4 Neutron flux

The neutron flux is defined to be the number of neutrons crossing unit area per second, the area being perpendicular to the neutron beam. It follows from eqn (1.6) and the remarks immediately after it that a simple expression for neutron flux in terms of the wave function $\psi(\mathbf{r})$ and speed $v$ is

$$\Phi = |\psi(\mathbf{r})|^2 v. \tag{1.8}$$

## 1.2 Spin and polarization

According to quantum mechanics, the component of angular momentum parallel to a given direction is quantized into discrete values separated by $\hbar$. The neutron has spin quantum number $s_n = \frac{1}{2}$, and so the allowed values of spin angular momentum are $+\frac{1}{2}\hbar$ and $-\frac{1}{2}\hbar$. These two states, known colloquially as 'spin-up' and 'spin-down', represent the two polarization states of the neutron. In practical situations the quantization direction is usually defined by a magnetic field.

The concept of polarization extends easily to a neutron beam. If the spin states of the individual neutrons in a beam are random then the beam is *unpolarized*, but if one spin state predominates then the beam is *polarized*. The polarization of the beam is described by a vector $\mathbf{P}$ whose components $P_\alpha$ ($\alpha = x, y, z$) are given by

$$P_\alpha = \frac{n_\alpha - n_{\bar{\alpha}}}{n_\alpha + n_{\bar{\alpha}}}, \qquad (1.9)$$

where $n_\alpha$ and $n_{\bar{\alpha}}$ are the numbers of 'up' and 'down' neutrons with respect to the $\alpha$ direction. From (1.9) it can be seen that the projection of $\mathbf{P}$ along any given direction lies in the range $-1$ to $+1$. The extremes ($-1$ and $+1$) correspond to beams in which all the neutrons are in the same spin state. An unpolarized beam has $\mathbf{P} = 0$.

Despite the fact that the neutron is uncharged it nevertheless carries a magnetic dipole moment.[8] The neutron's magnetic moment is given by

[8]A consequence of the neutron's internal quark structure.

$$\boldsymbol{\mu}_n = -2\gamma\mu_N\mathbf{s}_n, \qquad (1.10)$$

where $\mu_N = 5.051 \times 10^{-27}\,\mathrm{J\,T^{-1}}$ is the nuclear magneton, and $\gamma = 1.913$. In a magnetic field of flux density $\mathbf{B}$ the magnetic moment has potential energy

$$V_B = -\boldsymbol{\mu}_n \cdot \mathbf{B}, \qquad (1.11)$$

(the Zeeman interaction), and experiences a torque $\mathbf{T}$ given by

$$\mathbf{T} = \boldsymbol{\mu}_n \times \mathbf{B}. \qquad (1.12)$$

Hence, if $\boldsymbol{\mu}_n \parallel \mathbf{B}$ then $\boldsymbol{\mu}_n$ is unaffected by the field, reflecting that the neutron is in a spin eigenstate — either up or down. In general, however, $\boldsymbol{\mu}_n$ precesses around $\mathbf{B}$ at the Larmor angular frequency

$$\omega_L = \gamma_L B, \qquad (1.13)$$

where $\gamma_L = 2\gamma\mu_N/\hbar = 2\pi \times 2.916 \times 10^7\,\mathrm{rad\,s^{-1}T^{-1}}$ is the *gyromagnetic ratio* of the neutron. To gauge the importance of this we note that the magnetic moment of a thermal neutron travelling at $2{,}200\,\mathrm{ms^{-1}}$ in Earth's magnetic field, $B_E = 5 \times 10^{-5}\,\mathrm{T}$, will precess through one complete revolution in about $0.7\,\mathrm{m}$. This means that fields comparable with Earth's field can have a significant depolarizing effect. To prevent depolarization, it is essential to maintain a guide field whose strength is well in excess of Earth's field (or indeed any other stray fields) all along the neutron path.

## 1.3   Production of neutrons

In ordinary condensed matter, neutrons are bound in the nucleus of atoms. Neutrons can be released from the nucleus in several different nuclear reactions. The experimental discovery of the neutron by Chadwick in 1932 employed one such reaction.[9] Chadwick fired alpha ($\alpha$) particles[10] from the radioactive decay of polonium at a target of beryllium (Be) to produce neutrons by the following ($\alpha$, n) reaction:

$$^9\mathrm{Be} + \alpha \;\;\rightarrow\;\; {}^{12}\mathrm{C} + \mathrm{n}.$$

Neutron sources based on ($\alpha$, n) reactions but with radium in place of polonium as the $\alpha$ emitter were used in the first demonstrations of neutron diffraction from crystals,[11] and are still used as portable devices for calibrating neutron detectors. A typical such source might produce $10^8\,\mathrm{n\,s^{-1}}$, giving a neutron flux at a distance of $1\,\mathrm{m}$ of $10^3\,\mathrm{cm^{-2}\,s^{-1}}$. This is several orders of magnitude too small to be useful for contemporary neutron scattering research.

Below we briefly describe the principal methods used to produce neutrons for research.

### 1.3.1   Reactor sources

Today, two distinct methods are employed to produce neutrons at high-flux neutron scattering facilities. One of these is *nuclear fission*, specifically neutron-induced fission in a nuclear reactor. This is the process by which a heavy nucleus such as $^{235}\mathrm{U}$ absorbs a neutron and subsequently splits into two or more lighter nuclei accompanied by the release of an average of 2.4 neutrons per fission with typical kinetic energies of about $2\,\mathrm{MeV}$ per neutron (Fig. 1.4).

The reason why neutrons are released in fission can be understood from the chart in Fig. 1.3 which plots the number of neutrons against the number of protons for the long-lived isotopes of the chemical elements. The chart shows that in stable nuclei the number of neutrons equals or exceeds the number of protons, and that the neutron/proton ratio increases with increasing atomic number $Z$. Hence, neutrons will be released when a heavy nucleus splits into stable lighter nuclei.

The core of a nuclear reactor containing the fissile fuel elements is arranged so that the neutrons released in one fission have a high probability of inducing at least one additional fission by collision with other

[9]James Chadwick (1891–1974) won the 1935 Nobel Prize in Physics for his discovery of the neutron (Chadwick, 1932).

[10]The alpha particle is the nucleus of a helium atom, comprising two protons and two neutrons.

[11]The first experiments were performed independently by von Halban jr. and Preiswerk (1936), and Mitchell and Powers (1936).

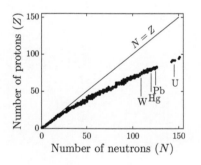

**Fig. 1.3** Chart of nuclides (Segrè chart) with half-lives greater than one million years. Nuclei important for neutron production are indicated. Information extracted from the Chart of Nuclides database maintained by the National Nuclear Data Center, http://www.nndc.bnl.gov/chart/.

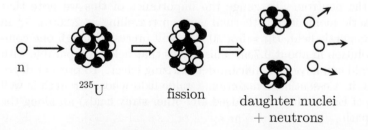

**Fig. 1.4** Neutron-induced fission of the $^{235}\mathrm{U}$ nucleus.

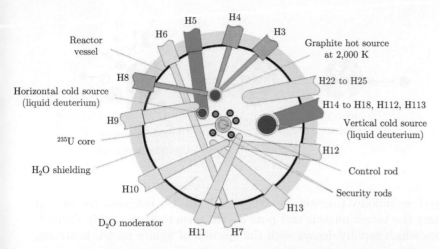

**Fig. 1.5** Schematic of the high-flux reactor at the Institut Laue–Langevin in Grenoble, France. Labels of the form 'H*n*' indicate beam tubes and guides. (Image credit: the Institut Laue–Langevin.)

nuclei. This in turn induces another fission, and so on, causing a *chain reaction*. Neutron beam reactors are designed so that the majority of the excess neutrons which do not participate in fission reactions escape from the core. To be suitable for scattering experiments these neutrons must be slowed down to reduce their kinetic energy from ∼MeV to less than 1 eV. This takes place in a *moderator*, which is a medium that slows the neutrons down by repeated collisions with the nuclei of the moderator substance. After many collisions the neutrons reach thermal equilibrium, whereupon their energy distribution is Maxwellian (Fig. 1.6) and their mean kinetic energy is determined by the temperature of the moderator. Examples of actual moderator substances are liquid hydrogen at 25 K, liquid water at 300 K, and solid graphite at 2,400 K. For obvious reasons, neutrons emerging from these moderators are termed cold, thermal, and hot, respectively. On emerging from the moderator the neutrons are transported along beam tubes and guides to the neutron scattering instrumentation.

A diagram showing the key components of the research reactor at the Institut Laue–Langevin (ILL) is provided in Fig. 1.5. The reactor operates at an output power of 58 MW and produces a steady flux of neutrons in the moderator region of $1.5 \times 10^{15}$ cm$^{-2}$ s$^{-1}$. There are a number of cold, thermal, and hot moderators, and the neutrons from these are delivered to over forty instruments designed for a wide range of scattering and fundamental physics experiments.

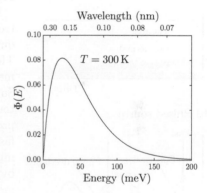

**Fig. 1.6** The flux per unit energy from a moderator at a temperature of 300 K. The curve is a normalized Maxwellian distribution describing the effusion of neutrons from the moderator:

$$\Phi(E) = \frac{2\pi E}{(\pi k_{\mathrm{B}} T)^{3/2}} \exp(-E/k_{\mathrm{B}} T).$$

## 1.3.2  Spallation sources

The other method used to produce neutrons for condensed matter research is *spallation*.[12] Nuclear spallation is the process by which light nuclear fragments are ejected from a heavy nucleus as a result of bombardment by high-energy (∼1 GeV) protons. A synchrotron, cyclotron, or linear accelerator is used to produce pulses of protons that travel at close to the speed of light. The spallation target is made from a heavy

[12]The name comes from *spall*, a word used in the mining industry for a chip or flake ejected when rock is mechanically crushed.

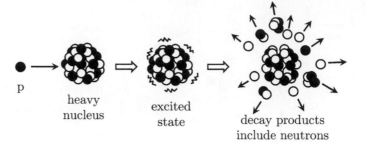

p    heavy
nucleus    excited
state    decay products
include neutrons

**Fig. 1.7** Proton-induced spallation.

metal, commonly tungsten, mercury, or lead. On collision, the proton enters the target nucleus and puts it into a short-lived, highly excited, state which rapidly decays with the ejection of lighter nuclei, neutrons, and other fundamental particles (Fig. 1.7). The de-excitation process can also result in fission. Approximately thirty neutrons are produced per incident proton, and their average energy is around 20–30 MeV. The neutrons are slowed down in moderators similar to those used in reactor sources before being transported down beam tubes to the instruments. Figure 1.9 shows the layout of the ISIS spallation neutron source.

In contrast to reactor sources, which produce a constant flux of neutrons, spallation sources produce neutrons in a train of pulses whose frequency varies between 10 and 60 Hz depending on the facility, Fig. 1.8. The majority of spallation sources have a pulse duration of ~100 μs immediately after the moderator (an exception is the European Spallation Source (ESS) which employs 3 ms pulses). The neutrons emerging from the moderator have a spread of energies and hence speeds, but as they propagate along a beamline they disperse in space and time, with the faster neutrons arriving before the slower ones. One can therefore determine the speed of the neutrons (and hence their energy and wavelength) by measuring their flight time over a known distance. Neutron instruments at spallation sources employ a variety of time-of-flight techniques to make optimum use of the pulse structure of the source for different types of measurement, so although the time-average neutron flux at a pulsed spallation source is often less than that of a high-flux reactor, experiments are usually completed in about the same amount of time.

(a) Continuous source

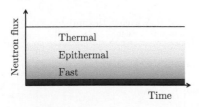

(b) Pulsed source

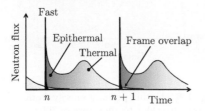

**Fig. 1.8** Time structure of the neutron flux on (a) a continuous and (b) a pulsed source. On a continuous source, the flux of fast, epithermal, and thermal neutrons is constant, whereas on a pulsed source, neutrons of different energy arrive at different times (epithermal neutrons are neutrons that have been slowed down by collisions but have not yet reached thermal equilibrium). Some neutrons from the low-energy tail of the $n^{th}$ pulse can overlap with the high-energy part of the $(n+1)^{th}$ pulse. This is called *frame overlap*.

Although reactor sources are normally continuous and spallation sources pulsed, there do exist counter-examples. The IBR-2 reactor at the Joint Institute for Nuclear Research in Dubna, Russia, has two reflectors, one of which rotates at twice the frequency of the other. When the reflectors coincide near the core there is a reactor power pulse. This mechanism results in neutron pulses of duration approximately 0.2 ms and frequency 5 Hz which feed a range of different time-of-flight instruments. Conversely, the SINQ neutron scattering facility at the Paul Scherrer Institut, Switzerland, is fed by a proton cyclotron that operates at a frequency of 50 MHz. The frequency is sufficiently high that the neutron pulses produced by spallation in the lead target fully overlap with the adjacent pulses, resulting in a continuous beam.

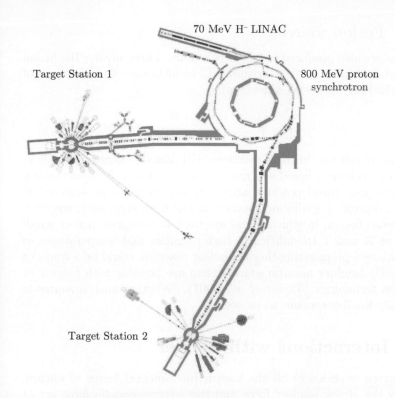

70 MeV H⁻ LINAC

Target Station 1

800 MeV proton
synchrotron

Target Station 2

**Fig. 1.9** Schematic of the ISIS spallation neutron and muon source at Harwell–Oxford, U.K. (Image credit: STFC ISIS Neutron and Muon Source.)

### 1.3.3   Compact accelerator-driven sources

Most neutron beam experiments require a high neutron flux, and can be performed only at large-scale facilities operating at nuclear reactors or spallation sources. On the other hand, there has been growth in the development of relatively low power accelerator-driven sources which operate at energies below 100 MeV. At these energies, spallation of a heavy nucleus is no longer the most efficient process for neutron production. Instead, the best neutron yields are obtained in (p, n) or (d, n) nuclear reactions with light nuclei such as $^7$Li and $^9$Be. For example,

$$^9\mathrm{Be} + \mathrm{p} \;\rightarrow\; {}^9\mathrm{B} + \mathrm{n}.$$

These *compact accelerator-driven neutron sources* (CANS) have a neutron yield typically four or five orders of magnitude lower than that of reactor or spallation sources but require significantly less shielding, making them physically smaller and hence more efficient in transporting the neutrons from target to sample. As a result, the usable flux can be within one or two orders of magnitude of that available at high-flux facilities, making CANS a practical alternative for techniques that do not always require the highest flux, such as diffraction, small-angle scattering, reflectometry, and imaging, as well as for training and test purposes.

Another approach under consideration is neutron production by high-power lasers. Intense laser light can generate charged particles which subsequently produce neutrons via nuclear reactions. If viable, this technology could open up the possibility of 'table-top' neutron sources.

### 1.3.4   Fusion sources

Neutrons are also produced in fusion reactions. These involve the fusion of ions of deuterium (D) and tritium (T) to form one of the isotopes of helium (He) together with a neutron:

$$D + D \quad \rightarrow \quad {}^3\text{He} + \text{n}$$
$$D + T \quad \rightarrow \quad {}^4\text{He} + \text{n}.$$

Neutrons are released in these reactions with kinetic energies $E \sim \text{MeV}$. Table-top accelerator-based neutron sources based on DT fusion (known as *neutron generators*) produce more neutrons than $(\alpha, \text{n})$ sources but less than reactor or spallation sources. It has been suggested, however, that inertial fusion, in which lasers are used to compress a very small capsule of D and T to sufficiently high densities and temperatures so as to initiate a propagating thermonuclear reaction, could be a route to significantly brighter neutron sources than are possible with reactor or spallation technology (Taylor *et al.*, 2007). Whether such a source is practically feasible remains to be seen.

## 1.4   Interactions with matter

The neutron experiences all the known fundamental forces of nature, but only the *strong nuclear force* and the *electromagnetic force* are of direct relevance to the study of condensed matter. The gravitational attraction between particles is so small as to be utterly negligible, and the weak nuclear force is only relevant insofar as it causes $\beta$–decay and thereby limits the lifetime of the neutron.

### 1.4.1   The nuclear interaction

Most of the present chapter is devoted to the principles of neutron scattering from atomic nuclei via the strong nuclear force, the force responsible for binding protons and neutrons together in the nucleus. Although unimaginably strong, the strong force only acts over a very short distance of order $10^{-15}$ m. The neutron–nucleus interaction depends on the number of neutrons and protons in the nucleus and also on the combined spin of the neutron–nucleus system.

Because the range of the strong force is so short the probability that a neutron will be scattered as it passes by a nucleus is extremely small. The separation between atomic nuclei in a solid or liquid is usually greater than $10^{-10}$ m, and so the neutron will in all probability travel deep into the interior of a sample, typically $10^{-2}$ m (1 cm), before encountering the force field of a nucleus. This high degree of penetration means that surface effects can usually be neglected, and that neutrons sample a large volume of the target. Moreover, neutrons deposit so little energy in a sample that they cause no physical damage and do not ionize the atoms. Hence, neutron scattering is non-destructive and is a bulk probe of matter.

For comparison, X-rays penetrate typically $10^{-4}$ to $10^{-7}$ m depending on the X-ray energy and on the atomic number of the elements present, and electron beams usually only penetrate a few atomic layers beneath the surface. Deep penetration is one of the great strengths of neutron scattering, but the associated low scattering rate can be a problem. Large samples or high neutron fluxes are often essential.

### 1.4.2 The electromagnetic interaction

Being neutral particles, neutrons do not scatter via the Coulomb interaction. They do experience electromagnetic forces though, principally on account of the neutron's intrinsic magnetic dipole moment — see Table 1.1 and eqn (1.10). The magnetic dipole moment allows the neutron to scatter from magnetic flux variations in the target, and thus provides the neutron with its ability to probe magnetism. Sources of magnetic scattering include the orbital motion and intrinsic spin of unpaired electrons, and any non-uniformity in the magnetization of a sample. The magnetic interaction depends on the relative orientation of the magnetic moment of the neutron and the scattering field, and hence an analysis of scattering processes according to the polarization state of the incident and scattered neutron beams can often yield extra information on the magnetic state of the scattering system.

## 1.5 Scattering kinematics

Figure 1.10(a) depicts a generic scattering event. A neutron with initial wavevector $\mathbf{k_i}$ and energy $E_i$ is scattered through an angle $\phi$ into a final state with wavevector $\mathbf{k_f}$ and energy $E_f$. The vector

$$\mathbf{Q} = \mathbf{k_i} - \mathbf{k_f} \tag{1.14}$$

is known as the *scattering vector*, and the triangle formed by the vectors $\mathbf{k_i}$, $\mathbf{k_f}$, and $\mathbf{Q}$ is called the *scattering triangle*, Fig. 1.10(b). From eqns (1.1) and (1.2) it is apparent that $\hbar\mathbf{Q}$ is the amount of momentum transferred from the neutron to the sample, and

$$\hbar\omega = E_i - E_f \tag{1.15}$$

is the amount of energy transferred to the sample.[13]

The scattering process is said to be *elastic* if $E_i = E_f$, i.e. the neutron's energy remains unchanged, and *inelastic* if $E_i \neq E_f$. From the cosine rule, and eqn (1.3),

$$Q^2 = k_i^2 + k_f^2 - 2k_i k_f \cos\phi \tag{1.16}$$

$$= \frac{2m_n}{\hbar^2}\left\{ E_i + E_f - 2(E_i E_f)^{\frac{1}{2}}\cos\phi \right\}, \tag{1.17}$$

where $Q = |\mathbf{Q}|$, etc.

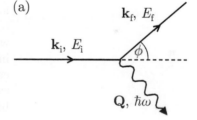

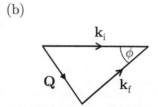

**Fig. 1.10** (a) Representation of a scattering event. (b) The corresponding scattering triangle.

[13] $\mathbf{Q}$ and $\omega$ are often referred to as the *momentum transfer* and *energy transfer* respectively, the constant $\hbar$ being omitted for brevity.

For elastic scattering, $|\mathbf{k}_i| = |\mathbf{k}_f| = k$ and so the scattering triangle is isosceles. Hence,

$$Q = 2k\sin(\phi/2) = \frac{4\pi}{\lambda}\sin(\phi/2) \qquad \text{(elastic scattering)}. \qquad (1.18)$$

For the remainder of this chapter, and throughout the whole of Chapter 2, we will be concerned only with elastic scattering. Inelastic scattering will be introduced in Chapter 3.

## 1.6    Scattering from nuclei

### 1.6.1    Isolated bound nucleus: the scattering length

In this section we shall consider the scattering of neutrons by an isolated nucleus which is rigidly bound to its surroundings. The rigidity constraint means that neutrons cannot gain or lose kinetic energy on scattering.

Consider the incident neutrons as a plane wave travelling in the $z$ direction, and suppose the nucleus is fixed at the origin. The effect of the neutron–nucleus interaction is to produce a weak spherical wave radiating from the nucleus. This process is represented (in two dimensions) in Fig. 1.11. The scattered wave is spherical for three reasons. First, the strong nuclear force is isotropic. Second, the range of the nuclear force is much shorter than any neutron wavelength we are ever going to encounter in the study of condensed matter, so interference effects within the nucleus can be ignored. Third, the wave representing the incident neutrons is virtually unaffected by the interaction. This last condition is equivalent to our earlier statement that, owing to the short range of the nuclear force, most neutrons pass by the nucleus without being deflected. The production of spherical waves as a result of a localized perturbation is like the effect of a stone thrown into still water. The stone strongly distorts the water surface in the immediate vicinity of the impact, but further away the disturbance quickly takes the form of circular outgoing waves.[14]

If the incident neutron plane wave is represented by the wave function (1.5), then the wave function for the scattered neutrons can be written

$$\psi_{\text{sc}}(\mathbf{r}) = -\frac{b}{r}A\exp(ikr), \qquad (1.19)$$

where $r$ is the distance from the centre of the nucleus and $b$ is a parameter that controls the amplitude of the outgoing spherical wave relative to the incident plane wave. The minus sign means that if $b$ is positive then there is a phase change of $\pi$ on scattering from the nucleus, which is what is often observed in the reflection of classical waves. For example, sound waves in air undergo a phase change of $\pi$ on reflection from the ground.

The constant $b$ has dimensions of length, and is known as the *bound nuclear scattering length*. It is a characteristic of the nucleus, and varies

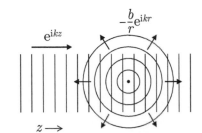

**Fig. 1.11** Interaction of a neutron plane wave of unit amplitude with a nucleus to produce a spherical scattered wave. The lines represent the wave crests.

[14]In the formal theory, the scattered wave is expressed as a sum of partial waves with different angular momentum quantum numbers $l$. At low energies and large distances from a localized target region the scattered wave is dominated by the spherical component ($l = 0$). This is called the $S$-wave scattering.

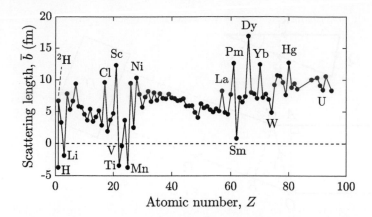

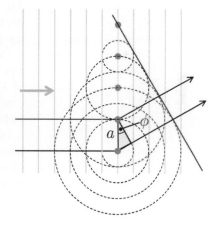

**Fig. 1.12** Bound coherent nuclear scattering lengths for the natural elements and deuterium ($^2$H). Some outlying points are labelled with the chemical symbol. The data can be found in Appendix A.

both across the periodic table and among the isotopes of a given element. It also depends on the spin state of the neutron–nucleus system. $b$ is found to be positive for most nuclei, but it can also be negative or even complex. Figure 1.12 plots the isotopically averaged scattering lengths of the natural elements against atomic number. There is a gradual increase with atomic number, but also considerable scatter.

## 1.6.2 Scattering from systems of bound nuclei

We now take a step towards an expression for the scattering from a real sample. Instead of an isolated bound nucleus, as in the previous section, consider a system made up of a collection of many bound nuclei not necessarily all identical. When the specimen is illuminated with neutrons, each nucleus becomes an independent source of spherical waves. These waves overlap, and the total scattered wave at any point in space is a superposition of the scattered waves from each individual nucleus.

We first consider a simple example of superposition, and then tackle the general case.

**Fig. 1.13** First order of diffraction from a line of nuclei with separation $a$. At large distances, the spherical wavefronts from each nucleus are approximately planar, and interfere constructively to give a plane wave at an angle $\phi$ to the incident beam, where $\lambda = a \sin \phi$.

### A one-dimensional diffraction grating

A beam of neutrons of wavelength $\lambda$ is incident normally on a line of $N$ equally spaced identical nuclei, as shown in Fig. 1.13. Adjacent nuclei are separated by a distance $a$. Spherical scattered waves emerge from each of the nuclei, and we consider the superposition of these waves at the position of a detector. The detector is sufficiently far away that the beams from each nucleus to the detector may be considered parallel, all making an angle $\phi$ to the incident beam direction.

In superposing all the scattered waves we must take into account that the total distance from source to detector varies slightly according to the position of each nucleus. By geometry, the path difference for the waves scattered from adjacent nuclei is $a \sin \phi$, corresponding to a number of wavelengths equal to $(a \sin \phi)/\lambda$. In the special case when the path difference is a whole number $p$ of wavelengths all the scattered waves

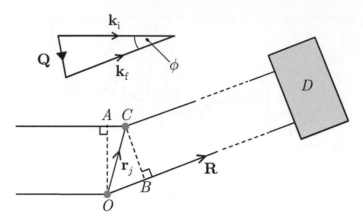

**Fig. 1.14** Scattering geometry for a general configuration of nuclei. The scattering triangle for elastic scattering is shown at the top.

emerge from the grating in phase and so reinforce one another. The result is a wave whose amplitude is $N$ times larger than that from a single nucleus. This effect is known as *constructive interference*, and the condition for it to happen is

$$p\lambda = a \sin \phi. \tag{1.20}$$

The construction in Fig. 1.13 shows how spherical waves from adjacent nuclei combine to produce a plane wave when the path difference is one wavelength $(p = 1)$.

Constructive interference occurs only at very specific angles given by eqn (1.20). At other angles the path difference is not an integral multiple of $\lambda$, and there is a high degree of cancellation when all the waves are combined. The resultant wave amplitude is then comparable in magnitude to that from an individual nucleus.

In summary, the scattering from the diffraction grating contains intense peaks at specific angles separated by intervals of relatively low intensity. This type of scattering pattern is characteristic of all periodic structures.

### General scattering system of bound nuclei

Ultimately, we want to be able to calculate the count rate measured by a detector for neutron diffraction from an arbitrary distribution of nuclei. Let $\mathbf{R}$ be the position vector of the detector $D$ relative to an origin $O$ which we locate for convenience on one of the nuclei. Let the position vector of the $j$th nucleus be $\mathbf{r}_j$ with respect to $O$. The arrangement is shown in Fig. 1.14.

The wave that arrives at $D$ via nucleus $j$ travels a distance $AC$ more and $OB$ less than the wave from $O$, corresponding to a phase difference $\delta_j = k(AC - OB)$, with $k = 2\pi/\lambda$. In all practical situations $|\mathbf{r}_j| \ll |\mathbf{R}|$, and so the rays $OD$ and $CD$ are virtually parallel to one another. It is convenient then to express the phase difference in terms of the wavevectors of the incident and scattered neutrons, so that

$$\delta_j = \mathbf{k_i} \cdot \mathbf{r}_j - \mathbf{k_f} \cdot \mathbf{r}_j$$
$$= \mathbf{Q} \cdot \mathbf{r}_j, \tag{1.21}$$

where $\mathbf{Q}$ is the scattering vector — see eqn (1.14).

Inclusion of $\delta_j$ from eqn (1.21) into the exponential term of eqn (1.19) enables us to express the total scattered wave amplitude at the detector position for an incident wave of unit amplitude as

$$\Psi_{\mathrm{sc}}(\mathbf{R}) = \exp(ikR) \sum_j \frac{-b_j}{|\mathbf{R} - \mathbf{r}_j|} \exp(i\mathbf{Q} \cdot \mathbf{r}_j). \tag{1.22}$$

The summation extends over all nuclei in the sample, and $|\mathbf{R} - \mathbf{r}_j|$ is the distance from nucleus $j$ to the detector. To a very good approximation $|\mathbf{R} - \mathbf{r}_j| \approx |\mathbf{R}|$ since, as noted above, $|\mathbf{r}_j| \ll |\mathbf{R}|$. With this approximation, and the fact that intensity is given by $|\Psi_{\mathrm{sc}}(\mathbf{R})|^2$ as discussed previously, the neutron count rate measured in the detector is proportional to the function

$$I(\mathbf{Q}) = \left| \sum_j b_j \exp(i\mathbf{Q} \cdot \mathbf{r}_j) \right|^2. \tag{1.23}$$

Equation (1.23) shows us that the scattering intensity is a function of the amount of momentum transferred to the sample, rather than of the initial or final states of the neutron.

In principle, eqn (1.23) enables us to calculate the scattering from any system providing we know the locations and types of all the nuclei. This task may at first sight appear formidable, but it can be greatly simplified in most cases by exploiting special properties of the scattering system. One such property is periodicity, and to illustrate this we will now evaluate $I(\mathbf{Q})$ for the diffraction grating considered above.

## $I(\mathbf{Q})$ for a one-dimensional diffraction grating

We refer again to Fig. 1.13, and let the position vectors of the nuclei be $\mathbf{r}_n = n\mathbf{a}$, where $\mathbf{a}$ is the vector joining one nucleus to its neighbour immediately above. As $\mathbf{k_i}$ is perpendicular to $\mathbf{a}$ the phase difference associated with the $n$th nucleus is from (1.21) $\delta_n = -n\mathbf{k_f} \cdot \mathbf{a} = kna \sin \phi$. From eqn (1.23) the intensity is[15]

$$I(\mathbf{Q}) = \left| \sum_{n=0}^{N-1} b \exp(ikna \sin \phi) \right|^2$$
$$= b^2 \frac{\sin^2(\frac{1}{2} kNa \sin \phi)}{\sin^2(\frac{1}{2} ka \sin \phi)}$$
$$= b^2 \frac{\sin^2(\frac{1}{2} Q_x Na)}{\sin^2(\frac{1}{2} Q_x a)}. \tag{1.24}$$

[15]The sum of a geometric series is

$$\sum_{n=0}^{N-1} u^n = \frac{1 - u^N}{1 - u}.$$

Here, $u = \exp(ika \sin \phi)$. Note also that the component of $\mathbf{Q}$ along the line of nuclei is $Q_x = k \sin \phi$.

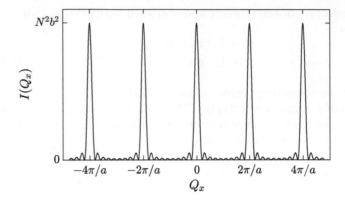

**Fig. 1.15** Diffracted intensity calculated from eqn (1.24) for a line of 10 nuclei separated by a distance $a$. The intensity is plotted as a function of $Q_x$, the component of the scattering vector parallel to the line of nuclei.

$I(\mathbf{Q})$ becomes very large whenever the denominator in eqn (1.24) vanishes, i.e. when $\frac{1}{2}ka\sin\phi = p\pi$, or $p\lambda = a\sin\phi$, where $p$ is an integer. This is the same condition for constructive interference as found previously, eqn (1.20), but now we have the benefit of the complete intensity function. Recast in terms of $\mathbf{Q}$ this condition reads $Q_x = 2p\pi/a$. The scattering does not vary with $Q_y$ or $Q_z$.

Figure 1.15 shows the intensity calculated for $N = 10$ plotted as a function of $Q_x$. Since $\sin\phi \to \phi$ as $\phi \to 0$ the height of the principal maxima observed in $I(\mathbf{Q})$ is seen from (1.24) to be $N^2b^2$. In the interval between the principal maxima $I(\mathbf{Q})$ has some small oscillations with a period of $2\pi/(Na)$ and an average intensity of order $b^2$.

Before finishing this section it is worth drawing attention to a very general feature of diffraction nicely illustrated by this simple model. We have seen that when expressed as a function of $Q_x$ the scattering exhibits, (1) sharp peaks separated by $2\pi/a$ associated with the distance $a$ between the nuclei, and (2) rapid oscillations with a period of $2\pi/D$, where $D = Na$ is the full length of the line of nuclei. The generalization can be summarized like this:

*Features in the object with a characteristic size $\Delta x$*
*will produce features in the scattering that vary*
*on a scale $\Delta Q_x \sim 2\pi/\Delta x$.*

The resemblance to Heisenberg's Uncertainty Principle (1.4) is not a coincidence.

### 1.6.3 Elastic scattering as a Fourier transform

Following on from the treatment presented above, we shall now show that the scattering intensity for a general system of bound nuclei given in (1.23) can be written as a Fourier transform.[16] This has both theoretical significance and practical value, since Fourier transforms have many useful properties.

We first define the *scattering length density*,

$$n_b(\mathbf{r}) = \sum_j b_j \delta(\mathbf{r} - \mathbf{r}_j), \qquad (1.25)$$

[16]The Fourier transform is introduced in Section B.3.

where the delta function $\delta(\mathbf{r}-\mathbf{r}_j)$ represents any sharply peaked function centred on $\mathbf{r} = \mathbf{r}_j$ with a volume of unity (see Section B.2). It follows that integration of $n_b(\mathbf{r})$ over a given volume tells us the amount of scattering length in the volume.[17] The $\delta$-function has the property that for any smooth function $f(\mathbf{r})$, the integral of $f(\mathbf{r})\delta(\mathbf{r}-\mathbf{r}_j)$ is equal $f(\mathbf{r}_j)$. Using this property we can write the sum that appears in eqn (1.23) as

$$\sum_j b_j \exp(\mathrm{i}\mathbf{Q} \cdot \mathbf{r}_j) = \int n_b(\mathbf{r}) \exp(\mathrm{i}\mathbf{Q} \cdot \mathbf{r}) \, \mathrm{d}^3\mathbf{r}. \tag{1.26}$$

In this form, the scattering amplitude is seen to be the Fourier transform of the scattering length density, and eqn (1.23) becomes

$$I(\mathbf{Q}) = \left| \int n_b(\mathbf{r}) \exp(\mathrm{i}\mathbf{Q} \cdot \mathbf{r}) \, \mathrm{d}^3\mathbf{r} \right|^2. \tag{1.27}$$

The formulation of elastic neutron scattering in terms of a Fourier transform draws a direct parallel with the treatment of interference phenomena in other scattering techniques, including optics (Fraunhofer diffraction), non-resonant X-ray scattering, and electron diffraction.

[17]The use of a $\delta$-function to represent the spatial distribution of a nucleus is justified by the extremely short range of the nuclear potential, and will be employed again in Section 4.1.1 to represent the neutron–nucleus interaction potential in the quantum-mechanical theory of scattering.

## 1.7   Cross-sections

### 1.7.1   Scattering cross-sections

An extremely useful feature of neutron scattering is the ability to determine absolute scattering intensities, i.e. the actual number of counts per second that would be recorded in the detector. The expressions for $I(\mathbf{Q})$ derived in the previous section do not allow us to do this because in deriving them we missed out some constants of proportionality. We now obtain these constants.

The easiest way to calculate absolute intensities is to express the scattering as a *cross-section*. As the name implies, a cross-section is a kind of area. It can be thought of as the effective area of a target that the neutrons must hit in order for a particular scattering process to occur. The larger the cross-section, the higher the scattering probability.

The simplest cross-section is the *total scattering cross-section* which measures the liklihood of neutrons being scattered by a target irrespective of any changes in direction or energy during the scattering process. The total scattering cross-section is defined by

$$\sigma_{\mathrm{s}} = \frac{(\text{total number of neutrons scattered per second})}{\Phi_0}, \tag{1.28}$$

where $\Phi_0$ is the incident neutron flux given by eqn (1.8).

To illustrate the meaning of the cross-section let us calculate $\sigma_{\mathrm{s}}$ for the isolated, bound nucleus we considered in section 1.2. From eqns (1.5), (1.8) and (1.19), the incident flux is $|A|^2 v$ and the scattered flux at a distance $r$ from the nucleus is $|A|^2 b^2 v / r^2$. Eqn (1.28) requires the total scattering in all directions, and this can be obtained by integration of the scattered flux over a closed surface surrounding the nucleus. In the present case the scattering is isotropic, and so we take the closed surface to be a sphere of radius $r$. The integration then corresponds to multiplication by $4\pi r^2$, the surface area of the sphere. Hence, the total cross-section for a bound nucleus is,

$$\sigma_{\mathrm{s}} = \frac{4\pi r^2 \times |A|^2 b^2 v/r^2}{|A|^2 v}$$

$$= 4\pi b^2 \qquad \text{(bound nucleus)}. \qquad (1.29)$$

We can acquire some intuition for the total cross-section by using classical mechanics to calculate $\sigma_{\mathrm{s}}$ for a solid sphere. Let the radius of the sphere be $a$. The number of neutrons scattered by the sphere is just equal to the number that strike it, and as the sphere presents an area of $\pi a^2$ perpendicular to the neutron beam the number of neutrons scattered per second is $\pi a^2 \Phi_0$. Dividing by $\Phi_0$ as prescribed by eqn (1.28) we find

$$\sigma_{\mathrm{s}} = \pi a^2 \qquad \text{(classical solid sphere)}, \qquad (1.30)$$

which is just the cross-sectional area of the sphere, as could have been predicted.

The similarity between eqns (1.29) and (1.30) might tempt us to look upon the nucleus as a solid sphere of radius $2b$. Such a view would be over-simplistic, but there is some merit in regarding the scattering length as an effective range for the nuclear scattering potential.

Scattering lengths have magnitudes that are typically in the range $10^{-15}$ to $10^{-14}$ m, and so the total cross-section of a nucleus is of order $10^{-28}$ m$^2$. For this reason, scattering lengths are usually quoted in femtometres[18] (fm), and cross-sections in units of barns[19] (b), where

$$1\,\mathrm{fm} = 10^{-15}\,\mathrm{m}$$
$$1\,\mathrm{b} = 10^{-28}\,\mathrm{m}^2.$$

The total scattering cross-section represents all scattered neutrons, but in an experiment we are usually interested in the scattering in a given direction. Accordingly, we introduce a second scattering cross-section, the *differential scattering cross-section*, defined by

$$\frac{\mathrm{d}\sigma}{\mathrm{d}\Omega} = \frac{\text{(number of neutrons scattered per second into solid angle d}\Omega \text{ about a given direction)}}{\Phi_0 \times \mathrm{d}\Omega}. \qquad (1.31)$$

The meaning of solid angle is explained in Fig. 1.16. In eqn (1.31), $\mathrm{d}\Omega$ could be taken to be the solid angle subtended by a detector at the sample position. It would be more correct, however, to regard the differential cross-section as the limit of the right-hand side of eqn (1.31) as $\mathrm{d}\Omega$ tends to zero. This limit corresponds to a detector with infinitesimally small area over which there is no variation in the scattering intensity.

[18]The femtometre is an SI unit of length which is also called the fermi (1 femtometre = 1 fermi) in honour of the Nobel-prizewinning Italian physicist Enrico Fermi (1901–1954), whose name is attached to so many things.

[19]The name was coined by U.S. nuclear scientists during the Second World War. It is derived from the proverbial phrase 'couldn't hit the broad side of a barn', which refers to someone with a lousy aim. The point is that a cross-section of 1 b is actually enormous when compared with many of the sub-atomic reactions scientists attempt to study with particle accelerators.

As an example, the differential cross-section for elastic scattering from fixed nuclei is just the function $I(\mathbf{Q})$ given in eqn (1.23), i.e.

$$\frac{d\sigma}{d\Omega} = \left| \sum_j b_j \exp(i\mathbf{Q} \cdot \mathbf{r}_j) \right|^2 \qquad \text{(elastic scattering)}. \qquad (1.32)$$

This follows from eqns (1.5), (1.8), (1.22), and (1.31) if one expresses $d\Omega = dS/R^2$, where $R$ is the distance from the sample to a small area element $dS$ (e.g. the detector) perpendicular to the scattered neutrons. For the case of an isolated bound nucleus, eqn (1.32) becomes

$$\frac{d\sigma}{d\Omega} = b^2 \qquad \text{(bound nucleus)}, \qquad (1.33)$$

which is just the total cross-section (1.29) divided by $4\pi$, as expected.

When the detection system has energy analysis we can sort the scattered neutrons according to their final energies, recording how many are scattered into a given range of final energies. We can then study inelastic scattering. For this purpose we use the *partial differential cross-section*,

$$\frac{d^2\sigma}{d\Omega dE_f} = \frac{\left(\begin{array}{c}\text{number of neutrons scattered per second}\\ \text{into solid angle } d\Omega \text{ about a given direction}\\ \text{with final energy between } E_f \text{ and } E_f + dE_f\end{array}\right)}{\Phi_0 \times d\Omega \times dE_f}. \qquad (1.34)$$

### 1.7.2 Absorption cross-section

Scattering is not the only outcome of the neutron–nucleus interaction. The next most likely process is absorption of the neutron into the nucleus, leaving the nucleus in an excited state which may at some later time relax by emission of a gamma-ray photon.

The removal of neutrons from the beam via nuclear reactions needs to be taken into account if we wish to determine absolute intensities correctly. To quantify this loss we define the *absorption cross-section $\sigma_a$* in the same way as the total scattering cross-section, eqn (1.28), except that we substitute the absorption rate in place of the scattering rate:

$$\sigma_a = \frac{(\text{total number of neutrons absorbed per second})}{\Phi_0}. \qquad (1.35)$$

Neutron absorption cross-sections vary greatly from element to element, and also amongst the isotopes of a given element. Some examples are given in Table 1.2. The amount of absorption also depends on the energy of the neutron. Often, the absorption is small compared to the scattering, and so can be neglected. Certain isotopes, however, absorb neutrons very strongly, and when this occurs it means that the energy of the neutron is close to a resonance in the nuclear excitation spectrum (see Fig. 1.17). Examples of nuclei which show resonance behaviour with thermal neutrons are $^{10}$B, $^{113}$Cd, and $^{157}$Gd.

Strong absorption can be a problem for neutron scattering experiments, but strongly absorbing materials find great practical use. For

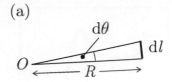

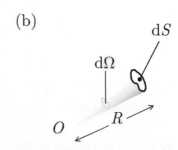

**Fig. 1.16** A solid angle is a three-dimensional analogue of an angle.
(a) An angle (in radians) is defined by $d\theta = dl/R$, and measures how large the line segment $dl$ appears when viewed from the point $O$. (b) A solid angle (in steradians) is defined by $d\Omega = dS/R^2$, and measures how large the area $dS$ appears from $O$.

**Table 1.2** Neutron absorption cross-sections of selected nuclei at a neutron energy of 25.3 meV. A more complete listing is given in Appendix A.

| nuclide | $\sigma_a$ (b) |
|---|---|
| $^1$H | 0.333 |
| $^3$He | 5,333 |
| $^4$He | 0.000 |
| $^6$Li | 940 |
| $^7$Li | 0.045 |
| $^{10}$B | 3,835 |
| $^{11}$B | 0.006 |
| $^{nat}$B | 767 |
| $^{nat}$O | 0.000 |
| $^{nat}$Sm | 5,922 |
| $^{nat}$Cd | 2,520 |
| $^{157}$Gd | 259,000 |
| $^{nat}$Gd | 49,700 |

$^{nat}$ natural isotopic mixture.

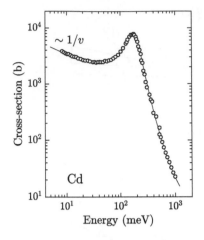

**Fig. 1.17** Energy dependence of the neutron cross-section of cadmium (Rainwater *et al.*, 1947; Brockhouse, 1953). The line is a fit to the single-level Breit–Wigner formula. A useful resource for nuclear data including plots of energy-dependent neutron cross-sections can be found at http://t2.lanl.gov/nis/data.shtml.

example, the nuclides $^3$He, $^6$Li, and $^{10}$B are used in neutron detectors and monitors, and pieces of Cd metal or $B_4C$ are often used to cover sample appendages in order to reduce stray scattering from the sample environment. Gd-impregnated paint is also used for this purpose.

When the neutron energy is well below a resonance the absorption cross-section is commonly observed to vary inversely with the neutron's speed. This so-called '$1/v$ law' comes about physically because the chance of capture by the nucleus increases in proportion to the time spent in the vicinity of the nuclear potential, which varies as $1/v$ or equivalently as $1/E^{1/2}$. Values of $\sigma_a$ are usually tabulated for neutrons of energy 25.3 meV, but using the $1/v$ law one can obtain the absorption at other energies by multiplying by a factor $(25.3/E)^{1/2}$, with $E$ expressed in meV. Fig. 1.17 illustrates the resonance and $1/v$ law features in the energy-dependent absorption cross-section of cadmium.

### 1.7.3    Complex scattering lengths

The process of neutron absorption also has an effect on the nuclear scattering amplitude. Usually this effect is small and can be neglected. Near an absorption resonance, however, the scattering amplitude can vary appreciably with neutron energy.

In Chapter 9, we will show that the presence of absorption requires us to express the scattering length as a complex number, $b = b' - \mathrm{i}b''$, with $b'$ and $b''$ the real and (minus) the imaginary parts of $b$, respectively. This means that the amplitude of the scattered wave in eqn (1.19) is now $|b|$ instead of $b$, and that there is an additional phase difference of $\tan^{-1}(b''/b')$ between the incident and scattered waves.

Both the real and imaginary parts of $b$ depend on neutron energy, and we will discover in Chapter 9 that $b''$ is related to the absorption cross-section by

$$b'' = \frac{k}{4\pi}\,\sigma_a, \qquad (1.36)$$

where $k = 2\pi/\lambda$ is the magnitude of the neutron wavevector.

### 1.7.4    Mean free path and transmission

The further a neutron beam travels through a material the more its intensity decreases as scattering and absorption processes remove neutrons from the beam. We can calculate the attenuation as a function of distance as follows. For simplicity, let there be only one type of atom, and let the number density of these atoms is $n$. In a short distance $\mathrm{d}x$ there are $n\,\mathrm{d}x$ atoms per unit area, and if the area of the neutron beam is $A$ then the number of atoms in the beam is $nA\,\mathrm{d}x$. Suppose the total scattering cross-section per atom is $\sigma_s$, then from eqn (1.28) the total number of neutrons scattered per second out of the beam within the length $\mathrm{d}x$ is $\Phi(x)\sigma_s nA\,\mathrm{d}x$, where $\Phi(x)$ is the flux at point $x$. This change in the number of neutrons in the beam can also be expressed in terms of the differential of the flux as $-A\,\mathrm{d}\Phi(x)$, where the minus sign reflects that the flux is decreasing with $x$. Including absorption as well as scattering we arrive at a differential equation for the flux,

$$\frac{d\Phi}{dx} = -\Phi(x)n(\sigma_s + \sigma_a). \tag{1.37}$$

Integration of eqn (1.37) leads to the well-known Beer–Lambert law,

$$\Phi(x) = \Phi(0)\exp\{-n(\sigma_s + \sigma_a)x\}. \tag{1.38}$$

The distance over which the flux decreases by a factor $1/e \approx 0.37$ is known as the *mean free path* $\lambda$. From eqn (1.38),

$$\lambda = \frac{1}{n(\sigma_s + \sigma_a)}. \tag{1.39}$$

The *transmission* $T$ of a sample is the fraction of the incident neutrons that survive the passage through the sample without being scattered or absorbed. If the path length through the sample is $d$ then from (1.38) the transmission is[20]

$$T = \exp\{-n(\sigma_s + \sigma_a)d\}. \tag{1.40}$$

If more than one type of atom is present then eqns (1.37)–(1.40) can be generalized by the replacement of $n(\sigma_s + \sigma_a)$ by $\sum_j n_j\{\sigma_s(j) + \sigma_a(j)\}$, where $n_j$ is the number density of atom type $j$. Equivalently, one can identify some convenient grouping of atoms and use the number density and cross-sections for this grouping.

# 1.8 Coherent and incoherent scattering

In Section 1.6.1 we introduced the scattering length $b$ as the amplitude of the scattered wave emanating from the nucleus. We also mentioned that $b$ varies from isotope to isotope and depends on the relative spin orientations of the neutron and nucleus. At first sight, therefore, it seems that we need information on the isotope and spin state of *every* nucleus in order to calculate the scattering cross-section, e.g. eqn (1.32).

Fortunately, nature comes to the rescue in most cases of practical interest by ensuring that the distribution of isotopes and nuclear spin orientations varies randomly from nucleus to nucleus. This randomness simplifies the calculation immensely. To see this, let us first consider the isotopic variation of $b$, neglecting for the time being the spin dependence.

## 1.8.1 Isotopic dependence of scattering length

Suppose that the nucleus at site $j$ in a sample has several different isotopes (see Fig. 1.18), and that the probability $p_i$ of a particular isotope $i$ is proportional to its natural abundance. This means that $p_i$ does not depend on which isotopes are to be found on the sites immediately adjacent to $j$ or, for that matter, on any other sites. In other words, the distribution of isotopes is random and uncorrelated.

[20]Reflections at the sample surfaces are negligible unless the neutron beam makes a glancing angle with the surface, see Chapter 9.

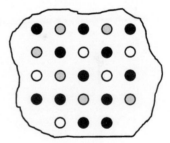

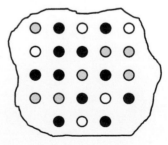

**Fig. 1.18** Random distribution of isotopes in two identical samples containing one type of atom with three different isotopes (black, grey, and white circles).

Now consider a second sample of the same material, identical to the first in every way except in the detail of which isotope occurs on each site. As explained above, the distribution of isotopes is governed only by their natural abundance, and so the arrangement of isotopes among the sites will almost certainly be different from the first sample. Even so, if the number of nuclei is large then we would expect the scattering from both samples to be virtually indistinguishable. This is because scattering takes place at random positions in the sample,[21] so after many scattering events the neutrons probe all parts of the sample and in effect perform an average over the isotopic composition of the sample.

Carrying this reasoning one stage further we assert that the scattering from one particular sample of a material is, to a very good approximation, the same as the scattering averaged over an ensemble of many such samples differing only in the arrangement of isotopes among the sites. This permits us to write eqn (1.32) as

$$\frac{d\sigma}{d\Omega} = \sum_{jk} \overline{b_j b_k} \, \exp\{i\mathbf{Q} \cdot (\mathbf{r}_k - \mathbf{r}_j)\}, \tag{1.41}$$

in which the squared modulus has been written out as a product two factors, one the complex conjugate of the other, and we have assumed for simplicity that the scattering lengths are real. The bar signifies the average over the natural isotopic distribution. As there are no correlations between sites we can express the average for 'self' $(j = k)$ terms and 'cross' $(j \neq k)$ terms in the summation as

$$\begin{aligned} \overline{b_j b_k} &= \overline{b_j^2} & (j = k), \\ &= \overline{b}_j \, \overline{b}_k & (j \neq k), \end{aligned} \tag{1.42}$$

where the mean and mean squared scattering lengths are obtained from the probabilities for each isotope in the usual way:

$$\begin{aligned} \overline{b} &= \sum_i p_i b_i \\ \overline{b^2} &= \sum_i p_i b_i^2. \end{aligned} \tag{1.43}$$

We can now separate the self and cross terms in eqn (1.41):

$$\begin{aligned} \frac{d\sigma}{d\Omega} &= \sum_{j \neq k} \overline{b_j b_k} \, \exp\{i\mathbf{Q} \cdot (\mathbf{r}_k - \mathbf{r}_j)\} + \sum_j \overline{b_j^2} \\ &= \sum_{jk} \overline{b}_j \overline{b}_k \, \exp\{i\mathbf{Q} \cdot (\mathbf{r}_k - \mathbf{r}_j)\} + \sum_j (\overline{b_j^2} - \overline{b}_j^2). \end{aligned} \tag{1.44}$$

To arrive at eqn (1.44) we have added self terms of the form $\overline{b}_j^2$ into the first summation so as to extend it to include all $j$ and $k$, and then to compensate we have subtracted these same terms from the second summation so that the overall result is the same.

[21]More precisely, scattering probes a sub-volume of the sample over which the neutron wave is coherent. For a bulk sample this 'scattering volume' is usually much smaller than the volume of the sample itself, so each neutron probes a different part of the sample.

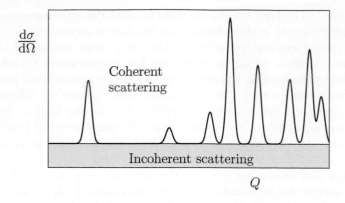

$\dfrac{\mathrm{d}\sigma}{\mathrm{d}\Omega}$

Coherent
scattering

Incoherent scattering

$Q$

**Fig. 1.19** Schematic plot to illustrate the coherent and incoherent components of the differential cross-section.

The division of the cross-section in the manner of (1.44) proves to be useful for the interpretation of the scattering, and defines the *coherent* and *incoherent* differential cross-sections:

$$\left(\frac{\mathrm{d}\sigma}{\mathrm{d}\Omega}\right)_{\text{coh}} = \sum_{jk} \bar{b}_j\,\bar{b}_k\,\exp\{i\mathbf{Q}\cdot(\mathbf{r}_k - \mathbf{r}_j)\}$$

$$= \left|\sum_j \bar{b}_j\,\exp(i\mathbf{Q}\cdot\mathbf{r}_j)\right|^2 \tag{1.45}$$

$$\left(\frac{\mathrm{d}\sigma}{\mathrm{d}\Omega}\right)_{\text{inc}} = \sum_j (\overline{b_j^2} - \bar{b}_j{}^2). \tag{1.46}$$

The coherent cross-section depends on the positions of the atoms relative to one another, and therefore contains information on the structure of the sample. The incoherent cross-section, on the other hand, depends only on which atoms are present, irrespective of their arrangement. As far as an experiment is concerned, any variation in the intensity with $\mathbf{Q}$ due to interference effects derives from the coherent part of the scattering, while the incoherent scattering simply produces a 'flat background'.[22] The two contributions are illustrated schematically in Fig. 1.19.

Although the incoherent scattering is often a nuisance in experiments this is not always the case. In Chapter 5 we will show that inelastic incoherent scattering can provide important information on the dynamical properties of materials.

[22] In fact, there will be a weak $\mathbf{Q}$ dependence to the incoherent scattering due to the vibrational motion of the atoms (the Debye–Waller factor), as explained in Chapter 5.

## 1.8.2   Nuclear spin dependence of scattering length

The neutron has spin $s_{\text{n}} = \frac{1}{2}$. Many nuclei also possess an intrinsic spin. When a neutron interacts with a nucleus of spin $I$ the strength of the interaction depends on the combined spin of the system, which according to quantum mechanics can be either $I + \frac{1}{2}$ or $I - \frac{1}{2}$. Hence, the scattering length takes on two possible values which we denote by $b_+$ and $b_-$.

We now need to find out how to take into account the spin dependence of the scattering length in the coherent and incoherent cross-sections. To keep things easy, lets first consider the case of a single isotope of one type of atom. Unless we are at extremely low temperatures ($\lesssim$ mK), where nuclear spin ordering can occur, we can safely assume that there are no correlations between the nuclear spin orientations on different sites. The atoms exhibit a random variation of scattering lengths, just as in the previous section, except now the different scattering lengths originate from the randomness in the nuclear spin orientations rather than in the isotope distribution. Hence, we can separate the cross-section into coherent and incoherent parts in the same manner as eqns (1.45) and (1.46) except that to calculate $\bar{b}$ and $\overline{b^2}$ from eqns (1.43) we use the probabilities $p_+$ and $p_-$ for the two combined spins $I + \frac{1}{2}$ and $I - \frac{1}{2}$:

$$\bar{b} = p_+ b_+ + p_- b_-$$
$$\overline{b^2} = p_+ b_+^2 + p_- b_-^2. \tag{1.47}$$

What determines $p_+$ and $p_-$? Quantum mechanics tells us that a state with angular momentum quantum number $J$ has $2J + 1$ possible orientations of the angular momentum vector. Accordingly, there are $2(I + \frac{1}{2}) + 1 = 2I + 2$ orientations of spin $I + \frac{1}{2}$, and $2(I - \frac{1}{2}) + 1 = 2I$ orientations of spin $I - \frac{1}{2}$. The probabilities $p_+$ and $p_-$ are proportional to the number of possible orientations, and $p_+ + p_- = 1$. Hence,

$$p_+ = \frac{I+1}{2I+1}$$
$$p_- = \frac{I}{2I+1}. \tag{1.48}$$

Now we proceed to the general case, where the sample contains atoms which have more than one isotope *and* each isotope may have non-zero spin. If we assume that the distributions of isotopes and spins are random throughout the sample, as above, then we now need to average over both the isotope and spin distributions. From eqns (1.43), (1.47), and (1.48),

$$\bar{b} = \sum_i p_i \left\{ \frac{I_i+1}{2I_i+1} (b_+)_i + \frac{I_i}{2I_i+1} (b_-)_i \right\}$$
$$\overline{b^2} = \sum_i p_i \left\{ \frac{I_i+1}{2I_i+1} (b_+)_i^2 + \frac{I_i}{2I_i+1} (b_-)_i^2 \right\}. \tag{1.49}$$

These formulae enable us to work out $\bar{b}_j$ and $(\overline{b_j^2} - \bar{b}_j^2)$ for all sites $j$ in a sample as required for the calculation of the coherent and incoherent cross-sections from eqns (1.45) and (1.46). Should any of the scattering lengths be complex then all that changes in eqns (1.49) is that $(b_+)_i^2$ and $(b_-)_i^2$ are replaced by $|b_+|_i^2$ and $|b_-|_i^2$, and hence $\overline{b^2}$ becomes $\overline{|b|^2}$.

$|b|^2 = bb^* = b^*b$

## 1.8.3   Other sources of incoherence

The key point is that coherent scattering depends on the mean scattering lengths, whereas incoherent scattering is proportional to the variance of the scattering lengths. Without random fluctuations in the scattering length on a given site there is no incoherent scattering.

As we have seen, randomness in the scattering from the nucleus arises from the presence of different isotopes and from the random orientation of nuclear spins. Hence, we often use the terms *isotopic incoherence* and *spin incoherence*. But these are not the only possible sources of randomness. If a material contains two or more atoms of similar size and ionization state then it is possible for them to become scrambled. A well-known example is $\beta$-brass, which is an alloy of Cu and Zn. Below 740 K the atoms are ordered on a regular periodic array, but above 740 K the material becomes a random alloy, meaning there is an equal chance of finding either Cu or Zn on any site. As the nuclei of Cu and Zn have different scattering lengths this randomness gives rise to incoherent scattering. We call this *chemical incoherence*. If chemical incoherence is important then we must amend (1.49) to include an average over the relative abundance of each atomic species.

Another type of incoherent scattering is magnetic scattering of neutrons from a paramagnet. In an ideal paramagnet the atomic magnetic moments are randomly oriented, causing incoherent scattering in an analogous way to nuclear spin incoherence. Magnetic incoherent scattering is $Q$-dependent due to the magnetic form factor — see Chapter 6.

For completeness, it is worth noting that incoherent scattering is rarely encountered in X-ray and electron scattering experiments. This is because the interaction of X-rays and electrons with matter is dominated by charge scattering, and usually there are no fluctuations in charge density on a given site. Possible exceptions are chemical incoherence, and random variations in the ionization state of the atoms.

## 1.8.4   Tables of scattering lengths and cross-sections

In practice, it is rarely necessary to evaluate eqns (1.49) because detailed tabulations exist, including information on most of the isotopes. Such a compilation is given in Appendix A. The quantities tabulated include (i) $\bar{b}$, which is known as the *coherent bound scattering length*, (ii) the bound-atom total scattering cross-section $\sigma_s$ defined by[23]

$$\sigma_s = 4\pi \overline{|b|^2}, \tag{1.50}$$

and (iii) the absorption cross-section $\sigma_a$. Values are given for some individual isotopes as well as for the natural isotopic mix.

The total scattering cross-section is conventionally divided into coherent and incoherent parts, $\sigma_s = \sigma_{coh} + \sigma_{inc}$, defined by

$$\sigma_{coh} = 4\pi |\bar{b}|^2 \tag{1.51}$$

and

$$\sigma_{inc} = 4\pi (\overline{|b|^2} - |\bar{b}|^2). \tag{1.52}$$

[23] Expressions (1.50)–(1.52) are written for a general complex scattering length.

From the tabulated values of $\bar{b}$ and $\sigma_s$ one can use (1.50) and (1.52) to work out $\sigma_{inc}$ for individual isotopes, or for a mixture of isotopes. In the latter case one must use the isotopically averaged values of $\bar{b}$ and $\sigma_s$ to work out $\sigma_{inc}$ for the mixture because the values of $\sigma_{inc}$ for each isotope do not add linearly — see eqns (1.49). Providing there is no other source of incoherence one can, however, calculate the total incoherent scattering from a sample simply by adding up the incoherent scattering from each atom present, as may be seen from eqn (1.46).

### 1.8.5   Scattering properties of hydrogen

To illustrate the use of eqns (1.49)–(1.52) we will evaluate the various bound scattering cross-sections of the important element hydrogen.

Hydrogen has three isotopes, $^1$H (protium), $^2$H (deuterium), and $^3$H (tritium). The relative abundance is $^1$H (99.985%), $^2$H (0.015%), and a negligible amount of the radioactive isotope $^3$H. The $^1$H nucleus is just a proton, which is a spin-$\frac{1}{2}$ particle. When a neutron interacts with a proton the combined spin is either 0 or 1. These correspond to singlet and triplet states, respectively, since they have degeneracies of 1 and 3. The measured values of the singlet and triplet scattering lengths are

$$b_- = -47.5\,\text{fm} \quad \text{and} \quad b_+ = 10.85\,\text{fm}.$$

If we had 100% $^1$H then the sum over isotopes in eqns (1.49) would contain only one term, and the average over spin states with $I = \frac{1}{2}$ gives

$$\bar{b} = \tfrac{3}{4}b_+ + \tfrac{1}{4}b_- = -3.74\,\text{fm},$$

and

$$\overline{b^2} = \tfrac{3}{4}b_+^2 + \tfrac{1}{4}b_-^2 = 6.49\,\text{b}.$$

Hence, from eqns (1.50)–(1.52),

$$\left.\begin{array}{l} \sigma_s = 81.7\,\text{b} \\ \sigma_{coh} = \phantom{0}1.8\,\text{b} \\ \sigma_{inc} = 79.9\,\text{b} \end{array}\right\} {}^1\text{H}.$$

The $^2$H nucleus contains a proton and a neutron, and the total nuclear spin is $I = 1$. The measured scattering lengths are

$$b_- = 0.98\,\text{fm} \quad \text{and} \quad b_+ = 9.53\,\text{fm},$$

and a similar procedure to that above gives

$$\left.\begin{array}{l} \bar{b} = 6.67\,\text{fm} \\ \sigma_s = 7.6\,\text{b} \\ \sigma_{coh} = 5.6\,\text{b} \\ \sigma_{inc} = 2.0\,\text{b} \end{array}\right\} {}^2\text{H}.$$

Several features of the scattering from hydrogen are worth noting. Firstly, the scattering from a sample containing natural hydrogen displays an abnormally large incoherent component arising from the large difference between $b_+$ and $b_-$. This incoherent scattering is often a nuisance because it acts as a source of unwanted background. For this reason, neutron scatterers usually try to minimize the amount of hydrogen-containing substances in the neutron beam during their experiments.[24] Second, the incoherent cross-section of $^2$H is much smaller than that of $^1$H, and so another way to reduce the incoherent background is to prepare hydrogen-containing samples with $^2$H instead of $^1$H. Finally, the coherent scattering lengths of $^1$H and $^2$H are quite different, and indeed are of opposite sign. This means that the coherent scattering from certain parts of a sample can be enhanced through selective replacement of $^1$H by $^2$H, a process known as *isotopic labelling*. The ability to do isotopic labelling experiments is an extremely powerful feature of neutron scattering, and will be discussed further in Chapter 2.

[24] An exception is experiments which exploit the large incoherent cross-section of hydrogen to study dynamics — see Chapter 5.

### 1.8.6  Cross-sections for a free atom

Most of the time it is safe to assume that the atoms in systems studied by neutron scattering are bound.[25] This assumption will break down if the kinetic energy of the incident neutron is larger than the binding potential that holds the atoms in place, in which case the atoms will recoil under the impact from a neutron.

[25] This does not imply, however, that the atoms are necessarily stationary.

By transforming to the centre-of-mass frame one can show that the problem of a neutron in collision with a free atom is equivalent to one in which the atom is bound (i.e. has infinite mass) and the neutron has a reduced mass $\mu = m_\mathrm{n}M/(m_\mathrm{n} + M)$, where $M$ is the atomic mass. As we shall show in Section 4.1.1, the effective potential $V_\mathrm{N}(\mathbf{r})$ that describes the interaction of a nucleus with a bound atom depends on the ratio $b/m_\mathrm{n}$, and since $V_\mathrm{N}(\mathbf{r})$ must be independent of reference frame the scattering length for a free atom must be $b \times \mu/m_\mathrm{n}$. It follows that the scattering cross-sections for free and bound atoms are related by

$$\sigma_\mathrm{s}(\text{free}) = \left(\frac{A}{A+1}\right)^2 \sigma_\mathrm{s}(\text{bound}), \qquad (1.53)$$

where $A = M/m_\mathrm{n}$ is essentially the relative atomic mass. The effect is large for hydrogen ($A = 1$), but declines rapidly with atomic number.

As an example, Fig. 1.20 shows the total cross-section of H in gaseous 1,3-butadiene ($C_4H_6$) at room temperature, determined via (1.40) from transmission measurements. For neutron energies above about $1\,\mathrm{eV}$ the nuclei recoil freely and the data approach the cross-section for free protons, which from (1.53) with $A = 1$ is $\sigma_\mathrm{s}(\text{free}) = \sigma_\mathrm{s}(\text{bound})/4 = 20.4\,\mathrm{b}$. Below $1\,\mathrm{eV}$ the cross-section increases, and if the H atoms were stationary then the cross-section would tend towards $\sigma_\mathrm{s}(\text{bound}) = 81.7\,\mathrm{b}$. In reality, the low-energy cross-section per proton exceeds $\sigma_\mathrm{s}(\text{bound})$ owing to the effects of thermally excited intra-molecular vibrations and the translational motion of the molecules in the gas.

In the case of nuclear absorption, the cross-section does not depend on the mass of the neutron and so

$$\sigma_\mathrm{a}(\text{free}) = \sigma_\mathrm{a}(\text{bound}). \qquad (1.54)$$

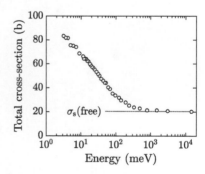

**Fig. 1.20** Total cross-section per proton of H in $C_4H_6$ at room temperature as a function of neutron energy. (After Melkonian, 1949.)

# Chapter summary

- The basic properties of the neutron have been introduced, including their wave–particle properties and spin.
- The principal methods for neutron production have been covered, especially reactor and spallation sources.
- Neutrons interact with nuclei via the strong nuclear force and with unpaired electrons via electromagnetic forces.
- In a scattering process,

$$\mathbf{Q} = \mathbf{k}_i - \mathbf{k}_f$$
$$\hbar\omega = E_i - E_f,$$

where $\hbar\mathbf{Q}$ is the momentum transfer and $\hbar\omega$ is the energy transfer.
- Scattering is elastic if $\hbar\omega = 0$, and inelastic if $\hbar\omega \neq 0$.
- The nuclear scattering length $b$ is the (complex) amplitude of the spherical neutron wave scattered from a nucleus. It varies from isotope to isotope, and depends on the relative orientations of the neutron and nuclear spins.
- Some nuclei absorb neutrons with energies below $\sim 1\,\mathrm{eV}$ very strongly. These include $^3\mathrm{He}$, $^6\mathrm{Li}$, $^{10}\mathrm{B}$, $^{113}\mathrm{Cd}$, and $^{157}\mathrm{Gd}$, which are used in neutron detectors and shielding materials.
- The probability of scattering or absorption is represented by cross-sections (total, differential, double differential, absorption), which represent the effective area that the neutrons must hit for the process to occur.
- Neutron scattering cross-sections can be divided into coherent and incoherent parts. For elastic scattering from a collection of bound nuclei, these are

$$\left(\frac{d\sigma}{d\Omega}\right)_{\mathrm{coh}} = \sum_{jk} \overline{b}_j^* \,\overline{b}_k \, \exp\{i\mathbf{Q} \cdot (\mathbf{r}_k - \mathbf{r}_j)\}$$
$$\left(\frac{d\sigma}{d\Omega}\right)_{\mathrm{inc}} = \sum_j \frac{(\sigma_{\mathrm{inc}})_j}{4\pi}.$$

The coherent elastic scattering contains inter-site interference effects and gives information on structure. The incoherent elastic scattering gives a featureless background.
- Nuclear coherent elastic scattering can also be expressed as the Fourier transform of the scattering length density, $n_b(\mathbf{r})$:

$$\left(\frac{d\sigma}{d\Omega}\right)_{\mathrm{coh}} = \left| \int n_b(\mathbf{r}) \exp(i\mathbf{Q} \cdot \mathbf{r}) \, d^3\mathbf{r} \right|^2 .$$

# Further reading

More details about neutron production, including a brief history, can be found in Willis and Carlile (2009).

A review of compact accelerator-driven neutron sources has been given by Anderson *et al.* (2016).

# Exercises

(1.1) Obtain the following handy formulae for neutrons:

$$E = \frac{0.818}{\lambda^2} = 0.02072k^2,$$

where $E$ is in units of meV, $\lambda$ is in nm, and $k$ is in nm$^{-1}$.

(1.2) Calculate the energy (in meV) of (a) a neutron of wavelength 0.2 nm and (b) a photon of wavelength 0.2 nm.

(1.3) At pulsed sources, the neutron's time-of-flight $t$ measured over a known path length $L$ is used to determine its wavelength $\lambda$. Show that

$$\lambda = \frac{ht}{m_{\mathrm{n}}L},$$

where $h$ is Planck's constant.

(1.4) In a partially polarized neutron beam, 90% of the neutrons have spin 'up' and 10% have spin 'down'. What is the beam polarization $P$?

(1.5) For the one-dimensional diffraction grating shown in Fig. 1.13, show the $Q_x = k \sin \phi$, where the $x$-axis is along the line of nuclei.

(1.6) Show that the coherent differential cross-section for a rigid, homonuclear, diatomic molecule is

$$\left(\frac{\mathrm{d}\sigma}{\mathrm{d}\Omega}\right)_{\mathrm{coh}} = 4\bar{b}^2 \cos^2(\tfrac{1}{2}Qd\cos\theta),$$

where $d$ is the separation of the atoms and $\theta$ is the angle between $\mathbf{Q}$ and the axis of the molecule [the cross-section for a gas of diatomic molecules with random orientations is obtained in Section 2.5.1].

(1.7) Calculate the mean free path for neutrons of energy 35 meV in (a) aluminium, (b) water, and (c) cadmium.

(1.8) The table below gives the experimental values of the natural abundance ($p$), spin ($I$) and scattering lengths ($b_{\pm}$ in fm) of the individual isotopes ($i$) of nickel. With the exception of $^{61}$Ni, the isotopes have zero spin. Calculate the values of $\bar{b}$ and $\sigma_{\mathrm{inc}}$ for a natural nickel sample.

| $i$ | $p_i$ | $I_i$ | $(b_+)_i$ | $(b_-)_i$ |
|---|---|---|---|---|
| $^{58}$Ni | 0.683 | 0 | 14.4 | 14.4 |
| $^{60}$Ni | 0.261 | 0 | 2.8 | 2.8 |
| $^{61}$Ni | 0.011 | 3/2 | 4.6 | 12.6 |
| $^{62}$Ni | 0.036 | 0 | $-8.7$ | $-8.7$ |
| $^{64}$Ni | 0.009 | 0 | $-0.4$ | $-0.4$ |

(1.9) Work out the composition of an $H_2O/D_2O$ mixture with zero average coherent scattering cross-section (the so-called *semi-transparent mixture*).

# Diffraction in the Static Approximation

2

Neutron diffraction, or coherent elastic scattering,[1] is used to obtain information on the structure of materials at the atomic scale. In this chapter we shall derive expressions for the coherent elastic scattering cross-section for the most frequently occurring forms of condensed matter. For simplicity, we shall assume that the structures are perfectly rigid so that the neutron energy before and after scattering is exactly the same. This approximation is known as the *static approximation*.[2] Hence, our starting point will be the coherent differential cross-section in the form derived in Chapter 1,

$$\left(\frac{\mathrm{d}\sigma}{\mathrm{d}\Omega}\right)_{\mathrm{coh}} = \sum_{jk} \overline{b}_j^* \, \overline{b}_k \, \exp\{\mathrm{i}\mathbf{Q} \cdot (\mathbf{r}_k - \mathbf{r}_j)\}. \tag{2.1}$$

The neglect of atomic motion is acceptable since our aim is to bring out the principal features of the diffraction from different types of structure. The reader should bear in mind, however, that atoms in condensed matter are never truly stationary even at absolute zero temperature. Dynamic effects will be considered in Chapters 3 and 5.

Crystal structure determination is one of the most important applications of neutron scattering, and the first part of this chapter is concerned with diffraction from crystalline materials. After this, we explore what happens when the periodicity is removed, for example due to reduced dimensionality or when materials have only short-range order, such as in liquids and glasses. Finally we consider diffraction from isolated and condensed molecules.

## 2.1 Crystals

In order to develop the theory of neutron diffraction from crystals we need to be familiar with the elementary concepts of crystallography.

---

[1]The terms *diffraction* and *coherent elastic scattering* are used synonymously.
[2]The validity and implications of the static approximation are discussed in more detail in Sections 3.8.4 and 5.3.4.

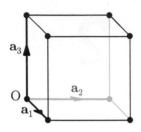

**Fig. 2.1** Lattice translation vectors for the primitive cubic lattice.

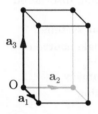

**Fig. 2.2** Lattice translation vectors for the primitive tetragonal lattice.

[3]Auguste Bravais (1811–1863).

[4]An earlier enumeration by Moritz Frankenheim had come up with fifteen distinct lattice types, but Bravais showed that two of Frankenheim's lattices were in fact equivalent.

### 2.1.1 Crystal lattice

A crystal is a periodic structure built from a large number of identical units that stack together to fill space. The underlying periodicity is represented by a *lattice*, which is a periodic array of points such that each point is in an identical environment. A lattice may be generated mathematically by

$$\mathbf{l} = n_1\mathbf{a}_1 + n_2\mathbf{a}_2 + n_3\mathbf{a}_3, \qquad (2.2)$$

where $\mathbf{l}$ represents the position vector of a lattice point with respect to some arbitrarily chosen origin which is also a lattice point, $\mathbf{a}_1$, $\mathbf{a}_2$, and $\mathbf{a}_3$ are three non-coplanar vectors known as *primitive lattice vectors* or *lattice translation vectors*, and $n_1$, $n_2$, and $n_3$ are integers.

A simple type of lattice is the primitive cubic lattice, for which $\mathbf{a}_1$, $\mathbf{a}_2$, and $\mathbf{a}_3$ are mutually perpendicular and all of the same length, Fig. 2.1. Such a lattice has a high degree of symmetry. As well as having the translational symmetry expressed by eqn (2.2) the primitive cubic lattice also has the same rotational and reflection symmetries as a cube.

We can generate other lattice types by varying the lengths of the primitive lattice vectors and/or the angles between them. For example, starting with a primitive cubic lattice, if we make the length of $\mathbf{a}_3$ different from that of $\mathbf{a}_1$ and $\mathbf{a}_2$ we would have a *primitive tetragonal* lattice, Fig. 2.2. It follows that cubic and tetragonal lattices do not share all the same symmetry elements. For example, cubic lattices have 4-fold rotational symmetry about axes parallel to $\mathbf{a}_1$, $\mathbf{a}_2$ and $\mathbf{a}_3$ passing through the lattice points, whereas tetragonal lattices have a 4-fold axis parallel to $\mathbf{a}_3$ but only 2-fold axes parallel to $\mathbf{a}_1$ and $\mathbf{a}_2$. This example illustrates how symmetry can be used to characterize different lattice types.

The first correct enumeration of all possible lattice types was done by Bravais[3] in around 1848. He showed that there are only fourteen distinct types of lattice in three dimensions.[4] These are known as the *Bravais lattices*. In two dimensions there are only five distinct lattice types. The principal features of the Bravais lattices are given in Table 2.1.

### 2.1.2 Motif

The arrangement of atoms associated with a lattice point is known as the *motif* or *basis*. As every lattice point is in an identical environment, a knowledge of the lattice type together with the motif specifies the crystal structure completely.

### 2.1.3 Unit cell

A *unit cell* is a fundamental building block of a crystal. When many identical unit cells are stacked together they fill space completely and generate the whole crystal.

For a given crystal structure there are many different choices of unit cell. The smallest of these contain only one lattice point and are known as *primitive* unit cells. By 'contain only one lattice point' we mean that

**Table 2.1** Principal features of the Bravais lattices. The fourteen lattices are grouped into seven lattice systems by their conventional unit cell.

| Lattice system | Conventional unit cell | Lattice symbol[a] | Number of lattice points in conventional unit cell |
|---|---|---|---|
| triclinic | $a \neq b \neq c$ <br> $\alpha \neq \beta \neq \gamma$ | P | 1 |
| monoclinic | $a \neq b \neq c$ <br> $\alpha = \gamma = 90°; \beta > 90°$ | P <br> C(A) | 1 <br> 2 |
| orthorhombic | $a \neq b \neq c$ <br> $\alpha = \beta = \gamma = 90°$ | P <br> C(A,B) <br> I <br> F | 1 <br> 2 <br> 2 <br> 4 |
| tetragonal | $a = b \neq c$ <br> $\alpha = \beta = \gamma = 90°$ | P <br> I | 1 <br> 2 |
| rhombohedral[b] | $a = b \neq c$ <br> $\alpha = \beta = 90°, \gamma = 120°$ | R | 3 |
| hexagonal[b] | $a = b \neq c$ <br> $\alpha = \beta = 90°, \gamma = 120°$ | P | 1 |
| cubic | $a = b = c$ <br> $\alpha = \beta = \gamma = 90°$ | P <br> I <br> F | 1 <br> 2 <br> 4 |

[a] Symbols: P  primitive
  I  body-centred
  F  all-face-centred
  C(A,B)  base-centred on the C (or A, or B) faces
  R  rhombohedrally centred hexagonal cell containing 3 lattice points
    with fractional coordinates $0, 0, 0;\ \frac{2}{3}, \frac{1}{3}, \frac{1}{3};\ \frac{1}{3}, \frac{2}{3}, \frac{2}{3}$

[b] A rhombohedral cell with $a = b = c$, $\alpha = \beta = \gamma \neq 90°$ is also used. For the rhombohedral lattice the cell is primitive, but for the hexagonal lattice it contains three lattice points with fractional coordinates $0, 0, 0;\ \frac{2}{3}, \frac{1}{3}, \frac{1}{3};\ \frac{1}{3}, \frac{2}{3}, \frac{2}{3}$.

the number of unit cells in the crystal is equal to the number of lattice points.

Primitive unit cells do not always have a regular shape, and sometimes it is possible to construct a larger unit cell with a simpler geometry. This latter type of unit cell is called a *conventional* unit cell, and is often employed in preference to a primitive cell to make calculations easier.

It is often the case in practice that the unit cell, whether primitive or conventional, has the shape known as a *parallelepiped* shown in Fig. 2.3. The cell has basis vectors **a**, **b**, and **c** (not necessarily the same as the primitive lattice vectors $\mathbf{a_1}$, $\mathbf{a_2}$, $\mathbf{a_3}$) with angles of $\alpha$, $\beta$, $\gamma$ between them, as indicated. The lengths of **a**, **b**, **c**, together with the angles $\alpha$, $\beta$, $\gamma$, are known as the *cell parameters*. The cell volume is given by

$$v_0 = \mathbf{a} \cdot \mathbf{b} \times \mathbf{c} = \mathbf{b} \cdot \mathbf{c} \times \mathbf{a} = \mathbf{c} \cdot \mathbf{a} \times \mathbf{b}. \qquad (2.3)$$

The use of a conventional unit cell is illustrated in Fig. 2.4 for the two-dimensional body-centred rectangular lattice generated by primitive lattice vectors $\mathbf{a_1}$ and $\mathbf{a_2}$. Two primitive unit cells are shown, together

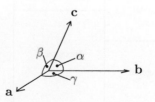
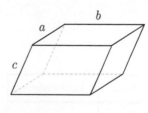

**Fig. 2.3** Geometry and cell parameters for a parallelepiped-shaped unit cell.

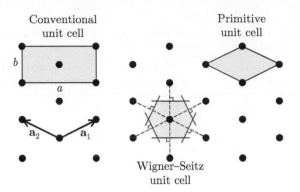

**Fig. 2.4** The two-dimensional body-centred rectangular lattice, showing two different primitive unit cells and the conventional rectangular (non-primitive) unit cell.

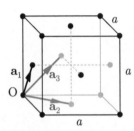

**Fig. 2.5** Lattice translation vectors for the face-centred cubic lattice. The conventional cubic unit cell is also shown.

[5]The origin of a unit cell can be located anywhere, but it is usually placed on a lattice point for convenience.

| Crystal system | Lattice system |
|---|---|
| Trigonal | Rhombohedral |
| | Hexagonal |
| Hexagonal | |

**Fig. 2.6** Lattice types for the trigonal and hexagonal crystal systems.

with the rectangular conventional unit cell of dimensions $a \times b$. The latter is seen to contain 2 lattice points. One of the primitive unit cells shown is the *Wigner–Seitz cell*. This cell is defined to be the area (volume in three dimensions) enclosed by the set of planes that perpendicularly bisect the vectors from the origin to the neighbouring lattice points. The Wigner–Seitz cell has the same symmetry as the lattice, and is best known in the context of wave propagation in crystalline solids where it is used to define the first Brillouin zone.

A well-known example of a three-dimensional lattice whose conventional cell is not primitive is the face-centred cubic (cubic F) lattice. The conventional cell is a cube of side $a$ $(= b = c)$, whereas the primitive lattice vectors connect a lattice point at the corner of the cube to lattice points at the centres of each of the nearest faces, as shown in Fig. 2.5.[5]

### 2.1.4  Crystal symmetry

In the classification of crystals we refer to two types of symmetry, *point symmetry* and *translational symmetry*. Point symmetry operations are geometric transformations that leave at least one point fixed. The set of point symmetries that transform an object back onto itself forms a *point group*. The symmetry operations in a point group can include only rotations, reflections, inversion, and combinations of these. If the object is a crystal then the point group symmetries must be compatible with the translational symmetry of the lattice. The 14 Bravais lattices permit only 1-, 2-, 3-,- 4-, and 6-fold rotational symmetries, and restrict the number of distinct crystallographic point groups to 32 in total.

Point symmetry is used to classify crystals into *crystal systems*. There are seven crystal systems in total, six of which have the same names as the corresponding Bravais lattices: triclinic, monoclinic, orthorhombic, tetragonal, hexagonal, and cubic. For example, cubic crystals always have one of the three cubic lattice types: cubic P (primitive), cubic I (body-centred), or cubic F (face-centred), see Table 2.1. The seventh crystal system is the trigonal system. Rather confusingly, trigonal crystals can have either the rhombohedral lattice or the hexagonal lattice, as represented in Fig. 2.6. The trigonal and hexagonal systems differ in that trigonal crystals have one 3-fold or $\bar{3}$-fold axis of symmetry whereas hexagonal crystals have one 6-fold or $\bar{6}$-fold axis of symmetry.[6]

The overall symmetry of a crystal includes both the lattice translations and the point symmetries. In addition, two other types of symmetry operation may be present, the screw and glide operations. These are combinations of a translation through a fraction of a lattice vector with either a rotation (screw) or a reflection (glide). The symmetry groups formed from the complete set of crystal symmetries are called *space groups* or *Fedorov groups*. There are 230 distinct space group types, 73 of which do not contain screw or glide operations (the *symmorphic* space groups) and 157 that do (the *non-symmorphic* space groups).

### 2.1.5   Atomic coordinates, directions, and planes

There are widely used conventions for the notation to describe the geometry of crystals, and we shall adhere to them in this book.

The coordinates of an atom in the unit cell are given as fractions of the cell parameters, so $x, y, z$ (no parentheses) represents an atom displaced from the origin by the vector $x\mathbf{a} + y\mathbf{b} + z\mathbf{c}$. For example, an atom right in the centre of the unit cell would have fractional coordinates $\frac{1}{2}, \frac{1}{2}, \frac{1}{2}$.

Crystallographic directions are indicated by a vector written in square brackets as $[uvw]$ corresponding to $u\mathbf{a} + v\mathbf{b} + w\mathbf{c}$, and crystallographic planes are represented by their *Miller indices*. The Miller indices $(hkl)$ describe a set of parallel planes such that if one of the planes passes through the origin $O$ then an adjacent plane intersects the $\mathbf{a}$, $\mathbf{b}$ and $\mathbf{c}$ axes at $a/h$, $b/k$ and $c/l$, respectively (see Fig. 2.7).

### 2.1.6   The reciprocal lattice

For any real-space lattice there exists a dual lattice in wavevector space known as the *reciprocal lattice*. In the simplest terms, the reciprocal lattice is the Fourier transform of the lattice, and is obtained from the solutions to the equation (see Section B.4.1)

$$\exp(i\mathbf{Q} \cdot \mathbf{l}) = 1 \text{ for all } \mathbf{l}. \tag{2.4}$$

We shall denote a $\mathbf{Q}$ vector that satisfies (2.4) by $\mathbf{G}$, and in Section B.4.1 we show that the set of all $\mathbf{G}$ vectors may be generated by

$$\mathbf{G} = m_1\mathbf{a}_1^* + m_2\mathbf{a}_2^* + m_3\mathbf{a}_3^*, \tag{2.5}$$

where $m_1$, $m_2$ and $m_3$ are integers and $\mathbf{a}_1^*$, $\mathbf{a}_2^*$ and $\mathbf{a}_3^*$ are basis vectors in $\mathbf{Q}$-space which depend on the primitive lattice vectors $\mathbf{a}_1$, $\mathbf{a}_2$ and $\mathbf{a}_3$ through the relations

$$\mathbf{a}_1^* = \frac{2\pi}{v_0} \mathbf{a}_2 \times \mathbf{a}_3, \quad \mathbf{a}_2^* = \frac{2\pi}{v_0} \mathbf{a}_3 \times \mathbf{a}_1, \quad \mathbf{a}_3^* = \frac{2\pi}{v_0} \mathbf{a}_1 \times \mathbf{a}_2. \tag{2.6}$$

Here $v_0 = \mathbf{a}_1 \cdot \mathbf{a}_2 \times \mathbf{a}_3$, etc, is the volume of the primitive unit cell of the real-space lattice. It follows from (2.6) that

$$\mathbf{a}_1 \cdot \mathbf{a}_1^* = \mathbf{a}_2 \cdot \mathbf{a}_2^* = \mathbf{a}_3 \cdot \mathbf{a}_3^* = 2\pi,$$

and

$$\mathbf{a}_1 \cdot \mathbf{a}_2^* = \mathbf{a}_1 \cdot \mathbf{a}_3^* = \mathbf{a}_2 \cdot \mathbf{a}_1^* = \ldots = 0. \tag{2.7}$$

[6]The symbol $\bar{n}$ denotes an *improper* rotation, which is a proper rotation $n$ followed by the inversion operation $\bar{1}$.

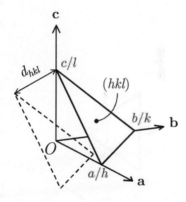

**Fig. 2.7** Representation of crystal planes by Miller indices $(hkl)$. The inter-planar spacing is denoted by $d_{hkl}$.

It is apparent from eqn (2.2) that (2.5) defines a lattice in $\mathbf{Q}$-space, and as $\mathbf{Q}$-space is often referred to as *reciprocal space* the lattice is called the reciprocal lattice. Similarly, the $\mathbf{G}$ vectors are called *reciprocal lattice vectors*.

The reciprocal lattice is a very useful concept in many areas of solid state science.[7] It is a fundamental property of any periodic structure, and arises naturally in the description of phenomena that involve waves propagating in a crystal. The reciprocal lattice is linked directly to the lattice in 'real' or 'direct' space via eqns (2.6), so even if it might at first sight appear something of a mathematical abstraction the reciprocal lattice is straightforward to construct in practice.

[7]The reciprocal lattice concept was first used to interpret X-ray diffraction patterns by Paul Peter Ewald (1888–1985) in 1913. At the same time, he introduced the sphere of diffraction now known as the Ewald sphere (see Section 10.2.4).

### 2.1.7 Reciprocal lattice units

A general wavevector in reciprocal space,

$$\mathbf{Q} = u_1 \mathbf{a}_1^* + u_2 \mathbf{a}_2^* + u_3 \mathbf{a}_3^*, \tag{2.8}$$

is often represented by the shorthand $(u_1, u_2, u_3)$. The dimensionless coefficients $u_1$, $u_2$, and $u_3$ are the components of $\mathbf{Q}$ in what are known as *reciprocal lattice units*, or r.l.u. for short. For example, $(\frac{1}{2}, \frac{1}{2}, \frac{1}{2})$ expresses the vector $\frac{1}{2}\mathbf{a}_1^* + \frac{1}{2}\mathbf{a}_2^* + \frac{1}{2}\mathbf{a}_3^*$ in r.l.u. When $u_1$, $u_2$, and $u_3$ are integers, $(u_1, u_2, u_3)$ represents a reciprocal lattice vector.

## 2.2 Diffraction from crystals

### 2.2.1 Bragg's Law

The idea that the atoms in crystals lie on planes is central to our understanding of diffraction. Consider a set of crystallographic planes with Miller indices $(hkl)$, and let $d_{hkl}$ be the spacing between adjacent planes (Fig. 2.7). We can imagine that the planes act like mirrors, reflecting a small fraction of the neutrons incident on them. The reflection process is *specular*, i.e. the angle of incidence and the angle of reflection are the same, as shown in Fig. 2.8.

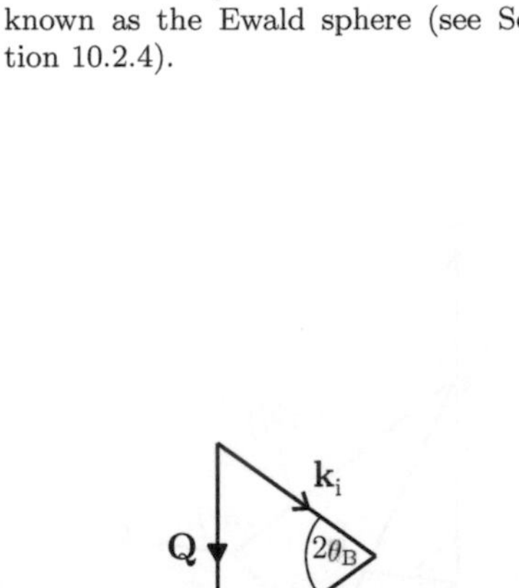

**Fig. 2.8** Bragg reflection from a set of parallel planes. Length $AO = OB = d_{hkl} \sin\theta_{\mathrm{B}}$, so the path difference $AOB$ between reflections from adjacent planes is $2d_{hkl} \sin\theta_{\mathrm{B}}$. Top: scattering triangle.

Applying the same approach as was used with the one-dimensional diffraction grating in Section 1.6.2, we search for the condition that the reflected waves interfere constructively to produce a large resultant amplitude. This occurs when the path difference for waves reflected from adjacent planes is a whole number $n$ of wavelengths, and from Fig. 2.8 the condition is

$$n\lambda = 2d_{hkl} \sin\theta_{\mathrm{B}}, \tag{2.9}$$

where $\theta_{\mathrm{B}}$ is the angle that the incident/reflected beams make with the planes (the *Bragg angle*). The value of $n$ is known as the *order* of diffraction, for example $n = 2$ corresponds to second order. Equation (2.9) is the well-known *Bragg Law* of diffraction[8] which provides us with the condition that must be satisfied if we are to observe a (Bragg) peak in the diffracted intensity.

[8]William L. Bragg (1890–1971) derived the equation now known as Bragg's law in 1912, while still a research student at Cambridge University. He collaborated with his father, William H. Bragg (1862–1942), to develop X-ray diffraction techniques for solving crystal structures. The father-and-son team shared the Nobel Prize in Physics in 1915.

## 2.2.2  Laue formulation of diffraction

Bragg's law allows us to determine the interplanar spacings in a crystal from a set of observed Bragg peak angles. From these data one can deduce the lattice type and the unit cell dimensions. Bragg's law does not, however, predict the intensity of the Bragg peaks, and because of this we have only limited information on the arrangement of atoms in the unit cell. To overcome this limitation we now proceed to evaluate the coherent elastic differential cross-section for a crystal. We continue to assume rigidly bound atoms.

In evaluating the cross-section we need to perform a sum over every atom (nucleus, to be precise) in the crystal. Since all unit cells are identical the simplest way to do this is first to sum over the atoms in one unit cell, and then to sum over all the unit cells.

Let the primitive unit cells in a crystal be labelled by $l = 1, 2, \ldots, N$ where $N$ is the total number of unit cells in the crystal (assumed very large), and let us specify the location of the origin of a particular primitive unit cell by a lattice vector $\mathbf{l}$, as in eqn (2.2). The position of the $d$th atom in the $l$th unit cell is given by

$$\mathbf{r}_{ld} = \mathbf{l} + \mathbf{d}, \tag{2.10}$$

where $\mathbf{d}$ is the displacement relative to the origin of the unit cell.

The differential cross-section (2.1) may be written as

$$\left(\frac{d\sigma}{d\Omega}\right)_{\text{coh}} = \left| \sum_j \bar{b}_j \exp(i\mathbf{Q} \cdot \mathbf{r}_j) \right|^2$$

$$= \left| \sum_{ld} \bar{b}_d \exp\{i\mathbf{Q} \cdot (\mathbf{l} + \mathbf{d})\} \right|^2$$

$$= \left| \sum_l \exp(i\mathbf{Q} \cdot \mathbf{l}) \right|^2 \left| \sum_d \bar{b}_d \exp(i\mathbf{Q} \cdot \mathbf{d}) \right|^2. \tag{2.11}$$

The last line of (2.11) has been written as the product of two factors, the first involving a sum over the lattice vectors, and the second a sum over the atoms within one unit cell. We shall consider the first factor here, and the second factor in Section 2.2.4.

The sum contained in the first factor in (2.11) represents the Fourier transform of the lattice. This sum can be interpreted with the aid of the reciprocal lattice concept introduced in Section 2.1.6, and is essentially zero except when $\mathbf{Q}$ is one of the reciprocal lattice vectors $\mathbf{G}$ when it becomes equal to $N$. In the present context we shall express $\mathbf{G}$ as

$$\mathbf{G} = h\mathbf{a}_1^* + k\mathbf{a}_2^* + l\mathbf{a}_3^*, \tag{2.12}$$

where $h$, $k$, $l$ are integers and $\mathbf{a}_1^*$, $\mathbf{a}_2^*$, $\mathbf{a}_3^*$ are the basis vectors of the reciprocal lattice, which are related to the primitive lattice vectors $\mathbf{a}_1$, $\mathbf{a}_2$, $\mathbf{a}_3$ by eqns (2.6).

To summarize: a necessary condition for strong diffraction from a crystal is that $\mathbf{Q}$ coincides with a reciprocal lattice vector (2.12), i.e.

$$\mathbf{Q} = \mathbf{G}. \tag{2.13}$$

[9]Max von Laue (1879–1960), was a German physicist who, together with Paul Knipping and Walter Friedrich, first demonstrated the diffraction of X-rays from crystals. Laue was awarded the 1914 Nobel Prize in Physics for this work.

Equation (2.13) is the *Laue equation*, or *Laue condition*.[9] Physically, the Laue equation represents the condition that waves scattered from equivalent atoms in each unit cell are in phase and so interfere constructively.

### 2.2.3   Equivalence of Bragg and Laue formulations

We now show that the Laue equation represents the same condition as Bragg's law. To do this we need two properties that link reciprocal lattice vectors to crystal planes.

(1)   The reciprocal lattice vector $\mathbf{G}$ is perpendicular to the planes with Miller indices $(hkl)$.

(2)   The length of $\mathbf{G}$ is equal to $2\pi$ divided by the spacing $d_{hkl}$ between the $(hkl)$ planes, i.e.

$$|\mathbf{G}| = 2\pi/d_{hkl}. \tag{2.14}$$

The reader is invited to prove these two results in Exercise 2.1.

Let us now take the modulus of both sides of the Laue equation, eqn (2.13), and substitute eqn (2.14) for $|\mathbf{G}|$:

$$|\mathbf{Q}| = |\mathbf{G}|$$
$$= 2\pi/d_{hkl}. \tag{2.15}$$

From the scattering triangle shown in Fig. 2.8 it is easy to show for the case of elastic scattering ($|\mathbf{k_i}| = |\mathbf{k_f}|$) that

$$|\mathbf{Q}| = (4\pi/\lambda)\sin\theta_{\mathrm{B}}, \tag{2.16}$$

and by combining eqns (2.15) and (2.16) we arrive at

$$\lambda = 2d_{hkl}\sin\theta_{\mathrm{B}}, \tag{2.17}$$

which is Bragg's Law as given in eqn (2.9) for the special case of $n = 1$.

The treatment of higher orders of diffraction in the Laue formulation is conceptually different from that in the Bragg formulation. Consider the following example. Suppose we are measuring the second-order Bragg reflection from the (100) planes of atoms in a crystal. This means in reality that there are two wavelengths path difference between waves scattered from adjacent planes of atoms, as shown in Fig. 2.9.

Now let us consider the set of planes with Miller indices (200). By definition, the (200) planes have half the spacing of the (100) planes, and so are not true crystallographic planes since only half of the (200) planes actually contain atoms. Nevertheless, if we pretend that *all* the (200) planes reflect neutrons then it is evident from Fig. 2.9 that first-order diffraction from the (200) planes would be equivalent to second-order diffraction from the (100) planes.

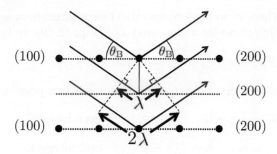

**Fig. 2.9** Miller indices and higher orders of diffraction. Second-order diffraction from the (100) planes is equivalent to first-order diffraction from the (200) planes.

The reciprocal lattice is infinite in extent, and so all possible values of $h$, $k$ and $l$ are permitted even though the corresponding planes may not all contain atoms. In this way, the Laue equation naturally includes higher orders, and the connection with the Bragg formulation is that the $n$th order of diffraction from the $(hkl)$ crystallographic planes is equivalent to the first order of diffraction from a set of hypothetical planes with Miller indices $(nh, nk, nl)$ corresponding to the reciprocal lattice vector $\mathbf{G} = nh\mathbf{a}_1^* + nk\mathbf{a}_2^* + nl\mathbf{a}_3^*$.

## 2.2.4  The structure factor

Let us suppose that the crystal is aligned on a diffractometer such that the scattering vector $\mathbf{Q}$ satisfies the Laue equation (2.13) for a particular $\mathbf{G} = h\mathbf{a}_1^* + k\mathbf{a}_2^* + l\mathbf{a}_3^*$ corresponding to the planes with Miller indices $(hkl)$, whereupon we expect to observe a large intensity of scattered neutrons in the detector. To find out exactly how large the intensity is, we must consider the second factor in eqn (2.11). Writing $\mathbf{d} = x_d\mathbf{a}_1 + y_d\mathbf{a}_2 + z_d\mathbf{a}_3$, and making use of eqns (2.7) to simplify the exponential, we can express the second factor of eqn (2.11) as $|F_N(\mathbf{G})|^2$, where

$$F_N(\mathbf{G}) = \sum_d \bar{b}_d \exp(i\mathbf{G} \cdot \mathbf{d}) \tag{2.18}$$

$$= \sum_d \bar{b}_d \exp\{2\pi i(hx_d + ky_d + lz_d)\}. \tag{2.19}$$

The function $F_N(\mathbf{G})$ is known as the *nuclear unit-cell structure factor*, and is the sum of the amplitudes of the waves scattered from the nuclei of each atom in the unit cell for the case when $\mathbf{Q} = \mathbf{G}$. Formally, the structure factor is the Fourier transform of the coherent scattering length density in one unit cell.[10] As the lattice sum in (2.11) is the same for all $\mathbf{G}$, the intensity (i.e. cross-section) of a Bragg peak is seen to be proportional to $|F_N(\mathbf{G})|^2$.

As an example, let us calculate the Bragg peak structure factors for the well-known ionic salt cesium chloride (CsCl). The crystal structure of CsCl has a primitive cubic lattice. For convenience, we locate the Cs ions on the lattice points and let these be the corners of the cubic unit cell (Fig. 2.10). The basis then consists of Cs at $0, 0, 0$ and Cl at $\frac{1}{2}, \frac{1}{2}, \frac{1}{2}$. Because the unit cell is primitive there is only one Cs and one Cl ion

[10]See Section 1.6.3.

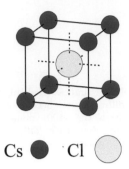

Cs ● · Cl ◯

**Fig. 2.10** Crystal structure and cubic unit call of CsCl.

per unit cell, so there is no need to specify the coordinates of any other atoms. Substituting these for $x_d$, $y_d$, and $z_d$ in eqn (2.19) we find

$$F_{\mathrm{N}}(\mathbf{G}) = \bar{b}_{\mathrm{Cs}} + \bar{b}_{\mathrm{Cl}} \exp\{i\pi(h + k + l)\}.$$

As $h$, $k$, and $l$ are integers, the structure factor has two possible values:

$$F_{\mathrm{N}}(\mathbf{G}) = \bar{b}_{\mathrm{Cs}} + \bar{b}_{\mathrm{Cl}} \quad (h + k + l = \text{even integer}),$$

$$= \bar{b}_{\mathrm{Cs}} - \bar{b}_{\mathrm{Cl}} \quad (h + k + l = \text{odd integer}).$$

The practical consequence of this result is that the intensities of the Bragg peaks measured from CsCl will be proportional to $(\bar{b}_{\mathrm{Cs}} + \bar{b}_{\mathrm{Cl}})^2$ or $(\bar{b}_{\mathrm{Cs}} - \bar{b}_{\mathrm{Cl}})^2$ depending upon the parity of $h + k + l$. For example, the ratio of the intensity of the (200) Bragg peak to that of the (111) is $(\bar{b}_{\mathrm{Cs}} + \bar{b}_{\mathrm{Cl}})^2/(\bar{b}_{\mathrm{Cs}} - \bar{b}_{\mathrm{Cl}})^2 = (5.42 + 9.58)^2/(5.42 - 9.58)^2 = 13.0$.

### 2.2.5   Non-primitive unit cells

If we choose to use a non-primitive unit cell to describe a crystal structure, then the crystallographic plane defined by Miller indices $(hkl)$ is not in general the same as the $(hkl)$ plane defined on a primitive cell. It is important, therefore, to specify the cell on which the diffraction peaks are indexed.

Furthermore, if a non-primitive cell is used, not all $(hkl)$ correspond to reciprocal lattice points. To see this, consider a unit cell whose shape is a parallelepiped (Fig. 2.3). The unit cells form an array whose periodicity can be described by a lattice with basis vectors **a**, **b**, and **c**. If the unit cell is not primitive then the unit cell lattice is a sublattice of the crystal lattice. If we now construct a 'reciprocal lattice' from the vectors $h\mathbf{a}^* + k\mathbf{b}^* + l\mathbf{c}^*$, where

$$\mathbf{a}^* = \frac{2\pi}{v_0}\,\mathbf{b}\times\mathbf{c}, \qquad \mathbf{b}^* = \frac{2\pi}{v_0}\,\mathbf{c}\times\mathbf{a}, \qquad \mathbf{c}^* = \frac{2\pi}{v_0}\,\mathbf{a}\times\mathbf{b}, \qquad (2.20)$$

and $v_0$ is the unit cell volume, eqn (2.3), then only a subset of the points so-defined will belong to the true reciprocal lattice of the crystal. The remaining points are not reciprocal lattice points, and so the Laue condition (2.13) is not satisfied at these points. The corresponding $hkl$ reflections are said to be *systematically absent*.

Put another way, any reciprocal lattice vector may be expressed as

$$\mathbf{G} = h\mathbf{a}^* + k\mathbf{b}^* + l\mathbf{c}^*, \qquad (2.21)$$

[11] Because **a**, **b**, and **c** are not necessarily the same as the lattice basis vectors $\mathbf{a}_1$, $\mathbf{a}_2$, and $\mathbf{a}_3$.

but for non-primitive cells only certain sets of integers $h$, $k$ and $l$ are allowed.[11] The rules which govern the allowed integers are called *selection rules*, and are particular to the lattice type and choice of unit cell. One can find the selection rules by calculating the structure factor of the set of lattice points contained in the non-primitive unit cell.

As an example, we calculate the structure factor of sodium chloride (NaCl). NaCl has a face-centred cubic lattice, and we choose to locate the Na ions at the lattice points (Fig. 2.11). The conventional cubic unit cell is non-primitive, and contains four lattice points, located at $0, 0, 0$; $\frac{1}{2}, \frac{1}{2}, 0$; $\frac{1}{2}, 0, \frac{1}{2}$; and $0, \frac{1}{2}, \frac{1}{2}$. The basis comprises a single pair of Na and Cl ions, and the Cl may be taken to be displaced from the Na by $\mathbf{c}/2$. Substituting the coordinates of the four Na and four Cl ions in the unit cell into eqn (2.19) we obtain

$$F_{\mathrm{N}}(\mathbf{G}) = \{1 + \mathrm{e}^{\mathrm{i}\pi(h+k)} + \mathrm{e}^{\mathrm{i}\pi(h+l)} + \mathrm{e}^{\mathrm{i}\pi(k+l)}\} \, (\overline{b}_{\mathrm{Na}} + \overline{b}_{\mathrm{Cl}} \mathrm{e}^{\mathrm{i}\pi l}).$$

The first factor is the structure factor of the four lattice points in the unit cell. This factor is non-zero if the integers $h$, $k$, and $l$ are all even or all odd, which are the selection rules for a face-centred lattice. Hence, reflections with mixed parity Miller indices, e.g. 100, are systematically absent. The second factor is the structure factor of the basis. Overall,

$$F_{\mathrm{N}}(\mathbf{G}) = 4(\overline{b}_{\mathrm{Na}} + \overline{b}_{\mathrm{Cl}}) \qquad (h, k, l \text{ all even}),$$

$$= 4(\overline{b}_{\mathrm{Na}} - \overline{b}_{\mathrm{Cl}}) \qquad (h, k, l \text{ all odd}),$$

$$= 0 \qquad (\text{otherwise}).$$

The occurrence of systematic absences can be understood physically as follows. Taking the 100 reflection of a crystal with a face-centred lattice as an example, Fig. 2.12, we note that half-way between two adjacent (100) scattering planes there is another plane with the same density of atoms as on the (100) planes. The scattered waves from these interleaved planes are of the same amplitude as those from the (100) planes but are half a wavelength out of phase. This causes destructive interference and results in zero total amplitude for the 100 reflection.

## 2.2.6 The phase problem

In general, the structure factor is a complex number which can be written

$$F_{\mathrm{N}}(\mathbf{G}) = f(\mathbf{G})\mathrm{e}^{\mathrm{i}\phi(\mathbf{G})},$$

where the modulus $f$ and phase $\phi$ are real numbers. Both $f$ and $\phi$ contain information about the atomic positions. If we could measure the modulus and phase of structure factors we could directly compute the positions of the atoms in the unit cell without any prior knowledge of the structure simply by performing a Fourier transform on a complete set of measured structure factors. Unfortunately, this is not possible because the diffracted intensity is proportional to $|F_{\mathrm{N}}|^2$, and

$$|F_{\mathrm{N}}|^2 = f\mathrm{e}^{\mathrm{i}\phi} \times f\mathrm{e}^{-\mathrm{i}\phi} = f^2.$$

So all information about the phase $\phi$ of the structure factor is lost. This is known as the *phase problem* in crystallography.

Methods do exist to get around the phase problem. For example, one can vary the nuclear scattering lengths at one or more sites in the structure, either by making chemical or isotopic substitutions which preserve

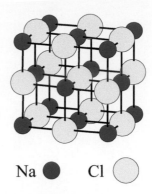

**Fig. 2.11** Crystal structure and conventional cubic unit call of NaCl.

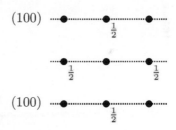

**Fig. 2.12** Plan view of the face-centred cubic lattice looking down the $c$-axis. The fractions indicate fractional heights in the cubic unit cell. Two adjacent (100) planes are marked.

the crystal structure, or by exploiting the energy dependence of the scattering length in the vicinity of a strong nuclear absorption resonance. However, the vast majority of crystal structures are solved by refinement of a model from an initial 'guess' at the atomic positions until the diffracted intensities calculated from the model give the best agreement with the measured intensities.

### 2.2.7    Friedel's law

Friedel's law states that $|F_N(-\mathbf{G})|^2 = |F_N(\mathbf{G})|^2$. It applies when either (i) the crystal has inversion symmetry, or (ii) the nuclear scattering lengths are real. In the latter case, we see from (2.18) that

$$F_N(-\mathbf{G}) = \sum_d \bar{b}_d \exp(-i\mathbf{G} \cdot \mathbf{d}) = F_N^*(\mathbf{G}). \qquad (2.22)$$

Hence,

$$\begin{aligned} |F_N(-\mathbf{G})|^2 &= F_N(-\mathbf{G})F_N^*(-\mathbf{G}) \\ &= F_N^*(\mathbf{G})F_N(\mathbf{G}) \\ &= |F_N(\mathbf{G})|^2. \end{aligned} \qquad (2.23)$$

Reflections related by the inversion operation in reciprocal space are called *Friedel pairs*. Result (2.23) tells us that the intensities of a Friedel pair are the same, which means that the distribution of Bragg peak intensities in reciprocal space has inversion symmetry whether or not the crystal has inversion symmetry. Friedel's law is violated for non-centrosymmetric crystals if the neutron energy is close to a nuclear absorption resonance, because in that case the scattering length takes the complex form $b = b' - ib''$ (Section 1.7.3), and so (2.22) does not hold. Violation of Friedel's law can be exploited to find the handedness of chiral crystals or the polarity in polar crystals.

### 2.2.8    Summary: the differential cross-section

The scattering from crystalline solids consists of sharp (Bragg) peaks in very specific directions described by Bragg's Law, eqn (2.9), or equivalently, by the Laue condition, eqn (2.13). The intensity of a Bragg peak is proportional to $|F_N(\mathbf{G})|^2$, where $F_N(\mathbf{G})$ is the nuclear unit-cell structure factor, eqn (2.19).

There now remains one loose end to tie up. In Section 2.2.2 we argued that the first factor in eqn (2.11) leads to sharp Bragg peaks in the cross-section, but we did not give a mathematical expression for it. In Section B.4.2 it is shown that the sum over lattice vectors can be written

$$\left| \sum_l \exp(i\mathbf{Q} \cdot \mathbf{l}) \right|^2 = N\frac{(2\pi)^3}{v_0} \sum_{\mathbf{G}} \delta(\mathbf{Q} - \mathbf{G}). \qquad (2.24)$$

Here, $\delta(\mathbf{Q} - \mathbf{G})$ is a Dirac delta function, which is a sharply peaked function centred at $\mathbf{Q} = \mathbf{G}$ with a volume of unity when integrated

(a) Transverse

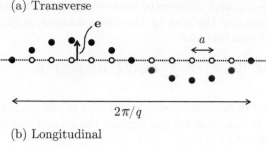

$$2\pi/q$$

(b) Longitudinal

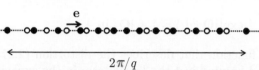

$$2\pi/q$$

**Fig. 2.13** Two types of superstructural distortion on a one-dimensional atomic chain. (a) Transverse ($\mathbf{e} \perp \mathbf{q}$). (b) Longitudinal ($\mathbf{e} \parallel \mathbf{q}$). The open and filled circles represent the undistorted and distorted positions, respectively.

over all components of $\mathbf{Q}$ (Section B.2). These delta functions describe mathematically the Bragg peaks that are observed whenever the Laue equation (2.13) is satisfied.

Inserting eqn (2.24) into (2.11) we finally arrive at the general expression for the coherent differential cross-section for elastic scattering from a periodic crystal composed of rigidly bound nuclei:

$$\left(\frac{d\sigma}{d\Omega}\right)_{\text{coh}} = N\frac{(2\pi)^3}{v_0}\sum_{\mathbf{G}}|F_{\text{N}}(\mathbf{G})|^2\delta(\mathbf{Q}-\mathbf{G}), \qquad (2.25)$$

where $F_{\text{N}}(\mathbf{G})$ is the structure factor of the reflection measured at $\mathbf{Q} = \mathbf{G}$, given in eqn (2.19).

## 2.2.9   Periodic distortion of a crystal

Various physical mechanisms exist by which atoms in a crystal can become slightly displaced from their mean positions. If the displacement pattern is periodic then it produces a *superstructure*, which generates diffraction peaks at $\mathbf{Q}$ positions corresponding to the Fourier components of the modulation. These are known as *superstructure peaks*.

To illustrate the main properties of the diffraction from superstructures we consider a crystal that undergoes a small sinusoidal distortion. After the distortion we assume that all the atoms in the $l$th unit cell have a displacement $\mathbf{u}_l$ relative to their original positions,[12] so that the position vector of the $d$th atom in the $l$th unit cell becomes [c.f. eqn (2.10)]

$$\mathbf{r}_{ld} = \mathbf{l} + \mathbf{u}_l + \mathbf{d}, \qquad (2.26)$$

where,

$$\mathbf{u}_l = \mathbf{e}\sin(\mathbf{q}\cdot\mathbf{l}). \qquad (2.27)$$

The 'polarization' vector $\mathbf{e}$ describes the amplitude and direction of the atomic displacements, and the wavevector $\mathbf{q}$ specifies the direction and period ($= 2\pi/|\mathbf{q}|$) of the sinusoidal modulation, see Fig. 2.13.

[12]We make the assumption that the displacement is the same for all atoms within a unit cell for simplicity. In general, the amplitude, phase and direction of the atomic displacements could vary from atom to atom within the unit cell.

To calculate the coherent differential cross-section we substitute (2.26) into (2.1), and factorize the sum in the same manner as (2.11). The summation over $l$ now becomes

$$\sum_l \exp\{i\mathbf{Q} \cdot (\mathbf{l} + \mathbf{u}_l)\} = \sum_l \exp(i\mathbf{Q} \cdot \mathbf{l}) \exp(i\mathbf{Q} \cdot \mathbf{u}_l). \qquad (2.28)$$

Distortions in crystals are usually very small, and it will usually be found that $\mathbf{Q} \cdot \mathbf{u}_l \ll 1$. In this case we can expand the second exponential on the right-hand side of (2.28):

$$\exp(i\mathbf{Q} \cdot \mathbf{u}_l) = 1 + i\mathbf{Q} \cdot \mathbf{u}_l + \cdots. \qquad (2.29)$$

[13] From Euler's formula,

$$\sin x = (e^{ix} - e^{-ix})/2i.$$

Writing the sine function that describes the distortion (2.27) in exponential form[13] and keeping just the linear term in the expansion (2.29) we can approximate the right-hand side of (2.28) by

$$\sum_l \left[ \exp(i\mathbf{Q} \cdot \mathbf{l}) + \frac{\mathbf{Q} \cdot \mathbf{e}}{2} \left[ \exp\{i(\mathbf{Q} + \mathbf{q}) \cdot \mathbf{l}\} - \exp\{i(\mathbf{Q} - \mathbf{q}) \cdot \mathbf{l}\} \right] \right]. \qquad (2.30)$$

From eqn (B.46) we see that the first lattice sum in (2.30) results in an array of sharply peaked functions centred on the reciprocal lattice vectors $\mathbf{G}$, and the second and third sums produce arrays of smaller peaks at $\mathbf{G} - \mathbf{q}$ and $\mathbf{G} + \mathbf{q}$, respectively. For an infinite lattice the peaks are sharp and do not overlap, so when the modulus of (2.30) is squared to calculate the scattering cross-section we can ignore the cross terms and write the self terms as delta functions, exactly as was done in (2.24). The final expression for the cross-section when $\mathbf{Q} \cdot \mathbf{e} \ll 1$ is

$$\left( \frac{d\sigma}{d\Omega} \right)_{\mathrm{coh}} \simeq N \frac{(2\pi)^3}{v_0} \sum_{\mathbf{G}} \left[ |F_{\mathrm{N}}(\mathbf{G})|^2 \delta(\mathbf{Q} - \mathbf{G}) \right.$$
$$\left. + \tfrac{1}{4}(\mathbf{Q} \cdot \mathbf{e})^2 |F_{\mathrm{N}}(\mathbf{G} \mp \mathbf{q})|^2 \delta(\mathbf{Q} \pm \mathbf{q} - \mathbf{G}) \right]. \qquad (2.31)$$

Comparing (2.25) and (2.31) we see that both contain the same set of principal Bragg peaks centred on the reciprocal lattice vectors $\mathbf{G}$, but for the distorted structure the principal Bragg peaks are flanked by satellites displaced by wavevectors $\pm\mathbf{q}$ from the reciprocal lattice vectors, see Fig. 2.14. These satellites are the superstructure peaks arising from the periodicity of the modulation. The structure factors that determine the intensities of the superstructure peaks must be evaluated at $\mathbf{G} \pm \mathbf{q}$, but if $\mathbf{q}$ is small then $F_{\mathrm{N}}(\mathbf{G} \pm \mathbf{q})$ is usually similar to $F_{\mathrm{N}}(\mathbf{G})$.

The factor $(\mathbf{Q} \cdot \mathbf{e})^2$ associated with the superstructure peaks means their intensity increases as $Q^2$ and depends on the angle between $\mathbf{Q}$ and $\mathbf{e}$. When $\mathbf{Q} \perp \mathbf{e}$ the superstructure peaks have zero intensity. Physically, this is because in Bragg diffraction $\mathbf{Q}$ is perpendicular to the planes, and displacements parallel to Bragg planes do not introduce any extra phase difference between the waves scattered from adjacent planes. The $Q^2$ variation reflects the fact that at larger $Q$ the degree of constructive interference between the scattered waves increases.

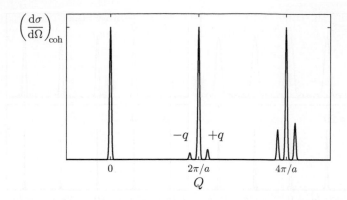

**Fig. 2.14** Diffraction intensity from a one-dimensional superstructure such as one of those shown in Fig. 2.13

As well as the superstructure peaks at $\pm\mathbf{q}$ it is common to observe weaker satellites at $\pm 2\mathbf{q}$, $\pm 3\mathbf{q}$, etc, relative to $\mathbf{G}$. These are known as *harmonics*, and they appear for two different reasons. First, because distortions in real materials are not always perfectly sinusoidal, and so $\mathbf{u}_l$ may contain Fourier components that are multiples of $\pm\mathbf{q}$. These higher-order Fourier components produce peaks at harmonic positions, and the relative intensities of these peaks provides information on the shape of the distortion. Second, because higher-order terms in the expansion (2.29) also give rise to peaks at harmonics of $\pm\mathbf{q}$, as can be seen by treating the quadratic term in the same way as we did for the linear term.[14] Harmonics generated this way become more important as the size of the distortion increases, but are always present at some level even when the distortion is perfectly sinusoidal.

The method described in this section can easily be extended to other types of periodic distortion. The particular case we have worked through here, of a sinusoidal distortion, will provide useful insights into the cross-section for scattering from vibrations in crystals, obtained in Chapter 5. Crystal vibrations are just dynamic sinusoidal distortions of the crystal, so several of the same features crop up. In particular, the polarization vector and $Q^2$ variation of the superstructure peak intensity will be encountered again in the one-phonon scattering cross-section.

[14]Note that
$$\sin^2 x = \tfrac{1}{2} - \tfrac{1}{4}(\mathrm{e}^{2\mathrm{i}x} + \mathrm{e}^{-2\mathrm{i}x}).$$

## 2.3 Disorder in crystalline materials

### 2.3.1 Short-range order

So far in this chapter we have dealt with periodic structures that extend infinitely in all directions. Infinity is unimaginably large, and in reality the theory generally works well for finite-sized systems as long as they are not too small. What counts as too small depends on the resolution of the diffractometer, but the lower limit is typically in the range $10^1$–$10^3$ unit cells in each direction. If the structure is smaller than this threshold size in one or more direction then the Bragg peaks will no longer be 'resolution-limited' and the scattering will contain additional features relating to the size and shape of the scattering system.

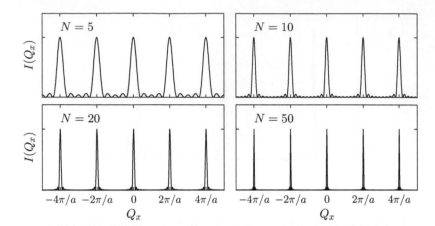

**Fig. 2.15** Diffracted intensity for the one-dimensional diffraction grating with different numbers $N$ of nuclei.

The one-dimensional diffraction grating provides a good illustration of this finite-size effect. Fig. 2.15 shows the diffracted intensity calculated from eqn (1.24) for several gratings with different numbers of nuclei. As $N$ is reduced the main diffraction peaks become broader and decrease in height ($\propto N^2$) while the subsidiary peaks increase in size.

This example is representative of the very general phenomenon of peak broadening due to *short-range order*. Periodic structures that are ordered over only a short distance have diffraction peaks whose intrinsic widths (i.e. the width of the peak in the absence of any broadening due to the resolution of the diffractometer) vary inversely with the range of the ordering. The peaks are no longer well described by the delta function scattering cross-section of eqn (2.24), but instead are spread out in reciprocal space. Coherent scattering which is not highly localized in reciprocal space is called *diffuse scattering*.

Speaking generally, diffuse scattering arises whenever there is a deviation from perfect periodic order. As well as finite-size effects there are many other sources of disorder in real materials. Examples are dislocations, stacking faults, twin boundaries, vacancies, impurities, and substitutional disorder. In the following sub-sections we describe a number of concepts connected with diffuse scattering from crystals.

### 2.3.2   The correlation length

It is often possible to describe disorder in condensed matter with a model in which the average coherence of the crystalline order decays exponentially with distance. This model applies when the periodicity of the scattering system is interrupted by occasional blemishes, such as dislocations or twin boundaries, which occur at random and have the effect of subdividing the system into many differently sized domains. Inside one domain the order is perfect, but on crossing from one domain to the next the periodicity is interrupted randomly.

We recall from eqn (2.1) that the coherent differential cross-section contains the complex amplitude $\exp\{i\mathbf{Q} \cdot (\mathbf{r}_k - \mathbf{r}_j)\}$ summed over pairs of atoms $j$ and $k$. The basic assumption of the exponential decay model is that if we take a walk through the crystal then the chance of encountering a domain boundary is the same for each step we take, regardless of how far we have already travelled. This idea is analogous to that used to derive the exponential decay law for the attenuation of a neutron beam in a sample (Section 1.7.4). Thus, if $dx/\xi$ is the probability of encountering a blemish in the interval $dx$ then the probability of travelling between atoms $j$ and $k$ without encountering a blemish is $\exp(-|\mathbf{r}_k - \mathbf{r}_j|/\xi)$. The parameter $\xi$ is known as the *correlation length*. If atoms $j$ and $k$ are in the same crystalline domain then their contribution to the differential cross-section is exactly as set out in eqn (2.1). But if they are in different domains then their positions are uncorrelated and on average the scattering from this pair of atoms will vanish, the average being taken over an ensemble of systems, in the same sense as described in Section 1.8.1. Once the ensemble average has been performed, the effect of the exponential correlation probability is simply to replace the factor $\bar{b}_j^* \bar{b}_k$ in eqn (2.1) by $\bar{b}_j^* \bar{b}_k \exp(-|\mathbf{r}_k - \mathbf{r}_j|/\xi)$.

For illustration, we evaluate the cross-section for the one-dimensional diffraction grating introduced in Section 1.6.2 but this time include an exponential correlation probability. For simplicity, we assume the line of atoms is infinitely long so that the sum over $k$ is virtually the same for any $j$.[15] With these modifications, eqn (2.1) becomes (for real $\bar{b}$)

$$\left(\frac{d\sigma}{d\Omega}\right)_{\text{coh}} = N\bar{b}^2 \sum_{n=-\infty}^{\infty} \exp(-|n|\,a/\xi)\exp(iQ_x na), \qquad (2.32)$$

where $Q_x$ is the component of $\mathbf{Q}$ parallel to the line of nuclei. Evaluation of the summation involves some interesting algebraic manipulations, but not ones we wish to become involved with here. The determined reader is referred to Exercise 2.6. The result is

$$\left(\frac{d\sigma}{d\Omega}\right)_{\text{coh}} = N\bar{b}^2 \frac{2\pi}{a} \sum_{p=-\infty}^{\infty} \frac{1/(\pi\xi)}{(Q_x - 2\pi p/a)^2 + 1/\xi^2}$$

$$= N\bar{b}^2 \frac{2\pi}{a} \sum_{p=-\infty}^{\infty} L(Q_x - 2\pi p/a, 1/\xi), \qquad (2.33)$$

where $L(x - x_0, \Gamma)$ denotes a normalized *Lorentzian* function centred at $x = x_0$ with full width at half maximum $2\Gamma$, see eqn (B.5). The scattering comprises a sum of Lorentzians centred on $Q_x = 2\pi p/a$, where $p$ is an integer. From (2.33),

$$\Gamma = \frac{1}{\xi}, \qquad (2.34)$$

i.e. the Lorentzian half width at half maximum is the reciprocal of the correlation length. Fig. 2.16 plots the scattering for the case $\xi = 3a$.

We can easily extend this idea into three dimensions. As an example, consider a layered crystal that has perfect periodicity within the layers,

[15] This approximation is valid when the broadening due to the finite size of the system is much less than the broadening due to the correlation length. In other words, when $Na \gg \xi$.

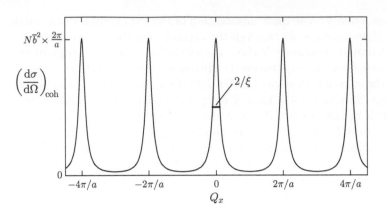

**Fig. 2.16** Diffuse scattering from a one-dimensional diffraction grating with correlation length $\xi = 3a$.

but has imperfect periodicity described by the exponential correlation model perpendicular to the layers. The scattering would consist of peaks centred about the reciprocal lattice points $\mathbf{G}$ but these peaks would be elongated along the direction perpendicular to the layers, like a mountain with two long ridges descending from the summit in opposite directions. The condition for us to observe scattering is now less stringent than in eqn (2.25). All we require is that $\mathbf{Q}$ lies somewhere on one of these ridges. Scans perpendicular to the ridges will contain very sharp peaks, whereas scans along the ridges will show diffuse scattering with maxima whenever $\mathbf{Q} = \mathbf{G}$.

It would be useful to have a simple mathematical expression analogous to eqn (2.25) to describe this kind of diffuse scattering for a general three-dimensional crystal with more than one atom in the unit cell. Unfortunately the evaluation of the cross-section is no longer straightforward because the lack of long-range periodicity means that the summation over all nuclei does not factorize in the manner of eqn (2.11). In many systems of interest, however, the correlation length is much larger than the unit cell dimensions and hence the diffuse scattering is relatively narrow. In this case, the coherent cross-section is to a good approximation given by eqn (2.25) with Lorentzians in place of the delta functions. For example, the cross-section for the layered crystal described above is given in this approximation by

$$\left(\frac{d\sigma}{d\Omega}\right)_{\text{coh}} = N\frac{(2\pi)^3}{v_0}\sum_{\mathbf{G}}|F_{\text{N}}(\mathbf{G})|^2\delta(\mathbf{Q}_{2\text{d}} - \mathbf{G}_{2\text{d}})L(Q_z - G_z, 1/\xi), \quad (2.35)$$

where $Q_z$ and $\mathbf{Q}_{2\text{d}}$ are the components of $\mathbf{Q}$ parallel and perpendicular to the ridge direction, and similarly for $G_z$ and $\mathbf{G}_{2\text{d}}$. The two-dimensional delta function $\delta(\mathbf{Q}_{2\text{d}} - \mathbf{G}_{2\text{d}})$ represents a sharply peaked function that integrates to unity over the $Q_xQ_y$ plane. When $\xi \to \infty$ the Lorentzians also become delta functions, and the cross-section reverts to (2.25).

### 2.3.3   Two- and one-dimensional systems

The other extreme limit is when $\xi \to 0$, indicating a complete lack of coherence. If this occurs in just one direction then we have in effect a two-dimensional (2d) crystal with perfect periodicity within a plane, but no correlations between the atoms in different planes. This limit is not often encountered in real crystal structures, but there is an extensive class of layered materials that exhibit essentially 2d magnetic ordering. The absence of any correlations perpendicular to the plane means that the diffuse ridges become totally flat, and so the scattering takes the form of rods passing through points of the 2d reciprocal lattice. One of these is illustrated in Fig. 2.17(a), and the corresponding expression for the cross-section is

$$\left(\frac{d\sigma}{d\Omega}\right)_{\text{coh}} = N\frac{(2\pi)^2}{s_0}\sum_{\mathbf{G}_{2d}}|F(\mathbf{G}_{2d})|^2\delta(\mathbf{Q}_{2d} - \mathbf{G}_{2d}) \quad \text{(2d)}, \quad (2.36)$$

where $F(\mathbf{G}_{2d})$ is now the structure factor for a 2d periodic structure and $s_0$ is the area of the unit cell.

By an obvious extension of these arguments we can also deduce that the diffuse scattering from a 1d periodic structure consists of a set of parallel planes in reciprocal space separated by the reciprocal lattice vector for the 1d periodicity. The scattering is depicted in Fig. 2.17(b), and the cross-section is

$$\left(\frac{d\sigma}{d\Omega}\right)_{\text{coh}} = N\frac{2\pi}{a}\sum_{G_{1d}}|F(G_{1d})|^2\delta(Q_{1d} - G_{1d}) \quad \text{(1d)}, \quad (2.37)$$

where $a$, $F(G_{1d})$, $Q_{1d}$, and $G_{1d}$ are the 1d analogues of the corresponding quantities in eqn (2.36).

### 2.3.4 Point defects

The presence of a point defect (e.g. a vacancy, interstitial, impurity, or substitutional atom)[16] in a crystal causes the surrounding atoms to be displaced from their regular sites. The scattering due to the defect, which is known as *Huang scattering*, depends upon the difference between the distorted and undistorted crystal as well as on the defect atom or vacancy itself.

For illustration, consider an impurity atom with real scattering length $b_2$ in a host consisting of atoms with real scattering length $b_1$ arranged on a Bravais lattice — see Fig. 2.18. We choose the origin to be at the impurity, and represent the static distortion field around the impurity by the vectors $\mathbf{u}_l$ ($\mathbf{u}_0 = 0$) which give the displacements of atoms belonging to lattice sites $l$ from their equilibrium positions.

The summation needed to evaluate the coherent differential cross-section (2.1) can be written

$$\sum_l b_l \exp\{i\mathbf{Q}\cdot(\mathbf{l} + \mathbf{u}_l)\} = b_1\sum_l \exp(i\mathbf{Q}\cdot\mathbf{l}) + F_{\text{dis}}(\mathbf{Q}), \quad (2.38)$$

where

$$F_{\text{dis}}(\mathbf{Q}) = b_1\sum_l \exp(i\mathbf{Q}\cdot\mathbf{l})\{\exp(i\mathbf{Q}\cdot\mathbf{u}_l) - 1\} + b_2 - b_1 \quad (2.39)$$

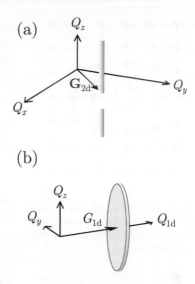

**Fig. 2.17** Illustrations of the distribution of diffuse scattering in reciprocal space from (a) two-dimensional and (b) one-dimensional structures. In (a) $\mathbf{Q}_{2d} = (\mathbf{Q}_x, \mathbf{Q}_y)$, and in (b) $Q_{1d} = Q_x$.

[16]A vacancy is a regular atomic site in a crystal that is unoccupied. An interstitial atom is an atom that does not occupy a regular atomic site of the crystal structure. Substitutional disorder, or 'site-mixing', means that atoms of one type partially occupy regular crystallographic sites belonging to atoms of a different type, both atomic species being intrinsic to the material.

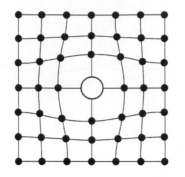

**Fig. 2.18** Distortion of the crystal structure around a substitutional impurity. The difference in size of the atoms is greatly exaggerated.

[17]Note that $\mathbf{u}(\mathbf{q})$ is real because a Bravais lattice has inversion symmetry.

[18]The distinction between liquids and glasses is that in a liquid the atoms are free to roam around, whereas in glassy solids the atoms are frozen into a particular configuration which does not change on the timescale of the experiment. Hence, liquids 'flow' whereas glassy solids do not.

is the structure factor of the distortion. The first term in (2.38) is identical with that for the undistorted crystal, and gives an array of delta functions centred at the reciprocal lattice vectors. The cross-section for the complete system is the squared modulus of (2.38),

$$\left(\frac{d\sigma}{d\Omega}\right)_{\text{coh}} = N\frac{(2\pi)^3}{v_0}b_1^2 \sum_{\mathbf{G}} \delta(\mathbf{Q}-\mathbf{G}) + |F_{\text{dis}}(\mathbf{Q})|^2. \tag{2.40}$$

We have neglected the cross-terms as they are non-zero only at the positions of the delta functions.

If the displacements are sufficiently small then $\mathbf{Q}\cdot\mathbf{u}_l \ll 1$. In this case we can use the approximation $\exp(i\mathbf{Q}\cdot\mathbf{u}_l) - 1 \simeq i\mathbf{Q}\cdot\mathbf{u}_l$, and the convolution theorem (see Section B.3.2), to show that

$$F_{\text{dis}}(\mathbf{Q}) \simeq ib_1 \sum_{\mathbf{G}} \mathbf{Q}\cdot\mathbf{u}(\mathbf{q}) + b_2 - b_1,$$

where $\mathbf{q} = \mathbf{Q}-\mathbf{G}$ and $\mathbf{u}(\mathbf{q})$ is the Fourier transform of $\mathbf{u}(\mathbf{r})$, a continuous distortion field that describes the displacements $\mathbf{u}_l$ around the impurity. Hence,[17]

$$|F_{\text{dis}}(\mathbf{Q})|^2 \simeq b_1^2 \left[\sum_{\mathbf{G}} \{\mathbf{Q}\cdot\mathbf{u}(\mathbf{q})\}^2 + (b_2/b_1 - 1)^2\right]. \tag{2.41}$$

We see that the Huang scattering is a diffuse feature under the Bragg peaks. The shape of the Huang scattering is the same at each Bragg peak, but the intensity increases as $Q^2$. By measuring $\mathbf{u}(\mathbf{q})$ one can obtain $\mathbf{u}(\mathbf{r})$, which provides information on the structure of the defect.

### 2.3.5 Thermal diffuse scattering

Thermal diffuse scattering (TDS) is another process which produces broad features in the diffraction from crystals. TDS arises from inelastic scattering from phonons which is unavoidably recorded due to the imperfect energy resolution of the diffractometer. As TDS is a consequence of crystal dynamics we shall not discuss it further here, but instead refer the reader to Section 10.2.3.

## 2.4  Liquids and disordered solids

As already explained, diffuse scattering is a consequence of short-range order. The breakdown of long-range order is at its most extreme in liquids and disordered solids (i.e. glasses).[18] In an ideal liquid there are no correlations between the relative positions of atoms that are far apart. The overall distribution of atoms, however, is not entirely random. On the contrary, neighbouring atoms in a liquid are highly correlated as a result of the way in which the atoms pack together. These correlations can extend for several interatomic distances, and it is this local structure and the associated dynamics that are of interest.

## 2.4.1 Spherical averaging

For isotropic or randomly oriented samples, such as liquids, glasses, and powdered crystals, the observed diffraction represents an average over all possible orientations of the system. An orientational average can be performed with the help of Fig. 2.19. The differential cross-section (2.1) contains the factor $\exp\{i\mathbf{Q} \cdot (\mathbf{r}_k - \mathbf{r}_j)\} = \exp(iQr_{jk}\cos\theta)$, where $r_{jk} = |\mathbf{r}_{jk}| = |\mathbf{r}_k - \mathbf{r}_j|$ and $\theta$ is the angle between $\mathbf{r}_{jk}$ and $\mathbf{Q}$. This factor takes the same value for all directions of $\mathbf{r}_{jk}$ which lie on a cone of semi-angle $\theta$. The fraction of directions between $\theta$ and $\theta + d\theta$ is $\frac{1}{2}\sin\theta\,d\theta$, so the spherical average is

$$\langle\exp\{i\mathbf{Q} \cdot (\mathbf{r}_k - \mathbf{r}_j)\}\rangle = \int_0^\pi \exp(iQr_{jk}\cos\theta)\tfrac{1}{2}\sin\theta\,d\theta$$

$$= \frac{-1}{2iQr_{jk}}\left[\exp(iQr_{jk}\cos\theta)\right]_0^\pi$$

$$= \frac{\sin Qr_{jk}}{Qr_{jk}} \equiv j_0(Qr_{jk}). \qquad (2.42)$$

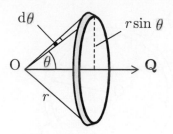

**Fig. 2.19** Calculation of the fraction of directions between polar angles $\theta$ and $\theta + d\theta$ for spherical averaging. The shaded strip lives on the surface of a sphere of radius $r$. The area of the strip is $2\pi r^2 \sin\theta\,d\theta$, which is a fraction $\frac{1}{2}\sin\theta\,d\theta$ of the surface area $4\pi r^2$ of the sphere.

$j_0(x)$ is the first spherical Bessel function, also known as the sinc function.

Hence, the spherically averaged differential cross-section (2.1) is

$$\left(\frac{d\sigma}{d\Omega}\right)_{\mathrm{coh}} = \sum_j |\bar{b}_j|^2 + \sum_{j\neq k} \bar{b}_j^* \bar{b}_k \frac{\sin(Qr_{jk})}{Qr_{jk}}, \qquad (2.43)$$

where for convenience we have separated terms with $j = k$ and $j \neq k$. Further configurational averaging of the second term will be necessary if the positions of atoms $j$ and $k$ are not fixed with respect to one another.

## 2.4.2 Pair distribution function

### One-component systems

The structure of a disordered material is often described by the *pair distribution function*, $g(r)$. For a one-component system, i.e. a system with only one type of particle, $g(r)$ is defined by

$$n_0 g(r) 4\pi r^2 dr = dN(r), \qquad (2.44)$$

where $dN(r)$ is the average number of particles between concentric spheres of radius $r$ and $r + dr$ centred on a particle, and $n_0$ is the bulk density of the material, i.e. the average number of particles per unit volume.[19] Note that $r$ rather than $\mathbf{r}$ appears in (2.44) because a disordered material is *isotropic*, i.e. the properties of the material are the same in all directions.

The definition of $g(r)$ is illustrated in Fig. 2.20. At large $r$, $g(r) \to 1$. This is because as $r$ increases the number of atoms in the shell volume increases, so the number of atoms per unit volume in the shell tends towards the bulk density. We see, therefore, that $g(r)$ is the factor by which the average density of particles at a distance $r$ from a given

[19]Sometimes the factor $n_0$ is absorbed into the definition of $g(r)$.

(a)

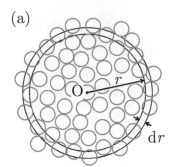

(b)

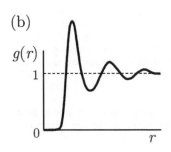

**Fig. 2.20** The pair distribution function $g(r)$ for a monatomic system. (a) The volume of a shell bounded by spheres of radius $r$ and $r + dr$ is $4\pi r^2 dr$. The number of atoms whose centres fall within this spherical shell is $n_0 g(r) 4\pi r^2 dr$. (b) Sketch of $g(r)$ as a function of $r$.

[20] For liquids, an additional factor is that the atoms are mobile, so the configuration of the system is constantly changing. During the course of a measurement, therefore, the system passes through many different configurations, and a configurational average is performed with the passage of time.

particle varies from the bulk density. Note that the origin particle does not count towards $g(r)$, so if the particles have a strong repulsive core then $g(0) = 0$.

We shall now show that diffraction from a disordered material is related to $g(r)$ by a Fourier transform, with the consequence that $g(r)$ can be obtained directly from experiment without the phase problem encountered in crystal structure solution (Section 2.2.6). To see this it is useful to introduce a more general, anisotropic pair distribution function $g(\mathbf{r})$. Still considering a one-component system, we define

$$n_0 g(\mathbf{r}) = \frac{1}{N} \sum_{j \neq k} \langle \delta(\mathbf{r} - \mathbf{r}_k + \mathbf{r}_j) \rangle$$
$$= \sum_{k \neq 0} \langle \delta(\mathbf{r} - \mathbf{r}_k + \mathbf{r}_0) \rangle. \tag{2.45}$$

The angular brackets denote an ensemble average over configurations of the system, and $N$ is the total number of particles. The second line follows because the ensemble average makes the summation over $k$ the same for any $j$, so instead of summing over $j$ we might as well fix $j = 0$ and multiply by $N$. Integration of the right-hand side of (2.45) over a small volume $dV$ gives the number of particles in $dV$, so $g(\mathbf{r})$ is the factor by which the average density at a displacement $\mathbf{r}$ from the origin particle varies from the bulk density.

The coherent differential cross-section for a rigid system is given by eqn (2.1), but for a large enough system we can argue in the same spirit as in Section 1.8.1 that the cross-section for one particular sample is for all practical purposes the same as the average cross-section for many similar samples. Systems with disorder do not have a unique configuration, and so the ensemble average includes many different configurations with essentially the same energy. The justification here is that the neutrons probe correlations within a *scattering volume* which is usually much smaller than the physical size of the sample as a whole. The spatial configuration of the atoms within a scattering volume may vary from place to place in the sample, and if the number scattering volumes in the entire sample is large enough then it is reasonable to assume that a representative set of all configurations is sampled. It is a good approximation, therefore, to perform an ensemble average over all possible configurations,[20] and we write the coherent differential cross-section

$$\left( \frac{d\sigma}{d\Omega} \right)_{coh} = \left\langle \sum_{jk} \bar{b}_j^* \bar{b}_k \exp\{i\mathbf{Q} \cdot (\mathbf{r}_k - \mathbf{r}_j)\} \right\rangle, \tag{2.46}$$

with the angular brackets again denoting the ensemble (configurational) average. Usually, the structure of a material is not influenced by the distribution of isotopes, and in this case the scattering lengths can be excluded from the ensemble average. Hence, for a material containing only one chemical species,

$$\left( \frac{d\sigma}{d\Omega} \right)_{coh} = |\bar{b}|^2 \sum_{jk} \langle \exp\{i\mathbf{Q} \cdot (\mathbf{r}_k - \mathbf{r}_j)\} \rangle. \tag{2.47}$$

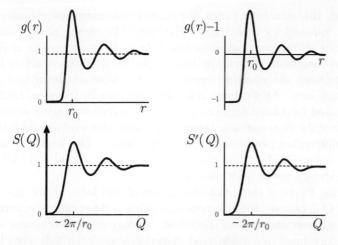

**Fig. 2.21** The left panels depict the typical form of the pair distribution function $g(r)$ and structure factor $S(Q)$ for a one-component liquid. The right panels show the functions $g(r) - 1$ and $S'(Q)$, which are related via eqn (2.51). $S'(Q)$ is the same as $S(Q)$ but without the $\delta$-function at $Q = 0$.

Using property (B.7) of the delta function we can express the cross-section as

$$\left(\frac{d\sigma}{d\Omega}\right)_{\text{coh}} = |\bar{b}|^2 \sum_{jk} \int \langle \delta(\mathbf{r} - \mathbf{r}_k + \mathbf{r}_j) \rangle \exp(i\mathbf{Q} \cdot \mathbf{r}) \, d^3\mathbf{r}$$

$$= N|\bar{b}|^2 S(\mathbf{Q}), \tag{2.48}$$

where, from (2.45),

$$S(\mathbf{Q}) = 1 + n_0 \int g(\mathbf{r}) \exp(i\mathbf{Q} \cdot \mathbf{r}) \, d^3\mathbf{r}. \tag{2.49}$$

The function $S(\mathbf{Q})$ is known as the *structure factor*.[21] Apart from an additive constant ($= 1$), $S(\mathbf{Q})$ is the Fourier transform of the pair distribution function. The aforementioned constant originates from the $j = k$ terms in the double summation in (2.48), which are explicitly excluded from the definition (2.45) of $g(\mathbf{r})$.

Because $g(\mathbf{r}) \to 1$ as $\mathbf{r} \to \infty$, the Fourier transform of $g(\mathbf{r})$ contains a delta function at $\mathbf{Q} = 0$. Physically, this represents scattering in the forward direction. This scattering can never be measured, but to avoid the inelegance of a singularity an alternative definition of the structure factor is frequently adopted,

$$S'(\mathbf{Q}) = 1 + n_0 \int \{g(\mathbf{r}) - 1\} \exp(i\mathbf{Q} \cdot \mathbf{r}) \, d^3\mathbf{r}. \tag{2.50}$$

$S'(\mathbf{Q})$ and $S(\mathbf{Q})$ are identical apart from the delta function at $\mathbf{Q} = 0$.

Returning to disordered materials, for which $g(\mathbf{r})$ is isotropic, we can integrate (2.50) over the angular coordinates using eqn (2.42) to give

$$S'(Q) = 1 + \frac{4\pi n_0}{Q} \int_0^\infty \{g(r) - 1\} r \sin(Qr) \, dr. \tag{2.51}$$

Figure 2.21 contains schematic illustrations of $g(r)$, $S(Q)$, $g(r) - 1$, and $S'(Q)$ for a typical disordered material containing one type of particle, and Fig. 2.22 shows experimental results for molten aluminium.

[21]Or, sometimes, the *static structure factor*. Confusingly, the term *structure factor* as used here refers to a scattering *intensity*, whereas in crystallography it represents a scattering *amplitude*, see eqn (2.19).

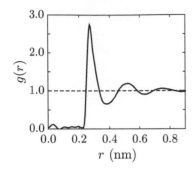

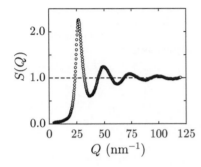

**Fig. 2.22** Measured structure factor $S(Q)$ and derived pair distribution function $g(r)$ for molten aluminium at a temperature of 980 K (Dahlborg *et al.*, 2013. Data courtesy of U. Dahlborg and M. Calvo-Dahlborg.)

In general, the structure factor for disordered materials exhibits a strong peak followed by decaying oscillations. The strong peak, called the *first sharp diffraction peak*, appears at $Q \simeq 2\pi/r_0$, where $r_0$ is the mean nearest-neighbour separation, and originates from constructive interference between the neutron waves scattered from an atom and its nearest neighbours. As we have remarked previously (Section 1.6.2), atoms separated by a fixed distance produce coherent scattering around the wavevector $2\pi/$(inter-atomic separation), and this explains why the first sharp diffraction peak appears at $Q \simeq 2\pi/r_0$. The width of this peak will be determined by how close the distances to second and higher nearest neighbours are to multiples of $r_0$.

The limiting forms of the diffraction at small and large $Q$ are also of interest. As $Q$ decreases, the coherent scattering is determined by correlations over greater and greater distances. Widely separated atoms are virtually uncorrelated in a disordered material, and so the diffracted intensity drops as $Q \to 0$. The scattering at $Q = 0$ does not entirely vanish though. This is because all materials are to some extent compressible, and hence there exist long-range density fluctuations which scatter neutrons at small $Q$. This property will be discussed in Section 2.4.4. In the opposite limit, $Q \to \infty$, the coherent scattering becomes insensitive to interatomic correlations, and only the 'self' terms, i.e. $j = k$ in eqn (2.47), contribute to $S(Q)$. Hence, $S(\infty) = 1$. Finally, we recall that the total observed diffraction is a sum of coherent and incoherent scattering, the latter of which (if non-zero) adds a flat (i.e. $Q$-independent) background onto the patterns shown in Fig. 2.21.

**Multicomponent systems**

For systems containing more than one chemical species it is often useful to describe the correlations and scattering in terms of partial, or species-dependent, functions. Consider a system containing several species such that $c_\alpha = N_\alpha/N$ is the concentration of species $\alpha$, where $N = \sum_\alpha N_\alpha$ remains the total number of atoms and $n_0 = N/V$ the bulk particle density. We now decompose the differential cross-section into 'self' and 'difference' terms,[22]

$$\left(\frac{d\sigma}{d\Omega}\right) = \sum_j \overline{|b_j|^2} + \sum_{j \neq k} \overline{b}_j^* \overline{b}_k \langle \exp\{i\mathbf{Q} \cdot (\mathbf{r}_k - \mathbf{r}_j)\}\rangle$$

$$= N \sum_\alpha c_\alpha \overline{|b_\alpha|^2} + N \sum_{\alpha\beta} c_\alpha c_\beta \overline{b}_\alpha^* \overline{b}_\beta \{S_{\alpha\beta}(\mathbf{Q}) - 1\}, \qquad (2.52)$$

where $S_{\alpha\beta}(\mathbf{Q})$ are the *partial structure factors*,[23]

$$S_{\alpha\beta}(\mathbf{Q}) - 1 = \frac{1}{Nc_\alpha c_\beta} \sum_{j_\alpha \neq k_\beta} \langle \exp\{i\mathbf{Q} \cdot (\mathbf{r}_{k_\beta} - \mathbf{r}_{j_\alpha})\}\rangle. \qquad (2.53)$$

The notation $j_\alpha$ in the summations is used to indicate the $j$th atom of type $\alpha$. The constraint $j_\alpha \neq k_\beta$ is only relevant for $\alpha = \beta$. The

[22]As opposed to the usual separation into coherent and incoherent terms.

[23]This is the Faber–Ziman definition of the partial structure factors. Other definitions are also in use (see Keen, 2001). Note that $S_{\alpha\beta} = S_{\beta\alpha}$ since the atoms are treated as static.

partial structure factors are related by Fourier transform to the partial pair distribution functions $g_{\alpha\beta}$,

$$S_{\alpha\beta}(\mathbf{Q}) = 1 + n_0 \int g_{\alpha\beta}(\mathbf{r}) \exp(i\mathbf{Q} \cdot \mathbf{r}) \, d^3\mathbf{r}, \qquad (2.54)$$

and

$$n_0 g_{\alpha\beta}(\mathbf{r}) = \frac{1}{N c_\alpha c_\beta} \sum_{j_\alpha \neq k_\beta} \langle \delta(\mathbf{r} - \mathbf{r}_{k_\beta} + \mathbf{r}_{j_\alpha}) \rangle. \qquad (2.55)$$

The partial pair distribution functions have the same interpretation as before, except they are specific to a given atom type. In particular, $g_{\alpha\beta}(\mathbf{r})$ represents the factor by which the density of particles of type $\beta$ at a displacement $\mathbf{r}$ from a particle of type $\alpha$ differs from the average density of particles of type $\beta$.[24]

From definitions (2.54) and (2.55), and the discussion in the previous section, it is easy to see that $g_{\alpha\beta}(\infty) = 1$. Hence, functions $S'_{\alpha\beta}(\mathbf{Q})$ can be defined exactly as in (2.50) to avoid the delta function at $\mathbf{Q} = 0$. Note also that $S_{\alpha\beta}(\infty) = S'_{\alpha\beta}(\infty) = 1$, and that for a single species $(c_\alpha = 1, c_\beta = 0)$ $S_{\alpha\beta}(\mathbf{Q})$ as defined in (2.52) reduces to $S(\mathbf{Q})$ in (2.48).

A knowledge of all the partial pair distribution functions amounts to a complete description of the structure of a multicomponent system. One of the great strengths of neutron scattering is that by measuring several samples with different isotopic compositions one can determine the partial structure factors, and hence the partial pair distribution functions. The method works because, as seen in eqn (2.52) above, the coherent scattering is just a weighted summation of partial structure factors, the weighting factors being simple products of the concentrations and coherent scattering lengths of the different species. For example, all three partial structure factors $S_{AA}$, $S_{BB}$, and $S_{AB}$ of a binary fluid containing components $A$ and $B$ can be obtained by simple arithmetic from measurements of three chemically identical samples with different isotopic compositions — see Fig. 2.23 and Exercise 2.7.

### 2.4.3   Density correlations and diffraction

In this section we provide an alternative view of neutron diffraction by showing that the cross-section can be interpreted in terms of density fluctuations.

**One-component systems**

We can represent the spatial variation in the particle density by

$$n(\mathbf{r}) = \sum_j \delta(\mathbf{r} - \mathbf{r}_j). \qquad (2.56)$$

The function $n(\mathbf{r})$ is extremely 'spiky', but it does correctly represent the density of particles because integration of $n(\mathbf{r})$ over any volume of space yields the number of particles in that volume.

[24]For an isotropic system we can define $g_{\alpha\beta}(r)$ in a form analogous to (2.44),

$$c_\beta n_0 g_{\alpha\beta}(r) 4\pi r^2 \, dr = dN_{\alpha\beta},$$

where $dN_{\alpha\beta}$ is the number of particles of type $\beta$ between distances $r$ and $r + dr$ from a particle of type $\alpha$.

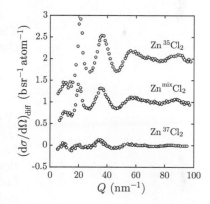

**Fig. 2.23** Diffraction from samples of molten $ZnCl_2$ prepared with different ratios of the isotopes $^{35}Cl$ and $^{37}Cl$ — 99.3:0.7 (top); 67.7:32.3 (middle); 2.7:97.3 (bottom). The quantity plotted is the 'difference' differential cross-section, i.e. the second term in eqn (2.52), and the top two curves have been offset vertically for clarity. (After Biggin and Enderby, 1981.)

We anticipate that density fluctuations may be correlated, and we characterize these correlations by the function

$$G(\mathbf{r}) = \frac{1}{N} \int \langle n(\mathbf{r}')n(\mathbf{r}' + \mathbf{r}) \rangle \, d^3\mathbf{r}'. \tag{2.57}$$

$G(\mathbf{r})$ is called the *density–density correlation function*, and is constructed from the product of the densities at two points in the material separated by $\mathbf{r}$, averaged over all configurations of the entire system.

From (2.56),

$$G(\mathbf{r}) = \frac{1}{N} \sum_{jk} \int \langle \delta(\mathbf{r}' - \mathbf{r}_j)\delta(\mathbf{r}' + \mathbf{r} - \mathbf{r}_k) \rangle \, d^3\mathbf{r}'$$

$$= \frac{1}{N} \sum_{jk} \langle \delta(\mathbf{r} - \mathbf{r}_k + \mathbf{r}_j) \rangle$$

$$= n_0 g(\mathbf{r}) + \delta(\mathbf{r}). \tag{2.58}$$

The last line follows from (2.45), and shows that $G(\mathbf{r})$ and $g(\mathbf{r})$ are closely related. Hence, from (2.49),

$$S(\mathbf{Q}) = \int G(\mathbf{r}) \exp(i\mathbf{Q} \cdot \mathbf{r}) \, d^3\mathbf{r}. \tag{2.59}$$

We see, therefore, that for single-component systems, neutron diffraction measures the Fourier transform of the density–density correlation function. In the next chapter we will show that in the Born approximation the formulation of neutron scattering theory in terms of correlation functions is completely general. Also, in Chapter 5 we will find that inelastic scattering from nuclei probes correlations in time as well as position.

**Multicomponent systems**

As a generalization of (2.57)–(2.59) for multicomponent systems we may define the partial density–density correlation functions

$$G_{\alpha\beta}(\mathbf{r}) = \frac{1}{Nc_\alpha c_\beta} \int \langle \hat{n}_\alpha(\mathbf{r}')\hat{n}_\beta(\mathbf{r}' + \mathbf{r}) \rangle \, d^3\mathbf{r}'$$

$$= \frac{1}{Nc_\alpha c_\beta} \sum_{j_\alpha, k_\beta} \langle \delta(\mathbf{r} - \mathbf{r}_{k_\beta} + \mathbf{r}_{j_\alpha}) \rangle, \tag{2.60}$$

so that the partial static structure factors in (2.52) are given by

$$S_{\alpha\beta}(\mathbf{Q}) - 1 = \int G_{\alpha\beta}(\mathbf{r}) \exp(i\mathbf{Q} \cdot \mathbf{r}) \, d^3\mathbf{r} - \frac{1}{c_\alpha}\delta_{\alpha\beta}, \tag{2.61}$$

where $\delta_{\alpha\beta}$ is the Kronecker delta,[25] $c_\alpha = N_\alpha/N$ is the concentration of species $\alpha$, and $N = \sum_\alpha N_\alpha$ is the total number of particles.

[25]Not to be confused with the Dirac delta function. The Kronecker delta is defined as

$$\delta_{\alpha\beta} = \begin{cases} 1 & \text{if } \alpha = \beta \\ 0 & \text{if } \alpha \neq \beta. \end{cases}$$

Leopold Kronecker was a nineteenth century mathematician, who is quoted as having said 'God made the integers, everything else is the work of man'.

## Scattering length density correlations

We can also extend the notion of density–density correlations to include the scattering lengths of the nuclei. This is accomplished by defining the *scattering length density* as in eqn (1.25),

$$n_b(\mathbf{r}) = \sum_j \overline{b}_j \delta(\mathbf{r} - \mathbf{r}_j), \tag{2.62}$$

and the correlation function for the scattering length density,

$$G_b(\mathbf{r}) = \frac{1}{N} \int \langle n_b(\mathbf{r}') n_b(\mathbf{r}' + \mathbf{r}) \rangle \, \mathrm{d}^3 \mathbf{r}'. \tag{2.63}$$

It follows that $G_b(\mathbf{r})$ is related to the coherent cross-section by

$$\left( \frac{\mathrm{d}\sigma}{\mathrm{d}\Omega} \right)_{\mathrm{coh}} = N \int G_b(\mathbf{r}) \exp(\mathrm{i}\mathbf{Q} \cdot \mathbf{r}) \, \mathrm{d}^3 \mathbf{r}. \tag{2.64}$$

This shows quite generally that neutrons probe the scattering-length-weighted density–density correlation function.[26]

## 2.4.4   The zero-Q limit and bulk fluctuations

In this section we consider the scattering at vanishingly small $\mathbf{Q}$. As noted in Section 2.4.2, the coherent cross-section contains a delta function at $\mathbf{Q} = 0$ which we never observe experimentally because it is inseparable from the unscattered neutron beam. What we are interested in here, therefore, is not the scattering exactly at $\mathbf{Q} = 0$, but the limit of the scattering as $\mathbf{Q} \to 0$.

We consider first a one-component system, for which we shall find that the scattering depends on fluctuations in particle density. To show this, we write the scattering in terms of the difference between the local particle density and the average particle density,

$$\Delta n(\mathbf{r}) = n(\mathbf{r}) - n_0. \tag{2.65}$$

Here, $n_0 = N/V$ is the bulk density, i.e. the total number of particles in the sample divided by the total volume of the sample, and is a constant. As discussed in Section 2.4.2, neutrons probe correlations within a scattering volume usually much smaller than the physical size of the sample as a whole. The average density within a scattering volume $v$,

$$n = \frac{1}{v} \int_v n(\mathbf{r}) \, \mathrm{d}^3 \mathbf{r}, \tag{2.66}$$

may vary throughout the sample, in which case the bulk density is an ensemble average of $n$ over many scattering volumes:

$$n_0 = \langle n \rangle. \tag{2.67}$$

Substituting (2.65) into (2.48), (2.57), and (2.59) we obtain

$$\left( \frac{\mathrm{d}\sigma}{\mathrm{d}\Omega} \right)_{\mathrm{coh}} = |\overline{b}|^2 \int \int \langle \{ \Delta n(\mathbf{r}') \Delta n(\mathbf{r}' + \mathbf{r}) + n_0^2 \} \rangle \exp(\mathrm{i}\mathbf{Q} \cdot \mathbf{r}) \, \mathrm{d}^3 \mathbf{r}' \, \mathrm{d}^3 \mathbf{r}. \tag{2.68}$$

[26] By comparing (2.64) with (2.52) and (2.54) we can also write $G_b(\mathbf{r})$ in terms of partial pair distribution functions:

$$G_b(\mathbf{r}) = \sum_{\alpha\beta} c_\alpha c_\beta \overline{b}_\alpha^* \overline{b}_\beta n_0 g_{\alpha\beta}(\mathbf{r}) + \left( \sum_\alpha c_\alpha |\overline{b}_\alpha|^2 \right) \delta(\mathbf{r}).$$

Both integrals extend over the same scattering volume, and the omitted terms containing $\langle \Delta n(\mathbf{r}') \rangle$ and $\langle \Delta n(\mathbf{r}' + \mathbf{r}) \rangle$ vanish after performing the integration and ensemble average. This follows from (2.65)–(2.67). The $n_0^2$ term in (2.68) gives rise to a term proportional to $\delta(\mathbf{Q})$. This is the same delta function we subtracted off $S(\mathbf{Q})$ to define $S'(\mathbf{Q})$ in (2.50), and we do not see it if we extrapolate the scattering back to zero $\mathbf{Q}$. Hence,

$$\lim_{\mathbf{Q} \to 0} \left( \frac{d\sigma}{d\Omega} \right)_{\text{coh}} = |\bar{b}|^2 \int \int \langle \Delta n(\mathbf{r}') \Delta n(\mathbf{r}' + \mathbf{r}) \rangle \, d^3\mathbf{r}' \, d^3\mathbf{r}. \qquad (2.69)$$

If we substitute $\mathbf{r}'' = \mathbf{r}' + \mathbf{r}$, and integrate over $\mathbf{r}''$ instead of $\mathbf{r}$, then

$$\lim_{\mathbf{Q} \to 0} \left( \frac{d\sigma}{d\Omega} \right)_{\text{coh}} = |\bar{b}|^2 \int \int \langle \Delta n(\mathbf{r}') \Delta n(\mathbf{r}'') \rangle \, d^3\mathbf{r}' \, d^3\mathbf{r}''$$

$$= |\bar{b}|^2 \left\langle \left\{ \int \Delta n(\mathbf{r}') \, d^3\mathbf{r}' \right\}^2 \right\rangle$$

$$= |\bar{b}|^2 \langle (\Delta n)^2 \rangle v^2 \qquad (2.70)$$

The last line follows from (2.65) and (2.66), and shows that the zero-$\mathbf{Q}$ limit of the scattering is proportional to the mean squared fluctuation in the particle density on a scale of the scattering volume.

The significance of this result is that $\langle (\Delta n)^2 \rangle$ is directly related to the compressibility of the material, which is measurable by bulk techniques. To establish the relationship let the scattering volume contain a fixed number of particles. Density fluctuations about the average density $n_0$ convert to volume fluctuations about the average volume $v_0$ via $\Delta n = -(n_0/v_0)\Delta v$, so that

$$\langle (\Delta n)^2 \rangle v_0^2 = n_0^2 \langle (\Delta v)^2 \rangle. \qquad (2.71)$$

We can regard the particles in $v$ to be a small subsystem in equilibrium with the surrounding system, which is maintained at fixed temperature $T$ and pressure $p$. Standard thermodynamic arguments can then be used to show that

$$\langle (\Delta v)^2 \rangle = k_{\text{B}} T v_0 \kappa_T, \qquad (2.72)$$

where $\kappa_T$ is the bulk isothermal compressibility, defined by

$$\kappa_T = -\frac{1}{V} \left( \frac{\partial V}{\partial p} \right)_T. \qquad (2.73)$$

The compressibility measures 'squashiness'. Equations (2.71) and (2.72) tell us that the more squashy a material is, the greater is the amplitude of the density fluctuations, which also accords with our intuition.

Equations (2.70)–(2.72) show that the cross-section for volume $v_0 \approx v$ is proportional to $v_0$, so to obtain the cross-section for the sample as a whole we simply replace $v_0$ by $V$. This means that the final result does not depend on the size of the scattering volume. Since $N = n_0 V$,

$$\lim_{\mathbf{Q} \to 0} \left( \frac{d\sigma}{d\Omega} \right)_{\text{coh}} = N|\bar{b}|^2 S'(0) = N|\bar{b}|^2 n_0 k_{\text{B}} T \kappa_T, \qquad (2.74)$$

whereupon,[27]

$$S'(0) = n_0 k_B T \kappa_T. \tag{2.75}$$

[27]Remember that $S'(\mathbf{Q})$ is the same as $S(\mathbf{Q})$ apart from the delta function at $\mathbf{Q} = 0$ in the latter — see eqn (2.50).

For systems containing more than one species, a more general treatment would show that the zero-$\mathbf{Q}$ scattering depends on fluctuations in composition as well as density.

Most solids and condensed fluids have a relatively small compressibility, and so the coherent scattering almost vanishes as $\mathbf{Q} \to 0$. An exception to this is in the vicinity of phase transitions, where density fluctuations can become substantial. A gas, on the other hand, is highly compressible. From the ideal gas equation of state $p = n_0 k_B T$ the compressibility (2.73) is found to be $\kappa_T = 1/(n_0 k_B T)$, so that

$$S'(0) = 1 \qquad [\text{ideal gas}]. \tag{2.76}$$

Physically this reflects that the particles in an ideal gas have no volume and do not interact, i.e. the particles are uncorrelated.

## 2.5   Molecular systems

Suppose we have a disordered material composed not of single atoms but of rigid molecules, such as liquid ammonia ($NH_4$). The summation over pairs of atoms in eqn (2.46) can then be split into *intra-molecular* and *inter-molecular* terms, according to whether $j$ and $k$ belong to the same or to different molecules:

$$\left(\frac{d\sigma}{d\Omega}\right)_{\text{coh}} = \left( \sum_{\substack{jk \\ \text{intra}- \\ \text{molecule}}} + \sum_{\substack{jk \\ \text{inter}- \\ \text{molecule}}} \right) \bar{b}_j^* \bar{b}_k \, \langle \exp\{i\mathbf{Q}\cdot(\mathbf{r}_k - \mathbf{r}_j)\}\rangle. \tag{2.77}$$

Evaluation of the *inter-* term requires information about the way the molecules fit together to form the local structure of the material, and considerations analogous to those discussed in the previous sections apply. The intra-molecular summation is somewhat easier to calculate since it depends only on the orientation and internal structure of the individual molecules.

In order to discuss the intra- and inter-molecular terms separately, let us for simplicity consider a molecular fluid with only one type of molecule, and define the functions

$$P(\mathbf{Q}) = \frac{1}{|B|^2} \sum_{\substack{jk \\ \text{intra}- \\ \text{molecule}}} \bar{b}_j^* \bar{b}_k \, \langle \exp\{i\mathbf{Q}\cdot(\mathbf{r}_k - \mathbf{r}_j)\}\rangle \tag{2.78}$$

and

$$S_{12}(\mathbf{Q}) = \frac{1}{N|B|^2} \sum_{\substack{jk \\ \text{inter}- \\ \text{molecule}}} \bar{b}_j^* \bar{b}_k \, \langle \exp\{i\mathbf{Q}\cdot(\mathbf{r}_k - \mathbf{r}_j)\}\rangle \tag{2.79}$$

where,

$$B = \sum_{\substack{j \\ \text{intra}- \\ \text{molecule}}} \bar{b}_j. \tag{2.80}$$

The summations in (2.78) and (2.79) are now over all atoms in a single molecule and a single pair of molecules, respectively. The constant $B$ is the coherent scattering length of the molecule as a whole, and normalizes the function $P(\mathbf{Q})$ to unity at $\mathbf{Q} = 0$. From (2.77)–(2.80) we can now write

$$\left(\frac{\mathrm{d}\sigma}{\mathrm{d}\Omega}\right)_{\mathrm{coh}} = N|B|^2\{P(\mathbf{Q}) + S_{12}(\mathbf{Q})\}, \qquad (2.81)$$

where $N$ is now the total number of molecules in the sample. For a sample containing molecules of several different types the above formalism can easily be generalized by the use of species-dependent functions, as employed to describe multicomponent systems in Section 2.4.2.

### 2.5.1  The form factor

The intra-molecular scattering function $P(\mathbf{Q})$ depends on the size, shape and orientation of the molecules, and is called the *form factor*. It is what we would measure if the molecules were completely independent of one another, e.g. in an ideal gas. Neutron diffraction measurements on molecules in the gaseous phase are possible, but difficult. Of greater practical relevance is that there are occasions when it is possible to determine the intra-molecular scattering even when the sample is a condensed phase. This is often the goal of low $Q$ diffraction (generally known as *small-angle scattering*, or SANS) which, as discussed in Section 2.6, is designed to study the size and internal structure of large objects.

A number of different methods can be used to calculate form factors from (2.78) depending on the nature of the system under investigation. Here we illustrate the basic idea with a very simple system, the homonuclear diatomic molecule. Section 2.6 contains calculations of $P(\mathbf{Q})$ for more complex systems.

For simple molecules it is convenient to calculate the form factor starting from the expression

$$P(\mathbf{Q}) = \frac{1}{|B|^2}\left\langle |A(\mathbf{Q})|^2 \right\rangle, \qquad (2.82)$$

where[28]

$$A(\mathbf{Q}) = \sum_j \bar{b}_j \, \exp(\mathrm{i}\mathbf{Q} \cdot \mathbf{r}_j). \qquad (2.83)$$

[28]The function $A(\mathbf{Q})$ defined in (2.83) is the same $A(\mathbf{Q})$ as used in expressions for unpolarized neutron scattering elsewhere in this book.

Equations (2.82) and (2.83) are equivalent to (2.78), $B$ being the scattering length of the particle as a whole. The summation in (2.83) is over the scattering centres within the scattering particle. This formulation is useful when the positions of the atoms within the scattering particles are well defined. The ensemble average then takes account only of variations in the orientation of the particles.

Homonuclear diatomic molecules, such as $N_2$, contains two identical atoms bonded together. The molecule has a dumbbell shape. We shall assume that individual molecules are randomly oriented and that there are no correlations between the molecules. We shall also assume, as elsewhere in this chapter, that the atoms are entirely rigid.

Consider first a molecule whose axis is aligned at an angle $\theta$ to the direction of the scattering vector $\mathbf{Q}$ (Fig. 2.24). Let the origin $O$ be at the centre of mass of a molecule. From eqn (2.78) and (2.80) the contribution from this molecule to the form factor is (for real $\bar{b}$)

$$|A(\mathbf{Q})|^2 = \bar{b}^2 \left\{ 2 + \exp(iQd\cos\theta) + \exp(-iQd\cos\theta) \right\}$$

$$= 2\bar{b}^2 \left\{ 1 + \cos(Qd\cos\theta) \right\}, \tag{2.84}$$

where $d$ is the distance between the two atoms in the molecule.

We now average over an ensemble of molecules taking into account their random orientations. To do this we use the fact that the fraction of all directions represented by angles $\theta$ in the range $\theta$ to $\theta+d\theta$ is $\frac{1}{2}\sin\theta\,d\theta$. The orientational average, therefore, is performed by multiplication of the right-hand side of eqn (2.84) by $\frac{1}{2}\sin\theta\,d\theta$, and integration over $\theta$ from 0 to $\pi$. With $B = 2\bar{b}$, the result for the form factor is

$$P(Q) = \left\{ \frac{1}{2} + \frac{\sin(Qd)}{2Qd} \right\}. \tag{2.85}$$

This expression can also be obtained directly from the orientational-averaged cross-section (2.43) together with (2.80)–(2.81). The $Q$ dependence of the form factor (2.85) is plotted in Fig. 2.25.

## 2.5.2 Inter-molecular interference

The inter-molecular term $S_{12}(\mathbf{Q})$ in the cross-section (2.81) depends both on the internal structure of the molecules and on how the molecules as a whole fit together. In general, these properties are not separable. However, if the molecules are such that the way they pack together does not depend on their shape then we can describe inter-molecular correlations with a molecular structure factor and a molecular pair distribution function, analogous to those of a simple liquid (Section 2.4.2). Here we will show how this can be done.

First, we label the molecules by an index $l$, $1 \leq l \leq N$, where $N$ is the total number of molecules. We can now write the position vector of the $d$th atom (strictly speaking the $d$th nucleus) in the $l$th molecule as

$$\mathbf{r}_{ld} = \mathbf{l} + \mathbf{d}_l, \tag{2.86}$$

where $\mathbf{l}$ is the position vector of the centre of mass of the $l$th molecule, and $\mathbf{d}_l$ is the displacement of the $d$th atom from the centre of mass. In terms of these coordinates the inter-molecular term (2.79) in the coherent cross-section (2.81) becomes

$$S_{12}(\mathbf{Q}) = \frac{1}{N|B|^2} \left\langle \sum_{l \neq l'} \exp\{i\mathbf{Q} \cdot (\mathbf{l}' - \mathbf{l})\} \sum_{dd'} \bar{b}_d^* \bar{b}_{d'} \exp\{i\mathbf{Q} \cdot (\mathbf{d}_{l'}' - \mathbf{d}_l)\} \right\rangle. \tag{2.87}$$

This expression for $S_{12}(\mathbf{Q})$ is completely general, and cannot be developed further without assumptions. For the particular case of near-spherical molecules it is reasonable to assume that inter-molecular forces

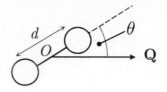

**Fig. 2.24** The coordinate system used to calculate the form factor for a homonuclear diatomic molecule.

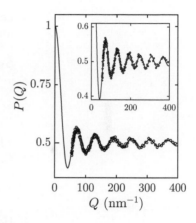

**Fig. 2.25** Form factor for a homonuclear diatomic molecule, eqn (2.85), plotted together with data for nitrogen ($N_2$) gas at 25°C and 35 atm pressure. The insert is an enlargement of the central part of the plot. The curve is calculated for an internuclear separation of 0.107 nm. (After Page and Powles, 1975.)

are not affected by the orientations of the molecules. Hence,

$$S_{12}(\mathbf{Q}) = \frac{1}{|B|^2}\{S_{\rm cm}(\mathbf{Q}) - 1\}\left\langle \sum_{dd'} \bar{b}_d^* \bar{b}_{d'} \exp\{i\mathbf{Q}\cdot(\mathbf{d}'-\mathbf{d})\}\right\rangle, \quad (2.88)$$

where $S_{\rm cm}(\mathbf{Q})$ is the molecular structure factor, which describes inter-molecular correlations in the same way as the structure factor $S(\mathbf{Q})$ does for simple liquids. Thus, we can express $S_{\rm cm}(\mathbf{Q})$ in terms of a molecular pair distribution function $g_{\rm cm}(\mathbf{r})$ with the same physical meaning as explained in Section 2.4.2, only with molecules in place of atoms.

If we further neglect correlations between displacements $\mathbf{d}$ and $\mathbf{d}'$, i.e. we assume that the relative orientations of different molecules are uncorrelated, as would be expected for near-spherical molecules, then

$$S_{12}(\mathbf{Q}) = \frac{1}{|B|^2}\{S_{\rm cm}(\mathbf{Q}) - 1\}\left\langle \sum_d \bar{b}_d^* \exp(-i\mathbf{Q}\cdot\mathbf{d})\right\rangle\left\langle \sum_{d'} \bar{b}_{d'} \exp(i\mathbf{Q}\cdot\mathbf{d}')\right\rangle$$

$$= \frac{1}{|B|^2}\{S_{\rm cm}(\mathbf{Q}) - 1\}\,|\langle A(\mathbf{Q})\rangle|^2\,, \quad (2.89)$$

where $A(\mathbf{Q})$ is defined in (2.83). For spherical molecules, $|\langle A(\mathbf{Q})\rangle|^2 = \langle|A(\mathbf{Q})|^2\rangle = |B|^2 P(Q)$, so that the cross-section (2.81) simplifies to

$$\left(\frac{d\sigma}{d\Omega}\right)_{\rm coh} = N|B|^2 P(Q) S_{\rm cm}(Q). \qquad [\text{spherical molecules}] \quad (2.90)$$

Hence, the scattering factorizes into intra-molecular and inter-molecular terms, which means that the molecular structure factor (and hence the molecular pair distribution function) can be extracted from the data provided the molecular form factor is known.

## 2.6   Small-angle neutron scattering

Low $Q$ neutron scattering, which is usually called small-angle neutron scattering[29] (SANS), is used to determine the form factors of large particles of typical size 1–100 nm. More generally, SANS probes the global (i.e. large scale) structure of a scattering object rather than its atomic-scale details. It has become a very important tool in polymer and colloid science, in materials science, and in biology, and is beginning to have an impact in nanotechnology.

When viewed on a macroscopic scale, a SANS scattering system can be regarded as built from composite units containing many atoms each, or even as a continuous object. This coarse-grained description is valid because interference effects in the scattering from the global structure of a scattering particle of characteristic size $R$ appear when $Q \sim 1/R$, well below the typical range of scattering vectors $Q \sim 1/d$ where interference from correlations at the atomic scale show up (here $d$ is the typical separation of the atoms).

In the following sections we calculate the low $Q$ form factors of several types of large object. The examples are chosen to illustrate different methods that can be used to calculate $P(\mathbf{Q})$.

[29]Small angles are needed to achieve the required low $Q$ unless very large wavelengths are available.

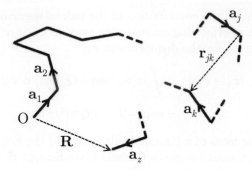

**Fig. 2.26** Random walk model of flexible polymer chain. Each chain segment has length $a$.

## 2.6.1   Flexible linear polymer

Polymers are large molecules constructed from many identical small molecular units called monomers. If we know how these units are connected together then we can in principle calculate the form factor from (2.78), averaging over all orientations and conformations of the molecule. However, the calculation involves very large numbers of atoms, and it is hardly practical to obtain a formula applicable for all $\mathbf{Q}$. Instead, we will find an approximate formula valid for small $\mathbf{Q}$. To do this we ignore individual atoms and instead take the monomers to be the fundamental scattering objects, with each monomer having a scattering length $b_{\mathrm{m}}$ equal to the sum of the coherent scattering lengths of the nuclei in the monomer.

We model the polymer by a linear chain of $z$ independent segments, each of length $a$, which represent the monomers. The segments are attached to one another by infinitely flexible joints so that the orientations of any pair of segments are uncorrelated (Fig. 2.26). This model is equivalent to the famous *random walk* problem in statistics. One property of a random walk is that the mean squared distance travelled after $z$ steps is

$$\langle R^2 \rangle = \langle (\mathbf{a}_1 + \mathbf{a}_2 + \ldots + \mathbf{a}_z)^2 \rangle$$

$$= za^2, \tag{2.91}$$

where the vectors $\mathbf{a}_i$ represent the individual steps of the walk (i.e. the polymer segments). The cross terms $\langle \mathbf{a}_i \cdot \mathbf{a}_j \rangle$ vanish because the steps are uncorrelated. It can be shown that the probability distribution function $w(\mathbf{R})$ for the vector $\mathbf{R}$ joining the start to the finish of the walk tends towards a Gaussian distribution as $z$ tends to infinity:

$$w(\mathbf{R}) = \left( \frac{3}{2\pi za^2} \right)^{\frac{3}{2}} \exp\left( -\frac{3R^2}{2za^2} \right). \tag{2.92}$$

Here, $w(\mathbf{R}) \, \mathrm{d}^3\mathbf{R}$ is the probability that if the walk starts at the origin then it finishes within the volume $\mathrm{d}^3\mathbf{R}$ centred on $\mathbf{R}$. The value of $\langle R^2 \rangle$ calculated from (2.92) agrees with that given in (2.91).

The vector $\mathbf{r}_{jk} = \mathbf{r}_k - \mathbf{r}_j$ joining any two segments $j$ and $k$ follows the same distribution (2.92) except $\mathbf{r}_{jk}$ replaces $\mathbf{R}$ and $|k - j|$ replaces $z$.

Hence, we can evaluate the contribution of the pair of segments $j$ and $k$ to the form factor by performing an ensemble average using $w(\mathbf{r}_{jk})\,\mathrm{d}^3\mathbf{r}_{jk}$ as the weighting factor for the displacement $\mathbf{r}_{jk}$:

$$\langle\exp\{i\mathbf{Q}\cdot(\mathbf{r}_k-\mathbf{r}_j)\}\rangle = \int w(\mathbf{r}_{jk})\exp(i\mathbf{Q}\cdot\mathbf{r}_{jk})\,\mathrm{d}^3\mathbf{r}_{jk}$$

$$= \exp(-|k-j|Q^2a^2/6). \tag{2.93}$$

The integral has the form of a Fourier transform, and the Fourier transform of a Gaussian function is also a Gaussian (see Section B.3.3). From (2.78), the form factor for a flexible polymer is[30]

$$
\begin{aligned}
P(Q) &= \frac{1}{z^2}\sum_{j=1}^{z}\sum_{k=1}^{z}\exp(-|k-j|Q^2a^2/6) \\
&= \frac{1}{z^2}\left\{z + 2\sum_{j=k+1}^{z}\sum_{k=1}^{z-1}x^{(j-k)}\right\} \qquad [\,x=\exp(-Q^2a^2/6)\,] \\
&= \frac{1}{z^2}\left\{z + 2\sum_{m=1}^{z-1}\sum_{l=1}^{m}x^l\right\} \qquad [\,l=j-k \ \text{and} \ m=z-k\,] \\
&\simeq \frac{2}{u^2}\left\{\exp(-u)+u-1\right\}, \tag{2.94}
\end{aligned}
$$

where $u = zQ^2a^2/6 = Q^2R_{\mathrm{g}}^2$. The parameter $R_{\mathrm{g}}$ is the *radius of gyration* of the polymer, see (2.104). The algebra required to reach the last line involves the summation of geometric series. We have neglected terms of order $u/z$, which are small providing $Q \ll 1/a$. The form factor (2.94) is plotted in Fig. 2.27, together with measurements on polyethylene chains under conditions where the polymer is almost ideally Gaussian and intermolecular interference can be neglected.

[30]Expression (2.94) for the form factor of a flexible polymer was first obtained by Peter Debye (1947).

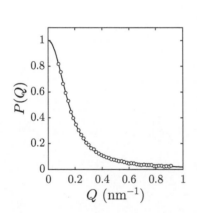

**Fig. 2.27** Form factor of a Gaussian linear polymer. The data are SANS measurements of a polyethylene sample of molecular weight $55\,\mathrm{kg\,mol^{-1}}$ in dilute solution in deuterated diphenyl at $125°C$, and the line is a fit to the Debye form factor eqn (2.94) with radius of gyration $R_{\mathrm{g}} = 10.26\,\mathrm{nm}$. (Adapted from Boothroyd *et al.*, 1989.)

## 2.6.2   Continuous particles: the sphere

A useful simplification for macroscopic scattering bodies is to replace the summation in (2.83) by an integral. This amounts to the replacement of a distribution of coherent scattering lengths $\overline{b}_j$ located at point-like nuclei by a smooth function $n_b(\mathbf{r})$, defined in eqn (2.62), which describes the locally averaged scattering length density. For the same reason as discussed at the beginning of Section 2.6, this procedure is valid providing $Q$ is significantly smaller than reciprocal of the interatomic separation.

With this simplification, eqn (2.83) becomes

$$A(\mathbf{Q}) = \int n_b(\mathbf{r})\exp(i\mathbf{Q}\cdot\mathbf{r})\,\mathrm{d}^3\mathbf{r}. \tag{2.95}$$

We see that $A(\mathbf{Q})$ is the Fourier transform of the scattering length density of the particle.

To illustrate this method let us obtain the form factor for a spherical particle of radius $R$ with a constant scattering length density $n_b$. We

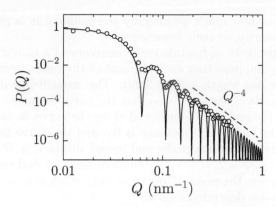

**Fig. 2.28** The form factor of a uniform sphere, eqn (2.97), together with SANS measurements on polystyrene latex spheres of average radius 72 nm in dilute salt solution. The dashed represents Porod's law $P(Q) \sim Q^{-4}$, which describes the scattering from randomly oriented sharp interfaces. (Data courtesy of Rennie *et al.*, 2013.)

express the volume integral in spherical polar coordinates, taking the polar axis along $\mathbf{Q}$:

$$
\begin{aligned}
A(Q) &= n_b \int_0^{2\pi} \int_0^{\pi} \int_0^R \exp(iQr\cos\theta)\, r^2 \sin\theta\, d\phi\, d\theta\, dr \\
&= 4\pi n_b \int_0^R \frac{\sin(Qr)}{Qr}\, r^2\, dr \\
&= \frac{4\pi n_b}{Q^3}\{\sin(QR) - QR\cos(QR)\}.
\end{aligned}
\tag{2.96}
$$

The form factor $P(Q)$ for the sphere, obtained from (2.96), (2.82), and $B = (4/3)\pi R^3 n_b$, is

$$
P(Q) = \frac{9}{u^6}(\sin u - u\cos u)^2, \qquad [u = QR], \tag{2.97}
$$

and is plotted Fig. 2.28. The function has a series of sharp minima and can be used to obtain an accurate value of $R$ from experimental measurements on spherical particles. In practice, the effects of polydispersity and instrumental resolution will cause the minima to be broader and less deep than in the ideal case, as can be seen in Fig. 2.28.

### 2.6.3 Fractals

Fractals are structures that 'look' the same on all length scales.[31] They have been found in many different guises in the natural world, ranging from coastlines and clouds to Romanesco broccoli. A variety of different condensed matter systems exhibit fractal behaviour, amongst which are flexible polymers. As we will now show, the scattering from fractals is characterized by a power-law scaling in $Q$.

For illustration, Fig. 2.29 shows a simple geometrical fractal known as the Koch curve. It is constructed from the fundamental unit shown at the top, which is made from four line segments of equal length. Four of these units are then assembled in the same pattern as the fundamental unit, to make a composite object with sixteen line segments. This procedure is repeated over and over again. The resulting object looks the

[31]The term *fractal* was coined in 1975 by the French mathematician Benoit Mandelbrot (1924–2010).

[32] Obviously there are physical limits. A fractal structure cannot continue below the atomic length scale or exceed the size of the Universe. We are usually content to call an object fractal if it is self-similar over several orders of magnitude of scale.

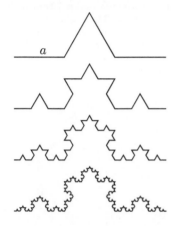

**Fig. 2.29** The Koch curve, an example of a fractal. The curve consists of many copies of the fundamental four-segment motif, and looks the same on all length scales.

same irrespective of how much we magnify the image of it, a property known as *self-similarity*, or *scale invariance*.[32]

An important property of fractals is the existence of a characteristic *fractal dimension*. Suppose that each segment of the fundamental unit of the Koch curve has length $a$ (Fig. 2.29). The straight-line distance between the two ends of the unit is $3a$, but the curve distance, i.e. the distance between the end points measured along the curve, is $4a$. After the second step the end-to-end distance is $9a$, and the curve length is $16a$. So, after the $n$th iteration the end-to-end distance is $R = 3^n a$ and the curve distance is $L = 4^n a$. By eliminating $n$ we find that $L = a(R/a)^{\ln 4/\ln 3}$. Hence, the mass $M$ of the object, which is proportional to $L$, has a power-law dependence on $R$:

$$M \propto R^{\ln 4/\ln 3}. \tag{2.98}$$

Such a power-law relation between $M$ and $R$ is a general feature of fractals, and is characterized by the fractal dimension $D$, defined by

$$M \propto R^D. \tag{2.99}$$

For the Koch curve, $D = \ln 4/\ln 3 \simeq 1.26$, and for a Gaussian polymer $D = 2$, as can be seen from eqn (2.91)

Suppose we perform a neutron diffraction experiment from a fractal object. What would be the form factor? To answer this question we consider the scattering cross-section in the form given by (2.64), i.e. in terms of a correlation function. We assume the scattering length density is proportional to the mass density, so that the form factor is proportional to the Fourier transform of the density–density correlation function $G(\mathbf{r})$, given by (2.57). We also assume the sample contains an ensemble of fractals in random orientations. This means $G(\mathbf{r}) \to G(r)$, $P(\mathbf{Q}) \to P(Q)$, and we can perform the angular part of the Fourier transform as in (2.42), to give

$$P(Q) \propto \int G(r) \frac{\sin(Qr)}{Qr} r^2 \, \mathrm{d}r. \tag{2.100}$$

From (2.57) we see that $G(r)$ describes the average density at a distance $r$ from a point in the system. We can obtain this by considering concentric spheres of radius $r$ and $r + \mathrm{d}r$. The volume enclosed between these spheres is $4\pi r^2 \mathrm{d}r$, and the mass contained in this volume is $M(r+\mathrm{d}r) - M(r)$. Hence, from (2.99),

$$G(r) \propto \lim_{\mathrm{d}r \to 0} \frac{(r + \mathrm{d}r)^D - r^D}{4\pi r^2 \, \mathrm{d}r}$$

$$\propto r^{D-3}. \tag{2.101}$$

Substituting (2.101) into (2.100) we obtain

$$P(Q) \propto \int_{r_0}^{r_\infty} r^{D-3} \frac{\sin(Qr)}{Qr} r^2 \, dr$$

$$\propto \frac{Q^{3-D}}{Q^3} \int_{u_0}^{u_\infty} u^{D-2} \sin u \, du \qquad [\,u = Qr\,]$$

$$\propto Q^{-D} \qquad [\,1/r_\infty \ll Q \ll 1/r_0\,]. \tag{2.102}$$

Here, $r_0$ and $r_\infty$ are the lower and upper length scales between which fractal behaviour applies. The power law obtained for $P(Q)$ in (2.102) is valid within the corresponding range of $Q$ indicated. The SANS data shown in Fig. 2.30 for a silica aerogel exhibits this type of scaling behaviour over two decades of $Q$. Silica aerogels are highly porous networks of connected silica particles, and under the right preparation conditions a fractal network is formed.

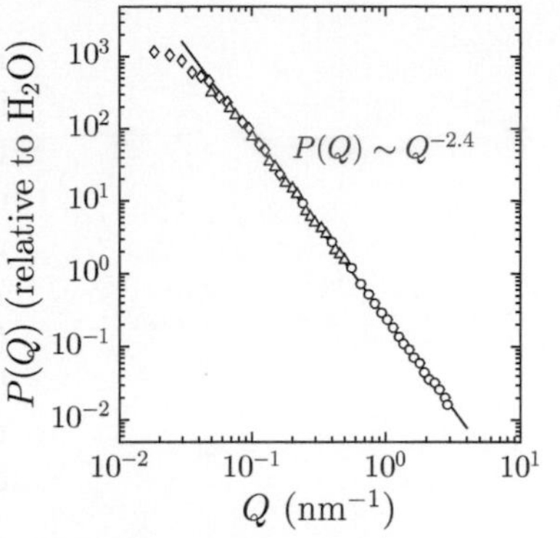

The concept of fractals is very useful in describing the conformation of flexible polymers. The random-walk polymer described in the previous section is a good example. From (2.94) it can be seen that $P(Q) \propto Q^{-2}$ in the range $1/\langle R^2 \rangle^{1/2} \ll Q \ll 1/a$, where $\langle R^2 \rangle^{1/2}$ is the typical size of the polymer chain and $a$ is the size of a monomer. Thus, within these limits a Gaussian flexible polymer can be regarded as a fractal with dimension $D = 2$, as already noted above. Polymers in dilute solution in good solvents are also self-similar, but are less compact than Gaussian polymers and have $D = 5/3$.

**Fig. 2.30** SANS from a self-similar silica aerogel of density $95 \, \mathrm{kgm^{-3}}$ (after Wacher *et al.*, 1988). The straight line is a fit to the scaling form $P(Q) \sim Q^{-D}$ for a mass fractal.

### 2.6.4   The Guinier approximation

When $Q$ is very small the particle form factor takes on a particularly simple approximate form. Assuming the particles are randomly oriented and the scattering lengths real,[33]

$$P(Q) \simeq \exp(-\tfrac{1}{3} Q^2 R_g^2). \tag{2.103}$$

Here, $R_g$ is the *radius of gyration* $R_g$ of the particle, defined by[34]

$$R_g^2 = \frac{1}{B} \int_0^\infty |\mathbf{r} - \mathbf{r}_{\mathrm{cm}}|^2 n_b(\mathbf{r}) \, d^3\mathbf{r}, \tag{2.104}$$

where $B$ is given in eqn (2.80) and $\mathbf{r}_{\mathrm{cm}}$ is the centre of mass of the particle. Equation (2.103), which is derived in Exercise 2.9, is a good approximation when $Q^2 R_g^2 < 1$ and the scattering particles are independent, i.e. the inter-particle interference term $S_{12}(\mathbf{Q}) = 0$.

The Guinier approximation is useful because it can be used to obtain a model-independent value for $R_g$. The usual procedure is to plot $\ln\{I(Q)\}$ vs $Q^2$, where $I(Q)$ is the measured intensity. In the limit as $Q \to 0$ the plot becomes a straight line with gradient $-R_g^2/3$.

[33] If the scattering lengths are complex then $R_g^2$ is replaced by $\mathrm{Re}\{R_g^2\}$ in (2.103).

[34] The standard definition of $R_g$ in mechanics has $\rho(\mathbf{r})/M$ instead of $n_b(\mathbf{r})/B$, where $\rho(\mathbf{r})$ is the mass density within the particle and $M$ is its total mass. The two definitions are the same if the composition of the particle is such that $n_b(\mathbf{r})$ is proportional $\rho(\mathbf{r})$.

## 2.6.5   Particles in continuous media: contrast

Let us now consider the practical scenario in which the particle of interest is embedded in another medium, such as a solvent. We will show here that the low $Q$ scattering from the embedded particle depends on the difference between the scattering length densities of the particle and medium. This difference is known as *contrast*.

For simplicity, we shall assume that the scattering particle is isolated (so that there is no inter-particle interference) and that its size is much larger than the atomic scale. We shall also assume that the medium surrounding the scattering particle is continuous when viewed on a length scale comparable to the size of the scattering particle.

These assumptions allow us to describe the particle and medium by continuous scattering length densities $n_{b_\mathrm{p}}(\mathbf{r})$ and $n_{b_0}$, respectively, and employ the integral form of the coherent cross-section (1.27),

$$\left(\frac{\mathrm{d}\sigma}{\mathrm{d}\Omega}\right)_{\mathrm{coh}} = \left| \int_V n_b(\mathbf{r})\,\exp(\mathrm{i}\mathbf{Q}\cdot\mathbf{r})\,\mathrm{d}^3\mathbf{r} \right|^2 . \qquad (2.105)$$

The integration extends over the whole sample volume $V$, and can be split into separate terms corresponding to the particle and medium. Thus,

$$\int_V n_b(\mathbf{r})\,\mathrm{e}^{\mathrm{i}\mathbf{Q}\cdot\mathbf{r}}\,\mathrm{d}^3\mathbf{r} = \int_{V_\mathrm{p}} n_{b_\mathrm{p}}(\mathbf{r})\,\mathrm{e}^{\mathrm{i}\mathbf{Q}\cdot\mathbf{r}}\,\mathrm{d}^3\mathbf{r} + \int_{V_0} n_{b_0}\,\mathrm{e}^{\mathrm{i}\mathbf{Q}\cdot\mathbf{r}}\,\mathrm{d}^3\mathbf{r}$$

$$= \int_{V_\mathrm{p}} \left\{ n_{b_\mathrm{p}}(\mathbf{r}) - n_{b_0} \right\} \mathrm{e}^{\mathrm{i}\mathbf{Q}\cdot\mathbf{r}}\,\mathrm{d}^3\mathbf{r} + \int_V n_{b_0}\,\mathrm{e}^{\mathrm{i}\mathbf{Q}\cdot\mathbf{r}}\,\mathrm{d}^3\mathbf{r},$$
$$(2.106)$$

where $V_\mathrm{p}$ and $V_0$ are the volumes of the particle and medium, respectively. The second integral in the second line of eqn (2.106) is proportional to $\delta(\mathbf{Q})$ (see Section B.2.3) and so only contributes to the scattering at zero $\mathbf{Q}$ where it is inseparable from the unscattered beam. The measurable part of the coherent cross-section is therefore

$$\left(\frac{\mathrm{d}\sigma}{\mathrm{d}\Omega}\right)_{\mathrm{coh}} = \left| \int_{V_\mathrm{p}} \left\{ n_{b_\mathrm{p}}(\mathbf{r}) - n_{b_0} \right\} \exp(\mathrm{i}\mathbf{Q}\cdot\mathbf{r})\,\mathrm{d}^3\mathbf{r} \right|^2 . \qquad (2.107)$$

As an example, the intra-particle term in the coherent cross-section for a sample containing $N$ spherical particles of radius $R$ and constant scattering length density $n_{b_\mathrm{p}}$, dissolved in a solvent, would be, from (2.81) and (2.107),

$$\left(\frac{\mathrm{d}\sigma}{\mathrm{d}\Omega}\right)_{\mathrm{coh}} = N \left(\frac{4\pi R^3}{3}\right)^2 |n_{b_\mathrm{p}} - n_{b_0}|^2 P(Q), \qquad (2.108)$$

where $P(Q)$ is the form factor for a sphere, eqn (2.97).

Similarly, the intra-molecular cross-section for a solution containing $N$ flexible polymers, each with $z$ monomers, is found to be

$$\left(\frac{\mathrm{d}\sigma}{\mathrm{d}\Omega}\right)_{\mathrm{coh}} = N z^2 |b_m - \beta b_0|^2 P(Q), \qquad (2.109)$$

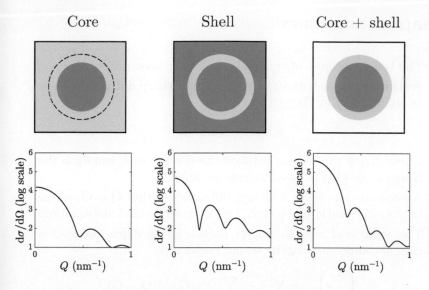

**Fig. 2.31** Contrast variation of different parts of a core-and-shell particle. Left: shell and solvent are contrast-matched. Middle: core and solvent are contrast-matched. Right: no contrast-matching. The simulations are done for a mean core radius of 10 nm, a shell thickness of 4 nm, and a Gaussian core size distribution with standard deviation 0.8 nm.

where $b_m$ and $b_0$ are the scattering lengths of a monomer and a solvent molecule, respectively, $\beta = v_m/v_0$ is the ratio of the volume of a monomer to the volume of a solvent molecule, and $P(Q)$ is the form factor for the polymer.

These two examples serve to illustrate that the intra-molecular scattering depends upon the contrast, i.e. the difference between the scattering length density of the particle and that of the solvent.

If the contrast should happen to be zero then there will be no coherent scattering from the dissolved particles. In this case we say the particle and solvent are *contrast matched*. This opens up a number of interesting possibilities. For example, suppose a solution contains two types of solute particles, call them A and B. If we engineer the system so that the contrast between the B particles and the solvent is zero, making the B particles 'invisible', then we can study the A particles on their own. Alternatively, if the solute particle is built from two distinct units each with a different scattering length density then one can study one of the units on its own by contrast matching the other unit with the solvent.

Figure (2.31) provides an illustration of the method of contrast matching. In this example the solute particle consists of a uniform spherical core coated with an outer shell of a different material. By contrast-matching the solvent to either the shell (left) or the core (middle) one can study the core (left) or shell (middle) in isolation. Without contrast-matching (right) the scattering comes from both parts of the composite particle.

The simplest way to tune contrast in practice is to use a mixture of two chemically identical species with different isotopic compositions. A mixture of protonated and deuterated molecules is often used for this purpose (see Section 1.8.5).

# Chapter summary

- The basic principles of crystallography have been reviewed.
- Scattering from crystals takes the form of sharp diffracted beams at angles given by Bragg's law,

$$n\lambda = 2d_{hkl}\sin\theta_{\mathrm{B}},$$

where $d_{hkl}$ is the spacing between the $(hkl)$ planes, and $\theta_{\mathrm{B}}$ is the Bragg angle ($2\theta_{\mathrm{B}}$ is the scattering angle.)

- A necessary condition for Bragg diffraction is that $\mathbf{Q} = \mathbf{G}$, where $\mathbf{G}$ is a reciprocal lattice vector. This is known as the Laue equation.

- The differential cross-section for nuclear Bragg diffraction from a crystal composed of rigidly bound nuclei is

$$\left(\frac{\mathrm{d}\sigma}{\mathrm{d}\Omega}\right)_{\mathrm{coh}} = N\frac{(2\pi)^3}{v_0}\sum_{\mathbf{G}}|F_{\mathrm{N}}(\mathbf{G})|^2\delta(\mathbf{Q}-\mathbf{G}),$$

where $F_{\mathrm{N}}(\mathbf{G})$ is the structure factor,

$$F_{\mathrm{N}}(\mathbf{G}) = \sum_d \overline{b}_d\,\exp(\mathrm{i}\mathbf{G}\cdot\mathbf{d}).$$

- The pair distribution function for a single-component fluid is

$$g(\mathbf{r}) = \frac{1}{n_0}\sum_{k\neq 0}\langle\delta(\mathbf{r}-\mathbf{r}_k+\mathbf{r}_0)\rangle.$$

$g(\mathbf{r})$ is the factor by which the average particle density at a displacement $\mathbf{r}$ from the origin particle varies from the bulk density.

- The (static) structure factor is related to $g(\mathbf{r})$ by

$$S(\mathbf{Q}) = 1 + n_0\int\{g(\mathbf{r})-1\}\exp(\mathrm{i}\mathbf{Q}\cdot\mathbf{r})\,\mathrm{d}^3\mathbf{r}, \qquad (\mathbf{Q}\neq 0).$$

- The density–density correlation function is defined as

$$G(\mathbf{r}) = \frac{1}{N}\int\langle n(\mathbf{r}')n(\mathbf{r}'+\mathbf{r})\rangle\,\mathrm{d}^3\mathbf{r}'.$$

- Partial or species-dependent functions $S_{\alpha\beta}(\mathbf{Q})$, $g_{\alpha\beta}(\mathbf{r})$ and $G_{\alpha\beta}(\mathbf{r})$ are used for multicomponent systems.

- $S(\mathbf{Q})$ is the Fourier transform of $G_b(\mathbf{r})$, the correlation function for the scattering length density $n_b(\mathbf{r})$.

- The form of $S(\mathbf{Q})$ for molecular systems has been given, and the principles of small-angle scattering have been described.

# Further reading

The International Tables for Crystallography (http://www.iucr.org) contain everything you need to know about crystal structure, crystal symmetry, crystal structure determination, and more besides. Burns and Glazer (2013) is a pedagogical introduction to crystal structure and symmetry, and Sivia (2011) is likewise for scattering theory, especially elastic scattering. A detailed account of disorder and diffuse scattering in crystalline

materials is given in Nield and Keen (2000). Theories of liquids are described fully by Egelstaff (1992) and by Hansen and McDonald (2013), and neutron and X-ray diffraction studies of liquids and glasses are reviewed by Fischer *et al.* (2006). Small-angle neutron scattering from polymers is covered in more detail by Higgins and Benoit (1994), and the theory of SANS by certain fractal systems is described by Teixeira (1988).

# Exercises

(2.1) Show (a) that the reciprocal lattice vector $\mathbf{G}$ is perpendicular to the planes with Miller indices $(hkl)$, and (b) that $|\mathbf{G}| = 2\pi/d_{hkl}$, where $d_{hkl}$ is the interplanar spacing.

(2.2) Show that diffraction from crystals having a body-centred lattice type is allowed when $h+k+l = 2n$, where $n$ is an integer. Derive the equivalent selection rule for the C base-centred lattice type. Explain the relation between the selection rule and the reciprocal lattice.

(2.3) The crystal structure of silicon (Si) has a face-centred cubic lattice and a basis consisting of Si atoms at $0,0,0$ and $\frac{1}{4},\frac{1}{4},\frac{1}{4}$. Draw the conventional unit cell in plan view, and show that Bragg reflections with $h+k+l = 4n+2$ are systematically absent.

(2.4) This exercise is based on a landmark experiment by C. G. Shull *et al.*, *Phys. Rev.* **73**, 842 (1948).

X-ray diffraction from sodium hydride (NaH) had established that the Na atoms are arranged on a face-centred cubic lattice. The H atoms were thought to be displaced from the Na atoms either by $\frac{1}{4},\frac{1}{4},\frac{1}{4}$ or by $\frac{1}{2},\frac{1}{2},\frac{1}{2}$, to form the zinc blende (ZnS) structure or the sodium chloride (NaCl) structure, respectively. To distinguish these models, Shull *et al.* performed a neutron powder diffraction measurement. The intensity of the 111 Bragg peak was found to be much larger than the intensity of the 200 peak. Deduce which of the two structure models for NaH is correct.

(2.5) Silver chloride (AgCl) has a face-centred cubic lattice with cell parameter $a = 0.556$ nm. The structure is to be investigated with neutrons of wavelength $\lambda = 0.236$ nm.

On a scale of $2\,\text{cm}:10\,\text{nm}^{-1}$ draw the $(hhl)$ section through the reciprocal lattice, i.e. the plane containing all reciprocal lattice points of the form $(hhl)$, where the Miller indices refer to the conventional unit cell. Label reciprocal lattice points that have indices in the range $0 \le h \le 3$ and $0 \le l \le 4$. Calculate the magnitudes of the neutron wavevector and the reciprocal lattice vector $\mathbf{G}_{220} = 2\mathbf{a}^* + 2\mathbf{b}^*$. Construct the scattering triangle for diffraction from the (220) planes. What is the angle $\phi$ between $\mathbf{k}_i$ and $\mathbf{k}_f$? What is the relation between $\phi$ and the Bragg angle $\theta_B$? What is the angle between $\mathbf{k}_i$ and the $c$ axis of the crystal?

(2.6) [Harder]. Derive eqn (2.33) from (2.32).

Hint: write (2.32) in the form

$$h(Q_x) = \int_{-\infty}^{\infty} f(x)g(x)\exp(iQ_x x)\,dx,$$

where

$$f(x) = \sum_{n=-\infty}^{\infty} \delta(x - na)$$
$$g(x) = \exp(-|x|/\xi),$$

and then use the convolution theorem (B.27) to

write

$$h(Q_x) = \frac{1}{2\pi} \int_{-\infty}^{\infty} F(Q')G(Q_x - Q') \, dQ',$$

with $F$ and $G$ the Fourier transforms of $f$ and $g$.

(2.7) Equation (2.52) expresses the differential cross-section for a multicomponent fluid in terms of the partial structure factors. Calculate the weights $c_\alpha c_\beta \bar{b}_\alpha^* \bar{b}_\beta$ associated with each partial structure factor of molten $ZnCl_2$ for the three samples measured by Biggin and Enderby, 1981. Solve the equations to obtain the three partial structure factors in terms of the differential cross-sections. Considering the data shown in Fig. 2.23, describe any difficulties that might be encountered in attempting to obtain accurate partial structure factors.

(2.8) Starting from the form factor for a sphere, eqn (2.97), and the differential cross-section for a dilute dispersion of spherical particles, eqn (2.108), derive Porod's law,

$$\left(\frac{d\sigma}{d\Omega}\right)_{\text{coh}} \simeq \frac{2\pi(\Delta n_b)^2 S}{Q^4} \quad (Q \gg 1/R),$$

where $\Delta n_b$ is the contrast between the scattering length density of the sphere and solvent, and $S$ is the total surface area of the spheres.

(2.9) Derive Guinier's approximation, eqn (2.103), for $QR_g \ll 1$. Start from

$$P(Q) = \frac{1}{B^2} \left\langle \left| \int n_b(\mathbf{r}) \exp(i\mathbf{Q} \cdot \mathbf{r}) \, d^3\mathbf{r} \right|^2 \right\rangle_\Omega,$$

which comes from eqns (2.82) and (2.95) (the angle brackets $\langle \cdots \rangle_\Omega$ indicate an orientational average), and by expanding the exponential function as

$$\exp(i\mathbf{Q} \cdot \mathbf{r}) \simeq 1 + i\mathbf{Q} \cdot \mathbf{r} - \frac{1}{2}(\mathbf{Q} \cdot \mathbf{r})^2 + \cdots .$$

show that

$$P(Q) \simeq 1 - \frac{1}{B} \int n_b(\mathbf{r}) \, \langle (\mathbf{Q} \cdot \mathbf{r})^2 \rangle_\Omega \, d^3\mathbf{r}.$$

Next, average $(\mathbf{Q} \cdot \mathbf{r})^2$ over all orientations of $\mathbf{Q}$ to obtain $\langle (\mathbf{Q} \cdot \mathbf{r})^2 \rangle_\Omega = Q^2 r^2 / 3$. Finally, taking the origin to be at the centre of mass in eqn (2.104), i.e. $\mathbf{r}_{\text{cm}} = 0$, show that

$$P(Q) \simeq 1 - \frac{Q^2 R_g^2}{3} \simeq \exp(-\frac{1}{3}Q^2 R_g^2).$$

(2.10) The table below gives SANS data in the Guinier regime for a dilute solution of a jellyfish protein. Determine the radius of gyration of the protein molecules.

| $Q(\text{nm}^{-1})$ | $I(Q)$ | $Q(\text{nm}^{-1})$ | $I(Q)$ |
|---|---|---|---|
| 0.209 | 0.536 | 0.308 | 0.492 |
| 0.226 | 0.534 | 0.332 | 0.476 |
| 0.245 | 0.524 | 0.359 | 0.467 |
| 0.264 | 0.518 | 0.388 | 0.444 |
| 0.285 | 0.499 | 0.419 | 0.434 |

# Kinematical Theory of Scattering

**3**

The previous two chapters described a geometrical method to calculate the scattering cross-section for general structures built from rigid (i.e. static) nuclei. That approach employs the classical theory of waves as used to describe simple interference phenomena in other fields such as physical optics, and is both intuitive and mathematically straightforward. A disadvantage, however, is that it does not readily extend to more general scattering problems, such as magnetic scattering and inelastic scattering from non-rigid structures.

In this chapter we shall outline a more powerful theory of neutron scattering, the so-called *kinematical* theory, which employs quantum-mechanical perturbation theory. Though still not the most general scattering theory, the kinematical theory is sufficient for the vast majority of condensed matter studies, and only becomes inapplicable when strong multiple scattering effects are present. In such cases, the *dynamical theory* described in Chapter 9 is superior.

## 3.1 The partial differential cross-section

The task we face is to derive a general formula for the partial differential cross-section, defined in eqn (1.34) of Chapter 1. To recap, this quantity describes the probability that a neutron with initial wavevector $\mathbf{k}_i$ is scattered by the sample into a final state whose wavevector lies within a small interval about $\mathbf{k}_f$. The acceptable interval includes wavevectors whose direction falls within the range of solid angle $d\Omega$ centred on the direction of $\mathbf{k}_f$, and whose magnitude corresponds to a final energy $E_f$ that lies within the range $dE_f$ centred on $E_f = \hbar^2 k_f^2/(2m_n)$.

For the sake of generality, we shall include from the outset the dependence of the cross-section on the spin of the neutron. We recall that the neutron is a spin-$\frac{1}{2}$ particle, and therefore has two polarization states. These correspond to the spin being parallel or antiparallel to a quantization direction usually defined by an external magnetic field. During the scattering process the polarization state of the neutron may change, or the quantization direction for the scattered neutrons may differ from that of the incident neutrons. We wish to be able to calculate the cross-section for all such processes.

Denoting the initial and final polarization states of the neutron by $\sigma_i$ and $\sigma_f$ respectively (the precise meaning of these labels will be revealed in Chapter 4), we may write the partial differential cross-section as[1]

$$\left(\frac{\mathrm{d}^2\sigma}{\mathrm{d}\Omega\mathrm{d}E_f}\right)_{(\mathbf{k}_i,\sigma_i)\to(\mathbf{k}_f,\sigma_f)}. \tag{3.1}$$

To avoid an unwieldy notation, we shall hereafter omit the subscript $(\mathbf{k}_i,\sigma_i)\to(\mathbf{k}_f,\sigma_f)$.

By writing the partial differential cross-section in the form (3.1) we emphasize that it relates to the change in state of the neutron. The real interest, however, is in the intrinsic properties of the sample. What we must do now, therefore, is determine what the experimentally measured cross-section tells us about the scattering system.

When a neutron is scattered from state $(\mathbf{k}_i,\sigma_i)$ to state $(\mathbf{k}_f,\sigma_f)$ there is an associated transition in the scattering system from an initial state $\lambda_i$ to a final state $\lambda_f$. For this process to occur the total energy must be conserved, i.e.

$$E_i + E_{\lambda_i} = E_f + E_{\lambda_f}, \tag{3.2}$$

where $E_{\lambda_i}$ and $E_{\lambda_f}$ are the initial and final energies of the scattering system.

To include the possibility that there may be more than one transition that can bring about the change $(\mathbf{k}_i,\sigma_i) \to (\mathbf{k}_f,\sigma_f)$ while conserving energy we sum over all such transitions, weighting each contribution by the probability $p_{\lambda_i}$ for the initial state $\lambda_i$.[2]

The weights $p_{\lambda_i}$ come from the statistical distribution of initial states, which is usually the result of the sample being in thermal equilibrium with its surroundings. By incorporating this weighted summation into eqn (1.34) we can express the partial differential cross-section (3.1) as

$$\left(\frac{\mathrm{d}^2\sigma}{\mathrm{d}\Omega\mathrm{d}E_f}\right) = \lim_{\mathrm{d}\Omega,\,\mathrm{d}E_f\to0} \frac{\sum_{\lambda_i} p_{\lambda_i} \sum_{\lambda_f} R_{i\to f}}{\Phi_0\,\mathrm{d}\Omega\,\mathrm{d}E_f}, \tag{3.3}$$

where $\Phi_0$ is the flux of neutrons incident on the sample, and $R_{i\to f}$ is the transition rate for $(\mathbf{k}_i,\sigma_i,\lambda_i) \to (\mathbf{k}_f,\sigma_f,\lambda_f)$ including all processes in which the final neutron wavevector lies in the interval defined by $(\mathrm{d}\Omega,\mathrm{d}E_f)$ centred on $\mathbf{k}_f$ and which satisfy energy conservation.

## 3.2   The Born approximation

To continue the development of eqn (3.3) we need to evaluate the transition rate $R_{i\to f}$. This can be accomplished by means of time-dependent perturbation theory. The starting point is an incident plane wave with wavevector $\mathbf{k}_i$ which, all of a sudden, experiences a stationary interaction potential represented by $V(\mathbf{r})$. For neutron scattering from condensed matter, $V(\mathbf{r})$ includes terms from the nuclear interaction and from the interaction with electromagnetic fields. The explicit forms for these interactions, some of which depend on the initial and final spin states of the neutron, are given in Chapter 4.

---

[1] There is no need to include an interval $\mathrm{d}\sigma_f$ of final polarization states because the neutron polarization states are discrete and small in number.

[2] We are assuming that the state of the scattering system (i.e. the sample) varies in space and time, so after many scattering events the neutrons measure an average over the statistical distribution of initial states.

The effect of the interaction potential is to scatter the incident neutrons into different plane wave[3] and spin states, with an associated change in the state of the target. The key assumption here is that $V$ is weak, by which we mean that the state function representing the final state of the neutron + target system is hardly any different from that for the incident state. This is known as the *first Born approximation*.

Amongst the final states there may be some having target state and neutron spin equal to $(\lambda_f, \sigma_f)$, and at the same time having neutron wavevector within the desired interval around $\mathbf{k}_f$. Since $E_i$, $E_{\lambda_i}$ and $E_{\lambda_f}$ are all fixed, the requirement of energy conservation, eqn (3.2), means that $R_{i \to f}$ vanishes unless $E_f = E_i + E_{\lambda_i} - E_{\lambda_f}$. Standard quantum-mechanical perturbation theory in the Born approximation can be used to show that the number of transitions per unit time into these states is given by an expression known as *Fermi's Golden Rule*:[4]

$$R_{i \to f} = \frac{2\pi}{\hbar} |\langle f| V |i\rangle|^2 g(E_f) \frac{d\Omega}{4\pi}, \quad \text{if } E_f = E_i + E_{\lambda_i} - E_{\lambda_f},$$

$$= 0 \qquad \text{otherwise.} \qquad (3.4)$$

In this expression, $g(E_f)$ is the density of states function defined such that $g(E_f)dE_f$ is the number of neutron plane wave states in $\mathbf{k}$-space with energies between $E_f$ and $E_f + dE_f$, the factor $d\Omega/(4\pi)$ is the fraction of all directions that lie in the range $d\Omega$, and the Dirac notation for the matrix element is shorthand for the inner product[5]

$$\langle f| V |i\rangle = \int \Psi_f^* V \Psi_i d\tau, \qquad (3.5)$$

where $\Psi_i$ and $\Psi_f$ are the initial and final state functions for the neutron + target, and $d\tau$ represents the composite volume element for all integration variables.

To evaluate $g(E_f)$ we use the device known as *box normalization*, in which we imagine the neutrons and scattering system to be enclosed in a large box. The neutron wave functions $\psi_{\mathbf{k}}(\mathbf{r})$ are normalized plane wave solutions of the time-independent Schrödinger equation,[6] and are given by

$$\psi_{\mathbf{k}}(\mathbf{r}) = \frac{1}{V_0^{\frac{1}{2}}} \exp(i\mathbf{k} \cdot \mathbf{r}), \qquad (3.6)$$

where $V_0$ is the volume of the box. We impose periodic boundary conditions, and this restricts the wavevector components to discrete values $k_x = n(2\pi/L_x)$, etc, where $n$ is an integer and $L_x$ is the size of the box in the $x$ direction. The allowed wavevector states form a lattice of points in $\mathbf{k}$-space with a volume $(2\pi)^3/V_0$ per point, and so the density of allowed states in $\mathbf{k}$-space is

$$g(\mathbf{k}) = \frac{V_0}{(2\pi)^3}. \qquad (3.7)$$

The density of states function $g(E)$ is related to $g(\mathbf{k})$ by

$$g(E)dE = 4\pi k^2 g(\mathbf{k})dk, \qquad (3.8)$$

[3] Actually, the neutrons are scattered into spherical wave states, but the distinction between plane and spherical waves is unimportant at large distances from the target where the curvature of the spherical wavefronts is negligible.

[4] See Section C.1.9.

[5] See Section C.1.3.

[6] See eqn (C.22).

in which the factor $4\pi k^2 \mathrm{d}k$ is the volume in **k**-space containing all states with wavevectors between $k$ and $k + \mathrm{d}k$. Neutrons used in condensed matter experiments are non-relativistic, and so the relationship between energy and wavevector is that given in eqn (1.3),

$$E = \frac{\hbar^2 k^2}{2m_\mathrm{n}}. \tag{3.9}$$

Combining eqns (3.7), (3.8) and (3.9) one finds

$$g(E) = \frac{V_0}{2\pi^2}\frac{m_\mathrm{n}}{\hbar^2}k. \tag{3.10}$$

Returning to eqn (3.4), we introduce a mathematical trick to get round the awkwardness of the energy conservation condition. Let us define a function $f(E_\mathrm{f})$ with the property that $f(E_\mathrm{f}) = 1$ if $E_\mathrm{f}$ lies in the small energy interval $\mathrm{d}E_\mathrm{f}$ centred on $E_\mathrm{i} + E_{\lambda_\mathrm{i}} - E_{\lambda_\mathrm{f}}$, and $f(E_\mathrm{f}) = 0$ otherwise. We then write eqn (3.4) as

$$R_{\mathrm{i}\to\mathrm{f}} = \frac{2\pi}{\hbar}|\langle \mathrm{f}|\,V\,|\mathrm{i}\rangle|^2 g(E_\mathrm{f})\frac{\mathrm{d}\Omega}{4\pi}f(E_\mathrm{f}). \tag{3.11}$$

By introducing the 'filter' function $f(E_\mathrm{f})$ like this we have slightly relaxed the conservation of energy condition since we allow final neutron energies within a range $\mathrm{d}E_\mathrm{f}$ to survive. This inexactness, however, need not worry us as we shall eventually take the limit $\mathrm{d}E_\mathrm{f} \to 0$, at which point energy conservation is once again satisfied.

In order to obtain the sought-after expression for the partial differential cross-section we will substitute eqns (3.10) and (3.11) into eqn (3.3). In doing so, however, the function $f(E_\mathrm{f})$ will need to be divided by $\mathrm{d}E_\mathrm{f}$. The ratio $f(E_\mathrm{f})/\mathrm{d}E_\mathrm{f}$ is a very narrow 'top hat' function with a width $\mathrm{d}E_\mathrm{f}$ and a height $1/\mathrm{d}E_\mathrm{f}$. As $\mathrm{d}E_\mathrm{f} \to 0$ the width of this function tends to zero but its area remains unity. The reader will immediately recognize this to be a delta function:[7]

[7]See Section B.2.

$$\lim_{\mathrm{d}E_\mathrm{f}\to 0} f(E_\mathrm{f})/\mathrm{d}E_\mathrm{f} = \delta(E_{\lambda_\mathrm{f}} - E_{\lambda_\mathrm{i}} - \hbar\omega), \tag{3.12}$$

where $\hbar\omega = E_\mathrm{i} - E_\mathrm{f}$ is the *energy transfer* (see Section 1.5), i.e. the amount of energy transferred from the neutron to the target.

The last quantity we need in eqn (3.3) is the incident neutron flux $\Phi_0$. Neutron flux is discussed in Section 1.1.4 and given by eqn (1.8). The wave function for the incident neutrons has the form of (3.6), and the speed of the (non-relativistic) incident neutrons is $v = \hbar k_\mathrm{i}/m_\mathrm{n}$. Therefore,

$$\Phi_0 = \frac{1}{V_0}\frac{\hbar k_\mathrm{i}}{m_\mathrm{n}}. \tag{3.13}$$

Bundling eqns (3.10)–(3.13) into (3.3) we obtain

$$\frac{\mathrm{d}^2\sigma}{\mathrm{d}\Omega \mathrm{d}E_\mathrm{f}} = V_0^2\frac{k_\mathrm{f}}{k_\mathrm{i}}\left(\frac{m_\mathrm{n}}{2\pi\hbar^2}\right)^2\sum_{\lambda_\mathrm{i}}p_{\lambda_\mathrm{i}}\sum_{\lambda_\mathrm{f}}|\langle \mathrm{f}|\,V\,|\mathrm{i}\rangle|^2\delta(E_{\lambda_\mathrm{f}} - E_{\lambda_\mathrm{i}} - \hbar\omega). \tag{3.14}$$

We note that the cross-section is independent of the volume $V_0$ of the hypothetical box, as we should expect, because the factor $V_0^2$ shown explicitly in eqn (3.14) is cancelled by a factor $1/V_0^2$ that arises in the squared matrix element $|\langle f| V |i\rangle|^2$ due to the plane-wave normalization prefactor $1/V_0^{\frac{1}{2}}$ included in eqn (3.6). To avoid the appearance of $V_0$ altogether we shall henceforth omit this prefactor and instead use the form $\exp(i\mathbf{k}\cdot\mathbf{r})$ for the neutron wave functions in the matrix element, which we shall write as $\langle \mathbf{k}_f\sigma_f\lambda_f| V |\mathbf{k}_i\sigma_i\lambda_i\rangle$.

We proceed by performing the integral over the spatial coordinates of the neutron contained in the matrix element:[8]

$$\langle \mathbf{k}_f| V |\mathbf{k}_i\rangle = \int \exp(-i\mathbf{k}_f\cdot\mathbf{r})V(\mathbf{r})\exp(i\mathbf{k}_i\cdot\mathbf{r})\,\mathrm{d}^3\mathbf{r}$$
$$= \int V(\mathbf{r})\exp(i\mathbf{Q}\cdot\mathbf{r})\,\mathrm{d}^3\mathbf{r}$$
$$= V(\mathbf{Q}), \tag{3.15}$$

[8]The spin part of the matrix element is independent of the spatial part, and will be considered in Chapter 4.

where $V(\mathbf{Q})$ is the Fourier transform of the interaction potential. The result that neutron scattering depends on the Fourier transform of $V(\mathbf{r})$ is an inherent property of the Born approximation which, as noted earlier, assumes that both the incident and scattered waves are plane waves.

Putting (3.15) in (3.14) we arrive at the central result of this section, which is a master equation for the partial differential cross-section:

$$\frac{\mathrm{d}^2\sigma}{\mathrm{d}\Omega\mathrm{d}E_f} = \frac{k_f}{k_i}\left(\frac{m_n}{2\pi\hbar^2}\right)^2\sum_{\lambda_i}p_{\lambda_i}\sum_{\lambda_f}|\langle\sigma_f\lambda_f| V(\mathbf{Q}) |\sigma_i\lambda_i\rangle|^2\delta(E_{\lambda_f}-E_{\lambda_i}-\hbar\omega). \tag{3.16}$$

## 3.3 Elastic scattering

Let us pause for a moment and compare eqn (3.16) with the results obtained in Chapter 2 for the special case of a rigid scattering target.

If the atoms are rigidly bound then the neutron does not change the state of the target in the scattering process. More technically, the perturbing operator $V$ does not act on the target eigenstates. In this case the matrix element in eqn (3.16) factorizes:

$$\langle\sigma_f\lambda_f| V(\mathbf{Q}) |\sigma_i\lambda_i\rangle = \langle\sigma_f| V(\mathbf{Q}) |\sigma_i\rangle\langle\lambda_f|\lambda_i\rangle$$
$$= \langle\sigma_f| V(\mathbf{Q}) |\sigma_i\rangle. \tag{3.17}$$

The second line follows because for a rigid target $\lambda_i = \lambda_f$, and the target eigenfunctions are orthonormal:

$$\langle\lambda_f|\lambda_i\rangle = 1 \quad (\lambda_i = \lambda_f),$$
$$= 0 \quad (\lambda_i \neq \lambda_f). \tag{3.18}$$

Equation (3.16) then further simplifies: (i) the summation over $\lambda_f$ contains only the term $\lambda_f = \lambda_i$, (ii) the summation over $\lambda_i$ disappears because $\sum_{\lambda_i} p_{\lambda_i} = 1$, and (iii) $E_{\lambda_i} = E_{\lambda_f}$ in the delta function. With these simplifications eqn (3.16) becomes

$$\frac{d^2\sigma}{d\Omega dE_f} = \frac{k_f}{k_i} \left( \frac{m_n}{2\pi\hbar^2} \right)^2 |\langle \sigma_f | V(\mathbf{Q}) | \sigma_i \rangle|^2 \delta(\hbar\omega). \qquad (3.19)$$

The delta function in (3.19) means that the cross-section vanishes for final neutron energies that are different from the initial energy. In other words, the scattering from a rigid structure is totally elastic, as expected. The delta function singularity in energy is so sharp (infinitely so) that any attempt to measure inelastic scattering would be limited by the experimental energy resolution. A more useful quantity, therefore, is the differential cross-section which, as readers will recall, is what is measured in a detector that records all scattered neutrons regardless of their energy. Integration of eqn (3.19) over $E_f$ leads to

$$\frac{d\sigma}{d\Omega} = \left( \frac{m_n}{2\pi\hbar^2} \right)^2 |\langle \sigma_f | V(\mathbf{Q}) | \sigma_i \rangle|^2. \qquad (3.20)$$

## 3.4  Correlation functions

One slightly discouraging aspect of the master equation (3.16) is the presence of the delta function. The scattering cross-section is often a smooth function of energy, so why should it contain something as hideous and unphysical as a delta function?

In this section we show how the delta function can be eliminated and the cross-section written in a way that is more intuitive and often easier to calculate from a microscopic model. The procedure, which is principally due to Van Hove,[9] will lead us to a formulation of neutron scattering in terms of various *correlation functions* of operators associated with the scattering system. These correlation functions are equilibrium properties of the system, and can be calculated by well-established methods in quantum and statistical mechanics.

The development involves certain formal manipulations which will be introduced one at a time in the following sections.

### 3.4.1  Integral representation of the delta function

The removal of the delta function is made possible by the identity[10]

$$\delta(E_{\lambda_f} - E_{\lambda_i} - \hbar\omega) = \frac{1}{2\pi\hbar} \int_{-\infty}^{\infty} \exp\{i(E_{\lambda_f} - E_{\lambda_i})t/\hbar\} \exp(-i\omega t) \, dt.$$

Using this, we can write the product of the squared matrix element and the delta function in eqn (3.16) as

$$|\langle f| V |i \rangle|^2 \delta(E_{\lambda_f} - E_{\lambda_i} - \hbar\omega)$$

$$= \frac{1}{2\pi\hbar} \int_{-\infty}^{\infty} \langle i| V^\dagger |f \rangle \langle f| V |i \rangle \exp\{i(E_{\lambda_f} - E_{\lambda_i})t/\hbar\} \exp(-i\omega t) \, dt,$$

$$(3.21)$$

[9]Léon Van Hove (1924–1990), Belgian theoretical physicist who showed that inelastic neutron scattering in the Born approximation can be expressed in terms of space- and time-dependent pair correlation functions (Van Hove, 1954).

[10]See Section B.2.3.

where $V$ and its Hermitian adjoint operator $V^\dagger$ are related by[11]

$$\langle i | V^\dagger | f \rangle = \langle f | V | i \rangle^*.$$

[11] See Section C.1.5.

Here, we have re-introduced the shorthand $\langle f|V|i \rangle \equiv \langle \sigma_f \lambda_f | V(\mathbf{Q}) | \sigma_i \lambda_i \rangle$, and used the fact that the matrix elements are independent of time to take them inside the time integral.

## 3.4.2 Introduction of Heisenberg operators

We now show how the integrand in (3.21) can be made more compact by use of the so-called Heisenberg representation of the operators[12] (see Section C.1.10). If $A$ is an operator corresponding to a physical observable then the time-dependent Heisenberg operator $\hat{A}(t)$ is constructed from $A$ as follows:[13]

$$\hat{A}(t) = \exp(i\mathcal{H}t/\hbar)\, A\, \exp(-i\mathcal{H}t/\hbar), \tag{3.22}$$

where $\mathcal{H}$ is the Hamiltonian for the target, and $t$ is time. The exponential function of $\mathcal{H}$ represents the power series expansion

$$\exp x = 1 + x + \frac{x^2}{2!} + \dots, \tag{3.23}$$

with $x = \pm i\mathcal{H}t/\hbar$.

To proceed, we need to consider the action of the operator $\exp(\pm i\mathcal{H}t/\hbar)$ on the state functions $|i\rangle$ and $|f\rangle$. Consider first the action on an eigenstate $|\lambda\rangle$ of the target obeying

$$\mathcal{H} | \lambda \rangle = E_\lambda | \lambda \rangle, \tag{3.24}$$

where $E_\lambda$ is the corresponding eigenvalue. It follows from (3.23) and repeated application of (3.24) that

$$\exp(-i\mathcal{H}t/\hbar) | \lambda \rangle = \exp(-E_\lambda t/\hbar) | \lambda \rangle. \tag{3.25}$$

The same relation holds if we replace $|\lambda\rangle$ by $|i\rangle$ or $|f\rangle$ because $\mathcal{H}$ acts only on the component of the state functions pertaining to the target.

Considering the Heisenberg operator corresponding to $V$, we now see from (3.25) that

$$\begin{aligned}
\langle f | V | i \rangle \exp\{i(E_{\lambda_f} - E_{\lambda_i})t/\hbar\} &= \langle f | \exp(i\mathcal{H}t/\hbar)\, V\, \exp(-i\mathcal{H}t/\hbar) | i \rangle \\
&= \langle f | \hat{V}(t) | i \rangle,
\end{aligned} \tag{3.26}$$

and hence (3.21) may be written

$$|\langle f | V | i \rangle|^2 \delta(E_{\lambda_f} - E_{\lambda_i} - \hbar\omega) = \frac{1}{2\pi\hbar} \int_{-\infty}^{\infty} \langle i | \hat{V}^\dagger | f \rangle \langle f | \hat{V}(t) | i \rangle \exp(-i\omega t)\, dt. \tag{3.27}$$

Here we have used the shorthand $\hat{V}^\dagger$ for $\hat{V}^\dagger(0)$.

[12] In brief, the usual Schrödinger picture associates physical observables with operators that are constant in time, and builds all the time dependence into the state function $|\Psi(t)\rangle$. In the Heisenberg picture, on the other hand, all the time dependence is contained in the operators by means of the construction given in eqn (3.22), while the state functions are independent of time. The expectation value of a physical observable represented by an operator $A$ is $\langle \Psi(t)|A|\Psi(t)\rangle$ in the Schrödinger formulation and $\langle \Psi|\hat{A}(t)|\Psi\rangle$ in the Heisenberg picture.

[13] The 'hat' over the $A$ in $\hat{A}(t)$ is used throughout this book to distinguish a Heisenberg time-dependent operator from the operator that represents the physical observable in the Schrödinger formulation. Note that $\hat{A}(0) \equiv A$.

### 3.4.3 Summation over target eigenstates

The final step in formulating the scattering cross-section (3.16) in terms of a correlation function of the sample is to perform the summation over final states of the target, and to average over the population of initial states. This procedure involves only the parts of the state functions $|i\rangle$ and $|f\rangle$ that depend on the target variables, i.e. $|\lambda_i\rangle$ and $|\lambda_f\rangle$. It is therefore convenient to suppress the $|\mathbf{k}\sigma\rangle$ parts of the state functions for the time being. We do this by defining the (Heisenberg) operator

$$\hat{A}(t) = \left(\frac{m_n}{2\pi\hbar^2}\right) \langle\sigma_f| \hat{V}(\mathbf{Q},t) |\sigma_i\rangle, \qquad (3.28)$$

and its Hermitian conjugate

$$\hat{A}^\dagger(t) = \left(\frac{m_n}{2\pi\hbar^2}\right) \langle\sigma_i| \hat{V}^\dagger(\mathbf{Q},t) |\sigma_f\rangle. \qquad (3.29)$$

The $\mathbf{Q}$ dependence of $\hat{A}(t)$ has been suppressed for now to simplify the notation, and the constant $m_n/2\pi\hbar^2$ is included for later convenience.

Since the target eigenstates form a complete set we can use the closure relation (C.19) to contract the product of matrix elements:

$$\sum_{\lambda_f}\langle\lambda_i| \hat{A}^\dagger |\lambda_f\rangle\langle\lambda_f| \hat{A}(t) |\lambda_i\rangle = \langle\lambda_i| \hat{A}^\dagger\hat{A}(t) |\lambda_i\rangle. \qquad (3.30)$$

The quantity that enters into the cross-section is then the average of (3.30) over the initial states, and we will employ the standard notation

$$\langle\hat{A}^\dagger\hat{A}(t)\rangle \equiv \sum_{\lambda_i} p_{\lambda_i}\langle\lambda_i| \hat{A}^\dagger\hat{A}(t) |\lambda_i\rangle \qquad (3.31)$$

to indicate this statistical average. If the target is in thermal equilibrium, which is usually the case, then $\langle\ldots\rangle$ represents a thermal average. In this case,

$$p_{\lambda_i} = \frac{1}{Z} \exp(-\beta E_{\lambda_i}), \qquad (3.32)$$

where,

$$Z = \sum_{\lambda_i} \exp(-\beta E_{\lambda_i}) \qquad (3.33)$$

is the partition function, and $\beta = 1/(k_B T)$.

### 3.4.4 New formulation for the cross-section

We can now bring together all the results of the previous three subsections. Starting from the Born approximation for the partial differential scattering cross-section, eqn (3.16), we represented the delta function as a time integral in (3.21), introduced a Heisenberg operator in (3.26) to remove the complex exponential containing the target eigenvalues, and performed the summation over final target states and average over initial target states in (3.30) and (3.31) respectively. Defining $\hat{A}$ in terms of the interaction potential via (3.28) we reach the desired expression for the cross-section:

$$\frac{d^2\sigma}{d\Omega dE_f} = \frac{k_f}{k_i}\frac{1}{2\pi\hbar}\int_{-\infty}^{\infty}\langle\hat{A}^\dagger\hat{A}(t)\rangle\,\exp(-i\omega t)\,dt. \quad (3.34)$$

The function $\langle\hat{A}^\dagger\hat{A}(t)\rangle$ is known as a *correlation function*, and the neutron cross-section is thus seen to be proportional to the time Fourier transform of a correlation function. The significance of this is that the operator $\hat{A}$ represents a physical property of the scattering system, and the correlation function describes how that physical property varies with time. To be precise, the correlation function measures how the value of a physical quantity relates to its value at a time $t$ later. If the sample is in equilibrium then the time dependence is a result of thermal and quantum fluctuations, and the time Fourier transform of the correlation function is then the *spectrum of spontaneous fluctuations* of the quantity represented by $\hat{A}$.

## 3.4.5   The response function and intermediate function

The time Fourier transform of a correlation function is known as a *response function*. From (3.28) and (3.29), the correlation function is

$$\langle\hat{A}^\dagger\hat{A}(t)\rangle = \left(\frac{m_n}{2\pi\hbar^2}\right)^2\langle\,\langle\sigma_i|\,\hat{V}^\dagger(\mathbf{Q})\,|\sigma_f\rangle\langle\sigma_f|\,\hat{V}(\mathbf{Q},t)\,|\sigma_i\rangle\,\rangle, \quad (3.35)$$

and the corresponding response function[14] is

$$S(\mathbf{Q},\omega) = \frac{1}{2\pi\hbar}\int_{-\infty}^{\infty}\langle\hat{A}^\dagger\hat{A}(t)\rangle\,\exp(-i\omega t)\,dt \quad (3.36)$$

$$= \frac{1}{2\pi\hbar}\int_{-\infty}^{\infty}I(\mathbf{Q},t)\,\exp(-i\omega t)\,dt, \quad (3.37)$$

where we have introduced

$$I(\mathbf{Q},t) = \langle\hat{A}^\dagger\hat{A}(t)\rangle. \quad (3.38)$$

It now becomes clear that the dependence of $S(\mathbf{Q},\omega)$ on the neutron wavelengths before and after scattering and on the scattering angle appears only in the quantities $\mathbf{Q}$ and $\omega$, the momentum and energy transfer, respectively (see Section 1.5). This becomes clear when we use (3.36) to write the scattering cross-section (3.34) as

$$\frac{d^2\sigma}{d\Omega dE_f} = \frac{k_f}{k_i}\,S(\mathbf{Q},\omega). \quad (3.39)$$

The appeal of (3.39) is that it factorizes the cross-section into a part that depends on the particular neutron wavelengths used in the experiment (the $k_f/k_i$ factor), and a part that depends only on the physical properties of the system under investigation (the response function).

In the context of scattering theory, the correlation function $I(\mathbf{Q},t)$ usually goes by the name *intermediate scattering function*, or *intermediate function* for short, reflecting the direct relation (via Fourier transform) between it and the scattering function $S(\mathbf{Q},\omega)$.

[14] $S(\mathbf{Q},\omega)$ is also known as the *scattering function*, *dynamical structure factor* or *scattering law*. Readers should be alert to the fact that definitions of $S(\mathbf{Q},\omega)$ are often given with different multiplicative constants in front. Sometimes, for example, parameters associated with the interaction potential are taken outside, and often a factor $N$ (the total number of scattering units) appears in front. What is important, however, is the exclusion of the $k_f/k_i$ factor since this makes $S(\mathbf{Q},\omega)$ a property of the sample and not dependent on the neutron-related details of how the experiment was performed.

## 3.5    Properties of correlation and response functions

### 3.5.1    Properties of the correlation function

Remember: $\hat{A}^\dagger$ is short for $\hat{A}^\dagger(0)$.

We first quote four useful properties possessed by $\langle \hat{A}^\dagger \hat{A}(t) \rangle$. The first three identities are proved in Section C.1.11.

(1) If the system is in thermal equilibrium then the correlation function should not depend on the origin of time. In other words,

$$\langle \hat{A}^\dagger \hat{A}(t) \rangle = \langle \hat{A}^\dagger(t_0) \hat{A}(t_0 + t) \rangle. \tag{3.40}$$

(2) A related identity is

$$\langle \hat{A}^\dagger \hat{A}(t) \rangle = \langle \hat{A} \hat{A}^\dagger(-t + i\hbar\beta) \rangle, \tag{3.41}$$

where $\beta = 1/(k_B T)$.

(3) The complex conjugate of the correlation function is

$$\langle \hat{A}^\dagger \hat{A}(t) \rangle^* = \langle \hat{A}^\dagger(t) \hat{A} \rangle. \tag{3.42}$$

(4) The final property we need is the relationship between the correlation functions for $\mathbf{Q}$ and $-\mathbf{Q}$. We observe firstly that $V(\mathbf{r})$ is a Hermitian operator, i.e. $V^\dagger(\mathbf{r}) = V(\mathbf{r})$. This is because the interaction potential is a physical observable.[15] Hence, from eqn (3.15),

[15]See Section C.1.5.

$$V(-\mathbf{Q}) = V^\dagger(\mathbf{Q}). \tag{3.43}$$

To within a constant, the operator $\hat{A}(t)$ is the matrix element of $\hat{V}(\mathbf{Q}, t)$ between initial and final spin states of the neutron, eqn (3.28), so that $\hat{A}(-\mathbf{Q}, t) \neq \hat{A}^\dagger(\mathbf{Q}, t)$ unless

$$\langle \sigma_f | \hat{V}(\mathbf{Q}, t) | \sigma_i \rangle = \langle \sigma_i | \hat{V}(\mathbf{Q}, t) | \sigma_f \rangle. \tag{3.44}$$

Condition (3.44) is trivially satisfied if $\sigma_i = \sigma_f$, and there are some other cases where the scattering does not depend on the neutron spin states, for example scattering from randomly oriented nuclear spins, and magnetic scattering from a paramagnet. In general, however, the relation analogous to (3.43) for $\hat{A}$ is

$$\hat{A}(-\mathbf{Q}, t)_{\sigma_f \to \sigma_i} = \hat{A}^\dagger(\mathbf{Q}, t)_{\sigma_i \to \sigma_f}, \tag{3.45}$$

so that, with the $\mathbf{Q}$ dependence of $\hat{A}$ shown explicitly, the $-\mathbf{Q}$ correlation function can be written

$$\langle \hat{A}^\dagger(-\mathbf{Q}) \hat{A}(-\mathbf{Q}, t) \rangle_{\sigma_f \to \sigma_i} = \langle \hat{A}(\mathbf{Q}) \hat{A}^\dagger(\mathbf{Q}, t) \rangle_{\sigma_i \to \sigma_f}. \tag{3.46}$$

The correlation function for *unpolarized* neutron scattering contains an average over all initial and final spin states of the neutron, in which case

$$\langle \hat{A}^\dagger(-\mathbf{Q}) \hat{A}(-\mathbf{Q}, t) \rangle_{\text{unpol}} = \langle \hat{A}(\mathbf{Q}) \hat{A}^\dagger(\mathbf{Q}, t) \rangle_{\text{unpol}}. \tag{3.47}$$

## 3.5.2   Response function is real and positive

Because the response function is effectively the scattering probability it must be real and positive. The truth of this assertion is evident from eqns (3.16) and (3.39), since all quantities in the former expression are real and positive. The reader can verify the realness of $S(\mathbf{Q}, \omega)$ directly from the definition of the response function as the Fourier transform of the correlation function — see Exercise 3.1.

## 3.5.3   The Principle of Detailed Balance

An inelastic scattering process involves a change in the state of the sample from $\lambda_i$ to $\lambda_f$, a change in the neutron spin from $\sigma_i$ to $\sigma_f$, and a transfer of momentum and energy from the neutron to the sample. Suppose, now, that the sample is initially in the state $\lambda_f$. Then we could measure the same transition but in reverse: $\lambda_f \to \lambda_i$ instead of $\lambda_i \to \lambda_f$. This time the neutron spin would change from $\sigma_f$ to $\sigma_i$, and momentum and energy would transfer from the sample back to the neutron. The 'forward' and 'backward' processes are represented by the response functions $S(\mathbf{Q}, \omega)_{\sigma_i \to \sigma_f}$ and $S(-\mathbf{Q}, -\omega)_{\sigma_f \to \sigma_i}$ respectively, and for a sample in thermal equilibrium there exists a particularly simple relation between these response functions known as the *Principle of Detailed Balance*.

Before deriving this relation let us be more explicit about what the forward and backward processes involve. When a neutron converts a system to an excited state which is higher in energy than the initial state we call this a *neutron energy-loss* or *Stokes* process (also referred to colloquially as 'down-scattering'). In an energy-loss process, momentum and energy are transferred from the neutron to the system, thereby creating an excitation. The corresponding *neutron energy-gain* (or anti-Stokes, or 'up-scattering') process is when a pre-existing excitation is annihilated in the scattering event, thus transferring momentum and energy from the system to the neutron. For a neutron energy-gain process to occur the system must be at a temperature greater than absolute zero so that a thermal population of excitations exists prior to the scattering event. Neutron energy-loss processes, on the other hand, can occur at any temperature including absolute zero. Fig. 3.1 provides a pictorial illustration of neutron energy-loss and energy-gain processes.

Now let us establish the relationship between the response functions $S(\mathbf{Q}, \omega)_{\sigma_i \to \sigma_f}$ and $S(-\mathbf{Q}, -\omega)_{\sigma_f \to \sigma_i}$. From (3.36), (3.46) and (3.41),

$$
\begin{aligned}
S(-\mathbf{Q}, -\omega)_{\sigma_f \to \sigma_i} &= \frac{1}{2\pi\hbar} \int_{-\infty}^{\infty} \langle \hat{A}^\dagger(-\mathbf{Q})\hat{A}(-\mathbf{Q}, t) \rangle_{\sigma_f \to \sigma_i} \exp(\mathrm{i}\omega t)\, \mathrm{d}t \\
&= \frac{1}{2\pi\hbar} \int_{-\infty}^{\infty} \langle \hat{A}(\mathbf{Q})\hat{A}^\dagger(\mathbf{Q}, t) \rangle_{\sigma_i \to \sigma_f} \exp(\mathrm{i}\omega t)\, \mathrm{d}t \\
&= \frac{1}{2\pi\hbar} \int_{-\infty}^{\infty} \langle \hat{A}^\dagger(\mathbf{Q})\hat{A}(\mathbf{Q}, -t + \mathrm{i}\hbar\beta) \rangle_{\sigma_i \to \sigma_f} \exp(\mathrm{i}\omega t)\, \mathrm{d}t \\
&= \frac{1}{2\pi\hbar} \int_{-\infty}^{\infty} \langle \hat{A}^\dagger(\mathbf{Q})\hat{A}(\mathbf{Q}, t') \rangle_{\sigma_i \to \sigma_f} \exp(-\mathrm{i}\omega t' - \beta\hbar\omega)\, \mathrm{d}t',
\end{aligned}
$$

Fig. 3.1 Diagrams showing processes in which the neutron (a) creates and (b) annihilates an excitation having a momentum $\hbar \mathbf{Q}_0$ and an energy $\hbar\omega_0$. The momentum and energy transferred from the neutron to the system in each case is (a) $(\mathbf{Q},\omega) = (\mathbf{Q}_0,\omega_0)$ (neutron energy-loss), and (b) $(\mathbf{Q},\omega) = (-\mathbf{Q}_0, -\omega_0)$ (neutron energy-gain).

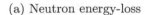
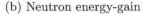
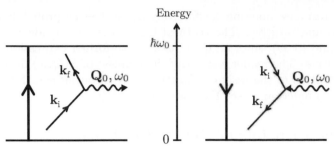

(a) Neutron energy-loss  (b) Neutron energy-gain

where $t' = -t + \mathrm{i}\hbar\beta$ and $\beta = 1/(k_B T)$, so that,

$$S(-\mathbf{Q}, -\omega)_{\sigma_f \to \sigma_i} = \exp(-\beta\hbar\omega)\, S(\mathbf{Q}, \omega)_{\sigma_i \to \sigma_f}. \qquad (3.48)$$

This expression is known as the *Principle of Detailed Balance* (or just 'detailed balance', for short).

In the case of unpolarized neutron scattering (Section 4.9), or if the scattering potential is spin-independent in the sense of (3.44), then we need not be specific about the initial and final neutron spin states in eqn (3.48). Another special case is when the scattering potential is symmetric in $\mathbf{Q}$, in which case we do not have to worry about changing the sign of $\mathbf{Q}$.

Physically, detailed balance reflects (i) that in thermal equilibrium the population of states with energy $\hbar\omega$ above the ground state is proportional to the Boltzmann factor $\exp(-\beta\hbar\omega)$, and (ii) that the scattering probability is proportional to the population of the initial state irrespective of whether we are creating or destroying the excitation — see comments leading to eqn (3.3).

Detailed balance has important consequences for experiment. It tells us that we can measure the same excitation in neutron energy-loss or in neutron energy-gain, but in the latter case the scattering function is smaller by a factor $\exp(-\beta\hbar\omega)$. This could be an important consideration if $\hbar\omega > k_B T$. Figure 3.2 provides an illustration of this property. The intensity penalty for energy-gain scattering is partially offset by the $k_f/k_i$ term in the cross-section, eqn (3.39), which is greater than unity for energy-gain and less than unity for energy-loss.

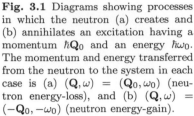
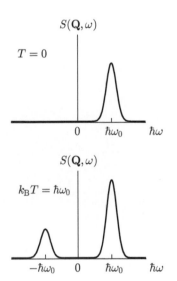

Fig. 3.2 Intensity of energy-loss and energy-gain scattering for a transition with energy $\hbar\omega_0$. The upper figure is for a temperature $T = 0$, and the lower figure is for $k_B T = \hbar\omega_0$.

## 3.6 Static and dynamic correlations

In this section we will show how consideration of the limiting behaviour of the correlation function at long times allows us to separate the elastic and inelastic components of the cross-section.

The analysis rests on the assumption that any dynamical correlation in a real system will, after a sufficiently long time, eventually decay to zero.[16] This means that after an infinite amount of time has elapsed only the static correlations of the system will remain.

[16]Otherwise we would have perpetual motion.

Consider the following analogy: two dancers A and B are located in different rooms so they cannot see each other (Fig. 3.3). At $t = 0$, they begin to perform the same dance. They dance about a fixed point on the floor, but they are not provided with any music or beat to keep them in time. Initially they remain quite well coordinated with one another, but as time passes slight irregularities in their technique and sense of timing, and their mutual concealment, mean that the two dancers become increasingly out of step. From time to time they must inevitably stop to rest, and this further contributes to the loss of synchronization.

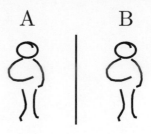

**Fig. 3.3** Static and dynamic correlations. Dancers A and B dance the same sequence of steps but have no way of knowing what each other is doing. Their movements are initially well correlated, but become increasingly less so as time elapses.

Suppose we ask the following. After a time $t$ has elapsed, how well can we predict where B is in her dance routine if all we know is what A was doing at the start of the exercise? If $t$ were small (compared with the cycle of the dance) then the answer would be 'very well', but for large $t$ we would have very little confidence at all in our prediction. The only thing we do know for certain is that A and B stay approximately the same distance apart. In other words, after a long interval of time there remains a correlation between A and B's average positions, but little or no correlation between their movements.

Returning to neutron scattering, we recognize that the correlation function $\langle \hat{A}^\dagger \hat{A}(t) \rangle$ can be separated into a dynamic part which tends to zero as $t$ tends to infinity, like the synchronization of the dancers movements, and a static part equal to what is left at $t = \infty$, like the separation of the dancers. If the system is ergodic, i.e. the thermal average of $\hat{A}(t)$ is independent of $t$, so that $\langle \hat{A}(\infty) \rangle = \langle \hat{A}(0) \rangle = \langle \hat{A} \rangle$, then the static part is given by

$$\langle \hat{A}^\dagger \hat{A}(\infty) \rangle = \langle \hat{A}^\dagger \rangle \langle \hat{A} \rangle$$
$$= |\langle \hat{A} \rangle|^2. \qquad (3.49)$$

Hence, from (3.36), (3.49) and (B.16), the response function is given by

$$S(\mathbf{Q}, \omega) = \frac{1}{2\pi\hbar} \int_{-\infty}^{\infty} \langle \hat{A}^\dagger \hat{A}(t) \rangle \exp(-i\omega t) \, dt$$

$$= |\langle \hat{A} \rangle|^2 \delta(\hbar\omega) + \widetilde{S}(\mathbf{Q}, \omega), \qquad (3.50)$$

where $\widetilde{S}(\mathbf{Q}, \omega)$ is the component of the response function that derives from dynamic correlations. The delta function in the first term of (3.50) expresses the fact that the scattering associated with the static part of the correlation function is purely elastic. All the inelastic scattering from the system is contained in $\widetilde{S}(\mathbf{Q}, \omega)$. The static and dynamic components of $S(\mathbf{Q}, \omega)$ are illustrated in Fig. 3.4.

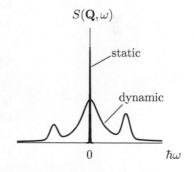

**Fig. 3.4** Schematic of the response function $S(\mathbf{Q}, \omega)$. The elastic component arises from static correlations, and the inelastic component arises from dynamic correlations.

## 3.7   The Fluctuation–Dissipation theorem

A general property of weak experimental probes is that the size of the response of the system is proportional to the size of the stimulus. The constant of proportionality is known as a linear response coefficient. In this section we describe how the neutron scattering response function relates to a linear response coefficient known as the *generalized susceptibility*. This relationship provides an alternative way to connect the results of a neutron scattering experiment with a microscopic theory of the scattering system. A more complete account of linear response theory is given in Appendix D.

A simple way to look at the neutron scattering process is as follows. The neutron couples weakly to the sample via an interaction potential, establishing a frequency- and wavevector-dependent force to which the sample responds. If the amplitude of the response is large when probed at a particular wavevector $\mathbf{Q}$ and angular frequency $\omega$ then there will be a corresponding enhancement in the scattering response $S(\mathbf{Q}, \omega)$. In other words, a large $S(\mathbf{Q}, \omega)$ means that the system reacts strongly when perturbed at $\mathbf{Q}$ and $\omega$. It follows that the system will be particularly susceptible to fluctuations away from equilibrium whose $\mathbf{Q}$ and $\omega$ correspond to large $S(\mathbf{Q}, \omega)$.

Therefore, there is a direct link between the scattering intensity and the size of the spontaneous fluctuations. This link is expressed in a very simple relationship known as the *Fluctuation–Dissipation theorem*.

To be specific, let the perturbation acting on the system be represented by a position- and time-dependent 'force' $f(\mathbf{r}, t)$. Without loss of generality, let us consider just one Fourier component of $f(\mathbf{r}, t)$,

$$f(\mathbf{Q}, \omega) \exp\{\mathrm{i}(\omega t - \mathbf{Q} \cdot \mathbf{r})\}. \tag{3.51}$$

We suppose that the force couples linearly to a physical property of the system which we will call the 'displacement', $a(\mathbf{r}, t)$. The relationship between the Fourier components of the force and displacement can be written

$$a(\mathbf{Q}, \omega) = \chi(\mathbf{Q}, \omega) f(\mathbf{Q}, \omega), \tag{3.52}$$

where

$$\chi(\mathbf{Q}, \omega) = \chi'(\mathbf{Q}, \omega) - \mathrm{i}\chi''(\mathbf{Q}, \omega) \tag{3.53}$$

is a complex function known as the *generalized susceptibility*. As usual, the use of a complex function allows us to describe the phase difference between force and displacement as well as their relative amplitudes. The existence of a phase difference implies dissipation of energy in the system, and the the amount of energy dissipated is governed by the *absorptive* part $\chi''$ of the generalized susceptibility.

When we say the response is *linear* in the force we mean that if the force doubles then so does the displacement. In other words, $\chi(\mathbf{Q}, \omega)$ does not depend on the amplitude of the force. The linear approximation is usually valid if the perturbation is very weak, which is the case for neutron scattering in the Born approximation.

The fact that neutron scattering produces a linear response in systems is one of the great strengths of the technique because linear response theory provides us with many simple relationships between various correlation and response functions of systems in thermal equilibrium (see Appendix D). In particular, it can be used to write the inelastic part of the response function in terms of the absorptive part of the generalized susceptibility, generating the equation that embodies the Fluctuation–Dissipation theorem as far as neutron scattering is concerned,[17]

$$\widetilde{S}(\mathbf{Q},\omega) = \{1 + n(\omega)\}\frac{1}{\pi}\chi''(\mathbf{Q},\omega), \qquad (3.54)$$

where

$$n(\omega) = \frac{1}{\exp(\beta\hbar\omega) - 1} \qquad (3.55)$$

is the Planck distribution function, which is shown in Fig. 3.5. The factor $\{1 + n(\omega)\}$ is usually referred to as the *detailed balance factor*, or simply, the *temperature factor*.

Whether one chooses to calculate $\widetilde{S}(\mathbf{Q},\omega)$ from the dynamic part of the correlation function, eqn (3.50), or from the generalized susceptibility via (3.54) is largely a matter of convenience. Some models, e.g. harmonic approximation for quantized lattice vibrations (Chapter 5), lead naturally to the time correlation function, whereas others, e.g. the band model of magnetic dynamics (Section 8.6), are more amenable to the calculation of the generalized susceptibility. The generalized susceptibility is particularly useful in describing magnetic scattering, though not exclusively so.

[17]A derivation of eqn (3.54) is given in Sections D.3–D.4.

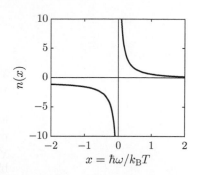

**Fig. 3.5** The Planck distribution describes the statistics of systems of particles with zero chemical potential, such as photons or phonons. It is a special case of the Bose–Einstein distribution. $n(\omega)$ gives the thermal population of a state at an energy $\hbar\omega$ above the ground state.

## 3.8 Sum rules, limiting cases, and approximations

Given unlimited beam time, the surest way to gain the maximum information about the static and dynamic properties of a system would be to measure $S(\mathbf{Q},\omega)$ over as large a range of $\mathbf{Q}$ and $\omega$ as is practical. Often, however, we can learn about specific properties of the system just by measuring certain limits of the response function or by performing summations over parts of it. As well as being a quick way of gaining information, this approach can also serve as a way of checking neutron scattering results against data from other techniques or against strict requirements imposed by theory. In Section 3.6 we encountered one such example: the infinite time limit of the correlation function, which corresponds to static correlations and gives rise to purely elastic scattering. In this section we consider several other cases where a clear physical interpretation can be obtained from either a limit or an integral of a correlation or response function.

### 3.8.1 Uniform static susceptibility

The generalized susceptibility, eqns (3.52) and (3.53), possesses the property that its reactive and absorptive parts are related to one another by the Kramers–Kronig transformation.[18] A special case is the expression for $\chi'(\mathbf{Q},0)$, the reactive part of the zero-frequency susceptibility. From

[18]See Section D.5.

(D.41) and (3.54),

$$\chi'(\mathbf{Q}, 0) = \frac{1}{\pi} \int_{-\infty}^{\infty} \frac{\chi''(\mathbf{Q}, \omega)}{\omega} \, d\omega$$

$$= \int_{-\infty}^{\infty} \frac{\widetilde{S}(\mathbf{Q}, \omega)}{\omega} \{1 - \exp(-\beta\hbar\omega)\} \, d\omega. \qquad (3.56)$$

$\chi'(\mathbf{Q}, 0)$ represents the in-phase response of the system to a spatially-varying perturbation with a wavevector $\mathbf{Q}$.

The $\mathbf{Q} = 0$ limit of (3.56) is of particular significance since the 'uniform' or 'bulk' static susceptibility $\chi'(0,0)$ is accessible to other experimental techniques. In the case of the magnetism, for example, $\chi'(0,0)$ is proportional to the isothermal zero-field magnetic susceptibility $\chi_{\mathrm{M}} = (M/H)_{H=0}$, where $M$ is the magnetization and $H$ is the applied field.[19] $\chi_{\mathrm{M}}$ is the quantity measured in a laboratory magnetometer. Another example is the case of a classical monatomic fluid, for which $\chi'(0,0) = n\kappa_T$ (see Exercise 3.4), where $\kappa_T = -(1/V)(\partial V/\partial p)_T$ is the isothermal compressibility, $n$ is the number of atoms per unit volume, $V$ is the volume and $p$ the pressure.

Hence, eqn (3.56) provides a direct connection between the neutron inelastic scattering spectrum and data obtained by other techniques. This link is useful as a check of consistency. Because a neutron spectrum is necessarily measured over a finite range of energy we can use (3.56) to assess whether there is significant spectral weight outside the energy range over which we have measured.

Equation (3.56) is an example of what is known as a *sum rule*, meaning here that, irrespective of the functional form of $\widetilde{S}(\mathbf{Q}, \omega)$, the integral in (3.56) is constrained by theory to yield a definite answer, namely the static susceptibility of the system. Further examples of sum rules are to be found in the following sections.

### 3.8.2   Spectral-weight function

From (3.56), it is readily seen that the dynamical part of the response function can be written

$$\widetilde{S}(\mathbf{Q}, \omega) = \omega\{1 + n(\omega)\}\chi'(\mathbf{Q}, 0)F(\mathbf{Q}, \omega), \qquad (3.57)$$

where

$$\int_{-\infty}^{\infty} F(\mathbf{Q}, \omega) \, d\omega = 1. \qquad (3.58)$$

The function $F(\mathbf{Q}, \omega)$ is known as the *spectral-weight function*. Because it is normalized to unity it allows us to compare the intrinsic line shapes of peaks in energy spectra. It is also useful when we do not have a theoretical form for $\widetilde{S}(\mathbf{Q}, \omega)$. In these cases it is often found that simple phenomenological spectral-weight functions, such as the Gaussian and Lorentzian (see Sections B.1.1 and B.1.2), provide a good description of peaks found in neutron energy spectra. The functions $F(\mathbf{Q}, \omega)$ and $\widetilde{S}(\mathbf{Q}, \omega)$ are compared in Fig. 3.6.

[19]See Section 6.4 and Section D.9.

$F(\mathbf{Q}, \omega)$

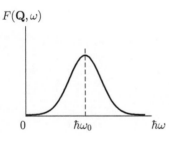

$\widetilde{S}(\mathbf{Q}, \omega)$

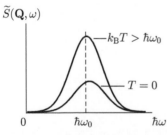

**Fig. 3.6** Gaussian spectral-weight function $F(\mathbf{Q}, \omega)$ and the corresponding dynamical response function $\widetilde{S}(\mathbf{Q}, \omega)$ at two temperatures. Note the asymmetry in $\widetilde{S}(\mathbf{Q}, \omega)$ due to the $\omega\{1 + n(\omega)\}$ factor in (3.57). The asymmetry is largest when $k_{\mathrm{B}}T < \hbar\omega_0$.

### 3.8.3 Structure factor, $S(\mathbf{Q})$

Integration of $S(\mathbf{Q}, \omega)$ over $\hbar\omega$ yields the *structure factor*:[20]

$$S(\mathbf{Q}) = \int_{-\infty}^{\infty} S(\mathbf{Q}, \omega)\,\mathrm{d}(\hbar\omega). \qquad (3.59)$$

From (3.36), with $\hat{A}^{\dagger} \equiv \hat{A}^{\dagger}(0)$,

$$
\begin{aligned}
S(\mathbf{Q}) &= \frac{1}{2\pi\hbar} \int_{-\infty}^{\infty} \int_{-\infty}^{\infty} \langle \hat{A}^{\dagger}(0)\hat{A}(t) \rangle \exp(-\mathrm{i}\omega t)\,\mathrm{d}t\,\mathrm{d}(\hbar\omega) \\
&= \int_{-\infty}^{\infty} \langle \hat{A}^{\dagger}(0)\hat{A}(t) \rangle\,\delta(t)\,\mathrm{d}t \\
&= \langle \hat{A}^{\dagger}(0)\hat{A}(0) \rangle \\
&= \langle |\hat{A}|^2 \rangle. \qquad (3.60)
\end{aligned}
$$

Between the first and second lines we interchanged the order of integration and used the integral representation of the delta function (B.16).

From (3.60) it can be seen that $S(\mathbf{Q})$ measures the *equal-time*, or *instantaneous*, correlation function $\langle \hat{A}^{\dagger}(0)\hat{A}(0) \rangle$. In other words, we can think of $S(\mathbf{Q})$ as providing a snapshot of the state of the system at $t = 0$. For a system in thermal equilibrium $\langle \hat{A}^{\dagger}(0)\hat{A}(0) \rangle = \langle \hat{A}^{\dagger}(t)\hat{A}(t) \rangle$, from property (3.40), and so $S(\mathbf{Q})$ is the same irrespective of the particular time at which we take the snapshot.

It is important to appreciate that $S(\mathbf{Q})$ does not simply describe elastic scattering. $S(\mathbf{Q})$ actually represents the sum of *all* correlations, both static and dynamic. To understand this point it is helpful to look at the form of the response function given in eqn (3.50). There, we split $S(\mathbf{Q}, \omega)$ into two parts, representing static and dynamic correlations respectively. The static part, call it $S_{\mathrm{el}}(\mathbf{Q}, \omega)$, contains a delta function in $\hbar\omega$, so that if we integrate over $\hbar\omega$ we obtain

$$S_{\mathrm{el}}(\mathbf{Q}) = |\langle \hat{A}(\mathbf{Q}) \rangle|^2, \qquad (3.61)$$

where for clarity we show the $\mathbf{Q}$ dependence of $\hat{A}$ explicitly. On the other hand, from (3.50) and (3.60),

$$
\begin{aligned}
S(\mathbf{Q}) &= S_{\mathrm{el}}(\mathbf{Q}) + \tilde{S}(\mathbf{Q}) \\
&= \langle |\hat{A}(\mathbf{Q})|^2 \rangle. \qquad (3.62)
\end{aligned}
$$

Therefore, $S_{\mathrm{el}}(\mathbf{Q})$ is the squared modulus of the average of $\hat{A}$, whereas $S(\mathbf{Q})$ is the average of the squared modulus of $\hat{A}$.

Taking as an example the case of scattering from atomic nuclei, $S(\mathbf{Q})$ measures the instantaneous positions of the nuclei, whereas $S_{\mathrm{el}}(\mathbf{Q})$ measures the time-averaged position of the nuclei. The usefulness of $S(\mathbf{Q})$ arises in describing systems such as liquids which are correlated when viewed over short times, but which have no permanent order. In such

[20]Here, $S(\mathbf{Q})$ is *not* shorthand for $S(\mathbf{Q}, \omega = 0)$. Sometimes $S(\mathbf{Q})$ is referred to as the *static structure factor*, but this is misleading because it actually represents both static and dynamic correlations.

cases, the equal-time correlation function can be calculated from a series of snap-shots of the system generated in a simulation that includes a model for the interactions between the particles, and the results compared directly with an experimental measurement of $S(\mathbf{Q})$.

### 3.8.4   Static approximation

An experiment that measures $S(\mathbf{Q})$ is called a *total scattering* measurement. The proper way to do this is to measure $S(\mathbf{Q}, \omega)$ as a function of $\omega$ at a fixed $\mathbf{Q}$, and then to perform the integration in (3.59) after the experiment. If, however, the scattering extends over only a small range of energy transfer relative to the incident neutron energy then an alternative approach is to use a detector that does not have energy analysis. In this case the detector performs the integration directly. This method is not exact, but has the advantage of being very much quicker than the first. The *static approximation* is the name given to the assumption that this method provides a good estimate of $S(\mathbf{Q})$.

To understand the limits of validity of the static approximation, consider a typical experimental setup in which a monochromatic beam of neutrons of energy $E_i$ is scattered from a sample into a detector at angle $\phi$ to the incident beam direction. Suppose that the detector records all scattered neutrons, regardless of their final energy $E_f$, and for simplicity assume the detection efficiency does not vary with neutron energy. The smallest possible $E_f$ is zero, and so the detector records

$$\frac{\mathrm{d}\sigma}{\mathrm{d}\Omega} = \int_0^\infty \left( \frac{\mathrm{d}^2\sigma}{\mathrm{d}\Omega\mathrm{d}E_f} \right)_\phi \mathrm{d}E_f, \qquad (3.63)$$

where the $\phi$ subscript indicates that the integral over $E_f$ is to be performed at constant scattering angle. The energy transfer is $\hbar\omega = E_i - E_f$, so with $E_i$ fixed, $\mathrm{d}E_f = -\mathrm{d}(\hbar\omega)$. Hence, from the definition of the response function (3.39) we can write (3.63) as

$$\frac{\mathrm{d}\sigma}{\mathrm{d}\Omega} = \int_{-\infty}^{E_i} \frac{k_f}{k_i} S(\mathbf{Q}, \omega)_\phi \, \mathrm{d}(\hbar\omega). \qquad (3.64)$$

The static approximation makes the assumption that the right-hand side of (3.64) is equal to $S(\mathbf{Q})$, which from (3.59) is manifestly not true. Firstly, the upper limit of integration is $E_i$ rather than $\infty$, and second, both $k_f$ and $\mathbf{Q}$ vary along the integration path — see eqns (1.16) and (1.17). This latter point is illustrated by the scattering triangle in Fig. 3.7, which shows how $\mathbf{Q}$ varies with $E_f = \hbar^2 k_f^2/(2m_n)$ when the scattering angle $\phi$ and $E_i$ are fixed.

For the static approximation to be valid we require either that the total amount of inelastic scattering is small compared with the elastic scattering, or that the inelastic scattering does not extend in energy beyond $E_i$ so as to make the upper limit of integration in (3.64) effectively infinity. We also require that $k_f/k_i \simeq 1$ and that $\mathbf{Q}$ does not change appreciably during the integration. If the energy spectrum of the sample has a cut-off $\hbar\omega_c$ then these conditions are met providing $E_i \gg \hbar\omega_c$, which means that $|\Delta\mathbf{Q}| \ll k_f$ (i.e. $k_i \simeq k_f$).

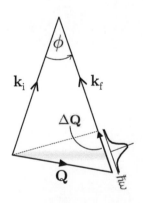

**Fig. 3.7** Scattering into an energy-integrating detector at a fixed angle $\phi$ to the incident beam. Both $k_f$ and $\mathbf{Q}$ vary as the energy integration is performed over the distribution of neutrons scattered by the sample into the detector. The static approximation applies when $|\Delta\mathbf{Q}| \ll k_f$.

The static approximation usually holds well for Bragg scattering at low temperatures, but its validity must be carefully evaluated for diffuse scattering measurements, such as studies of liquids and disordered solids. Various methods have been developed to calculate inelasticity corrections to measured structure factors obtained in the static approximation, and these are described in Section 10.2.7.

### 3.8.5 Bulk fluctuations and the classical limit

When $k_BT \gg \hbar\omega$, quantum effects are negligible and the Planck distribution (3.55) approximates to $k_BT/\hbar\omega$. In this limit, eqn (3.57) becomes

$$\widetilde{S}(\mathbf{Q}, \omega) \simeq \frac{k_BT}{\hbar} \chi'(\mathbf{Q}, 0) F(\mathbf{Q}, \omega) \qquad (k_BT \gg \hbar\omega). \tag{3.65}$$

This is the classical limit of the inelastic response function. If, further, $k_BT \gg \hbar\omega_c$, where $\hbar\omega_c$ is the cut-off in the energy spectrum, then (3.65) is applicable throughout the entire energy spectrum and we see immediately from the normalization condition (3.58) of the spectral-weight function that

$$\widetilde{S}(\mathbf{Q}) \simeq k_BT\chi'(\mathbf{Q}, 0) \qquad (k_BT \gg \hbar\omega_c). \tag{3.66}$$

This is the structure factor for a classical system without static correlations. An example of such a system is a classical fluid. Following the discussion in Section 3.8.1, we see that for a classical monatomic fluid

$$\widetilde{S}(0) = k_BT n\kappa_T \qquad \text{(classical monatomic fluid),} \tag{3.67}$$

where $n$ is the number of atoms per unit volume and $\kappa_T$ is the isothermal compressibility.

The $\mathbf{Q} = 0$ limit of $\widetilde{S}(\mathbf{Q})$ relates directly to bulk (i.e. long-wavelength) fluctuations in $A$, the physical observable represented by the operator $\hat{A}$. This can be seen from (3.61) and (3.62), according to which

$$\widetilde{S}(0) = \langle|\hat{A}(0)|^2\rangle - |\langle\hat{A}(0)\rangle|^2$$

$$= \langle|\Delta\hat{A}|^2\rangle, \tag{3.68}$$

where $\Delta\hat{A} = \hat{A}(0) - \langle\hat{A}(0)\rangle$. We see that $\widetilde{S}(0)$ is the variance of the spectrum of bulk fluctuations in $A$.

### 3.8.6 Energy moments of $S(\mathbf{Q}, \omega)$

The energy moments of $S(\mathbf{Q}, \omega)$ are defined by

$$S_n(\mathbf{Q}) = \int_{-\infty}^{\infty} (\hbar\omega)^n S(\mathbf{Q}, \omega)\, \mathrm{d}(\hbar\omega). \tag{3.69}$$

The zeroth moment is the structure factor $S(\mathbf{Q})$ already discussed in Section 3.8.3. The higher moments ($n \geq 1$) can be obtained from experiment if $S(\mathbf{Q}, \omega)$ is measurable over the full range of $\omega$.

The energy moments are useful in a number of ways. For some systems the expressions for $S_n(\mathbf{Q})$ depend in a simple way on basic physical properties of the sample. In these cases an experimental determination of one or more of the energy moments can provide direct information on the target (see Section 5.5.2 for an example). Conversely, if the moments of the scattering system can be calculated from the known properties of the system then they provide a good way to check an experimental measurement of $S(\mathbf{Q},\omega)$. If no complete theoretical description of the target exists then one can sometimes obtain a rough idea of the general form of $S(\mathbf{Q},\omega)$ from approximate calculations of the $n \geq 1$ moments.

We now show how to obtain an expression for $S_n(\mathbf{Q})$ in terms of derivatives of the correlation function. First, we invert the Fourier transform in (3.36) and (3.37) to give

$$\langle \hat{A}^{\dagger}(\mathbf{Q})\hat{A}(\mathbf{Q},t) \rangle = I(\mathbf{Q},t) = \int_{-\infty}^{\infty} S(\mathbf{Q},\omega)\exp(\mathrm{i}\omega t)\,\mathrm{d}(\hbar\omega). \qquad (3.70)$$

Next, we differentiate both sides of this equation with respect to $t$:

$$\frac{\partial}{\partial t}I(\mathbf{Q},t) = \mathrm{i}\int_{-\infty}^{\infty} \omega S(\mathbf{Q},\omega)\exp(\mathrm{i}\omega t)\,\mathrm{d}(\hbar\omega). \qquad (3.71)$$

By setting $t = 0$ and multiplying through by $\hbar/\mathrm{i}$ we see that the right-hand side of (3.71) becomes the first moment $S_1(\mathbf{Q})$. Repeating this procedure $n$ times we find

$$S_n(\mathbf{Q}) = \left(\frac{\hbar}{\mathrm{i}}\right)^n \left\{ \frac{\partial^n}{\partial t^n}I(\mathbf{Q},t) \right\}_{t=0}. \qquad (3.72)$$

## 3.9   Examples of correlation and response functions

As has been repeatedly emphasized in this chapter, correlation functions are an extremely powerful way to represent neutron scattering, and it is worth our while devoting some time towards acquiring a better feeling for correlation functions and what they mean. Here we look at some simple examples.

### 3.9.1   Classical damped harmonic oscillator

The damped harmonic oscillator is a useful and familiar model for many physical systems. It is also an example of a model for which it is very easy to derive the generalized susceptibility, and hence the neutron scattering response function via the Fluctuation–Dissipation theorem (3.54).

We consider a particle of mass $m$ in a harmonic potential (e.g. a mass suspended by a spring). When displaced a distance $x$ from its equilibrium position the particle experiences a restoring force $-m\omega_0^2 x$, where $\omega_0$ is the angular frequency of the undamped oscillator. We assume a frictional damping force of the form $-m\gamma \mathrm{d}x/\mathrm{d}t$, where $\gamma$ is a constant.

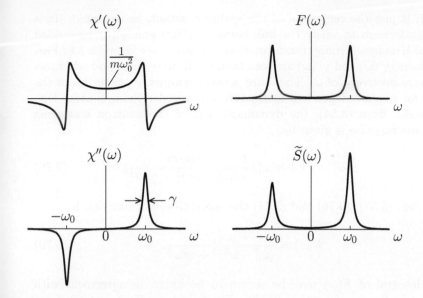

**Fig. 3.8** Linear response functions for a damped harmonic oscillator. The curves are calculated from eqns (3.76)–(3.79) with $\gamma/\omega_0 = 0.15$. A temperature $T = 2\hbar\omega_0/k_B$ is used for $\widetilde{S}(\omega)$.

The equation of motion for this particle when subjected to an external force $f(t)$ is

$$m\frac{d^2x}{dt^2} + m\gamma\frac{dx}{dt} + m\omega_0^2 x = f(t). \qquad (3.73)$$

This is a linear equation, and so we can use linear response theory to derive the various correlation and response functions.

Let us consider a single Fourier component $f(\omega)\exp(i\omega t)$ of $f(t)$, and look for a steady-state solution of the form $x(t) = a(\omega)\exp(i\omega t)$. Substituting $f(t)$ and $x(t)$ into (3.73) we find

$$\{-m\omega^2 + im\gamma\omega + m\omega_0^2\}\, a(\omega)\exp(i\omega t) = f(\omega)\exp(i\omega t),$$

so that

$$a(\omega) = \frac{f(\omega)/m}{\omega_0^2 - \omega^2 + i\omega\gamma}. \qquad (3.74)$$

From (3.52), the generalized susceptibility is

$$\chi(\omega) = \frac{a(\omega)}{f(\omega)}$$

$$= \frac{1/m}{\omega_0^2 - \omega^2 + i\omega\gamma}, \qquad (3.75)$$

and so from (3.53) the real and (minus) the imaginary parts of $\chi(\omega)$ are

$$\chi'(\omega) = \frac{(\omega_0^2 - \omega^2)/m}{(\omega_0^2 - \omega^2)^2 + \gamma^2\omega^2}$$

$$\chi''(\omega) = \frac{\gamma\omega/m}{(\omega_0^2 - \omega^2)^2 + \gamma^2\omega^2}. \qquad (3.76)$$

These are plotted in Fig. 3.8. The static susceptibility is seen to be

$$\chi'(0) = \frac{1}{m\omega_0^2}, \qquad (3.77)$$

[21] One can also, with somewhat more labour, verify the more general formulation of the Kramers–Kronig transformation (D.41).

which is just the reciprocal of the spring constant, as expected. It is straightforward to verify the link between $\chi''(\omega)$ and $\chi'(0)$ prescribed by the Kramers–Kronig transformation[21] (3.56) — see Exercise 3.6. Furthermore, $\chi'(\omega)$ and $\chi''(\omega)$ are seen from (3.76) to be even and odd functions, respectively, of $\omega$. These are general symmetry properties of the $\chi(\omega)$ for a spin-independent interaction, and are derived in Section D.8.

Finally, from (3.54), the dynamical part of the neutron scattering response function is given by

$$\widetilde{S}(\omega) = \{1 + n(\omega)\}\frac{1}{\pi}\frac{\gamma\omega/m}{(\omega_0^2 - \omega^2)^2 + \gamma^2\omega^2}, \qquad (3.78)$$

and from (3.57), (3.76) and (3.77) the spectral-weight function is

$$F(\omega) = \frac{\gamma\omega_0^2/\pi}{(\omega_0^2 - \omega^2)^2 + \gamma^2\omega^2}. \qquad (3.79)$$

The integral of $F(\omega)$ may be shown to be unity, in agreement with the normalization condition (3.58). The functions $F(\omega)$ and $\widetilde{S}(\omega)$ are plotted in Fig. 3.8. Note that when the damping is small ($\gamma \ll \omega_0$) the full width at half maximum of the peaks is approximately $\gamma$.

### 3.9.2   Lorentzian $\chi''(\omega)$ function

A typical form for the generalized susceptibility that describes the dynamics of weakly interacting condensed matter systems is

$$\chi(\omega) = \frac{1}{2\omega_0}\left(\frac{1}{\omega_0 + \omega - i\gamma} + \frac{1}{\omega_0 - \omega + i\gamma}\right). \qquad (3.80)$$

This expression provides an approximate description of many types of lightly damped excitations, such as anharmonic phonons. The absorptive part of (3.80) is

$$\chi''(\omega) = \frac{\gamma}{2\omega_0}\left[\frac{1}{(\omega_0 - \omega)^2 + \gamma^2} - \frac{1}{(\omega_0 + \omega)^2 + \gamma^2}\right], \qquad (3.81)$$

which consists of two Lorentzian functions, one with positive amplitude centred at $\omega_0$ and the other with negative amplitude centred at $-\omega_0$, see Fig. 3.9. When the damping is small ($\gamma \ll \omega_0$) the full width at half maximum of the peaks is approximately $2\gamma$.

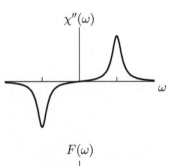

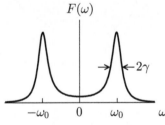

**Fig. 3.9** Lorentzian $\chi''(\omega)$, eqn (3.81), and corresponding spectral function calculated for $\gamma/\omega_0 = 0.15$.

### 3.9.3   Lorentzian spectral function

Another useful lineshape function derives from a phenomenological model in which the relaxation function (see Section D.6) is chosen to have the form of a damped oscillation,

$$R(t) = \exp(-\gamma|t|)\cos(\omega_0 t). \qquad (3.82)$$

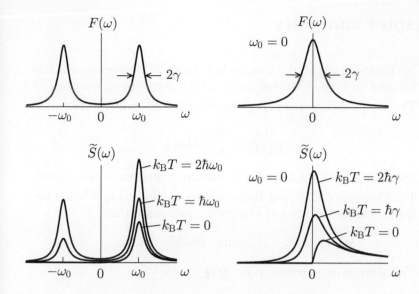

**Fig. 3.10** Lorentzian spectral function $F(\omega)$ defined by eqn (3.83) and corresponding scattering function $\widetilde{S}(\omega)$. The left panels show an inelastic response ($\omega_0 \neq 0$) calculated for $\gamma/\omega_0 = 0.15$, and the right panels show a quasielastic response ($\omega_0 = 0$). The scattering function is plotted for three different temperatures as indicated.

This function is constructed to satisfy a general symmetry requirement that $R(-t) = R^*(t)$. The corresponding spectral-weight function, proportional to the Fourier transform of $R(t)$ [see eqn (D.52)], is given by

$$F(\omega) = \frac{\gamma/2\pi}{(\omega_0 - \omega)^2 + \gamma^2} + \frac{\gamma/2\pi}{(\omega_0 + \omega)^2 + \gamma^2}, \qquad (3.83)$$

from which

$$\chi''(\omega) = \chi'(0)\frac{\gamma\omega}{2}\left[\frac{1}{(\omega - \omega_0)^2 + \gamma^2} + \frac{1}{(\omega + \omega_0)^2 + \gamma^2}\right]. \qquad (3.84)$$

Figure (3.10) shows $F(\omega)$ from (3.83) for oscillators with $\omega_0 \neq 0$ and $\omega_0 = 0$. The corresponding scattering functions are shown at three different temperatures.

The Lorentzian spectral function has been used to describe the lineshape of $f$-electron spectra in metals, e.g. heavy fermion and intermediate valence compounds, where the damping is caused by interactions between the localized $f$ states and conduction electrons (Becker *et al.*, 1977; Holland-Moritz *et al.*, 1982). Note, however, that (3.84) is not physically realistic at large $\omega$ because when $\omega \gg \omega_0$, $\chi''(\omega) \sim \omega^{-1}$, which means that the integral of $\chi''(\omega)$ diverges.[22] Therefore, if (3.84) is to be used over a wide energy range one needs to modify the high-energy behaviour so that the integral remains finite, as required by the structure factor sum rule (3.59).

[22]Mathematically, this is because the $R(t)$ given in (3.82) is not analytic at $t = 0$.

# Chapter summary

- We have introduced the kinematical theory of scattering, which is founded on Fermi's Golden Rule and the Born approximation.
- The partial differential scattering cross-section may be written

$$\frac{\mathrm{d}^2\sigma}{\mathrm{d}\Omega\mathrm{d}E_\mathrm{f}} = \frac{k_\mathrm{f}}{k_\mathrm{i}}S(\mathbf{Q},\omega),$$

  where $S(\mathbf{Q},\omega)$ is the response or scattering function of the system.

- The Principle of Detailed Balance gives the following relation between neutron energy-loss and energy-gain scattering:

$$S(-\mathbf{Q},-\omega)_{\sigma_\mathrm{f}\to\sigma_\mathrm{i}} = \exp(-\beta\hbar\omega)\, S(\mathbf{Q},\omega)_{\sigma_\mathrm{i}\to\sigma_\mathrm{f}}.$$

- Two alternative expressions for $S(\mathbf{Q},\omega)$ are

$$S(\mathbf{Q},\omega) = \left(\frac{m_\mathrm{n}}{2\pi\hbar^2}\right)^2 \sum_{\lambda_\mathrm{i}} p_{\lambda_\mathrm{i}} \sum_{\lambda_\mathrm{f}} |\langle\sigma_\mathrm{f}\lambda_\mathrm{f}|\, V(\mathbf{Q})\, |\sigma_\mathrm{i}\lambda_\mathrm{i}\rangle|^2 \delta(E_{\lambda_\mathrm{f}} - E_{\lambda_\mathrm{i}} - \hbar\omega)$$

$$= \frac{1}{2\pi\hbar}\int_{-\infty}^{\infty} \langle\hat{A}^\dagger(\mathbf{Q})\hat{A}(\mathbf{Q},t)\rangle\, \exp(-\mathrm{i}\omega t)\,\mathrm{d}t,$$

  where $\hat{A}(\mathbf{Q},t)$ is the Heisenberg operator representing the Fourier transform of the interaction potential, including the initial and final spin states of the neutron, eqn (3.28).

- $\langle\hat{A}^\dagger(\mathbf{Q})\hat{A}(\mathbf{Q},t)\rangle = I(\mathbf{Q},t)$ is called the intermediate scattering function and describes space- and time-dependent correlations in the property represented by $\hat{A}$.

- $S(\mathbf{Q},\omega)$ can be divided into an elastic part $S_\mathrm{el}$ due to static correlations, and an inelastic part $\widetilde{S}$ due to dynamic correlations. These may be written

$$S_\mathrm{el}(\mathbf{Q},\omega) = |\langle\hat{A}(\mathbf{Q})\rangle|^2\delta(\hbar\omega)$$

$$\widetilde{S}(\mathbf{Q},\omega) = \{1 + n(\omega)\}\frac{1}{\pi}\chi''(\mathbf{Q},\omega).$$

  The second expression says that inelastic scattering is proportional to $\chi''(\mathbf{Q},\omega)$, the absorptive part of the generalized susceptibility. This relation comes from the Fluctuation–Dissipation theorem.

- The structure factor

$$S(\mathbf{Q}) = \int_{-\infty}^{\infty} S(\mathbf{Q},\omega)\,\mathrm{d}(\hbar\omega) = \langle|\hat{A}(\mathbf{Q})|^2\rangle$$

  measures equal-time (instantaneous) correlations.

- The static approximation is the assumption that $S(\mathbf{Q})$ is measured by a detector that does not have energy analysis.

# Further reading

Sears (1989) provides an account of scattering theory that goes beyond the Born approximation.

# Exercises

(3.1) By taking the complex conjugate of eqn (3.36), and making use of (3.40) and (3.42), show that $S^*(\mathbf{Q}, \omega) = S(\mathbf{Q}, \omega)$ and hence that the response function is real.

(3.2) Neutron scattering is used to measure a transition in which the final state of the system is at an energy $\hbar\omega_0 = 5\,\mathrm{meV}$ above the ground state. Calculate the ratio of the response function for neutron energy-gain scattering to that for neutron energy-loss scattering at temperatures of 10, 50 and 200 K.

(3.3) The absorptive part of the generalized susceptibility has the symmetry property [see eqn (D.76)] $\chi''(-\mathbf{Q}, -\omega)_{\sigma_f \to \sigma_i} = -\chi''(\mathbf{Q}, \omega)_{\sigma_i \to \sigma_f}$. Use this to derive the general expression for the detailed balance principle, eqn (3.48), from the Fluctuation–Dissipation theorem (3.54).

(3.4) Use results (2.75) and (3.67) to show that for a classical monatomic fluid $\chi'(0,0) = n\kappa_T$, where $\kappa_T$ is the bulk isothermal compressibility.

(3.5) The scattering function for a nucleus of mass $M$ in a classical ideal gas of atoms at temperature $T$ is

$$S(\mathbf{Q}, \omega) = \left(\frac{\beta}{4\pi E_r}\right)^{1/2} \exp\left\{-\frac{\beta}{4E_r}(\hbar\omega - E_r)^2\right\},$$

where $E_r = \hbar^2 Q^2 / 2M$ is the recoil energy and $\beta = 1/k_B T$. This result can be obtained from (5.153) by the substitution $\langle p_z \rangle^2 = 2MK/3 = Mk_B T$, where $K = 3k_B T/2$ is the mean kinetic energy per atom. Show from (3.69) that the first and second energy moments of $S(\mathbf{Q}, \omega)$ are

$$S_1(Q) = E_r, \qquad S_2(Q) = 2E_r k_B T + E_r^2.$$

(3.6) Verify that the generalized susceptibility for a damped harmonic oscillator, eqn (3.76), satisfies the Kramers–Kronig transformation for zero frequency, eqn (3.56).

# The Interaction Potential

# 4

Neutron scattering depends on the potential $V(\mathbf{r})$ that describes the neutron–matter interaction. In Chapter 3 we found that in the Born approximation the scattering cross-section contains the matrix element

$$\langle \sigma_f \lambda_f | V(\mathbf{Q}) | \sigma_i \lambda_i \rangle,$$

where $\lambda_i$ and $\lambda_f$ are the initial and final states of the scattering system, $\sigma_i$ and $\sigma_f$ are the initial and final spin states of the neutron, and

$$V(\mathbf{Q}) = \int V(\mathbf{r}) \exp(i\mathbf{Q} \cdot \mathbf{r}) \, d^3\mathbf{r} \tag{4.1}$$

is the Fourier transform of $V(\mathbf{r})$.

In this chapter we shall derive the form of the interaction potential $V(\mathbf{r})$ for all the important types of neutron–matter interaction encountered in practice, and we shall determine $V(\mathbf{Q})$ in each case. By considering the spin dependence of $V(\mathbf{Q})$ we shall obtain general expressions for neutron polarization analysis.

## 4.1 Nuclear interaction

### 4.1.1 The Fermi pseudopotential

The form of $V(\mathbf{r})$ for the nuclear interaction is complicated, and no complete description exists. We do know, however, that it is a scalar interaction with a very short range compared with the wavelength of neutrons used to study condensed matter, and so a bound nucleus will scatter neutrons isotropically.[1] Isotropic scattering can be described by a single parameter $b$, the scattering length, which determines the amplitude of the spherical scattered wave via eqn (1.19).

[1] See Section 1.6.1.

If the scattered wave depends on only one parameter, then a single parameter is also sufficient to describe the nuclear interaction potential. The simplest one-parameter function that can be used to model a very short-range potential is

$$V_N(\mathbf{r}) = a\,\delta(\mathbf{r}), \tag{4.2}$$

where $a$ is a constant, and the origin is taken to be at the nucleus.

The parameter $a$ is linked to the scattering length $b$ in a very simple way. We can find this connection by comparing the two expressions we have obtained for the differential scattering cross-section for a single bound nucleus. These are eqn (3.20), derived from the first Born approximation incorporating the interaction potential, and eqn (1.33), derived from a spherical scattered wave with amplitude $b/r$.

Substituting (4.2) into eqn (3.20) we obtain

$$\frac{d\sigma}{d\Omega} = \left(\frac{m_\mathrm{n}}{2\pi\hbar^2}\right)^2 |\langle \mathbf{k}_\mathrm{f}\sigma_\mathrm{f}| \, a \, \delta(\mathbf{r}) \, |\mathbf{k}_\mathrm{i}\sigma_\mathrm{i}\rangle|^2. \qquad (4.3)$$

If we assume for now that the potential is independent of spin then the state vectors $|\mathbf{k}\sigma\rangle$ can be factorized into spin and spatial parts,

$$\langle \mathbf{k}_\mathrm{f}\sigma_\mathrm{f}| \, a \, \delta(\mathbf{r}) \, |\mathbf{k}_\mathrm{i}\sigma_\mathrm{i}\rangle = a \, \langle\sigma_\mathrm{f}|\sigma_\mathrm{i}\rangle \int \exp(-\mathrm{i}\mathbf{k}_\mathrm{f}\cdot\mathbf{r})\delta(\mathbf{r})\exp(\mathrm{i}\mathbf{k}_\mathrm{i}\cdot\mathbf{r})\,\mathrm{d}^3\mathbf{r}.$$

The volume integral containing the delta function equates to unity (Section B.2), and $\langle\sigma_\mathrm{f}|\sigma_\mathrm{i}\rangle = 1$ since $\sigma_\mathrm{i} = \sigma_\mathrm{f}$ for a spin-independent potential. Hence, the cross-section is

$$\frac{d\sigma}{d\Omega} = \left(\frac{m_\mathrm{n}}{2\pi\hbar^2}\right)^2 a^2. \qquad (4.4)$$

For a single bound nucleus $d\sigma/d\Omega = b^2$ [eqn (1.33)], and so[2]

$$a = \frac{2\pi\hbar^2}{m_\mathrm{n}} \, b. \qquad (4.5)$$

In deriving this relationship we proceeded as if $a$ and $b$ were real quantities, but the result is equally valid if they are complex. From eqns (4.5) and (4.2), the effective interaction potential for a bound nucleus is now given in terms of the (complex) scattering length by

$$V_\mathrm{N}(\mathbf{r}) = \frac{2\pi\hbar^2}{m_\mathrm{n}} \, b \, \delta(\mathbf{r}) \qquad (4.6)$$

This potential is known as the *Fermi pseudopotential*. The term *pseudopotential* is applied because in deriving eqn (4.6) we used Fermi's Golden Rule despite the fact that first-order perturbation theory is not applicable close to the nucleus where the true $V_\mathrm{N}(\mathbf{r})$ is very large. On the other hand, at large distances from the nucleus the Born approximation is valid because the incident neutron wave is only weakly perturbed by the nuclear interaction and the scattered wave is spherical. Therefore, the Fermi pseudopotential is an artificial expression for $V_\mathrm{N}(\mathbf{r})$ constructed to give the correct isotropic scattering when used in formulae derived in the Born approximation. It saves us from having to worry about what actually happens in the vicinity of the nucleus.

In summary: the Fermi pseudopotential should not be regarded as a model for the actual interaction potential, but when used in the Born approximation it gives the same limiting form (i.e. a long way from the nucleus) for the scattered wave as would a more more realistic interaction potential in a more general scattering theory.

### 4.1.2   Scattering length operator

As discussed in Section 1.8.2, nuclei possessing spin exhibit two different values for the scattering length, $b_+$ and $b_-$, depending on whether the

[2]By convention, the positive root is chosen so as to make the experimental values of $b$ positive for the vast majority of nuclides. A positive scattering length corresponds to a repulsive pseudopotential, whereas the actual neutron–nucleus interaction is strongly attractive. This emphasizes the distinction between the pseudopotential and the true potential.

neutron spin is parallel or antiparallel to the nuclear spin. A convenient way to build this spin dependence into $V_N$ is to define an operator $b$ that acts on the combined neutron and nuclear state vector. The eigenvalues of this operator are $b_+$ and $b_-$, i.e.

$$b\,|+\rangle = b_+|+\rangle, \ \text{ and } \ b\,|-\rangle = b_-|-\rangle, \qquad (4.7)$$

where $|+\rangle$ and $|-\rangle$ are states of the combined neutron–nucleus system.

To find a representation for $b$, let $\mathbf{t} = \mathbf{I}+\mathbf{s_n}$ be an operator representing the total spin of the nucleus and neutron, $\mathbf{I}$ and $\mathbf{s_n}$ being the operators for the nuclear and neutron spins, respectively. The states $|+\rangle$ and $|-\rangle$ are eigenstates of the operators $\mathbf{t}^2$, $\mathbf{I}^2$ and $\mathbf{s}_n^2$ with eigenvalues given by[3]

$$\mathbf{t}^2\,|\pm\rangle = t_\pm(t_\pm + 1)\,|\pm\rangle$$
$$\mathbf{I}^2\,|\pm\rangle = I(I + 1)\,|\pm\rangle \qquad (4.8)$$
$$\mathbf{s}_n^2\,|\pm\rangle = \tfrac{1}{2}(\tfrac{1}{2} + 1)\,|\pm\rangle,$$

where $t_+ = I + \frac{1}{2}$ and $t_- = I - \frac{1}{2}$. These operators are related by

$$\mathbf{t}^2 = \mathbf{I}^2 + \mathbf{s}_n^2 + 2\mathbf{s_n} \cdot \mathbf{I}, \qquad (4.9)$$

so that from (4.8) and (4.9)

$$\mathbf{s_n} \cdot \mathbf{I}\,|+\rangle = \tfrac{1}{2}I\,|+\rangle, \ \text{ and } \ \mathbf{s_n} \cdot \mathbf{I}\,|-\rangle = -\tfrac{1}{2}(I+1)\,|-\rangle. \qquad (4.10)$$

If we now write

$$b = A + B\mathbf{s_n} \cdot \mathbf{I}, \qquad (4.11)$$

then it is easy to show from eqns (4.7) and (4.10) that

$$A = \frac{1}{2I + 1}\{(I + 1)b_+ + Ib_-\}, \qquad (4.12)$$

$$B = \frac{2}{2I + 1}(b_+ - b_-). \qquad (4.13)$$

Here, $A$ is the scattering length averaged over nuclear spin orientations — see Section 1.8.2.

[3]See Sections C.1.14 and C.1.15 for a summary of the properties of angular momentum in quantum mechanics.

### 4.1.3   Fourier transform of $V_N(\mathbf{r})$

From (4.6), the nuclear pseudopotential interaction operator for a system of many individual nuclei may be written

$$V_N(\mathbf{r}) = \frac{2\pi\hbar^2}{m_n} \sum_j b_j\, \delta\{\mathbf{r} - \mathbf{r}_j\}, \qquad (4.14)$$

where $\mathbf{r}$ is the neutron coordinate and $\mathbf{r}_j$ is the position vector of the $j$th nucleus. The Fourier transform of (4.14) is

$$V_N(\mathbf{Q}) = \frac{2\pi\hbar^2}{m_n} \sum_j b_j\, \exp\{i\mathbf{Q} \cdot \mathbf{r}_j\}, \qquad (4.15)$$

so that

$$\langle \sigma_{\mathrm{f}} | \, \hat{V}_{\mathrm{N}}(\mathbf{Q}, t) \, | \sigma_{\mathrm{i}} \rangle = \frac{2\pi\hbar^2}{m_{\mathrm{n}}} \sum_j \hat{\beta}_j(t) \, \exp\{i\mathbf{Q} \cdot \hat{\mathbf{r}}_j(t)\}, \tag{4.16}$$

where

$$\hat{\beta}_j(t) = \langle \sigma_{\mathrm{f}} | \, \hat{b}_j(t) \, | \sigma_{\mathrm{i}} \rangle. \tag{4.17}$$

Hence, from (3.35) and (3.38) the nuclear correlation function reads

$$I(\mathbf{Q}, t) = \sum_{jk} \langle \exp(-i\mathbf{Q} \cdot \hat{\mathbf{r}}_j) \, \hat{\beta}_j^\dagger \, \hat{\beta}_k(t) \, \exp\{i\mathbf{Q} \cdot \hat{\mathbf{r}}_k(t)\} \rangle. \tag{4.18}$$

The symbols $\hat{b}(t)$, $\hat{\mathbf{r}}(t)$, etc, denote Heisenberg operators, and $\hat{\mathbf{r}} \equiv \hat{\mathbf{r}}(0)$.

## 4.1.4   Coherent and incoherent cross-sections

Equation (4.18) can be simplified in several possible ways depending on the system under investigation and the experimental conditions.

(i) *No correlation between nuclear and target states*

The first simplification is to assume that the motion of the atoms is decoupled from the internal states of the nuclei (as represented by their scattering lengths). In other words, the force between any two atoms is assumed to be independent of which isotopes or spin orientations are found on the two nuclei.

This assumption allows us to separate (4.18) into a product of two factors, one describing correlations between the scattering lengths and the other describing correlations between the positions of the nuclei:

$$I(\mathbf{Q}, t) = \sum_{jk} \langle \hat{\beta}_j^\dagger \, \hat{\beta}_k(t) \rangle \, \langle \exp(-i\mathbf{Q} \cdot \hat{\mathbf{r}}_j) \, \exp\{i\mathbf{Q} \cdot \hat{\mathbf{r}}_k(t)\} \rangle. \tag{4.19}$$

The above simplification is almost always valid for condensed matter systems. Two cases where it breaks down are, firstly, when there are very strong quantum-mechanical exchange correlations in a system of identical nuclei, and second, when the difference in mass between isotopes causes local modes of oscillation that differ significantly in frequency. The reason the former can have a measurable effect on the scattering is that the symmetries of the spin and spatial wave functions of identical particles (nuclei in this case) are intrinsically coupled, as required by the Pauli Exclusion Principle. Therefore, for example, the rotational motion of a homonuclear diatomic molecule is dependent on the combined spin state of its two nuclei.

(ii) *Random isotope and nuclear spin distributions*

A further simplification is to assume that the different isotopes and/or spin states of the nuclei are distributed randomly among the sites for a given atomic species. Cases do exist where this assumption fails, but they are rare. An obvious exception is an isotopically labelled system.[4]

[4]Isotopic labelling is the process by which one part of a system is deliberately isotopically enriched, i.e. a particular isotope is in greater abundance in that part than elsewhere in the sample. The most common case is the selective replacement of protium ($^1$H) by deuterium ($^2$H) in one component of a hydrogen-containing sample.

Another example is the symmetry constraint on the nuclear spin and coordinate wave functions discussed in the previous section. A third case is nuclear spin ordering in a crystal. The latter only occurs at milliKelvin temperatures or lower, so is hardly ever important (unless, of course, one has set out to study it — see Section 7.7).

In most cases of interest, simplifications (i) and (ii) *both* apply. This means that after carrying out the factorization in eqn (4.19) we can replace the scattering length correlation function by a simple product and perform an ensemble average over the random distribution of isotopes and/or nuclear spin states, exactly as described in Sections 1.8.1 and 1.8.2. In doing so, we make the further assumption that the spin dynamics of the nucleus is sufficiently slow compared with the interaction time of the neutron that the nuclear spin orientations can be treated as static, which means that $\hat{\beta}_k(t)$ can be replaced by $\hat{\beta}_k(0) \equiv \beta_k$. Hence,

$$\overline{\beta_j^\dagger \beta_k} = \overline{|\beta_j|^2} \quad (j = k),$$
$$= \overline{\beta_j^\dagger}\,\overline{\beta_k} \quad (j \neq k), \tag{4.20}$$

where the bar signifies the isotope and/or nuclear spin average.

Inclusion of averages (4.20) in (4.19) and separation of terms in the summation in the manner of eqn (1.44) allows us to write

$$I_{\mathrm{coh}}(\mathbf{Q}, t) = \sum_{jk} \overline{\beta_j^\dagger}\,\overline{\beta_k} \langle \exp(-i\mathbf{Q} \cdot \hat{\mathbf{r}}_j) \exp\{i\mathbf{Q} \cdot \hat{\mathbf{r}}_k(t)\} \rangle, \tag{4.21}$$

and

$$I_{\mathrm{inc}}(\mathbf{Q}, t) = \sum_j (\overline{|\beta_j|^2} - |\overline{\beta}_j|^2)\langle \exp(-i\mathbf{Q} \cdot \hat{\mathbf{r}}_j) \exp\{i\mathbf{Q} \cdot \hat{\mathbf{r}}_j(t)\} \rangle. \tag{4.22}$$

These functions represent the spin-dependent coherent and incoherent nuclear scattering, respectively. Generalizing the account in Section 1.8, we remark, first, that nuclear scattering arises from correlations between the positions of nuclei at different times, and second, that coherent scattering includes correlations between both the same and different nuclei, whereas incoherent scattering arises only from self correlations.

In Section 4.5 we shall determine how the coherent and incoherent scattering depends on the neutron spin state before and after scattering, and in Section 4.9 we consider nuclear scattering of unpolarized neutrons.

## 4.2   Magnetic interaction

The neutron couples to microscopic electromagnetic fields via its magnetic dipole moment. Compared with the strong nuclear force, the electromagnetic interactions are much weaker and their range much longer. The Born approximation is again applicable because the interaction causes only a small perturbation to the incoming neutron wave, but the extended range of the interaction means that an accurate form for the interaction potential is required. We cannot get away with a pseudopotential this time.

There is more than one channel for electromagnetic interactions, but by far the strongest is the one usually called the *magnetic interaction*,

$$V_{\mathrm{M}}(\mathbf{r}) = -\boldsymbol{\mu}_{\mathrm{n}} \cdot \mathbf{B}(\mathbf{r}), \qquad (4.23)$$

corresponding to the potential energy of the neutron dipole moment $\boldsymbol{\mu}_{\mathrm{n}}$ in the local magnetic flux density (or induction) $\mathbf{B}(\mathbf{r})$. In order to calculate the cross-section from a microscopic electronic model we will need to develop expressions for $\mathbf{B}(\mathbf{r})$. We lay the foundations for this in the following section.

### 4.2.1 Flux density from unpaired electrons

The magnetic flux density from electrons in a medium is the sum of two contributions, one from the intrinsic spin of the electrons and the other from electronic currents due to the motion of the electrons. The latter is commonly referred to as the 'orbital' contribution, but this can be misleading since electrons in solids are not necessarily bound to atoms.

The spin contribution from a single electron can be written

$$\mathbf{B}_{\mathrm{S}}(\mathbf{r}) = -2\mu_{\mathrm{B}}\frac{\mu_0}{4\pi}\, \nabla \times \left(\frac{\mathbf{s} \times \mathbf{r}}{r^3}\right), \qquad (4.24)$$

and, for non-relativistic velocities,[5] the current contribution is

$$\mathbf{B}_{\mathrm{L}}(\mathbf{r}) = -2\mu_{\mathrm{B}}\frac{\mu_0}{4\pi}\frac{1}{\hbar}\frac{\mathbf{p} \times \mathbf{r}}{r^3}. \qquad (4.25)$$

$\mathbf{s}$ and $\mathbf{p}$ are the spin and momentum of the electron, respectively, and $\mu_{\mathrm{B}} = e\hbar/(2m_{\mathrm{e}}) = 9.27 \times 10^{-24}\,\mathrm{JT}^{-1}$ is the Bohr magneton. Equation (4.24) comes from the standard expression for the flux density of a magnetic dipole moment $\boldsymbol{\mu}$ (see, for example, Griffiths, 2017),

$$\mathbf{B}(\mathbf{r}) = \frac{\mu_0}{4\pi}\, \nabla \times \left(\frac{\boldsymbol{\mu} \times \mathbf{r}}{r^3}\right). \qquad (4.26)$$

In the present case the magnetic dipole moment is that of the electron, which is related to its intrinsic spin by

$$\boldsymbol{\mu}_{\mathrm{e}} = -g_{\mathrm{e}}\mu_{\mathrm{B}}\mathbf{s}, \qquad (4.27)$$

where $g_{\mathrm{e}} = 2.0023$ is the electron spin $g$-factor. From now on we shall take $g_{\mathrm{e}} = 2$ for simplicity.

Expressions (4.24) and (4.25) provide two distinct terms in the magnetic interaction potential (4.23), which thus takes the form[6]

$$\begin{aligned} V_{\mathrm{M}}(\mathbf{r}) &= -\boldsymbol{\mu}_{\mathrm{n}} \cdot \{\mathbf{B}_{\mathrm{S}}(\mathbf{r}) + \mathbf{B}_{\mathrm{L}}(\mathbf{r})\} \\ &= 2\gamma\mu_{\mathrm{N}}\,\mathbf{s}_{\mathrm{n}} \cdot \{\mathbf{B}_{\mathrm{S}}(\mathbf{r}) + \mathbf{B}_{\mathrm{L}}(\mathbf{r})\}. \end{aligned} \qquad (4.28)$$

[5]The magnetic flux density of a charge $q$ moving with non-relativistic velocity $\mathbf{v}$ is given by

$$\mathbf{B}(\mathbf{r}) = \frac{\mu_0}{4\pi}\frac{q\mathbf{v} \times \mathbf{r}}{r^3},$$

which reduces to (4.25) for $q = -e$ and $\mathbf{v} = \mathbf{p}/m_{\mathrm{e}}$.

[6]Recall eqn (1.10) relating the neutron spin and magnetic moment:

$$\boldsymbol{\mu}_{\mathrm{n}} = -2\gamma\mu_{\mathrm{N}}\mathbf{s}_{\mathrm{n}},$$

where $\mu_{\mathrm{N}}$ is the nuclear magneton and $\gamma = 1.913$.

## 4.2.2   Fourier transform of $V_\mathrm{M}(\mathbf{r})$

We continue to treat the spin and current terms in $V_\mathrm{M}$ separately. We consider a system of many electrons, and obtain the contribution from the $j$th electron. This electron has displacement $\mathbf{r}_j$ and spin $\mathbf{s}_j$. The Fourier transform of the spin flux density (4.24) is given by

$$\{\mathbf{B}_\mathrm{S}(\mathbf{Q})\}_j = -2\mu_\mathrm{B}\frac{\mu_0}{4\pi}\int \nabla \times \left(\frac{\mathbf{s}_j \times \mathbf{R}}{R^3}\right) \exp(i\mathbf{Q}\cdot\mathbf{r})\,\mathrm{d}^3\mathbf{r}, \qquad (4.29)$$

where $\mathbf{R} = \mathbf{r} - \mathbf{r}_j$. We use the identity (B.32)

$$\nabla \times \left(\frac{\mathbf{s} \times \mathbf{R}}{R^3}\right) = \frac{1}{2\pi^2}\int \hat{\mathbf{q}} \times (\mathbf{s} \times \hat{\mathbf{q}})\exp(i\mathbf{q}\cdot\mathbf{R})\,\mathrm{d}^3\mathbf{q},$$

where $\hat{\mathbf{q}}$ denotes the unit vector in the direction of $\mathbf{q}$, to write (4.29) as

$$\{\mathbf{B}_\mathrm{S}(\mathbf{Q})\}_j = -2\mu_\mathrm{B}\frac{\mu_0}{8\pi^3}\int\int \hat{\mathbf{q}} \times (\mathbf{s} \times \hat{\mathbf{q}})\exp\{i(\mathbf{Q}+\mathbf{q})\cdot\mathbf{R}\}\,\mathrm{d}^3\mathbf{R}\,\mathrm{d}^3\mathbf{q}.$$

Integration over $\mathbf{R}$ gives the same result as integration over $\mathbf{r}$ since the integral extends over all space and $\mathbf{r}_j$ is constant. Making use of the identity (B.15)

$$\int \exp\{i(\mathbf{Q}+\mathbf{q})\cdot\mathbf{R}\}\,\mathrm{d}^3\mathbf{R} = (2\pi)^3\delta(\mathbf{Q}+\mathbf{q}),$$

we obtain

$$\{\mathbf{B}_\mathrm{S}(\mathbf{Q})\}_j = -2\mu_\mathrm{B}\mu_0\{\hat{\mathbf{Q}} \times (\mathbf{s}_j \times \hat{\mathbf{Q}})\}\exp(i\mathbf{Q}\cdot\mathbf{r}_j). \qquad (4.30)$$

Next, the flux density due to the electron current. The Fourier transform of (4.25) is

$$\{\mathbf{B}_\mathrm{L}(\mathbf{Q})\}_j = -2\mu_\mathrm{B}\frac{\mu_0}{4\pi}\frac{1}{\hbar}\int \frac{\mathbf{p}_j \times \mathbf{R}}{R^3}\exp(i\mathbf{Q}\cdot\mathbf{r})\,\mathrm{d}^3\mathbf{r}$$

$$= -2\mu_\mathrm{B}\mu_0\frac{i}{\hbar Q}(\mathbf{p}_j \times \hat{\mathbf{Q}})\exp(i\mathbf{Q}\cdot\mathbf{r}_j). \qquad (4.31)$$

To do the integral, change the integration variable to $\mathbf{R} = \mathbf{r} - \mathbf{r}_j$ and use identity (B.30)

$$\int \frac{\mathbf{R}}{R^3}\exp(i\mathbf{Q}\cdot\mathbf{R})\,\mathrm{d}^3\mathbf{R} = \frac{4\pi i}{Q}\hat{\mathbf{Q}}.$$

Finally, combining the spin and orbital terms for all the electrons in the system we obtain from (4.28), (4.30) and (4.31),

$$V_\mathrm{M}(\mathbf{Q}) = -4\gamma\mu_0\mu_\mathrm{B}\mu_\mathrm{N}\,\mathbf{s}_\mathrm{n}\cdot\sum_j \{\hat{\mathbf{Q}} \times (\mathbf{s}_j \times \hat{\mathbf{Q}}) + \frac{i}{\hbar Q}(\mathbf{p}_j \times \hat{\mathbf{Q}})\}\exp(i\mathbf{Q}\cdot\mathbf{r}_j).$$

$$(4.32)$$

The magnetic scattering strength can be estimated from the numerical constants outside the summation in (4.32) and the factor $m_\mathrm{n}/(2\pi\hbar^2)$

which appears squared in front the cross-section (3.16). Multiplying these factors together, we find

$$2\mu_0 \frac{e\hbar}{2m_{\mathrm{e}}} \gamma \frac{e\hbar}{2m_{\mathrm{p}}} \frac{m_{\mathrm{n}}}{2\pi\hbar^2} = \gamma r_0,$$

where[7]

$$r_0 = \frac{\mu_0 e^2}{4\pi m_{\mathrm{e}}} = 2.818 \times 10^{-15}\,\mathrm{m} \qquad (4.33)$$

[7]The proton and neutron rest masses differ by about 1 part in $10^3$, but for present purposes we take $m_{\mathrm{p}} = m_{\mathrm{n}}$.

is the classical radius of the electron. The factor $\gamma r_0$ is similar in size to a typical nuclear scattering length, and this shows that the strengths of nuclear and magnetic scattering are comparable.

### 4.2.3   Magnetization and the magnetic interaction

We will now show that the right-hand side of (4.32) can be written in terms of a spatially varying microscopic magnetization $\mathbf{M}(\mathbf{r})$. The aim of this exercise is to express the neutron cross-section in terms of a quantity that has some intuitive meaning.

Before starting there are two issues which need clarification.[8] The first is that from Maxwell's equations there is an indeterminacy in the definition of $\mathbf{M}(\mathbf{r})$ due to a gauge freedom. This is because the physical quantity that appears in Maxwell's field equations for media, and upon which the neutron cross-section depends, is the electron current density $\mathbf{j}(\mathbf{r})$, which is related to the magnetization by $\mathbf{j} = \nabla \times \mathbf{M}$. Therefore, the transformation $\mathbf{M}' = \mathbf{M} + \nabla g$, where $g(\mathbf{r})$ is any continuous scalar function, gives the same $\mathbf{j}(\mathbf{r})$.[9] The choice of gauge has no impact on the scattering cross-section, but as we shall find shortly the gauge freedom means that neutron scattering can provide no information on the component of $\mathbf{M}$ parallel to $\mathbf{Q}$.

[8]An excellent discussion of these points is provided by Hirst (1997).

[9]From the identify $\nabla \times \nabla g(\mathbf{r}) = 0$.

The second point concerns the physical interpretation of magnetization. Because the electron spin is point-like the magnetization arising from spin can readily be understood in terms of spin density, which is a well-defined quantity. However, the physical meaning of the magnetization arising from electron currents is not so clear. In the case of atomic magnetism (electrons tightly bound to atoms) the electrons have bound orbits about a fixed origin (the nucleus) and one can define an orbital magnetization in terms of the density of orbital angular momentum. However, the electron current magnetization in the case when the electrons are not bound is not so obvious.

Following these remarks, let us explicitly separate the magnetization into spin and orbital (current) parts,

$$\mathbf{M}(\mathbf{r}) = \mathbf{M}_{\mathrm{S}}(\mathbf{r}) + \mathbf{M}_{\mathrm{L}}(\mathbf{r}). \qquad (4.34)$$

We define the spin magnetization as the density of electron spin magnetic moments. From (4.27),

$$\mathbf{M}_{\mathrm{S}}(\mathbf{r}) = -2\mu_{\mathrm{B}} \sum_j \delta(\mathbf{r} - \mathbf{r}_j)\mathbf{s}_j. \qquad (4.35)$$

The Fourier transform of $\mathbf{M}_S(\mathbf{r})$ is

$$\mathbf{M}_S(\mathbf{Q}) = \int \mathbf{M}_S(\mathbf{r}) \, \exp(\mathrm{i}\mathbf{Q} \cdot \mathbf{r}) \, \mathrm{d}^3\mathbf{r}$$

$$= -2\mu_\mathrm{B} \sum_j \mathbf{s}_j \, \exp(\mathrm{i}\mathbf{Q} \cdot \mathbf{r}_j). \qquad (4.36)$$

Hence,

$$\sum_j \hat{\mathbf{Q}} \times (\mathbf{s}_j \times \hat{\mathbf{Q}}) \exp(\mathrm{i}\mathbf{Q} \cdot \mathbf{r}_j) = -\frac{1}{2\mu_\mathrm{B}} \hat{\mathbf{Q}} \times \{\mathbf{M}_S(\mathbf{Q}) \times \hat{\mathbf{Q}}\}. \qquad (4.37)$$

The orbital current term in (4.32) is similarly related to an appropriately-defined orbital magnetization:

$$\sum_j \frac{\mathrm{i}}{\hbar Q} (\mathbf{p}_j \times \hat{\mathbf{Q}}) \exp(\mathrm{i}\mathbf{Q} \cdot \mathbf{r}_j) = -\frac{1}{2\mu_\mathrm{B}} \hat{\mathbf{Q}} \times \{\mathbf{M}_L(\mathbf{Q}) \times \hat{\mathbf{Q}}\}. \qquad (4.38)$$

This relation can be derived directly from a microscopic definition of the orbital magnetization (see Steinsvoll *et al.*, 1967). Substituting (4.37) and (4.38) into (4.32) we obtain the following expression,[10]

$$V_\mathrm{M}(\mathbf{Q}) = 2\gamma\mu_0\mu_\mathrm{N} \, \mathbf{s}_\mathrm{n} \cdot \mathbf{M}_\perp(\mathbf{Q}), \qquad (4.39)$$

where

$$\mathbf{M}_\perp(\mathbf{Q}) = \hat{\mathbf{Q}} \times \{\mathbf{M}(\mathbf{Q}) \times \hat{\mathbf{Q}}\} \qquad (4.40)$$

is the transverse magnetization, i.e. the projection of the total magnetization $\mathbf{M} = \mathbf{M}_S + \mathbf{M}_L$ onto the plane perpendicular to $\mathbf{Q}$, see Fig. 4.1. An alternative expression to (4.40) is (we omit the $\mathbf{Q}$ dependence of $\mathbf{M}$)

$$\mathbf{M}_\perp = \mathbf{M} - (\mathbf{M} \cdot \hat{\mathbf{Q}})\hat{\mathbf{Q}}, \qquad (4.41)$$

which is easily derived via the standard vector triple product identity. For later reference, we also note the following identity:

$$\mathbf{M}_\perp^\dagger \cdot \mathbf{M}_\perp = \{\mathbf{M}^\dagger - (\mathbf{M}^\dagger \cdot \hat{\mathbf{Q}})\hat{\mathbf{Q}}\} \cdot \{\mathbf{M} - (\mathbf{M} \cdot \hat{\mathbf{Q}})\hat{\mathbf{Q}}\}$$

$$= \mathbf{M}^\dagger \cdot \mathbf{M} - (\mathbf{M}^\dagger \cdot \hat{\mathbf{Q}})(\mathbf{M} \cdot \hat{\mathbf{Q}})$$

$$= \sum_{\alpha\beta} (\delta_{\alpha\beta} - \hat{Q}_\alpha \hat{Q}_\beta) M_\alpha^\dagger M_\beta. \qquad (4.42)$$

$\hat{\mathbf{Q}}$ is a unit vector in the direction of $\mathbf{Q}$.

[10]The vector $\mathbf{M}_\perp(\mathbf{Q})$ is proportional to the *magnetic interaction vector* $\mathbf{q} = \hat{\mathbf{Q}} \times \{\hat{\mathbf{M}}(\mathbf{Q}) \times \hat{\mathbf{Q}}\}$ first introduced by Halpern and Johnson (1939). Later accounts of magnetic neutron scattering theory, e.g. Lovesey (1984*b*) and Squires (1977), tend to define a magnetic interaction vector $\mathbf{Q}$ (not to be confused with the scattering vector) which in our notation is equal to $-\mathbf{M}(\mathbf{Q})/(2\mu_\mathrm{B})$. We prefer to write (4.39) and (4.40) in the form given here because $\mathbf{M}$ has a more intuitive meaning (at least for bound electrons) and, more pragmatically, because it avoids having two quite different quantities denoted by the same symbol $\mathbf{Q}$.

$$\mathbf{a} \times (\mathbf{b} \times \mathbf{c}) = (\mathbf{a} \cdot \mathbf{c})\mathbf{b} - (\mathbf{a} \cdot \mathbf{b})\mathbf{c}$$

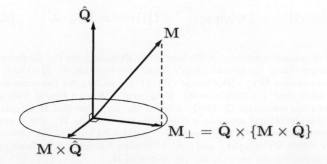

**Fig. 4.1** Diagram showing the relationship between the directions of the vectors $\hat{\mathbf{Q}}$, $\mathbf{M}$ and $\mathbf{M}_\perp = \hat{\mathbf{Q}} \times \{\mathbf{M} \times \hat{\mathbf{Q}}\}$. The vector $\mathbf{M} \times \hat{\mathbf{Q}}$ is perpendicular to both $\hat{\mathbf{Q}}$ and $\mathbf{M}$, so $\hat{\mathbf{Q}} \times \{\mathbf{M} \times \hat{\mathbf{Q}}\}$ is the projection of $\mathbf{M}$ onto the plane perpendicular to $\hat{\mathbf{Q}}$.

The significance of this formulation is that it shows that magnetic scattering only 'sees' the component of the magnetization perpendicular to $\mathbf{Q}$. The longitudinal magnetization is not accessible to experiment.[11]

### 4.2.4   Interaction with atomic electric fields

The next largest electromagnetic interactions arise from the interaction of the neutron dipole moment with atomic electric fields and with the nuclear magnetic dipole moment (if present). As we shall see, these interactions are smaller than $V_M$ by a factor of $m_e/m_n \sim 1/2000$, and for this reason they rarely come into play in conventional experiments designed to study the properties of condensed matter.[12]

The interaction with an electrostatic field $\mathbf{E}$ can be written

$$V_E(\mathbf{r}) = V_{SO}(\mathbf{r}) + V_F(\mathbf{r}), \tag{4.43}$$

where

$$V_{SO}(\mathbf{r}) = -\frac{1}{m_n c^2}\, \boldsymbol{\mu}_n \cdot (\mathbf{E} \times \mathbf{p}) \tag{4.44}$$

and

$$V_F(\mathbf{r}) = -\frac{\hbar\,\mu_n}{2m_n c^2}\, \nabla \cdot \mathbf{E}. \tag{4.45}$$

The term $V_{SO}$ is known as the *spin–orbit* interaction. It arises because the motion of the neutron through an electric field generates a magnetic field in the rest frame of the neutron given by (Griffiths, 2017)

$$\mathbf{B} = \frac{1}{c^2}\mathbf{E} \times \mathbf{v}. \tag{4.46}$$

This field produces a magnetic interaction in the rest frame of the neutron, and expression (4.44) for $V_{SO}$ follows immediately from eqns (4.46) and (4.23), and the non-relativistic momentum $\mathbf{p} = m_n\mathbf{v}$.

The term $V_F$ in (4.43), known as the Foldy interaction (Foldy, 1958), is due to the existence of internal electric fields within the neutron. These internal fields arise from the quark structure of the neutron, and also give the neutron its magnetic dipole moment.

In the case of elastic scattering, the Fourier transforms of $V_{SO}$ and $V_F$ are found to be

$$V_{SO}(\mathbf{Q}) = -2\gamma\mu_0\mu_B\mu_N\frac{m_e}{m_n}n(\mathbf{Q})\,\mathrm{i}\cot(\phi/2)\,\mathbf{s}_n \cdot \hat{\mathbf{z}}, \tag{4.47}$$

[12]A comprehensive discussion of these smaller electromagnetic terms has been given by Sears (1986).

[11]This feature of magnetic neutron scattering arises directly from the fundamental laws of electromagnetism. To see this, apply Maxwell's equation $\nabla \cdot \mathbf{B}(\mathbf{r}) = 0$ to a single Fourier component $\mathbf{B}(\mathbf{r}) = \mathbf{B}(\mathbf{Q})\exp(-\mathrm{i}\mathbf{Q}\cdot\mathbf{r})$. By writing out the components of $\nabla$ one can show that $\nabla \cdot \mathbf{B}(\mathbf{r}) = -\mathrm{i}\mathbf{Q} \cdot \mathbf{B}(\mathbf{Q})\exp(-\mathrm{i}\mathbf{Q} \cdot \mathbf{r})$. Hence, Maxwell's equation imposes the constraint $\mathbf{Q} \cdot \mathbf{B}(\mathbf{Q}) = 0$, i.e. the component of $\mathbf{B}(\mathbf{Q})$ parallel to $\mathbf{Q}$ is zero. This shows that neutrons scatter from variations in $\mathbf{B}_\perp$, the magnetic flux density perpendicular to $\mathbf{Q}$. In general, $\mathbf{B} = \mu_0(\mathbf{H} + \mathbf{M})$, but in the absence of free currents $\mathbf{H} = 0$ so $\mathbf{B}_\perp = \mu_0\mathbf{M}_\perp$ and eqn (4.39) then follows directly from (4.23) without the need to consider the microscopic source of $\mathbf{B}$.

and

$$V_{\mathrm{F}}(\mathbf{Q}) = \gamma\mu_0\mu_{\mathrm{B}}\mu_{\mathrm{N}}\frac{m_{\mathrm{e}}}{m_{\mathrm{n}}}n(\mathbf{Q}), \tag{4.48}$$

where $n(\mathbf{Q})$ is the Fourier transform of the electron density, $\phi$ is the scattering angle, and $\hat{\mathbf{z}}$ is a unit vector in the direction of $\mathbf{k}_{\mathrm{f}} \times \mathbf{k}_{\mathrm{i}}$, i.e. perpendicular to the scattering plane. For derivations of (4.47) and (4.48) the interested reader is referred to Squires (1977). Hence,

$$V_{\mathrm{E}}(\mathbf{Q}) = \gamma\mu_0\mu_{\mathrm{B}}\mu_{\mathrm{N}}\frac{m_{\mathrm{e}}}{m_{\mathrm{n}}}n(\mathbf{Q})\{1 - 2\mathrm{i}\cot(\phi/2)\,\mathbf{s}_{\mathrm{n}}\cdot\hat{\mathbf{z}}\}. \tag{4.49}$$

Comparing (4.49) with (4.32) we see that the interaction with electric fields is $\sim m_{\mathrm{e}}/m_{\mathrm{n}}$ times smaller than the magnetic interaction.

## 4.3  Spin dependence

To handle experiments with polarized neutrons we need to find general expressions for the scattering of neutrons from a specified initial spin state $\sigma_{\mathrm{i}}$ to a specified final spin state $\sigma_{\mathrm{f}}$. This requires the evaluation of the matrix element

$$\langle\sigma_{\mathrm{f}}|\,V(\mathbf{Q})\,|\sigma_{\mathrm{i}}\rangle. \tag{4.50}$$

In general, the interaction potential is the sum of the nuclear and all the electromagnetic terms. Looking at the expressions for $V(\mathbf{Q})$ for the various interaction potentials considered here, i.e. the nuclear interaction (4.11) and (4.15), the magnetic interaction (4.32) or (4.39), and the weaker spin–orbit and Foldy interactions (4.49), we see that there are two types of term: a spin-independent term, and a term containing the scalar product of the neutron magnetic moment with a vector field associated with the scattering system. Hence, we can write the potential in the general form[13]

[13]The factor $2\pi\hbar^2/m_{\mathrm{n}}$ cancels in the cross-section (3.16).

$$V(\mathbf{Q}) = \frac{2\pi\hbar^2}{m_{\mathrm{n}}}\{C(\mathbf{Q}) + 2\mathbf{s}_n\cdot\mathbf{D}(\mathbf{Q})\}, \tag{4.51}$$

where $C(\mathbf{Q})$ contains all the spin-independent terms, and $\mathbf{D}(\mathbf{Q})$ all the vector terms. For convenience, the terms contributing to $C(\mathbf{Q})$ and $\mathbf{D}(\mathbf{Q})$ for each type of interaction are gathered together in Table 4.1.

**Table 4.1** The terms $C(\mathbf{Q})$ and $\mathbf{D}(\mathbf{Q})$ in eqn (4.51) for each type of interaction.

| | $C(\mathbf{Q})$ | $\mathbf{D}(\mathbf{Q})$ |
|---|---|---|
| Nuclear | $\sum_j A_j \exp(\mathrm{i}\mathbf{Q}\cdot\mathbf{r}_j)$ | $\frac{1}{2}\sum_j B_j\mathbf{I}_j\exp(\mathrm{i}\mathbf{Q}\cdot\mathbf{r}_j)$ |
| Magnetic | $0$ | $\frac{\gamma r_0}{2\mu_{\mathrm{B}}}\mathbf{M}_\perp(\mathbf{Q})$ |
| Electric | $\frac{\gamma r_0}{2}\frac{m_{\mathrm{e}}}{m_{\mathrm{n}}}n(\mathbf{Q})$ | $-\frac{\gamma r_0}{2}\frac{m_{\mathrm{e}}}{m_{\mathrm{n}}}n(\mathbf{Q})\mathrm{i}\cot(\phi/2)\hat{\mathbf{z}}$ |

To work out the matrix elements (4.50) for a potential of the form (4.51) all we need are the values of the amplitudes $\langle\sigma_f|\sigma_i\rangle$ and matrix elements $\langle\sigma_f|\mathbf{s}_n|\sigma_i\rangle$ for the chosen initial and final spin states. The method for calculating these matrix elements is presented in Section C.1.15. To describe all cases we need to remember that the neutron has a spin of $\frac{1}{2}$, and so for any chosen quantization direction there are two eigenstates, corresponding to spin parallel ('up') or antiparallel ('down') to the quantization direction. With three orthogonal directions $x$, $y$ and $z$ this gives a total of 36 $(\sigma_i, \sigma_f)$ combinations.

The amplitudes $\langle\sigma_f|\sigma_i\rangle$ are listed in Table C.1, and the matrix elements $\langle\sigma_f|S_x|\sigma_i\rangle$, $\langle\sigma_f|S_y|\sigma_i\rangle$ and $\langle\sigma_f|S_z|\sigma_i\rangle$ are in Table C.2.

## 4.4   Polarization analysis

The term *polarization analysis* is used to describe experiments in which different components of the scattered neutron polarization are measured. Such experiments are used to measure the partial spin-state cross-sections, i.e. cross-sections for scattering from a specific $\sigma_i$ to a specific $\sigma_f$. These partial cross-sections can contain information about the scattering system which is not accessible if unpolarized neutrons are employed. Indeed, the use of polarization analysis is sometimes the only way to discriminate unambiguously between different sources of scattering. In this section we shall show how the partial spin-state cross-sections depend on correlations between the various terms that appear in $V(\mathbf{Q})$, with particular emphasis on structural and magnetic correlations.

To make these correlations explicit we shall work with the intermediate scattering function[14]

[14]In eqn (4.52) the outer angular brackets denote the thermal average.

$$I(\mathbf{Q},t) = \left(\frac{m_n}{2\pi\hbar^2}\right)^2 \Big\langle \langle\sigma_i|\,\hat{V}^\dagger(\mathbf{Q})\,|\sigma_f\rangle\langle\sigma_f|\,\hat{V}(\mathbf{Q},t)\,|\sigma_i\rangle \Big\rangle, \qquad (4.52)$$

rather than the scattering cross-section. Apart from the factor $k_f/k_i$, the cross-section is just the time Fourier transform of $I(\mathbf{Q},t)$ [see eqns (3.37)–(3.39)], so the reader can easily convert the expressions obtained in the rest of this chapter into scattering intensities.

Substituting the general form of the interaction defined by (4.51) into (4.52) we obtain

$$I^{\sigma_i\sigma_f}(t) = \Big\langle \langle\sigma_i|\,\hat{C}^\dagger + 2\mathbf{s}_n\cdot\hat{\mathbf{D}}^\dagger\,|\sigma_f\rangle\langle\sigma_f|\,\hat{C}(t) + 2\mathbf{s}_n\cdot\hat{\mathbf{D}}(t)\,|\sigma_i\rangle \Big\rangle$$

$$= \Big\langle \{\langle\sigma_i|\sigma_f\rangle\hat{C}^\dagger + 2\langle\sigma_i|\mathbf{s}_n|\sigma_f\rangle\cdot\hat{\mathbf{D}}^\dagger\}$$

$$\times\{\langle\sigma_f|\sigma_i\rangle\hat{C}(t) + 2\langle\sigma_f|\mathbf{s}_n|\sigma_i\rangle\cdot\hat{\mathbf{D}}(t)\} \Big\rangle. \qquad (4.53)$$

The initial and final spin states of the neutron are indicated as superscripts on $I(\mathbf{Q},t)$, and for simplicity we do not show explicitly the $\mathbf{Q}$ dependence of the functions. Note that $\hat{C}(t)$ and $\hat{\mathbf{D}}(t)$ are now Heisenberg operators. Eqn (4.53) and the coefficients given in Tables C.1 and C.2

enable us to work out the intermediate scattering function (and hence the cross-section) for any combination of initial and final spin states.

As an example, consider the particular case $|z\rangle \rightarrow |x\rangle$. From (4.53) and Tables C.1 and C.2 the scattering in this channel is given by

$$I^{zx}(t) = \left\langle \{\tfrac{1}{\sqrt{2}}\hat{C}^{\dagger} + \tfrac{1}{\sqrt{2}}[\hat{D}_x^{\dagger} - i\hat{D}_y^{\dagger} + \hat{D}_z^{\dagger}]\} \right.$$
$$\left. \times \{\tfrac{1}{\sqrt{2}}\hat{C}(t) + \tfrac{1}{\sqrt{2}}[\hat{D}_x(t) + i\hat{D}_y(t) + \hat{D}_z(t)]\} \right\rangle$$

$$= \tfrac{1}{2}\langle \hat{C}^{\dagger}\hat{C}(t) \rangle$$
$$+ \tfrac{1}{2}\left\langle \hat{C}^{\dagger}\{\hat{D}_x(t) + i\hat{D}_y(t) + \hat{D}_z(t)\} + \{\hat{D}_x^{\dagger} - i\hat{D}_y^{\dagger} + \hat{D}_z^{\dagger}\}\hat{C}(t) \right\rangle$$
$$+ \tfrac{1}{2}\left\langle \{\hat{D}_x^{\dagger} - i\hat{D}_y^{\dagger} + \hat{D}_z^{\dagger}\}\{\hat{D}_x(t) + i\hat{D}_y(t) + \hat{D}_z(t)\} \right\rangle .$$

In the following sections we will use (4.53) to treat a number of cases of practical interest.

## 4.5   Uniaxial polarization analysis

If the neutron spin quantization axis immediately after the sample is the same as that just before the sample then we can only analyze the scattered neutron polarization in one direction, i.e. parallel to the incident neutron polarization. Although not the most general type of polarized neutron setup, uniaxial polarization analysis (also known as longitudinal polarization analysis) has a number of very useful applications, one of the most important of which is in providing an unambiguous separation of magnetic and structural (non-magnetic) scattering, as we show below.

Suppose the neutron spin quantization axis is the $x$ direction. The $x$ component of neutron spin then has two stationary states, with the spin either parallel or antiparallel to $x$. If the scattering process is spin-dependent then a proportion of the neutron spins may 'flip' on scattering, i.e. change from $+x$ to $-x$, or vice versa. For obvious reasons, people speak of non-spin-flip (NSF) processes ($xx$ and $\bar{x}\bar{x}$) and spin-flip (SF) processes ($x\bar{x}$ and $\bar{x}x$) when performing uniaxial polarization analysis, where $\bar{x}$ is shorthand for $-x$. The corresponding spin-state cross-sections are therefore described by the four functions $I^{xx}(t), I^{\bar{x}\bar{x}}(t), I^{x\bar{x}}(t)$ and $I^{\bar{x}x}(t)$. From (4.53) and Tables C.1 and C.2 we find

$$I^{xx}(t) = \left\langle \hat{C}^{\dagger}\hat{C}(t) + \hat{C}^{\dagger}\hat{D}_x(t) + \hat{D}_x^{\dagger}\hat{C}(t) + \hat{D}_x^{\dagger}\hat{D}_x(t) \right\rangle$$

$$I^{\bar{x}\bar{x}}(t) = \left\langle \hat{C}^{\dagger}\hat{C}(t) - \hat{C}^{\dagger}\hat{D}_x(t) - \hat{D}_x^{\dagger}\hat{C}(t) + \hat{D}_x^{\dagger}\hat{D}_x(t) \right\rangle$$

$$I^{x\bar{x}}(t) = \left\langle \hat{D}_y^{\dagger}\hat{D}_y(t) + \hat{D}_z^{\dagger}\hat{D}_z(t) + i\{\hat{D}_y^{\dagger}\hat{D}_z(t) - \hat{D}_z^{\dagger}\hat{D}_y(t)\} \right\rangle$$

$$I^{\bar{x}x}(t) = \left\langle \hat{D}_y^{\dagger}\hat{D}_y(t) + \hat{D}_z^{\dagger}\hat{D}_z(t) - i\{\hat{D}_y^{\dagger}\hat{D}_z(t) - \hat{D}_z^{\dagger}\hat{D}_y(t)\} \right\rangle . \quad (4.54)$$

It can be seen that the scalar term $C$ gives only NSF scattering, whereas the vector term $\mathbf{D}$ produces NSF scattering from the component along the neutron spin quantization direction and SF scattering from the perpendicular components.

We will now use these general expressions for uniaxial polarization analysis to consider two common types of system: first, one in which the scattering is purely nuclear in origin, and second, one in which both nuclear and magnetic scattering are present.

### 4.5.1   Nuclear scattering

There are three types of correlation we need to consider in (4.54). For nuclear scattering these take the form (see Table 4.1)

$$\langle \hat{C}^\dagger \hat{C}(t) \rangle = \sum_{jk} \langle A_j^* A_k \exp(-i\mathbf{Q} \cdot \hat{\mathbf{r}}_j) \exp\{i\mathbf{Q} \cdot \hat{\mathbf{r}}_k(t)\} \rangle \tag{4.55}$$

$$\langle \hat{C}^\dagger \hat{D}_\alpha(t) \rangle = \frac{1}{2} \sum_{jk} \langle A_j^* B_k \exp(-i\mathbf{Q} \cdot \hat{\mathbf{r}}_j) I_{k\alpha} \exp\{i\mathbf{Q} \cdot \hat{\mathbf{r}}_k(t)\} \rangle \tag{4.56}$$

$$\langle \hat{D}_\alpha^\dagger \hat{D}_\beta(t) \rangle = \frac{1}{4} \sum_{jk} \langle B_j^* B_k \exp(-i\mathbf{Q} \cdot \hat{\mathbf{r}}_j) I_{j\alpha} I_{k\beta} \exp\{i\mathbf{Q} \cdot \hat{\mathbf{r}}_k(t)\} \rangle. \tag{4.57}$$

We shall assume that the distribution of isotopes and nuclear spin states is random among the sites and that there are no correlations between nuclear potential and nuclear coordinate states (see Section 4.1.4). The independence of nuclear potential and coordinate states allows us to take the $A$ and $B$ amplitudes and the $I_{j\alpha}$ operators outside the correlation functions. Random nuclear spins allows three further simplifications: (i) that $\langle \hat{C}^\dagger \hat{D}_\alpha(t) \rangle = 0$, (ii) that all the $I_{j\alpha} I_{k\beta}$ terms with $j \neq k$ or $\alpha \neq \beta$ vanish, and (iii) that the $I_{j\alpha} I_{j\alpha}$ terms can be replaced by $\frac{1}{3} I_j(I_j + 1)$, since for random spins

$$\langle I_x^2 \rangle = \langle I_y^2 \rangle = \langle I_z^2 \rangle = \tfrac{1}{3} I(I+1). \tag{4.58}$$

Applying these simplifications to the correlation functions (4.55)–(4.57) and substituting the results in (4.54) we obtain the following expressions for the NSF and SF intermediate scattering functions:

$$I_{\text{NSF}} = I_{\text{coh}}^{\text{N}} + I_{\text{inc}}^{\text{iso}} + \tfrac{1}{3} I_{\text{inc}}^{\text{sp}} \tag{4.59}$$

$$I_{\text{SF}} = \tfrac{2}{3} I_{\text{inc}}^{\text{sp}} \tag{4.60}$$

where

$$I_{\text{coh}}^{\text{N}} = \sum_{jk} \overline{A_j^*}\, \overline{A}_k \langle \exp(-i\mathbf{Q} \cdot \hat{\mathbf{r}}_j) \exp\{i\mathbf{Q} \cdot \hat{\mathbf{r}}_k(t)\} \rangle \tag{4.61}$$

$$I_{\text{inc}}^{\text{iso}} = \frac{1}{4\pi} \sum_j \left(\sigma_{\text{inc}}^{\text{iso}}\right)_j \langle \exp(-i\mathbf{Q} \cdot \hat{\mathbf{r}}_j) \exp\{i\mathbf{Q} \cdot \hat{\mathbf{r}}_j(t)\} \rangle \tag{4.62}$$

$$I_{\text{inc}}^{\text{sp}} = \frac{1}{4\pi} \sum_j \left(\sigma_{\text{inc}}^{\text{sp}}\right)_j \langle \exp(-i\mathbf{Q} \cdot \hat{\mathbf{r}}_j) \exp\{i\mathbf{Q} \cdot \hat{\mathbf{r}}_j(t)\} \rangle \tag{4.63}$$

and

$$(\sigma_{\text{inc}}^{\text{iso}})_j = 4\pi(\overline{|A_j|^2} - |\overline{A}_j|^2) \tag{4.64}$$

$$(\sigma_{\text{inc}}^{\text{sp}})_j = \pi\overline{|B_j|^2}I_j(I_j + 1). \tag{4.65}$$

The single-atom incoherent cross-sections $\sigma_{\text{inc}}^{\text{iso}}$ and $\sigma_{\text{inc}}^{\text{sp}}$ defined in (4.64) and (4.65) represent the two types of nuclear incoherent scattering, the first due to the random distribution of isotopes (isotopic incoherence) and the second due to the random nuclear spin states (spin incoherence). The bar above terms on the right-hand sides of (4.61), (4.64) and (4.65) indicates an average over the distribution of isotopes.[15] To simplify later expressions we will introduce the operator

$$\hat{N}(\mathbf{Q}, t) = \sum_j \overline{b}_j \exp\{i\mathbf{Q} \cdot \hat{\mathbf{r}}_j(t)\} \tag{4.66}$$

to represent the nuclear coherent scattering amplitude for random nuclear spins, so that

$$I_{\text{coh}}^{\text{N}} = \langle \hat{N}^\dagger(\mathbf{Q})\hat{N}(\mathbf{Q}, t)\rangle. \tag{4.67}$$

In (4.66) and subsequent expressions we have denoted the spin-averaged nuclear scattering length by $b$ rather than $A$, in line with common usage [c.f. (4.12) and (1.49)].

The preceding analysis leads to the following important properties of nuclear scattering: coherent scattering and isotopic incoherent scattering do not change the spin state of the neutron, whereas spin incoherent scattering does. In the case of spin incoherence, two-thirds of the scattered neutrons have their spins flipped whereas one-third are not flipped.

These results can alternatively be stated in terms of the polarizations of the incident and scattered neutron beams. For nuclear coherent scattering and isotopic incoherent scattering $\mathbf{P}_{\text{f}} = \mathbf{P}_{\text{i}}$, whereas for nuclear spin incoherent scattering $\mathbf{P}_{\text{f}} = -\frac{1}{3}\mathbf{P}_{\text{i}}$. When both isotopic and spin incoherence are present $\mathbf{P}_{\text{f}}$ lies somewhere between $-\frac{1}{3}\mathbf{P}_{\text{i}}$ and $\mathbf{P}_{\text{i}}$.

In a classic experiment, Moon, Riste, and Koehler (1969) tested the predictions of (4.59) and (4.60) by measuring the NSF and SF incoherent elastic scattering from samples of nickel and vanadium. The results

[15] $A$ is spin-independent, and the nuclear spin average has already been applied to the term containing $B$.

(a)

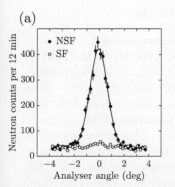

(b)

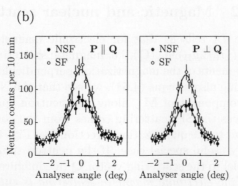

**Fig. 4.2** Uniaxial polarization analysis of nuclear incoherent scattering. (a) Isotopic incoherent scattering from nickel. (b) Nuclear-spin incoherent scattering from vanadium. The proportions of NSF and SF scattering satisfy the predictions of eqns (4.59) and (4.60). In (b) the results are independent of the direction of $\mathbf{P}$ relative to $\mathbf{Q}$ because the nuclear spins are random. (After Moon *et al.*, 1969.)

are reproduced in Fig. 4.2. Nickel has several isotopes, but the most abundant all have $I = 0$. The incoherent scattering is therefore almost entirely isotopic in origin, and should give virtually all NSF scattering. Vanadium, on the other hand, has only one abundant isotope, and the nuclear spin of this isotope is not zero. In this case, the NSF incoherent scattering is expected to be close to one-half the intensity of the SF incoherent scattering. Both of these predictions are confirmed by the results shown in Fig. 4.2.

We conclude this section with a couple of comments on the usefulness of polarization analysis for nuclear scattering. First of all, since nuclear coherent scattering does not change the polarization state of the neutron no new information on structural correlations can be obtained by performing polarization analysis. However, one potentially useful application of polarization analysis is to separate nuclear coherent from nuclear incoherent scattering in systems where the spin incoherence dominates the isotopic incoherence. In such cases, the coherent scattering is given by $I_{\text{NSF}} - \frac{1}{2}I_{\text{SF}}$, as can be seen from (4.59) and (4.60). An obvious case is hydrogen-containing systems, since the proton has a spin incoherent cross-section an order of magnitude larger than for any other nucleus (see Section 1.8.5). This method has been applied to analyze data from non-crystalline polymers and biological materials, Fig. 4.3, for which the coherent scattering tends to be diffusive (see Section 2.6) and difficult to separate from the incoherent scattering by other methods.

Second, we note that eqns (4.59) and (4.60) also show that polarization analysis can be used to separate isotopic and spin incoherent scattering when nuclear coherent scattering is absent. This could in principle be used as a way to obtain information on the single-particle dynamics of individual atomic species in a multi-component system, since from (4.62) and (4.63), the incoherent scattering is a sum over the self correlation functions for each species weighted by the single-atom cross-section. The weighting factors for each species will in general be different for isotopic and spin incoherence, so the self correlation function for species with dominant isotopic or spin incoherence could be measured. Use of different isotopes could also be employed to selectively enhance the signal from different species.

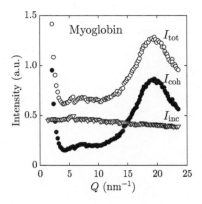

**Fig. 4.3** Separation of coherent and incoherent nuclear scattering from a solution of myoglobin in $D_2O$. (After Gaspar *et al.*, 2010.)

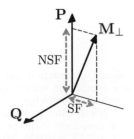

**Fig. 4.4** The components of $\mathbf{M}_\perp$ parallel and perpendicular to $\mathbf{P}$ give rise to NSF and SF scattering, respectively, where $\mathbf{P}$ represents the neutron spin quantization direction.

### 4.5.2  Magnetic and nuclear scattering

Table 4.1 tells us that for electronic magnetic scattering $C = 0$ and $\mathbf{D} = (\gamma r_0 / 2\mu_{\mathbf{B}})\mathbf{M}_\perp(\mathbf{Q})$, i.e. magnetic scattering depends on the Fourier components of the magnetization perpendicular to the scattering vector. Looking also at eqns (4.54), we see that the NSF scattering comes from the component of $\mathbf{M}_\perp$ along the neutron spin quantization direction, whereas the SF scattering comes from the components of $\mathbf{M}_\perp$ perpendicular to the quantization direction (see Fig. 4.4). This presents us with a very useful special case. If the neutron spin quantization direction is parallel to the scattering vector $\mathbf{Q}$ (a configuration we denote by $\mathbf{P} \parallel \mathbf{Q}$) then *the electronic magnetic scattering is entirely spin-flip*. Therefore, the $\mathbf{P} \parallel \mathbf{Q}$ configuration makes it possible to separate electronic mag-netic scattering (entirely SF) from nuclear coherent scattering (entirely NSF).

To be specific, we choose a Cartesian coordinate system in which the scattering vector $\mathbf{Q}$ is parallel to the $x$ axis, so that $\mathbf{M}_{\perp} = (0, M_y, M_z)$. We consider three orthogonal configurations of the neutron spin quan-tization direction, (i) $\mathbf{P} \parallel \mathbf{Q}$, (ii) $\mathbf{P} \perp \mathbf{Q}$ with $\mathbf{P}$ along the $y$ axis, and (iii) $\mathbf{P} \perp \mathbf{Q}$ with $\mathbf{P}$ along the $z$ axis. The configurations are represented in Fig. 4.5. From (4.54), the scattering in the NSF and SF channels in each case is given by

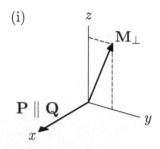

(i) $\mathbf{P} \parallel \mathbf{Q}$, i.e. $\mathbf{P} \parallel x$

$$I_{\text{NSF}}^x = I_{\text{coh}}^N + I_{\text{inc}}^{\text{iso}} + \tfrac{1}{3} I_{\text{inc}}^{\text{sp}} \tag{4.68}$$

$$I_{\text{SF}}^x = \tfrac{2}{3} I_{\text{inc}}^{\text{sp}} + I_{yy}^M + I_{zz}^M \pm J_{yz}^M, \tag{4.69}$$

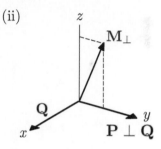

(ii) $\mathbf{P} \perp \mathbf{Q}$ with $\mathbf{P} \parallel y$

$$I_{\text{NSF}}^y = I_{\text{coh}}^N + I_{\text{inc}}^{\text{iso}} + \tfrac{1}{3} I_{\text{inc}}^{\text{sp}} \pm I_y^{\text{NM}} + I_{yy}^M \tag{4.70}$$

$$I_{\text{SF}}^y = \tfrac{2}{3} I_{\text{inc}}^{\text{sp}} + I_{zz}^M, \tag{4.71}$$

(iii) $\mathbf{P} \perp \mathbf{Q}$ with $\mathbf{P} \parallel z$

$$I_{\text{NSF}}^z = I_{\text{coh}}^N + I_{\text{inc}}^{\text{iso}} + \tfrac{1}{3} I_{\text{inc}}^{\text{sp}} \pm I_z^{\text{NM}} + I_{zz}^M \tag{4.72}$$

$$I_{\text{SF}}^z = \tfrac{2}{3} I_{\text{inc}}^{\text{sp}} + I_{yy}^M, \tag{4.73}$$

In the above, $I_{yy}^M$ and $I_{zz}^M$ are just the $yy$ and $zz$ magnetic correlations,

$$I_{yy}^M = \left(\frac{\gamma r_0}{2\mu_B}\right)^2 \langle \hat{M}_y^{\dagger}(\mathbf{Q}) \hat{M}_y(\mathbf{Q}, t) \rangle, \quad I_{zz}^M = \left(\frac{\gamma r_0}{2\mu_B}\right)^2 \langle \hat{M}_z^{\dagger}(\mathbf{Q}) \hat{M}_z(\mathbf{Q}, t) \rangle, \tag{4.74}$$

and in (4.69) $J_{yz}^M$ is the scattering from correlations between orthogonal components of $\mathbf{M}_{\perp}$ and $\mathbf{M}_{\perp}^*$ (known as magnetic chiral scattering), i.e.

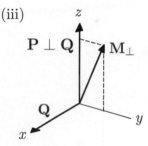

**Fig. 4.5** Definition of the three configu-rations of uniaxial polarization analysis given in the text.

$$J_{yz}^M = \left(\frac{\gamma r_0}{2\mu_B}\right)^2 \mathrm{i} \left\{ \langle \hat{M}_y^{\dagger}(\mathbf{Q}) \hat{M}_z(\mathbf{Q}, t) \rangle - \langle \hat{M}_z^{\dagger}(\mathbf{Q}) \hat{M}_y(\mathbf{Q}, t) \rangle \right\}. \tag{4.75}$$

The $+$ and $-$ signs in front of $J_{yz}^M$ in (4.69) are for the $x\bar{x}$ and $\bar{x}x$ SF channels, respectively. The terms $I_y^{\text{NM}}$ and $I_z^{\text{NM}}$ in (4.70) and (4.72) represent scattering from correlations between the nuclear coordinate states and one component of the magnetization,

$$I_{\alpha}^{\text{NM}} = \left(\frac{\gamma r_0}{2\mu_B}\right) \left\{ \langle \hat{N}^{\dagger}(\mathbf{Q}) \hat{M}_{\alpha}(\mathbf{Q}, t) \rangle + \langle \hat{M}_{\alpha}^{\dagger}(\mathbf{Q}) \hat{N}(\mathbf{Q}, t) \rangle \right\}, \tag{4.76}$$

where $\hat{N}(\mathbf{Q}, t)$ is the nuclear coherent scattering amplitude, eqn (4.66). The $+$ and $-$ signs in (4.70) are for the $yy$ and $\bar{y}\bar{y}$ NSF channels, re-spectively. Similarly in (4.72) for the $zz$ and $\bar{z}\bar{z}$ NSF channels.

Equations (4.68) and (4.69) reiterate our statement that when $\mathbf{P} \parallel \mathbf{Q}$ all the electronic magnetic scattering is in the SF channel and all the coherent nuclear scattering is in the NSF channel. They also show that if we want to isolate the magnetic scattering entirely then we need a way to separate it from the nuclear spin incoherent scattering $I_{\mathrm{inc}}^{\mathrm{sp}}$, two-thirds of which appears in the SF channel. In many cases, $I_{\mathrm{inc}}^{\mathrm{sp}}$ can be treated as a featureless background under the magnetic signal of interest and fitted to a straight line. A more certain method, however, is to measure $I_{\mathrm{SF}}$ with both $\mathbf{P} \parallel \mathbf{Q}$ and $\mathbf{P} \perp \mathbf{Q}$ (either $\mathbf{P} \parallel y$ or $\mathbf{P} \parallel z$) and subtract the two signals. The difference $I_{\mathrm{SF}}^{\parallel} - I_{\mathrm{SF}}^{\perp}$ contains purely magnetic scattering, as is clear from (4.69), (4.71), and (4.73), with the assurance that any background signals have been eliminated in the subtraction.

More generally, eqns (4.68)–(4.73) can be manipulated to obtain $I_{yy}^{\mathrm{M}}$, $I_{zz}^{\mathrm{M}}$, and $J_{yz}^{\mathrm{M}}$ separately, which may be necessary if the magnetic system is anisotropic. The difference between the $I_{\mathrm{SF}}^{x}$ signals measured in the $x\bar{x}$ and $\bar{x}x$ channels gives $J_{yz}^{\mathrm{M}}$, and knowing this we can extract $I_{yy}^{\mathrm{M}}$ from $I_{\mathrm{SF}}^{x} - I_{\mathrm{SF}}^{y}$ and $I_{zz}^{\mathrm{M}}$ from $I_{\mathrm{SF}}^{x} - I_{\mathrm{SF}}^{z}$. Similarly, $I_{y}^{\mathrm{NM}}$ can be determined from the two channels of $I_{\mathrm{NSF}}^{y}$, and likewise for $I_{z}^{\mathrm{NM}}$ from $I_{\mathrm{NSF}}^{z}$.

## 4.6    XYZ polarization analysis

The techniques for separating magnetic and nuclear scattering discussed in the previous section depend upon aligning the neutron polarization along directions parallel and perpendicular to the scattering vector. This requirement cannot be met on an instrument with a multidetector because each point on the detector corresponds to a different scattering vector. The so-called XYZ method described here is an implementation of uniaxial polarization analysis that enables magnetic scattering to be separated on a multidetector.[16]

The equations of XYZ polarization analysis are a generalization of those presented in the previous section. As before, we measure the SF and NSF channels with the polarization aligned along three orthogonal axes, but now we do not require $\mathbf{Q}$ to be along $x$. Instead, the axes are usually defined parallel to the axes of the Helmholtz coils which surround the sample, with $x$ and $y$ in the horizontal plane and $z$ vertically upwards to make a right-handed set (see Fig. 4.6). The vector $\mathbf{M}_{\perp}$ will be different for each point on the detector and will in general depend on all three components of $\mathbf{M}$. The expressions for SF and NSF scattering with $\mathbf{P}$ parallel to each of $x$, $y$ and $z$ are now

$$I_{\mathrm{NSF}}^{x} = I_{\mathrm{coh}}^{\mathrm{N}} + I_{\mathrm{inc}}^{\mathrm{iso}} + \tfrac{1}{3} I_{\mathrm{inc}}^{\mathrm{sp}} \pm I_{x}^{\mathrm{NM}} + I_{xx}^{\mathrm{M}} \tag{4.77}$$

$$I_{\mathrm{SF}}^{x} = \tfrac{2}{3} I_{\mathrm{inc}}^{\mathrm{sp}} + I_{yy}^{\mathrm{M}} + I_{zz}^{\mathrm{M}} \pm J_{yz}^{\mathrm{M}} \tag{4.78}$$

$$I_{\mathrm{NSF}}^{y} = I_{\mathrm{coh}}^{\mathrm{N}} + I_{\mathrm{inc}}^{\mathrm{iso}} + \tfrac{1}{3} I_{\mathrm{inc}}^{\mathrm{sp}} \pm I_{y}^{\mathrm{NM}} + I_{yy}^{\mathrm{M}} \tag{4.79}$$

$$I_{\mathrm{SF}}^{y} = \tfrac{2}{3} I_{\mathrm{inc}}^{\mathrm{sp}} + I_{xx}^{\mathrm{M}} + I_{zz}^{\mathrm{M}} \pm J_{zx}^{\mathrm{M}} \tag{4.80}$$

[16]For a review, see Stewart *et al.* (2009).

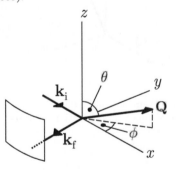

**Fig. 4.6** Geometry employed in XYZ polarization analysis. $\mathbf{Q}$ is the scattering vector, and the spherical polar angles $\theta$ and $\phi$ specify the direction of $\mathbf{Q}$ with respect to the Cartesian axes. Neutrons are shown scattering into a two-dimensional multidetector.

$$I_{\text{NSF}}^z = I_{\text{coh}}^{\text{N}} + I_{\text{inc}}^{\text{iso}} + \tfrac{1}{3}I_{\text{inc}}^{\text{sp}} \pm I_z^{\text{NM}} + I_{zz}^{\text{M}} \qquad (4.81)$$

$$I_{\text{SF}}^z = \tfrac{2}{3}I_{\text{inc}}^{\text{sp}} + I_{xx}^{\text{M}} + I_{yy}^{\text{M}} \pm J_{xy}^{\text{M}}. \qquad (4.82)$$

The nuclear and magnetic terms on the right-hand side of eqns (4.77)–(4.82) are defined in the same way as in (4.61)–(4.67) and (4.74)–(4.76) *except* that the magnetic terms are explicitly built from the components of $\mathbf{M}_\perp$, for example[17] [c.f. eqn (4.74)]

$$I_{yy}^{\text{M}} = \left(\frac{\gamma r_0}{2\mu_{\text{B}}}\right)^2 \langle \hat{M}_{\perp y}^\dagger \hat{M}_{\perp y}(t)\rangle. \qquad (4.83)$$

From eqn (4.41) it follows that[18]

$$\mathbf{M}_\perp =$$

$$\begin{pmatrix} M_x(1 - \cos^2\phi\sin^2\theta) - M_y\sin\phi\cos\phi\sin^2\theta - M_z\cos\phi\sin\theta\cos\theta \\ -M_x\sin\phi\cos\phi\sin^2\theta + M_y(1 - \sin^2\phi\sin^2\theta) - M_z\sin\phi\sin\theta\cos\theta \\ -M_x\cos\phi\sin\theta\cos\theta - M_y\sin\phi\sin\theta\cos\theta + M_z\sin^2\theta \end{pmatrix},$$

$$(4.84)$$

where $\theta$ and $\phi$ specify the direction of $\mathbf{Q}$ in spherical polar coordinates with respect to the Cartesian axes (see Fig. 4.6). Equation (4.84) makes it possible to write each of the polarization channels in (4.77)–(4.82) in terms of correlations between the components of $\mathbf{M}$. For the particular case of a one-dimensional multidetector in the horizontal plane (i.e. $\theta = \pi/2$) the expressions have been given previously by Schärpf and Capellmann (1993).[19]

Various manipulations of the different XYZ polarization channels can now be performed in order to separate magnetic and nuclear scattering, and to obtain information on different magnetic correlations. For a crystalline sample with anisotropic magnetic correlations, a further transformation of coordinates between the XYZ axes and the crystallographic axes would be required to write the correlations in terms of magnetization components along the crystal axes. One could then make a rigorous test of a theoretical model by calculating the different XYZ polarization signals for all points on the multidetector and comparing the calculations with experimental data.

The general expressions for each polarization channel in terms of the correlation functions are straightforward to derive but rather lengthy. For illustration, we consider the special case where the magnetic correlations are isotropic in the spectrometer coordinates, i.e. $\langle \hat{M}_x^\dagger \hat{M}_x(t)\rangle = \langle \hat{M}_y^\dagger \hat{M}_y(t)\rangle = \langle \hat{M}_z^\dagger \hat{M}_z(t)\rangle = \tfrac{1}{2}\langle \mathbf{M}_\perp^\dagger \cdot \mathbf{M}_\perp(t)\rangle$, and $\langle \hat{M}_x^\dagger \hat{M}_y(t)\rangle = 0$, etc. This applies, for example, to polycrystalline samples which are paramagnets or have certain types of magnetic order (e.g. collinear antiferromagnets), and to single-crystal paramagnets with negligible spin anisotropy. We also neglect the nuclear–magnetic interference terms like $I_x^{\text{NM}}$ which, if present, can always be determined from the difference between the $xx$ and $\bar{x}\bar{x}$ NSF channels. With these simplifications, the XYZ polarization

[17]The $\mathbf{P} \parallel \mathbf{Q} \parallel x$ configuration discussed in the previous section is a special case of the XYZ geometry, and for that case $M_{\perp x} = 0$, $M_{\perp y} = M_y$ and $M_{\perp z} = M_z$.

[18]In Cartesian coordinates, $\hat{\mathbf{Q}} = (\cos\phi\sin\theta, \sin\phi\sin\theta, \cos\theta)$, where $\theta$ and $\phi$ are the spherical polar angles.

[19]This paper also describes the implementation of XYZ polarization analysis on the D7 instrument at the Institut Laue–Langevin. D7 has a one-dimensional multidetector surrounding the sample in the horizontal plane.

channels (4.77)–(4.82) are given by (Ehlers *et al.*, 2013)

$$I_{\mathrm{NSF}}^x = I_{\mathrm{coh}}^{\mathrm{N}} + I_{\mathrm{inc}}^{\mathrm{iso}} + \tfrac{1}{3}I_{\mathrm{inc}}^{\mathrm{sp}} + \tfrac{1}{2}I^{\mathrm{M}}(1 - \cos^2\phi\sin^2\theta) \tag{4.85}$$

$$I_{\mathrm{SF}}^x = \tfrac{2}{3}I_{\mathrm{inc}}^{\mathrm{sp}} + \tfrac{1}{2}I^{\mathrm{M}}(1 + \cos^2\phi\sin^2\theta) \tag{4.86}$$

$$I_{\mathrm{NSF}}^y = I_{\mathrm{coh}}^{\mathrm{N}} + I_{\mathrm{inc}}^{\mathrm{iso}} + \tfrac{1}{3}I_{\mathrm{inc}}^{\mathrm{sp}} + \tfrac{1}{2}I^{\mathrm{M}}(1 - \sin^2\phi\sin^2\theta) \tag{4.87}$$

$$I_{\mathrm{SF}}^y = \tfrac{2}{3}I_{\mathrm{inc}}^{\mathrm{sp}} + \tfrac{1}{2}I^{\mathrm{M}}(1 + \sin^2\phi\sin^2\theta) \tag{4.88}$$

$$I_{\mathrm{NSF}}^z = I_{\mathrm{coh}}^{\mathrm{N}} + I_{\mathrm{inc}}^{\mathrm{iso}} + \tfrac{1}{3}I_{\mathrm{inc}}^{\mathrm{sp}} + \tfrac{1}{2}I^{\mathrm{M}}\sin^2\theta \tag{4.89}$$

$$I_{\mathrm{SF}}^z = \tfrac{2}{3}I_{\mathrm{inc}}^{\mathrm{sp}} + \tfrac{1}{2}I^{\mathrm{M}}(2 - \sin^2\theta). \tag{4.90}$$

where

$$I^{\mathrm{M}} = \left(\frac{\gamma r_0}{2\mu_{\mathrm{B}}}\right)^2 \langle \hat{\mathbf{M}}_\perp^\dagger \cdot \hat{\mathbf{M}}_\perp(t) \rangle. \tag{4.91}$$

Equations (4.85)–(4.90) can be combined in various ways, but to isolate the magnetic scattering one simple approach is to use the following:

$$I^{\mathrm{M}}(\tfrac{3}{2}\sin^2\theta - 1) = 2I_{\mathrm{NSF}}^z - I_{\mathrm{NSF}}^x - I_{\mathrm{NSF}}^y \tag{4.92}$$

$$= I_{\mathrm{SF}}^x + I_{\mathrm{SF}}^y - 2I_{\mathrm{SF}}^z. \tag{4.93}$$

Knowing the angle $\theta$ away from the vertical, one can use (4.92) and/or (4.93) to isolate magnetic scattering at any point on a two-dimensional multidetector.[20]

The XYZ polarization analysis method with a one-dimensional multidetector has been used for a number of years in studies of the diffuse scattering from short-range atomic and magnetic correlations in spin glasses, paramagnets, and disordered alloys.[21] The development of spin-polarized $^3$He filter cells for use as neutron spin polarizers and analysers makes it practical to implement XYZ polarization analysis with a two-dimensional scattering geometry as described above, and this opens up the possibility of performing polarization analysis on time-of-flight spectrometers with multidetectors covering large solid angles.

## 4.7 Polarized incident beam

Next we calculate the scattering cross-section without polarization analysis for a polarized incident beam.[22] Without polarization analysis the detector counts all scattered neutrons irrespective of their spin state, which is equivalent to summing the spin-up and spin-down states in the scattered beam for an arbitrarily chosen final spin quantization direction. To allow for the incident polarization, we weight the partial spin-state cross-sections by the fraction of neutrons with each initial spin state.

[20]Note, however, that with these combinations the strength of the magnetic signal decreases away from the horizontal plane, and vanishes entirely at $\theta = \sin^{-1}(2/3)^{\frac{1}{2}} = 54.7°$. A method involving additional measurements has been proposed to get round this problem (Ehlers *et al.*, 2013).

[21]For a review, see Stewart *et al.* (2000).

[22]Known as a *half-polarized* setup.

To proceed, let us take the incident polarization vector to be parallel to the $x$ axis, i.e. [see eqn (1.9)]

$$\mathbf{P}_{\mathrm{i}} = \begin{pmatrix} f_x - f_{\bar{x}} \\ 0 \\ 0 \end{pmatrix}, \tag{4.94}$$

where $f_x$ is the fraction of incident neutrons in the 'up' eigenstate with respect to the $x$ direction, and $\bar{x}$ is shorthand for $-x$. If we also choose the $x$ axis as the quantization direction for the sum over final spin states, then the scattering intensity for an incident beam with polarization vector (4.94) is given by[23]

$$I^x = f_x(I^{xx} + I^{x\bar{x}}) + f_{\bar{x}}(I^{\bar{x}x} + I^{\bar{x}\bar{x}}). \tag{4.95}$$

As explained above, we could equally well have chosen $y$ or $z$ for the scattered neutron quantization direction, but $x$ is convenient because it means we can use eqns (4.54) for uniaxial polarization analysis to write down the partial functions $I^{xx}$, $I^{x\bar{x}}$, etc. This way we obtain

$$I^x = (f_x + f_{\bar{x}})\langle \hat{C}^\dagger \hat{C}(t) \rangle + (f_x - f_{\bar{x}})\langle \hat{C}^\dagger \hat{D}_x(t) + \hat{D}_x^\dagger \hat{C}(t) \rangle$$

$$+ (f_x + f_{\bar{x}})\langle \hat{\mathbf{D}}^\dagger \cdot \hat{\mathbf{D}}(t) \rangle + \mathrm{i}(f_x - f_{\bar{x}})\langle \hat{D}_y^\dagger \hat{D}_z(t) - \hat{D}_z^\dagger \hat{D}_y(t) \rangle. \tag{4.96}$$

Using (4.94), and the fact that $(f_x + f_{\bar{x}}) = 1$, we can generalize (4.96) for an arbitrary $\mathbf{P}_{\mathrm{i}}$ as follows,

$$I(\mathbf{P}_{\mathrm{i}}) = \langle \hat{C}^\dagger \hat{C}(t) \rangle + \langle \hat{\mathbf{D}}^\dagger \cdot \hat{\mathbf{D}}(t) \rangle + \mathbf{P}_{\mathrm{i}} \cdot \langle \hat{C}^\dagger \hat{\mathbf{D}}(t) + \hat{\mathbf{D}}^\dagger \hat{C}(t) \rangle + \mathrm{i}\mathbf{P}_{\mathrm{i}} \cdot \langle \hat{\mathbf{D}}^\dagger \times \hat{\mathbf{D}}(t) \rangle. \tag{4.97}$$

The validity of this result can easily be confirmed by writing down the expressions corresponding to (4.96) for $\mathbf{P}_{\mathrm{i}}$ parallel to the $y$ and $z$ axes.

[23] As ever, the cross-section is actually the time Fourier transform of (4.95), multiplied by the factor $k_{\mathrm{f}}/k_{\mathrm{i}}$.

## 4.8  Zero-field spherical neutron polarimetry

In the standard uniaxial (longitudinal) polarization analysis technique, described in Section 4.5, a guide field is applied at the sample position and only the component of the neutron polarization parallel or antiparallel to the field direction is measured. For many problems this is adequate, but sometimes more information can be obtained if all the components of the final neutron polarization are measured, i.e. both the longitudinal and the transverse components. The technique for doing this is known as *spherical neutron polarimetry* (SNP), or alternatively *three-dimensional polarization analysis* or *vector polarization analysis*. To avoid depolarization of the neutron beam the sample is usually contained in a zero-field chamber.

Spherical neutron polarimetry can be used to measure the partial spin state cross-sections (4.53) directly, but the method more often employed is to measure the final polarization $\mathbf{P}_{\mathrm{f}}$ for a given direction of the incident polarization $\mathbf{P}_i$. Let us first suppose that $\mathbf{P}_i$ is parallel to the $x$ axis, i.e. $\mathbf{P}_i$ is given by eqn (4.94). In that case

$$\mathbf{P}_f = \frac{1}{I^x} \begin{pmatrix} f_x I^{xx} + f_{\bar{x}} I^{\bar{x}x} - f_x I^{x\bar{x}} - f_{\bar{x}} I^{\bar{x}\bar{x}} \\ f_x I^{xy} + f_{\bar{x}} I^{\bar{x}y} - f_x I^{x\bar{y}} - f_{\bar{x}} I^{\bar{x}\bar{y}} \\ f_x I^{xz} + f_{\bar{x}} I^{\bar{x}z} - f_x I^{x\bar{z}} - f_{\bar{x}} I^{\bar{x}\bar{z}} \end{pmatrix}, \qquad (4.98)$$

where $I^x$ is given by (4.96). With eqn (4.53), Tables C.1 and C.2, and some lengthy algebra, one can show that the three components of $\mathbf{P}_f$ are given by

$$I^x P_{f,x} = (f_x - f_{\bar{x}})\langle \hat{C}^\dagger \hat{C}(t) \rangle + \langle \hat{C}^\dagger \hat{D}_x(t) + \hat{D}_x^\dagger \hat{C}(t) \rangle$$
$$+ (f_x - f_{\bar{x}})\langle \hat{D}_x^\dagger \hat{D}_x(t) - \hat{D}_y^\dagger \hat{D}_y(t) - \hat{D}_z^\dagger \hat{D}_z(t) \rangle$$
$$+ \mathrm{i}\langle \hat{D}_z^\dagger \hat{D}_y(t) - \hat{D}_y^\dagger \hat{D}_z(t) \rangle$$

$$I^x P_{f,y} = \langle \hat{C}^\dagger \hat{D}_y(t) + \hat{D}_y^\dagger \hat{C}(t) \rangle + \mathrm{i}(f_x - f_{\bar{x}})\langle \hat{C}^\dagger \hat{D}_z(t) - \hat{D}_z^\dagger \hat{C}(t) \rangle$$
$$+ (f_x - f_{\bar{x}})\langle \hat{D}_y^\dagger \hat{D}_x(t) + \hat{D}_x^\dagger \hat{D}_y(t) \rangle + \mathrm{i}\langle \hat{D}_x^\dagger \hat{D}_z(t) - \hat{D}_z^\dagger \hat{D}_x(t) \rangle$$

$$I^x P_{f,z} = \langle \hat{C}^\dagger \hat{D}_z(t) + \hat{D}_z^\dagger \hat{C}(t) \rangle - \mathrm{i}(f_x - f_{\bar{x}})\langle \hat{C}^\dagger \hat{D}_y(t) - \hat{D}_y^\dagger \hat{C}(t) \rangle$$
$$+ (f_x - f_{\bar{x}})\langle \hat{D}_z^\dagger \hat{D}_x(t) + \hat{D}_x^\dagger \hat{D}_z(t) \rangle + \mathrm{i}\langle \hat{D}_y^\dagger \hat{D}_x(t) - \hat{D}_x^\dagger \hat{D}_y(t) \rangle.$$

$$(4.99)$$

The generalization of (4.99) for an arbitrary $\mathbf{P}_i$ can be written

$$I(\mathbf{P}_i)\mathbf{P}_f = \langle \hat{C}^\dagger \hat{C}(t) \rangle \mathbf{P}_i - \langle \hat{\mathbf{D}}^\dagger \cdot \hat{\mathbf{D}}(t) \rangle \mathbf{P}_i$$
$$+ \mathrm{i}\langle \hat{C}^\dagger \hat{\mathbf{D}}(t) - \hat{\mathbf{D}}^\dagger \hat{C}(t) \rangle \times \mathbf{P}_i + \langle \hat{\mathbf{D}}^\dagger \{\hat{\mathbf{D}}(t) \cdot \mathbf{P}_i\} + \{\hat{\mathbf{D}}^\dagger \cdot \mathbf{P}_i\}\hat{\mathbf{D}}(t) \rangle$$
$$+ \langle \hat{C}^\dagger \hat{\mathbf{D}}(t) + \hat{\mathbf{D}}^\dagger \hat{C}(t) \rangle - \mathrm{i}\langle \hat{\mathbf{D}}^\dagger \times \hat{\mathbf{D}}(t) \rangle, \qquad (4.100)$$

where $I(\mathbf{P}_i)$ is given by (4.97). Equation (4.100) is written such that the first line contains terms which do not change the polarization of the beam, the second line contains terms which cause rotation of the initial polarization, and the third line contains terms which create polarization irrespective of the polarization of the incident beam.[24]

Results (4.97) and (4.100), which give the scattering intensity and final polarization for any $\mathbf{P}_i$, are known as the *Blume–Maleev equations*.[25] A few observations can be made about these formulae.

[24]Equation (4.100) can be written in the form $\mathbf{P}_f = \mathsf{R}\mathbf{P}_i + \mathbf{P}_0$. The $3 \times 3$ matrix $\mathsf{R}$ has the effect of rotating and scaling the initial polarization vector, and the vector $\mathbf{P}_0$ is the created polarization. The components of $\mathsf{R}$ and $\mathbf{P}_0$ are contained in the matrices in eqn (4.104). The transpose of the first matrix in (4.104) (without the multiplicative factors for the rows) is equal to $I(\mathbf{P}_i)\mathsf{R}$, and the transpose of any row of the second matrix (again without the multiplicative factor) is $I(\mathbf{P}_i)\mathbf{P}_0$.

[25]A bit of history. The theory of polarized neutron scattering was first outlined by Halpern and Johnson (1939), and the first experiments with polarized neutrons were performed in 1959 on an instrument with a polarized incident beam (Nathans *et al.*, 1959). The general expressions (4.97) and (4.100) were derived independently by Blume (1963) and Maleev *et al.* (1963) using a density matrix formalism. In this book we have obtained the same results as Blume and Maleev *et al.* by explicit use of the neutron spin matrix elements. Our approach is more accessible to the general reader, and is similar to that followed by Moon *et al.* (1969) in their classic paper on uniaxial polarization analysis.

First, we verify the statements made in Section 4.5 about the final polarization for purely nuclear scattering. If the signal only contains nuclear coherent scattering and/or isotopic incoherent scattering, which are described by the first term in each of eqns (4.97) and (4.100), then it can be seen that $\mathbf{P}_f = \mathbf{P}_i$. If, on the other hand, the signal contains only nuclear spin incoherent scattering then the only non-zero terms are those of the form $\langle \hat{D}_\alpha^\dagger \hat{D}_\beta(t) \rangle$ with $\alpha = \beta$, and these terms are all equal to $\frac{1}{3} I_{inc}^{sp}$, where $I_{inc}^{sp}$ is given by (4.63). Hence, $\langle \hat{\mathbf{D}}^\dagger \cdot \hat{\mathbf{D}}(t) \rangle = I_{inc}^{sp}$, and $\langle \hat{\mathbf{D}}^\dagger \{ \hat{\mathbf{D}}(t) \cdot \mathbf{P}_i \} \rangle = \langle \{ \hat{\mathbf{D}}^\dagger \cdot \mathbf{P}_i \} \hat{\mathbf{D}}(t) \rangle = \frac{1}{3} I_{inc}^{sp} \mathbf{P}_i$, since only one component of the vector $\hat{\mathbf{D}}$ projects onto $\mathbf{P}_i$. Collecting these terms together, we obtain $\mathbf{P}_f = -\frac{1}{3} \mathbf{P}_i$. When all types of nuclear scattering are present the final polarization is given in the notation of (4.61)–(4.63) by

$$\mathbf{P}_f = \frac{I_{coh}^N + I_{inc}^{iso} - \frac{1}{3} I_{inc}^{sp}}{I_{coh}^N + I_{inc}^{iso} + I_{inc}^{sp}} \mathbf{P}_i, \qquad (4.101)$$

which lies somewhere between $-\frac{1}{3} \mathbf{P}_i$ and $\mathbf{P}_i$.

Second, for an isotropic paramagnet (see Exercise 4.5)

$$\mathbf{P}_f = -(\hat{\mathbf{Q}} \cdot \mathbf{P}_i) \hat{\mathbf{Q}}. \qquad (4.102)$$

In this case, the final polarization lies along $-\hat{\mathbf{Q}}$ and has a magnitude between 0 and 1, depending on the orientation of $\mathbf{P}_i$ relative to $\mathbf{Q}$. The reason why (4.102) differs from the result $\mathbf{P}_f = -\frac{1}{3} \mathbf{P}_i$ for nuclear spin incoherent scattering is that all spin components contribute in nuclear spin scattering, whereas only components perpendicular to $\mathbf{Q}$ contribute to electronic magnetic scattering.

The third remark concerns the terms in (4.100) which do not depend on $\mathbf{P}_i$, i.e. the $\hat{C}^\dagger \hat{\mathbf{D}}$ and $\hat{\mathbf{D}}^\dagger \hat{C}$ terms and the so-called *chiral* term $\hat{\mathbf{D}}^\dagger \times \hat{\mathbf{D}}(t)$. These terms can be non-zero only if there is a unique direction. They make it possible for an initially unpolarized beam ($\mathbf{P}_i = 0$) to acquire some degree of polarization after scattering. In magnetic systems, the $\hat{C}^\dagger \hat{\mathbf{D}}$-type terms describe interference between nuclear and magnetic scattering and create polarization along the direction of the magnetic interaction vector $\mathbf{M}_\perp$. A ferromagnet is an example where nuclear–magnetic interference is important since in a ferromagnet the nuclear and magnetic Bragg scattering occur at the same $\mathbf{Q}$. An example where the chiral term is non-zero is diffraction from a magnetic structure for which $\mathbf{M}_\perp \neq \mathbf{M}_\perp^*$. This arises when a magnetic structure does not have a centre of inversion symmetry, such as a helical magnetic structure. Because $\mathbf{M}_\perp$ and $\mathbf{M}_\perp^*$ are both perpendicular to $\mathbf{Q}$ the chiral term creates polarization along either $\mathbf{Q}$ or $-\mathbf{Q}$.

Fourth, if a magnetic structure does have a centre of inversion symmetry then $\mathbf{M}_\perp = \mathbf{M}_\perp^*$ and hence the chiral term must vanish. The Blume–Maleev equations now tell us that $I(\mathbf{P}_i) = (\gamma r_0 / 2 \mu_B)^2 |\mathbf{M}_\perp|^2$ and $\mathbf{P}_f = -\mathbf{P}_i + 2(\hat{\mathbf{M}}_\perp \cdot \mathbf{P}_i) \hat{\mathbf{M}}_\perp$. This vector relation is illustrated geometrically in Fig. 4.7. It can be seen that $\mathbf{P}_f$ and $\mathbf{P}_i$ are related by a rotation through 180° about the direction of $\mathbf{M}_\perp$.

Fifth, now turn to the first term on the second line of (4.100), which has the form $i \langle \hat{C}^\dagger \hat{\mathbf{D}}(t) - \hat{\mathbf{D}}^\dagger \hat{C}(t) \rangle \times \mathbf{P}_i$. In magnetic systems this term

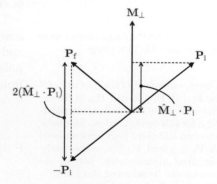

**Fig. 4.7** Relationship between $\mathbf{P}_f$ and $\mathbf{P}_i$ for magnetic diffraction from magnetic structures with a centre of symmetry ($\mathbf{M}_\perp = \mathbf{M}_\perp^*$).

corresponds to nuclear–magnetic interference. Two conditions must be satisfied for it to be non-zero: (i) magnetic and nuclear scattering must occur at the same $\mathbf{Q}$, and (ii) the phase difference between the magnetic and nuclear scattering must be neither 0° nor 180°. When these conditions are met the effect is to rotate the polarization towards $\mathbf{Q}$.

Finally, let us note that the terms on the right-hand side of the Blume–Maleev equations (4.97) and (4.100) are thermal averages. For example, if there are two or more magnetic domains in a magnetically ordered material then the scattering intensity from each domain must be averaged with respect to the domain populations, and the domain-averaged intensities then used to calculate $I(\mathbf{P_i})$ and $\mathbf{P_f}$.

### 4.8.1   The polarization matrix

[26]'Convention' is perhaps too strong a word. The polarization matrix can also be found in the literature defined as the transpose of that used here, i.e. with the first index for the scattered polarization and the second for the initial polarization.

The experimental implementation of SNP involves the measurement of three orthogonal components of $\mathbf{P_f}$ for each of three orthogonal directions of $\mathbf{P_i}$. These nine values are collected together into what is known as the *polarization matrix* P. By convention,[26] the element $\mathsf{P}_{\alpha\beta}$ of P is the $\beta$ component of the scattered neutron polarization for an incident beam polarized in the $\alpha$ direction ($\alpha, \beta = x, y, z$). This is determined experimentally from

$$\mathsf{P}_{\alpha\beta} = \frac{n_{\alpha\beta} - n_{\alpha\bar{\beta}}}{n_{\alpha\beta} + n_{\alpha\bar{\beta}}}, \tag{4.103}$$

where $n_{\alpha\beta}$ and $n_{\alpha\bar{\beta}}$ are the numbers of scattered neutrons (corrected for background) with spins parallel and antiparallel with respect to the $\beta$ direction. The polarization matrix can also be obtained with the incident polarization directed along the $-x$, $-y$ and $-z$ directions, giving a total of eighteen terms for both the $\mathsf{P}_{\alpha\beta}$ and $\mathsf{P}_{\bar{\alpha}\beta}$ matrices.

[27]Up to now, SNP measurements have only been possible in a plane.

[28]The measurement of the first row of P is made with $\mathbf{P_i} \parallel x$, and $P_{i,x}$ is the experimental value of the incident polarization for this configuration. Similarly, the second and third rows are measured with $\mathbf{P_i} \parallel y$ and $\mathbf{P_i} \parallel z$, respectively. In an ideal measurement, perfect polarization can be achieved along any direction, e.g. $P_{i,x} = 1$. In practice, the incident polarization is typically in the range 0.8–0.9.

The coordinates are usually defined so that the $x$ axis is parallel to the scattering vector $\mathbf{Q}$, the $z$ axis is perpendicular to the scattering plane,[27] and the $y$ axis makes up a right-handed set. With this convention the polarization matrix for magnetic and coherent nuclear scattering is found from (4.100) to be[28]

$$\mathsf{P} = \begin{pmatrix} I^N_{\text{coh}} - I^M_{yy} - I^M_{zz} & J^{NM}_z & -J^{NM}_y \\ -J^{NM}_z & I^N_{\text{coh}} + I^M_{yy} - I^M_{zz} & I^M_{yz} \\ J^{NM}_y & I^M_{yz} & I^N_{\text{coh}} - I^M_{yy} + I^M_{zz} \end{pmatrix} \begin{matrix} \times P_{i,x}/I^x \\ \times P_{i,y}/I^y \\ \times P_{i,z}/I^z \end{matrix}$$

$$+ \begin{pmatrix} -J^M_{yz} & I^{NM}_y & I^{NM}_z \\ -J^M_{yz} & I^{NM}_y & I^{NM}_z \\ -J^M_{yz} & I^{NM}_y & I^{NM}_z \end{pmatrix} \begin{matrix} \times 1/I^x \\ \times 1/I^y \\ \times 1/I^z \end{matrix}, \tag{4.104}$$

where

$$I_{yz}^{M} = \left(\frac{\gamma r_0}{2\mu_{B}}\right)^2 \left\{ \langle \hat{M}_y^\dagger(\mathbf{Q}) \hat{M}_z(\mathbf{Q},t)\rangle + \langle \hat{M}_z^\dagger(\mathbf{Q}) \hat{M}_y(\mathbf{Q},t)\rangle \right\} \quad (4.105)$$

$$J_{\alpha}^{NM} = \left(\frac{\gamma r_0}{2\mu_{B}}\right) \mathrm{i} \left\{ \langle \hat{N}^\dagger(\mathbf{Q}) \hat{M}_\alpha(\mathbf{Q},t)\rangle - \langle \hat{M}_\alpha^\dagger(\mathbf{Q}) \hat{N}(\mathbf{Q},t)\rangle \right\}, \quad (4.106)$$

($\alpha = y, z$), and the other symbols have been defined in (4.66)–(4.67) and (4.74)–(4.76). The expressions to the right of the matrices are multiplicative factors for the rows, and have in their denominators

$$\begin{aligned} I^x &= I_{coh}^N + I_{yy}^M + I_{zz}^M + J_{yz}^M P_{i,x} \\ I^\alpha &= I_{coh}^N + I_{yy}^M + I_{zz}^M + I_\alpha^{NM} P_{i,\alpha} \quad (\alpha = y, z), \end{aligned} \quad (4.107)$$

These are the expressions obtained from (4.97) for the scattering measured without polarization analysis when the incident beam is polarized along each of the axes.

Although the mathematics of SNP may appear rather formidable, the expressions are actually quite straightforward to use in practice. In Chapter 7 we will illustrate the various terms in the spin-dependent cross-section and polarization formulae by considering diffraction from some specific magnetic systems.

## 4.9   Unpolarized neutrons

Many experiments do not employ polarized neutrons, so in this section we derive the correlation functions for unpolarized neutron scattering.

By *unpolarized neutron scattering*, we mean that the incident beam contains an equal proportion of 'up' and 'down' spins, referred to any given quantization direction, and all scattered neutrons are counted regardless of their spin state. For simplicity let us take $x$ as the quantization direction both before and after scattering. To obtain the unpolarized-neutron correlation function we then need to average over $\sigma_i = +x$ and $-x$ with equal weighting and sum over $\sigma_f = +x$ and $-x$, i.e.

$$\begin{aligned} I_{unpol}(t) &= \sum_{\sigma_i} p_{\sigma_i} \sum_{\sigma_f} I^{\sigma_i \sigma_f}(t) \\ &= \frac{1}{2} \{ I^{xx}(t) + I^{\bar{x}\bar{x}}(t) + I^{x\bar{x}}(t) + I^{\bar{x}x}(t) \}, \end{aligned} \quad (4.108)$$

where $p_{+x} = p_{-x} = 1/2$.

The four terms required in (4.108) are given by eqns (4.54). Summing these up gives

$$I_{unpol}(t) = \langle \hat{C}^\dagger \hat{C}(t)\rangle + \langle \hat{\mathbf{D}}^\dagger \cdot \hat{\mathbf{D}}(t)\rangle. \quad (4.109)$$

This result is the same as that obtained when $\mathbf{P}_i$ is set to zero (i.e. the incident beam is unpolarized) in eqn (4.97), the cross-section for a polarized incident beam. From (4.109), it can be seen that unpolarized neutron scattering is not sensitive to correlations between $\hat{C}$ and $\hat{\mathbf{D}}$, or to correlations between different components of the vector $\hat{\mathbf{D}}$.

### 4.9.1   Magnetic and nuclear scattering

For the nuclear scattering, we make the usual assumptions of random isotope distributions and nuclear spin orientations, and lack of any correlation between nuclear states and the motion of the atoms (see Section 4.1.4). The consequences of these assumptions for the terms that appear in eqn (4.109) were discussed in Section 4.5. Accordingly, the intermediate scattering function for magnetic and nuclear scattering of unpolarized neutrons is found to be

$$I_{\text{unpol}} = I_{\text{coh}}^{\text{N}} + I_{\text{inc}}^{\text{N}} + I^{\text{M}}, \tag{4.110}$$

where

$$I_{\text{coh}}^{\text{N}} = \sum_{jk} \overline{b_j^* \, b_k} \, \langle \exp(-i\mathbf{Q} \cdot \hat{\mathbf{r}}_j) \, \exp\{i\mathbf{Q} \cdot \hat{\mathbf{r}}_k(t)\}\rangle \tag{4.111}$$

$$I_{\text{inc}}^{\text{N}} = \sum_{j} \frac{(\sigma_{\text{inc}})_j}{4\pi} \, \langle \exp(-i\mathbf{Q} \cdot \hat{\mathbf{r}}_j) \, \exp\{i\mathbf{Q} \cdot \hat{\mathbf{r}}_j(t)\}\rangle \tag{4.112}$$

$$I^{\text{M}} = \left(\frac{\gamma r_0}{2\mu_{\text{B}}}\right)^2 \langle \hat{\mathbf{M}}_\perp^\dagger(\mathbf{Q}) \cdot \hat{\mathbf{M}}_\perp(\mathbf{Q}, t)\rangle \tag{4.113}$$

$$= \left(\frac{\gamma r_0}{2\mu_{\text{B}}}\right)^2 \sum_{\alpha\beta} (\delta_{\alpha\beta} - \hat{Q}_\alpha \hat{Q}_\beta) \langle \hat{M}_\alpha^\dagger(\mathbf{Q}) \hat{M}_\beta(\mathbf{Q}, t)\rangle. \tag{4.114}$$

[29] Remember that $\hat{\mathbf{Q}}$ is a unit vector in the direction of $\mathbf{Q}$, and that the hat on $\mathbf{M}$ denotes a Heisenberg operator, not a unit vector.

In (4.114) we have used (4.41) and (4.42) to write $I^{\text{M}}$ in terms of vector components,[29] where $\alpha$, $\beta$ stand for $x$, $y$, $z$, and $\delta_{\alpha\beta}$ is the Kronecker delta.

The nuclear coherent scattering (4.111) is the same as eqn (4.61) since $\overline{A} = \overline{b}$, see (4.12) and (1.49). The nuclear incoherent scattering (4.112) is the sum of the isotopic and spin incoherence terms given in (4.62)–(4.65). Hence,

$$\frac{\sigma_{\text{inc}}}{4\pi} = \overline{|A|^2} - |\overline{A}|^2 + \tfrac{1}{4}\overline{|B|^2} I(I+1)$$

$$= \overline{|b|^2} - |\overline{b}|^2. \tag{4.115}$$

One can verify the equivalence of the two expressions on the right-hand side of (4.115) by substituting formulae (4.12) and (4.13) for the $A$ and $B$ coefficients and comparing the result with the expressions for the spin–isotope averages $\overline{b}$ and $\overline{b^2}$ given in (1.49).

# Chapter summary

- The Fermi pseudopotential

$$V_N(\mathbf{r}) = \frac{2\pi\hbar^2}{m_n} \, b \, \delta(\mathbf{r}),$$

  describes the interaction of neutrons with nuclei in the Born approximation. The scattering length operator $b$ represents the spin dependence of the nuclear potential.

- The Fourier transform of $V_N(\mathbf{r})$ for a system of nuclei is

$$\frac{m_n}{2\pi\hbar^2} \, V_N(\mathbf{Q}) = \sum_j b_j \, \exp(\mathrm{i}\mathbf{Q} \cdot \mathbf{r}_j).$$

- The magnetic interaction potential and its Fourier transform are

$$V_M(\mathbf{r}) = -\boldsymbol{\mu}_n \cdot \mathbf{B}(\mathbf{r})$$
$$\frac{m_n}{2\pi\hbar^2} \, V_M(\mathbf{Q}) = \frac{\gamma r_0}{\mu_B} \, \mathbf{s}_n \cdot \mathbf{M}_\perp(\mathbf{Q}),$$

  where $\gamma = 1.913$, $r_0 = 2.818\,\mathrm{fm}$, and $\mathbf{M}_\perp(\mathbf{Q}) = \hat{\mathbf{Q}} \times \{\mathbf{M}(\mathbf{Q}) \times \hat{\mathbf{Q}}\}$ is the projection of $\mathbf{M}(\mathbf{Q})$ onto the plane perpendicular to $\mathbf{Q}$. $\mathbf{M}(\mathbf{Q})$ is the Fourier transform of the magnetization.

- Scattering from a neutron spin state $\sigma_i$ to a final spin state $\sigma_f$ depends on the amplitudes $\langle \sigma_f | \sigma_i \rangle$ and matrix elements $\langle \sigma_f | \mathbf{s}_n | \sigma_i \rangle$ for a spin-$\frac{1}{2}$ particle, which are given in Tables C.1 and C.2.

- Neutron polarization is described by a vector $\mathbf{P}$ with components

$$P_\alpha = f_\alpha - f_{\bar{\alpha}},$$

  where $f_\alpha$ is the fraction of neutrons with spin parallel to the $\alpha$ direction ($\alpha = x, y, z$).

- Uniaxial polarization analysis uses a continuous guide field to make the neutron spin quantization axis the same before and after the sample. The scattered neutrons are sorted into spin-flip (SF) and non-spin-flip (NSF) scattering channels.

- Nuclear coherent and isotopic incoherent scattering are both NSF; nuclear spin incoherent scattering is 1/3 NSF and 2/3 SF.

- Components of $\mathbf{M}_\perp$ along $\mathbf{P}$ give NSF scattering, while components of $\mathbf{M}_\perp$ perpendicular to $\mathbf{P}$ give SF scattering. When $\mathbf{P} \parallel \mathbf{Q}$, electronic magnetic scattering is entirely SF.

- The Blume–Maleev equations (4.97) and (4.100) give the final polarization $\mathbf{P}_f$ for an arbitrary initial polarization $\mathbf{P}_i$.

- Spherical neutron polarimetry involves measurement of three orthogonal components of $\mathbf{P}_f$ for each of three orthogonal components of $\mathbf{P}_i$ with the sample in zero field. The results are expressed as the polarization matrix.

# Further reading

A number of excellent reviews on polarized neutron scattering and spherical neutron polarimetry are contained in Chatterji (2006). The earlier review by Hicks (1996) is also well worth consulting.

# Exercises

(4.1) In Section 4.1.2, the nuclear scattering length operator was written in the form $b = A + B\mathbf{s}_n \cdot \mathbf{I}$. Starting from the eigenvalue equations (4.7) and (4.10), derive the expressions for the parameters $A$ and $B$ given in (4.12) and (4.13).

(4.2) Measurements on a polarized beam instrument record 36,212 counts in the $+z$ channel and 1,405 counts in the $-z$ channel. What is the beam polarization and its statistical uncertainty? What is likely to be the main source of systematic error?

(4.3) Prove the following results for uniaxial polarization analysis from (4.68)–(4.73):

$$J_{yz}^{M} = \tfrac{1}{2}(I_{SF}^{x\bar{x}} - I_{SF}^{\bar{x}x})$$
$$I_{yy}^{M} = \tfrac{1}{2}(I_{SF}^{x\bar{x}} + I_{SF}^{\bar{x}x}) - I_{SF}^{y}$$
$$I_{y}^{NM} = \tfrac{1}{2}(I_{NSF}^{yy} - I_{NSF}^{\bar{y}\bar{y}}) \, .$$

(4.4) Show that the non-spin-flip and spin-flip intensities for polarized neutron scattering from an isotropic paramagnet are given by

$$I_{NSF} = \tfrac{1}{2}I^{M}\{1 - (\hat{\mathbf{Q}} \cdot \hat{\boldsymbol{P}})^{2}\}$$
$$I_{SF} = \tfrac{1}{2}I^{M}\{1 + (\hat{\mathbf{Q}} \cdot \hat{\boldsymbol{P}})^{2}\},$$

where $I^{M}$ is given by (4.91).
[Hint: Choose $\mathbf{P} \parallel x$ and apply eqns (4.85)–(4.86).]

(4.5) Starting from the Blume–Maleev equations (4.97) and (4.100), show that when scattering from an isotropic paramagnet the final polarization is given by eqn (4.102),

$$\mathbf{P}_f = -(\hat{\mathbf{Q}} \cdot \mathbf{P}_i)\hat{\mathbf{Q}} \, .$$

[Hint: Choose $\mathbf{Q} \parallel x$ and put $\mathbf{M}_\perp = (0, M_y, M_z)$.]
Show that in uniaxial polarization analysis this reduces to

$$P_f = -(\hat{\mathbf{Q}} \cdot \hat{\mathbf{P}}_i)^{2} P_i \, .$$

# Nuclear Scattering

**5**

Neutron scattering via the nuclear interaction makes it possible to probe the arrangements and dynamics of atoms and molecules in condensed matter. In this chapter, the kinematical theory described in Chapters 3 and 4 will be applied to calculate the elastic and inelastic nuclear scattering from ordered and disordered solids, liquids and complex molecular systems. In the case of diffraction, i.e. coherent elastic scattering, we shall find that some of the results obtained in Chapter 2 for stationary atoms require modification to take into account atomic motion. The bulk of the chapter, however, will be concerned with the calculation of cross-sections for inelastic scattering. These enable us to learn about the dynamics of individual particles and about the collective motions of atoms, such as crystal vibrations and molecular motions.

Calculations of nuclear scattering for virtually all systems of practical interest begin with the expressions for the coherent and incoherent double differential cross-sections,

$$\left(\frac{\mathrm{d}^2\sigma}{\mathrm{d}\Omega\mathrm{d}E_{\mathrm{f}}}\right)_{\mathrm{coh}} = \frac{k_{\mathrm{f}}}{k_{\mathrm{i}}}\frac{1}{2\pi\hbar}\sum_{jk}\overline{b}_j^*\,\overline{b}_k\int_{-\infty}^{\infty}\langle\exp(-\mathrm{i}\mathbf{Q}\cdot\hat{\mathbf{r}}_j)\exp\{\mathrm{i}\mathbf{Q}\cdot\hat{\mathbf{r}}_k(t)\}\rangle$$
$$\times\,\exp(-\mathrm{i}\omega t)\,\mathrm{d}t, \tag{5.1}$$

and

$$\left(\frac{\mathrm{d}^2\sigma}{\mathrm{d}\Omega\mathrm{d}E_{\mathrm{f}}}\right)_{\mathrm{inc}} = \frac{k_{\mathrm{f}}}{k_{\mathrm{i}}}\frac{1}{2\pi\hbar}\sum_{j}\frac{(\sigma_{\mathrm{inc}})_j}{4\pi}\int_{-\infty}^{\infty}\langle\exp(-\mathrm{i}\mathbf{Q}\cdot\hat{\mathbf{r}}_j)\exp\{\mathrm{i}\mathbf{Q}\cdot\hat{\mathbf{r}}_j(t)\}\rangle$$
$$\times\,\exp(-\mathrm{i}\omega t)\,\mathrm{d}t. \tag{5.2}$$

These expressions are for unpolarized neutrons,[1] and assume that the distribution of isotopes and/or nuclear spin states among the sites for a given atomic species is completely random. They also assume that the motions of the system are independent of the nuclear states (i.e. the dynamics do not depend on the isotopic or spin state of any given nucleus). If these assumptions are not all valid then one must revert to the general expressions for nuclear scattering, eqns (4.18)–(4.19) or (4.55)–(4.57).

Coherent scattering provides information on the relative positions of the atoms, and their collective dynamics. The coherent scattering length $\overline{b}_j$ that appears in (5.1) is an average over the natural distribution of isotopes and random nuclear spin states for the atomic species that occupies site $j$, as explained in Section 1.8. Incoherent scattering provides information on single particle dynamics and arises from the random variation of the nuclear scattering amplitude due to the random distribution

[1]See Section 4.9.

of isotopes and nuclear spin states among the sites. The single-atom incoherent cross-section $\sigma_{\text{inc}}$ in (5.2) is proportional to the statistical variance of the scattering length [see eqns (1.52) and (4.115)], and is made up of two components, one due to the random distribution of isotopes (isotopic incoherence) and the second due to the random nuclear spin states (spin incoherence).

Possible ways of exploiting polarized neutrons for nuclear scattering were discussed in Section 4.5. Briefly, polarized neutrons can be useful for separating coherent from incoherent scattering in systems where the spin incoherence dominates the isotopic incoherence, such as hydrogen-containing systems. It is also possible to use polarization analysis to separate the spin incoherent scattering from the isotopic incoherent scattering, which in principle allows one to measure single particle dynamics with a degree of selectively, i.e. to separate the single-particle dynamics of different species in a multi-component system. Generally speaking, though, polarization analysis is not commonly employed when studying purely nuclear scattering because it does not present many opportunities for measuring correlations that cannot be measured by unpolarized neutron scattering.[2] For this reason, we shall only concern ourselves in this chapter with unpolarized neutron scattering.

[2]By contrast, neutron polarization analysis of magnetic scattering provides access to a great deal more information than can be measured with unpolarized neutrons.

## 5.1   The harmonic oscillator

The harmonic oscillator is an extremely useful model for many different processes in matter. In this section we calculate the scattering cross-section for a nucleus in a harmonic (quadratic) potential.[3] The exercise will serve two purposes. Firstly, the cross-section is directly applicable in a number of cases of interest, especially the spectra of molecular vibrations. Second, the problem of the isolated quantum harmonic oscillator provides a simple introduction to the mathematical procedures needed to deal with the more general case of coupled harmonic oscillators, such as the vibrations of atoms in a crystal. The introduction of pseudo-boson (creation and annihilation) operators greatly simplifies the analysis of the harmonic oscillator, and analogous operator methods are frequently employed to calculate the scattering from various other types of cooperative excitations, such as spin waves.

[3]See Zemach and Glauber (1956).

The Hamiltonian for a mass $M$ in an isotropic harmonic potential with 'spring constant' $K$ is

$$\mathcal{H} = \frac{\mathbf{p}^2}{2M} + \tfrac{1}{2}M\omega_0^2\mathbf{r}^2, \qquad (5.3)$$

where $\omega_0 = (K/M)^{1/2}$ is the angular frequency of the classical harmonic oscillator described by the same Hamiltonian. Because the potential is isotropic, eqn (5.3) is separable into different Cartesian components, $\mathcal{H} = \mathcal{H}_x + \mathcal{H}_y + \mathcal{H}_z$, and so it will be sufficient to consider the harmonic oscillator in one dimension. The Hamiltonian for this is

$$\mathcal{H} = \frac{p^2}{2M} + \tfrac{1}{2}M\omega_0^2 x^2. \qquad (5.4)$$

Unlike the classical harmonic oscillator, whose energy varies continuously, the quantum oscillator has discrete energy levels given by[4]

$$E_n = (n + \tfrac{1}{2})\hbar\omega_0, \quad n = 0, 1, 2, \ldots \tag{5.5}$$

originating from the non-commutation of $x$ and $p$,

$$[x, p] = i\hbar. \tag{5.6}$$

Because the energy levels are equally spaced and characterized by the integers $n$, it is convenient to denote the corresponding eigenstates by the ket $|n\rangle$, the ground state being $|0\rangle$.[5] Hence,

$$\mathcal{H}|n\rangle = (n + \tfrac{1}{2})\hbar\omega_0\,|n\rangle. \tag{5.7}$$

The eigenstates are taken to be orthonormal, i.e. $\langle n|n'\rangle = \delta_{nn'}$.

## 5.1.1   Annihilation and creation operators

In order to derive useful properties of the harmonic operator, such as the scattering cross-section, we will need to evaluate matrix elements of operators between different eigenstates. One way to do this is to solve the Schrödinger equation for the system to obtain explicit wave functions for each $|n\rangle$ in terms of the coordinate $x$. In many cases, however, the algebra is greatly simplified by working in the so-called *occupation number representation*.

The principle of this approach is to keep to a minimum what needs to be known about the eigenstates. We do this by defining operators that transform one state of the system into another. The complete set of states can then be generated from the ground state. The operators that perform this function are defined by

$$a = \left(\frac{1}{2M\hbar\omega_0}\right)^{1/2}(M\omega_0 x + ip)$$

$$\tag{5.8}$$

$$a^\dagger = \left(\frac{1}{2M\hbar\omega_0}\right)^{1/2}(M\omega_0 x - ip),$$

and are called *annihilation* and *creation* operators. From (5.6) and (5.8) it is easy to show that $a$ and $a^\dagger$, which are Hermitian adjoints,[6] satisfy what are known as the Bose commutation relations

$$[a, a^\dagger] = 1$$

$$\tag{5.9}$$

$$[a, a] = [a^\dagger, a^\dagger] = 0,$$

and that the Hamiltonian (5.4) may be rewritten as

$$\mathcal{H} = (a^\dagger a + \tfrac{1}{2})\hbar\omega_0. \tag{5.10}$$

Comparing (5.10) and (5.7) one sees that the eigenvalues of the operator product $a^\dagger a$ are just the integers $n$, i.e.

$$a^\dagger a|n\rangle = n|n\rangle. \tag{5.11}$$

[4]Consult any textbook on quantum mechanics. Additional results on the quantum harmonic oscillator are given in Section C.1.12.

[5]Dirac's bra and ket notation is explained in Section C.1.3.

[6]See Section C.1.5.

With slightly more effort (see Exercise 5.3), it can be shown from (5.9) and (5.10) that

$$a|n\rangle = n^{\frac{1}{2}}|n-1\rangle \tag{5.12}$$

$$a^{\dagger}|n\rangle = (n+1)^{\frac{1}{2}}|n+1\rangle. \tag{5.13}$$

Hence, $a$ and $a^{\dagger}$ convert $|n\rangle$ into $|n-1\rangle$ and $|n+1\rangle$, respectively, which is why they are called *annihilation* and *creation* operators.[7] The corresponding Heisenberg operators have explicit time dependence[8]

$$\hat{a}(t) = a\exp(-\mathrm{i}\omega_0 t), \tag{5.14}$$

$$\hat{a}^{\dagger}(t) = a^{\dagger}\exp(\mathrm{i}\omega_0 t). \tag{5.15}$$

From (5.8), (5.14) and (5.15)

$$\hat{x}(t) = \left(\frac{\hbar}{2M\omega_0}\right)^{1/2}\{a\exp(-\mathrm{i}\omega_0 t) + a^{\dagger}\exp(\mathrm{i}\omega_0 t)\}. \tag{5.16}$$

### 5.1.2   Harmonic oscillator in thermal equilibrium

We need to consider a quantum harmonic oscillator in thermal equilibrium at temperature $T$. The probability $p_n$ of the oscillator being in state $|n\rangle$ with energy $E_n$ is given by the Boltzmann distribution

$$p_n = \frac{1}{Z}\exp(-\beta E_n), \tag{5.17}$$

where $\beta = 1/k_{\mathrm{B}}T$ and

$$
\begin{aligned}
Z &= \sum_{n=0}^{\infty}\exp(-\beta E_n) \\
&= \exp(-\tfrac{1}{2}\beta\hbar\omega_0)\sum_{n=0}^{\infty}\exp(-n\beta\hbar\omega_0) \\
&= \frac{\exp(-\tfrac{1}{2}\beta\hbar\omega_0)}{1 - \exp(-\beta\hbar\omega_0)}.
\end{aligned} \tag{5.18}
$$

Here, we have substituted for $E_n$ using (5.5) and evaluated the summation using the standard result for an infinite geometric series.[9]

From (5.10), the average energy of the oscillator is given by

$$\langle E\rangle = (\langle a^{\dagger}a\rangle + \tfrac{1}{2})\hbar\omega_0, \tag{5.19}$$

and from (5.11), (5.17), and (5.18)

$$
\begin{aligned}
\langle a^{\dagger}a\rangle = \langle n\rangle &= \sum_n np_n \\
&= \frac{\sum_{n=0}^{\infty}n\exp(-n\beta\hbar\omega_0)}{\sum_{n=0}^{\infty}\exp(-n\beta\hbar\omega_0)}.
\end{aligned} \tag{5.20}
$$

[7]The terms *lowering* and *raising* operators are also used.

[8]See Section C.1.12.

[9]The infinite geometric series sums to

$$\sum_{n=0}^{\infty}u^n = \frac{1}{1-u}, \quad (|u| < 1).$$

In (5.18), put $u = \exp(-\beta\hbar\omega_0)$.

The simplest way to evaluate the sums in (5.20) is note that the denominator is the same infinite geometric series as encountered in (5.18), and that the numerator is obtained by differentiation of this series with respect to $-\beta\hbar\omega_0$. The result is

$$\langle a^\dagger a \rangle = \frac{1}{\exp(\beta\hbar\omega_0) - 1} \equiv n(\omega_0). \tag{5.21}$$

With the aid of the Bose commutation relations (5.9) one can also show that

$$\langle aa^\dagger \rangle = \frac{1}{1 - \exp(-\beta\hbar\omega_0)} \equiv n(\omega_0) + 1. \tag{5.22}$$

In eqns (5.21) and (5.22), $n(\omega)$ is the Planck distribution defined previously in (3.55). In general, $n(\omega)$ gives the average occupation number of boson quasiparticles in a state with energy $\hbar\omega$. For the harmonic oscillator this corresponds to the expectation value of the level index $n$, and from (5.19) the average energy is then $\langle E \rangle = \{n(\omega_0) + \frac{1}{2}\}\hbar\omega_0$.

We shall require the thermal average of two other quantities:

$$\langle x^2 \rangle = \frac{\hbar}{2M\omega_0} \coth(\tfrac{1}{2}\beta\hbar\omega_0), \tag{5.23}$$

and

$$\langle \exp x \rangle = \exp(\tfrac{1}{2}\langle x^2 \rangle). \tag{5.24}$$

These two results are derived in Section C.1.12. Relation (5.24) is known as Bloch's identity.

## 5.1.3 Scattering cross-section

We now proceed to calculate the scattering cross-section for a nucleus in an isotropic harmonic potential. We make the assumption that the eigenstates of the harmonic oscillator are independent of the internal state of the nucleus (i.e. the particular isotope or spin state). The basic quantity of interest is then the correlation function in (4.19), which we write

$$\langle \exp(-i\mathbf{Q} \cdot \hat{\mathbf{r}}) \exp\{i\mathbf{Q} \cdot \hat{\mathbf{r}}(t)\} \rangle = \langle \exp U \exp V \rangle, \tag{5.25}$$

where

$$U = -i\mathbf{Q} \cdot \hat{\mathbf{r}}, \text{ and } V = i\mathbf{Q} \cdot \hat{\mathbf{r}}(t). \tag{5.26}$$

Here, $\hat{\mathbf{r}}(t)$ is the Heisenberg operator representing the displacement of the nucleus, and $\hat{\mathbf{r}}$ is shorthand for $\hat{\mathbf{r}}(0)$.

The scalar products in (5.26) make the correlation function only dependent on the component of the displacement parallel to $\mathbf{Q}$, and because the harmonic potential is assumed to be isotropic we can choose this direction to be $x$. The variables $U$ and $V$ are then proportional to the operators $\hat{x}$ and $\hat{x}(t)$ whose properties were explored in the previous two sections. In particular, the commutator $[U, V]$ is proportional to $[\hat{x}, \hat{x}(t)]$, which from (5.16) turns out to be a complex number (rather than an operator). This means we can use the Baker–Campbell–Hausdorff formula (see Section C.1.7) to write (5.25) in the form

$$\langle \exp U \exp V \rangle = \langle \exp(U + V) \rangle \, \exp(\tfrac{1}{2}[U, V]). \tag{5.27}$$

Now, $U + V$ is proportional to $\hat{x} - \hat{x}(t)$, and so Bloch's identity (5.24) applies to $\langle \exp(U + V) \rangle$, giving

$$\langle \exp(U + V) \rangle = \exp\{\tfrac{1}{2} \langle (U + V)^2 \rangle\}. \tag{5.28}$$

Substituting (5.28) into (5.27) and combining the exponentials we obtain

$$\langle \exp U \exp V \rangle = \exp\{\tfrac{1}{2} \langle (U + V)^2 + UV - VU \rangle\}$$
$$= \exp(\tfrac{1}{2}\langle U^2 \rangle + \tfrac{1}{2}\langle V^2 \rangle) \, \exp\langle UV \rangle$$
$$= \exp(-Q^2 \langle x^2 \rangle) \, \exp\{Q^2 \langle \hat{x}\hat{x}(t) \rangle\}. \tag{5.29}$$

In the last line we set $\langle x^2 \rangle = \langle \hat{x}^2(t) \rangle$, since the thermal average is independent of the origin of time. An expression for $\langle x^2 \rangle$ is given in (5.23).

The scattering cross-section is proportional to the response function[10] $S(\mathbf{Q}, \omega)$, which is in turn proportional to the time Fourier transform of the correlation function (5.29). If, for simplicity, we assume that the spin of the nucleus is randomly oriented, then from (3.36), (4.19) and (4.20) the response function for unpolarized neutrons is given by[11]

$$S(\mathbf{Q}, \omega) = \frac{1}{2\pi\hbar} \, \overline{|b|^2} \, \exp(-Q^2 \langle x^2 \rangle) \int_{-\infty}^{\infty} \exp\{Q^2 \langle \hat{x}\hat{x}(t) \rangle\} \, \exp(-\mathrm{i}\omega t) \, \mathrm{d}t. \tag{5.30}$$

The energy dependence of the scattering arises from the time integral in (5.30), which contains the correlation function $\langle \hat{x}\hat{x}(t) \rangle$. From (5.16),

$$\langle \hat{x}\hat{x}(t) \rangle = \frac{\hbar}{2M\omega_0} \langle aa^\dagger \exp(\mathrm{i}\omega_0 t) + a^\dagger a \exp(-\mathrm{i}\omega_0 t) \rangle$$
$$= \frac{\hbar}{2M\omega_0} \left[ \{n(\omega_0) + 1\} \exp(\mathrm{i}\omega_0 t) + n(\omega_0) \exp(-\mathrm{i}\omega_0 t) \right]. \tag{5.31}$$

On substituting (5.31) into (5.30) we are faced with a formidable-looking integral, the result of which can, however, be expressed rather concisely in terms of Bessel functions.[12] The result is

$$S(\mathbf{Q}, \omega) = \overline{|b|^2} \, \exp(-Q^2 \langle \hat{x}^2 \rangle) \, \exp(\tfrac{1}{2}\beta\hbar\omega) \sum_{n=-\infty}^{\infty} I_n(y) \, \delta(\hbar\omega - n\hbar\omega_0), \tag{5.32}$$

where $n$ is an integer, $I_n(y)$ is a modified Bessel function of the first kind,[13] and

$$y = \frac{\hbar Q^2}{2M\omega_0 \sinh(\tfrac{1}{2}\beta\hbar\omega_0)}. \tag{5.33}$$

The spectrum described by (5.32) contains a sequence of sharp peaks separated in energy by $\hbar\omega_0$. The $n$th peak corresponds to the transfer of $n$ quanta from the neutron to the oscillator ($n > 0$) or from the

---

[10] See Section 3.4.5:

$$\frac{\mathrm{d}^2\sigma}{\mathrm{d}\Omega \mathrm{d}E_{\mathrm{f}}} = \frac{k_{\mathrm{f}}}{k_{\mathrm{i}}} \, S(\mathbf{Q}, \omega).$$

[11] The bar in $\overline{|b|^2}$ indicates an average over nuclear spin orientations. We are assuming that the measurement involves averaging the scattering of many neutrons during which time the nuclear spin is constantly tumbling around. With only one nucleus it is meaningless to split the cross-section into coherent and incoherent parts since both measure the same thing. For polarized neutron scattering, replace $\overline{|b|^2}$ by $|\hat{\beta}|^2$ — see Section 4.1.4.

[12] See Zemach and Glauber (1956).

[13] Bessel functions of the first kind are defined for integer $n$ by the series

$$J_n(y) = \sum_{m=0}^{\infty} \frac{(-1)^m}{m! \, (n+m)!} \left(\frac{y}{2}\right)^{n+2m}.$$

The *modified* Bessel functions of the first kind are given for integer $n$ by

$$I_n(y) = \frac{J_n(\mathrm{i}y)}{\mathrm{i}^n},$$

so that $I_{-n}(y) = I_n(y)$.

(a)

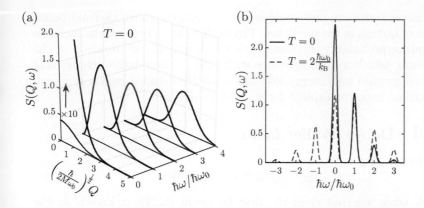

(b)

**Fig. 5.1** The response function eqn (5.32) of a quantum harmonic oscillator plotted (a) as a function of reduced energy transfer $\hbar\omega/\hbar\omega_0$ and reduced momentum transfer $(\hbar/2M\omega_0)^{\frac{1}{2}}Q$ at temperature $T = 0$, and (b) as a function of $\hbar\omega/\hbar\omega_0$ at fixed $(\hbar/2M\omega_0)^{\frac{1}{2}}Q = 1$ and two different temperatures. The simulations assume $|b|^2 = 1$ for simplicity.

oscillator to the neutron ($n < 0$). The $n = 0$ term corresponds to elastic scattering. From the property $I_{-n}(y) = I_n(y)$ it can be seen that the response function satisfies the detailed balance condition $S(-\mathbf{Q}, -\omega) = \exp(-\beta\hbar\omega)\,S(\mathbf{Q}, \omega)$, eqn (3.48).

Figure 5.1 illustrates the form of the response function (5.32). For $n > 0$ the scattering has a maximum as a function $Q$. The maximum shifts to higher $Q$ with increasing $n$. The neutron can only absorb energy from the target if the oscillator is in an excited state, and this is why the neutron energy-gain scattering ($\hbar\omega < 0$) is negligible when $k_B T \ll \hbar\omega_0$.

The vibrational spectrum of uranium nitride (UN), shown in Fig. 5.2, provides a fine illustration of the QHO model. The spectrum contains a series of peaks at equally spaced energies. These peaks correspond to the quantized vibrational energy levels of the light N atoms, each of which vibrates independently in an octahedron of heavy U atoms.

For later reference, it is instructive to examine the elastic ($n = 0$) and single-quantum ($n = \pm 1$) terms in $S(\mathbf{Q}, \omega)$. In (5.32) these are proportional to $I_0(y)$ and $I_{\pm 1}(y)$ respectively, but in the limit of small $Q$ we can obtain approximate expressions for them directly from eqn (5.30) by expanding the first exponential inside the integral:

$$\exp\{Q^2\langle\hat{x}\hat{x}(t)\rangle\} = 1 + Q^2\langle\hat{x}\hat{x}(t)\rangle + \frac{1}{2!}Q^4\langle\hat{x}\hat{x}(t)\rangle^2 + \ldots. \quad (5.34)$$

If we neglect all but the first two terms on the right-hand side then from (5.30), (5.31), and (B.16),

$$S(\mathbf{Q}, \omega) \approx \overline{|b|^2}\exp(-Q^2\langle x^2\rangle)$$
$$\times \left\{\delta(\hbar\omega) + \frac{\hbar Q^2}{2M\omega_0}[\{n(\omega_0) + 1\}\delta(\hbar\omega - \hbar\omega_0) + n(\omega_0)\delta(\hbar\omega + \hbar\omega_0)]\right\}. \quad (5.35)$$

This approximation is valid providing $\hbar Q^2 n(\omega_0)/(2M\omega_0) \ll 1$, which from (5.23) is essentially the same as the condition $Q^2\langle x^2\rangle \ll 1$. The three terms in the large bracket are seen to correspond to elastic scattering and inelastic scattering with exchange of a single quantum of energy.

(a)

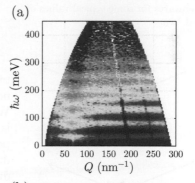

(b)

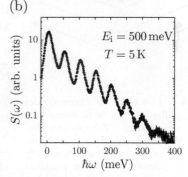

**Fig. 5.2** Spectrum of local vibrational modes in uranium nitride (Aczel *et al.*, 2012). (a) Intensity map as a function of $Q$ and energy transfer. (b) Scattering integrated over $Q$ plotted as a function of energy transfer. (Data courtesy of A. A. Aczel and S. E. Nagler, Oak Ridge National Laboratory.)

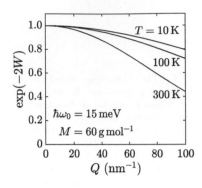

**Fig. 5.3** $Q$ dependence of the squared Debye–Waller factor, evaluated at three temperatures for some typical values of $M$ and $\hbar\omega_0$.

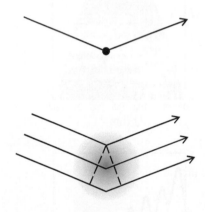

**Fig. 5.4** The effect of thermal and quantum fluctuations is to smear out the nucleus into a diffuse object (below). Waves scattered from different parts of the distribution are not exactly in phase (lower), leading to a reduction in intensity compared with an hypothetical stationary nucleus (upper).

The latter terms are seen to depend on $n(\omega_0)$, the mean thermal population of quanta in the oscillator. Physically, this is because the inelastic scattering probability increases with the amplitude of vibration of the oscillator, which in turn increases with $n(\omega_0)$. The relative strengths of the energy-gain and energy-loss terms satisfy detailed balance, as can be verified from the explicit form (5.21) of $n(\omega)$.

### 5.1.4 Debye–Waller factor

The factor

$$\exp(-\tfrac{1}{2}Q^2\langle x^2\rangle) = \exp(-W), \qquad (5.36)$$

which when squared gives the first factor in (5.29), is known as the *Debye–Waller factor*. From (5.23),

$$W = \frac{\hbar Q^2}{4M\omega_0}\coth(\tfrac{1}{2}\beta\hbar\omega_0). \qquad (5.37)$$

The Debye–Waller factor has a value between 0 and 1. It depends on temperature and on the scattering vector $Q$, as illustrated in Fig. 5.3, but not on the neutron energy transfer. The physical origin of the Debye–Waller factor is in the motion of the nucleus. Although the nucleus is point-like, the uncertainty in its position due to thermal and zero-point fluctuations means that the neutrons scatter from an apparently diffuse object. This is illustrated in Fig. 5.4. The scattered waves from different parts of a diffuse object are not exactly in phase, and when they are superposed there is a degree of cancellation. This reduces the amplitude of the scattered beam relative to that from a point-like object.

## 5.2  Crystalline solids

We recall from Chapter 2 that the periodicity of a crystalline solid is represented by a lattice, and that with each lattice point one may associate a primitive unit cell containing a collection of atoms. The translational symmetry means that the arrangement of atoms in each unit cell is identical.

Let the position of the $d$th atom in the primitive unit cell associated with lattice point $l$ be represented by

$$\mathbf{r}_{ld}(t) = \mathbf{l} + \mathbf{d} + \mathbf{u}_{ld}(t), \qquad (5.38)$$

where $\mathbf{l}$ is the vector from the origin to lattice point $l$, $\mathbf{d}$ connects the lattice point to the equilibrium position of the atom, and $\mathbf{u}_{ld}(t)$ is the displacement of the atom from its equilibrium position.

The coherent scattering cross-section (5.1) contains a summation over all pairs of nuclei, and from (5.38) this can be separated into a sum over all pairs of lattice points $l, l'$ and a sum over all pairs of atoms in the unit cell $d, d'$. Because the lattice is effectively infinite the sum over $l$ is the same for each $l'$, so we need only include one value of $l'$, say $l' = 0$, and multiply the cross-section by $N$, the number of lattice points. With this simplification the coherent scattering cross-section for unpolarized neutrons becomes

$$\left(\frac{\mathrm{d}^2\sigma}{\mathrm{d}\Omega\mathrm{d}E_\mathrm{f}}\right)_\mathrm{coh} = \frac{k_\mathrm{f}}{k_\mathrm{i}}\frac{N}{2\pi\hbar}\sum_l \exp(\mathrm{i}\mathbf{Q}\cdot\mathbf{l})\sum_{dd'}\bar{b}_{d'}^*\,\bar{b}_d\,\exp\{\mathrm{i}\mathbf{Q}\cdot(\mathbf{d}-\mathbf{d}')\}$$

$$\times\int_{-\infty}^{\infty}\langle\exp(-\mathrm{i}\mathbf{Q}\cdot\hat{\mathbf{u}}_{0d'})\exp\{\mathrm{i}\mathbf{Q}\cdot\hat{\mathbf{u}}_{ld}(t)\}\rangle\exp(-\mathrm{i}\omega t)\,\mathrm{d}t, \qquad (5.39)$$

and from (5.2) the incoherent cross-section for unpolarized neutrons is

$$\left(\frac{\mathrm{d}^2\sigma}{\mathrm{d}\Omega\mathrm{d}E_\mathrm{f}}\right)_\mathrm{inc} = \frac{k_\mathrm{f}}{k_\mathrm{i}}\frac{N}{2\pi\hbar}\sum_d\frac{(\sigma_\mathrm{inc})_d}{4\pi}$$

$$\times\int_{-\infty}^{\infty}\langle\exp(-\mathrm{i}\mathbf{Q}\cdot\hat{\mathbf{u}}_d)\exp\{\mathrm{i}\mathbf{Q}\cdot\hat{\mathbf{u}}_d(t)\}\rangle\exp(-\mathrm{i}\omega t)\,\mathrm{d}t. \qquad (5.40)$$

The factor $N$ appears in (5.40) because the 'self' correlation function in the integral is the same for every $l$.

To convert (5.39) and (5.40) into more manageable forms we repeat the steps from (5.25) to (5.29) used in the development of the harmonic oscillator. In the case of coherent scattering, we define

$$U = -\mathrm{i}\mathbf{Q}\cdot\hat{\mathbf{u}}_{0d'}, \quad\text{and}\quad V = \mathrm{i}\mathbf{Q}\cdot\hat{\mathbf{u}}_{ld}(t). \qquad (5.41)$$

Hence,

$$\langle\exp U\exp V\rangle = \exp\{-(W_d + W_{d'})\}\exp\langle UV\rangle, \qquad (5.42)$$

where

$$W_d = \tfrac{1}{2}\langle(\mathbf{Q}\cdot\hat{\mathbf{u}}_d)^2\rangle. \qquad (5.43)$$

The first exponential on the right-hand side of (5.42) is the joint Debye–Waller factor associated with atoms $d$ and $d'$ in the unit cell. Note that there is no dependence on $l$ because by translational symmetry the mean squared displacement of an atom at a particular crystallographic site is the same in all unit cells. The analysis for incoherent scattering is the same except that we put $d' = d$.

The problem that remains is to evaluate the time Fourier transform of the function $\exp\langle UV\rangle$. The easiest way to proceed is to expand the exponential as a power series in $\langle UV\rangle$ just as we did in (5.34) for the harmonic oscillator:

$$\exp\langle UV\rangle = 1 + \langle UV\rangle + \frac{1}{2!}\langle UV\rangle^2 + \dots \qquad (5.44)$$

The term in $\langle UV\rangle^n$ corresponds to exchange of $n$ quanta, referred to as $n$-phonon processes. When the condition $\langle UV\rangle \ll 1$ holds, as it generally will if $N$ is large, it will suffice to retain only the first few terms in the series. In the following sections we shall consider explicitly the elastic and one-phonon scattering, and we shall also comment on the form of the scattering from multi-phonon processes.

### 5.2.1   Elastic scattering

If we retain just the first term ($= 1$) in the series (5.44) then the time integral in (5.39) can be written (see Section B.2.3)

$$\frac{1}{2\pi\hbar} \int_{-\infty}^{\infty} \exp(-i\omega t)\, dt = \delta(\hbar\omega).$$
(5.45)

Hence, the cross-section contains purely elastic scattering, which means we can set $k_i = k_f$. Since no real detector can measure a $\delta$-function energy dependence it is more convenient to integrate over $E_f$ and describe the scattering by the differential cross-section. If, in addition, we use the results in Section B.4.2 to rewrite the sum over lattice vectors as

$$\sum_l \exp(i\mathbf{Q} \cdot \mathbf{l}) = \frac{(2\pi)^3}{v_0} \sum_{\mathbf{G}} \delta(\mathbf{Q} - \mathbf{G}),$$
(5.46)

where $\mathbf{G}$ is a reciprocal lattice vector and $v_0$ is the volume of the unit cell (2.3), then (5.39) becomes

$$\begin{aligned}
\left(\frac{d\sigma}{d\Omega}\right)_{\text{coh}}^{\text{el}} &= \int_0^{\infty} \left(\frac{d^2\sigma}{d\Omega dE_f}\right)_{\text{coh}} dE_f \\
&= N\frac{(2\pi)^3}{v_0} \sum_{\mathbf{G}} \delta(\mathbf{Q} - \mathbf{G}) \\
&\quad \times \sum_{dd'} \bar{b}_{d'}^* \bar{b}_d \exp(-W_d - W_{d'}) \exp\{i\mathbf{Q} \cdot (\mathbf{d} - \mathbf{d}')\} \\
&= N\frac{(2\pi)^3}{v_0} \sum_{\mathbf{G}} |F_N(\mathbf{G})|^2 \delta(\mathbf{Q} - \mathbf{G}),
\end{aligned}$$
(5.47)

where

$$F_N(\mathbf{G}) = \sum_d \bar{b}_d \exp(-W_d) \exp(i\mathbf{G} \cdot \mathbf{d})$$
(5.48)

is the nuclear unit-cell structure factor.

Glancing back at eqns (2.18) and (2.25), we see that elastic scattering from a crystal with mobile atoms is the same as that from a rigid crystal apart from the factor of $\exp(-W_d)$ in the structure factor. Also, Friedel's law (Section 2.2.7) still applies, since from (5.43) $W_d(-\mathbf{Q}) = W_d(\mathbf{Q})$. Hence, results derived for a rigid crystal can readily be adapted for a real crystal simply by inclusion of the Debye–Waller factor.

The corresponding expression for the incoherent differential cross-section derived from (5.40) is

$$\left(\frac{d\sigma}{d\Omega}\right)_{\text{inc}}^{\text{el}} = N \sum_d \frac{(\sigma_{\text{inc}})_d}{4\pi} \exp(-2W_d).$$
(5.49)

### 5.2.2   Atomic vibrations in crystals

The atoms in crystals vibrate about their equilibrium positions, and the vibrations on different sites are coupled by inter-atomic forces. When the amplitudes of vibration are small compared with the interatomic separations one can usually make the *harmonic approximation*, namely that the equilibrium position of the atoms is at the minimum of a quadratic potential. The coupled oscillations of the atoms in a harmonic crystal can then be described by a set of *normal modes*, states of motion in which all the atoms vibrate at the same frequency. Inelastic neutron scattering from atomic vibrations occurs via exchange of energy quanta with the normal modes of the crystal.

We start this section by deriving the classical vibration spectrum of a simple one-dimensional system. The analysis of this system will illustrate much of the phenomenology associated with crystal dynamics, and we shall outline how the method can be extended to arbitrary crystal structures. After that, we give the quantum treatment of crystal dynamics, which introduces the concept of *phonon* quasiparticles to describe the vibrational energy of the crystal in the quantum picture. Finally we show how the cross-section for neutron inelastic scattering from a crystal can be expressed in terms of creation and annihilation of phonons.

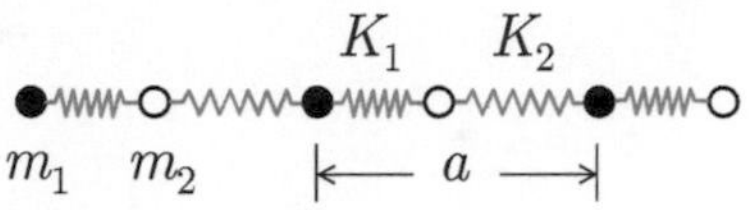

**Fig. 5.5** Diatomic chain of atoms.

## Diatomic chain

Consider a one-dimensional crystal composed of two different atoms arranged in pairs along a line, as shown in Fig. 5.5. The two atoms have masses $m_1$ and $m_2$ respectively, and are coupled by harmonic forces. The crystal can be generated from a one-dimensional lattice with lattice parameter $a$ and a two-atom basis. For simplicity, we locate the lattice points on the $m_1$ atoms.

We assume that the electron clouds move rigidly with the nucleus,[14] and consider nearest-neighbour forces only. The total energy (kinetic + elastic potential energy) of the chain is then given by

[14]This is known as the *adiabatic*, or *Born–Oppenheimer*, approximation.

$$E = \tfrac{1}{2}\sum_l (m_1\dot{u}_l^2 + m_2\dot{v}_l^2) + \tfrac{1}{2}\sum_l \{K_1(v_l-u_l)^2 + K_2(v_{l-1}-u_l)^2\}, \quad (5.50)$$

where $u_l$ and $v_l$ are the displacements from equilibrium of masses $m_1$ and $m_2$ in the $l$th primitive unit cell, and $K_1$ and $K_2$ are the spring constants (also known as force constants) connecting $m_1$ to the adjacent $m_2$ masses, as shown in Fig. 5.5.

The total energy of the system is a constant, and so the derivative of (5.50) with respect to any coordinate is zero. Hence, by differentiating with respect to $u_l$ and $v_l$ we obtain the equations of motion

$$m_1\ddot{u}_l = K_1(v_l - u_l) + K_2(v_{l-1} - u_l)$$
$$m_2\ddot{v}_l = K_2(u_{l+1} - v_l) + K_1(u_l - v_l). \quad (5.51)$$

We expect travelling waves to propagate on the chain, so we take as trial solutions[15]

[15]These trial solutions are basis states from which a general solution can be constructed by linear superposition — see eqns (5.62).

$$u_l(t) = \frac{1}{\sqrt{m_1}}\alpha_q \exp\{i(qla - \omega_q t)\}$$
$$v_l(t) = \frac{1}{\sqrt{m_2}}\beta_q \exp\{i(qla - \omega_q t)\}. \quad (5.52)$$

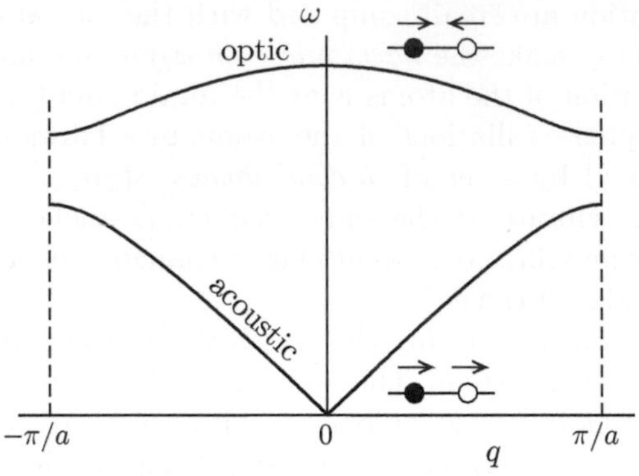

**Fig. 5.6** Dispersion relation for the diatomic chain of atoms. Only the first Brillouin zone is plotted. The relative displacements of the two masses for modes on the acoustic and optic branches near to $q = 0$ are shown.

Substitution of (5.52) into (5.51) leads to a pair of simultaneous linear equations for $\alpha_q$ and $\beta_q$ which must be satisfied if the travelling wave solutions are to satisfy the equations of motion. These simultaneous equations can be written concisely in matrix form as

$$
\begin{pmatrix} \frac{K_1+K_2}{m_1} & -\frac{K_1+K_2\exp(-iqa)}{\sqrt{m_1 m_2}} \\ -\frac{K_1+K_2\exp(iqa)}{\sqrt{m_1 m_2}} & \frac{K_1+K_2}{m_2} \end{pmatrix} \begin{pmatrix} \alpha_q \\ \beta_q \end{pmatrix} = \omega_q^2 \begin{pmatrix} \alpha_q \\ \beta_q \end{pmatrix},
$$
(5.53)

or

$$
\mathsf{D}_q \mathbf{x}_q = \omega_q^2 \mathbf{x}_q.
$$
(5.54)

The 2–by–2 matrix $\mathsf{D}_q$ is called the *dynamical matrix*. From (5.53), the transpose of $\mathsf{D}_q$ is seen to satisfy $\mathsf{D}_q^{\mathrm{T}} = \mathsf{D}_q^* = \mathsf{D}_{-q}$. A matrix such as $\mathsf{D}_q$ whose transpose equals its complex conjugate is an *Hermitian* matrix,[16] and it is well known that Hermitian matrices have real eigenvalues and orthogonal eigenvectors. From (5.54), the eigenvalues of $\mathsf{D}_q$ are seen to correspond to the squares of the angular frequencies of vibration. Solving the secular equation $|\mathsf{D}_q - \omega_q^2 \mathsf{I}| = 0$, where $\mathsf{I}$ is the 2–by–2 identity matrix, we find

[16]Note that the $1/\sqrt{m_{1,2}}$ prefactors in (5.52) are needed to ensure that $\mathsf{D}_q$ is Hermitian.

$$
\omega_{qj}^2 = \frac{(K_1 + K_2)(m_1 + m_2)}{2 m_1 m_2} +
$$

$$
(-1)^j \sqrt{\frac{(K_1+K_2)^2(m_1+m_2)^2}{4 m_1^2 m_2^2} - \frac{4 K_1 K_2}{m_1 m_2} \sin^2(\tfrac{1}{2} qa)}, \quad (5.55)
$$

where $j = 1, 2$ is an index for the two solutions.

The travelling wave solutions (5.52) are called *normal modes*. They are characterized by the fact that every atom vibrates with the same frequency. Equation (5.55) linking the wavevector and angular frequency of the normal modes is known as the *dispersion relation*. The dispersion relation for the diatomic chain is plotted in Fig. 5.6, and consists of two continuous curves, or *branches*, corresponding to $j = 1$ and 2. Both branches are periodic functions of $q$ with period $2\pi/a$.

To find the relative motion of the two atoms at a given point on the dispersion relation we need the eigenvectors. These can be obtained from (5.53) and (5.54) in the usual way, and may be written

$$\mathbf{x}_{qj} \equiv \mathbf{x}_s = \begin{pmatrix} e_{1s} \\ e_{2s} \end{pmatrix}, \tag{5.56}$$

where $s \equiv qj$ is an index labelling the normal modes, and

$$\frac{e_{2s}}{e_{1s}} = \frac{\beta_s}{\alpha_s} = \left(\frac{m_2}{m_1}\right)^{1/2} \frac{K_1 + K_2 \exp(iqa)}{K_1 + K_2 - m_2\omega_s^2}. \tag{5.57}$$

By convention, the eigenvectors are normalized to unity, so $e_{1s} = \alpha_s/|\mathbf{x}_s|$ and $e_{2s} = \beta_s/|\mathbf{x}_s|$. Granted this, the reader is invited to verify the orthonormality condition, $(\mathbf{x}_{qj})^* \cdot \mathbf{x}_{qj'} = \delta_{jj'}$ (Exercise 5.4).

Let us now examine the eigenvectors near the point $q = 0$. From (5.52) and (5.56) it can be seen that the relative amplitude and phase of the two masses is given by the ratio $e_{2s}/e_{1s}$ weighted by the factor $(m_1/m_2)^{1/2}$. From (5.55) and (5.57), as $q \to 0$,

$$\left(\frac{m_1}{m_2}\right)^{1/2} \frac{e_{2s}}{e_{1s}} \to \begin{cases} 1 & (j = 1), \\ -m_1/m_2 & (j = 2). \end{cases} \tag{5.58}$$

Hence, near $q = 0$ on the lower branch $(j = 1)$ the two masses in the unit cell oscillate in phase and with the same amplitude. This motion corresponds to ordinary elastic waves, i.e. sound waves, and for this reason the lower branch is called the *acoustic* branch. Sound waves have the property that the frequencies vanish as $q \to 0$. At $q = 0$ on the upper branch $(j = 2)$ the two masses oscillate in anti-phase, i.e. with a phase difference of $\pi$, the amplitude of mass $m_2$ being a factor $m_1/m_2$ times that of $m_1$. The frequency does not vanish as $q \to 0$, but instead tends to a constant. These vibrations are called *optic* modes.[17] The relative motions of the atoms at $q = 0$ are indicated on Fig. 5.6.

If the chain is of finite length then only certain discrete values of $q$ are allowed. Suppose there are $N$ pairs of atoms, so the length of the chain is $Na$. Let us connect the two ends of the chain together to form a ring[18] and impose periodic boundary conditions on the displacements:

$$u_{l+N}(t) = u_l(t). \tag{5.59}$$

Hence, from (5.52),

$$\exp(iqNa) = 1, \tag{5.60}$$

and so

$$q = \frac{2\pi n}{Na}, \tag{5.61}$$

where $n$ is an integer. We see that in one period $2\pi/a$ of the dispersion curve there are $N$ allowed values of $q$ and hence $2N$ normal modes (two modes per $q$, one on each branch). This includes all the modes since there are $2N$ equations of motion in (5.51). Hence, all distinct modes of vibration are represented in one period of the dispersion curve, and conventionally we choose the interval $-\pi/a \le q \le \pi/a$, known as the *first Brillouin zone*.[19]

[17]So-called because modes on the optic branch close to $q = 0$ can be observed in optical spectra providing the electronic charge on the atoms is different.

[18]Providing $N$ is large, this trick avoids end effects while retaining the essential physics of the one-dimensional chain.

[19]Léon Brillouin (1889–1969), French physicist.

The normal modes are particular solutions to the coupled-mass problem, but because the equations of motion are linear the most general solution for $u_l(t)$ and $v_l(t)$ is a superposition of normal modes, which we can write

$$u_l(t) = \frac{1}{\sqrt{Nm_1}} \sum_s e_{1s} \exp(iqla) Q_s(t)$$

$$v_l(t) = \frac{1}{\sqrt{Nm_2}} \sum_s e_{2s} \exp(iqla) Q_s(t), \qquad (5.62)$$

where[20]

$$Q_s(t) = A_s \exp(-i\omega_s t) + B_s \exp(i\omega_s t). \qquad (5.63)$$

[20]The eigenvector $(e_{1s}, e_{2s})$ is the same whether we have $\exp(i\omega_s t)$ or $\exp(-i\omega_s t)$ in (5.52).

The $Q_s(t)$ are known as *normal coordinates*. The fact that $u_l(t)$ and $v_l(t)$ are real, i.e. $u_l(t) = u_l^*(t)$ and $v_l(t) = v_l^*(t)$, means that

$$Q_{-s}(t) = Q_s^*(t)$$

$$= A_s^* \exp(i\omega_s t) + B_s^* \exp(-i\omega_s t), \qquad (5.64)$$

where $-s$ stands for $-qj$, and $\omega_{-s} = \omega_s$.

The complete set of normal coordinates (5.63) includes $4N$ unknown constants, because there are $N$ distinct values of $q$, two values of $j$, and $A_s$ and $B_s$ are complex constants constrained by (5.64). These constants are fixed by the initial conditions (e.g. the positions and velocities of the 2N masses at $t = 0$).

**Three-dimensional crystal**

Consider now a three-dimensional crystal with $N$ primitive unit cells each containing $p$ atoms. The procedure described for the diatomic chain is straightforward to generalize as follows:

(i) The displacement of the $d$th atom in the $l$th primitive unit is represented by the vector $\mathbf{u}_{ld}(t)$. The normal modes are plane waves, each having a wavevector $\mathbf{q}$.

(ii) The values of $\mathbf{q}$ allowed by periodic boundary conditions form a lattice of points in three-dimensional wavevector or reciprocal space.

(iii) There are $3p$ equations of motion ($p$ atoms, each with three displacement components). The dynamical matrix $\mathbf{D_q}$ is a $3p \times 3p$ Hermitian matrix, and therefore has real eigenvalues $\omega_{qj}^2$ and orthogonal eigenvectors $\mathbf{x}_{qj}$ ($j = 1, 2, \ldots, 3p$). The eigenvectors corresponding to the same $\mathbf{q}$ are orthonormal, i.e.

$$(\mathbf{x}_{qj})^* \cdot \mathbf{x}_{qj'} = \delta_{jj'}. \qquad (5.65)$$

The dynamical matrix satisfies

$$\mathbf{D_q^T} = \mathbf{D_q^*} = \mathbf{D_{-q}}, \qquad (5.66)$$

and so,

$$\mathbf{x}_{-\mathbf{q}j} = (\mathbf{x}_{\mathbf{q}j})^* \quad \text{and} \quad \omega_{-\mathbf{q}j} = \omega_{\mathbf{q}j}. \tag{5.67}$$

(iv) Each eigenvector contains $p$ complex vectors called *polarization vectors*:

$$\mathbf{x}_{\mathbf{q}j} = \mathbf{x}_s = \begin{pmatrix} \mathbf{e}_{1s} \\ \mathbf{e}_{2s} \\ \vdots \\ \mathbf{e}_{ps} \end{pmatrix}. \tag{5.68}$$

The polarization vector $\mathbf{e}_{ds}$ gives the amplitude,[21] direction and relative phase of the oscillation of atom $d$ in the normal mode specified by $s \equiv \mathbf{q}j$. From (5.67)

$$\mathbf{e}_{d-s} = (\mathbf{e}_{ds})^*, \tag{5.69}$$

where $-s$ stands for $-\mathbf{q}j$.

[21]The amplitude of vibration is actually proportional to $|\mathbf{e}_{ds}|/\sqrt{m_d}$.

(v) The dispersion relation is periodic in the reciprocal lattice, and has $3p$ branches. The first Brillouin zone is defined as the Wigner–Seitz unit cell in reciprocal space (see Section 2.1.3). By applying periodic boundary conditions one can show that the first Brillouin zone contains $N$ allowed values of $\mathbf{q}$, and hence contains all $3Np$ distinct normal modes.

(vi) In general, the atoms are in a state which can be expressed as a superposition of the normal mode displacements,

$$\mathbf{u}_{ld}(t) = \frac{1}{\sqrt{Nm_d}} \sum_s \mathbf{e}_{ds} \exp(i\mathbf{q} \cdot \mathbf{l}) Q_s(t), \tag{5.70}$$

where $Q_s(t)$ has the same form as (5.63). Because $\mathbf{u}_{ld}(t)$ is real and the polarization vectors satisfy (5.69),

$$Q_{-s}(t) = Q_s^*(t). \tag{5.71}$$

### 5.2.3 Quantization of normal modes

The total vibrational energy of a harmonic crystal, including coupling between all pairs of atoms (not just nearest neighbours), is

$$E = \tfrac{1}{2} \sum_{ld} m_d \dot{u}_{ld}^2 + \tfrac{1}{2} \times \tfrac{1}{2} \sum_{ld} \sum_{l'd'} \sum_{\alpha\beta} K_{ldl'd'}^{\alpha\beta} (u_{ld}^\alpha - u_{l'd'}^\beta)^2, \tag{5.72}$$

where $K_{ldl'd'}^{\alpha\beta}$ is the force constant between the $\alpha$ component of the displacement of atom $ld$ and the $\beta$ component of the displacement of atom $l'd'$. The extra factor of $1/2$ in front of the potential energy term corrects for double-counting in the summation.

Using eqns (5.65) to (5.71) we can write $E$ in terms of the normal coordinates. The result is (Exercise 5.6),

$$E = \tfrac{1}{2} \sum_s (\dot{Q}_s \dot{Q}_{-s} + \omega_s^2 Q_s Q_{-s}). \tag{5.73}$$

It can further be shown (Exercise 5.7) that the quantum-mechanical operators corresponding to the normal coordinates satisfy the commutation relation

$$[Q_s, \dot{Q}_{-s'}] = i\hbar \delta_{ss'}. \tag{5.74}$$

We recognize, e.g. from (5.6), that $Q_s$ and $\dot{Q}_{-s}$ represent canonically conjugate variables analogous to displacement and momentum respectively. We also see, e.g. by comparison with (5.3), that the Hamiltonian for the system, obtained from (5.73) by replacing the variables by operators, corresponds to a sum of harmonic oscillator Hamiltonians. Therefore, *each of the normal modes of a harmonic crystal behaves as an independent harmonic oscillator.*

From the treatment of the harmonic oscillator given earlier we expect that the energy in each normal mode should be quantized in units of $\hbar\omega_s$. To verify this we introduce normal mode annihilation and creation operators resembling (5.8):

$$a_s = \left( \frac{1}{2\hbar\omega_s} \right)^{1/2} (\omega_s Q_s + i\dot{Q}_s)$$

$$a_s^\dagger = \left( \frac{1}{2\hbar\omega_s} \right)^{1/2} (\omega_s Q_{-s} - i\dot{Q}_{-s}). \tag{5.75}$$

From (5.74) these operators may be seen to satisfy the Bose commutation relations

$$[a_s, a_{s'}^\dagger] = \delta_{ss'}, \tag{5.76}$$

and from (5.73) the Hamiltonian transforms into

$$\mathcal{H} = \sum_s (a_s^\dagger a_s + \tfrac{1}{2})\hbar\omega_s. \tag{5.77}$$

This is a sum of harmonic oscillator Hamiltonians, one for each normal mode, confirming that the normal modes have discrete energy levels. The quantum of energy $\hbar\omega_s$ associated with a normal mode is known as a *phonon*.

Finally, from (5.75),

$$Q_s = \left( \frac{\hbar}{2\omega_s} \right)^{1/2} (a_s + a_{-s}^\dagger). \tag{5.78}$$

Equations (5.70) and (5.78) enable us to replace all local displacement and momentum coordinates by operators for the annihilation and creation of phonons. This completes the transformation into the so-called *phonon representation*, in which the state of a harmonic crystal is described in terms of the phonon occupation number of each of the normal modes. We will now use this representation to calculate the neutron scattering cross-section for one-phonon and multi-phonon processes.

### 5.2.4   One-phonon coherent scattering

From (5.39), (5.41), and (5.44) the one-phonon coherent scattering cross-section can be written

$$
\left(\frac{\mathrm{d}^2\sigma}{\mathrm{d}\Omega\mathrm{d}E_\mathrm{f}}\right)_\mathrm{coh}^{\mathrm{1\,ph}} = \frac{k_\mathrm{f}}{k_\mathrm{i}}\, N \sum_l \exp(\mathrm{i}\mathbf{Q}\cdot\mathbf{l}) \sum_{dd'} \bar{b}_{d'}^* \, \bar{b}_d \, \exp\{\mathrm{i}\mathbf{Q}\cdot(\mathbf{d}-\mathbf{d}')\}
$$

$$
\times \frac{1}{2\pi\hbar}\int_{-\infty}^{\infty} \langle UV\rangle \exp(-\mathrm{i}\omega t)\,\mathrm{d}t, \tag{5.79}
$$

where,

$$
\langle UV\rangle = \langle \mathbf{Q}\cdot\hat{\mathbf{u}}_{0d'}\, \mathbf{Q}\cdot\hat{\mathbf{u}}_{ld}(t)\rangle
$$

$$
= \frac{1}{N\sqrt{m_d m_{d'}}} \sum_{ss'} \mathbf{Q}\cdot(\mathbf{e}_{d's'})^* \, \mathbf{Q}\cdot\mathbf{e}_{ds} \exp(\mathrm{i}\mathbf{q}\cdot\mathbf{l}) \langle \hat{Q}_{s'}^\dagger \hat{Q}_s(t)\rangle.
$$

$$
\tag{5.80}
$$

Here we have exploited the fact that $\hat{\mathbf{u}}$ is a Hermitian operator ($\hat{\mathbf{u}} = \hat{\mathbf{u}}^\dagger$), and used (5.70) and (5.71) to express the result in terms of the normal mode correlation function. From (5.78), the latter is given by

$$
\langle \hat{Q}_{s'}^\dagger \hat{Q}_s(t)\rangle = \left(\frac{\hbar^2}{4\omega_s\omega_s'}\right)^{1/2} \langle\, (\hat{a}_{s'}^\dagger + \hat{a}_{-s'})\{\hat{a}_s(t) + \hat{a}_{-s}^\dagger(t)\}\,\rangle. \tag{5.81}
$$

For the reasons given in the discussion before eqn (C.65), only products comprising one creation and one annihilation operator contribute to the correlation function. Furthermore, only the terms with $s = s'$ survive because the normal modes are orthogonal. Hence,

$$
\langle \hat{Q}_s^\dagger \hat{Q}_s(t)\rangle = \left(\frac{\hbar}{2\omega_s}\right) \langle\, \hat{a}_s^\dagger \hat{a}_s(t) + \hat{a}_{-s}\hat{a}_{-s}^\dagger(t)\,\rangle
$$

$$
= \left(\frac{\hbar}{2\omega_s}\right) [\, n(\omega_s)\exp(-\mathrm{i}\omega_s t) + \{n(\omega_{-s}) + 1\}\exp(\mathrm{i}\omega_{-s}t)\,].
$$

$$
\tag{5.82}
$$

To reach the second line, the time dependence of the operators is written explicitly, according to (5.14) and (5.15), and the $\langle\hat{a}_s^\dagger\hat{a}_s\rangle$ and $\langle\hat{a}_{-s}\hat{a}_{-s}^\dagger\rangle$ correlation functions are then expressed in terms of the average occupancy of the modes by (5.21) and (5.22).

Now let us put together (5.79)–(5.82). Using (5.45) and (5.46) to write the time integral and lattice sum as $\delta$-functions (see also Sections B.2.3 and B.4.2), and the fact that $\omega_{-s} = \omega_s$, see eqn (5.67), we find

$$
\left(\frac{\mathrm{d}^2\sigma}{\mathrm{d}\Omega\mathrm{d}E_\mathrm{f}}\right)_\mathrm{coh}^{\mathrm{1\,ph}} = \frac{k_\mathrm{f}}{k_\mathrm{i}}\,\frac{(2\pi)^3}{v_0} \sum_\mathbf{G} \sum_s |G_s(\mathbf{Q})|^2
$$

$$
\times \frac{1}{2\omega_s} [\, n(\omega_s)\delta(\mathbf{Q}+\mathbf{q}-\mathbf{G})\delta(\omega+\omega_s)
$$

$$
+ \{n(\omega_s)+1\}\delta(\mathbf{Q}-\mathbf{q}-\mathbf{G})\delta(\omega-\omega_s)\,], \tag{5.83}
$$

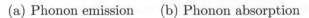
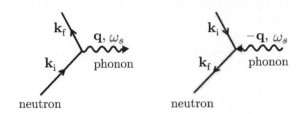

(a) Phonon emission    (b) Phonon absorption

**Fig. 5.7** Feynman diagrams to illustrate (a) phonon emission, and (b) phonon absorption processes in neutron scattering.

where $\mathbf{G}$ is a reciprocal lattice vector, and

$$G_s(\mathbf{Q}) = \sum_d \frac{\bar{b}_d}{\sqrt{m_d}}\,\mathbf{Q}\cdot\mathbf{e}_{ds}\,\exp(-W_d)\exp(i\mathbf{Q}\cdot\mathbf{d}) \tag{5.84}$$

is the *phonon structure factor* for mode $s$. The expression for $W_d$ is

$$W_d(\mathbf{Q}) = \frac{1}{2Nm_d}\sum_s \frac{\hbar}{2\omega_s}|\mathbf{Q}\cdot\mathbf{e}_{ds}|^2\coth(\beta\hbar\omega_s), \tag{5.85}$$

which comes from (5.43), (5.70) and (5.82).

The form of (5.83) warrants some interpretation. The $\delta$-functions mean that scattering only occurs when $\mathbf{Q}$ and $\omega$ jointly satisfy the conditions

$$\mathbf{Q} = \mathbf{G} \pm \mathbf{q} \quad \text{and} \quad \omega = \pm\omega_s, \tag{5.86}$$

where $\mathbf{q}$ is defined within the first Brillouin zone. The two conditions in (5.86) correspond to conservation of momentum[22] and energy in the scattering process. The conservation laws mean that on scattering, a neutron can emit a phonon of energy $\hbar\omega_s$ and wavevector $\mathbf{q}$, or can absorb a phonon of energy $\hbar\omega_s$ and wavevector $-\mathbf{q}$. The emission and absorption processes are represented diagrammatically in Fig. 5.7.

Since modes with $\mathbf{q}$ and $-\mathbf{q}$ have the same frequency ($\omega_s = \omega_{-s}$), if we were to scan the neutron energy transfer at a fixed $\mathbf{Q} = \mathbf{G} \pm \mathbf{q}$ then we would observe a pair of peaks, one at $\omega = -\omega_s$ and the other at $\omega = \omega_s$. The size of the neutron energy-gain peak ($\omega = -\omega_s$) is proportional to $n(\omega_s)$, and the size of the energy-loss peak ($\omega = \omega_s$) is proportional to $n(\omega_s) + 1$. Therefore, as noted in Section 5.1.3, the intensities of the energy-gain and energy-loss peaks are in the ratio $n(\omega_s)/\{n(\omega_s)+1\} = \exp(-\hbar\omega_s/k_BT)$ as prescribed by detailed balance. This means that when $k_BT \ll \hbar\omega_s$ the energy-gain peak is significantly smaller than the energy-loss peak, a consequence of the very few thermally activated phonons available from which neutrons can gain energy. At high temperatures ($k_BT \gg \hbar\omega_s$), $n(\omega_s)+1 \approx n(\omega_s) \approx k_BT/\hbar\omega_s \gg 1$. The energy-gain and energy-loss peaks are then almost the same size, and both are much larger than the energy-loss peak at low temperatures. Hence, a phonon measurement is often performed at elevated temperatures where the phonon population is large. This tactic must be used with care, however, since normal mode frequencies and widths can vary significantly with temperature.

[22]Note that no momentum is actually associated with the phonon itself. All of the momentum $\hbar\mathbf{Q}$ transferred from the neutron is absorbed by the crystal as a whole plus any objects attached to the crystal. The process is analogous to what happens when a pendulum is struck from the side — after transients have died down, all the momentum from the initial impulse is transmitted via the point of attachment to the rigid structure supporting the pendulum and then, if the structure is fixed to the ground, absorbed by the Earth. The average momentum of the oscillating pendulum itself is zero. The quantity $\hbar\mathbf{q}$ is in many ways analogous to a momentum, and is often called the *quasi-momentum* or *crystal momentum*, but in phonon collision processes quasi-momentum is conserved only to within an additive vector $\hbar\mathbf{G}$ — see eqn (5.86)

As well as ensuring conservation of momentum, the δ-function involving **Q** in (5.83) also contains the fact that the normal-mode dispersion relation is periodic in reciprocal space. This means that a particular normal mode $s = \mathbf{q}j$ can be measured in any Brillouin zone. It simply requires that **Q** differs from the reciprocal lattice vector **G** at the centre of the Brillouin zone by an amount **q**. The measurement of the same phonon in different Brillouin zones is illustrated in Fig. 5.8.

Reciprocal space periodicity allows us considerable experimental flexibility, but in planning a measurement one must be aware that the one-phonon scattering intensity can vary significantly from one Brillouin zone to another. This variation is described by the squared modulus of the phonon structure factor $G_s(\mathbf{Q})$ defined in eqn (5.84). The value of $G_s(\mathbf{Q})$ depends on three factors:

(i) The scalar product $\mathbf{Q} \cdot \mathbf{e}_{ds}$. This gives rise to a $Q^2$ variation in the scattering intensity, and so it is generally an advantage to perform measurements at large $Q$.[23] The scalar product also depends on the direction of **Q**, and one should ideally choose a Brillouin zone in which the direction of **Q** is predominantly parallel to the atomic displacements for the particular mode under investigation. This will depend on which branch of the dispersion relation the mode belongs.

(ii) The Debye–Waller factor, $\exp(-W_d)$. As discussed in Section 5.1.4, $\exp(-W_d)$ decreases with $Q$ and so moderates the aforementioned $Q^2$ variation. The effect can be significant for light atoms.

(iii) The phase factor, $\exp(i\mathbf{Q} \cdot \mathbf{d})$. This causes interference between waves scattered from atoms at different sites in the unit cell. It depends on both the magnitude and direction of **Q**, and on the crystal structure.

One can readily appreciate from this discussion that for crystals with more than one atom in the unit cell it is a great advantage to have a crystal dynamics model to predict how the scattering intensity from each phonon branch varies among the Brillouin zones.

Finally, it is instructive to consider the origin of the $1/\omega_s$ factor in the one-phonon cross-section (5.83). We note from (5.79) and (5.80) that the one-phonon scattering cross-section is proportional to the square of the amplitude of the normal mode coordinate. Considering normal modes as harmonic oscillators, we see from (5.19) and (5.21) that the mean vibrational energy in a mode $s$ is given by

$$\langle E_{n_s} \rangle = (\langle n_s \rangle + \tfrac{1}{2})\hbar\omega_s$$

For a harmonic oscillator $\langle E_{n_s} \rangle$ is also proportional to $\omega_s^2 \langle x_{n_s}^2 \rangle$. Therefore, for a fixed $\langle n_s \rangle$,

$$\langle x_{n_s}^2 \rangle \propto \frac{1}{\omega_s}.$$

This simple analysis reveals that the $1/\omega_s$ factor traces back to the fact that the higher frequency modes of vibration in a crystal have a smaller amplitude of vibration for a given phonon occupancy than the lower frequency modes.

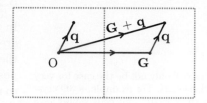

**Fig. 5.8** The same phonon mode with wavevector **q** can be measured in different Brillouin zones displaced by **G**.

[23] The $Q^2$ variation has the same physical origin as that in the diffraction from a static, sinusoidally modulated structure considered in Section 2.2.9.

## 5.2.5  Scattering intensity of a phonon

In the one-phonon scattering cross-section (5.83) the vibrational modes are represented by a set of discrete wavevectors $\mathbf{q}$. In reality, however, the allowed values of $\mathbf{q}$ are generally too close together to admit any possibility of resolving individual modes.[24] It is more practical, therefore, to represent the dispersion relation as a continuous function. This means that the summations over $\mathbf{G}$ and $\mathbf{q}$ can be replaced by an integral over the density of states,

$$\sum_{\mathbf{G}} \sum_{\mathbf{q}} \to \frac{V_0}{(2\pi)^3} \int d^3\mathbf{k},$$

where $\mathbf{k} = \mathbf{G} \pm \mathbf{q}$ spans all of reciprocal space, and the factor $V_0/(2\pi)^3 = Nv_0/(2\pi)^3$ is the density of states in three-dimensional wavevector space — see eqn (3.7). After integration over $\mathbf{k}$, the one-phonon scattering cross-section may be written

$$\left(\frac{d^2\sigma}{d\Omega dE_{\mathrm{f}}}\right)_{\mathrm{coh}}^{1\,\mathrm{ph}} = \frac{k_{\mathrm{f}}}{k_{\mathrm{i}}} N \sum_j |G_s(\mathbf{Q})|^2$$

$$\times \frac{1}{2\omega_s} \left[ n(\omega_s)\delta(\omega + \omega_s) + \{n(\omega_s) + 1\}\delta(\omega - \omega_s) \right]. \quad (5.87)$$

The summation is over the phonon branches but not any longer over $\mathbf{q}$. Instead, the phonon eigenvalues and eigenvectors are to be regarded as continuous and extended functions in wavevector space, with the periodicity of the reciprocal lattice. Therefore, $\omega_s$ now stands for $\omega_j(\mathbf{Q})$, and $\mathbf{e}_{ds}$ stands for $\mathbf{e}_{dj}(\mathbf{Q})$. The normal mode wavevector $\mathbf{q}$ is still given by $\mathbf{Q} = \mathbf{G} + \mathbf{q}$.

A traditional way to measure a phonon is to scan the neutron energy transfer $\hbar\omega$ across the phonon branch at a fixed value of $\mathbf{Q}$ (a constant-$\mathbf{Q}$ scan). If the data are corrected for the $k_{\mathrm{f}}/k_{\mathrm{i}}$ factor in the double differential cross-section then the integrated intensity $\mathcal{I}_{\mathrm{ph}}(\mathbf{Q})$ of the phonon peak (integrated over energy) is given by

$$\mathcal{I}_{\mathrm{ph}}(\mathbf{Q}) = N \frac{\hbar}{2\omega_s} |G_s(\mathbf{Q})|^2 \times \begin{cases} n(\omega_s) & \text{neutron energy-gain peak,} \\ n(\omega_s) + 1 & \text{neutron energy-loss peak.} \end{cases}$$
$$(5.88)$$

The factor of $\hbar$ comes from identity (B.8): $\delta(\omega - \omega_s) = \hbar\delta(\hbar\omega - \hbar\omega_s)$.

## 5.2.6  Absolute calibration from an acoustic phonon

The integrated intensity $\mathcal{I}_{\mathrm{ph}}(\mathbf{Q})$ depends on the directions, amplitudes and relative phases of the individual atomic vibrations through the polarization vectors contained in $G_s(\mathbf{Q})$, eqn (5.84). These typically vary with $\mathbf{Q}$ in a complicated way which needs to be calculated from a crystal dynamics model. One notable exception is the $\mathbf{q} \to 0$ limit of acoustic phonons, for which the phonon intensity simplifies considerably as we shall now show.

[24]This will only not be the case for very small crystals. For example, a 3D crystal with $N = 10^6$ unit cells would have about 50 $\mathbf{q}$ points along a line from the centre of the Brillouin zone to a Brillouin zone boundary. The linear dimension of a crystal of this size would typically be less than $100\,\mathrm{nm}$.

For acoustic modes, as $\mathbf{q} \to 0$ all the atoms vibrate in the same direction, and with the same amplitude and phase. This is shown explicitly for the diatomic chain in eqn (5.58). The amplitude of vibration of atom $d$ is proportional to $\mathbf{e}_{ds}/\sqrt{m_d}$, so for an acoustic mode the factor $(\mathbf{Q} \cdot \mathbf{e}_{ds})/\sqrt{m_d}$ which appears in $G_s(\mathbf{Q})$ is independent of $d$ and can be taken outside the summation in (5.84). If $\psi$ is the angle between $\mathbf{Q}$ and $\mathbf{e}_{ds}$, then

$$\left| \frac{\mathbf{Q} \cdot \mathbf{e}_{ds}}{\sqrt{m_d}} \right|^2 = \frac{Q^2 |\mathbf{e}_{ds}|^2 \cos^2 \psi}{m_d}$$

$$= \frac{Q^2 \cos^2 \psi}{M}, \tag{5.89}$$

where $M = \sum_d m_d$ is the total mass of atoms in the unit cell. To reach the second line of (5.89), multiply both sides of the first line by $m_d$ and sum over $d$. Note that $\sum_d |\mathbf{e}_{ds}|^2 = 1$ from the normalization of the eigenvectors as defined in (5.65) and (5.68). Hence, for a neutron energy-loss measurement of an acoustic phonon,

$$\lim_{\mathbf{q} \to 0} |G_s(\mathbf{Q})|^2 = \frac{|\mathbf{G}|^2 \cos^2 \psi}{M} \left| \sum_d \bar{b}_d \exp(-W_d) \exp(i\mathbf{G} \cdot \mathbf{d}) \right|^2$$

$$= \frac{|\mathbf{G}|^2 \cos^2 \psi}{M} |F_N(\mathbf{G})|^2, \tag{5.90}$$

where $F_N(\mathbf{G})$ is the nuclear structure factor of the Bragg peak at $\mathbf{Q} = \mathbf{G}$, eqn (5.48). Putting (5.90) into (5.88) we obtain,

$$\lim_{\mathbf{q} \to 0} \mathcal{I}_{\mathrm{ph}}(\mathbf{Q}) = N \frac{\hbar}{2\omega_s} \frac{|\mathbf{G}|^2 \cos^2 \psi}{M} |F_N(\mathbf{G})|^2 \{n(\omega_s) + 1\}. \tag{5.91}$$

Eqn (5.91) may be used to calibrate a spectrometer. One simply compares the measured integrated intensity of a small-$\mathbf{q}$ acoustic phonon with the value calculated from (5.91) to obtain the calibration constant. The method does not depend on a model for the crystal dynamics; only the crystal structure and the phonon frequency are required.

## 5.2.7 Longitudinal and transverse modes

In general, there is no simple relationship between the directions of the atomic displacements, represented by the $\mathbf{e}_{ds}$ vectors, and the direction of $\mathbf{q}$. If, however, all the atoms should happen to be vibrating along a direction parallel to $\mathbf{q}$ then this is called a *longitudinal mode*, and if all the atoms are vibrating perpendicularly to $\mathbf{q}$ then this is a *transverse mode*. These special cases can arise when $\mathbf{q}$ is along a high-symmetry direction in the crystal.

Take copper, for example, which has a cubic structure. The phonon dispersion relations along two of the high-symmetry directions are shown in Fig. 5.9. For $\mathbf{q} \parallel (100)$, the $\mathbf{e}_{ds}$ vectors are either all parallel to $\mathbf{q}$ (longitudinal) or all perpendicular to $\mathbf{q}$ (transverse). The transverse modes

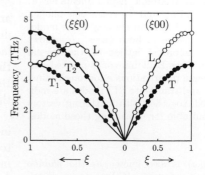

**Fig. 5.9** Phonon dispersion relations in copper along two high-symmetry directions. The two transverse modes are degenerate along $(\xi 00)$. The data and fits are from Svensson *et al.* (1967).

### (a) Longitudinal

### (b) Transverse

### (c)

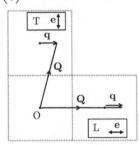

**Fig. 5.10** Atomic displacements for (a) longitudinal and (b) transverse phonon modes. The wavevector **q** of the modes is the same in both cases. (c) shows the reciprocal lattice of the crystal with three Brillouin zones marked out. Phonon polarizations **e** for longitudinal (L) and transverse (T) modes are indicated, together with scattering vectors **Q** suitable for measuring these modes.

[25] $g(\omega)$ is also known as the *phonon density of states*. If there are $N$ primitive unit cells in the crystal and $p$ atoms in the basis, then

$$\int_0^\infty g(\omega)\,\mathrm{d}\omega = 3Np.$$

are doubly degenerate because the two orthogonal directions for the polarization vectors, e.g. [010] and [001], are equivalent in cubic symmetry. For **q** ∥ (110), the polarization directions of the two transverse modes are not equivalent so the $T_1$ and $T_2$ modes are not degenerate.

The measurement of longitudinal and transverse modes provides useful insight into the way mode polarization affects the one-phonon scattering cross-section. A simple example is illustrated in Fig. 5.10. The crystal has a square lattice, and panels (a) and (b) show normal modes propagating in the (horizontal) $x$ direction with longitudinal and transverse polarizations, respectively. Both modes have the same **q**. If we want to obtain the maximum signal then the $\mathbf{Q} \cdot \mathbf{e}_{ds}$ factor in the one-phonon structure factor (5.84) dictates that we should choose the direction of **Q** parallel to the atomic displacements. With **q** ∥ $x$, if we choose **Q** ∥ $x$ then we obtain maximum scattering from the longitudinal mode and zero scattering from the transverse mode. Conversely, we should choose **Q** ∥ $y$ to measure the transverse mode **q** ∥ $x$. Examples of suitable configurations are shown in Fig. 5.10(c). Note that if **Q** is in the first Brillouin zone, i.e. $\mathbf{G} = 0$, then **Q** is always parallel to **q** and so it is impossible to measure transverse modes.

### 5.2.8  One-phonon incoherent scattering

To obtain the one-phonon incoherent cross-section we follow the same procedure as in Section 5.2.4, except that we start from (5.40) instead of (5.39) and we set $U = -\mathrm{i}\mathbf{Q} \cdot \mathbf{u}_d$ and $V = \mathrm{i}\mathbf{Q} \cdot \mathbf{u}_d(t)$. The result is

$$\left(\frac{\mathrm{d}^2\sigma}{\mathrm{d}\Omega\mathrm{d}E_\mathrm{f}}\right)_\mathrm{inc}^{1\,\mathrm{ph}} = \frac{k_\mathrm{f}}{k_\mathrm{i}} \sum_s \sum_d \frac{(\sigma_\mathrm{inc})_d}{4\pi} \frac{1}{m_d} |\mathbf{Q} \cdot \mathbf{e}_{ds}|^2 \exp(-2W_d)$$

$$\times \frac{1}{2\omega_s}\left[n(\omega_s)\delta(\omega+\omega_s) + \{n(\omega_s)+1\}\delta(\omega-\omega_s)\right]. \quad (5.92)$$

In contrast to the one-phonon coherent cross-section [eqn (5.83)], the one-phonon incoherent cross-section does not contain a $\delta$-function in **Q**, and neither does it contain any interference term due to the positions of different atoms in the unit cell, as appears in the one-phonon structure factor for coherent scattering, eqn (5.84). Hence, for a given energy the scattering varies continuously with **Q**, crossing smoothly from one Brillouin zone to another, and exhibits only a weak **Q** dependence from the polarization and Debye–Waller factors.

All modes having angular frequency $\omega$ contribute to the one-phonon incoherent scattering at a given **Q** and $\omega$, and for this reason it is useful to recast eqn (5.92) in terms of the frequency distribution of the normal modes. Specifically, we introduce the *vibrational density of states*[25] function $g(\omega)$, defined such that $g(\omega)\mathrm{d}\omega$ is the number of normal modes in the angular frequency range from $\omega$ to $\omega + \mathrm{d}\omega$. We can now replace the summation over the normal modes in (5.92) by an integral over $g(\omega)$,

$$\sum_s \cdots \rightarrow \int \cdots g(\omega)\,\mathrm{d}\omega,$$

so that[26]

$$\left(\frac{\mathrm{d}^2\sigma}{\mathrm{d}\Omega \mathrm{d}E_\mathrm{f}}\right)^{1\,\mathrm{ph}}_{\mathrm{inc}} = \frac{k_\mathrm{f}}{k_\mathrm{i}} \sum_d \frac{(\sigma_{\mathrm{inc}})_d}{4\pi} \frac{1}{m_d} \exp(-2W_d) \int_{-\infty}^{\infty} g(\omega') \langle |\mathbf{Q}\cdot\mathbf{e}_{ds}|^2 \rangle_{\omega'}$$

$$\times \frac{1}{2\omega'}\left[n(\omega')\delta(\omega+\omega') + \{n(\omega')+1\}\delta(\omega-\omega')\right]\mathrm{d}\omega'. \quad (5.93)$$

Here, $\langle |\mathbf{Q}\cdot\mathbf{e}_{ds}|^2\rangle_{\omega'}$ is the value of $|\mathbf{Q}\cdot\mathbf{e}_{ds}|^2$ averaged over all modes with angular frequency $\omega'$. The integration is routine, and the identity

$$n(-\omega) + 1 = -n(\omega), \quad (5.94)$$

which follows from (5.21) and (5.22), allows us to combine the energy-gain and energy-loss terms into one. Hence,

$$\left(\frac{\mathrm{d}^2\sigma}{\mathrm{d}\Omega \mathrm{d}E_\mathrm{f}}\right)^{1\,\mathrm{ph}}_{\mathrm{inc}} = \frac{k_\mathrm{f}}{k_\mathrm{i}} \sum_d \frac{(\sigma_{\mathrm{inc}})_d}{4\pi} \frac{1}{m_d} \exp(-2W_d) \langle |\mathbf{Q}\cdot\mathbf{e}_{ds}|^2\rangle_\omega$$

$$\times \frac{g(\omega)}{2\omega}\{n(\omega)+1\}. \quad (5.95)$$

As an example, consider the simplest possible system, a cubic crystal with one atom of mass $m$ per primitive unit cell. For this case it can be shown that $\langle |\mathbf{Q}\cdot\mathbf{e}_s|^2\rangle_\omega = \frac{1}{3}Q^2$, so that

$$\left(\frac{\mathrm{d}^2\sigma}{\mathrm{d}\Omega \mathrm{d}E_\mathrm{f}}\right)^{1\,\mathrm{ph}}_{\mathrm{inc}} = \frac{k_\mathrm{f}}{k_\mathrm{i}} \frac{\sigma_{\mathrm{inc}}}{4\pi} \frac{Q^2}{3m} \exp(-2W) \frac{g(\omega)}{2\omega}\{n(\omega)+1\},$$

and the phonon density of states can then be obtained directly from the one-phonon incoherent scattering cross-section. Figure 5.11 shows the phonon density of states of vanadium measured this way. Vanadium is a body-centred cubic metal and an almost pure incoherent scatterer.

It will always be preferable to learn about the vibrational dynamics of a material from the phonon dispersion relations rather than from the density of states. However, systematic measurements of phonon frequencies at many points in the Brillouin zone and the subsequent modelling can be time-consuming. Moreover a single crystal is required for such measurements, and sometimes a suitable crystal is simply not available. In these circumstances, some useful information can be obtained relatively quickly with a polycrystalline sample. A polycrystalline sample contains many small, randomly aligned, single-crystal grains, so the scattering is averaged over all orientations. If one can neglect the anisotropy in the Debye–Waller factor then after orientational averaging the factor $\langle |\mathbf{Q}\cdot\mathbf{e}_{ds}|^2\rangle_\omega$ becomes $\frac{1}{3}Q^2\langle |\mathbf{e}_{ds}|^2\rangle_\omega$, and the one-phonon incoherent cross-section is given by

$$\left(\frac{\mathrm{d}^2\sigma}{\mathrm{d}\Omega \mathrm{d}E_\mathrm{f}}\right)^{1\,\mathrm{ph}}_{\mathrm{inc}} \approx \frac{k_\mathrm{f}}{k_\mathrm{i}} \sum_d \frac{(\sigma_{\mathrm{inc}})_d}{4\pi} \frac{Q^2}{3m_d} \exp(-2W_d) \frac{g_d(\omega)}{2\omega}\{n(\omega)+1\}. \quad (5.96)$$

The function

$$g_d(\omega) = g(\omega)\langle |\mathbf{e}_{ds}|^2\rangle_\omega \quad (5.97)$$

is the *partial vibrational density of states*, which measures how much the motion of atom $d$ contributes towards the density of states.[27]

[26]For negative values of $\omega$ we define $g(-\omega) = g(\omega)$.

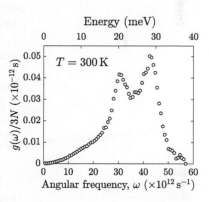

**Fig. 5.11** Phonon density of states of vanadium measured at room temperature. (After Delaire *et al.*, 2008, and Kresch, 2009.)

[27]From (5.65) and (5.68) it follows that $\sum_d |\mathbf{e}_{ds}|^2 = 1$, so $\sum_d g_d(\omega) = g(\omega)$.

### 5.2.9   Incoherent approximation

The preceding section did not touch upon how to separate the incoherent from the coherent scattering when both are present, as is usually the case. One possibility is to use polarization analysis, since a proportion of the neutron spins flip during nuclear spin incoherent scattering but not during nuclear coherent scattering[28] (see Section 4.5.1).

Alternatively, the need to isolate the incoherent scattering can be avoided if the measurements are performed at large scattering vector. When $Q \gg 2\pi/d$, where $d$ is the nearest-neighbour distance, one can ignore interatomic interference effects and treat each atom as scattering independently of the others. This is because coherence effects are washed out at large $Q$ due to the imperfect $Q$ resolution of any real measurement. As a result, the $j \neq k$ terms in the coherent cross-section eqn (5.1) can be neglected. Hence, at large $Q$, the nuclear scattering is described by the same expressions as for incoherent scattering except that $(\sigma_{\mathrm{inc}})_d$ is replaced by the single-atom total scattering cross-section $\sigma_{\mathrm{d}} = (\sigma_{\mathrm{coh}})_d + (\sigma_{\mathrm{inc}})_d$. This is called the *incoherent approximation*.

The incoherent approximation can be deployed to interpret one-phonon scattering at large $Q$ from polycrystalline samples. On a sample containing two or more different atomic species such a measurement does not yield the total phonon density of states, as is clear from (5.96). Rather, it gives an average over the partial densities of states weighted by several species-dependent factors. It is usual to introduce a *generalized phonon density of states* $G(\omega)$, defined by

$$G(\omega) = \sum_d \frac{\sigma_d}{4\pi} \frac{1}{m_d} \exp(-2W_d)\, g_d(\omega), \tag{5.98}$$

which can be calculated from a crystal dynamics model and compared with the results of a measurement via (5.96):

$$G(\omega) \propto \frac{\omega}{Q^2} \frac{1}{n(\omega)+1} \frac{k_{\mathrm{i}}}{k_{\mathrm{f}}} \left( \frac{\mathrm{d}^2\sigma}{\mathrm{d}\Omega\mathrm{d}E_{\mathrm{f}}} \right)^{1\,\mathrm{ph}}. \tag{5.99}$$

As the single-atom cross-sections appear as weighting factors in the definition (5.98) of $G(\omega)$ one can, in principle, deduce the partial vibrational density of states for each atomic species by performing a series of measurements on samples with different isotopic compositions. There is also an inverse proportionality to the atomic mass, so light atoms with large total cross-sections, such as hydrogen, will give strong scattering and will dominate whenever there is a peak in the corresponding $g_d(\omega)$. Therefore, a consideration of the ratios $\sigma_d/m_d$ can be a useful guide to the interpretation of features in the phonon spectrum.

As an example, the measured and calculated $G(\omega)$ for vanadium pentoxide ($V_2O_5$) is shown in Figure 5.12. The weighting factors $\sigma_d/m_d$ in $\mathrm{b\,amu^{-1}}$ are V: 0.100, O: 0.264, which means that $G(\omega)$ is dominated by oxygen vibrations. An *ab initio* model makes it possible to identify the underlying atomic displacements associated with features in the spectrum. For example, the peak centred on 124 meV derives from stretching modes of the shortest V–O bonds.

[28]This method only works in the absence of electronic magnetism, which also causes spin-flip scattering.

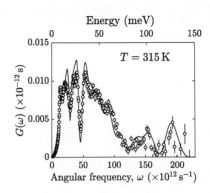

**Fig. 5.12** Generalized phonon density of states of $V_2O_5$ normalized to unity. The circles are experimental data, and the line is an *ab initio* density functional theory calculation. (Singh *et al.*, 2017. Data courtesy of R. Mittal.)

## 5.2.10   Multi-phonon scattering

Multi-phonon scattering occurs when the neutron couples to two or more phonons. It can be calculated from the higher-order terms in the expansion of $\exp\langle UV \rangle$ in eqn (5.44). Since $n$-phonon scattering derives from the $\langle UV \rangle^n$ term in the series, it follows that the $n$-phonon cross-sections vary in proportion to $Q^{2n}$. Hence, as $Q$ increases, higher-order processes become increasingly more significant.

Multi-phonon processes satisfy conservation laws analogous to (5.86). For example, the conditions for coherent two-phonon scattering are

$$\mathbf{Q} = \mathbf{G} \pm \mathbf{q}_1 \pm \mathbf{q}_2 \quad \text{and} \quad \omega = \pm\omega_{s_1} \pm \omega_{s_2}, \qquad (5.100)$$

The incoherent two-phonon cross-section has the same condition on energy but does not have the wavevector conservation law.

From the second of (5.100), it follows that if $\omega_{\max}$ is the highest angular frequency in the vibrational spectrum then two-phonon scattering will be observed up to $\omega = 2\omega_{\max}$. When $\omega \leq 2\omega_{\max}$ it is always possible to find combinations of two normal modes that satisfy (5.100). It follows that coherent two-phonon scattering occurs everywhere in reciprocal space and does not exhibit sharp peaks as a function of $\mathbf{Q}$ and $\omega$. There can, however, be enhancements in the coherent two-phonon scattering at certain points in $\mathbf{Q}$–$\omega$ space if a particularly large number of pairs of modes satisfy (5.100) at those points.

Generally, multi-phonon scattering appears as a relatively small and smoothly varying 'background' under the scattering of interest. Sometimes, however, it is useful to have an estimate of the multi-phonon scattering. A serious hindrance is that the computational effort required to calculate $n$-phonon scattering increases dramatically with $n$. Because multi-phonon scattering tends to be most important at large $Q$ the incoherent approximation (see Section 5.2.9) is usually invoked. Several approximate methods have been devised, and a discussion of these can be found in Marshall and Lovesey (1971).

## 5.2.11   Phonon damping

The one-phonon coherent and incoherent cross-sections (5.87) and (5.92) contain $\delta$-functions in $\omega$. This is because within the harmonic approximation the vibrational normal modes are non-interacting and have infinite lifetime. In real materials, however, there are a number of different mechanisms by which phonons can interact with their environment, causing spectral lineshapes to shift in frequency and broaden (phonon damping). Possible sources of damping include scattering from static disorder, phonon–phonon interactions due to anharmonicity, electron–phonon coupling in metals, and scattering from magnetic excitations. By studying how the damping varies throughout the Brillouin zone one can conduct a detailed examination of these mechanisms, which are relevant for understanding the thermal and transport properties of solids.

Providing the broadening is small compared with the phonon frequency one can retain the concept of quasiparticles and describe the

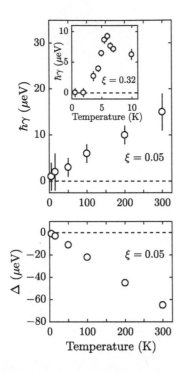

**Fig. 5.13** Phonon damping in lead measured by the NRSE technique implemented on a triple-axis spectrometer. Upper panel: linewidth, $\hbar\gamma$. Lower panel: peak shift, $\Delta$. The phonon modes are all on the $(\xi\xi0)\ T_1$ branch as indicated. The inset shows the effect of superconductivity on the low temperature linewidth. (Adapted from Habicht *et al.*, 2004, and Keller *et al.*, 2006.)

damping in terms of a generalized susceptibility of the form given in eqn (3.80), whose dissipative part is the difference of two Lorentzian functions, eqn (3.81). Practically, this means that the terms in square brackets $[\cdots]$ on the second lines of (5.87) and (5.92)–(5.93) are replaced by

$$\{n(\omega) + 1\}\left\{\frac{\gamma_s/\pi}{(\omega - \omega_s - \Delta_s)^2 + \gamma_s^2} - \frac{\gamma_s/\pi}{(\omega + \omega_s + \Delta_s)^2 + \gamma_s^2}\right\}.$$

The parameters $\Delta_s$ and $\gamma_s$ correspond to the frequency shift and broadening of a mode $s$ whose bare frequency (i.e. unrenormalized by interactions) is $\omega_s$. The peaks described by this lineshape have full width at half maximum of approximately $2\gamma_s$. The reciprocal half width $1/\gamma_s$ can be interpreted as the lifetime of the phonon. When $\Delta_s$ and $\gamma_s$ are zero, the lineshape reverts back to the undamped form.

Extraction of intrinsic phonon linewidths and frequency shifts is complicated by resolution (see Section 10.1.6). The neutron resonance spin echo (NRSE) technique has been developed specifically to enable the line widths of dispersive excitations to be determined with $\mu$eV energy resolution over the entire Brillouin zone (see the article by Keller *et al.* in Mezei *et al.*, 2003). Figure 5.13 shows phonon damping in lead as a function of temperature. The measurements were made by the NRSE technique. In general, the phonons broaden and soften with increasing temperature, but at low temperatures the behaviour is anomalous because lead is a superconductor. Below the superconducting transition temperature, which for lead is $T_c = 7.2\,\mathrm{K}$, a gap opens in the electronic density of states. Upon cooling, the damping of a phonon whose energy is smaller than the zero-temperature gap first increases and then decreases (see inset to Fig. 5.13). The increase is because there is a pile-up of electronic states either side of the gap, so when the phonon energy matches the gap energy there is more phase space for scattering than in the normal state. The subsequent decrease in damping at lower temperatures is because when the gap is larger than the phonon energy the phonon cannot excite electrons across the gap.

## 5.3    Correlation and response functions

As we have seen in the preceding sections, the excitations of systems with quadratic potentials are long-lived normal mode vibrations (harmonic waves in the case of a periodic crystal), and the scattering can be calculated directly from a model for the inter-atomic potentials. In materials with more complex dynamics, such as liquids, the excitations do not have such a simple interpretation and we need a more general approach. In Chapter 2, we saw how the static structure of disordered materials can be described by pair correlation functions, and we showed that the nuclear elastic scattering can be expressed in terms of the Fourier transform of these functions. Here, we show how this formalism can be generalized to describe inelastic scattering from mobile nuclei.

## 5.3.1   One-component systems

The time-dependent pair correlation function for a system composed of one type of scattering particle is defined by

$$G(\mathbf{r}, t) = \frac{1}{N} \int \langle \hat{n}(\mathbf{r}', 0) \hat{n}(\mathbf{r}' + \mathbf{r}, t) \rangle \, \mathrm{d}^3 \mathbf{r}', \qquad (5.101)$$

where

$$\hat{n}(\mathbf{r}, t) = \sum_j \delta\{\mathbf{r} - \hat{\mathbf{r}}_j(t)\} \qquad (5.102)$$

is the time-dependent particle density operator. Here, $\hat{\mathbf{r}}_j(t)$ is the Heisenberg operator representing the displacement of the $j$th nucleus. $G(\mathbf{r}, t)$ describes the correlation between the density at position $\mathbf{r}' + \mathbf{r}$ at time $t$ and the density at position $\mathbf{r}'$ at time $t = 0$, averaged over $\mathbf{r}'$ and over all configurations of the system.

We now show how $G(\mathbf{r}, t)$ is experimentally accessible to neutron inelastic scattering. We restrict the analysis to unpolarized neutrons. Because we are dealing with a single scattering species it will be convenient to write the double differential scattering cross-sections in the form

$$\left( \frac{\mathrm{d}^2 \sigma}{\mathrm{d}\Omega \mathrm{d}E_\mathrm{f}} \right)_\mathrm{coh} = \frac{k_\mathrm{f}}{k_\mathrm{i}} N |\bar{b}|^2 S(\mathbf{Q}, \omega) \qquad (5.103)$$

and

$$\left( \frac{\mathrm{d}^2 \sigma}{\mathrm{d}\Omega \mathrm{d}E_\mathrm{f}} \right)_\mathrm{inc} = \frac{k_\mathrm{f}}{k_\mathrm{i}} N \frac{\sigma_\mathrm{inc}}{4\pi} S_\mathrm{i}(\mathbf{Q}, \omega), \qquad (5.104)$$

where $S(\mathbf{Q}, \omega)$ is the dynamical response function for coherent scattering, and $S_\mathrm{i}(\mathbf{Q}, \omega)$ is the equivalent function for incoherent scattering. The response functions defined in (5.103) and (5.104) do not contain the nuclear scattering lengths, and therefore differ dimensionally from those used elsewhere. To be able to separate the scattering lengths from the response functions we have assumed that there is no correlation between the nuclear and target states and that the distribution of isotopes and nuclear spin orientations is random. The validity of these assumptions was discussed in Section 4.1.4.

From (5.1), the coherent response function is given by

$$S(\mathbf{Q}, \omega) = \frac{1}{2\pi\hbar} \frac{1}{N} \sum_{jk} \int \langle \exp(-\mathrm{i}\mathbf{Q} \cdot \hat{\mathbf{r}}_j) \exp\{\mathrm{i}\mathbf{Q} \cdot \hat{\mathbf{r}}_k(t)\} \rangle \exp(-\mathrm{i}\omega t) \, \mathrm{d}t,$$

$$(5.105)$$

where $\hat{\mathbf{r}}_j \equiv \hat{\mathbf{r}}_j(0)$. Now, from (5.102),

$$\sum_j \exp\{\mathrm{i}\mathbf{Q} \cdot \hat{\mathbf{r}}_j(t)\} = \int \hat{n}(\mathbf{r}, t) \exp(\mathrm{i}\mathbf{Q} \cdot \mathbf{r}) \, \mathrm{d}^3 \mathbf{r} \equiv \hat{n}(\mathbf{Q}, t), \qquad (5.106)$$

so (5.105) becomes[29]

$$S(\mathbf{Q}, \omega) = \frac{1}{2\pi\hbar} \frac{1}{N} \int \langle \hat{n}^\dagger(\mathbf{Q}, 0) \hat{n}(\mathbf{Q}, t) \rangle \exp(-\mathrm{i}\omega t) \, \mathrm{d}t. \qquad (5.107)$$

[29]From (5.106), $\hat{n}^\dagger(\mathbf{Q}, t) = \hat{n}(-\mathbf{Q}, t)$ since $\hat{\mathbf{r}}_j(t)$ is a Hermitian operator.

We can now write $S(\mathbf{Q}, \omega)$ in terms of $G(\mathbf{r}, t)$, because from (5.106,)

$$\langle \hat{n}^\dagger(\mathbf{Q}, 0) \hat{n}(\mathbf{Q}, t) \rangle$$

$$= \int \int \langle \hat{n}(\mathbf{r}', 0) \exp(-i\mathbf{Q} \cdot \mathbf{r}') \hat{n}(\mathbf{r}'', t) \exp(i\mathbf{Q} \cdot \mathbf{r}'') \rangle \, \mathrm{d}^3\mathbf{r}' \, \mathrm{d}^3\mathbf{r}''$$

$$= \int \int \langle \hat{n}(\mathbf{r}', 0) \hat{n}(\mathbf{r}' + \mathbf{r}, t) \rangle \exp(i\mathbf{Q} \cdot \mathbf{r}) \, \mathrm{d}^3\mathbf{r}' \, \mathrm{d}^3\mathbf{r} \quad [\mathbf{r}'' = \mathbf{r}' + \mathbf{r}]$$

$$= N \int G(\mathbf{r}, t) \exp(i\mathbf{Q} \cdot \mathbf{r}) \, \mathrm{d}^3\mathbf{r}. \tag{5.108}$$

Hence,

$$S(\mathbf{Q}, \omega) = \frac{1}{2\pi\hbar} \int \int G(\mathbf{r}, t) \exp\{i(\mathbf{Q} \cdot \mathbf{r} - \omega t)\} \, \mathrm{d}^3\mathbf{r} \, \mathrm{d}t. \tag{5.109}$$

This shows that $S(\mathbf{Q}, \omega)$ and $G(\mathbf{r}, t)$ are related by space and time Fourier transformation. From the inverse Fourier transform relations (Section B.3.1),

$$G(\mathbf{r}, t) = \frac{\hbar}{(2\pi)^3} \int \int S(\mathbf{Q}, \omega) \exp\{-i(\mathbf{Q} \cdot \mathbf{r} - \omega t)\} \, \mathrm{d}^3\mathbf{Q} \, \mathrm{d}\omega. \tag{5.110}$$

Hence, $G(\mathbf{r}, t)$ is obtainable from a measurement of the coherent scattering cross-section. An alternative expression for $G(\mathbf{r}, t)$, derived from (5.105) and (5.110), is

$$G(\mathbf{r}, t) = \frac{1}{(2\pi)^3} \frac{1}{N} \sum_{jk} \int \langle \exp(-i\mathbf{Q} \cdot \hat{\mathbf{r}}_j) \exp\{i\mathbf{Q} \cdot \hat{\mathbf{r}}_k(t)\} \rangle \exp(-i\mathbf{Q} \cdot \mathbf{r}) \, \mathrm{d}^3\mathbf{Q}. \tag{5.111}$$

In the same spirit, the response function for incoherent scattering,

$$S_\mathrm{i}(\mathbf{Q}, \omega) = \frac{1}{2\pi\hbar} \frac{1}{N} \sum_j \int \langle \exp(-i\mathbf{Q} \cdot \hat{\mathbf{r}}_j) \exp\{i\mathbf{Q} \cdot \hat{\mathbf{r}}_j(t)\} \rangle \exp(-i\omega t) \, \mathrm{d}t, \tag{5.112}$$

can be written

$$S_\mathrm{i}(\mathbf{Q}, \omega) = \frac{1}{2\pi\hbar} \int \int G_\mathrm{s}(\mathbf{r}, t) \exp\{i(\mathbf{Q} \cdot \mathbf{r} - \omega t)\} \, \mathrm{d}^3\mathbf{r} \, \mathrm{d}t, \tag{5.113}$$

where $G_\mathrm{s}(\mathbf{r}, t)$ is the 'self' time-dependent pair correlation function, given by

$$G_\mathrm{s}(\mathbf{r}, t) = \frac{1}{(2\pi)^3} \frac{1}{N} \sum_j \int \langle \exp(-i\mathbf{Q} \cdot \hat{\mathbf{r}}_j) \exp\{i\mathbf{Q} \cdot \hat{\mathbf{r}}_j(t)\} \rangle \exp(-i\mathbf{Q} \cdot \mathbf{r}) \, \mathrm{d}^3\mathbf{Q}. \tag{5.114}$$

We shall also encounter the *intermediate scattering function* $I(\mathbf{Q}, t)$, which is the spatial Fourier transform of $G(\mathbf{r}, t)$ and the $\omega$ Fourier transform of $S(\mathbf{Q}, \omega)$,[30]

[30]This definition of $I(\mathbf{Q}, t)$ differs from the one introduced in Section 3.4.5 in that it is normalized by $N$ and does not include the scattering lengths.

$$I(\mathbf{Q}, t) = \frac{1}{N} \sum_{jk} \langle \exp(-i\mathbf{Q} \cdot \hat{\mathbf{r}}_j) \exp\{i\mathbf{Q} \cdot \hat{\mathbf{r}}_k(t)\} \rangle \qquad (5.115)$$

$$= \int G(\mathbf{r}, t) \exp(i\mathbf{Q} \cdot \mathbf{r}) \, \mathrm{d}^3\mathbf{r} \qquad (5.116)$$

$$= \hbar \int S(\mathbf{Q}, \omega) \exp(i\omega t) \, \mathrm{d}\omega, \qquad (5.117)$$

and the *self intermediate scattering function* $I_s(\mathbf{Q}, t)$, which is similarly related to $G_s(\mathbf{r}, t)$ and $S_i(\mathbf{Q}, \omega)$.

## 5.3.2   Physical interpretation of $G(\mathbf{r}, t)$ and $G_s(\mathbf{r}, t)$

Because they are built from quantum-mechanical operators, any attempt to attach physical meaning to the time-dependent pair correlation functions needs to be approached with caution. Both $G(\mathbf{r}, t)$ and $G_s(\mathbf{r}, t)$ are, in general, complex functions whose imaginary parts derive from quantum properties of the system (see Exercise 5.8). The quantum effects arise because the operators $\exp(-i\mathbf{Q} \cdot \hat{\mathbf{r}}_j)$ and $\exp\{i\mathbf{Q} \cdot \hat{\mathbf{r}}_k(t)\}$ in (5.111) and (5.114) do not commute, except at $t = 0$. This non-commutation is because (i) the Heisenberg operator $\hat{\mathbf{r}}_k(t)$ contains the Hamiltonian for the system, which is a function of the position and momentum of each of the particles, and (ii) the momentum operator for the $j$th particle does not commute with the position operator $\hat{\mathbf{r}}_j$.

In the absence of quantum effects, however, $G(\mathbf{r}, t)$ and $G_s(\mathbf{r}, t)$ do admit to a simple interpretation. If we allow $\hat{\mathbf{r}}_k(t)$ and $\hat{\mathbf{r}}_j$ to commute then

$$G^{\mathrm{cl}}(\mathbf{r}, t) = \frac{1}{(2\pi)^3} \frac{1}{N} \sum_{jk} \int \langle \exp[-i\mathbf{Q} \cdot \{\hat{\mathbf{r}}_j - \hat{\mathbf{r}}_k(t) + \mathbf{r}\}] \rangle \, \mathrm{d}^3\mathbf{Q}$$

$$= \frac{1}{N} \sum_{jk} \langle \delta\{\mathbf{r} - \hat{\mathbf{r}}_k(t) + \hat{\mathbf{r}}_j\} \rangle$$

$$= \sum_k \langle \delta\{\mathbf{r} - \hat{\mathbf{r}}_k(t) + \hat{\mathbf{r}}_0\} \rangle. \qquad (5.118)$$

The last line follows because when all the nuclei are equivalent the sum over $k$ is the same for all $j$, so in place of the sum over $j$ we have chosen $j = 0$ and multiplied by $N$.

The final expression in (5.118) shows that under classical conditions $G(\mathbf{r}, t) \, \mathrm{d}^3\mathbf{r}$ represents the probability of finding a particle at time $t$ in the volume $\mathrm{d}^3\mathbf{r}$ at a displacement $\mathbf{r}$ away from a point occupied by a particle (the same or different) at $t = 0$. The equivalent expression for $G_s(\mathbf{r}, t)$ is

$$G_s^{\mathrm{cl}}(\mathbf{r}, t) = \langle \delta\{\mathbf{r} - \hat{\mathbf{r}}_0(t) + \hat{\mathbf{r}}_0\} \rangle, \qquad (5.119)$$

which means that under classical conditions $G_s(\mathbf{r}, t) \, \mathrm{d}^3\mathbf{r}$ represents the probability of finding a particle at time $t$ in the volume $\mathrm{d}^3\mathbf{r}$ at a displacement $\mathbf{r}$ away from a point occupied by *the same* particle at $t = 0$.

### 5.3.3   Classical approximation to $G(\mathbf{r}, t)$ and $S(\mathbf{Q}, \omega)$

The classical pair correlation functions $G^{\mathrm{cl}}(\mathbf{r}, t)$ and $G_{\mathrm{s}}^{\mathrm{cl}}(\mathbf{r}, t)$ are relevant because theoretical calculations for dense fluids, such as molecular dynamics simulations, are not usually carried out quantum-mechanically. For this reason it is useful to know under what circumstances the classical correlation functions are applicable, and to work out corrections that give a better approximation to the true scattering function.

The functions $G^{\mathrm{cl}}(\mathbf{r}, t)$ and $G_{\mathrm{s}}^{\mathrm{cl}}(\mathbf{r}, t)$ are real-valued, and for isotropic fluids in equilibrium they are symmetric functions of $\mathbf{r}$ and $t$. It follows from (5.109) and (5.113) that

$$S^{\mathrm{cl}}(-\mathbf{Q}, -\omega) = S^{\mathrm{cl}}(\mathbf{Q}, \omega) \quad \text{and} \quad S_{\mathrm{i}}^{\mathrm{cl}}(-\mathbf{Q}, -\omega) = S_{\mathrm{i}}^{\mathrm{cl}}(\mathbf{Q}, \omega), \quad (5.120)$$

which means that the classical response functions do not satisfy the detailed balance condition (3.48). The reason for this is that during a neutron scattering event the neutron exchanges energy and momentum with the target, taking the target briefly out of equilibrium. The initial recoil of an isolated nucleus is an example. Because the classical correlation functions are calculated from the *equilibrium* state of the liquid they do not include the recoil. By contrast, $G(\mathbf{r}, t)$ and $G_{\mathrm{s}}(\mathbf{r}, t)$ derive from quantum-mechanical scattering theory which correctly describes the dynamics of the target during the entire scattering process, at least in the Born approximation.

The classical correlation functions will apply when the characteristic length and time scales probed by the neutron greatly exceed the average distance and time needed for the target to relax back to equilibrium. To quantify these conditions we consider a single nucleus of mass $M$ in thermal equilibrium at temperature $T$. The de Broglie wavelength of the nucleus is given by $\lambda_{\mathrm{B}} = h/p$, where $p^2/(2M) \sim k_{\mathrm{B}}T$. Hence, $\lambda_{\mathrm{B}} \sim h/(Mk_{\mathrm{B}}T)^{1/2}$. The time for the nucleus to travel a distance $\lambda_{\mathrm{B}}$ is $t_{\mathrm{B}} \sim M\lambda_{\mathrm{B}}/p$. The neutron probes correlations on a length scale of order $2\pi/Q$ and on a time scale of order $2\pi/\omega$, so to avoid quantum effects we must have that $2\pi/Q \gg \lambda_{\mathrm{B}}$ and $2\pi/\omega \gg t_{\mathrm{B}}$. Hence, the scattering process will be classical if

$$Q \ll \frac{(Mk_{\mathrm{B}}T)^{\frac{1}{2}}}{\hbar} \quad \text{and} \quad \hbar\omega \ll k_{\mathrm{B}}T. \quad (5.121)$$

A simple procedure to extend the validity of the classical response functions is to make the approximation[31] (Schofield, 1960)

$$S(\mathbf{Q}, \omega) \approx \exp\left(\beta\hbar\omega/2\right) S^{\mathrm{cl}}(\mathbf{Q}, \omega), \quad (5.122)$$

$\beta = 1/k_{\mathrm{B}}T$, and similarly for $S_{\mathrm{i}}(\mathbf{Q}, \omega)$. The response function defined by (5.122) satisfies detailed balance by construction, rather than by introducing new physics into the classical description of the dynamics. It can be shown that the response function constructed this way is correct to first order in $\hbar$, so it should be a very good approximation close to classical conditions.

[31] The approximation in (5.122) is equivalent to replacing $t$ by $t - \mathrm{i}\hbar\beta/2$ in $G^{\mathrm{cl}}(\mathbf{r}, t)$.

### 5.3.4   Relation between $G(\mathbf{r}, 0)$ and $S(\mathbf{Q})$

As mentioned above, at $t = 0$ the quantum-mechanical operators in $G(\mathbf{r}, t)$ commute, so this is a special case when the classical expressions apply. By comparing (5.118) with (2.58) we see that

$$G(\mathbf{r}, 0) = G(\mathbf{r})$$
$$= n_0 g(\mathbf{r}) + \delta(\mathbf{r}), \tag{5.123}$$

where $G(\mathbf{r})$ is the density–density correlation function defined in Section 2.4.3, $n_0$ is the bulk density, and $g(\mathbf{r})$ is the pair distribution function defined in Section 2.4.2. Hence, from (2.59) the static structure factor is given by

$$S(\mathbf{Q}) = \int G(\mathbf{r}, 0) \exp(\mathrm{i}\mathbf{Q} \cdot \mathbf{r}) \, \mathrm{d}^3\mathbf{r}, \tag{5.124}$$

which, from (5.110) and (B.15), can also be written

$$S(\mathbf{Q}) = \int_{-\infty}^{\infty} S(\mathbf{Q}, \omega) \, \mathrm{d}(\hbar\omega). \tag{5.125}$$

Two important conclusions can be drawn from the foregoing. The first is that $S(\mathbf{Q})$ is determined by *instantaneous* density–density correlations. This is because $G(\mathbf{r}, 0)$ reflects correlations in density at different points in space *at the same time*. In effect, $S(\mathbf{Q})$ measures the average of many snapshots of the atomic positions.

The second conclusion is that expressions for $S(\mathbf{Q})$ derived with the assumption of rigid atoms, such as those in Chapter 2, can be used to interpret experiments on real systems with mobile atoms as long as the measurement integrates over all the inelastic scattering. This procedure is called the *static approximation*. The usual approach is to use a neutron detection system that records all scattered neutrons irrespective of their energy and assume this measures $S(\mathbf{Q})$ directly. Factors affecting the validity of the static approximation are discussed in Section 3.8.4.

### 5.3.5   Multicomponent systems

**Partial correlation and response functions**

One way to treat multicomponent systems is via partial, or species-dependent, correlation and response functions, as set out in Section 2.4.2. For a system of several species with concentrations $c_\alpha = N_\alpha/N$, where $N = \sum_\alpha N_\alpha$, we can write in place of (5.103),

$$\left(\frac{\mathrm{d}^2\sigma}{\mathrm{d}\Omega \mathrm{d}E_\mathrm{f}}\right)_\mathrm{coh} = \frac{k_\mathrm{f}}{k_\mathrm{i}} N \sum_{\alpha\beta} \bar{b}_\alpha^* \bar{b}_\beta c_\alpha c_\beta S_{\alpha\beta}(\mathbf{Q}, \omega), \tag{5.126}$$

where[32]

$$S_{\alpha\beta}(\mathbf{Q}, \omega) = \frac{1}{2\pi\hbar} \int\int G_{\alpha\beta}(\mathbf{r}, t) \exp\{\mathrm{i}(\mathbf{Q} \cdot \mathbf{r} - \omega t)\} \, \mathrm{d}^3\mathbf{r} \, \mathrm{d}t, \tag{5.127}$$

[32]In general, $S_{\alpha\beta} \neq S_{\beta\alpha}$. In fact, as shown in Section D.8, $S_{\alpha\beta}$ is the complex conjugate of $S_{\beta\alpha}$.

with

$$G_{\alpha\beta}(\mathbf{r},t) = \frac{1}{N} \int \langle \hat{n}_\alpha(\mathbf{r}',0)\hat{n}_\beta(\mathbf{r}'+\mathbf{r},t)\rangle \, d^3\mathbf{r}'. \tag{5.128}$$

The incoherent cross-section can be written similarly.

### Scattering length density

An alternative way of describing multicomponent systems is with correlation functions defined in terms of the scattering length density operator

$$\hat{n}_b(\mathbf{r},t) = \sum_j \overline{b}_j \delta\{\mathbf{r} - \hat{\mathbf{r}}_j(t)\}. \tag{5.129}$$

Hence,

$$G_b(\mathbf{r},t) = \frac{1}{N} \int \langle \hat{n}_b^\dagger(\mathbf{r}',0)\hat{n}_b(\mathbf{r}'+\mathbf{r},t)\rangle \, d^3\mathbf{r}', \tag{5.130}$$

and[33]

$$S(\mathbf{Q},\omega) = \frac{1}{2\pi\hbar} \int \int G_b(\mathbf{r},t) \exp\{i(\mathbf{Q}\cdot\mathbf{r} - \omega t)\} \, d^3\mathbf{r}\, dt$$

$$= \frac{1}{2\pi\hbar}\frac{1}{N} \int \langle \hat{n}_b^\dagger(\mathbf{Q},0)\hat{n}_b(\mathbf{Q},t)\rangle \, \exp(-i\omega t)\, dt. \tag{5.131}$$

This formalism emphasizes that coherent nuclear scattering actually arises from correlations in the scattering length density, rather than in the particle density. Furthermore, (5.131) shows that the scattering length density operator $\hat{n}_b(\mathbf{Q},t)$ is the operator $\hat{A}(\mathbf{Q},t)$ that represents the neutron–nucleus interaction and which appears in the general form of the response function, eqn (3.36).

The use of (5.129)–(5.131) is valid for unpolarized neutron scattering from systems with uncorrelated nuclear spins and whose internal nuclear states are not coupled to the motion of the atoms. For the completely general case we must replace the mean scattering length in (5.129) with $\hat{\beta}_j(t)$, the matrix element of the scattering length operator between initial and final spin states of the neutron — see (4.17). Hence,[34]

$$\hat{n}_\beta(\mathbf{r},t) = \sum_j \hat{\beta}_j(t)\delta\{\mathbf{r} - \hat{\mathbf{r}}_j(t)\}, \tag{5.132}$$

and $\hat{n}_\beta$ replaces $\hat{n}_b$ in (5.130) and (5.131).

## 5.4   Sum rules

An interesting, and sometimes useful, property of the coherent response function is that if all the inelastic and elastic scattering is added together and averaged over all $\mathbf{Q}$, then the result is a constant equal to the mean squared nuclear scattering length of the sample. This is called the *sum rule* for nuclear scattering. We can derive it from the general expression for the coherent response function for nuclear scattering from multicomponent systems,[35]

[33]The scattering lengths are now included in $S(\mathbf{Q},\omega)$, giving it dimensions of area.

[34]$\hat{n}_\beta(\mathbf{r},t)$ is not a Hermitian operator.

[35]The definition of $S(\mathbf{Q},\omega)$ in (5.133) is the same as that in (5.131). Both include the scattering lengths.

$$S(\mathbf{Q},\omega) = \frac{1}{2\pi\hbar}\frac{1}{N}\sum_{jk}\bar{b}_j^*\bar{b}_k\int_{-\infty}^{\infty}\langle\exp(-i\mathbf{Q}\cdot\hat{\mathbf{r}}_j)\exp\{i\mathbf{Q}\cdot\hat{\mathbf{r}}_k(t)\}\rangle\exp(-i\omega t)\,\mathrm{d}t.$$

(5.133)

Integrating over energy,

$$S(\mathbf{Q}) = \int_{-\infty}^{\infty} S(\mathbf{Q},\omega)\,\mathrm{d}(\hbar\omega)$$

$$= \frac{1}{N}\sum_{jk}\bar{b}_j^*\bar{b}_k\,\langle\exp\{i\mathbf{Q}\cdot(\hat{\mathbf{r}}_k - \hat{\mathbf{r}}_j)\}\rangle,$$

(5.134)

where we have used eqn (B.14) and the fact that $\hat{\mathbf{r}}_j$ and $\hat{\mathbf{r}}_k$ commute, since they both correspond to the same time ($t = 0$). We now average $S(\mathbf{Q})$ over $\mathbf{Q}$,

$$\langle S\rangle_{\mathbf{Q}} = \frac{1}{V_{\mathbf{Q}}}\int_{V_{\mathbf{Q}}} S(\mathbf{Q})\,\mathrm{d}^3\mathbf{Q},$$

(5.135)

where $V_{\mathbf{Q}}$ is an appropriate volume of $\mathbf{Q}$ space. In general this means integrating over the whole of $\mathbf{Q}$ space, but in practice it is usually sufficient to integrate up to some maximum $\mathbf{Q}$ above which $S(\mathbf{Q})$ does not exhibit any measurable variation. For periodic systems it may be sufficient to average over just a few Brillouin zones. For example, the scattering from a monatomic crystal with a primitive lattice has the same periodicity as the reciprocal lattice, so it is sufficient to average over one Brillouin zone.

If $\mathbf{r}_j \neq \mathbf{r}_k$ in (5.134) then the integral in (5.135) vanishes because of the oscillatory form of the integrand, whereas if $\mathbf{r}_j = \mathbf{r}_k$ then $S(\mathbf{Q})$ is independent of $\mathbf{Q}$. We make the assumption that the nuclei are sufficiently localized that they cannot coincide in position, so that $\mathbf{r}_j = \mathbf{r}_k$ implies $j = k$. Hence, we arrive at the sum rule for coherent nuclear scattering,

$$\langle S\rangle_{\mathbf{Q}} = \frac{1}{N}\sum_j |\bar{b}_j|^2.$$

(5.136)

The same procedure applied to the 'self' response function leads to the sum rules for incoherent nuclear scattering:

$$S_i(\mathbf{Q}) = \langle S_i\rangle_{\mathbf{Q}} = \frac{1}{N}\sum_j \frac{(\sigma_{\mathrm{inc}})_j}{4\pi}.$$

(5.137)

Sum rules can be used to see if a measurement has captured all the scattering, or to test the reliability of corrections applied to the data. An obvious requirement is to calibrate the scattering intensity so that the response function is obtained in absolute units (see Section 10.1.4). The response function is then integrated over energy and averaged over an appropriate volume of $\mathbf{Q}$ space. If the result does not agree with the predictions of the sum rule then it could indicate that the data have been inadequately corrected, or there may be scattering worth searching for outside the limits probed in the experiment.

## 5.5    Neutron Compton scattering

Traditional Compton scattering is the process by which high-energy photons scatter from electrons. This technique can be used to probe electron momentum distributions in atoms, molecules, and solids. *Neutron Compton scattering* (also known as *deep inelastic neutron scattering*) is an analogous technique that allows the experimental determination of atomic momentum distributions in condensed matter. It involves the scattering of high-energy neutrons from atomic nuclei, and works best at very large scattering vectors (typically $Q \geq 200\,\mathrm{nm}^{-1}$).

### 5.5.1    Impulse approximation

The interpretation of neutron Compton scattering is based on two assumptions, known collectively as the *impulse approximation*. The first is the incoherent approximation, which applies when the momentum transfer is very large, as discussed in Section 5.2.9. The incoherent approximation entails the neglect of inter-particle interference effects. The second assumption is that in the collision between a neutron and a target nucleus the nucleus recoils as if it were free. This is equivalent to neglecting the characteristic dynamics of the system in the final state of the nucleus.[36]

[36]Note, however, that the dynamics in the initial state are not neglected, and appear in the form of the equilibrium momentum density of the nucleus.

[37]If more than one type of atom is present then simply add the differential cross-sections for each atom type. This is because there are no inter-particle terms in the incoherent approximation.

We now obtain the differential cross-section in the impulse approximation for an ensemble of mobile atoms, such as a liquid. We shall restrict the calculation to a single type of atom for simplicity,[37] and use (5.103) and (5.104) to define the response functions. The incoherent approximation means that the coherent and incoherent response functions are the same. The simplest starting point is eqn (3.16), which together with (4.15) and (5.103) or (5.104) gives the following expression for the response function of a single nucleus (unpolarized neutrons),

$$S(\mathbf{Q},\omega) = S_\mathrm{i}(\mathbf{Q},\omega) = \sum_{\lambda_\mathrm{i}} p_{\lambda_\mathrm{i}} \sum_{\lambda_\mathrm{f}} |\langle \lambda_\mathrm{f}| \exp(i\mathbf{Q}\cdot\mathbf{r})|\lambda_\mathrm{i}\rangle|^2 \delta(E_{\lambda_\mathrm{f}} - E_{\lambda_\mathrm{i}} - \hbar\omega),$$

(5.138)

where $\mathbf{r}$ is the position vector of the nucleus.

As the nucleus is taken to be a free particle in the final state we adopt the momentum representation for the wave functions, writing

$$|\lambda_\mathrm{i}\rangle = \frac{1}{V^{\frac{1}{2}}} \exp(i\mathbf{q}\cdot\mathbf{r}) \equiv |\mathbf{q}\rangle, \quad \text{and} \quad |\lambda_\mathrm{f}\rangle = \frac{1}{V^{\frac{1}{2}}} \exp(i\mathbf{q}'\cdot\mathbf{r}) \equiv |\mathbf{q}'\rangle.$$

(5.139)

Here, $\mathbf{q}$ and $\mathbf{q}'$ are wavevectors of the nucleus with corresponding momenta $\mathbf{p} = \hbar\mathbf{q}$ and $\mathbf{p}' = \hbar\mathbf{q}'$. The $|\mathbf{q}\rangle$ can be regarded as a Fourier component of the initial state function of the nucleus, with associated probability $p_\mathbf{q} \equiv p_{\lambda_\mathrm{i}}$. Therefore, we rewrite (5.138) as

$$S(\mathbf{Q},\omega) = \sum_\mathbf{q} p_\mathbf{q} \sum_{\mathbf{q}'} |\langle \mathbf{q}'| \exp(i\mathbf{Q}\cdot\mathbf{r})|\mathbf{q}\rangle|^2 \delta(E_{\mathbf{q}'} - E_\mathbf{q} - \hbar\omega). \quad (5.140)$$

The matrix element in (5.140) takes the form

$$\langle \mathbf{q}'|\exp(\mathrm{i}\mathbf{Q}\cdot\mathbf{r})|\mathbf{q}\rangle = \frac{1}{V}\int \exp\{\mathrm{i}(\mathbf{Q}+\mathbf{q}-\mathbf{q}')\cdot\mathbf{r}\}\,\mathrm{d}^3\mathbf{r}, \qquad (5.141)$$

which is zero unless

$$\mathbf{Q}+\mathbf{q}-\mathbf{q}' = 0, \qquad (5.142)$$

in which case

$$\langle \mathbf{q}'|\exp(\mathrm{i}\mathbf{Q}\cdot\mathbf{r})|\mathbf{q}\rangle = 1. \qquad (5.143)$$

Condition (5.142) represents conservation of momentum in the collision between the neutron and the nucleus. This condition is satisfied by only one value of $\mathbf{q}'$ for a given $\mathbf{q}$ and $\mathbf{Q}$, so the summation over $\mathbf{q}'$ in (5.140) contains only one non-zero term (5.143). Furthermore,

$$E_{\mathbf{q}'} - E_{\mathbf{q}} = \frac{\hbar^2}{2M}(\mathbf{Q}+\mathbf{q})^2 - \frac{\hbar^2 q^2}{2M}$$

$$= E_{\mathrm{r}} + \frac{\hbar^2}{M}\mathbf{Q}\cdot\mathbf{q}. \qquad (5.144)$$

where $E_{\mathrm{r}} = \hbar^2 Q^2/2M$ is the recoil energy of a stationary nucleus of mass $M$ (Fig. 5.14). Hence,

$$S(\mathbf{Q},\omega) = \sum_{\mathbf{q}} p_{\mathbf{q}}\,\delta\!\left(E_{\mathrm{r}} + \frac{\hbar^2}{M}\mathbf{Q}\cdot\mathbf{q} - \hbar\omega\right)$$

$$= \int n(\mathbf{p})\,\delta\!\left(E_{\mathrm{r}} + \frac{\hbar}{M}\mathbf{Q}\cdot\mathbf{p} - \hbar\omega\right)\mathrm{d}^3\mathbf{p} \qquad (5.145)$$

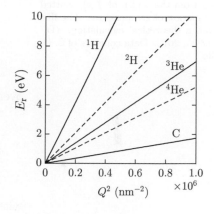

**Fig. 5.14** Recoil energy as a function of $Q^2$ for several light atoms.

In the second line we replaced the probability of discrete $\mathbf{q}$ vectors by the momentum distribution function $n(\mathbf{p})$, defined such that $n(\mathbf{p})\mathrm{d}^3\mathbf{p}$ is the fraction of atoms in the initial state with momenta in the volume $\mathrm{d}^3\mathbf{p}$ centred on $\mathbf{p}$.

When the scattering vector is large enough that $E_{\mathrm{r}} \gg (\hbar/M)\mathbf{Q}\cdot\mathbf{p}$ the scattering intensity, considered as a function of $\hbar\omega$ at fixed $\mathbf{Q}$, consists of a peak centred at $E_{\mathrm{r}}$ whose width depends on the magnitude of $\mathbf{Q}$ and on the distribution of atomic momenta along the direction of $\mathbf{Q}$. The dependence of the peak position on the mass of the target atom means that the scattering from different atomic species (or isotopes) can be separated.

## 5.5.2   The Compton profile

Let us take the $z$ axis to be along the direction of $\mathbf{Q}$, so that $\mathbf{Q}\cdot\mathbf{p} = Qp_z$, and define

$$J(p_z) = \int_{-\infty}^{\infty}\int_{-\infty}^{\infty} n(\mathbf{p})\,\mathrm{d}p_x\mathrm{d}p_y. \qquad (5.146)$$

We can now express the response function in terms of $J$ by integrating first over the $p_x$ and $p_y$ momentum components in (5.145), and then

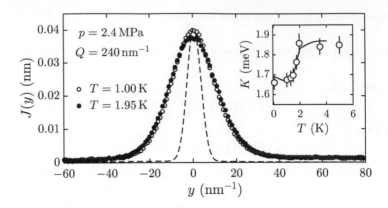

**Fig. 5.15** Compton profile $J(y)$ for liquid $^4$He measured at $Q = 240\,\text{nm}^{-1}$. The scaling variable $y$ is given in units with $\hbar = 1$. The broken line shows the experimental resolution. Inset: the mean kinetic energy per atom $K$ determined from the width of $J(y)$, plotted as a function of temperature. The line is from a theoretical calculation. (Diallo *et al.*, 2012. Data courtesy of S. O. Diallo.)

over $p_z$ using the properties of the delta function (see Section B.2):

$$S(\mathbf{Q},\omega) = \int J(p_z)\,\delta\!\left(E_\mathrm{r} + \frac{\hbar Q}{M}p_z - \hbar\omega\right)\mathrm{d}p_z$$

$$= \frac{M}{\hbar Q}J(y), \tag{5.147}$$

where

$$y = \frac{M}{\hbar Q}(\hbar\omega - E_\mathrm{r}). \tag{5.148}$$

The function $J(y)$, which gives the momentum distribution along the direction of $\mathbf{Q}$, is known as the Compton profile. It follows from (5.147) and (5.148) that $QS(\mathbf{Q},\omega)$ depends only on the variable $y$, and not on $Q$ and $\omega$ separately, a property known as *y scaling*.[38] The experimental observation of $y$ scaling is a good test of the validity of the impulse approximation.

[38]The scaling variable $y$ has dimensions of momentum.

For a system in which the target atom experiences anisotropic forces, such as an atom in a crystal, $J(y)$ will vary with the direction of $\mathbf{Q}$, and from this variation one can learn about the local bonding. For an isotropic system, such as a liquid, the direction of $\mathbf{Q}$ is immaterial.

A useful quantity is the mean atomic kinetic energy, $K = \langle p^2\rangle/2M$. For an isotropic system $\langle p^2\rangle = 3\langle p_z^2\rangle$, so $K$ can be obtained from the Compton profile as follows:

$$K = \frac{3}{2M}\int_{-\infty}^{\infty} y^2 J(y)\,\mathrm{d}y. \tag{5.149}$$

For an anisotropic system, $K$ represents the kinetic energy associated with atomic motion along $\mathbf{Q}$.

As an example, Fig. 5.15 shows measurements of the Compton profile and mean kinetic energy of liquid $^4$He at temperatures above and below the lambda transition, which occurs at $T_\lambda = 1.86\,\text{K}$ for the pressure $p = 2.4\,\text{MPa}$ (24 bar). Upon cooling through $T_\lambda$ there is a reduction in the width of $J(y)$ and a corresponding drop in $K$. These anomalies are associated with a Bose–Einstein condensation of the $^4$He atoms.

### 5.5.3 Gaussian Compton profile

A useful form for the Compton profile is

$$J(y) = \sqrt{\frac{3}{4\pi MK}}\, \exp\left(-\frac{3y^2}{4MK}\right). \qquad (5.150)$$

This is a Gaussian distribution (see Section B.1.1) with standard deviation $\sigma = \sqrt{2MK/3}$. For an isotropic system, $K$ is the mean kinetic energy as can easily be verified from (5.149). Figure 5.16 is a two-dimensional plot of $J(y)$ from (5.150) as a function of the experimental variables $Q$ and $\hbar\omega$. A constant-$Q$ cut through the $J(y)$ surface yields a Gaussian centred on $\hbar\omega = E_\mathrm{r}$ with a width proportional to $Q$.

The Gaussian is often a good model for the measured Compton profile. One system for which the Gaussian profile is rigorously correct is the harmonic oscillator. To show this we exploit some results derived in Section 5.1. According to the impulse approximation the nucleus is free after the collision, so we put $\hat{\mathbf{r}}(t) = \hat{\mathbf{r}} + \hat{\mathbf{p}}t/M$ in identity (5.27) and write[39]

$$\langle \exp(-i\mathbf{Q}\cdot\hat{\mathbf{r}})\exp\{i\mathbf{Q}\cdot\hat{\mathbf{r}}(t)\}\rangle = \left\langle \exp\left(\frac{it}{M}\mathbf{Q}\cdot\hat{\mathbf{p}}\right)\right\rangle \exp\left(\frac{i\hbar Q^2 t}{2M}\right). \qquad (5.151)$$

The position–momentum commutation relations, e.g. (5.6), were used to simplify the commutator in (5.27). The thermal average can be evaluated for the harmonic oscillator following the same procedure as used to derive Bloch's identity (5.24). The result is

$$\left\langle \exp\left(\frac{it}{M}\mathbf{Q}\cdot\hat{\mathbf{p}}\right)\right\rangle = \exp\left(-\frac{Q^2\langle p_z^2\rangle t^2}{2M^2}\right), \qquad (5.152)$$

where the $z$ axis is once again taken to be along $\mathbf{Q}$. After substituting (5.152) into (5.151) we take the time Fourier transform to obtain the response function

$$S(\mathbf{Q},\omega) = \frac{1}{2\pi\hbar}\int_{-\infty}^{\infty}\exp\left(-\frac{Q^2\langle p_z^2\rangle t^2}{2M^2}+\frac{i\hbar Q^2 t}{2M}\right)\exp(-i\omega t)\,\mathrm{d}t$$

$$= \frac{M}{\hbar Q}\sqrt{\frac{1}{2\pi\langle p_z^2\rangle}}\,\exp\left(-\frac{y^2}{2\langle p_z^2\rangle}\right), \qquad (5.153)$$

which, since $K = 3\langle p_z^2\rangle/2M$, contains the Gaussian profile of (5.150). We can replace $K$ with an explicit function of temperature because for a three-dimensional harmonic oscillator $K = \frac{3}{2}M\omega_0^2\langle z^2\rangle$, and $\langle z^2\rangle$ is given by (5.23). Hence,

$$K = \frac{3\hbar\omega_0}{4}\coth(\tfrac{1}{2}\beta\hbar\omega_0) \qquad [\beta = 1/(k_\mathrm{B}T)]. \qquad (5.154)$$

Another system that can be rigorously described by a Gaussian Compton profile is a classical (Boltzmann) gas of atoms in thermal equilibrium.

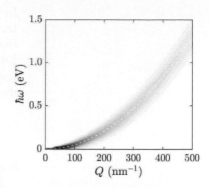

**Fig. 5.16** The Gaussian Compton profile eqn (5.150) plotted as a function of $Q$ and $\hbar\omega$. The profile is calculated for $M = 4\,\mathrm{u}$ ($^4$He) and mean kinetic energy $K = \frac{3}{2}k_\mathrm{B}T$ of a classical gas at $T = 100\,\mathrm{K}$. The broken white line is the recoil energy $E_\mathrm{r}$ of $^4$He.

[39] Equation (5.151) is not specific to the harmonic oscillator. It is a general expression for the nuclear correlation function in the impulse approximation.

In such a system the atoms do not interact with one another, and so are treated as free particles in both the initial and final states. This means that the impulse approximation holds under all conditions, and so the calculation is not restricted to large scattering vectors. We can derive the Gaussian Compton profile for a classical gas from the Boltzmann distribution, which for one momentum component $y$ may be written

$$J(y) = \left(\frac{1}{2\pi M k_{\mathrm{B}} T}\right)^{1/2} \exp\left(-\frac{y^2}{2M k_{\mathrm{B}} T}\right). \qquad (5.155)$$

For a classical gas $K = \frac{3}{2} k_{\mathrm{B}} T$, which takes us immediately to (5.150).

Finally, let us obtain $G(\mathbf{r}, t)$, the time-dependent pair correlation function for the Gaussian Compton profile. This requires the space and time Fourier transform of $S(\mathbf{Q}, \omega)$, as prescribed in (5.110). For the Gaussian profile (5.150) the integrals are relatively painless, and lead to

$$G(\mathbf{r}, t) = \left(\frac{1}{2\pi\sigma^2(t)}\right)^{3/2} \exp\left(-\frac{r^2}{2\sigma^2(t)}\right), \qquad (5.156)$$

where

$$\sigma^2(t) = \frac{2Kt}{3M}\left(t - \frac{3\mathrm{i}\hbar}{2K}\right). \qquad (5.157)$$

We find that $G(\mathbf{r}, t)$ is complex, which is a consequence of quantum effects, as discussed in Section 5.3.2. Because of the impulse approximation, the form of $G(\mathbf{r}, t)$ obtained here is generally only valid at short times. Therefore, to calculate $G(\mathbf{r}, t)$ valid for all $t$ for the harmonic oscillator one must Fourier transform the exact expression for $S(\mathbf{Q}, \omega)$ given in (5.32). An exception is the classical gas. In this case, the impulse approximation is exact and so eqns (5.156) and (5.157) hold for all $t$.

## 5.6 Dynamics of disordered and complex systems

For the remainder of this chapter we shall be considering the dynamics of systems without long-range crystalline order, including simple liquids and glasses, molecular fluids, and much of soft matter. The dynamics of such systems extends over a very wide range of time scales. We have already seen how neutron Compton scattering can be used to examine systems at the shortest time and distance scales, and to measure single-particle momentum distributions and mean kinetic energies (Section 5.5). At the opposite end of the spectrum (long times and distances) one finds centre-of-mass diffusion and molecular reorientation processes. These are discussed below in Section 5.8.

In between these two extremes one finds a variety of different dynamical modes, including bond stretching and bending modes, as well as localized vibrations, librations, and torsional oscillations of molecular sub-units. These processes, which are particular cases of optic modes, produce a series of inelastic peaks in neutron spectra. The measurement and analysis of these modes, called *vibrational spectroscopy*, provides a wealth of information on molecular structure and interactions, as well as being a way to identify and characterize complex systems through their vibrational 'fingerprint'.

## 5.7 Vibrational spectroscopy

Neutron vibrational spectroscopy is complementary to infra-red and Raman spectroscopy, techniques which also probe vibrational spectra but with photons instead of neutrons. The main advantages of neutrons over photons for vibrational spectra are (i) neutrons are not governed by the optical selection rules,[40] so all modes are measurable in principle, (ii) hydrogen-containing materials scatter very strongly and offer the possibility of isotopic labelling, (iii) neutrons can probe materials which are opaque to light, and (iv) neutron spectra are relatively easy to model — the validity of the Born approximation ensures that intensities are proportional to the amount of material in the beam and to the square of the mode amplitudes. The main drawbacks of neutrons are (i) that relatively large samples are required ($\sim$1 g), (ii) the energy resolution is generally not as good as with photon spectroscopies, and (iii) neutron scattering is not as readily accessible as infra-red and Raman spectroscopies, which are available in many laboratories. The complementarity of the three techniques is illustrated in Fig. 5.17, which compares the vibrational spectrum of an organic molecule as measured by infra-red, Raman, and neutron spectroscopy.

The dynamics of complex systems are made easier to unpick if the various types of motion present are uncoupled, or nearly so, which can happen if their frequencies are well separated. To see what simplifications arise from this assumption, consider a molecule whose motion can be decomposed into centre-of-mass translations, rigid rotations about the centre of mass, and atomic vibrations. For simplicity, we assume a hydrogen-containing molecule, so that the incoherent scattering from the protons dominates the contributions from other atoms. Similar to eqn (5.38) for the atoms in crystals, the position of the $j^{\text{th}}$ hydrogen atom may be described by the Heisenberg operator

$$\hat{\mathbf{r}}_j(t) = \hat{\mathbf{R}}(t) + \hat{\mathbf{d}}_j(t) + \hat{\mathbf{u}}_j(t), \tag{5.158}$$

where $\hat{\mathbf{R}}$ represents the position of the centre of mass of the molecule, $\hat{\mathbf{d}}_j$ gives the displacement of the mean position of atom $j$ relative to the centre of mass, and $\hat{\mathbf{u}}_j$ is the displacement of atom $j$ from its mean position. The time dependence of $\mathbf{d}$ comes from rigid rotations of the molecule, which are represented by an orientational degree of freedom. With the assumption that the translational, orientational, and vibrational degrees of freedom are uncoupled, the self intermediate scattering function[41] for atom $j$ may be factorized as

$$I_s(\mathbf{Q}, t) = I^{\text{cm}}(\mathbf{Q}, t) I^{\text{rot}}(\mathbf{Q}, t) I^{\text{vib}}(\mathbf{Q}, t), \tag{5.159}$$

[40]Infra-red (IR) and Raman processes are governed by selection rules which determine which vibrational modes are optically active. In a system with a centre of inversion symmetry, for example, odd parity modes are IR-active but even parity modes are IR-inactive. The reverse is true for Raman scattering.

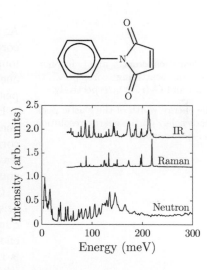

**Fig. 5.17** Vibrational spectrum of $N$–phenylmaleimide ($C_{10}H_7NO_2$) measured by infra-red (IR), Raman and neutron spectroscopy. (Parker, 2006. Data courtesy of S. F. Parker.)

[41]Obtained by substitution of (5.158) into eqn (5.115) with retention of the $j = k$ terms only.

where

$$I^{\mathrm{cm}}(\mathbf{Q},t) = \langle \exp(-\mathrm{i}\mathbf{Q}\cdot\hat{\mathbf{R}})\exp\{\mathrm{i}\mathbf{Q}\cdot\hat{\mathbf{R}}(t)\}\rangle, \qquad (5.160)$$

and the functions $I^{\mathrm{rot}}(\mathbf{Q},t)$ and $I^{\mathrm{vib}}(\mathbf{Q},t)$ are defined similarly for the rotational and vibrational correlations. There are no cross-correlations providing the different dynamical degrees of freedom are uncoupled. The incoherent scattering function $S_{\mathrm{i}}(\mathbf{Q},\omega)$ is the time Fourier transform of $I_{\mathrm{s}}(\mathbf{Q},t)$. Applying the convolution theorem (Section B.3.2) to (5.159), we may write $S_{\mathrm{i}}(\mathbf{Q},\omega)$ as

$$S_{\mathrm{i}}(\mathbf{Q},\omega) = S_{\mathrm{i}}^{\mathrm{cm}}(\mathbf{Q},\omega) * S_{\mathrm{i}}^{\mathrm{rot}}(\mathbf{Q},\omega) * S_{\mathrm{i}}^{\mathrm{vib}}(\mathbf{Q},\omega). \qquad (5.161)$$

Equations (5.160) and (5.161) provide useful starting points for data analysis providing separate models for the translational, rotational, and vibrational dynamics are available. Note that to account for the resolution of the spectrometer, the resolution function must also be convolved with (5.161) (see Section 10.1.6).

## 5.8   Quasielastic neutron scattering (QENS)

As explained in Section 3.6, *elastic* scattering arises from purely static correlations, i.e. correlations that remain in the system even after a very long time has elapsed.[42] Elastically scattered neutrons have the same energy before and after scattering, and give rise to a very sharp 'elastic peak' at $\omega = 0$.[43]

By definition, *inelastic* scattering is when there is a change in the energy of the neutrons on scattering, i.e. $\omega \neq 0$, and is caused by dynamic correlations. Inelastic scattering is usually associated with processes that leave the system in a state whose energy differs from that of the initial state.

*Quasielastic scattering* is a special case of inelastic scattering which arises from random rearrangements of the system due to thermal or quantum fluctuations between different configurations with the same energy. The energy transferred during a quasielastic scattering process is rapidly dissipated in the system, leaving it in an eventual final state with the same energy as the initial state. The quasielastic response consists of an asymmetric peak at the base of the elastic scattering. The energy-loss side of the quasielastic peak ($\omega > 0$) has a higher intensity than the energy-gain side ($\omega < 0$), as required by detailed balance, and the frequency width of the quasielastic peak is $\sim 1/\tau$, where $\tau$ is the characteristic time for configurational fluctuations of the system.[44]

Spectrometers capable of measuring quasielastic scattering require very good energy resolution. The backscattering spectrometer (see Section 10.3.4) is the type used most often, but the highest resolution is achieved with the neutron spin echo technique, described in Section 5.9.

The sorts of processes that give rise to quasielastic scattering include (i) diffusion of particles in liquids and solids, (ii) random jump processes

[42] Nuclear coherent and incoherent elastic scattering correspond to $G(\mathbf{r},\infty)$ and $G_{\mathrm{s}}(\mathbf{r},\infty)$, respectively.

[43] The energy width of the elastic peak depends solely on the energy resolution of the spectrometer.

[44] In accordance with Heisenberg's Uncertainty Principle, $\Delta E \Delta t \geq \hbar/2$.

between two or more equivalent molecular conformations caused either by thermal activation or by quantum-mechanical tunneling, and (iii) fluctuations in the direction of magnetic moments in disordered magnetic systems. In the following sections we cover a number of important examples from categories (i) and (ii).

## 5.8.1   Translational self-diffusion

In this section we are concerned with unrestricted *single-particle diffusion*, also called *self-diffusion*, which is probed by incoherent quasielastic neutron scattering. The single-particle dynamics is described by the self pair correlation function $G_s(\mathbf{r}, t)$, which is the space and time Fourier transform of the response function for incoherent scattering, eqn (5.113). We shall develop equations for $G_s(\mathbf{r}, t)$ for the case when the diffusing particle moves continuously

### Continuous diffusion

Macroscopically, the diffusion of particles in solids and liquids is driven by concentration gradients, but the mechanism of diffusion can be understood on a microscopic level in terms of a succession of random steps due to collisions. If the length and time intervals over which we observe the diffusing process are much greater than the average step size and inter-collision time, then we can describe the process by a continuous concentration profile $n(\mathbf{r}, t)$ which obeys Fick's first law,

$$\mathbf{j}(\mathbf{r}, t) = -D\nabla n(\mathbf{r}, t), \tag{5.162}$$

where $\mathbf{j}(\mathbf{r}, t)$ is the particle flux (the number of particles crossing unit area in unit time, the area being perpendicular to the direction of flow), $\nabla n$ is the concentration gradient,[45] and $D$ is the diffusion constant. If the number of particles is conserved then the concentration obeys the continuity equation,[46]

$$\frac{\partial n}{\partial t} = -\nabla \cdot \mathbf{j}. \tag{5.163}$$

Combining (5.162) and (5.163) we obtain Fick's second law,

$$\frac{\partial n}{\partial t} = D\nabla^2 n, \tag{5.164}$$

which assumes $D$ is independent of concentration.[47]

It will be recalled (Section 5.3.2) that $G_s^{cl}(\mathbf{r}, t)$ represents the probability of finding a particle at time $t$ in a unit volume at displacement $\mathbf{r}$ from a point occupied by the same particle at $t = 0$. It follows that $G_s^{cl}(\mathbf{r}, t)$ is the solution of (5.164) for the particular case when the initial condition is $n(\mathbf{r}, 0) = \delta(\mathbf{r})$. For an isotropic system the solution is[48]

$$G_s^{cl}(r, t) = \left(\frac{1}{4\pi D|t|}\right)^{\frac{3}{2}} \exp(-r^2/4D|t|), \tag{5.165}$$

[45] In Cartesian coordinates,
$$\nabla = \left(\frac{\partial}{\partial x}, \frac{\partial}{\partial y}, \frac{\partial}{\partial z}\right).$$

[46] The continuity equation is a mathematical statement of the fact that any change in the number of particles in a volume must be balanced by an equal number of particles flowing into the volume through its surface. This can be readily shown from eqn (5.163): integrate both sides over a volume and use the divergence theorem to convert the right-hand side into a surface integral.

[47] This is not always the case.

[48] The Laplacian for an isotropic system is $\nabla^2 = \frac{\partial^2}{\partial r^2} + \frac{2}{r}\frac{\partial}{\partial r}$.

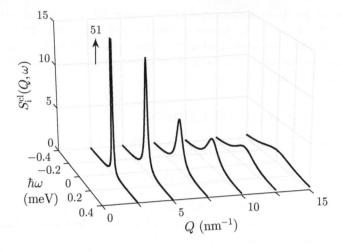

**Fig. 5.18** The Lorentzian incoherent response function for continuous self-diffusion, eqn (5.166). The diffusion constant used in the calculation is $D = 1.5 \times 10^{-9} \, \mathrm{m^2 \, s^{-1}}$, which is a typical value for self-diffusion in simple liquids.

which can be verified by direct substitution into (5.164). Fourier transformation of (5.165) over space and time coordinates yields

$$S_{\mathrm{i}}^{\mathrm{cl}}(Q, \omega) = \frac{1}{\pi} \frac{\hbar\Gamma(Q)}{(\hbar\omega)^2 + \{\hbar\Gamma(Q)\}^2}, \qquad (5.166)$$

where,

$$\Gamma(Q) = DQ^2. \qquad (5.167)$$

This is a Lorentzian function of energy centred on $\hbar\omega = 0$, with an energy width (half-width at half-maximum) of $\hbar\Gamma(Q)$. Figure 5.18 shows a two-dimensional plot of $S_{\mathrm{i}}^{\mathrm{cl}}(Q, \omega)$ from (5.166) and (5.167) as a function of $Q$ and $\omega$.

By using (5.122) we can obtain an approximation to the true incoherent response function:

$$S_{\mathrm{i}}(Q, \omega) \approx \exp\left(\beta\hbar\omega/2\right) \frac{1}{\pi} \frac{\hbar\Gamma(Q)}{(\hbar\omega)^2 + \{\hbar\Gamma(Q)\}^2} \qquad [\beta = 1/(k_{\mathrm{B}}T)]. \qquad (5.168)$$

The validity of this approximation was discussed in Section 5.3.3.

It is often difficult to separate the incoherent from the coherent scattering when both are present, but it can be done. One way is to perform the measurements at low $Q$ where the coherent structure factor is very small (see Section 2.4.2 and Fig. 2.22). A good example is the measurement of the self-diffusion constant of titanium, illustrated in Fig. 5.19. The QENS is dominated by the incoherent scattering ($\sigma_{\mathrm{inc}} = 2.87 \, \mathrm{b}$, $\sigma_{\mathrm{coh}} = 1.49 \, \mathrm{b}$) as long as $Q$ is well below the first maximum in $S(Q)$, which for titanium is at $26.5 \, \mathrm{nm}^{-1}$. For $Q$ below about $10 \, \mathrm{nm}^{-1}$ the inelastic response is well described by a Lorentzian function (5.166), and the half width agrees well with the prediction $\Gamma = DQ^2$ of the continuous diffusion model, eqn (5.166). Another method was used in the study of liquid argon by Sköld *et al.* (1972). These authors obtained the coherent scattering from measurements on liquid $^{36}\mathrm{Ar}$, an isotope with a purely coherent cross-section, and the incoherent scattering from

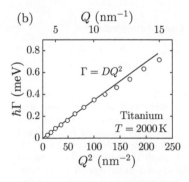

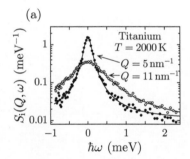

**Fig. 5.19** Incoherent quasielastic scattering from liquid titanium at a temperature of $T = 2000 \, \mathrm{K}$ (Meyer *et al.*, 2009). (a) $S_{\mathrm{i}}(Q, \omega)$ at two different $Q$ values. (b) Half-width at half-maximum $\hbar\Gamma$ obtained from fits to the continuous diffusion model eqn (5.166). The line is calculated from (5.167) with $D = 5.3 \times 10^{-9} \, \mathrm{m^2 s^{-1}}$. (Data courtesy of A. Meyer.)

a mixture of $^{36}$Ar and $^{40}$Ar for which the scattering is predominantly incoherent. The Lorentzian half width is plotted in Fig. 5.20 and agrees reasonably well with the continuous diffusion model, eqn (5.166).

Superficially, the continuous diffusion model seems to work well for liquid Ar, but there are discrepancies particularly for $Q > 30\,\text{nm}^{-1}$. There was also found to be scattering in excess of a Lorentzian for energies above $2\,\text{meV}$. These discrepancies occur because self-diffusion is influenced at shorter length and time scales by the local structure of the liquid. At very short time scales the recoil of the Ar atoms becomes important, and then the incoherent response should tend towards that given by (5.147) and (5.155) for a classical gas. Information on the detailed mechanism of diffusion in liquids requires a comparison of careful measurements of $S_i(Q, \omega)$ over a wide range of $Q$ and $\omega$ with microscopic theory or molecular dynamics simulations.

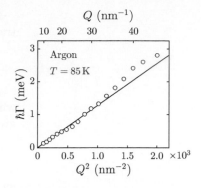

**Fig. 5.20** Half-width at half-maximum $\hbar\Gamma$ of incoherent quasielastic scattering from liquid argon obtained from fits to the continuous diffusion model eqn (5.166) (Sköld *et al.*, 1972). The line is calculated from (5.167) with $D = 1.94 \times 10^{-9}\,\text{m}^2\text{s}^{-1}$.

## Jump diffusion

The continuous diffusion model works well at small $Q$, but does not give any information on the mechanism of diffusion. The simplest model for the diffusion mechanism was developed by Chudley and Elliott (1961), who were interested in describing the effects of short-range order on diffusion in liquids. The model also lends itself well, perhaps better even, to the diffusion of atoms in crystalline solids. The assumptions of the Chudley–Elliott model are as follows. Diffusion takes place by a succession of random jumps between fixed sites. The diffusing atom remains on a site for a certain time $\tau$, after which it jumps to another site in a time much less than $\tau$. Sites that can be reached in a single jump form a spatial arrangement related to the local structure of the material. Interactions between different diffusing atoms are neglected, and diffusional and vibrational motions are assumed to be uncoupled.

For simplicity let us assume that the diffusion takes place on a Bravais lattice, so that all sites are equivalent, and that jumps can only take place to one of the $z$ nearest-neighbour sites.[49] The probability $P(\mathbf{r}, t)$ of finding the atom at a displacement $\mathbf{r}$ from the origin at time $t$ is then given by the rate equation

[49] It is straightforward to extend the model to non-Bravais lattices and energetically inequivalent sites — see Hempelmann (2000).

$$\frac{\partial P(\mathbf{r}, t)}{\partial t} = \frac{1}{z\tau} \sum_j \{P(\mathbf{r} + \mathbf{l}_j, t) - P(\mathbf{r}, t)\}, \quad (5.169)$$

where the $\mathbf{l}_j$ are vectors connecting a site to its nearest neighbours. The first term on the right-hand side is the rate of increase in the occupancy probability of site $\mathbf{r}$ due to jumps *onto* $\mathbf{r}$ from neighbouring sites, and the second term is the corresponding decrease in probability due to jumps *away from* $\mathbf{r}$. If we adopt the boundary condition

$$P(\mathbf{r}, 0) = \delta(\mathbf{r}), \quad (5.170)$$

then $P(\mathbf{r}, t)$ is just the classical self correlation function $G_s^{cl}(\mathbf{r}, t)$.

Equation (5.169) can most easily be integrated after a spatial Fourier transform, which converts it to

$$\frac{\partial I_s^{\mathrm{cl}}(\mathbf{Q},t)}{\partial t} = \frac{1}{z\tau} \sum_j \{\exp(-i\mathbf{Q}\cdot\mathbf{l}_j) - 1\}\, I_s^{\mathrm{cl}}(\mathbf{Q},t), \qquad (5.171)$$

where $I_s^{\mathrm{cl}}(\mathbf{Q},t)$ is the (classical) self intermediate scattering function, given by [c.f. (5.116)]

$$I_s^{\mathrm{cl}}(\mathbf{Q},t) = \int G_{\mathrm{s}}^{\mathrm{cl}}(\mathbf{r},t)\exp(i\mathbf{Q}\cdot\mathbf{r})\,\mathrm{d}^3\mathbf{r}. \qquad (5.172)$$

The boundary condition (5.170) becomes

$$I_s^{\mathrm{cl}}(\mathbf{Q},0) = 1, \qquad (5.173)$$

and the solution to (5.171) is, by inspection,

$$I_s^{\mathrm{cl}}(\mathbf{Q},t) = \exp(-\Gamma(\mathbf{Q})t), \qquad (5.174)$$

where,

$$\Gamma(\mathbf{Q}) = \frac{1}{z\tau}\sum_j \{1 - \exp(-i\mathbf{Q}\cdot\mathbf{l}_j)\}. \qquad (5.175)$$

Finally, by taking the time Fourier transform of (5.174) we obtain the incoherent response function,

$$S_{\mathrm{i}}^{\mathrm{cl}}(\mathbf{Q},\omega) = \frac{1}{\pi}\frac{\hbar\Gamma(\mathbf{Q})}{(\hbar\omega)^2 + \{\hbar\Gamma(\mathbf{Q})\}^2}. \qquad (5.176)$$

Comparing this response with the corresponding expression for the continuous diffusion model, eqn (5.166), we see that both are Lorentzian functions of energy but now the energy width $\hbar\Gamma(\mathbf{Q})$ of the quasielastic scattering is anisotropic (i.e. dependent on the direction of $\mathbf{Q}$). This is a consequence of the fixed geometry of the lattice of sites. In fact, to within a constant, $\Gamma(\mathbf{Q})$ is the Fourier transform of the arrangement of nearest-neighbour sites, as is evident from (5.175).

A simple application of the Chudley–Elliott model is to atoms diffusing on a primitive cubic lattice. Each lattice site has six nearest neighbours ($z = 6$) and the jump vectors $\mathbf{l}_j$ are $(\pm a, 0, 0)$, $(0, \pm a, 0)$ and $(0, 0, \pm a)$, where $a$ is the lattice constant. Hence, from (5.175),

$$\Gamma(\mathbf{Q}) = \frac{1}{3\tau}\{3 - \cos(Q_x a) - \cos(Q_y a) - \cos(Q_z a)\}. \qquad (5.177)$$

For small $Q$ we can use the expansion $\cos x \approx 1 - x^2/2$ to write $\Gamma(\mathbf{Q}) \approx Q^2 a^2/6\tau$. Hence, in this limit we recover the same form as that derived from the continuous diffusion model, and from (5.167) we find

$$D = \frac{a^2}{6\tau}. \qquad (5.178)$$

This provides a useful connection between the microscopic and macroscopic descriptions of diffusion.

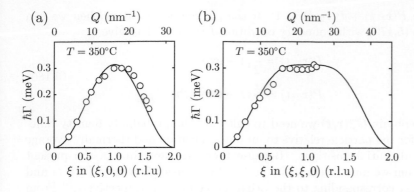

Fig. 5.21 Quasielastic widths extracted from data collected on a single crystal of $PdH_{0.03}$, with **Q** along (a) (100), (b) (110), Rowe *et al.* (1972). The line is calculated from the Chudley–Elliott model assuming the H atoms diffuse among the octahedral interstitial sites of the host, and with $\tau = 2.8\,\mathrm{ps}$.

Incoherent quasielastic neutron scattering has been used extensively to investigate hydrogen diffusion in metals. The large incoherent cross-section of hydrogen means that any coherent quasielastic scattering can be neglected. Corrections can be made for the incoherent scattering of the host metal from measurements on a hydrogen-free sample. A classic example is the diffusion of hydrogen in palladium. Palladium has a face-centred cubic (FCC) lattice, and H atoms are found to jump randomly among interstitial sites with octahedral symmetry. Figure 5.21 shows the measured quasielastic widths for a single crystal of Pd containing 3% H. The anisotropy in the widths is seen to agree well with the predictions of the Chudley–Elliott model (see Exercise 5.9), although the detailed mechanism of diffusion in this and other systems probably requires a more sophisticated model.

## 5.8.2 Spatially limited jump diffusion

We next consider diffusion among a restricted set of sites. This applies in cases where the diffusing particle is trapped within a region which is spatially or energetically separated from the rest of the system. The particle then jumps among the fixed number of sites in the trap. Examples of spatially limited diffusion are the rotational jump motion of whole molecules or of molecular side groups between different orientations with similar energies, and the tunneling dynamics of a molecule between two or more equivalent configurations (isomers).

The method of analysis is similar to that used to solve the Chudley–Elliott model in the previous section. In order to illustrate it, we consider the simple case of a particle jumping over an energy barrier between two sites $\mathbf{r}_1$ and $\mathbf{r}_2$ (Fig. 5.22). The mean residence time on the sites is $\tau$. Let the probability of finding the particle on site $\mathbf{r}_1$ at time $t$ be $P(\mathbf{r}_1, t)$, and similarly $P(\mathbf{r}_2, t)$ for site $\mathbf{r}_2$. These probabilities are given by rate equations like (5.169). For example,

$$\frac{\partial P(\mathbf{r}_1, t)}{\partial t} = \frac{1}{\tau}\{P(\mathbf{r}_2, t) - P(\mathbf{r}_1, t)\}$$

$$= \frac{1}{\tau}\{1 - 2P(\mathbf{r}_1, t)\}, \tag{5.179}$$

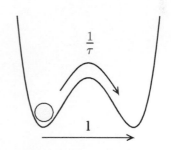

Fig. 5.22 Jump diffusion between two sites displaced by **l**.

since $P(\mathbf{r}_1,t) + P(\mathbf{r}_2,t) = 1$. If the particle starts at $\mathbf{r}_1$ then we can solve (5.179) with boundary condition $P(\mathbf{r}_1,0) = 1$ to give

$$P(\mathbf{r}_1,t) = \frac{1}{2}\{1 + \exp(-2t/\tau)\}$$

$$P(\mathbf{r}_2,t) = \frac{1}{2}\{1 - \exp(-2t/\tau)\}. \qquad (5.180)$$

To obtain $G_{\mathrm{s}}^{\mathrm{cl}}(\mathbf{r},t)$ we need to calculate the probability density function for the particle relative to an initial position and thermally average over all initial positions. Because there are two sites in the present problem we must average the probability density functions $P_1(\mathbf{r},t)$ and $P_2(\mathbf{r},t)$ corresponding to the origin at $\mathbf{r}_1$ and $\mathbf{r}_2$, respectively.[50] From (5.180) and the analogous probabilities for the case $P(\mathbf{r}_2,0) = 1$ one can easily see that the probability density functions are

$$P_1(\mathbf{r},t) = \frac{1}{2}\{1 + \exp(-2t/\tau)\}\delta(\mathbf{r}) + \frac{1}{2}\{1 - \exp(-2t/\tau)\}\delta(\mathbf{r} - \mathbf{l})$$

$$P_2(\mathbf{r},t) = \frac{1}{2}\{1 + \exp(-2t/\tau)\}\delta(\mathbf{r}) + \frac{1}{2}\{1 - \exp(-2t/\tau)\}\delta(\mathbf{r} + \mathbf{l}), \qquad (5.181)$$

where $\mathbf{l} = \mathbf{r}_2 - \mathbf{r}_1$. The self pair correlation function is then given by,

$$G_{\mathrm{s}}^{\mathrm{cl}}(\mathbf{r},t) = \frac{1}{2}\{P_1(\mathbf{r},t) + P_2(\mathbf{r},t)\}$$

$$= \frac{1}{2}\{1 + \exp(-2t/\tau)\}\delta(\mathbf{r})$$

$$+ \frac{1}{4}\{1 - \exp(-2t/\tau)\}\{\delta(\mathbf{r} - \mathbf{l}) + \delta(\mathbf{r} + \mathbf{l})\}. \qquad (5.182)$$

From (5.172), the self intermediate scattering function is found to be

$$I_s^{\mathrm{cl}}(\mathbf{Q},t) = A_0(\mathbf{Q}) + A_1(\mathbf{Q})\exp(-2t/\tau), \qquad (5.183)$$

where

$$A_0(\mathbf{Q}) = \frac{1}{2}\{1 + \cos(\mathbf{Q}\cdot\mathbf{l})\}$$

$$A_1(\mathbf{Q}) = \frac{1}{2}\{1 - \cos(\mathbf{Q}\cdot\mathbf{l})\}, \qquad (5.184)$$

and the Fourier transform gives the incoherent response function,

$$S_{\mathrm{i}}^{\mathrm{cl}}(\mathbf{Q},\omega) = A_0(\mathbf{Q})\delta(\hbar\omega) + A_1(\mathbf{Q})\frac{1}{\pi}\frac{\hbar\Gamma}{(\hbar\omega)^2 + (\hbar\Gamma)^2}, \qquad (5.185)$$

where $\Gamma = 2/\tau$.

For an isotropic sample, i.e. a powder or liquid, one must perform a spherical average over all possible orientations of $\mathbf{l}$. The result is that $A_0(\mathbf{Q})$ and $A_1(\mathbf{Q})$ are replaced by (see Section 2.4.1)

$$A_0(Q) = \frac{1}{2}\{1 + j_0(Ql)\}$$

$$A_1(Q) = \frac{1}{2}\{1 - j_0(Ql)\}, \qquad (5.186)$$

where $j_0(x) = \sin x / x$ is the first spherical Bessel function.[51]

[50] In the case of unrestricted diffusion on a Bravais lattice there was no need to average over all the initial positions because the lattice is infinite, and so the probability density function (5.169) is the same for all choices of origin.

[51] Also knows as the sinc function.

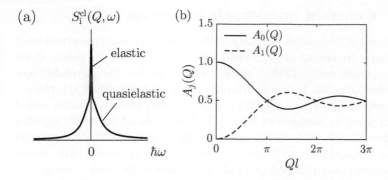

**Fig. 5.23** Incoherent scattering from single-particle jump diffusion on two sites a distance $l$ apart. (a) Elastic and quasielastic components shown schematically as a function of energy transfer. (b) $Q$ dependence of the spherically averaged elastic and quasielastic incoherent structure factors, eqns (5.186).

Comparing (5.185) with the equivalent expression for unrestricted jump diffusion, eqn (5.176), we see that both have a Lorentzian quasielastic component, but spatially limited diffusion has, in addition, a purely elastic component. The energy spectrum described by (5.185) is shown schematically in Fig. 5.23. The presence of an elastic component in the incoherent neutron scattering can therefore be taken as evidence for restricted motion of the diffusing particle, assuming, of course, that it does not come from the atoms making up the host structure.

Elastic scattering depends on the time-averaged particle occupancy of each site. The elastic component arises here because the diffusing particle is confined to wander back and forth between a restricted number of sites, so that after infinite time has elapsed the probability of finding the particle on each of the sites in the trap does not vanish, as it would for unrestricted diffusion. The coefficient $A_0(\mathbf{Q})$ of the delta function in (5.185) is called the *elastic incoherent structure factor* (EISF),[52] and is the squared modulus of the Fourier transform of the distribution of sites in the trap, normalized by the number of sites. For the two-site case this is evident from (5.184). For a general distribution of $N$ sites, each of which can accommodate the particle with equal probability, it can easily be shown[53] that the EISF is given by,

[52] Similarly, $A_1(\mathbf{Q})$ is the quasielastic incoherent structure factor.

[53] Put $t = \infty$ in (5.112) and follow the procedure in Section 3.6.

$$A_0(\mathbf{Q}) = \left| \frac{1}{N} \sum_j \langle \exp(i\mathbf{Q} \cdot \mathbf{r}_j) \rangle \right|^2, \qquad (5.187)$$

where the summation is over the $N$ sites that make up the trap.

From the sum rule for incoherent scattering, eqn (5.137), $S_i(\mathbf{Q})$ is independent of $\mathbf{Q}$, and so $A_0(\mathbf{Q}) + A_1(\mathbf{Q}) = 1$.

A good example of spatially limited diffusion is the dynamics of methyl ($CH_3$) side groups. Under certain conditions, the $CH_3$ groups can make random rotational jumps about the bond joining the C atom to the rest of the molecule. This bond is a threefold symmetry axis, so the $CH_3$ group diffuses between equivalent orientations by making $\pm 120°$ jumps. Figure 5.24 shows measurements of the EISF of $Si_8O_{12}(CH_3)_8$, which has a cage-like structure of Si and O with $CH_3$ groups bonded to Si atoms at the vertices of a cube. The experimental data are seen to be in good agreement with the 3-site jump model (see Exercise 5.10).

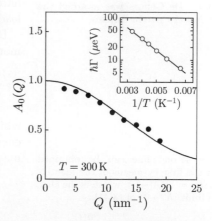

**Fig. 5.24** Elastic incoherent structure factor of $Si_8O_{12}(CH_3)_8$ (Jalarvo *et al.*, 2014). The line is calculated from the jump model with 120° rotations of the $CH_3$ groups. Insert: Arrhenius plot of the quasielastic width (HWHM) revealing activated behaviour with an activation energy of 53.1 meV. (Data courtesy of Mike Crawford.)

### 5.8.3   Coherent quasielastic scattering

The preceding discussion of quasielastic scattering was concerned with the diffusive dynamics of individual particles in liquids or solids as revealed by *incoherent* QENS. *Coherent* QENS probes the correlated motion of the particles. In the hydrodynamic limit (very small $Q$), the coherent response function $S(Q, \omega)$ consists of two inelastic peaks, called Brillouin lines, at energies of $\pm \hbar \omega_s(Q)$ due to sound waves,[54] together with a quasielastic peak due to collective thermal fluctuations. At larger $Q$ values, comparable with inverse inter-atomic distances, the sound waves become over-damped and only the quasielastic peak remains.[55]

The complexity of the many-body problem makes it difficult to develop simple models for coherent QENS, and so numerical methods are important in this field. *Molecular dynamics*, which involves computer simulations of the time evolution of systems containing many particles interacting via model inter-atomic potentials, has been widely used to calculate the cooperative motions of atoms and molecules in systems of interest to researchers in chemical physics, materials science, soft matter, and biology.

Despite the difficulties in describing correlated diffusive motions, one widely applicable result is that quasielastic linewidths narrow in the vicinity of the maximum in the structure factor $S(Q)$. This interesting phenomenon was first described by de Gennes (1959), and is known as *de Gennes narrowing*.[56] The usual explanation is that the maximum in $S(Q)$ is where the constructive interference is greatest and hence where the system is most correlated. The more correlated a system is, the harder it is to accommodate rearrangements and so the slower it decays, leading to a narrower quasielastic line.

De Gennes came to the same conclusion by considering a classical monatomic liquid and calculating the mean squared energy transfer

$$\overline{(\hbar\omega)^2} = \frac{S_2(\mathbf{Q})}{S(\mathbf{Q})}, \tag{5.188}$$

which is a measure of the quasielastic linewidth. $S_2(\mathbf{Q})$ is the second energy moment of $S(\mathbf{Q}, \omega)$, which from eqns (3.72), (5.106) and (5.115) may be written[57]

$$\begin{aligned}
S_2(\mathbf{Q}) &= \left(\frac{\hbar}{\mathrm{i}}\right)^2 \frac{1}{N} \left\{ \frac{\partial^2}{\partial t^2} \langle \hat{n}^\dagger(\mathbf{Q})\hat{n}(\mathbf{Q}, t) \rangle \right\}_{t=0} \\
&= \frac{\hbar^2}{N} \left\langle \frac{\partial \hat{n}}{\partial t}(\mathbf{Q}) \frac{\partial \hat{n}^\dagger}{\partial t}(\mathbf{Q}) \right\rangle \\
&= \frac{\hbar^2 Q^2}{N} \sum_{jk} \langle \dot{z}_j \dot{z}_k \exp\{\mathrm{i}Q(z_k - z_j)\} \rangle. \tag{5.189}
\end{aligned}$$

We have taken the $z$ coordinate to be along $\mathbf{Q}$, so that $\mathbf{Q} \cdot \mathbf{r}_j = Q z_j$, and $\dot{z}$ is shorthand for $\mathrm{d}z/\mathrm{d}t$. There is no correlation between the velocities of different particles, so only the $j = k$ terms are non-zero. Moreover,

[54]Sound waves are density waves, i.e. low-frequency longitudinal acoustic phonons, with wavelengths far in excess of inter-atomic distances.

[55]An exception is quantum fluids, such as liquid $^4$He, which support propagating modes to much higher $Q$ than do classical fluids.

[56]Pierre-Gilles de Gennes (1932–2007), French physicist who made important contributions in the fields of magnetism, superconductivity, and soft matter. De Gennes was awarded the 1991 Nobel Prize in Physics.

[57]The second line can be obtained from the first by means of the identity $\langle \hat{A}\hat{B}(t) \rangle = \langle \hat{B}\hat{A}(-t) \rangle$, which is the classical limit ($\hbar \to 0$) of (3.41):

$$\begin{aligned}
\frac{\partial^2}{\partial t^2} \langle \hat{n}^\dagger \hat{n}(t) \rangle &= \frac{\partial}{\partial t} \langle \hat{n}^\dagger \frac{\partial \hat{n}(t)}{\partial t} \rangle \\
&= \frac{\partial}{\partial t} \langle \frac{\partial \hat{n}}{\partial t} \hat{n}^\dagger(-t) \rangle \\
&= -\langle \frac{\partial \hat{n}}{\partial t} \frac{\partial \hat{n}^\dagger}{\partial t} \rangle.
\end{aligned}$$

In the last line we evaluated the time derivatives at $t = 0$ and used $\partial \hat{n}^\dagger(-t)/\partial t = -\partial \hat{n}^\dagger(t)/\partial t$.

from the equipartition theorem $\frac{1}{2}M\langle\dot{z}^2\rangle = \frac{1}{2}k_BT$, and therefore

$$\overline{(\hbar\omega)^2} = \frac{\hbar^2 Q^2 k_B T}{S(\mathbf{Q})M}. \tag{5.190}$$

This implies an energy narrowing at the maximum of $S(\mathbf{Q})$, consistent with our intuition. It also implies that the true quasielastic line shape must deviate from a Lorentzian function, because $\overline{(\hbar\omega)^2}$ is finite whereas the second moment of a Lorentzian is infinite.

Coherent quasielastic scattering can be studied experimentally on systems in which the cross-section for coherent scattering dominates that for incoherent scattering. A good example is liquid aluminium ($\sigma_{coh} = 1.495\,\text{b}$, $\sigma_{inc} = 0.008\,\text{b}$). Figure 5.25 illustrates the de Gennes narrowing of liquid aluminium at 943 K in the vicinity of the maximum of $S(Q)$ at $26.5\,\text{nm}^{-1}$.

## 5.8.4 High-resolution spectroscopy

We conclude this account of conventional quasielastic neutron scattering with some broader considerations. First, the QENS cross-section for real systems will contain Debye–Waller factors of the form $\exp(-Q^2\langle u^2\rangle)$, where $\langle u^2\rangle$ is some characteristic mean squared atomic displacement. Debye–Waller factors arise from random fluctuations in the atomic positions which spread out the average nuclear scattering potential (see Section 5.1.4). The effect on QENS is to reduce the intensity relative to that for point-like nuclei.

Second, QENS is not the only type of scattering that can be observed at very small energy transfers. The methyl group rotation described in Section 5.8.2 is a case in point. At sufficiently low temperatures, quantum-mechanical tunneling through the rotational potential barrier of the methyl group couples the individual proton wave functions and lifts the degeneracy of their ground states. The resulting energy splitting can be measured as a peak in the neutron spectrum. An example is shown in Fig. 5.26. At higher temperatures, coupling to phonons causes decoherence, and the behaviour crosses over to the classical stochastic jump diffusion described in Section 5.8.2.

Finally, it is worth mentioning that in QENS the elastic scattering is not, in reality, an infinitely narrow line in the energy spectrum. It is actually a peak with a width equal to the energy resolution of the spectrometer (often modelled as a Gaussian function — see Section 10.1.6). As a result, it is not always easy to separate the quasielastic scattering from the elastic peak unless the widths of both components are very different. Furthermore, the non-vanishing width of the elastic peak means that it includes scattering from motions on a timescale up to roughly the inverse of the frequency width as well as the truly elastic scattering. What appears elastic on one spectrometer, therefore, could turn out to be quasielastic if measured on another spectrometer with higher resolution. This has implications for measurement of the EISF. For example, quasielastic scattering due to reorientations of parts of a large molecule

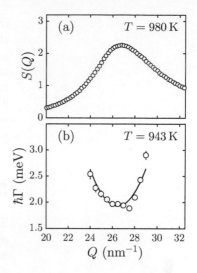

**Fig. 5.25** An example of de Gennes narrowing. (a) Structure factor $S(Q)$ of liquid aluminium, from Dahlborg *et al.* (2013) — see also Fig. 2.22. (b) Half width at half-maximum $\hbar\Gamma$ of coherent quasielastic scattering from liquid aluminium (Demmel *et al.*, 2011). The line is from a theoretical model. (Data courtesy of F. Demmel.)

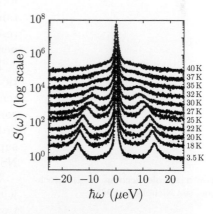

**Fig. 5.26** Tunneling spectra of 2-methylpyrazine with chloranilic acid at different temperatures (Rok *et al.*, 2018). The peaks at $\hbar\omega \neq 0$ are caused by a splitting of the ground state due to proton tunneling in the $CH_3$ groups. The curves are averaged over $Q$, and have been displaced vertically for clarity. (Data courtesy of Magdalena Rok.)

may be accompanied by what looks like elastic scattering. However, a slow translational diffusion of the centre of mass of the molecule would mean that the elastic peak is actually quasielastic. If the EISF associated with the reorientational motion is to be determined then one needs to carry out a proper integration of the (quasielastic) elastic peak over all energy at fixed $\mathbf{Q}$. The problems encountered with this are analogous to those discussed in connection with the static approximation — see Section 3.8.4.

## 5.9   Neutron spin echo spectroscopy

The measurement of very slow motions requires very high energy resolution. With conventional spectrometers the highest resolution available on current sources is $\sim 1\,\mu eV$, which corresponds to a timescale of $\sim 10^{-9}\,s$. This resolution is achieved with low-energy (cold) neutrons and filtering devices (monochromators/analysers or choppers) that select only a narrow band of energies from those incident upon them. An obvious disadvantage is that the flux of neutrons passed by the filter(s) decreases as the bandwidth decreases, and the count rate at the detector is correspondingly smaller.

Neutron spin echo (NSE) spectroscopy is an ingenious technique invented by Mezei (1972) to get around the flux limitation normally associated with high resolution. It uses the precession of a polarized beam of neutrons in a magnetic field to measure very precisely the change in wavelength of each scattered neutron in a way that does not depend on its initial wavelength. This means the incident neutron beam can contain a relatively large spread in wavelengths (typically 10%), which represents an enormous flux gain compared to conventional high-resolution neutron spectroscopy.

The NSE technique increases the dynamic range accessible by quasielastic neutron scattering up to $\sim 10^{-7}\,s$, which enables a wide range of systems to be studied, from the dynamics of polymer chains, glasses, and complex fluids, to the lifetimes of excitations in superfluid helium and ordered magnets, and the motion of magnetic domains and flux lines in superconductors. In this section we will give a short summary of the NSE technique.

The principle of NSE is as follows.[58] A beam of polarized neutrons with spins perpendicular to the beam direction passes down the axis of two identical solenoids placed one after the other. The currents in each solenoid generate axial magnetic fields which are the same in strength but which *in effect* point in opposite directions. The neutron spins precess around the magnetic fields in accordance with (1.12), and because the fields are in opposite directions in each solenoid the sense of precession is reversed. The field strength and flight time is the same in each solenoid, so the number of clockwise precessions in one solenoid is exactly matched by the number of anticlockwise precessions in the other. Hence, the neutrons exit the second solenoid with exactly the same spin

[58]For a simple account of the technique, see Nicholson (1981).

(a)

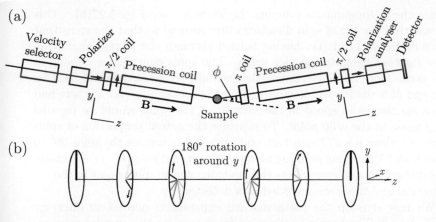

(b)

**Fig. 5.27** (a) The layout of a traditional neutron spin echo spectrometer. (b) Cartoon showing how the neutron spins de-phase in the first precession coil then re-phase in the second precession coil.

direction as when they entered the first solenoid. This is the *spin echo* effect. The echo condition is independent of the speed of the neutrons, so the beam can contain a range of wavelengths.

Now suppose the speed of the neutrons changes due to inelastic or quasielastic scattering from a sample placed between the two solenoids. The flight times in the two solenoids are now different, so the clockwise and anticlockwise precessions no longer balance. Instead of being parallel, the initial and final spin directions now differ by a phase angle which can be used to detect changes in energy in the sample. Practical values of field and neutron speed make energy changes as small as 10 neV observable. As we shall show below, after averaging over the full dynamic response of the sample the final polarization of the beam is proportional to the intermediate scattering function $I(\mathbf{Q}, t)$.

The practical layout of a traditional NSE spectrometer is sketched in Fig. 5.27(a). Neutrons from the source enter the spectrometer through a velocity selector which acts as a fairly coarse monochromator, giving typically 10% spread in the incident wavelength distribution. A polarizing supermirror after the velocity selector creates a polarized beam with the neutron spins aligned parallel to the beam direction, here defined to be the $z$ axis. A guide field pointing along $z$ maintains the neutron polarization until the neutrons reach a $\pi/2$ spin-turn coil. This device rotates the spins by 90° so that they emerge at right angles to the beam direction, along the $y$ axis say, and begin to precess around the $z$-axis field from the first precession solenoid. The number of precessions depends on the time in the solenoid, which means that the slower neutrons execute more turns than the faster ones. Hence, on exiting the solenoid the spins have 'fanned out' in the $xy$ plane.

The second precession solenoid is identical to the first, and is placed after the sample at a scattering angle $\phi$ to the incident beam direction. Contrary to how it was described earlier, the direction of the second precession field is not in reality the reverse of the first.[59] Instead, the same effect as a reversed field is achieved with a $\pi$ spin-turn coil placed between the two precession coils. The $\pi$ coil rotates the spins by 180° about the $y$ axis, so that the $x$ components of the spins reverse direction

[59] Otherwise there would be a region of zero field which could result in a loss of polarization.

while the $y$ components remains the same — see Fig. 5.27(b). This means that the fan of spin directions flips around so that the spins that were furthest ahead (i.e. having turned through the largest angle) are now furthest behind, and vice versa. The spins continue precessing in the same sense as they traverse the second precession solenoid, and by the end of it the fan will have closed up to produce an echo. If there had been no change in speed in the sample all the spins would be parallel to $y$ again at the echo point. To measure the actual proportion of spins pointing along $y$ a $\pi/2$ spin-turn coil placed here rotates the spins 90° in such a way that spins pointing along $y$ are rotated into the $+z$ direction. A supermirror analyser selects the neutrons whose spins point in the $+z$ direction and these are recorded in a detector.

We now develop the mathematical expressions needed to interpret NSE data.[60] The Larmor precession frequency for the angular momentum of a neutron in a steady magnetic field of flux density $B$ is $\omega_L = \gamma_L B$, where $\gamma_L = 2\gamma\mu_N/\hbar = 2\pi \times 2.916 \times 10^7\,\mathrm{rad\,s^{-1}\,T^{-1}}$ is the *gyromagnetic ratio* of the neutron (see Section 1.2). Over a path length $l$, the total precession angle of the spins is

$$\theta = \omega_L \tau = \frac{\gamma_L B l}{v}, \tag{5.191}$$

where $\tau$ is the flight time and $v$ is the speed of the neutron. In reality, $B$ is not constant over the path, so we must replace the product $Bl$ by the field integral

$$I_B = \int B\,\mathrm{d}l. \tag{5.192}$$

From the de Broglie equation (1.1) we can write $v = h/(m_n\lambda)$, where $h$ is Planck's constant, so that

$$\theta = \frac{\gamma_L m_n \lambda I_B}{h}. \tag{5.193}$$

After traversing the whole flight path the final beam polarization $P_z$ (equal to the $y$ component of polarization at the second $\pi/2$ coil) is given by

$$P_z = \cos(\theta_1 - \theta_2)$$
$$= \cos\{\frac{\gamma_L m_n}{h}(\lambda_i I_B^{(1)} - \lambda_f I_B^{(2)})\}, \tag{5.194}$$

where $\theta_1$ and $\theta_2$ are the precession angles in the two arms of the spectrometer, $\lambda_i$ and $\lambda_f$ are the initial and final neutron wavelengths, and $I_B^{(1)}$ and $I_B^{(2)}$ are the field integrals before and after the sample.

In NSE we are concerned with very small differences in the wavelengths ($\lambda_i \approx \lambda_f \approx \lambda$) and field integrals ($I_B^{(1)} \approx I_B^{(2)} \approx I_B$). To a very good approximation, therefore, we can write eqn (5.194) as

$$P_z = \cos\{\frac{\gamma_L m_n}{h}(I_B\Delta\lambda + \lambda\Delta I_B)\}, \tag{5.195}$$

where $\Delta\lambda = \lambda_f - \lambda_i$ and $\Delta I_B = I_B^{(2)} - I_B^{(1)}$.

[60]More detailed discussions of the theory can be found in Mezei (1980).

The next step is to average $P_z$ over the distribution of scattered neutron wavelengths. For this purpose, it is convenient to express $\Delta\lambda$ in terms of the energy transfer $\hbar\omega$. From (1.3) and (1.15)

$$\hbar\omega = \frac{h^2}{2m_n}\left(\frac{1}{\lambda_i^2} - \frac{1}{\lambda_f^2}\right)$$

$$\approx \frac{h^2}{m_n\lambda^3}\Delta\lambda. \tag{5.196}$$

Substitution of (5.196) into (5.195) allows us to write

$$P_z = \cos(\omega t + \alpha), \tag{5.197}$$

where

$$t = \frac{\gamma_L m_n^2 \lambda^3 I_B}{2\pi h^2} \quad \text{and} \quad \alpha = \frac{\gamma_L m_n \lambda \Delta I_B}{h}. \tag{5.198}$$

The parameter $t$ has dimensions of time, and is controlled by the neutron wavelength and the field integral.

To average $P_z$ over the distribution of final energies we need the scattering probability, which is given by the partial differential cross-section (3.39). Since $k_i \approx k_f$ this is essentially the response function $S(\mathbf{Q},\omega)$. Hence, the average over final energies can be written

$$P_z = \frac{\int S(\mathbf{Q},\omega)\cos(\omega t + \alpha)\,d(\hbar\omega)}{\int S(\mathbf{Q},\omega)\,d(\hbar\omega)}. \tag{5.199}$$

The practical limits of the integrals are set by the requirement to maintain good polarization in the precession coils, and this restricts the usable energy range to a small interval around zero energy transfer. If most of the spectral weight is in the quasielastic scattering contained within these limits, as is usually the case, then for the purpose of calculation the limits can be extended to $\pm\infty$ without much error. The denominator of (5.199) is then just the static structure factor $S(\mathbf{Q})$ — see (5.125). In the numerator, we can use a trigonometric identity to expand the cosine as $(\cos\omega t\cos\alpha - \sin\omega t\sin\alpha)$. To a good approximation, $S(\mathbf{Q},\omega)$ is the classical response function because $\hbar\omega \ll k_BT$, and so it is an even function of $\omega$ in the energy range probed by NSE — see Section 5.3.3. It follows that $S(\mathbf{Q},\omega)\sin\omega t$ is an odd function of $\omega$ and integrates to zero. With these considerations, (5.199) can be written

$cos(A+B) = \cos A\cos B - \sin A\sin B.$

$$P_z = \cos\alpha\frac{1}{S(\mathbf{Q})}\int S(\mathbf{Q},\omega)\cos\omega t\,d(\hbar\omega). \tag{5.200}$$

The integral is the real part of the intermediate scattering function $I(\mathbf{Q},t)$, which is identical with $S(\mathbf{Q})$ at $t=0$ — see (5.117). Hence,

$$P_z = I_0(\mathbf{Q},t)\cos\alpha, \tag{5.201}$$

where $I_0(\mathbf{Q},t)$ is the intermediate scattering function normalized to unity[61] at $t=0$.

[61] $I_0(\mathbf{Q},t) = I(\mathbf{Q},t)/S(\mathbf{Q})$ is often denoted by $S(\mathbf{Q},t)$ in the NSE literature.

Equation (5.201) shows that $I_0(\mathbf{Q}, t)$ can be determined from the amplitude of the oscillations in $P_z$ as $\alpha$ is varied (the echo amplitude). In practice, a small additional coil (the *phase coil*) is wound around the first precession solenoid, and a current sweep in this coil causes $\Delta I_B$ and hence $\alpha$ to be scanned very precisely. As mentioned earlier, $t$ is governed by $\lambda$ and $I_B$. In a traditional experiment $\lambda$ is fixed and the echo amplitude is measured at different values of $I_B$ set by the currents in the precession solenoids.[62] This yields $I_0(\mathbf{Q}, t)$ directly as a function of $t$.

[62]Time-of-flight NSE spectrometers have been constructed in which $\lambda$ varies at fixed $I_B$.

The analysis presented so far is correct for an ideal spectrometer. In practice, we must also take into account the distribution of wavelengths in the incident neutron beam and the distribution of paths (and hence field integrals) in the precession solenoids.

Let $f(\lambda)\,\mathrm{d}\lambda$ be the fraction of incident neutrons with wavelengths between $\lambda$ and $\lambda + \mathrm{d}\lambda$. The measured polarization is then the average of (5.201) over the wavelength distribution:

$$P_z = \int I_0(\mathbf{Q}, t) \cos\alpha \, f(\lambda) \, \mathrm{d}\lambda, \qquad (5.202)$$

Consider first the special case $t = 0$, whereupon $I_0(\mathbf{Q}, 0) = 1$. We see from (5.198) that $\alpha$ is proportional to the product $\Delta I_B \lambda$, and so from (5.202) the variation in $P_z$ with phase current (the echo) is the cosine Fourier transform of the wavelength distribution. When $t > 0$ this is no longer true since $I_0(\mathbf{Q}, t)$ depends on wavelength (through both $\mathbf{Q}$ and $t$), but it is approximately true as long as $I_0(\mathbf{Q}, t)$ varies slowly with wavelength over the range of wavelengths present, as is usually the case. A typical echo for a polychromatic neutron beam is illustrated in Fig. 5.28. The maximum in the envelope of the echo group occurs when $\Delta I_B = 0$, in which case

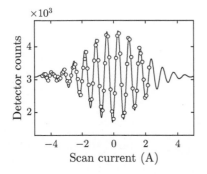

$\times 10^3$

**Fig. 5.28** A typical echo group recorded by scanning the phase current on the NSE spectrometer IN11C at the Institut Laue–Langevin. (Data courtesy of Peter Fouquet.)

$$(P_z)^{\mathrm{max}} = \int I_0(\mathbf{Q}, t) \, f(\lambda) \, \mathrm{d}\lambda. \qquad (5.203)$$

The procedure to obtain $(P_z)^{\mathrm{max}}$, therefore, is to measure $P_z$ as a function of the phase current and find the maximum of the echo.

The second issue we need to consider is the distribution of paths through the spectrometer, which exists because of the divergence of the neutron beam and the size of the sample. This means that the field integrals in the two arms of the spectrometer are not the same for all paths. To allow for this we write

$$\Delta I_B = \overline{\Delta I}_B + \delta I_B, \qquad (5.204)$$

where $\overline{\Delta I}_B$ is the average field integral and $\delta I_B$ is the deviation from average for a particular path. The maximum of the echo group will occur when $\overline{\Delta I}_B = 0$, but instead of setting $\alpha = 0$ as we did to reach (5.203) we must average $\cos\alpha$ over all paths to take into account that $\alpha$ depends on $\delta I_B$. So in place of (5.203) we write

$$(P_z)^{\mathrm{max}} = \int I_0(\mathbf{Q}, t) \, \langle\cos\alpha\rangle \, f(\lambda) \, \mathrm{d}\lambda, \qquad (5.205)$$

where the angular brackets denote an average over neutron paths. When $\delta I_B$ is small, $\cos \alpha$ does not vary much over the $\sim 10\%$ wavelength distribution and so it can be taken outside the integral. Therefore, to a good approximation,

$$(P_z)^{\text{max}} = \langle \cos \alpha \rangle \int I_0(\mathbf{Q}, t)\, f(\lambda)\, \mathrm{d}\lambda. \qquad (5.206)$$

The factor $\langle \cos \alpha \rangle$ is unknown, but it can easily be determined experimentally from a reference measurement on an elastic scattering sample.[63] For a purely elastic scatterer $I_0(\mathbf{Q}, t) = 1$, so that $(P_z)^{\text{max}} = \langle \cos \alpha \rangle$. Hence, the correction entails simple division of $(P_z)^{\text{max}}$ for the sample by $(P_z)^{\text{max}}$ for the reference elastic scatterer:

$$\frac{(P_z)^{\text{max}}}{(P_z)^{\text{max}}_{\text{elastic}}} = \int I_0(\mathbf{Q}, t)\, f(\lambda)\, \mathrm{d}\lambda. \qquad (5.207)$$

The effect of this correction is illustrated in Fig. 5.29

The above account of NSE spectroscopy assumes that the neutron polarization does not change on scattering from the sample, e.g. as for nuclear coherent and isotopic incoherent scattering. With nuclear spin incoherent scattering, on the other hand, the initial and final polarizations are related by $\mathbf{P}_{\text{f}} = -\frac{1}{3}\mathbf{P}_{\text{i}}$ [see Section 4.5.1 and eqn (4.101)], so when both coherent and spin incoherent scattering coexist the incoherent scattering subtracts from the coherent signal and the resulting polarization will depend on the relative amounts of each.

The polarization also changes in many magnetic scattering processes. In certain cases, such as isotropic paramagnets, the magnetic interaction rotates some or all of the incident polarization by $180°$ (see Exercise 4.5), thereby having the same effect as a $\pi$ spin-turn coil. To measure such processes, therefore, one simply removes the $\pi$ spin-turn coil from between the two precession coils. Without the $\pi$ coil, any nuclear scattering continues to de-phase in the second coil and does not produce an echo, so only the magnetic signal is detected. This way, NSE can be exploited to study a variety of magnetic phenomena, such as slow magnetic dynamics in spin glasses (see Mezei, 1980, and Pappas *et al.*, 2006).

To end this section, we list a few ongoing developments. We have already mentioned time-of-flight NSE spectroscopy. This is principally designed for use on spallation sources, which generate pulses containing a wide range of neutron wavelengths that can be separated by time-of-flight. A different variant is neutron resonance spin echo (NRSE), by which NSE resolution is achieved on measurements of propagating excitations, such as phonons and magnons (see Section 5.2.11). Spin echo small-angle neutron scattering (SESANS) has been developed to study structure on a length scale of $\sim 10\,\mu\text{m}$. To achieve this, the NSE precession coils have tilted faces and the polarized incident beam is split into two spatially separated beams, so that the precession phase after the second coil is proportional to the scattering angle. Finally, wide-angle NSE uses a novel magnetic field distribution and banks of multidetectors to enable simultaneous NSE measurements to be made over a wide range of scattering angles.

[63]Such as a glassy polymer. The reference elastic scatterer should ideally be the same shape and size as the real sample to replicate the same distribution of neutron paths.

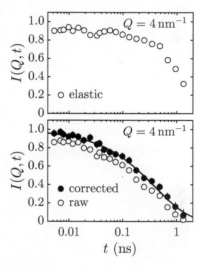

**Fig. 5.29** Resolution correction for NSE. Upper panel: spectrum of a glycerol sample at $T = 2\,\text{K}$, a good approximation to an elastic scatterer. Lower panel: spectrum of the same glycerol sample at $T = 300\,\text{K}$, before and after normalization to the elastic data. The line through the corrected data is a stretched exponential function (or Kohlrausch–Williams–Watts function) $f(t) = \exp\{-(t/\tau)^{\beta}\}$, with $\tau = 0.4\,\text{ns}$ and $\beta = 0.7$. The spectra were recorded on IN11C (Institut Laue–Langevin) with a wavelength of $0.55\,\text{nm}$. (Data courtesy of Peter Fouquet.)

# Chapter summary

- An overview has been given of the different types of structural dynamics found in condensed matter, and the associated neutron scattering cross-sections have been developed.

- In crystalline solids, the fundamental excitations are coupled atomic vibrations which in the harmonic approximation are described by a set of quantized normal modes (phonons). Inelastic neutron scattering occurs via exchange of energy quanta with the modes. The one-phonon scattering intensity varies as $Q^2$.

- The phonon dispersion relation $\omega(\mathbf{q})$ linking the angular frequency $\omega$ and wavevector $\mathbf{q}$ of the modes is periodic in the reciprocal lattice, i.e. $\omega(\mathbf{q} + \mathbf{G}) = \omega(\mathbf{q})$, where $\mathbf{G}$ is a reciprocal lattice vector.

- Elastic (Bragg) scattering from a crystal with mobile atoms is the same as for a rigid crystal apart from the inclusion of the Debye–Waller factor $\exp(-W_d)$ in the nuclear unit-cell structure factor,

$$F_{\mathrm{N}}(\mathbf{G}) = \sum_d \bar{b}_d \exp(-W_d) \exp(\mathrm{i}\mathbf{G} \cdot \mathbf{d}).$$

- The dynamics of non-crystalline materials, e.g. liquids, glasses, molecular fluids, soft matter, can be described by time-dependent van Hove pair correlation functions. For a one-component system

$$G(\mathbf{r}, t) = \frac{1}{N} \int \langle \hat{n}(\mathbf{r}', 0)\hat{n}(\mathbf{r}' + \mathbf{r}, t) \rangle \, \mathrm{d}^3 \mathbf{r}',$$

where $\hat{n}(\mathbf{r}, t) = \sum_j \delta\{\mathbf{r} - \hat{\mathbf{r}}_j(t)\}$ is the time-dependent particle density operator. The inelastic scattering response function is then

$$S(\mathbf{Q}, \omega) = \frac{1}{2\pi\hbar} \int \int G(\mathbf{r}, t) \exp\{\mathrm{i}(\mathbf{Q} \cdot \mathbf{r} - \omega t)\} \, \mathrm{d}^3 \mathbf{r} \, \mathrm{d}t.$$

- Approximate models often start from a classical version of $G(\mathbf{r}, t)$,

$$G^{\mathrm{cl}}(\mathbf{r}, t) = \sum_k \langle \delta\{\mathbf{r} - \hat{\mathbf{r}}_k(t) + \hat{\mathbf{r}}_0\} \rangle,$$

from which $S^{\mathrm{cl}}(\mathbf{Q}, \omega)$ can be calculated and the true response function approximated by $S(\mathbf{Q}, \omega) \simeq \exp(\beta\hbar\omega/2) S^{\mathrm{cl}}(\mathbf{Q}, \omega)$.

- Partial time-dependent correlation and response functions are defined for multicomponent systems.

- Single-particle recoil measurements by neutron Compton scattering probe momentum distributions and mean kinetic energies.

- Stretching, bending, and librational modes of molecular sub-units can be observed by vibrational spectroscopy.

- Quasielastic and neutron spin echo spectroscopy are used to measure slow, relaxational dynamics, such as centre-of-mass diffusion and molecular reorientations.

# Further reading

For an introduction to crystal dynamics, see Dove (2011). Two comprehensive reviews of neutron Compton scattering have been published by Andreani *et al.* (2005) and (2017). The technique of neutron vibrational spectroscopy is described fully in the book by Mitchell *et al.* (2005), and more details of the techniques and applications of quasielastic neutron scattering can be found in Bée (1988), Hempelmann (2000), and Garcia Sakai *et al.* (2012). Recent developments in neutron spin echo spectroscopy are described in Mezei *et al.* (2003), and applications of neutron spin echo in polymer science are reviewed by Richter *et al.* (2005).

# Exercises

(5.1) Starting from the Hamiltonian $\mathcal{H}$ for the harmonic oscillator, eqn (5.10), and the Bose commutation relations (5.9), show that $[\mathcal{H}, a^\dagger] = \hbar\omega_0 a^\dagger$. Hence, show that $a^\dagger|n\rangle$ is an energy eigenstate of $\mathcal{H}$ with eigenvalue $E_n + \hbar\omega_0$, and similarly $a|n\rangle$ is an eigenstate of $\mathcal{H}$ with eigenvalue $E_n - \hbar\omega_0$. These results demonstrate that $a^\dagger$ and $a$ are ladder operators which convert between energy eigenstates of the harmonic oscillator.

(5.2) Explain why the energy eigenvalues of the harmonic oscillator must be positive. Prove from (5.10) that the smallest energy eigenvalue is $E_0 = \frac{1}{2}\hbar\omega_0$, and hence that the energy eigenvalues are given by $E_n = (n + \frac{1}{2})\hbar\omega_0$.

(5.3) Show that eqns (5.12) and (5.13),

$$a|n\rangle = n^{\frac{1}{2}}|n-1\rangle$$
$$a^\dagger|n\rangle = (n+1)^{\frac{1}{2}}|n+1\rangle,$$

can be obtained from the results of Exercise 5.1, together with eqns (5.9) and (5.11).

(5.4) Show from (5.57) that the eigenvectors $\mathbf{x}_{qj}$ of the normal modes of the diatomic chain satisfy the orthonormality condition $(\mathbf{x}_{qj})^* \cdot \mathbf{x}_{qj'} = \delta_{jj'}$.

(5.5) (a) Show that the speed of sound on the diatomic chain in Section 5.2.2 is given by

$$v_{\mathrm{s}} = \sqrt{\frac{K_1 K_2 a^2}{(K_1 + K_2)(m_1 + m_2)}}.$$

(b) Show that the angular frequency of the optic mode at $q = 0$ is

$$\omega_0^{\mathrm{op}} = \sqrt{\frac{(K_1 + K_2)(m_1 + m_2)}{m_1 m_2}}.$$

(5.6) [Harder] The vibrational energy of a crystal (5.72) is the sum of the kinetic energy $T$ and the potential energy $V$. Show that

$$T = \frac{1}{2}\sum_s \dot{Q}_s \dot{Q}_{-s}$$
$$V = \frac{1}{2}\sum_s \omega_s^2 Q_s Q_{-s}.$$

Hint: For $T$, use (5.70) to substitute for $\dot{u}_{ld}$, then apply (B.50), (5.65), (5.68) and (5.69) to simplify. For $V$, write the potential energy term in (5.72) as

$$V = \frac{1}{2}\sum_{ld}\sum_{l'd'}\sum_{\alpha\beta} V_{ldl'd'}^{\alpha\beta} u_{ld}^\alpha u_{l'd'}^\beta, \qquad (5.208)$$

and argue that $V_{ldl'd'}^{\alpha\beta}$ does not depend on the actual positions of cells $l$ and $l'$, but only on the relative displacement $\mathbf{R} = \mathbf{l} - \mathbf{l}'$. Then use (5.70) to replace $u_{ld}^\alpha$ and $u_{l'd'}^\beta$ with normal mode operators, and show that

$$V = \frac{1}{2}\sum_{dd'}\sum_{\alpha\beta}\sum_{\mathbf{q}}\sum_{\mathbf{R}} \frac{V_{Rdd'}^{\alpha\beta}}{\sqrt{m_d m_{d'}}} \mathrm{e}^{-i\mathbf{q}\cdot\mathbf{R}}$$
$$\times \sum_{jj'} e_{d\mathbf{q}j}^\alpha e_{d'-\mathbf{q}j'}^\beta Q_{\mathbf{q}j} Q_{-\mathbf{q}j'}. \qquad (5.209)$$

Obtain the equation of motion from (5.208), and hence show that

$$\sum_{\mathbf{R}} \frac{V_{Rdd'}^{\alpha\beta}}{\sqrt{m_d m_{d'}}} \mathrm{e}^{-i\mathbf{q}\cdot\mathbf{R}} = D_{dd'\mathbf{q}}^{\alpha\beta},$$

where $D^{\alpha\beta}_{dd'\mathbf{q}}$ is an element of the dynamical matrix $\mathbf{D_q}$. Finally, write (5.209) in terms of the eigenvectors $\mathbf{x_{q}}_j$ of $\mathbf{D_q}$, where $\mathbf{D_q}\mathbf{x_{q}}_j = \omega^2_{\mathbf{q}j}\mathbf{x_{q}}_j$, to obtain the desired form for $V$.

(5.7) Show from the crystal Lagrangian $L = T - V$ that $P_s = \dot{Q}_{-s}$ is the momentum conjugate to $Q_s$, and hence that
$$[Q_s, \dot{Q}_{-s'}] = i\hbar\delta_{ss'}.$$

(5.8) Show from (5.110) that $G^*(\mathbf{r}, t) = G(-\mathbf{r}, -t)$. Use this result together with the stationarity property of correlation functions (3.40) to prove that
$$\mathrm{Im}\{G(\mathbf{r}, t)\} = \frac{1}{2\,\mathrm{i}N}\int\langle\,[\hat{n}(\mathbf{r'}, 0), \hat{n}(\mathbf{r'} + \mathbf{r}, t)]\,\rangle\,\mathrm{d}^3\mathbf{r'}.$$

Hence, argue that the imaginary part of $G(\mathbf{r}, t)$ derives from quantum properties of the system.

(5.9) Hydrogen atoms in palladium diffuse by jumping among the octahedral interstitial sites (Section 5.8.1). The nearest-neighbour jump vectors are $\pm\frac{1}{2}(a, a, 0)$, $\pm\frac{1}{2}(a, -a, 0)$, $\pm\frac{1}{2}(a, 0, a)$, $\pm\frac{1}{2}(a, 0, -a)$, $\pm\frac{1}{2}(0, a, a)$ and $\pm\frac{1}{2}(0, a, -a)$, where $a$ is the lattice constant. Show that the quasielastic width $\Gamma$ in the Chudley–Elliott model, eqn (5.175), for $\mathbf{Q}$ along high-symmetry directions is given by
$$\Gamma = \frac{4}{3\tau}\sin^2(\pi\xi/2), \qquad\qquad (\xi, 0, 0),$$
$$= \frac{1}{3\tau}\left[\sin^2(\pi\xi) + 4\sin^2(\pi\xi/2)\right], \quad (\xi, \xi, 0),$$
$$= \frac{1}{\tau}\sin^2(\pi\xi), \qquad\qquad\qquad (\xi, \xi, \xi).$$

where $\mathbf{Q} = (Q_x, Q_y, Q_z)$ with $Q_x = 2\pi\xi/a$, etc.

(5.10) Jump diffusion among three equivalent sites which form an equilateral triangle of side $d$ can be described by three rate equations,
$$\frac{\partial P(\mathbf{r}_i, t)}{\partial t} = \frac{1}{\tau}\{-2P(\mathbf{r}_i, t) + P(\mathbf{r}_j, t) + P(\mathbf{r}_k, t)\},$$

obtained by cyclic interchange of the site indices $i$, $j$ and $k$. Show that the spherically averaged elastic and quasielastic incoherent structure factors are given by
$$A_0(Q) = \frac{1}{3}\{1 + 2j_0(Qd)\}$$
$$A_1(Q) = \frac{2}{3}\{1 - j_0(Qd)\},$$

and that the quasielastic component of the incoherent scattering function is a Lorentzian with half-width at half-maximum $3\hbar/\tau$.

(5.11) A neutron spin echo spectrometer has a maximum field integral of $I_B = 0.25\,\mathrm{Tm}$. Consider neutrons of wavelength $\lambda_1 = 0.6\,\mathrm{nm}$ and $\lambda_2 = 1.2\,\mathrm{nm}$. For each wavelength, calculate (a) the number of spin rotations that take place in one of the precession coils, (b) the spin echo time $t$, and (c) the fractional energy resolution $\Delta(\hbar\omega)/E_\mathrm{i}$.

# Magnetic Scattering: General Properties

**6**

The experimental discovery of antiferromagnetism, made by neutron diffraction (Shull and Smart, 1949), was a landmark in the history of both neutron scattering and magnetism. Not only did it reveal one of the most interesting and important magnetic ground states, predicted some years earlier by Louis Néel,[1] but it also showed that neutron scattering could be used to probe magnetic materials on an atomic scale to gain information about electronic states and about the interactions between individual magnetic moments. Since then, neutron scattering has made an immense contribution in the field of magnetism, especially to the development of theoretical models.

The present chapter and the following two are concerned with the use of neutron scattering as a tool to learn about magnetic structure and dynamics. We begin by establishing the basic features of the magnetic scattering cross-section, and introduce the dipole approximation, which is a very convenient way to simplify the magnetic scattering operator. The remainder of the chapter is concerned with the calculation of magnetic form factors, and with sum rules for magnetic scattering.

Following on, in Chapter 7 we describe magnetic diffraction (i.e. elastic magnetic scattering), which is the technique to probe spatial arrangements of ordered magnetic moments, and Chapter 8 is concerned entirely with magnetic neutron spectroscopy (i.e. inelastic magnetic scattering), the technique for measuring magnetic excitations. Sections C.3 and C.4 include a summary of the magnetism of atoms and ions, including the effects of the ligand (crystal) field in solids.

[1]Louis Néel (1904–2000) shared the 1970 Nobel Prize in Physics with Hannes Alfvén for his pioneering studies of magnetism in solids.

## 6.1 Magnetic scattering cross-section

The partial differential neutron scattering cross-section in the first Born approximation is given by the master equation (Section 3.2)

$$\frac{\mathrm{d}^2\sigma}{\mathrm{d}\Omega \mathrm{d}E_f} = \frac{k_f}{k_i}\left(\frac{m_n}{2\pi\hbar^2}\right)^2 \sum_{\lambda_i} p_{\lambda_i} \sum_{\lambda_f} \left|\langle\sigma_f\lambda_f|\, V(\mathbf{Q})\, |\sigma_i\lambda_i\rangle\right|^2 \delta(E_{\lambda_f} - E_{\lambda_i} - \hbar\omega),$$

(6.1)

where $V(\mathbf{Q})$ is the Fourier transform of the interaction potential between the neutron and the scattering system. Labels $\lambda_i$, $\lambda_f$ and $\sigma_i$, $\sigma_f$ represent initial and final states of the target and the neutron spin, respectively.

[2]We neglect the smaller electromagnetic terms discussed in Section 4.2.4.

For magnetic materials, the interaction potential contains a nuclear term and a magnetic term,[2]

$$V(\mathbf{Q}) = V_N(\mathbf{Q}) + V_M(\mathbf{Q}). \tag{6.2}$$

Because the matrix element of $V(\mathbf{Q})$ is squared in (6.1) the cross-section can be separated into three terms, one that is purely nuclear, one that is purely magnetic, and the cross term, which gives interference between nuclear and magnetic scattering. This chapter is primarily concerned with the scattering that is purely magnetic. Pure nuclear scattering, however, is ever present, and the question of how to distinguish nuclear and magnetic scattering is addressed in Sections 4.5.2 and 6.2.2. The nuclear–magnetic interference term vanishes for unpolarized neutrons as it is linear in the neutron polarization (see Section 4.9), but useful information can be derived from it with polarized neutrons (Section 7.6).

Focusing on the pure magnetic scattering, we recall from Chapter 4 (Section 4.2) that the magnetic potential is

$$V_M(\mathbf{Q}) = -\boldsymbol{\mu}_n \cdot \mathbf{B}(\mathbf{Q}) = -\mu_0\boldsymbol{\mu}_n \cdot \mathbf{M}_\perp(\mathbf{Q}), \tag{6.3}$$

with

$$\mathbf{M}_\perp(\mathbf{Q}) = \hat{\mathbf{Q}} \times \{\mathbf{M}(\mathbf{Q}) \times \hat{\mathbf{Q}}\}. \tag{6.4}$$

$\boldsymbol{\mu}_n = -2\gamma\mu_N\mathbf{s}_n$

[3]Note that due to a gauge freedom $\mathbf{M}(\mathbf{Q})$ is not uniquely defined, see Section 4.2.3. The magnetic interaction vector $\mathbf{M}_\perp(\mathbf{Q})$, on the other hand, is uniquely defined.

[4]The $(\gamma r_0/2\mu_B)^2$ factor is taken out of $S(\mathbf{Q},\omega)$ to avoid having to carry around a lot of constants.

Here, $\boldsymbol{\mu}_n$ is the magnetic moment of the neutron, $\mathbf{B}(\mathbf{Q})$ is the Fourier transform of the inhomogeneous magnetic flux density (or magnetic induction) in the sample, and $\mathbf{M}(\mathbf{Q})$ is the Fourier transform of the corresponding magnetization.[3] $\mathbf{M}_\perp$ is the *magnetic interaction vector*, which is the component of $\mathbf{M}$ perpendicular to the scattering vector $\mathbf{Q}$, and $\hat{\mathbf{Q}}$ is a unit vector in the direction of $\mathbf{Q}$. By combining (6.1) and (6.3) one can write the magnetic scattering cross-section in the form[4]

$$\frac{d^2\sigma}{d\Omega dE_f} = \frac{k_f}{k_i}\left(\frac{\gamma r_0}{2\mu_B}\right)^2 S(\mathbf{Q},\omega), \tag{6.5}$$

where $\gamma = 1.913$, $r_0 = \mu_0 e^2/(4\pi m_e) = 2.818 \times 10^{-15}$ m, and

$$S(\mathbf{Q},\omega) = \sum_{\lambda_i} p_{\lambda_i} \sum_{\lambda_f} \left|2\langle\sigma_f|\mathbf{s}_n|\sigma_i\rangle \cdot \langle\lambda_f|\mathbf{M}_\perp(\mathbf{Q})|\lambda_i\rangle\right|^2 \delta(E_{\lambda_f} - E_{\lambda_i} - \hbar\omega) \tag{6.6}$$

is the magnetic response function. The matrix elements of $\mathbf{s}_n$ (the neutron spin operator[5]) and $\mathbf{M}_\perp$ separate because $\mathbf{s}_n$ acts only on the neutron spin part of the wave function and $\mathbf{M}_\perp$ acts only on the electronic part of the wave function. By selecting which initial and final neutron spin states to measure one can probe different components of $\mathbf{M}_\perp$ and hence isolate the partial magnetic response functions

[5]Section C.1.15 describes how to evaluate the matrix elements of $\mathbf{s}_n$, and the values are given in Table C.2.

$$S_{\alpha\beta}(\mathbf{Q},\omega) = \sum_{\lambda_i} p_{\lambda_i} \sum_{\lambda_f} \langle\lambda_i|M_\alpha^\dagger(\mathbf{Q})|\lambda_f\rangle\langle\lambda_f|M_\beta(\mathbf{Q})|\lambda_i\rangle\delta(E_{\lambda_f} - E_{\lambda_i} - \hbar\omega). \tag{6.7}$$

This is the basic principle of neutron polarization analysis, a full account of which is given in Section 4.4.

The partial magnetic response function (6.7) may alternatively be expressed as the time Fourier transform of a correlation function (Section 3.4),

$$S_{\alpha\beta}(\mathbf{Q}, \omega) = \frac{1}{2\pi\hbar} \int_{-\infty}^{\infty} \langle \hat{M}_\alpha^\dagger(\mathbf{Q}) \hat{M}_\beta(\mathbf{Q}, t) \rangle \exp(-\mathrm{i}\omega t)\, \mathrm{d}t. \qquad (6.8)$$

A magnetic pair correlation function can also be defined:

$$S_{\alpha\beta}(\mathbf{Q}, \omega) = \frac{N}{2\pi\hbar} \int_{-\infty}^{\infty} \int \Gamma_{\alpha\beta}(\mathbf{r}, t) \exp\{\mathrm{i}(\mathbf{Q} \cdot \mathbf{r} - \omega t)\}\, \mathrm{d}^3\mathbf{r}\, \mathrm{d}t, \qquad (6.9)$$

with

$$\Gamma_{\alpha\beta}(\mathbf{r}, t) = \frac{1}{N} \int \langle \hat{M}_\alpha(\mathbf{r}', 0) \hat{M}_\beta(\mathbf{r}' + \mathbf{r}, t) \rangle\, \mathrm{d}^3\mathbf{r}'. \qquad (6.10)$$

$\Gamma_{\alpha\beta}(\mathbf{r}, t)$ measures correlations between the magnetization at different points in space and time, and is the magnetic analogue of the pair correlation function $G(\mathbf{r}, t)$ introduced in Section 5.3. The normalization factor $N$ is included purely for convenience. The choice of $N$ does not affect the cross-section since it cancels, but it could for example be the number of unpaired electrons, or the number of magnetic atoms. One can show that (6.9)–(6.10) is the same as (6.8) by the same procedure as in (5.108).

In the case of unpolarized neutrons, the magnetic response function (6.6) reduces to (see Section 4.9)[6]

$$S(\mathbf{Q}, \omega) = \sum_{\lambda_\mathrm{i}} p_{\lambda_\mathrm{i}} \sum_{\lambda_\mathrm{f}} |\langle \lambda_\mathrm{f}| \mathbf{M}_\perp(\mathbf{Q}) |\lambda_\mathrm{i}\rangle|^2 \delta(E_{\lambda_\mathrm{f}} - E_{\lambda_\mathrm{i}} - \hbar\omega) \qquad (6.11)$$

$$= \sum_{\alpha\beta} (\delta_{\alpha\beta} - \hat{Q}_\alpha \hat{Q}_\beta) S_{\alpha\beta}(\mathbf{Q}, \omega). \qquad (6.12)$$

> [6] In eqn (4.42) it was shown that
>
> $$\mathbf{M}_\perp^\dagger \cdot \mathbf{M}_\perp = \sum_{\alpha\beta} (\delta_{\alpha\beta} - \hat{Q}_\alpha \hat{Q}_\beta) M_\alpha^\dagger M_\beta.$$
>
> For $\alpha \neq \beta$, only the symmetric combination $S_{\alpha\beta} + S_{\beta\alpha}$ appears in $S(\mathbf{Q}, \omega)$.

## 6.1.1 Magnetic correlations and generalized susceptibility

Following the approach in Section 3.6, we can write the magnetic response as the sum of static and dynamic contributions,

$$S_{\alpha\beta}(\mathbf{Q}, \omega) = \langle M_\alpha(-\mathbf{Q}) \rangle \langle M_\beta(\mathbf{Q}) \rangle \delta(\hbar\omega) + \tilde{S}_{\alpha\beta}(\mathbf{Q}, \omega). \qquad (6.13)$$

The first term corresponds to elastic scattering, and is usually presented in terms of the differential cross-section. For unpolarized neutrons this takes the form[7]

$$\frac{\mathrm{d}\sigma}{\mathrm{d}\Omega} = \left(\frac{\gamma r_0}{2\mu_\mathrm{B}}\right)^2 |\langle \mathbf{M}_\perp(\mathbf{Q}) \rangle|^2. \qquad (6.14)$$

> [7] For use with polarized neutrons, one simply includes the matrix elements of $\mathbf{s}_\mathrm{n}$ that are shown in eqn (6.6).

The dynamic part of (6.13) describes the inelastic scattering, and is related through the *Fluctuation–Dissipation* theorem to the imaginary part of the *generalized susceptibility*:[8]

> [8] See Sections 3.7 and D.2.

$$\tilde{S}_{\alpha\beta}(\mathbf{Q}, \omega) = \{1 + n(\omega)\} \frac{1}{\pi} \chi_{\beta\alpha}''(\mathbf{Q}, \omega), \qquad (6.15)$$

$$n(\omega) = \frac{1}{\exp(\beta\hbar\omega) - 1}$$

where $n(\omega)$ is the Planck distribution. For unpolarized neutrons,

$$\widetilde{S}(\mathbf{Q}, \omega) = \sum_{\alpha\beta} (\delta_{\alpha\beta} - \hat{Q}_\alpha \hat{Q}_\beta)\{1 + n(\omega)\}\frac{1}{\pi}\chi''_{\alpha\beta}(\mathbf{Q}, \omega). \qquad (6.16)$$

The generalized susceptibility $\chi_{\alpha\beta}(\mathbf{Q}, \omega)$ represents the magnetic response of the system in the $\alpha$ direction to a weak field that is applied in the $\beta$ direction and which varies sinusoidally in space and time.

### 6.1.2   Magnetic and structural correlations

It is common to assume that magnetic scattering of neutrons takes place without change in the vibrational motion of the atoms. This is often a good approximation, but not always. In this section we shall clarify this approximation and at the same time develop the cross-section one step closer to a form suitable for the interpretation of experiments.

We assume that the magnetization is sufficiently localized that it can be written as the sum of the magnetization of each ion, i.e.

$$\mathbf{M}(\mathbf{r}) = \sum_j \mathbf{M}_j(\mathbf{r} - \mathbf{r}_j). \qquad (6.17)$$

The Fourier transform of this $\mathbf{M}(\mathbf{r})$ is

$$\mathbf{M}(\mathbf{Q}) = \int \sum_j \mathbf{M}_j(\mathbf{r} - \mathbf{r}_j) \exp(\mathrm{i}\mathbf{Q} \cdot \mathbf{r}) \, \mathrm{d}^3\mathbf{r}$$

$$= \sum_j \left\{ \int \mathbf{M}_j(\mathbf{r}') \exp(\mathrm{i}\mathbf{Q} \cdot \mathbf{r}') \, \mathrm{d}^3\mathbf{r}' \right\} \exp(\mathrm{i}\mathbf{Q} \cdot \mathbf{r}_j) \qquad [\mathbf{r}' = \mathbf{r} - \mathbf{r}_j]$$

$$= \sum_j \mathbf{M}_j(\mathbf{Q}) \exp(\mathrm{i}\mathbf{Q} \cdot \mathbf{r}_j), \qquad (6.18)$$

where $\mathbf{M}_j(\mathbf{Q})$ is the Fourier transform of the magnetization of ion $j$. On substituting (6.18) into (6.8) we obtain

$$S_{\alpha\beta}(\mathbf{Q}, \omega) = \frac{1}{2\pi\hbar} \int_{-\infty}^{\infty} I_{\alpha\beta}(\mathbf{Q}, t) \exp(-\mathrm{i}\omega t) \, \mathrm{d}t, \qquad (6.19)$$

where $I_{\alpha\beta}(\mathbf{Q}, t)$ is a partial intermediate scattering function defined by

$$I_{\alpha\beta}(\mathbf{Q}, t) = \sum_{jk} \left\langle \exp(-\mathrm{i}\mathbf{Q} \cdot \hat{\mathbf{r}}_j) \hat{M}^\dagger_{\alpha j}(\mathbf{Q}) \hat{M}_{\beta k}(\mathbf{Q}, t) \exp\{\mathrm{i}\mathbf{Q} \cdot \hat{\mathbf{r}}_k(t)\} \right\rangle. \qquad (6.20)$$

If the magnetic correlations are decoupled from the structural correlations then the thermal average in (6.20) can be factorized to give

$$I_{\alpha\beta}(\mathbf{Q}, t) = \sum_{jk} \left\langle \exp(-\mathrm{i}\mathbf{Q} \cdot \hat{\mathbf{r}}_j) \exp\{\mathrm{i}\mathbf{Q} \cdot \hat{\mathbf{r}}_k(t)\} \right\rangle \left\langle \hat{M}^\dagger_{\alpha j}(\mathbf{Q}) \hat{M}_{\beta k}(\mathbf{Q}, t) \right\rangle. \qquad (6.21)$$

This simplification is valid when the magnetic dynamics are independent of the vibrational motion of the atoms.[9]

[9]This is often the case, but there are exceptions. One example is the coupling between magnon and phonon modes of the same symmetry, which leads to hybridized modes with both magnetic and vibrational character. Another example is the *dynamic Jahn–Teller effect*, which describes the coupling between lattice vibrations and the orbital fluctuations of an ion with an orbitally-degenerate ground state.

Despite the assumption of decoupled magnetic and atomic dynamics, the magnetic cross-section (6.21) still depends on atomic degrees of freedom. Physically, this is because the magnetic moments are attached to atoms which are always moving. Going one step further, we can separate each of the two correlation functions in (6.21) into a static part and a dynamic part (see Sections 3.6 and 6.1.1), making four possible magnetic scattering processes:

$$\langle \text{static}\rangle\langle \text{static}\rangle \quad \langle \text{static}\rangle\langle \text{dynamic}\rangle \quad \langle \text{dynamic}\rangle\langle \text{static}\rangle \quad \langle \text{dynamic}\rangle\langle \text{dynamic}\rangle,$$
$$(6.22)$$

where the first factor of each product represents the structural correlation function and the second factor the magnetic correlation function. The first two of these products correspond to magnetic scattering that is elastic in the structure, i.e. scattering which does not affect the atomic motion. This is the scattering which is usually of most interest in studies of magnetic systems.

The scattering associated with the third product in (6.22) is elastic in the magnetic part and inelastic in the structure, i.e. neutrons create or destroy phonons through the magnetic interaction without change to the magnetic system. It can only occur in magnetically ordered systems, and is usually called *magnetovibrational scattering*.[10] To calculate it, one retains only the static part of the magnetic intermediate scattering function, $\langle \hat{M}_{\alpha j}^{\dagger}(\mathbf{Q})\hat{M}_{\beta k}(\mathbf{Q}, t = \infty)\rangle = \langle M_{\alpha j}(-\mathbf{Q})\rangle\langle M_{\beta k}(\mathbf{Q})\rangle$. The magnetovibrational scattering occurs at the same $\mathbf{Q}$ and $\omega$ as the nuclear scattering, but the intensity is different because the phonon structure factor, c.f. eqn (5.84), contains the static magnetic amplitude $\langle M_{\alpha j}(\mathbf{Q})\rangle$ in place of the nuclear coherent scattering length $\bar{b}_j$ (as well as some different constants), and so the magnetovibrational and nuclear phonon scattering intensities are not proportional to one another.

The last product in (6.22) corresponds to inelastic scattering in both the magnetic and structural systems. This scattering is generally rather featureless because the spectra of magnetic and phonon excitations tend to be broad, and so the many different combinations of magnetic and phonon excitations are spread over a wide range of $\mathbf{Q}$ and $\omega$. An exception is when the system is a paramagnet (see Section 6.2.4). The magnetic scattering from a paramagnet is quasielastic, and providing the quasielastic width is narrow compared with the phonon frequencies the double-inelastic term in (6.22) will measure the phonon spectrum in much the same way as does magnetovibrational scattering.

### 6.1.3   Static lattice approximation

From now on we shall neglect the third and fourth terms in (6.22) and proceed to an expression for the part of the magnetic scattering that is elastic in the structure. The task is quite easy because the structural correlation function, i.e. the factor in (6.21) containing $\mathbf{r}_j$ and $\mathbf{r}_k$, also appears in the nuclear scattering cross-section, which we have treated

[10]The equivalent nuclear scattering process, in which neutrons interact via the nucleus and create or destroy magnetic excitations without change to the structure, is also possible in principle but requires the nuclear interaction potential to depend on the magnetic state of the system. Such a dependence could be produced by a hyperfine coupling between the nuclear and electronic moments, but as the hyperfine coupling is extremely weak the effect would be very small.

previously in Section 5.2. Following the same approach, we put

$$\mathbf{r}(t) = \mathbf{R} + \mathbf{u}(t), \tag{6.23}$$

where $\mathbf{u}$ is the displacement of an atom relative to its mean position $\mathbf{R}$, and write

$$\langle \exp(-i\mathbf{Q} \cdot \hat{\mathbf{u}}_j) \exp\{i\mathbf{Q} \cdot \hat{\mathbf{u}}_k(t)\} \rangle$$
$$= \exp(-W_j - W_k)\left\{ 1 + \langle UV \rangle + \frac{1}{2!}\langle UV \rangle^2 + \dots \right\}. \tag{6.24}$$

The first term in the curly brackets ($= 1$) corresponds to static correlations, and the remaining terms of the form $\langle UV \rangle^n$ describe $n$-phonon processes. The exponential in front of the curly brackets is the Debye–Waller factor, with $W_j = \frac{1}{2}\langle (\mathbf{Q} \cdot \hat{\mathbf{u}}_j)^2 \rangle$. If we retain just the static part of (6.24), then from (6.19), (6.21) and (6.23) the partial magnetic response function becomes

$$S_{\alpha\beta}(\mathbf{Q}, \omega) = \sum_{jk} \exp(-W_j - W_k) \exp\{i\mathbf{Q} \cdot (\mathbf{R}_k - \mathbf{R}_j)\}$$
$$\times \frac{1}{2\pi\hbar} \int_{-\infty}^{\infty} \left\langle \hat{M}_{\alpha j}^{\dagger}(\mathbf{Q}) \hat{M}_{\beta k}(\mathbf{Q}, t) \right\rangle \exp(-i\omega t)\, dt. \tag{6.25}$$

Equation (6.25), which assumes localized magnetic moments and no scattering-induced change in the atomic motion, is the usual starting point for interpreting magnetic scattering.

## 6.2 Dipole approximation

### 6.2.1 Spin and orbital magnetization

The magnetization generated by the electrons has two sources, (i) the dipole field created by the intrinsic spin of the electrons, and (ii) charge currents associated with the orbital motion of the electrons. In the first detailed treatment of magnetic neutron scattering, given by Halpern and Johnson (1939) following initial studies by Bloch (1936) and Schwinger (1937), the analysis was restricted to the case of atoms with zero orbital magnetic moment. The theory was extended to atoms with both spin and orbital moments by Trammell (1953), and the matrix elements in the cross-section were presented in a more compact form using Racah tensors by Johnston (1966), and others (Lovesey, 1969, Johnston and Rimmer, 1969).[11] A little later, Stassis and Deckman (1975) demonstrated that magnetic neutron scattering could be formulated as a sum of magnetic and electric multipole moments (dipole, quadrupole, etc), analogous to the standard approach used for magnetic photon scattering.

The task of calculating the matrix elements of $\mathbf{M}(\mathbf{Q})$ when both spin and orbital magnetization are present can be extremely demanding, especially for atoms with two or more unpaired electrons. Fortunately,

[11] A comprehensive account of the magnetic neutron scattering matrix elements in the Racah formulation is given in Balcar and Lovesey (1989).

the formulae simplify greatly in the typical domain in which magnetic neutron scattering experiments are performed, and this makes it feasible to apply the technique to study a wide range of problems in magnetism.

In the following sections we describe these simplifications, discuss when it is reasonable to use them, and derive the corresponding approximate forms for the neutron cross-section. The *dipole approximation*, as these simplifications are known (Section 6.2.2), is adequate for the vast majority of investigations on magnetic order and dynamics. General expressions for the matrix elements of $\mathbf{M}(\mathbf{Q})$ for localized electrons are given in Section 6.3.2. These are sometimes needed, for example when accurate intensity calculations are required, or when detailed information is sought on atomic magnetization distributions, or for the interpretation of high-energy atomic multiplet spectra (see Section 8.1.2).

## 6.2.2   Dipole approximation form factor

The approximation applies when $|\mathbf{Q}|$ is smaller than the reciprocal of the atomic radius, although it is often quite satisfactory for larger $|\mathbf{Q}|$. We shall neglect any contribution to the magnetization from filled shells and assume atomic-like orbitals for the unpaired electrons. We further assume that the initial and final states belong to the same $l^n$ configuration. The latter constraint has two implications. First, all the unpaired electrons have the same radial wave function.[12] Second, only terms in the interaction potential with even parity contribute to the scattering, i.e. odd-order magnetic and even-order electric multipoles.[13]

When $|\mathbf{Q}|$ is small the scattering is dominated by the magnetic dipole term, and it can be shown that, to a good approximation, $\mathbf{M}(\mathbf{Q})$ for a single atom can be written

$$\mathbf{M}(\mathbf{Q}) \simeq -2\mu_B \big[ \langle j_0(Q) \rangle \mathbf{S} + \tfrac{1}{2} \{ \langle j_0(Q) \rangle + \langle j_2(Q) \rangle \} \mathbf{L} \big], \tag{6.26}$$

where $\mathbf{S}$ and $\mathbf{L}$ are operators for the total spin and total orbital angular momentum, and $\langle j_n(Q) \rangle$ denotes the radial integral (see Fig. 6.1)

$$\langle j_n(Q) \rangle = \int_0^\infty j_n(Qr) r^2 R^2(r)\, dr, \tag{6.27}$$

in which $j_n(Qr)$ is a spherical Bessel function[14] of order $n$.

When $Q = 0$, $\langle j_0 \rangle = 1$ and $\langle j_2 \rangle = 0$, so the right-hand side of (6.26) becomes $-\mu_B(\mathbf{L} + 2\mathbf{S})$, which is just the magnetic dipole moment operator $\boldsymbol{\mu}$ — see Section C.3. For this reason, approximation (6.26) is known as the *dipole approximation*.[15]

Physically, the dipole approximation corresponds to the neglect of any directional variation of the spin and orbital magnetization within the atom, i.e. the spin and orbital magnetization vectors have unique directions (not necessarily the same) and the spatial distribution of magnetization is isotropic (see Fig. 6.2). This is a good approximation at small $Q$, specifically when $Q < 1/R$, $R$ being the typical orbital radius, because in this regime the scattering is not sensitive to fine details in the shape of the atomic orbitals.

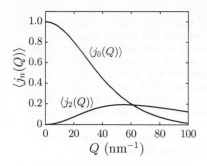

**Fig. 6.1** Radially averaged spherical Bessel functions $\langle j_0(Q) \rangle$ and $\langle j_2(Q) \rangle$. The curves are calculated for $Co^{2+}$, but are qualitatively similar for other magnetic ions.

[12] The radial wave function $R(r)$ is the solution of the one-electron radial Schrödinger equation, see Section C.2.1. The normalization condition is $\int_0^\infty r^2 R^2(r)\, dr = 1$.

[13] The multipole expansion of the scattering potential is described in Section C.5.

[14] Spherical Bessel functions appear in the solution of the wave equation in spherical polar coordinates. They can be evaluated from the relation
$$j_n(x) = (-x)^n \left( \frac{1}{x} \frac{d}{dx} \right)^n \frac{\sin x}{x}.$$

[15] The name is slightly misleading because when $Q \neq 0$ the right-hand side of (6.26) is not identical with the magnetic dipole part of $\mathbf{M}(\mathbf{Q})$ (see Section 6.3.1). For non-zero $\mathbf{Q}$, the dipole approximation is actually the approximation that results from retaining only the terms in $\mathbf{M}(\mathbf{Q})$ that depend on the magnitude but not the direction of $\mathbf{Q}$.

**Fig. 6.2** Illustration of the distribution of spin and orbital magnetization assumed in the dipole approximation. The size of the arrows represents the magnitude and direction of the magnetization. The patterns are equivalent to a single magnetic dipole moment spread out over a spherically symmetric electron cloud, but the radial dependences of the spin and orbital contributions are not the same.

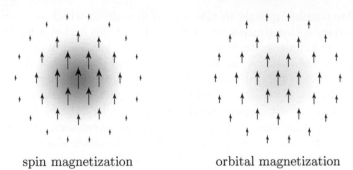

spin magnetization       orbital magnetization

[16]See Section C.4 for a discussion of the crystal field interaction and its energy scale relative to other interactions in magnetic ions.

The dipole approximation for $\mathbf{M}(\mathbf{Q})$ simplifies for weak and intermediate crystal fields:[16]

(a) *Weak crystal fields.* When the crystal field interaction is smaller than the spin–orbit coupling and intra-orbital Coulomb energy, which typically applies for $f$ electrons, the total angular momentum $\mathbf{J}$ is well defined. Within a $\mathbf{J}$ level one can replace $\mathbf{L}$ by $g_L\mathbf{J}$ and $2\mathbf{S}$ by $g_S\mathbf{J}$, where $g_L$ and $g_S$ are given in (C.133). With these substitutions, (6.26) becomes

$$\mathbf{M}(\mathbf{Q}) \simeq -g_J\mu_{\mathrm{B}}f(Q)\mathbf{J}. \qquad (6.28)$$

Here, $g_J$ is the Landé $g$-factor, eqn (C.131),

$$g_J = 1 + \frac{J(J+1) - L(L+1) + S(S+1)}{2J(J+1)}. \qquad (6.29)$$

with $L$, $S$, and $J$ the angular momentum quantum numbers, and

$$f(Q) = \langle j_0(Q)\rangle + \frac{2 - g_J}{g_J}\langle j_2(Q)\rangle \qquad (6.30)$$

is the *dipole magnetic form factor*, which in this approximation is a scalar function of $Q$ normalized to unity at $Q = 0$. Magnetic form factors are discussed in more detail in Section 6.3 below.

(b) *Intermediate crystal fields.* When the crystal field interaction is larger than the spin–orbit coupling but smaller than the intra-orbital Coulomb energy (typically applies to $3d$ electrons), the form taken by the dipole approximation depends on whether or not the orbital angular momentum of the ground state would be fully quenched by the crystal field *in the absence of spin–orbit coupling*. If not, then the spin and orbital angular momenta need to be treated independently and eqn (6.26) is the appropriate form for the dipole approximation. If it would be fully quenched, as is usually the case, then the observed moment equates to the spin moment plus a small additional orbital component either parallel or antiparallel to the spin moment induced by the non-zero spin–orbit coupling. The size of the orbital moment is given by how much the observed $g$-factor[17] differs from the spin-only value of 2. Hence, $\mathbf{L}$ can be

[17]See Section C.4.

replaced by $(g-2)\mathbf{S}$ and (6.26) can be written

$$\mathbf{M}(\mathbf{Q}) \simeq -g\mu_{\mathrm{B}} f(Q)\mathbf{S}, \qquad (6.31)$$

where the dipole form factor is now

$$f(Q) = \langle j_0(Q)\rangle + \left(\frac{g-2}{g}\right)\langle j_2(Q)\rangle. \qquad (6.32)$$

Whichever expression is more appropriate, be it (6.26), (6.28) or (6.31), the most useful property of the dipole approximation is that it simplifies the cross-section to the point where the only matrix elements that need be evaluated are those of an angular momentum operator of some kind. These are usually straightforward to calculate and have a simple physical interpretation. An additional property is that the $Q$ dependence of the matrix element is entirely contained in the dipole magnetic form factor, and is easy to determine since tabulations of $\langle j_0(Q)\rangle$ and $\langle j_2(Q)\rangle$ exist for virtually all ions of interest.[18]

Almost without exception, magnetic form factors decrease monotonically with increasing $Q$, and this property provides a practical way to identify a magnetic signal in neutron scattering: *Magnetic scattering is strong at small $Q$ and weak at large $Q$.* This is particularly useful for separating nuclear and magnetic inelastic scattering, because while magnetic scattering decreases with $Q$ according to $f^2(Q)$, nuclear one-phonon scattering increases as $Q^2$ (see Sections 5.1.3, 5.2.4 and 5.2.8).

### 6.2.3 Cross-section in the dipole approximation

To illustrate the simplifications permitted by the dipole approximation, consider a system which has only one type of magnetic ion. From (6.12) and (6.25), the unpolarized magnetic response function for this case is

$$S(\mathbf{Q},\omega) = f^2(Q)\exp(-2W)\sum_{\alpha\beta}(\delta_{\alpha\beta} - \hat{Q}_\alpha\hat{Q}_\beta)S_{\alpha\beta}(\mathbf{Q},\omega), \qquad (6.33)$$

where $S_{\alpha\beta}(\mathbf{Q},\omega)$ is now a *reduced* partial response function which excludes the magnetic form factor and Debye–Waller factor.[19] For intermediate crystal fields such that (6.31) applies, we use (6.32) to calculate $f(Q)$, and $S_{\alpha\beta}(\mathbf{Q},\omega)$ becomes[20]

$$S_{\alpha\beta}(\mathbf{Q},\omega) = g^2\mu_{\mathrm{B}}^2\sum_{jk}\exp\{i\mathbf{Q}\cdot(\mathbf{R}_k - \mathbf{R}_j)\}$$

$$\times \sum_{\lambda_i,\lambda_f} p_{\lambda_i}\langle\lambda_i|S_j^\alpha|\lambda_f\rangle\langle\lambda_f|S_k^\beta|\lambda_i\rangle\delta(E_{\lambda_f} - E_{\lambda_i} - \hbar\omega) \quad (6.34)$$

$$= g^2\mu_{\mathrm{B}}^2\sum_{jk}\exp\{i\mathbf{Q}\cdot(\mathbf{R}_k - \mathbf{R}_j)\}$$

$$\times \frac{1}{2\pi\hbar}\int_{-\infty}^{\infty}\langle\hat{S}_j^\alpha\hat{S}_k^\beta(t)\rangle\exp(-i\omega t)\,\mathrm{d}t. \qquad (6.35)$$

[18]See eqn (6.52) and associated references.

[19]In this respect, $S_{\alpha\beta}(\mathbf{Q},\omega)$ in (6.33) differs from that in (6.12), but we use the same symbol to avoid introducing a new notation. The two definitions of $S_{\alpha\beta}(\mathbf{Q},\omega)$ are the same when $\mathbf{Q}=0$.

[20]Note that the spin operator is Hermitian, i.e. $S^\dagger = S$.

In the weak field case, we use (6.30) for $f(Q)$, and replace $g$ by $g_J$ and $S_j^\alpha$ by $J_j^\alpha$ in (6.35).

If the magnetic atoms form a lattice, then the double summation in (6.35) can be replaced by one sum over the lattice vectors $\mathbf{l} = \mathbf{R}_k - \mathbf{R}_j$ and a second sum over $\mathbf{l}' = \mathbf{R}_j$. Since $\mathbf{l}$ and $\mathbf{l}'$ are both lattice vectors, each term in the sum over $\mathbf{l}'$ is the same and we can write $\langle \hat{S}_{\mathbf{l}'}^\alpha \hat{S}_{\mathbf{l}'+\mathbf{l}}^\beta(t) \rangle = \langle \hat{S}_0^\alpha \hat{S}_1^\beta(t) \rangle$. Hence,

$$ S_{\alpha\beta}(\mathbf{Q}, \omega) = g^2 \mu_\mathrm{B}^2 \frac{N}{2\pi\hbar} \sum_\mathbf{l} \exp(\mathrm{i}\mathbf{Q} \cdot \mathbf{l}) \int_{-\infty}^{\infty} \langle \hat{S}_0^\alpha \hat{S}_1^\beta(t) \rangle \exp(-\mathrm{i}\omega t)\, \mathrm{d}t, \tag{6.36} $$

where $N$ is the number of magnetic atoms. We see that the scattering from a lattice of spins is the space and time Fourier transform of the spin–spin correlation function. The inverse relation is

$$ \langle \hat{S}_0^\alpha \hat{S}_1^\beta(t) \rangle = \frac{\hbar}{N g^2 \mu_\mathrm{B}^2} \frac{v_0}{(2\pi)^3} \iint S_{\alpha\beta}(\mathbf{Q}, \omega) \exp\{-\mathrm{i}(\mathbf{Q} \cdot \mathbf{l} - \omega t)\}\, \mathrm{d}^3\mathbf{Q}\, \mathrm{d}\omega, \tag{6.37} $$

where the $\mathbf{Q}$ integration is performed over a primitive unit cell of the reciprocal lattice, which has a volume of $(2\pi)^3/v_0$.

If there are different magnetic atoms present then the $g^2$, $f^2(Q)$ and $\exp(-2W)$ factors in (6.33)–(6.35) need to be written as products $g_j g_k$, $f_j(Q) f_k(Q)$ and $\exp(-W_j) \exp(-W_k)$, and moved inside the summation over $j$ and $k$.

### 6.2.4 Cross-section for a local-moment paramagnet

The simplest system to which we can apply the above analysis is a local-moment paramagnet. In an ideal paramagnet there are no interactions between the magnetic moments, so there are no spatial magnetic correlations. This means that the scattering is incoherent, and the $\mathbf{Q}$ dependence of $S(\mathbf{Q}, \omega)$ comes entirely from the magnetic form factor $f(Q)$ and the Debye–Waller factor. For simplicity, we shall assume the form of the dipole approximation given by (6.31), and neglect any anisotropy in the $g$-factor.

As there are no magnetic interactions it costs no energy to change the directions of the moments. Hence, we can put $E_{\lambda_\mathrm{f}} = E_{\lambda_\mathrm{i}}$ in the reduced partial response function (6.34) and perform the summation over $\lambda_\mathrm{f}$ by the closure relation, eqn (C.19), to give

$$ S_{\alpha\beta}(\mathbf{Q}, \omega) = \mu_\mathrm{B}^2 \sum_{jk} g_j g_k \exp\{\mathrm{i}\mathbf{Q} \cdot (\mathbf{R}_k - \mathbf{R}_j)\} \langle S_j^\alpha S_k^\beta \rangle\, \delta(\hbar\omega). \tag{6.38} $$

After integration over $\hbar\omega$ this becomes

$$ S_{\alpha\beta}(\mathbf{Q}) = \mu_\mathrm{B}^2 \sum_{jk} g_j g_k \exp\{\mathrm{i}\mathbf{Q} \cdot (\mathbf{R}_k - \mathbf{R}_j)\} \langle S_j^\alpha S_k^\beta \rangle. \tag{6.39} $$

We see that the scattering depends on the equal-time correlation function $\langle S_j^\alpha S_k^\beta \rangle$, so even though there is no change in the neutron energy, the scattering is strictly quasielastic not elastic (see Section 3.8.3). Physically, this is because in a paramagnet the orientation of the spins is not static, but fluctuates randomly in time due to thermal agitation.

For a paramagnet there are no correlations between different spin components, and no correlations between spins on different sites. Hence,

$$
\begin{aligned}
\langle S_j^\alpha S_k^\beta \rangle &= 0 && (j \neq k) \\
&= \tfrac{1}{3} S_j (S_j + 1)\, \delta_{\alpha\beta} && (j = k).
\end{aligned}
\tag{6.40}
$$

The second line comes from $\langle \mathbf{S}^2 \rangle = S(S+1)$ and the fact that each component $S^\alpha$ of $\mathbf{S}$ is equivalent. With (6.40), the orientation factor in the response function for unpolarized neutrons (6.33) becomes

$$
\sum_{\alpha\beta} (\delta_{\alpha\beta} - \hat{Q}_\alpha \hat{Q}_\beta)\, \delta_{\alpha\beta} = \sum_\alpha (1 - \hat{Q}_\alpha^2) = 2,
\tag{6.41}
$$

and so

$$
S(\mathbf{Q}) = \frac{2}{3}\mu_{\mathrm{B}}^2 \sum_j g_j^2 f_j^2(Q) \exp(-2W_j) S_j(S_j + 1).
\tag{6.42}
$$

An ideal paramagnet can be realized if a small concentration of magnetic ions is dispersed in a non-magnetic host. Normally, however, spins on different sites are coupled by magnetic interactions which introduce correlations. Systems without long-range magnetic order but with short-range magnetic correlations are called *cooperative paramagnets*.

# 6.3  Magnetic form factors

In general, the magnetic form factor is a suitably normalized function that describes the $\mathbf{Q}$ dependence of the matrix element

$$
\langle \lambda_{\mathrm{f}} | \, \mathbf{M}(\mathbf{Q}) \, | \lambda_{\mathrm{i}} \rangle
$$

for a single atom or ion. When $\lambda_{\mathrm{i}}$ and $\lambda_{\mathrm{f}}$ are the same or have the same energy, i.e. the scattering is elastic or quasielastic, the magnetic form factor is, to within a constant, the Fourier transform of the intra-atomic magnetization. In this respect, it is just like other form factors encountered in scattering theory — see Sections 2.5 and 2.6 — all of which are Fourier transforms (or squared Fourier transforms[21]) of the scattering potential for an isolated scattering object (in the present case, a magnetic atom or ion).

The dipole magnetic form factors (6.30) and (6.32) encountered in the previous section are 'cheap-and-cheerful' approximations that can be used when high accuracy is not required. In what follows we present more general calculations of magnetic form factors, first for free ions and then for ions bound in solids. The expressions are strictly applicable only for elastic or quasielastic scattering but are usually also suitable for inelastic scattering ($E_{\mathrm{i}} \neq E_{\mathrm{f}}$) providing the angular momentum quantum numbers $J$, $L$, and $S$ are the same in the initial and final states. We discuss the inelastic form factor for more general cases in Section 6.3.2.

[21] Convention varies on the usage of the term *form factor*. Sometimes it is the Fourier transform, and other times it is the squared modulus of the Fourier transform. Normal practice in magnetic scattering is to use the former, and this is the convention we shall adopt, although sometimes we will find the squared modulus and then take the square root.

### 6.3.1 Free-ion magnetic form factors

A free ion is an ion not subject to any external interactions. In an ensemble of free magnetic ions every orientation of the atomic moment will occur with equal probability. We assume that the ions are immobile and that the electronic state of the ions does not change during scattering. The scattering from such ions is elastic, and from (6.5) and (6.11) the differential cross-section for unpolarized neutrons is given by

$$\frac{d\sigma}{d\Omega} = \left(\frac{\gamma r_0}{2\mu_B}\right)^2 \sum_{\lambda_i} p_{\lambda_i} \sum_{\lambda_f} |\langle \lambda_f | \mathbf{M}_\perp(\mathbf{Q}) | \lambda_i \rangle|^2. \qquad (6.43)$$

As the atomic moments are randomly aligned, we know from the outset that the scattering must be spatially isotropic, i.e. dependent on the magnitude but not the direction of $\mathbf{Q}$. Hence, the cross-section per ion can be written in the form

$$\frac{d\sigma}{d\Omega} \propto f^2(Q), \qquad (6.44)$$

which together with the normalization $f(0) = 1$ defines the free-ion magnetic form factor $f(Q)$.

We can obtain the coefficient of proportionality in (6.44) by evaluating (6.43) for $Q = 0$. We recall that $\mathbf{M}(\mathbf{Q})$ is the spatial Fourier transform of the magnetization density, and so $\mathbf{M}(0) = \boldsymbol{\mu}$, the operator representing the total magnetic moment of the ion. Hence, for $Q = 0$ the double summation in (6.43) is the expectation value of $\mu_\perp^2$ averaged over all directions. This is just $(2/3)\langle \mu^2 \rangle$,[22] whereupon

$$\frac{d\sigma}{d\Omega}(0) = \left(\frac{\gamma r_0}{2\mu_B}\right)^2 \frac{2}{3}\langle \mu^2 \rangle. \qquad (6.45)$$

We assume that the $LS$-coupling scheme applies for the free ion,[23] and that the ground state is a pure level, i.e. we ignore any admixed excited levels in the ground state. From eqn (C.130), the squared magnetic moment is given by $\langle \mu^2 \rangle = g_J^2 \mu_B^2 \langle J^2 \rangle = g_J^2 \mu_B^2 J(J+1)$, and so from (6.44) and (6.45),

$$\frac{d\sigma}{d\Omega} = (\gamma r_0)^2 \frac{g_J^2 J(J+1)}{6} f^2(Q). \qquad (6.46)$$

The general expression for $f^2(Q)$ for a shell with orbital quantum number $l$ in the $LS$-coupling scheme has been derived by several different methods,[24] and may be written

$$f^2(Q) = \frac{6}{g_J^2 J(J+1)} \sum_K \frac{3}{K+1} \{A(K-1,K) + B(K-1,K)\}^2. \qquad (6.47)$$

The $A$ and $B$ coefficients correspond to orbital and spin scattering, respectively, and are given in Section C.6. They depend on the angular momentum quantum numbers $l$, $S$, $L$ and $J$, and are linear combinations

[22]The moments are randomly oriented, so $\langle \mu_x^2 \rangle = \langle \mu_y^2 \rangle = \langle \mu_z^2 \rangle = \langle \mu^2 \rangle/3$. If we define the $z$ axis to be parallel to $\mathbf{Q}$ then $\langle \mu_\perp^2 \rangle = \langle \mu_x^2 \rangle + \langle \mu_y^2 \rangle = (2/3)\langle \mu^2 \rangle$.

[23]See Section C.2.2

[24]Trammell (1953); Johnston (1966); Stassis and Deckman (1975).

**Table 6.1** $a_K$, $b_K$ and $b'_K$ coefficients required to calculate the free-ion magnetic form factors in the ground state of the trivalent lanthanides, eqns (6.47)–(6.49). Note that $b'_7 = 0$.

| Ion | $^{2S+1}L_J$ | $a_1$ | $a_3$ | $a_5$ | $b_1$ | $b'_1$ | $b_3$ | $b'_3$ | $b_5$ | $b'_5$ | $b_7$ |
|---|---|---|---|---|---|---|---|---|---|---|---|
| $Ce^{3+}$ | $^2F_{5/2}$ | −1.1269 | −0.5216 | −0.2173 | 0.2817 | −0.2254 | 0.2087 | −0.5216 | 0.1087 | −1.0866 | 0 |
| $Pr^{3+}$ | $^3H_4$ | −1.7889 | 0 | 0.3078 | 0.5963 | −0.1723 | 0.2204 | 0.3005 | −0.1539 | 0.6707 | −0.1433 |
| $Nd^{3+}$ | $^4I_{9/2}$ | −2.1106 | 0.3476 | −0.1568 | 0.9045 | −0.0640 | 0.0869 | 0.1791 | −0.1046 | −0.8186 | 0.2323 |
| $Pm^{3+}$ | $^5I_4$ | −2.0870 | 0.3236 | −0.1281 | 1.1926 | 0.0632 | −0.0809 | −0.1667 | 0.0854 | 0.6684 | −0.1428 |
| $Sm^{3+}$ | $^6H_{5/2}$ | −1.6903 | 0 | 0.0856 | 1.4086 | 0.1628 | −0.1507 | −0.2055 | 0.0428 | −0.1866 | 0 |
| $Eu^{3+}$ | $^7F_0$ | 0 | 0 | 0 | 0 | 0 | 0 | 0 | 0 | 0 | 0 |
| $Gd^{3+}$ | $^8S_{7/2}$ | 0 | 0 | 0 | −2.6458 | 0 | 0 | 0 | 0 | 0 | 0 |
| $Tb^{3+}$ | $^7F_6$ | −1.0801 | −0.4020 | −0.1164 | −2.1603 | −0.1200 | 0.4020 | −0.1096 | 0.3492 | −0.0448 | 0.1819 |
| $Dy^{3+}$ | $^6H_{15/2}$ | −1.7743 | 0 | 0.2763 | −1.7743 | −0.1183 | 0.3677 | 0.1337 | −0.3685 | 0.1771 | −0.5675 |
| $Ho^{3+}$ | $^5I_8$ | −2.1213 | 0.3595 | −0.1738 | −1.4142 | −0.0471 | 0.1438 | 0.0981 | −0.2607 | −0.3342 | 1.0114 |
| $Er^{3+}$ | $^4I_{15/2}$ | −2.1292 | 0.3677 | −0.1842 | −1.0646 | 0.0473 | −0.1471 | −0.1003 | 0.2763 | 0.3543 | −1.1349 |
| $Tm^{3+}$ | $^3H_6$ | −1.8002 | 0 | 0.3492 | −0.7201 | 0.1200 | −0.4020 | −0.1462 | 0.4656 | −0.2238 | 0.9096 |
| $Yb^{3+}$ | $^2F_{7/2}$ | −1.1339 | −0.5471 | −0.2770 | −0.3780 | 0.1260 | −0.5471 | 0.1492 | −0.8311 | 0.1066 | −1.4750 |

of the $\langle j_n(Q) \rangle$ radial integrals defined in eqn (6.27). The summation index $K$ runs over odd integers from 1 to $2l - 1$ for $A$, and 1 to $2l + 1$ for $B$.[25] Only certain $\langle j_n(Q) \rangle$ appear for each $K$. These are given by the following expressions:

$$A(K-1, K) = a_K \big( \langle j_{K-1} \rangle + \langle j_{K+1} \rangle \big), \qquad (6.48)$$

$$B(K-1, K) = b_K \langle j_{K-1} \rangle + b'_K \langle j_{K+1} \rangle, \qquad (6.49)$$

where $a_K$, $b_K$, and $b'_K$ are numerical coefficients, and $b'_K = 0$ when $K = 2l + 1$. Here, and hereafter, we do not display the $Q$ dependence of $\langle j_n(Q) \rangle$ in order to simplify the notation. The magnetic dipole component of (6.47) is contained in the terms with $K = 1$. If just the dipole terms are kept then the form factor is very nearly (but not exactly) the same as the *dipole approximation* expression given in (6.30).[26]

Tables of the $a_K$, $b_K$, and $b'_K$ coefficients have been presented in several different places.[27] For ease of reference, Table 6.1 lists the $a_K$, $b_K$, and $b'_K$ coefficients for the ground state terms $^{2S+1}L_J$ of all the trivalent lanthanide ions. Similar data could also be given for $d$ electron systems, but the strong effect of the crystal field in solids tends to make the $d$ orbitals, and hence the form factor, highly anisotropic. In such cases the free-ion (isotropic) form factor is no longer suitable for accurate work, and as an approximation one might as well use the dipole form factor, eqn (6.32), which is simpler to compute. The exact method to calculate anisotropic form factors caused by crystal or magnetic fields is described in Section 6.3.2.

[25] The $A(K-1, K)$ and $B(K-1, K)$ coefficients originate from magnetic multipoles of odd rank $K$. Coefficients of the form $B(K, K)$ also exist in the magnetic neutron scattering cross-section [see eqns (6.70) and (6.73)] and originate from electric multipoles of even rank $K$. The latter do not contribute to the elastic free-ion form factor.

[26] The two expressions are identical when $Q = 0$, but the $b'_K$ coefficient is neglected in the standard dipole approximation so the $\langle j_2 \rangle$ term is different from that in the exact expression.

[27] Lander and Brun (1970); Balcar and Lovesey (1989); Osborn *et al.* (1991); Lovesey (2015).

To illustrate the above formalism we calculate the free-ion form factor of $Nd^{3+}$, whose magnetic ground state $^4I_{9/2}$ has quantum numbers $l = 3$, $L = 6$, $S = \frac{3}{2}$, $J = \frac{9}{2}$, from which $g_J = \frac{8}{11}$. From eqn (6.47) and Table 6.1, the exact free-ion form factor is found to be

$$f^2(Q) = \langle j_0\rangle^2 + 3.606\langle j_0\rangle\langle j_2\rangle + 3.316\langle j_2\rangle^2 + 0.157\langle j_2\rangle\langle j_4\rangle$$
$$+ 0.111\langle j_4\rangle^2 + 0.117\langle j_4\rangle\langle j_6\rangle + 0.227\langle j_6\rangle^2, \qquad (6.50)$$

and from (6.30), the dipole approximation form factor is given by

$$f_d^2(Q) = \langle j_0\rangle^2 + 3.500\langle j_0\rangle\langle j_2\rangle + 3.063\langle j_2\rangle^2. \qquad (6.51)$$

The different radial integrals $\langle j_n\rangle$, $n \leq 6$, have been calculated using relativistic Dirac–Fock atomic wave functions for most of the lanthanide (Freeman and Desclaux, 1979) and actinide ions (Desclaux and Freeman, 1978), and using non-relativistic Hartree–Fock wave functions for the $d$ ions (Watson and Freeman, 1961). Brown has provided a convenient analytic approximation to the $\langle j_n\rangle$ using the following parameterization:

$$\langle j_0\rangle = A\exp(-as^2) + B\exp(-bs^2) + C\exp(-cs^2) + D$$
$$\langle j_n\rangle = As^2\exp(-as^2) + Bs^2\exp(-bs^2) + Cs^2\exp(-cs^2) + Ds^2,$$
$$(n = 2, 4, 6), \qquad (6.52)$$

where $s = Q/4\pi$. The coefficients in (6.52) are tabulated for most of the important magnetic ions (for 3$d$, 4$d$, 4$f$ and 5$f$ ions, see Brown, 2004, and Lisher and Forsyth, 1971; for 5$d$ ions, see Kobayashi *et al.*, 2011).

In Fig. 6.3 we plot the free-ion magnetic form factor of $Nd^{3+}$ calculated with the appropriate combination of $\langle j_n\rangle$ functions required in the exact expression (6.50) and in the dipole approximation (6.51). Also shown is the magnetic dipole part of the exact form factor, which in this case is seen to be almost indistinguishable from the dipole approximation and the exact calculation.

This worked example demonstrates that the dipole approximation provides a very good approximation to the magnetic form factor for free ions. In the next section, however, we will show that the dipole approximation can have significant limitations for ions in solids due to the asymmetry in the atomic magnetization.

Figure 6.4 shows the free-ion magnetic form factors $f(Q)$ for some other representative ions. Note that the $f(Q)$ for 4$f$ ions are more extended in $Q$ than those of $d$ ions. This is because, on average, 4$f$ electronic orbitals are closer to the nucleus than are $d$ orbitals. Similarly, for a given electronic configuration, the $f(Q)$ for $d$ ions in different rows of the periodic table contract in the order 3$d$ to 4$d$ to 5$d$, in inverse relation to the radii of their $d$ orbitals. The peak in the form factor of $Sm^{3+}$ near $Q = 50\,\text{nm}^{-1}$ is an interference effect due to the non-monotonic radial variation of the magnetization, which is a consequence of the near-cancellation of the spin and orbital moments in $Sm^{3+}$ and the difference in the radial dependence of the spin and orbital magnetization.[28]

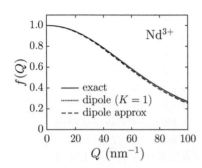

**Fig. 6.3** Free-ion magnetic form factor of $Nd^{3+}$. The full line is calculated from the exact expression (6.50), and the dashed line is calculated from the dipole approximation (6.51). The dotted line is the dipole part of the exact form factor (the $K = 1$ term in (6.47)).

[28]The ground state of $Sm^{3+}$ has only a small net magnetic dipole moment because the spin and orbital moments have almost the same magnitude and point in opposite directions. As a result, the coefficient of the $\langle j_0\rangle$ term in $f(Q)$, which is associated with the total magnetic moment, is relatively small compared with the coefficient of the $\langle j_2\rangle$ term, which originates almost entirely from orbital magnetic scattering.

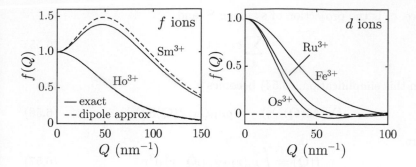

**Fig. 6.4** Free-ion magnetic form factors of selected $f$ and $d$ ions. Left: the magnetic form factor of $Ho^{3+}$ is typical of $4f$ ions, whereas that of $Sm^{3+}$ has a peak due to the near cancellation of the spin and orbital magnetization. Right: results for $Fe^{3+}$ ($3d^5$), $Ru^{3+}$ ($4d^5$) and $Os^{3+}$ ($5d^5$), which are from different rows in the periodic table but have the same outer $d$ electronic configuration.

## 6.3.2 Anisotropic magnetic form factors

Unlike the free-ion form factor, form factors for ions bound in solids are not generally isotropic because the ligand field acting on the electrons creates an aspherical charge density and hence an aspherical magnetization distribution. Spherical symmetry can also be broken by magnetic or exchange fields. If accurate intensities are required then the scattering must be calculated as a function of the orientation as well as the magnitude of $\mathbf{Q}$.

### Spin-only scattering

We first discuss scattering due to spin only, which applies to ions whose ground state term has $L = 0$, such as $Eu^{2+}$ and $Gd^{3+}$, and to ions whose orbital angular momentum is quenched, i.e. many $d$ electron systems. The reason for considering this special case first, apart from the fact that many systems of interest fall into this category, is that the absence of orbital angular momentum leads to a simplification in the structure of the theory similar to that achieved with the dipole approximation.

To proceed, we write the magnetization operator for spin-only ions as

$$\mathbf{M}(\mathbf{r}) = -g\mu_B \sum_j \delta(\mathbf{r} - \mathbf{r}_j)\mathbf{s}_j, \qquad (6.53)$$

where $\mathbf{s}_j$ is the spin[29] of the $j$th unpaired electron. We assume that the total spin $\mathbf{S} = \sum_j \mathbf{s}_j$ is the same in the initial and final state and that the spatial wave function is unchanged on scattering. Hence, the matrix element of $\mathbf{M}(\mathbf{Q})$, eqn (6.43), can be written

$$\langle \lambda_f | \mathbf{M}(\mathbf{Q}) | \lambda_i \rangle = -g\mu_B \langle \lambda_f | \int \sum_j \delta(\mathbf{r} - \mathbf{r}_j)\mathbf{s}_j \exp(i\mathbf{Q} \cdot \mathbf{r}) \, d^3\mathbf{r} | \lambda_i \rangle. \quad (6.54)$$

To simplify this expression, we note that when $\mathbf{S}$ is unchanged on scattering the operator $\mathbf{s}_j$ can be replaced inside the matrix element by $c_j \mathbf{S}$, where $c_j$ is a scalar. This result,[30] which is sometimes called the *Projection Theorem*, is a special case of the Wigner–Eckart Theorem. Physically, it means that when a set of spins $\mathbf{s}_j$ are coupled to form a resultant $\mathbf{S}$ the time average of each $\mathbf{s}_j$ follows the same direction as $\mathbf{S}$.

[29]More precisely, the *effective* spin — see eqn (C.143) and the discussion that precedes it.

[30]A formal proof can be found in Condon and Shortley (1935).

Hence, $c_j$ is the projection of $\mathbf{s}_j$ along $\mathbf{S}$, and

$$\sum_j c_j = 1. \tag{6.55}$$

With this simplification, (6.54) becomes

$$\langle \lambda_f | \mathbf{M}(\mathbf{Q}) | \lambda_i \rangle = -g\mu_B \langle \lambda_f | f(\mathbf{Q})\mathbf{S} | \lambda_i \rangle, \tag{6.56}$$

where,

$$f(\mathbf{Q}) = \int \rho_s(\mathbf{r}) \exp(i\mathbf{Q} \cdot \mathbf{r}) \, d^3\mathbf{r} \tag{6.57}$$

is the spin-only form factor, and

$$\rho_s(\mathbf{r}) = \sum_j \delta(\mathbf{r} - \mathbf{r}_j)c_j \tag{6.58}$$

is the spin density projected along the direction of $\mathbf{S}$. Note that $f(\mathbf{Q})$ is correctly normalized.[31]

With the aid of (6.56) we see that $\mathbf{M}(\mathbf{Q})$ reduces to

$$\mathbf{M}(\mathbf{Q}) = -g\mu_B f(\mathbf{Q})\mathbf{S}. \tag{6.59}$$

This is the same form as found in the dipole approximation (6.31), but here the form factor is the Fourier transform of the *total* spin density distribution, not just the dipole part. When all the unpaired electrons are equivalent the spin density $\rho_s(\mathbf{r})$ is proportional to the charge density of the unpaired electrons. This is not the case for ions with both spin and orbital angular momentum because the orbital current density has a different spatial distribution from the spin density.

Let us emphasize once again that the form of (6.59), like the dipole approximation for $\mathbf{M}(\mathbf{Q})$, is particularly simple in that it contains the product of two factors, one of which (the form factor) contains all the $\mathbf{Q}$ dependence of the scattering, and the other (the spin operator $\mathbf{S}$) acts only on the spin part of the scattering system. This means that neutron scattering from a system of spin-only ions directly probes spin–spin correlations, and the expressions in Section 6.2.3 apply.

We will now calculate the magnetic form factor for some simple cases of spin-only ions in crystals to illustrate the effect of aspherical charge (spin) distributions on the scattering. We begin by considering the form factor for one-electron orbitals in a ligand field, for which a relatively simple formula exists (Weiss and Freeman, 1959).

To derive the formula, we use the fact that for spin-only ions the magnetic form factor is the Fourier transform of the charge density,

$$f(\mathbf{Q}) = \int |\psi(\mathbf{r})|^2 \exp(i\mathbf{Q} \cdot \mathbf{r}) \, d^3\mathbf{r}. \tag{6.60}$$

In the absence of a ligand (crystal) field, the one-electron spatial wave functions may be written in separable form

$$\psi(\mathbf{r}) = R_{nl}(r)Y_{l,m}(\theta, \phi), \tag{6.61}$$

[31] From (6.57), (6.58) and (6.55),

$$f(0) = \int \rho_s(\mathbf{r}) \, d^3\mathbf{r}$$

$$= \int \sum_j \delta(\mathbf{r} - \mathbf{r}_j)c_j \, d^3\mathbf{r}$$

$$= \sum_j c_j$$

$$= 1.$$

where $R_{nl}(r)$ is the radial wave function and $Y_{l,m}(\theta, \phi)$ is the angular wave function, which is a spherical harmonic of rank $l$ and order $m$ — see Section C.2.1. The effect of a ligand field is to mix different one-electron states. Hence, eqn (6.60) contains terms like

$$f_{ij}(\mathbf{Q}) = \int R_i(r)R_j(r)Y_i^*(\theta, \phi)Y_j(\theta, \phi) \exp(i\mathbf{Q} \cdot \mathbf{r}) \, d^3\mathbf{r}, \qquad (6.62)$$

where $i$ and $j$ represent the set of quantum numbers $(nlm)$ for the one-electron states. The integral can be evaluated in spherical coordinates by use of a standard expansion of $\exp(i\mathbf{Q} \cdot \mathbf{r})$ in terms of products of spherical harmonics and Bessel functions.[32] The result may be written

$$f_{ij}(\mathbf{Q}) = \sqrt{4\pi} \sum_k i^k (2k+1)^{1/2} \langle j_k \rangle_{ij} \, c^k(l_i m_i, l_j m_j) Y^*_{k, m_i - m_j}(\beta, \alpha), \qquad (6.63)$$

where $\langle j_k \rangle_{ij}$ is defined by the radial integral (6.27) generalized to include two different radial wave functions, and $\beta$ and $\alpha$ are spherical angles that specify the direction of $\mathbf{Q}$ (Fig. 6.5). The $c^k(l_i m_i, l_j m_j)$ coefficients are integrals of products of three spherical harmonics and have been tabulated for various combinations of one-electron states (Condon and Shortley, 1935; Condon and Odabaşi, 1980). The coefficients for $d$ states ($l_i = l_j = 2$) are listed in Table 6.2 for convenience.

For illustration, let us work out the form factor for a single electron (or hole) in a $3d_{x^2-y^2}$ orbital. This is a good approximation to the hole state found in transition-metal ions with configuration $3d^9$, such as $Cu^{2+}$, in an octahedral ligand field with a small axial elongation The angular part of the wave function for the $d_{x^2-y^2}$ orbital is given by

$$d_{x^2-y^2} = \frac{1}{\sqrt{2}}(Y_{2,2} + Y_{2,-2}). \qquad (6.64)$$

Figure 6.6(a) shows the associated charge density of this orbital relative to the octahedral environment. Combining (6.64) and (6.61) and substituting in (6.60) we can write the magnetic form factor

$$f(\mathbf{Q}) = \frac{1}{2}(f_{22} + f_{-2-2} + f_{2-2} + f_{-22}), \qquad (6.65)$$

where $f_{22}$ denotes the function $f_{ij}(\mathbf{Q})$ in (6.63) with $m_i = 2$ and $m_j = 2$. In addition, for $3d$ orbitals we have that $n_i = n_j = 3$ and $l_i = l_j = 2$. With the aid of the Table 6.2, we find that

$$f(\mathbf{Q}) = \sqrt{4\pi} \left[ \langle j_0 \rangle Y_{0,0} + \tfrac{2\sqrt{5}}{7} \langle j_2 \rangle Y_{2,0} + \tfrac{1}{7} \langle j_4 \rangle Y_{4,0} + \tfrac{\sqrt{70}}{14} \langle j_4 \rangle (Y_{4,4} + Y^*_{4,4}) \right]$$

$$= \langle j_0 \rangle + \tfrac{5}{7}(3\cos^2\beta - 1)\langle j_2 \rangle$$
$$+ \tfrac{3}{56}(35\cos^4\beta - 30\cos^2\beta + 35\sin^4\beta\cos 4\alpha + 3)\langle j_4 \rangle. \qquad (6.66)$$

At $\mathbf{Q} = 0$, only the $\langle j_0 \rangle$ term is non-zero so $f(\mathbf{Q})$ is isotropic, but for $Q > 0$ it becomes increasingly anisotropic as $\langle j_0 \rangle$ decays with $Q$ while $\langle j_2 \rangle$ and $\langle j_4 \rangle$ grow. The anisotropy of the $d_{x^2-y^2}$ form factor is illustrated in Fig. 6.6(b), which plots an isosurface of $f(\mathbf{Q})$ in $\mathbf{Q}$-space.

[32] $\exp(i\mathbf{Q} \cdot \mathbf{r}) =$
$$4\pi \sum_{kq} i^k j_k(Qr) Y^*_{k,q}(\beta, \alpha) Y_{k,q}(\theta, \phi),$$
where $(\beta, \alpha)$ and $(\theta, \phi)$ specify the directions of $\mathbf{Q}$ and $\mathbf{r}$, respectively.

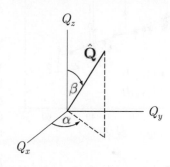

**Fig. 6.5** The spherical angles $\beta$ and $\alpha$ used to define the direction of $\mathbf{Q}$.

**Table 6.2** $c^k(l_i m_i, l_j m_j)$ coefficients required in eqn (6.63) to calculate the magnetic form factor for one-electron states. In the table here are listed only the coefficients for the particular case $l_i = l_j = 2$.

| | | | $k$ | |
| $m_i$ | $m_j$ | 0 | 2 | 4 |
| --- | --- | --- | --- | --- |
| $\pm 2$ | $\pm 2$ | 1 | $-2/7$ | $1/21$ |
| $\pm 2$ | $\pm 1$ | 0 | $\sqrt{6}/7$ | $-\sqrt{5}/21$ |
| $\pm 2$ | 0 | 0 | $-2/7$ | $\sqrt{15}/21$ |
| $\pm 1$ | $\pm 1$ | 1 | $1/7$ | $-4/21$ |
| $\pm 1$ | 0 | 0 | $1/7$ | $\sqrt{30}/21$ |
| 0 | 0 | 1 | $2/7$ | $2/7$ |
| $\pm 2$ | $\mp 2$ | 0 | 0 | $\sqrt{70}/21$ |
| $\pm 2$ | $\mp 1$ | 0 | 0 | $-\sqrt{35}/21$ |
| $\pm 1$ | $\mp 1$ | 0 | $-\sqrt{6}/7$ | $-\sqrt{40}/21$ |

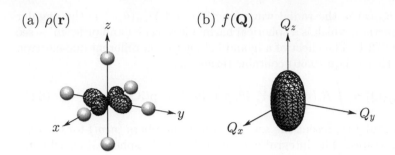

(a) $\rho(\mathbf{r})$        (b) $f(\mathbf{Q})$

**Fig. 6.6** Charge density and anisotropic form factor of the one-electron $d_{x^2-y^2}$ orbital. (a) Surface of constant charge density relative to an octahedral arrangement of anions. (b) Surface in $\mathbf{Q}$-space corresponding to $f(\mathbf{Q}) = 0.6$, calculated from eqn (6.66).

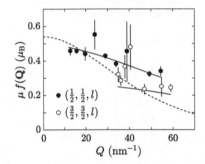

**Fig. 6.7** Anisotropic form factor of $Cu^{2+}$ in $YBa_2Cu_3O_{6.15}$ scaled by the ordered moment (Shamoto *et al.*, 1993). Filled circles are from the $(\frac{1}{2}, \frac{1}{2}, l)$ magnetic reflections, and open circles from the $(\frac{3}{2}, \frac{3}{2}, l)$ series. The solid lines are the $3d_{x^2-y^2}$ form factor for $Cu^{2+}$ calculated from eqn (6.66), and the broken line is the dipole form factor calculated from (6.32).

[33] Also known as Wigner, or vector-coupling, coefficients. There are several different notations, for example $\langle SM_S LM_L | JM_J \rangle$ is often written $\langle SLM_S M_L | JM_J \rangle$ or simply $\langle M_S M_L | JM_J \rangle$. Note also that the order of the indices in the coefficient matters, because $\langle SM_S LM_L | JM_J \rangle = (-1)^{L+S-J} \langle LM_L SM_S | JM_J \rangle$.

The anisotropy of the form factor can have a significant effect on scattering intensities. As an example, Fig. 6.7 compares the measured magnetic form factor of $Cu^{2+}$ in antiferromagnetically ordered $YBa_2Cu_3O_{6.15}$ with the calculated $3d_{x^2-y^2}$ and dipole approximation (isotropic) form factors (Shamoto *et al.*, 1993). The data exhibit considerable anisotropy consistent with the calculated anisotropic form factor.

### Spin and orbital scattering

When both spin and orbital angular momentum are present it is no longer possible to represent $\mathbf{M}(\mathbf{Q})$ in the form (6.59), i.e. as the product of a form factor and an angular momentum operator. If we wish to go beyond the dipole approximation then we must directly evaluate the matrix element of $\mathbf{M}(\mathbf{Q})$.

We assume the unpaired electrons are localized and belong to a pure $l^n$ configuration. It is usually convenient to use the $SM_S LM_L$ basis for $d$ electrons and the $SLJM_J$ basis for $f$ electrons (see Section C.4). These two representations are connected by the unitary transformation

$$|SM_S LM_L\rangle = \sum_{J,M_J} \langle SM_S LM_L | JM_J \rangle |SLJM_J\rangle,$$

$$|SLJM_J\rangle = \sum_{M_S,M_L} \langle JM_J | SM_S LM_L \rangle |SM_S LM_L\rangle,$$

(6.67)

where $\langle SM_S LM_L | JM_J \rangle = \langle JM_J | SM_S LM_L \rangle$ are Clebsch–Gordan coefficients.[33] If we choose the $SLJM_J$ basis, the scattering amplitude will contain matrix elements of the form (see Section 6.1)

$$\langle SLJM_J | \mathbf{M}_\perp(\mathbf{Q}) | S'L'J'M_J' \rangle.$$

The method to calculate these matrix elements is given in detail elsewhere (Johnston, 1966; Lovesey, 1984b; Balcar and Lovesey, 1989). It turns out that the expressions are simplest if the operators are written in spherical components, defined by

$$M_{\pm 1} = \mp \frac{1}{\sqrt{2}} (M_x \pm iM_y), \qquad M_0 = M_z,$$

(6.68)

so that

$$M_x = \frac{1}{\sqrt{2}} (M_{-1} - M_{+1}), \qquad M_y = \frac{i}{\sqrt{2}} (M_{-1} + M_{+1}).$$

(6.69)

After some intimidating algebra, the matrix element of the $q$th spherical component of $\mathbf{M}_\perp(\mathbf{Q})$ ($q = 0, \pm 1$) may be expressed as

$$\langle SLJM_J| M_{\perp,q}(\mathbf{Q}) |S'L'J'M'_J\rangle = -2\mu_B\sqrt{4\pi} \sum_{\widetilde{K},K} \{A(\widetilde{K},K) + B(\widetilde{K},K)\}$$

$$\times \langle KQJ'M'_J|JM_J\rangle\langle\widetilde{K}\widetilde{Q}KQ|1q\rangle Y_{\widetilde{K},\widetilde{Q}}(\hat{\mathbf{Q}}), \qquad (6.70)$$

with the following conditions from the Clebsch–Gordan coefficients:

$$Q = M_J - M'_J, \quad \widetilde{Q} = q - M_J + M'_J, \quad |Q| \leq K, \quad |\widetilde{Q}| \leq \widetilde{K}. \quad (6.71)$$

The $A(\widetilde{K},K)$ and $B(\widetilde{K},K)$ coefficients correspond to orbital (current) and spin scattering, respectively. These are the same coefficients that appear in (6.47)–(6.49), and are given explicitly in Section C.6. They are only non-zero in the following cases:

$$\begin{aligned} A(\widetilde{K},K): &\quad K = 1,3,...,2l-1, &\quad \widetilde{K} = K \pm 1, \\ B(\widetilde{K},K): &\quad K = 1,3,...,2l+1, &\quad \widetilde{K} = K \pm 1, &\qquad (6.72) \\ B(K,K): &\quad K = 2,4,...,2l. \end{aligned}$$

In addition, $A(K \pm 1, K) = 0$ when $S \neq S'$, and $B(K,K) = 0$ when $S = S'$, $L = L'$ and $J = J'$.

The anisotropy of the matrix element (6.70) is contained in the functions $Y_{\widetilde{K},\widetilde{Q}}(\hat{\mathbf{Q}}) = Y_{\widetilde{K},\widetilde{Q}}(\beta,\alpha)$, which are spherical harmonics of rank $\widetilde{K}$ and order $\widetilde{Q}$. The spherical harmonics are functions of the direction of $\hat{\mathbf{Q}}$, which is specified by the spherical angles $\beta$ and $\alpha$ (see Fig. 6.5).

We also give the corresponding expression to (6.70) for the operator $\mathbf{M}(\mathbf{Q})$. In doing so, we remark that $\mathbf{M}(\mathbf{Q})$ is not uniquely defined, in the sense that any function proportional to $\mathbf{Q}$ can be added to $\mathbf{M}(\mathbf{Q})$ without changing the neutron scattering cross-section (see Section 4.2.3). This allows us some freedom to select an expression for the matrix element of $\mathbf{M}(\mathbf{Q})$ which simplifies our calculations. The following is one such expression, chosen to remove the $A(K+1,K)$ and $B(K+1,K)$ terms:

$$\langle SLJM_J| M_q(\mathbf{Q}) |S'L'J'M'_J\rangle = -2\mu_B\sqrt{4\pi} \sum_K \langle KQJ'M'_J|JM_J\rangle$$

$$\times \left[ \frac{2K+1}{K+1}\{A(K-1,K) + B(K-1,K)\}\langle K-1\widetilde{Q}KQ|1q\rangle Y_{K-1,\widetilde{Q}}(\hat{\mathbf{Q}}) \right.$$

$$\left. + B(K,K)\langle K\widetilde{Q}KQ|1q\rangle Y_{K,\widetilde{Q}}(\hat{\mathbf{Q}}) \right].$$

$$(6.73)$$

The same restrictions as given in (6.71) and (6.72) apply, except that $|\widetilde{Q}| \leq K - 1$ in the second line of (6.73), and $|\widetilde{Q}| \leq K$ in the third line.

If the initial and final states are obtained in the $SM_SLM_L$ basis, then one can convert to the $SLJM_J$ basis using (6.67) and then use (6.70) or (6.73) to calculate the matrix elements. Alternatively, one can make

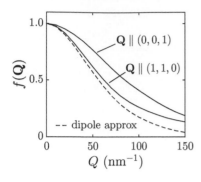

**Fig. 6.8** Anisotropic magnetic form factor of $Ce^{3+}$ in $CePd_2Si_2$. The broken line is the standard dipole approximation, eqn (6.30).

direct use of closed-form expressions given in Lovesey (1984b) and Balcar and Lovesey (1989) for the matrix elements of $\mathbf{M}$ in the $SM_SLM_L$ basis.

To illustrate the use of these formulae, we calculate the anisotropic paramagnetic form factor of $Ce^{3+}$ in $CePd_2Si_2$ (Rotter and Boothroyd, 2009). The tetragonal ligand field splits the $J = 5/2$ ground state level into three doublets, the wave function of the ground state doublet being $a|\pm5/2\rangle - b|\mp3/2\rangle$ in terms of $|M_J\rangle$ components. Experiments show that $a \simeq 0.53$ and $b \simeq 0.85$ (van Dijk *et al.*, 2000). As we are interested in paramagnetic scattering the initial and final angular momentum quantum numbers are the same ($L = L'$, $S = S'$ and $J = J'$), and hence there are no $B(K,K)$ coefficients. The required $A(K-1,K)$ and $B(K-1,K)$ coefficients for $Ce^{3+}$ are given in Table 6.1, and the radially averaged spherical Bessel functions $\langle j_n \rangle$ can be calculated as a function of $Q$ from the parameters given in Lisher and Forsyth (1971). The matrix elements of $\mathbf{M(Q)}$ are straightforward to calculate from (6.73), and are combined into a scattering intensity for a particular $\mathbf{Q}$ direction via eqns (6.11) and (6.12), then square-rooted and normalized to 1 at $Q = 0$.

Results for two directions of $\mathbf{Q}$ are shown in Fig. 6.8. It can be seen that $f(\mathbf{Q})$ for $\mathbf{Q} \parallel (0,0,1)$ exceeds that for $\mathbf{Q} \parallel (1,1,0)$ by about 30% over much of the displayed $Q$ range, and in both directions $f(\mathbf{Q})$ is appreciably greater than the approximate dipole form factor, eqn (6.30). For accurate work, therefore, the anisotropy needs to be taken into account.

### 6.3.3 Magnetic vs atomic form factors

The magnetic form factor for elastic neutron scattering is analogous to the atomic form factor that appears in X-ray diffraction. They differ, however, in which electrons are involved. X-rays scatter from all the electrons and so the form factor is the Fourier transform of the total charge density of the atom,[34] whereas magnetic neutron scattering involves just the electrons that contribute to the magnetization. The latter are in unfilled shells and tend to be furthest from the nucleus. Hence, the spatial variation of the charge and magnetization densities are different, and the $Q$ dependence of the form factors for the two techniques are also different. Figure 6.9 compares the (X-ray) atomic form factor with the (neutron) magnetic form factor for the $Fe^{3+}$ free ion. The magnetic form factor falls more rapidly with $Q$ than the atomic form factor because the distribution of magnetic electrons is more spatially extended than the average electron density in the ion.

Another difference between magnetic neutron and X-ray scattering is that the orbital and spin angular momentum enter into the X-ray scattering amplitude in a different way. For non-resonant X-ray scattering,

$$f(\mathbf{Q}) \sim \mathbf{M}_\perp^L(\mathbf{Q}) \cdot \mathbf{P}_L + \mathbf{M}^S(\mathbf{Q} \cdot \mathbf{P}_S, \tag{6.74}$$

where $\mathbf{M}^L$ and $\mathbf{M}^S$ are Fourier transforms of the orbital and spin magnetization, respectively, and $\mathbf{P}_L$ and $\mathbf{P}_S$ are factors which depend in distinct ways on the X-ray polarization. By performing measurements

[34]Here we are referring to the dominant Thomson scattering process, in which electrons are accelerated by the electric field of the X-rays and emit electromagnetic dipole radiation.

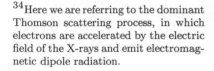
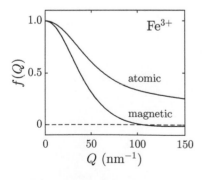

**Fig. 6.9** Atomic and magnetic free-ion form factors of $Fe^{3+}$.

with different X-ray polarizations it is possible to separate the spin and orbital magnetization.

### 6.3.4   Multipolar order

In general, the charge and current distributions of the unpaired electrons in solids can be represented by a series of electric and magnetic multipoles (see Section C.5). Usually when we speak about magnetic order we mean order of the magnetic dipole moments. However, higher-order multipoles can also exist either as secondary order parameters to dipolar order, or even as primary order parameters in the absence of dipole order when they are often called 'hidden' order parameters. Primary order parameters involving non-dipolar multipoles are believed to exist in several systems, including La-doped $CeB_6$, $NpO_2$, and $URu_2Si_2$.

As we saw in Sections 6.3.1–6.3.2, neutrons scatter from magnetic multipoles of odd rank and electric multipoles of even rank. The latter process, however, requires a change in one of the quantum numbers $L$, $S$, $J$, and so cannot be observed in elastic scattering. Scattering from magnetic multipoles of odd rank $K$ depends on the radially averaged spherical Bessel functions $\langle j_{K-1} \rangle$ and $\langle j_{K+1} \rangle$. Dipole ($K = 1$) scattering, if present, dominates at small $Q$ due to the $\langle j_0 \rangle$ term which starts from a value of 1 at $Q = 0$ and decreases with $Q$. All other $\langle j_n \rangle$ ($n > 0$) are zero at $Q = 0$ and increase with $Q$, reaching a maximum at some non-zero $Q$ before decreasing again (see Fig. 6.1). The scattering from non-dipolar multipoles does not depend on $\langle j_0 \rangle$, so is largest in the $Q$ range where the associated $\langle j_n \rangle$ functions have their maxima. The intensity of non-dipolar magnetic scattering is typically much smaller than that of dipolar scattering.

In principle, therefore, neutron diffraction can be used to probe order parameters involving magnetic multipoles higher than dipoles, but due to the small scattering cross-sections such measurements are likely to be challenging even in the absence of dipolar order. The relative insensitivity of neutrons to non-dipolar multipoles contrasts with the technique of resonant X-ray magnetic scattering performed at synchrotrons, which can separate the contribution from different multipoles by their distinct dependence on the X-ray scattering geometry and polarization.

## 6.4   Sum rules for magnetic scattering

Various sum rules can be obtained by integration of the magnetic response function $S(\mathbf{Q}, \omega)$ and generalized susceptibility $\chi(\mathbf{Q}, \omega)$ over $\mathbf{Q}$ or $\omega$, or both. Several general results are given in Section 3.8. Here we apply some of these results to magnetic scattering, and extend them in a couple of cases.

We start by summarizing the key relations for magnetic scattering. The magnetic response function is related to the double differential cross-section by eqn (6.5),

$$\frac{\mathrm{d}^2\sigma}{\mathrm{d}\Omega\mathrm{d}E_\mathrm{f}} = \frac{k_\mathrm{f}}{k_\mathrm{i}}\left(\frac{\gamma r_0}{2\mu_\mathrm{B}}\right)^2 S(\mathbf{Q},\omega), \tag{6.75}$$

where $(\gamma r_0/2)^2 = 7.27 \times 10^{-30}\,\mathrm{m}^2 = 72.7\,\mathrm{mb}$. A prerequisite to apply sum rules is that the scattering has been calibrated and expressed in absolute units. The double differential cross-section is usually given in units of $\mathrm{mb\,sr}^{-1}\mathrm{meV}^{-1}\mathrm{f.u.}^{-1}$, where f.u. stands for 'formula unit', for example a chemical formula, or a unit cell, or a single magnetic ion.

The magnetic response function $S(\mathbf{Q},\omega)$ contains the partial response function[35]

$$S_{\alpha\beta}(\mathbf{Q},\omega) = \frac{1}{2\pi\hbar}\int_{-\infty}^{\infty}\langle \hat{M}_\alpha^\dagger(\mathbf{Q})\hat{M}_\beta(\mathbf{Q},t)\rangle\,\exp(-\mathrm{i}\omega t)\,\mathrm{d}t, \tag{6.76}$$

whose dynamical part is related to the absorptive part of the generalized susceptibility by the Fluctuation–Dissipation theorem[36]

$$\widetilde{S}_{\alpha\beta}(\mathbf{Q},\omega) = \{1 + n(\omega)\}\frac{1}{\pi}\chi''_{\beta\alpha}(\mathbf{Q},\omega). \tag{6.77}$$

For unpolarized neutrons,

$$S(\mathbf{Q},\omega) = \sum_{\alpha\beta}(\delta_{\alpha\beta} - \hat{Q}_\alpha\hat{Q}_\beta)S_{\alpha\beta}(\mathbf{Q},\omega). \tag{6.78}$$

### 6.4.1  Uniform static susceptibility

The term *uniform static susceptibility* refers to $\chi'_{\alpha\beta}(0,0)$, the $\mathbf{Q} = 0$ and $\omega = 0$ limit of the reactive part of the generalized magnetic susceptibility.[37] It can be related to the dynamical magnetic response function via (6.77) and the Kramers–Kronig transformation eqn (D.41):

$$\chi'_{\beta\alpha}(0,0) = \int_{-\infty}^{\infty}\frac{\widetilde{S}_{\alpha\beta}(0,\omega)}{\omega}\{1 - \exp(-\beta\hbar\omega)\}\,\mathrm{d}\omega, \tag{6.79}$$

where $\widetilde{S}_{\alpha\beta}(0,\omega)$ is the dynamical part of the partial response function (6.76) extrapolated to $\mathbf{Q} = 0$.

An alternative, but equivalent relation (see Section D.6) is the inverse of (6.79),

$$\widetilde{S}_{\alpha\beta}(0,\omega) = \omega\{1 + n(\omega)\}\chi'_{\beta\alpha}(0,0)F_{\beta\alpha}(\omega), \tag{6.80}$$

where $F_{\alpha\beta}(\omega)$ is a spectral-weight function that has unit normalization,

$$\int_{-\infty}^{\infty}F_{\alpha\beta}(\omega)\,\mathrm{d}\omega = 1. \tag{6.81}$$

The usefulness of (6.79)–(6.80) lies in the fact that $\chi'(0,0)$ is proportional to the bulk magnetic susceptibility $\chi_\mathrm{M}$. These relations allow magnetometry measurements of $\chi_\mathrm{M}$ to be used as a consistency check on neutron spectra, e.g. to confirm that all magnetic spectral weight

---

[35] Note that $S_{\alpha\beta}(\mathbf{Q},\omega)$ defined here is the same as that in (6.12), but differs from the 'reduced' response function (6.33) which excludes the magnetic form factor and Debye–Waller factor. The two definitions are the same at $\mathbf{Q} = 0$.

[36] See Section D.7.

$$n(\omega) = \frac{1}{\exp(\beta\hbar\omega) - 1},$$

where $\beta = 1/k_\mathrm{B}T$.

[37] At zero frequency there is no dissipation, so $\chi'_{\alpha\beta}(\mathbf{Q},0) = \chi_{\alpha\beta}(\mathbf{Q},0)$.

is accounted for in the measured energy range and that non-magnetic backgrounds are correctly estimated. To do this in practice, we need the conversion between units of generalized susceptibility and units of magnetic susceptibility as used in electromagnetism. The relation, derived in Section D.9, is

$$\chi'(0,0) = \frac{V}{\mu_0}\chi_{\mathrm{M}} \quad \text{(SI)} \tag{6.82}$$

$$= \frac{10}{N_{\mathrm{A}}}\chi_{\mathrm{M}} \quad \text{(Gauss cgs)} \tag{6.83}$$

In (6.82), $\chi_{\mathrm{M}}$ is expressed in SI units[38] and is a dimensionless quantity, and $V$ is in m$^3$. In (6.83), $\chi_{\mathrm{M}}$ is in Gaussian cgs units, i.e. emu mol$^{-1}$ or equivalently cm$^3$ mol$^{-1}$. The particular volume $V$ to be used in (6.82) is determined by the normalization of the scattering intensity. For instance, if $\widetilde{S}_{\alpha\beta}(\mathbf{Q},\omega)$ is normalized per formula unit then $V$ is the volume of the formula unit,[39] and mol$^{-1}$ means per mole of formula units.

A simple example of how such a consistency check can work is provided by a study of the magnetic behaviour of the Yb ions in polycrystalline YbNi$_2$B$_2$C (Boothroyd *et al.*, 2003). For unpolarized neutron scattering from a powder sample, eqns (6.78) and (6.80) give

$$\widetilde{S}(0,\omega) = \omega\{1 + n(\omega)\}2\chi'(0,0)_{\mathrm{av}}\, F(\omega) \tag{6.84}$$

where $\chi_{\mathrm{av}} = \frac{1}{3}(\chi_{xx} + \chi_{yy} + \chi_{zz})$ is the powder-averaged, uniform static susceptibility. Equation (6.84), with a three-peak spectral-weight function, was fitted to the calibrated spectrum measured at $T = 5\,\mathrm{K}$, with $\chi'(0,0)_{\mathrm{av}}$ treated as a variable parameter. A good fit was achieved with $\chi'(0,0)_{\mathrm{av}} = 4.0 \times 10^{-3}\,\mathrm{meV}^{-1}$ per Yb ion. This converts via (6.83) to a magnetic susceptibility of $0.039\,\mathrm{emu\,(mol\ Yb)}^{-1}$ in Gaussian cgs units, which agrees well with the value $0.04\,\mathrm{emu\,(mol\ Yb)}^{-1}$ measured at $5\,\mathrm{K}$ by SQUID magnetometry.

## 6.4.2    Local susceptibility

The local, or $\mathbf{Q}$-averaged, susceptibility is defined as

$$\chi''(\omega) = \frac{\int \chi''(\mathbf{Q},\omega)\,\mathrm{d}^3\mathbf{Q}}{\int \mathrm{d}^3\mathbf{Q}}. \tag{6.85}$$

In principle, the integration is over the whole of $\mathbf{Q}$ space, but for periodic systems it can be sufficient to integrate over a small number of Brillouin zones corresponding to the periodicity of the scattering intensity in reciprocal space, providing account is taken of the fact that only fluctuating components perpendicular to $\mathbf{Q}$ are measured. The significance of $\chi''(\omega)$ is that it measures local magnetic correlations, and corresponds to a weighted density of magnetic modes, as we now show.

Suppose that the magnetization $\mathbf{M}(\mathbf{r})$ is sufficiently localized that it can be expressed as $\mathbf{M}(\mathbf{r}) = \sum_j \mathbf{M}_j(\mathbf{r} - \mathbf{r}_j)$, where $\mathbf{M}_j(\mathbf{r})$ is associated with atom $j$. Hence, from eqn (6.18), $\mathbf{M}(\mathbf{Q}) = \sum_j \mathbf{M}_j(\mathbf{Q}) \exp(i\mathbf{Q}\cdot\mathbf{r}_j)$.

[38] Système International d'unités.

[39] Note that $V\chi_{\mathrm{M}}$ (SI) is the magnetic moment induced on a formula unit per unit applied field.

Suppose also that $\mathbf{M}_j(\mathbf{Q}) = f_j(\mathbf{Q})\,\boldsymbol{\mu}_j$, where $f_j(\mathbf{Q})$ and $\boldsymbol{\mu}_j$ are the magnetic form factor and magnetic moment, respectively, for site $j$. This factorization is valid in the dipole approximation, e.g. eqns (6.28) and (6.31), or for spin-only scattering, eqn (6.59). For simplicity, we shall assume there is only one magnetic species in a single environment, so that $f_j(\mathbf{Q}) = f(\mathbf{Q})$ and $W_j(\mathbf{Q}) = W(\mathbf{Q})$.

With these assumptions, the response function for unpolarized neutrons (6.25) simplifies to[40]

[40] $\boldsymbol{\mu}$ is a Hermitian operator: $\boldsymbol{\mu}^\dagger = \boldsymbol{\mu}$.

$$\frac{S(\mathbf{Q},\omega)}{f^2(\mathbf{Q})\exp\{-2W(\mathbf{Q})\}} = \sum_{j,k} \exp\{i\mathbf{Q}\cdot(\mathbf{R}_k - \mathbf{R}_j)\}$$
$$\times \frac{1}{2\pi\hbar}\int_{-\infty}^{\infty} \langle \hat{\boldsymbol{\mu}}_{\perp j}\cdot\hat{\boldsymbol{\mu}}_{\perp k}(t)\rangle\,\exp(-i\omega t)\,\mathrm{d}t. \tag{6.86}$$

Let us now average both sides of (6.86) over $\mathbf{Q}$, in the manner of (6.85). As a shorthand, we use $\langle\ldots\rangle_{\mathbf{Q}}$ to indicate the $\mathbf{Q}$ average, and define

$$S(\omega) = \left\langle \frac{S(\mathbf{Q},\omega)}{f^2(\mathbf{Q})\exp\{-2W(\mathbf{Q})\}} \right\rangle_{\mathbf{Q}}. \tag{6.87}$$

Only the terms with $j = k$ on the right-hand side of (6.86) survive the $\mathbf{Q}$ averaging, because when $j \neq k$ the average tends to zero due to the oscillatory nature of the exponential factor. Therefore,

$$S(\omega) = \frac{1}{2\pi\hbar}\sum_j \int_{-\infty}^{\infty} \langle \hat{\boldsymbol{\mu}}_{\perp j}\cdot\hat{\boldsymbol{\mu}}_{\perp j}(t)\rangle_{\mathbf{Q}}\,\exp(-i\omega t)\,\mathrm{d}t$$
$$= \frac{N}{2\pi\hbar}\frac{2}{3}\int_{-\infty}^{\infty} \langle \hat{\boldsymbol{\mu}}\cdot\hat{\boldsymbol{\mu}}(t)\rangle\,\exp(-i\omega t)\,\mathrm{d}t, \tag{6.88}$$

where $N$ is the number of magnetic ions. Hereafter, we take $N = 1$ so that the scattering is normalized per magnetic ion.

To make the connection with the local susceptibility we $\mathbf{Q}$-average eqn (6.77), and together with (6.78) and (6.87) this gives for unpolarized neutrons

$$\tilde{S}(\omega) = \{1 + n(\omega)\}\frac{1}{\pi}\sum_{\alpha\beta}\left\langle (\delta_{\alpha\beta} - \hat{Q}_\alpha\hat{Q}_\beta)\frac{\chi''_{\alpha\beta}(\mathbf{Q},\omega)}{f^2(\mathbf{Q})\exp\{-2W(\mathbf{Q})\}} \right\rangle_{\mathbf{Q}}$$
$$= \{1 + n(\omega)\}\frac{1}{\pi}\frac{2}{3}\sum_\alpha \chi''_{\alpha\alpha}(\omega), \tag{6.89}$$

where $\chi''_{\alpha\alpha}(\omega)$ is the *reduced* local susceptibility, defined as in (6.85) but now normalized by the form factor and Debye–Waller factor.[41] In the case of an isotropic local susceptibility, i.e. $\chi''_{xx} = \chi''_{yy} = \chi''_{zz} = \chi''$,

[41] In performing the average over $\mathbf{Q}$ in (6.89) we again use the fact that the oscillatory part of $\chi''_{\alpha\beta}(\mathbf{Q},\omega)$ comes from interference effects which average to zero over the entirety of $\mathbf{Q}$ space.

$$\tilde{S}(\omega) = \{1 + n(\omega)\}\frac{2}{\pi}\chi''(\omega) \qquad \text{(isotropic)}. \tag{6.90}$$

To relate the reduced local susceptibility to the self correlation function we must isolate the dynamic part of (6.88). This is done by subtraction of the infinite time limit of the the correlation function (see

Section 3.6). Comparing the result with (6.89), we find

$$\chi''_{\alpha\alpha}(\omega) = \frac{1}{2\hbar}\{1 - \exp(-\beta\hbar\omega)\} \int_{-\infty}^{\infty} \left\{ \langle \hat{\mu}_\alpha \hat{\mu}_\alpha(t) \rangle - \langle \mu_\alpha \rangle^2 \right\} \exp(-i\omega t)\, dt$$

$$= \pi\{1 - \exp(-\beta\hbar\omega)\} \sum_{\lambda_i} p_{\lambda_i} \sum_{\lambda_f} |\langle \lambda_f | \mu_\alpha | \lambda_i \rangle|^2 \delta(E_{\lambda_f} - E_{\lambda_i} - \hbar\omega)$$

$$(6.91)$$

where $\beta = 1/(k_B T)$ and $\langle \mu_\alpha \rangle$ is the component of the ordered moment in the $\alpha$ direction. In the second line of (6.91) we only include the final states with $E_{\lambda_f} \neq E_{\lambda_i}$ so as to exclude the elastic scattering. In this form it can be seen that apart from the temperature-dependent pre-factor, $\chi''_{\alpha\alpha}(\omega)$ is constructed from the density of excited modes weighted by their transition probabilities from the ground state.

## 6.4.3 Total moment sum rule

Let us go one step further and integrate the unpolarized, $\mathbf{Q}$-averaged, reduced response function (6.88) over energy. Putting $N = 1$, we find

$$\int_{-\infty}^{\infty} S(\omega)\, d(\hbar\omega) = \frac{1}{2\pi\hbar}\frac{2}{3} \int_{-\infty}^{\infty} \int_{-\infty}^{\infty} \langle \hat{\boldsymbol{\mu}} \cdot \hat{\boldsymbol{\mu}}(t) \rangle \exp(-i\omega t)\, dt\, d(\hbar\omega)$$

$$= \frac{2}{3} \int_{-\infty}^{\infty} \langle \hat{\boldsymbol{\mu}} \cdot \hat{\boldsymbol{\mu}}(t) \rangle \delta(t)\, dt$$

$$= \frac{2}{3} \langle \boldsymbol{\mu}^2 \rangle. \qquad (6.92)$$

We have used the integral representation of the $\delta$-function (B.14) to go from the first to the second line. Equation (6.92) is called the *total moment sum rule*. It tells us that if we integrate $S(\omega)$ over energy then we should obtain the mean squared local moment of the magnetic ion. For a spin–only ion, for example, $\boldsymbol{\mu} = -g\mu_B \mathbf{S}$ and so[42]

$$\langle \boldsymbol{\mu}^2 \rangle = g^2 \mu_B^2 S(S+1). \qquad (6.93)$$

The sum rule for $S(\omega)$ expressed by (6.92) contains both the elastic and the inelastic scattering, and $\langle \boldsymbol{\mu}^2 \rangle$ accordingly includes both the ordered moment and the fluctuating moment. If one only integrates over the inelastic scattering then we have the sum rule for $\tilde{S}(\omega)$,

$$\int_{-\infty}^{\infty} \tilde{S}(\omega)\, d(\hbar\omega) = \frac{2}{3}(\langle \boldsymbol{\mu}^2 \rangle - \langle \boldsymbol{\mu} \rangle^2), \qquad (6.94)$$

and gives the square of the fluctuating moment. For a spin-only ion this amounts to $\frac{2}{3}g^2\mu_B^2 S$. The elastic part of the sum rule is then $\frac{2}{3}g^2\mu_B^2 S^2$.

These sum rules are useful for finding out whether a measurement has captured all the magnetic scattering, or to test the reliability of corrections applied to the data. If the energy integral does not agree with the sum rule then there could be magnetic scattering outside the limits probed in the experiment, or it might indicate that the data have been inadequately corrected for the non-magnetic scattering.

[42] For the particular case of $S = 1/2$, a sum rule applies separately to each component of spin:

$$\int_{-\infty}^{\infty} S_{\alpha\alpha}(\omega)\, d(\hbar\omega) = \tfrac{1}{4}g^2\mu_B^2$$

for $\alpha = x, y, z$ — for a proof, see Lorenzana *et al.* (2005) and Exercise 6.5.

# Chapter summary

- The basic theory of magnetic scattering has been presented. We have neglected (i) nuclear–magnetic interference, (ii) coupled magnetic and structural correlations, and (iii) magnetovibrational scattering, i.e. excitation of lattice modes via the magnetic interaction.

- We have defined a magnetic response function by,

$$\frac{\mathrm{d}^2\sigma}{\mathrm{d}\Omega\mathrm{d}E_\mathrm{f}} = \frac{k_\mathrm{f}}{k_\mathrm{i}} \left(\frac{\gamma r_0}{2\mu_\mathrm{B}}\right)^2 S(\mathbf{Q},\omega),$$

  where $(\gamma r_0/2)^2 = 7.27 \times 10^{-30}\,\mathrm{m}^2 = 72.7\,\mathrm{mb}$.

- The response function contains the partial response functions

$$S_{\alpha\beta}(\mathbf{Q},\omega) = \frac{1}{2\pi\hbar} \int_{-\infty}^{\infty} \langle \hat{M}_\alpha^\dagger(\mathbf{Q})\hat{M}_\beta(\mathbf{Q},t)\rangle \exp(-\mathrm{i}\omega t)\,\mathrm{d}t,$$

  where $\mathbf{M}(\mathbf{Q})$ is the Fourier transform of the magnetization.

- The magnetic scattering operator is $\mathbf{M}_\perp = \hat{\mathbf{Q}} \times \mathbf{M} \times \hat{\mathbf{Q}}$.

- The Fluctuation–Dissipation theorem relates the dynamical part of $S_{\alpha\beta}(\mathbf{Q},\omega)$ to the absorptive part of the generalized susceptibility:

$$\tilde{S}_{\alpha\beta}(\mathbf{Q},\omega) = \{1 + n(\omega)\}\frac{1}{\pi}\chi_{\beta\alpha}''(\mathbf{Q},\omega).$$

- For unpolarized neutrons,

$$S(\mathbf{Q},\omega) = \sum_{\alpha\beta}(\delta_{\alpha\beta} - \hat{Q}_\alpha\hat{Q}_\beta)S_{\alpha\beta}(\mathbf{Q},\omega).$$

- For localized (atomic) electrons, $\mathbf{M}(\mathbf{Q})$ can be expanded in a series of electromagnetic multipoles.

- When $Q$ is small, the magnetic dipole term dominates allowing a simplification known as the dipole approximation. When orbital angular momentum is quenched, $\mathbf{M}(\mathbf{Q}) \simeq -g\mu_\mathrm{B}f(Q)\mathbf{S}$, where where $f(Q)$ is the free-ion dipole magnetic form factor eqn (6.32). For weak crystal fields, $g \to g_J$, $\mathbf{S} \to \mathbf{J}$ and $f(Q)$ is given by (6.30).

- The shape of atoms bound in solids is aspherical due to interactions with ligand and exchange fields. For spin-only ions, $\mathbf{M}(\mathbf{Q}) = -g\mu_\mathrm{B}f(\mathbf{Q})\mathbf{S}$, where $f(\mathbf{Q})$ is an anisotropic form factor. General formulae are given for ions with both spin and orbital angular momenta.

- Magnetic neutron scattering satisfies various sum rules which we obtained by integrating $S(\mathbf{Q},\omega)$ and $\chi(\mathbf{Q},\omega)$ over $\mathbf{Q}$ or $\omega$, or both.

# Further reading

A concise summary of magnetic neutron scattering, including the application of polarized neutrons in studies of magnetism, is given by Enderle (2014). A number of excellent reviews on the application of neutrons to investigate magnetic materials can be found in Chatterji (2006).

Multipolar order and techniques available to interrogate it are reviewed by Hotta (2006) and Santini *et al.* (2009). The theory of X-ray magnetic scattering is reviewed by Altarelli (2006).

# Exercises

(6.1) Consider a state with well-defined $L$, $S$ and $J$. Show that the expectation values of the operators $\mathbf{L} \cdot \mathbf{J}$ and $\mathbf{S} \cdot \mathbf{J}$ are given by

$$2\langle \mathbf{L} \cdot \mathbf{J} \rangle = J(J+1) + L(L+1) - S(S+1)$$
$$2\langle \mathbf{S} \cdot \mathbf{J} \rangle = J(J+1) - L(L+1) + S(S+1).$$

Hence, show that the magnetic moment operator $\boldsymbol{\mu} = -\mu_{\rm B}(\mathbf{L} + 2\mathbf{S})$ is represented by $\boldsymbol{\mu} = -g_J \mu_{\rm B} \mathbf{J}$, where $g_J$ is given by eqn (6.29).

(6.2) Calculate for $Sm^{3+}$ the spin, orbital and total free-ion ordered magnetic moments, and the coefficients of the $\langle j_0 \rangle$ and $\langle j_2 \rangle$ terms in the magnetic form factor $f(Q)$. Hence, confirm that the peak in $f(Q)$ near $Q = 50\,\mathrm{nm}^{-1}$ (Fig. 6.4) originates from orbital scattering.

(6.3) The angular part of the wave function of the one-electron $d_{3z^2-r^2}$ orbital is $Y_{2,0}$. Use (6.63) and the coefficients for $m_i = m_j = 0$ in Table 6.2 to show that the $d_{3z^2-r^2}$ magnetic form factor is given by

$$f(\mathbf{Q}) = \langle j_0 \rangle - \tfrac{5}{7}(3\cos^2\beta - 1)\langle j_2 \rangle$$
$$+ \tfrac{9}{28}(35\cos^4\beta - 30\cos^2\beta + 3)\langle j_4 \rangle.$$

(6.4) Show that

$$\langle \hat{\boldsymbol{\mu}}_\perp \cdot \hat{\boldsymbol{\mu}}_\perp(t) \rangle_{\mathbf{Q}} = \frac{2}{3}\langle \hat{\boldsymbol{\mu}} \cdot \hat{\boldsymbol{\mu}}(t) \rangle,$$

as used in eqn (6.88), where the subscript $\mathbf{Q}$ denotes an average over all $\mathbf{Q}$, and $\boldsymbol{\mu}$ is the magnetic moment operator for a local moment.

(6.5) Consider the energy integral of $S_{\alpha\alpha}(\omega)$, the $\mathbf{Q}$-averaged, reduced partial response function for a single spin component. Adapt eqn (6.92) to this case, and hence prove that when $S = 1/2$ the following sum rule applies for each spin component:

$$\int_{-\infty}^{\infty} S_{\alpha\alpha}(\omega)\,\mathrm{d}(\hbar\omega) = \tfrac{1}{4}g^2 \mu_{\rm B}^2 \quad (\alpha = x, y, z).$$

Why do separate sum rules not apply to each spin component when $S > 1/2$?

# Magnetic Diffraction

**7**

Magnetic order can be found in a great many classes of material, including insulators, metals, semiconductors and superconductors, and in structural forms as diverse as crystals, glasses, molecular solids, polymers, and artificial nanostructures. The determination of magnetic structures is extremely important for understanding the magnetic properties of materials and for testing the validity of theoretical models.

The bulk of this chapter, contained in Sections 7.1 to 7.4, is concerned with magnetic structure determination by neutron diffraction. The variety of magnetic structures in real materials is extremely diverse, and some ingenuity may be needed to establish a magnetic ordering pattern uniquely. Neutron polarization analysis (Section 7.3) is a particularly powerful weapon which the experimentalist can deploy. Magnetism is usually associated with electrons, but ordering of magnetic moments attached to nuclei has also been observed by neutron diffraction, as described briefly in Section 7.7.

Some systems of coupled magnetic moments do not exhibit static long-range magnetic order, but instead have short-range magnetic correlations. There can be various reasons for this, including chemical or structural disorder, frustration, and even quantum fluctuations. Magnetic correlations also occur in the vicinity of phase transitions, e.g. from a paramagnet to a phase with long-range magnetic order. Short-range magnetic correlations are characterized by magnetic diffuse scattering, which is discussed in Section 7.5

As well as probing correlations between moments on different sites, neutron diffraction can also be used to determine the spatial distribution of the magnetization on an atomic scale through accurate measurements of magnetic form factors. This can reveal useful information about electronic structure and bonding. The method is outlined in Section 7.6.

## 7.1 Magnetic structures

### 7.1.1 Basic ideas

A magnetic structure is a spatially ordered arrangement of magnetic dipole moments, or *spins*.[1] For a magnetic structure to develop, there needs to be some form of magnetic coupling which favours a particular alignment of the moments with respect to one another. Generally, this coupling is provided by exchange interactions, although dipole–dipole forces can also drive magnetic order at low temperatures.

[1] Ordering of higher electric and magnetic multipoles can also occur, either concurrent with or at a different temperature from magnetic dipole ordering. The discussion here refers solely to magnetic dipole ordering.

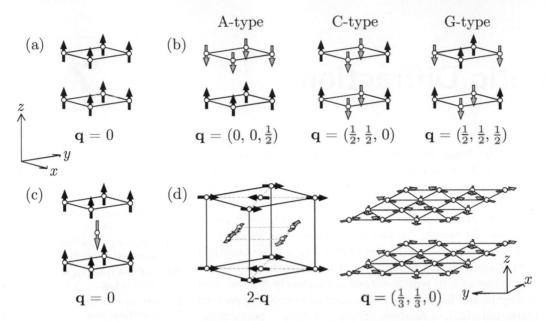

**Fig. 7.1** Illustrations of some magnetic structures found in magnetic crystals. (a) Ferromagnet. (b) Three different types of antiferromagnet. (c) Ferrimagnet. (d) Non-collinear magnetic structures. The propagation vector **q** is given for each structure. The 2-**q** structure in (d) has propagation vectors $\mathbf{q}_1 = (1, 0, 0)$ and $\mathbf{q}_2 = (0, 1, 0)$ for the $x$ and $y$ components of spin, respectively.

The simplest type of magnetic order is *ferromagnetism*. In a ferromagnet all the moments align parallel to one another, Fig. 7.1(a). This means that the magnetic lattice is the same as the crystal lattice, and so the magnetic and structural diffraction peaks coincide.

Another common type of magnetic order is *antiferromagnetism*. In an antiferromagnet, half the moments point in one direction and half point in the opposite direction. The moments are arranged spatially so that neighbouring moments are antiparallel in at least one direction. An antiferromagnet can be regarded as two interpenetrating ferromagnets whose magnetic moments exactly compensate one another. Figure 7.1(b) shows three types of antiferromagnetic structure. They differ with respect to the number of parallel and antiparallel nearest neighbours. The so-called G-type antiferromagnet is sometimes known as the Néel state.[2]

Crystals with magnetic atoms in two inequivalent sites coupled antiferromagnetically can form structures like the one shown in Fig. 7.1(c). This type of structure differs from an antiferromagnet in that the moments on each inequivalent site not only point in opposite directions but also have different sizes, which gives the material a net magnetic moment. Néel came up with this type of magnetic order to explain the properties of some ferrites. He called it *ferrimagnetism*.

Magnetic structures in which the moments are aligned either parallel or antiparallel to one another, such as those just described, are termed *collinear*. Structures in which moments can be found at angles other than 0° or 180° to one another are *non-collinear*. A couple of examples of non-collinear structures are given in Fig. 7.1(d).

[2]Louis Néel (1904–2000) was the first to suggest the possibility of magnetic structures with antiparallel neighbouring moments, and coined the term 'antiferromagnetism' for this type of structure.

## 7.1.2 Magnetic unit cell

The period of a magnetic structure is always equal to or greater than the period of the crystal lattice. If the magnetic and crystal lattice periods are related by an integer or a rational fraction then the magnetic structure is *commensurate* with the crystal lattice. It can then be convenient to describe the magnetic structure in terms of a *magnetic unit cell* which, when repeated in all directions, generates the complete magnetic structure. The magnetic unit cell of a ferromagnet [Fig. 7.1(a)] is the same as the chemical unit cell, but the magnetic unit cells of the antiferromagnets shown in Fig. 7.1(b) are double the chemical cells in one, two, and three directions, respectively, for the A-type, C-type and G-type structures. In commensurate magnetic structures one can often identify two or more *sublattices*. A magnetic sublattice describes the periodicity of any subset of the magnetic ions whose moments have translational symmetry. For example, each of the antiferromagnetic structures in Fig. 7.1(b) has two sublattices.

When a magnetic unit cell can be defined, one can describe the complete magnetic structure by specifying the magnetic lattice type and the magnitudes and directions of all the magnetic moments in the magnetic unit cell. Just as for crystal structures (Section 2.1.4), magnetic structures can be classified according to the set of symmetry operations that leave the structure invariant. In the magnetic case, however, there is need for a new anti-symmetry operation R which reverses the sign of the magnetic moment at each point in space without acting on the space coordinates. The R operator has the same effect as a time-reversal operation. The complete set of magnetic space groups (Shubnikov groups) is obtained by combining the crystallographic (Fedorov) space groups with the anti-symmetry operation. When this is done, a total of 1651 magnetic space groups is found.

A magnetic unit cell cannot be defined if the periodicity of the magnetic structure is not commensurate with the chemical lattice. Such structures are said to be *incommensurate*. Structures exist which have sinusoidal modulations in one or more spin component and these are frequently incommensurate. Some examples are shown in Fig. 7.2. The first is a collinear structure with a longitudinal modulation of the spins. This type of structure is often called a *spin-density wave*.[3] The second is a transverse helix,[4] in which the moments rotate through a fixed angle in a plane perpendicular to the line joining adjacent atoms along the axis of the helix. The third is an in-plane helix or cycloid, in which the modulation direction lies in the plane of rotation of the spins, and the fourth is conical helix in which the spins rotate on the surface of a cone.

## 7.1.3 Propagation vectors

Let us assume that a magnetic structure can be represented by a set of ordered moments associated with atomic sites. The periodicity of the magnetic structure implies a well-defined relationship between the

[3] *Spin density* is a property of metals, so only itinerant magnets can have spin-density wave order.

[4] Helical structures are sometimes called spirals. This term, however, is inappropriate because 'spiral' refers to a curve whose radius changes as it winds around an axis. In a helix, the radius remains constant. To be absolutely clear: a helical magnetic structure is *not* a spiral.

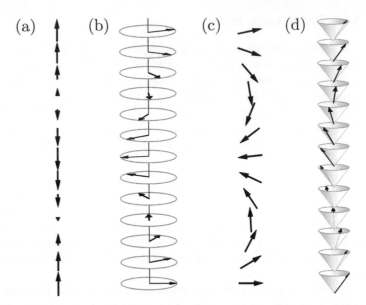

**Fig. 7.2** Some examples of incommensurate magnetic structures. (a) Longitudinal spin-density wave. (b) Transverse helix. (c) Cycloid. (d) Conical helix.

magnitudes and directions of ordered moments at equivalent positions in different chemical unit cells. This relationship can be expressed as a Fourier series. If $\boldsymbol{\mu}_{ld}$ is the magnetic moment on the $d$th site in the $l$th chemical unit cell associated with lattice vector $\mathbf{l}$, then

$$\boldsymbol{\mu}_{ld} = \sum_{\mathbf{q}} \mathbf{m}_{\mathbf{q},d} \exp(-\mathrm{i}\mathbf{q} \cdot \mathbf{l}). \tag{7.1}$$

The Fourier amplitude $\mathbf{m}_{\mathbf{q},d}$ is a complex vector which describes the amplitude and phase of each component of the ordered moment on site $d$. The lattice periodicity means that Fourier coefficients with wavevectors that differ by a reciprocal lattice vector are the same, i.e. $\mathbf{m}_{\mathbf{q}} = \mathbf{m}_{\mathbf{q}+\mathbf{G}}$.[5] Therefore, the summation can be restricted to wavevectors $\mathbf{q}$ inside the first Brillouin zone of the chemical lattice.[6] Magnetic moments are real, i.e. $\boldsymbol{\mu}^* = \boldsymbol{\mu}$, and from (7.1) this means that if a Fourier component $\mathbf{q}$ is non-zero then so is the component $-\mathbf{q}$, and that $\mathbf{m}_{-\mathbf{q}} = \mathbf{m}_{\mathbf{q}}^*$.

The principal wavevectors $\mathbf{q}$ in the Fourier decomposition (7.1) of a magnetic structure are called the *ordering wavevectors* or *propagation vectors*.[7] The vast majority of magnetic structures in relatively simple compounds can be described by a single propagation vector. For example, the ferromagnetic and ferrimagnetic structures in Fig. 7.1(a) and (c) both have propagation vector $\mathbf{q} = (0,0,0)$, because the magnetic moments are ordered in the same way in every unit cell. The magnetic moments in the antiferromagnetic structures in Fig. 7.1(b) reverse direction in adjacent chemical cells along the $z$ axis (A-type), $x$ and $y$ axes (C-type), and $x$, $y$ and $z$ axes (G-type). The propagation vectors (in r.l.u.) are therefore $\mathbf{q} = (0,0,\frac{1}{2})$, $\mathbf{q} = (\frac{1}{2},\frac{1}{2},0)$, and $\mathbf{q} = (\frac{1}{2},\frac{1}{2},\frac{1}{2})$, respectively.

Magnetic structures which have more than one propagation vector are called *multi-$\mathbf{q}$* structures. The first of the non-collinear structures

---

[5]To see this, write the inverse Fourier series $\mathbf{m}_{\mathbf{q}} = \frac{1}{N}\sum_l \boldsymbol{\mu}_l \exp(\mathrm{i}\mathbf{q}\cdot\mathbf{l})$, so that

$$\mathbf{m}_{\mathbf{q}+\mathbf{G}} = \frac{1}{N}\sum_l \boldsymbol{\mu}_l \exp\{\mathrm{i}(\mathbf{q}+\mathbf{G})\cdot\mathbf{l}\}$$

$$= \mathbf{m}_{\mathbf{q}}$$

since $\exp(\mathrm{i}\mathbf{G}\cdot\mathbf{l}) = 1$ by definition of the reciprocal lattice (see Section B.4.1).

[6]The first Brillouin zone is the Wigner–Seitz unit cell (see Section 2.1.3) centred on the origin of reciprocal space.

[7]Fourier components $\pm\mathbf{q}$ correspond to a single propagation vector $\mathbf{q}$.

in Fig. 7.1(d) is a *two*-$\mathbf{q}$ (2-$\mathbf{q}$) structure because it is described by two distinct propagation vectors, one for the $x$ components of the ordered moments and one for the $y$ components.

The structures shown in Fig. 7.2 all have a sinusoidal modulation in the vertical ($z$) direction. If we call the wavelength of the modulation $\lambda_m$, then the Fourier decomposition, eqn (7.1), of the structures has components $\pm\mathbf{q}$, where $\mathbf{q} = (0, 0, c/\lambda_m)$ is a propagation vector.[8] If the modulation of the moments along the propagation direction is not perfectly sinusoidal, for example if there were some squaring or bunching of the sinusoid, then the Fourier decomposition of the structure contains higher harmonics of the propagation vector, i.e. integer multiples of $\mathbf{q}$.

## 7.1.4   Magnetic domains

When magnetic order develops from the paramagnetic phase there is often more than one choice of equivalent magnetic structure that the system can adopt. This comes about because the symmetry of the ordered magnetic phase is generally lower than the symmetry of the paramagnetic phase (i.e. the symmetry of the crystal structure in the absence of magnetism). Take, for example, the two structures shown in Fig. 7.3(a). These are related by a 90° rotation about the $z$ axis, which is a symmetry element of the underlying square lattice but not a symmetry element of the magnetic structure itself. Both structures have identical energy, and there is no reason for the system to favour one over the other. We call these equivalent structures *magnetic domains*.

What tends to happen in practice is that a bulk sample will break up into many spatially separated microscopic regions each of which is a pure magnetic domain. The observed scattering intensity from such a sample will be an average over all the domains weighted by the population of each domain. The domain populations can be unequal if a macroscopic perturbation, such as a magnetic, electric, or strain field, is present which breaks the microscopic symmetry of the crystal so as to favour some magnetic domains over others.

Magnetic domains are classified according to the types of symmetry element that are lost in magnetic ordering: (a) configuration ($K$) domains (translational symmetry), (b) 180° domains (time-reversal symmetry), (c) orientation or spin ($S$) domains (rotational symmetry), (d) chiral domains (inversion symmetry). Examples are given in Fig. 7.3.

*Configuration domains* have propagation vectors whose directions are related by the symmetry elements of the paramagnetic phase, the so-called 'star' of $\mathbf{q}$. For example, in Fig. 7.3(a) the lattice has 4-fold rotational symmetry about the $z$ axis, so the propagation vector $\mathbf{q}_1 = (\frac{1}{2}, 0, 0)$ is related to the wavevectors $\mathbf{q}_2 = (0, \frac{1}{2}, 0)$, $\mathbf{q}_3 = (-\frac{1}{2}, 0, 0)$ and $\mathbf{q}_4 = (0, -\frac{1}{2}, 0)$. However, since $\mathbf{q}_1$ and $\mathbf{q}_3$ are connected by a reciprocal lattice vector, and $\mathbf{q}_2$ and $\mathbf{q}_4$ likewise, there are only two inequivalent configurational domains as shown in the figure.

Magnetic ordering corresponds to time-reversal symmetry breaking.[9] For any magnetic structure, there is an equivalent one in which all the

[8]The conical structure in Fig 7.2(d) has a second propagation vector, $\mathbf{q} = 0$, associated with the $z$ component of spin.

[9]Reversing the direction of time reverses the direction of the circulating currents which form the magnetic moments. A reversal of the magnetic moments is a symmetry operation for a paramagnet but not for a magnetically ordered material.

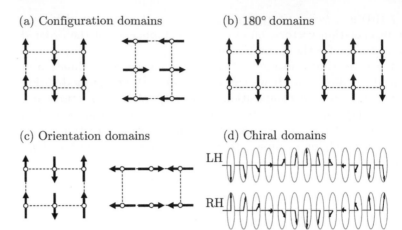

**Fig. 7.3** Examples of different types of magnetic domains.

magnetic moments are reversed. These two related structures are called *180° domains*. Fig. 7.3(b) shows two 180° domains.

*Orientation domains* have the same propagation vector but differ in the direction of $\mathbf{m}_{\mathbf{q},d}$, i.e. in the direction of the magnetic moments in the Fourier component with wavevector $\mathbf{q}$. The structures in Fig. 7.3(c) are orientation domains.

*Chiral domains* occur when a magnetic structure lacking a centre of symmetry forms in a crystal whose chemical structure has a centre of symmetry. In such cases, each domain with propagation vector $\mathbf{q}$ comes in two forms differing in their chirality, or handedness.[10] Right- and left-handed structures are mirror images of one another. An example is the helical structure in Fig. 7.2(b), which is described mathematically by

$$\boldsymbol{\mu}_l = \mu(\cos \mathbf{q} \cdot \mathbf{l}, \pm \sin \mathbf{q} \cdot \mathbf{l}, 0), \tag{7.2}$$

where $\mathbf{q} = (0, 0, q)$ r.l.u. and $\mathbf{q} \cdot \mathbf{l} = 2\pi q n_3$, $n_3 =$ integer as in (2.2). The two signs correspond to the right- and left-handed helices, as shown in Fig. 7.3(d), which are related by the inversion operation[11] $\mathbf{l} \to -\mathbf{l}$.

### 7.1.5   Representation analysis

As we have seen, symmetry plays an important role in the description of magnetic structures. Group theory, which is the mathematical language of symmetry, can be of great help in magnetic structure determination. The theory[12] makes use of *irreducible representations* (irreps) to describe how magnetic structures transform under the symmetry operations of the crystal space group and to enumerate the possible magnetic structures that are compatible with the symmetry of the spin Hamiltonian for a given $\mathbf{q}$. It writes the Fourier coefficients $\mathbf{m}_{\mathbf{q},d}$ as linear combinations of the irrep basis vectors, which allows one to fit the coefficients of the basis vectors rather than the $\mathbf{m}_{\mathbf{q},d}$. This has the advantage that the number of parameters to fit is much smaller and only symmetry-allowed structures are obtained. We will not describe the method here as there are a number of complete descriptions and software packages available.[13]

[10]The term *chiral* was introduced by Lord Kelvin, and is based on the Ancient Greek word $\chi\epsilon\iota\rho$ for hand. In case you wanted to know.

[11]Note that the direction of an individual magnetic moment is invariant under inversion.

[12]The theory was developed in detail by Bertaut (1968), although the use of group representations to describe magnetic structures was first introduced in an earlier paper by Alexander (1962).

[13]Widely used programs for symmetry analysis and data fitting include SARAh (http://fermat.chem.ucl.ac.uk/spaces/willsgroup/software/), BasIreps, which is integrated into the FullProf suite (https://www.ill.eu/sites/fullprof/), and JANA (http://jana.fzu.cz/).

## 7.2   Unpolarized neutron diffraction

With unpolarized neutrons, the total scattering from a magnetic substance is just the sum of the nuclear and magnetic scattering intensities.
There is no nuclear–magnetic interference because this term is linear
in the neutron polarization and therefore vanishes when averaged over
random neutron spin orientations (see Section 6.1).

Magnetic neutron diffraction (i.e. elastic magnetic scattering) is described by the static component of eqn (6.25), which for a single magnetic
domain may be written

$$\frac{\mathrm{d}\sigma}{\mathrm{d}\Omega} = \left(\frac{\gamma r_0}{2\mu_B}\right)^2 |\langle \mathbf{M}_\perp(\mathbf{Q})\rangle|^2$$

$$= \left(\frac{\gamma r_0}{2\mu_B}\right)^2 \left|\sum_j \langle \mathbf{M}_{\perp j}(\mathbf{Q})\rangle \exp(-W_j)\exp(\mathrm{i}\mathbf{Q}\cdot\mathbf{R}_j)\right|^2, \quad (7.3)$$

where $(\gamma r_0/2)^2 = 72.65$ mb, and

$$\mathbf{M}_{\perp j}(\mathbf{Q}) = \hat{\mathbf{Q}} \times \{\mathbf{M}_j(\mathbf{Q}) \times \hat{\mathbf{Q}}\}. \quad (7.4)$$

The summation is over all atoms carrying a magnetic moment, and
$\mathbf{M}_j(\mathbf{Q})$ is the Fourier transform of the magnetization associated with
site $j$. $\mathbf{M}_\perp$ is the component of $\mathbf{M}$ perpendicular to the scattering vector
$\mathbf{Q}$, and $\hat{\mathbf{Q}}$ is a unit vector in the direction of $\mathbf{Q}$. The Debye–Waller factor
$\exp(-W_j)$ takes into account atomic vibrations which have the effect of
reducing the intensity relatively to a perfectly rigid structure.

We will henceforth assume that the dipole approximation applies, either in the form of (6.28) or (6.31), which allows us to write

$$\langle \mathbf{M}_j(\mathbf{Q})\rangle = \boldsymbol{\mu}_j f_j(Q), \quad (7.5)$$

so that

$$\mathbf{M}(\mathbf{Q}) = \sum_j \boldsymbol{\mu}_j f_j(Q)\exp(-W_j)\exp(\mathrm{i}\mathbf{Q}\cdot\mathbf{R}_j) \quad (7.6)$$

where $f_j(Q)$ is the magnetic form factor and $\boldsymbol{\mu}_j$ the thermal average
of the magnetic moment of atom $j$. How to calculate $f_j(Q)$, and the
circumstances under which (7.5) applies, have been discussed in detail
in Sections 6.2.2 and 6.3. When the dipole approximation is inadequate
one can calculate $\langle \mathbf{M}_j(\mathbf{Q})\rangle$ from eqn (6.59), (6.70) or (6.73).[14]

[14]See also Rotter and Boothroyd (2009).

### 7.2.1   Structures described by a magnetic unit cell

In the case of magnetic structures which can be described by a magnetic
unit cell (see Section 7.1.2) we can exploit the commensurate periodicity
of the magnetic structure with respect to the crystal lattice to simplify
eqn (7.3). Following the steps in Section 5.2 we obtain the magnetic
analogue of the nuclear coherent elastic scattering cross-section (5.47),

$$\frac{\mathrm{d}\sigma}{\mathrm{d}\Omega} = N_m \frac{(2\pi)^3}{v_m}\sum_{\mathbf{G}_m} |\mathbf{F}_{M\perp}(\mathbf{Q})|^2 \delta(\mathbf{Q}-\mathbf{G}_m), \quad (7.7)$$

where

$$\mathbf{F}_{\mathrm{M}\perp}(\mathbf{Q}) = \hat{\mathbf{Q}} \times \{\mathbf{F}_{\mathrm{M}}(\mathbf{Q}) \times \hat{\mathbf{Q}}\}, \tag{7.8}$$

and

$$\mathbf{F}_{\mathrm{M}}(\mathbf{Q}) = p \sum_d \boldsymbol{\mu}_d f_d(Q) \exp(-W_d) \exp(\mathrm{i}\mathbf{Q} \cdot \mathbf{d}) \tag{7.9}$$

is the *magnetic structure factor*.[15] Here we have defined $p = \gamma r_0 / 2\mu_{\mathrm{B}}$.

In (7.7), $N_{\mathrm{m}}$ is the number of magnetic unit cells in the crystal, $v_{\mathrm{m}}$ is the volume of the magnetic unit cell, and the $\mathbf{G}_{\mathrm{m}}$ are reciprocal lattice vectors for the magnetic lattice. The summation in (7.9) is over the atoms in the magnetic unit cell. It is often convenient to use the identity (see Section 4.9)

$$|\mathbf{F}_{\mathrm{M}\perp}|^2 = \sum_{\alpha\beta} (\delta_{\alpha\beta} - \hat{Q}_\alpha \hat{Q}_\beta)(F_{\mathrm{M}}^*)_\alpha (F_{\mathrm{M}})_\beta. \tag{7.10}$$

These expressions show that the elastic scattering from a simple magnetic structure consists of a set of sharp peaks (magnetic Bragg peaks) associated with the magnetic reciprocal lattice vectors. The intensity of the Bragg peaks is governed in the main by three factors:[16] (i) the arrangement of the magnetic moments in the unit cell, (ii) the magnetic form factors, and (iii) the orientation of the scattering vector with respect to the ordered moments.

### Ferromagnet

Consider a simple ferromagnet with one magnetic atom per unit cell and moments aligned along the $z$ direction, as shown in Fig. 7.1(a). The summation in (7.9) has only one term and the ordered moment is given by $\boldsymbol{\mu} = (0, 0, \mu)$. The elastic magnetic scattering cross-section can easily be obtained from (7.7)–(7.10) and is given by

$$\frac{\mathrm{d}\sigma}{\mathrm{d}\Omega} = N \frac{(2\pi)^3}{v_0} (1 - \hat{Q}_z^2) |\mathbf{F}_{\mathrm{M}}(\mathbf{Q})|^2 \sum_{\mathbf{G}} \delta(\mathbf{Q} - \mathbf{G}), \tag{7.11}$$

where

$$\mathbf{F}_{\mathrm{M}}(\mathbf{Q}) = p \, \boldsymbol{\mu} f(Q) \exp(-W). \tag{7.12}$$

For a ferromagnet, the magnetic and chemical unit cells are the same ($N_{\mathrm{m}} = N$, $v_{\mathrm{m}} = v_0$) so the magnetic and structural Bragg peaks coincide at the reciprocal lattice vectors of the chemical lattice. The total intensity in the Bragg peaks is therefore the sum of the nuclear and magnetic intensities.[17] The factor $(1 - \hat{Q}_z^2)$ is sometimes called the *orientation factor*, and is equal to $\sin^2\theta$, where $\theta$ is the angle between $\mathbf{Q}$ and $\boldsymbol{\mu}$. This factor arises when the magnetization is projected onto the plane perpendicular to $\mathbf{Q}$ with eqn (7.10).

The cross-section (7.11) applies to a single-domain ferromagnet in which all the moments are aligned parallel to $z$. However, real ferromagnetic materials typically contain many domains, each of which is made up of a large number of atoms with magnetic moments aligned parallel to

[15]To within the constant $p$, the structure factor $\mathbf{F}_{\mathrm{M}}(\mathbf{Q})$ is the Fourier transform of the magnetization $\mathbf{M}(\mathbf{r})$ in the magnetic unit cell. This is different from $\mathbf{M}(\mathbf{Q})$, which is the Fourier transform of the magnetization in the entire crystal.

[16]The Debye–Waller factor is normally close to unity and varies only weakly with $Q$ at the small $Q$ values and low temperatures typically used to measure magnetic diffraction.

[17]As noted earlier, the nuclear–magnetic interference term averages to zero for unpolarized neutrons — see Section 6.1.

one another, the direction of alignment varying from domain to domain. Magnetic anisotropy tends to favour certain crystallographic directions ('magnetic easy axes'), and in the absence of a symmetry-breaking external perturbation, such as a magnetic field, domains with moments along all symmetry-equivalent directions will usually be present (Fig. 7.4).

The scattering from a multi-domain sample is the sum of the scattering from each individual domain. As an example, consider a cubic ferromagnet with $x$, $y$ and $z$ as the magnetic easy axes. This gives rise to three types of orientation domain. If these are present in equal proportions then the only modification to the cross-section (7.11) is to replace the orientation factor with its domain average, which is

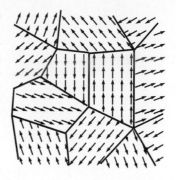

**Fig. 7.4** Schematic domain pattern in an unmagnetized ferromagnet.

$$\frac{1}{3}\sum_\alpha (1 - \hat{Q}_\alpha^2) = \frac{2}{3}, \quad (\alpha = x, y, z). \qquad (7.13)$$

The orientation factor is now independent of the direction of $\mathbf{Q}$.

Since the magnetic and nuclear Bragg peaks occur at the same points in reciprocal space for a ferromagnet, an obvious question is how to separate the nuclear and magnetic contributions. There are several ways. One is to perform measurements at temperatures above and below the Curie temperature $T_C$, which is the magnetic ordering temperature for a ferromagnet. Below $T_C$ the Bragg peaks contain both nuclear and magnetic contributions, whereas above $T_C$ only the nuclear component remains. The nuclear scattering usually varies only weakly with temperature (through the Debye–Waller factor) so a subtraction of the intensity measured above $T_C$ from that measured below $T_C$ usually provides a good separation of the magnetic and nuclear scattering.

A second method is to perform measurements in a magnetic field strong enough to overcome the magnetocrystalline anisotropy and align all the moments along the field direction. If the field is applied parallel to $\mathbf{Q}$ then the orientation factor vanishes and the elastic cross-section contains no magnetic scattering. The difference between the intensity measured with the field perpendicular and parallel to $\mathbf{Q}$ is then purely magnetic scattering. The expected variation is seen in the data for magnetite ($Fe_3O_4$) shown in Fig. 7.5. Magnetite is actually a ferrimagnet ($T_C = 858\,\mathrm{K}$), but the variation of diffracted intensity with field direction is the same as for a ferromagnet. In general, the 111 Bragg reflection of $Fe_3O_4$ is both nuclear and magnetic in origin. From (7.11), when the magnetic moments are fully aligned parallel to $\mathbf{Q}$ there is no magnetic scattering and the intensity is given by $|F_N|^2$, whereas when the moments are aligned perpendicular to $\mathbf{Q}$ there is both nuclear and magnetic intensity, $|F_N|^2 + |F_M|^2$. In zero field, the sample is unmagnetized and from (7.13) the domain-averaged intensity is then $|F_N|^2 + \frac{2}{3}|F_M|^2$.

Perhaps the most direct way to separate magnetic and non-magnetic scattering from a ferromagnet is to use polarized neutrons. Such measurements must be made in a magnetic field strong enough to align all the moments along the field direction, otherwise the neutron beam will depolarize in the spatially varying fields due to domains in the sample. Polarized neutron diffraction is discussed in Section 7.3.

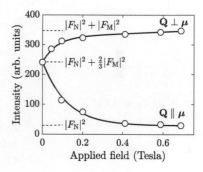

**Fig. 7.5** Intensity of the 111 reflection of $Fe_3O_4$ as a function of field applied parallel and perpendicular to $\mathbf{Q}$. At the highest field the magnetization directions of the individual domains become aligned with the field. (Adapted from Shull *et al.*, 1951.)

## Other collinear structures

In a collinear magnetic structure all the ordered moments point either parallel or antiparallel to a single direction. If $\boldsymbol{\eta}$ is a unit vector in this direction, then the elastic cross-section (7.7) may be written

$$\frac{d\sigma}{d\Omega} = N_m \frac{(2\pi)^3}{v_m}(1 - \hat{Q}_\eta^2)|F_M(\mathbf{Q})|^2 \sum_{\mathbf{G}_m} \delta(\mathbf{Q} - \mathbf{G}_m), \qquad (7.14)$$

where $\hat{Q}_\eta = \hat{\mathbf{Q}} \cdot \boldsymbol{\eta}$ is the component of $\hat{\mathbf{Q}}$ along $\boldsymbol{\eta}$, and the magnetic structure factor is now represented by the scalar function

$$F_M(\mathbf{Q}) = p \sum_d \sigma_d \mu_d f_d(Q) \exp(-W_d) \exp(i\mathbf{Q} \cdot \mathbf{d}), \qquad (7.15)$$

where $\sigma_d = +1$ or $-1$ depending on whether the moment at site $d$ is parallel or antiparallel to $\boldsymbol{\eta}$. To obtain the cross-section for an actual sample one must average over the cross-sections for all equivalent magnetic domains.

Consider the two-sublattice antiferromagnets shown in Fig. 7.1(b). For the A-type antiferromagnet, the magnetic unit cell has dimensions $a \times b \times 2c$ and contains two magnetic moments, both of magnitude $\mu$ but oppositely aligned, located at sites with fractional coordinates 0,0,0 and 0,0,1. From (7.15), it follows that the magnetic structure factor is

$$F_M(\mathbf{Q}) = p\mu f(Q) \exp(-W)\{1 - \exp(i\mathbf{Q} \cdot \mathbf{c})\}. \qquad (7.16)$$

Because the magnetic unit cell is doubled along the $z$ axis the magnetic reciprocal lattice vectors may be written

$$\mathbf{G}_m = h\mathbf{a}^* + k\mathbf{b}^* + l\mathbf{c}^* \quad \text{and} \quad h\mathbf{a}^* + k\mathbf{b}^* + (l + \tfrac{1}{2})\mathbf{c}^*, \qquad (7.17)$$

or $(h, k, l)$ and $(h, k, l+\tfrac{1}{2})$ in reciprocal lattice units, where $h$, $k$, and $l$ are integers and $\mathbf{a}^*$, $\mathbf{b}^*$, and $\mathbf{c}^*$ are the basis vectors for the reciprocal lattice of the chemical lattice (see Sections 2.1.6 and 2.2.2). The magnetic structure factor need only be evaluated at $\mathbf{Q} = \mathbf{G}_m$ because of the $\delta$-function in (7.14), and $F_M(\mathbf{G}_m)$ contains the factor

$$1 - \exp(i\mathbf{G}_m \cdot \mathbf{c}) = 0 \quad \text{for } (h, k, l)$$
$$= 2 \quad \text{for } (h, k, l + \tfrac{1}{2}). \qquad (7.18)$$

This shows that although both $(h, k, l)$ and $(h, k, l + \tfrac{1}{2})$ are magnetic reciprocal lattice vectors, magnetic Bragg scattering is only found at the $(h, k, l + \tfrac{1}{2})$ positions, see Fig. 7.6. Physically, this is because neutrons reflected from adjacent $(00l)$ atomic planes have a $\pi$ phase shift when undergoing magnetic scattering because the magnetic moments on adjacent planes point in opposite directions in this particular antiferromagnetic structure. The $\pi$ phase shift causes destructive interference. By contrast, nuclear Bragg scattering is found at the $(h, k, l)$ positions since these are reciprocal lattice vectors of the chemical lattice.

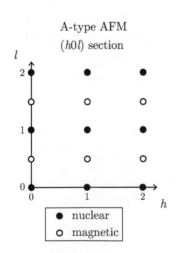

A-type AFM
$(h0l)$ section

**Fig. 7.6** $(h0l)$ section through the reciprocal lattice of an A-type antiferromagnet, showing the positions of the allowed nuclear and magnetic Bragg reflections.

Similar results are obtained for the other types of two-sublattice antiferromagnet shown in Fig. 7.1(b), as illustrated in Fig. 7.7. For the C-type structure the $\mathbf{G}_\mathrm{m}$ are of the form $(h, k, l)$ and $(h + \frac{1}{2}, k + \frac{1}{2}, l)$, and $F_\mathrm{M}(\mathbf{G}_\mathrm{m})$ contains the factor

$$1 - \exp(\mathrm{i}\mathbf{G}_\mathrm{m} \cdot \mathbf{a}) = 0 \quad \text{for } (h, k, l)$$
$$= 2 \quad \text{for } (h + \tfrac{1}{2}, k + \tfrac{1}{2}, l), \qquad (7.19)$$

while for the G-type structure the $\mathbf{G}_\mathrm{m}$ are $(h, k, l)$ and $(h + \frac{1}{2}, k + \frac{1}{2}, l + \frac{1}{2})$, and the factor in $F_\mathrm{M}(\mathbf{G}_\mathrm{m})$ is

$$1 - \exp(\mathrm{i}\mathbf{G}_\mathrm{m} \cdot \mathbf{a}) = 0 \quad \text{for } (h, k, l)$$
$$= 2 \quad \text{for } (h + \tfrac{1}{2}, k + \tfrac{1}{2}, l + \tfrac{1}{2}). \qquad (7.20)$$

These results easily generalize to two-sublattice ferrimagnets that have the same periodicities and therefore the same magnetic reciprocal lattice vectors $\mathbf{G}_\mathrm{m}$ as the two-sublattice antiferromagnets just considered, e.g. Fig. 7.1(c). If the unequal moments on the two sublattices are written $\mu(1 + \alpha)$ and $-\mu(1 - \alpha)$, $|\alpha| < 1$, then $F_\mathrm{M}(\mathbf{G}_\mathrm{m})$ is the same as for the antiferromagnets except that the factors analogous to those in (7.18)–(7.20) take the values $2\alpha$ and $2$, instead of $0$ and $2$. Therefore, magnetic scattering now appears at ferromagnetic/nuclear Bragg peak positions $(h, k, l)$ as well as at the antiferromagnetic positions.

The fact that neutron diffraction from two-sublattice antiferromagnets and ferrimagnets appears at half-odd-integer positions in reciprocal space can present a practical difficulty for measurements. If the crystal is studied on a diffractometer that employs a crystal monochromator then the neutron beam will usually contain higher-order harmonics of the fundamental neutron wavelength $\lambda$. If the magnetic structure has twice the period of the crystal structure then diffraction of the even-order harmonics ($\lambda/2$, $\lambda/4$, ...) from the crystal structure appears at the same positions in reciprocal space as the diffraction of the fundamental wavelength ($\lambda$) from the magnetic structure, potentially obscuring the magnetic signal. For example, magnetic diffraction of neutrons of wavelength $\lambda$ from an A-type antiferromagnet gives rise to a peak at $(0, 0, \frac{1}{2})$; nuclear diffraction of neutrons of wavelength $\lambda/2$ from the (001) atomic planes also results in a peak at $(0, 0, \frac{1}{2})$. To avoid confusing magnetic and higher-order nuclear scattering one can (i) place filters in the neutron beam to remove the higher-order contamination,[18] (ii) measure the temperature dependence of the peak and see whether it vanishes above the magnetic ordering temperature, or (iii) use neutron polarization analysis.

## Non-collinear structures

When the magnetic unit cell contains a non-collinear arrangement of ordered moments we must use the general expressions (7.7)–(7.10) to calculate the magnetic diffraction intensities. As an example, we consider the magnetic structure of the Mn spins in hexagonal $R\mathrm{MnO}_3$, where $R$ can be Sc, Y, Ho, Er, Tm, Yb or Lu.

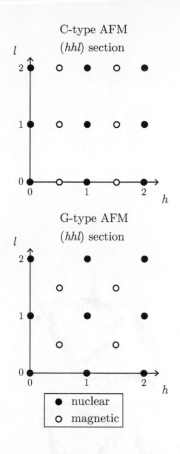

**Fig. 7.7** ($hhl$) section through the reciprocal lattices of C-type and G-type antiferromagnets, showing the positions of the allowed nuclear and magnetic Bragg reflections.

[18] See the account of neutron beam filters in Section 10.1.8.

**Table 7.1** Magnetic structure data for $RMnO_3$. The Mn positions are given in fractional coordinates relative to the unit cell, with $x_{Mn} \simeq \frac{1}{3}$. The magnetic moments are defined by their components along the orthonormal axes $\mathbf{e}_1, \mathbf{e}_2, \mathbf{e}_3$ (Fig. 7.8) and normalized to unit length. The upper and lower signs correspond to the $\alpha$ and $\beta$ structures, respectively.

| Mn position | | | Magnetic moment | | |
|---|---|---|---|---|---|
| $x$ | $y$ | $z$ | $\mu_1$ | $\mu_2$ | $\mu_3$ |
| $x_{Mn}$ | $0$ | $0$ | $\cos(\pi/6 - \psi)$ | $-\sin(\pi/6 - \psi)$ | $0$ |
| $0$ | $x_{Mn}$ | $0$ | $-\sin\psi$ | $\cos\psi$ | $0$ |
| $-x_{Mn}$ | $-x_{Mn}$ | $0$ | $-\cos(\pi/6 + \psi)$ | $-\sin(\pi/6 + \psi)$ | $0$ |
| $-x_{Mn}$ | $0$ | $1/2$ | $\pm\cos(\pi/6 - \psi)$ | $\mp\sin(\pi/6 - \psi)$ | $0$ |
| $0$ | $-x_{Mn}$ | $1/2$ | $\mp\sin\psi$ | $\pm\cos\psi$ | $0$ |
| $x_{Mn}$ | $x_{Mn}$ | $1/2$ | $\mp\cos(\pi/6 + \psi)$ | $\mp\sin(\pi/6 + \psi)$ | $0$ |

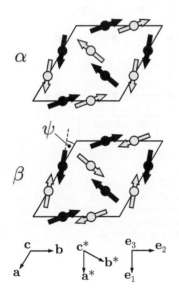

**Fig. 7.8** Possible arrangements of the Mn spins in $RMnO_3$ (after Brown and Chatterji, 2006). The figures show projections down the $c$ axis of the magnetic order in one unit cell of the crystal structure. Only the Mn atoms are shown. Black and grey symbols denote Mn atoms with coordinates $z = 0$ and $z = \frac{1}{2}$, respectively. The relationship between the primitive lattice vectors $\mathbf{a}, \mathbf{b}, \mathbf{c}$, the reciprocal lattice basis vectors $\mathbf{a}^*, \mathbf{b}^*, \mathbf{c}^*$, and the orthonormal axes $\mathbf{e}_1, \mathbf{e}_2, \mathbf{e}_3$ is shown.

The crystal structure of $RMnO_3$ contains near-perfect triangular networks of Mn atoms. The magnetic structure has a propagation vector of $\mathbf{q} = 0$, as evidenced by the observation of magnetic Bragg reflections with integer $hkl$. This means that the magnetic and chemical unit cells are the same. The unit cell contains six Mn atoms all of which carry magnetic moments, three in the basal plane ($z = 0$) and three at a fractional height $z = \frac{1}{2}$. The ordered moments on the Mn are all the same size and are oriented at 120° to the neighbouring moments within the same layer. This structure can propagate along the $c$ axis in two ways consistent with symmetry. Either the moments on Mn atoms at $-x, -y, \frac{1}{2}$ are parallel to those at $x, y, 0$, or else they are antiparallel. These possibilities are labelled $\alpha$ and $\beta$, and are shown in Fig. 7.8. The in-plane angle of the moments is not constrained by symmetry, and we let the moments point at a general angle $\psi$ to the axis on which the Mn lies. Details of the Mn positions and moment directions for the $\alpha$ and $\beta$ structures are given in Table 7.1.

The relation between the unit cell axes and the reciprocal lattice basis vectors is obtained from (2.20) and is shown in Fig. 7.8. It is convenient to express these vectors in terms of an orthonormal basis $\mathbf{e}_1, \mathbf{e}_2, \mathbf{e}_3$ in which $\mathbf{e}_1 \parallel \mathbf{a}^*$, $\mathbf{e}_3 \parallel \mathbf{c}^*$, and $\mathbf{e}_2$ chosen to make a right-handed set (Fig. 7.8):

$$\mathbf{a} = (\tfrac{\sqrt{3}}{2}, -\tfrac{1}{2}, 0)\, a, \qquad \mathbf{b} = (0, 1, 0)\, a, \qquad \mathbf{c} = (0, 0, 1)\, c,$$
$$\mathbf{a}^* = (\tfrac{2}{\sqrt{3}}, 0, 0)\, 2\pi/a, \qquad \mathbf{b}^* = (\tfrac{1}{\sqrt{3}}, 1, 0)\, 2\pi/a, \qquad \mathbf{c}^* = (0, 0, 1)\, 2\pi/c.$$

The reciprocal lattice vector corresponding to the $hkl$ reflection is then $\mathbf{G} = h\mathbf{a}^* + k\mathbf{b}^* + l\mathbf{c}^*$, and the structure factor can be obtained directly from eqn (7.9). Since the expression for general $hkl$ is rather lengthy we take for the sake of illustration reflections of the type $h0l$. The component of the magnetic structure factor along $\mathbf{e}_3$ is zero because the moments lie in the $ab$ plane, and the components along $\mathbf{e}_1$ and $\mathbf{e}_2$ are found to be (see Exercise 7.4)

$$\frac{F_{M_1}}{\Phi} = (C-1)\sin\psi + i\sqrt{3}S\cos\psi \pm (-1)^l\left[(C-1)\sin\psi - i\sqrt{3}S\cos\psi\right]$$

$$\frac{F_{M_2}}{\Phi} = (1-C)\cos\psi + i\sqrt{3}S\sin\psi \pm (-1)^l\left[(1-C)\cos\psi - i\sqrt{3}S\sin\psi\right],$$
(7.21)

where $C = \cos(2\pi h x_{Mn})$, $S = \sin(2\pi h x_{Mn})$ and $\Phi = p\mu f(G)\exp(-W)$.
The upper and lower signs are for the $\alpha$ and $\beta$ structures, respectively.

The intensity can now be calculated from (7.10). The result is

$$|\mathbf{F}_{M\perp}|^2 = |F_{M_1}|^2\cos^2\phi + |F_{M_2}|^2,$$
(7.22)

where $\phi$ is the angle between $\mathbf{G}$ and the $c$ axis, and $\cos\phi = lc^*/G$ with
$G = 2\pi\sqrt{4h^2/3a^2 + l^2/c^2}$.

For the particular case $x_{Mn} = 1/3$, which is very close to the value
found in $RMnO_3$,

$$|\mathbf{F}_{M\perp}|^2 = p^2\mu^2 f^2(G)\exp(-2W)\frac{9}{4}\Big[\{1 \pm (-1)^l\}^2(\sin^2\psi\cos^2\phi + \cos^2\psi)$$

$$+ \{1 \mp (-1)^l\}^2(\cos^2\psi\cos^2\phi + \sin^2\psi)\Big], \quad (h \neq 3n), \quad (7.23)$$

where $n$ is an integer, and $|\mathbf{F}_{M\perp}|^2 = 0$ for $h = 3n$.

Interestingly, eqn (7.23) reveals that an $\alpha$ structure with angle $\psi$ has
exactly the same magnetic diffraction intensities as a $\beta$ structure with
$\psi \pm \pi/2$. This property, known as *homometry*, means that the structural
pairs $\alpha(\psi)$ and $\beta(\psi \pm \pi/2)$ cannot be distinguished in a conventional
neutron diffraction experiment.[19]

## 7.2.2  General magnetic structures

To obtain the elastic cross-section for a general magnetic structure we
use (7.1) to express $\boldsymbol{\mu}_{ld}$, the magnetic moment at the $d$th site in the
chemical unit cell associated with lattice vector $\mathbf{l}$, as a Fourier series and
substitute this into (7.3). After performing the lattice vector sums we
find

$$\frac{d\sigma}{d\Omega} = N\frac{(2\pi)^3}{v_0}\sum_{\mathbf{G}}\sum_{\mathbf{q}}|\mathbf{F}_{M\perp}(\mathbf{Q})|^2\delta(\mathbf{Q} - \mathbf{q} - \mathbf{G}),$$
(7.24)

where

$$\mathbf{F}_M(\mathbf{Q}) = p\sum_d \mathbf{m}_{\mathbf{q},d}\, f_d(Q)\exp(-W_d)\exp(i\mathbf{Q}\cdot\mathbf{d}),$$
(7.25)

with $p = \gamma r_0/2\mu_B$. This shows that inside each Brillouin zone of the
chemical reciprocal lattice there exists a set of superstructure magnetic
Bragg peaks, one for each distinct wavevector $\mathbf{q}$ in the Fourier expansion
of the magnetic structure (7.1). These peaks are located at points in
reciprocal space where $\mathbf{Q} = \mathbf{G} + \mathbf{q}$.

[19]When $\mathbf{q} = 0$, homometric pairs can
in principle be distinguished by neutron
polarimetry via the nuclear–magnetic
interference scattering — see Brown
and Chatterji (2006).

For commensurate magnetic structures, the magnetic diffraction intensity calculated from (7.24) and (7.25) is the same as that obtained previously by the magnetic unit cell method. For example, the A-type antiferromagnet in Fig. 7.1(b) has one atom in the chemical unit cell, at $\mathbf{d} = (0,0,0)$, and the magnetic order is described by a single propagation vector $\mathbf{q} = (0,0,\frac{1}{2})$ r.l.u. The condition $\mathbf{Q} = \mathbf{G}+\mathbf{q}$ gives magnetic peaks at $(h,k,l+\frac{1}{2})$ positions in reciprocal space. For a given magnetic peak the value of $|\mathbf{F}_{\mathrm{M}}|^2$ calculated from (7.25) is a factor of 4 smaller than that from (7.16), but because the magnetic unit cell is twice the chemical unit cell the ratio $N/v_0$ in (7.24) is a factor of 4 larger than the ratio $N_{\mathrm{m}}/v_{\mathrm{m}}$ which appears in (7.14). Therefore, the diffracted intensities calculated from (7.24) and (7.14) are the same.

### Incommensurate magnetic structures

To illustrate the use of (7.24) and (7.25) for more general structures we calculate the diffracted intensity from (i) a longitudinal spin-density wave (LSDW), (ii) a transverse helix, and (iii) a cycloid. These structures are shown in Fig. 7.2(a)–(c).

(i) A LSDW with propagation wavevector $\mathbf{q} = (0,0,q)$ r.l.u. has magnetic moments which vary with position according to

$$\boldsymbol{\mu}_l = \mu(0,0,\cos\{\mathbf{q}\cdot\mathbf{l}+\phi\}),  \tag{7.26}$$

where $\phi$ is the phase of the LSDW, see Fig. 7.2(a). We have dropped the $d$ suffix as there is only one atom per unit cell in this example. The Fourier expansion of (7.26) has components with wavevectors $+\mathbf{q}$ and $-\mathbf{q}$, and the respective Fourier coefficients are

$$\mathbf{m}_{\mathbf{q}} = \tfrac{1}{2}\mu(0,0,\mathrm{e}^{-\mathrm{i}\phi}) \quad\text{and}\quad \mathbf{m}_{-\mathbf{q}} = \tfrac{1}{2}\mu(0,0,\mathrm{e}^{\mathrm{i}\phi}).$$

On substituting these into (7.10), (7.24) and (7.25) we obtain

$$\frac{\mathrm{d}\sigma}{\mathrm{d}\Omega} = N\frac{(2\pi)^3}{v_0}f^2(Q)\mathrm{e}^{-2W}(1-\hat{Q}_z^2)\,p^2\mu^2\frac{1}{4}\sum_{\mathbf{G}}\delta(\mathbf{Q}-\mathbf{G}\pm\mathbf{q}).  \tag{7.27}$$

(ii) The magnetic structure of a transverse helix with $\mathbf{q} = (0,0,q)$ is

$$\boldsymbol{\mu}_l = \mu(\cos\{\mathbf{q}\cdot\mathbf{l}+\phi\},\pm\sin\{\mathbf{q}\cdot\mathbf{l}+\phi\},0),  \tag{7.28}$$

Fig. 7.2(b), and (iii) that of a cycloid with $\mathbf{q} = (0,0,q)$ and moments rotating in the $xz$ plane, Fig. 7.2(c), is

$$\boldsymbol{\mu}_l = \mu(\cos\{\mathbf{q}\cdot\mathbf{l}+\phi\},0,\pm\sin\{\mathbf{q}\cdot\mathbf{l}+\phi\}).  \tag{7.29}$$

The $\pm$ signs correspond to the sense of rotation of the moments in the helix/cycloid. With the analogous Fourier decompositions of (7.28) and (7.29) it is straightforward to show that the elastic cross-sections for the helix and cycloid are the same as that for the LSDW, eqn (7.27), except that the orientation factor $(1-\hat{Q}_z^2)$ is replaced by $(1+\hat{Q}_z^2)$ for the transverse helix and by $(1+\hat{Q}_y^2)$ for the cycloid (see Exercise 7.5).

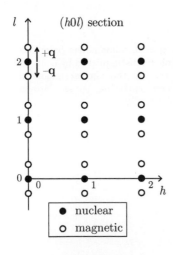

$l$      $(h0l)$ section

● nuclear
○ magnetic

**Fig. 7.9** $(h0l)$ section through the reciprocal lattice of an incommensurate magnetic structure with propagation vector $\mathbf{q} = (0,0,q)$, see Fig. (7.2). The crystal lattice is assumed to have orthogonal axes. The positions of the allowed nuclear reflections and magnetic satellites are shown.

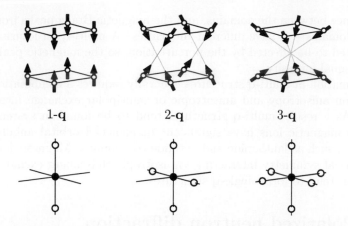

**Fig. 7.10** Set of longitudinal multi-**q** magnetic structures associated with the propagation vector $(0, 0, \frac{1}{2})$ on a primitive cubic lattice. The diagrams below illustrate the positions of the magnetic Bragg reflections (open circles) in reciprocal space relative to a reciprocal lattice point (filled circle). The relative intensities of the Bragg peaks will depend on the orientation factor.

The magnetic Bragg peaks for each of the incommensurate structures considered here occur at the same positions in reciprocal space, namely at $\mathbf{Q} = \mathbf{G} \pm \mathbf{q}$. In other words, each nuclear Bragg peak is flanked by two magnetic satellites, one on either side, as illustrated in Fig. 7.9. The difference between the diffraction intensities for each structure stems purely from the orientation factors.

## Multi-q structures

The conical structure shown in Fig. 7.2(d) is an example of a two-**q** structure, and has distinct propagation vectors $\mathbf{q}_1 = (0, 0, q)$ and $\mathbf{q}_2 = (0, 0, 0)$. The $\mathbf{q}_1$ component gives magnetic diffraction peaks at the positions shown in Fig. 7.9, and the $\mathbf{q}_2$ magnetic peaks coincide with the nuclear scattering peaks from the crystal structure.

Antiferromagnetism on a cubic lattice is interesting because it can often be difficult to distinguish between a multi-**q** structure and the existence of magnetic domains. Take, for example, the set of longitudinal antiferromagnetic structures on a primitive cubic lattice shown in Fig. 7.10. Longitudinal means that $\mathbf{m_q}$ is parallel to $\mathbf{q}$, e.g. the single-**q** (1-**q**) structure has propagation vector $\mathbf{q}_1 = (0, 0, \frac{1}{2})$ and spins pointing along $\mathbf{q}_1$. A single domain of this structure would give magnetic Bragg peaks at $\mathbf{G} \pm \mathbf{q}_1$, as indicated. The non-collinear 2-**q** structure is equivalent to a coherent superposition of two collinear 1-**q** structures with propagation vectors $\mathbf{q}_1$ and $\mathbf{q}_2 = (\frac{1}{2}, 0, 0)$, and would produce four magnetic peaks around each $\mathbf{G}$, as shown. However, this diffraction pattern is indistinguishable from that of an incoherent superposition of two equally populated domains with collinear 1-**q** order along $\mathbf{q}_1$ and $\mathbf{q}_2$. Similarly, the diffraction from the 3-**q** structure is indistinguishable from three equally populated 1-**q** domains.

The usual method to distinguish between multi-**q** and multiple 1-**q** domain structures is to apply an external perturbation, such as a magnetic field, electric field, or mechanical stress, in order to break the cubic symmetry macroscopically. If the sample contains multiple domains then on cooling through the magnetic transition the perturbation should create an imbalance between the domains, and the magnetic Bragg peaks from different domains will have different intensities. A multi-**q** structure is not expected to be affected by the perturbation, so the magnetic peaks will have equal intensity.

The formation of multi-**q** structures generally requires a combination of single-ion anisotropy and anisotropic or multipolar exchange interactions. As a result, multi-**q** structures tend to be found in systems where the magnetic ions have significant unquenched orbital angular momentum, such as lanthanide and actinide compounds. Magnetic ions with spherical symmetry interacting via isotropic Heisenberg exchange interactions tend to form single-**q** structures.

## 7.3   Polarized neutron diffraction

The use of polarized neutrons adds an extra dimension to neutron diffraction studies on magnetic materials. One important advantage is the ability to obtain an unambiguous separation of magnetic and nuclear scattering. Another is the ability to resolve different directional components of magnetic structure factors. In the study of non-collinear magnetic structures it is sometimes impossible to distinguish between two or more different viable magnetic structures without polarized neutrons.

In the following sections we will describe the different ways in which polarized neutron diffraction can be performed and the corresponding information that can be obtained. A full account of the theory of polarized neutron scattering was given in Section 4.4.

### 7.3.1   Uniaxial polarization analysis

Uniaxial polarization analysis (also known as longitudinal polarization analysis) is currently the most widely available implementation of neutron polarization analysis. In this method the polarization of the scattered neutrons is analysed parallel to the same axis along which the incident neutrons are polarized. This allows the measurement of four partial scattering cross-sections, corresponding to the two possible neutron spin states relative to the polarization axis ('up' or '+', and 'down' or '−') before and after scattering. The up-to-up (++) and down-to-down (−−) cross-sections are the non-spin-flip (NSF) channels, and the up-to-down (+−) and down-to-up (−+) cross-sections are the spin-flip (SF) channels.

The equipment needed to perform uniaxial polarization analysis was developed in the late 1960s by Moon *et al.* (1969). A generic setup is shown in Fig. 7.11. The main requirement is a continuous magnetic field over the neutron flight path through the sample. The strength of the field must be large enough to maintain the neutron polarization, and if the direction of the field changes before or after the sample then it must do so adiabatically, i.e. sufficiently smoothly to prevent depolarization of the neutron beam. The polarizer and analyser can be one of several devices: (i) a ferromagnetic crystal such as Heusler alloy ($Cu_2MnAl$)

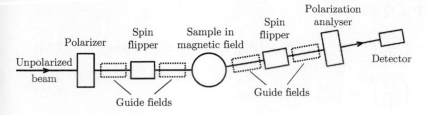

**Fig. 7.11** Schematic of an instrument for uniaxial polarization analysis.

which Bragg-scatters only one neutron spin state (see Section 7.3.2), (ii) a *supermirror* made from a graded sequence of alternating ferromagnetic (e.g. Fe) and non-magnetic (e.g. Si) layers which reflects only one neutron spin state, see Fig. 9.18, or (iii) a cell containing polarized $^3$He gas which transmits only one neutron spin state. Spin turning devices, or *flippers*, placed before and after the sample allow the direction of the neutron spins to be reversed. The magnetic field at the sample is typically provided by a system of Helmholtz coils which allows the neutron polarization to be rotated into any direction at the sample position. Guide fields maintain the neutron polarization between the polarizer and the sample, and between the sample and analyser.

The experimental procedure normally involves measurement of the four partial spin-state cross-sections (or a subset of them) with the neutron polarization **P** either parallel or perpendicular to the scattering vector **Q**, followed by manipulations to determine the desired magnetic terms. Corrections may be needed to compensate for the non-ideal polarization of the instrument and for background signals — see Section 10.1.7.

Following the usual convention, we define a set of Cartesian coordinates with $x$ parallel to **Q**, $y$ in the plane of incidence (the plane containing $\mathbf{k}_i$ and $\mathbf{k}_f$), and $z$ perpendicular to the plane of incidence, as shown in Fig. 7.12. The partial cross-sections describing nuclear and magnetic scattering for **P** parallel to each of the axes $x$, $y$, and $z$ were derived in Section 4.5, and are given by eqns (4.68)–(4.73). For elastic scattering these may be written:

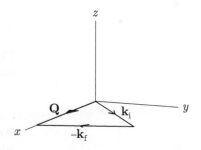

**Fig. 7.12** The standard (Blume–Maleev) polarization axis system.

(i) $\mathbf{P} \parallel \mathbf{Q}$ (i.e. $\mathbf{P} \parallel x$)

$$\left(\frac{d\sigma}{d\Omega}\right)_x^{\pm\pm} = S_{\text{coh}}^{\text{N}} + S_{\text{inc}}^{\text{iso}} + \tfrac{1}{3}S_{\text{inc}}^{\text{sp}} + B_{\text{NSF}} \qquad (7.30)$$

$$\left(\frac{d\sigma}{d\Omega}\right)_x^{\pm\mp} = \tfrac{2}{3}S_{\text{inc}}^{\text{sp}} + S_{yy}^{\text{M}} + S_{zz}^{\text{M}} \pm T_{yz}^{\text{M}} + B_{\text{SF}}, \qquad (7.31)$$

(ii) $\mathbf{P} \perp \mathbf{Q}$ with $\mathbf{P} \parallel y$

$$\left(\frac{d\sigma}{d\Omega}\right)_y^{\pm\pm} = S_{\text{coh}}^{\text{N}} + S_{\text{inc}}^{\text{iso}} + \tfrac{1}{3}S_{\text{inc}}^{\text{sp}} \pm S_y^{\text{NM}} + S_{yy}^{\text{M}} + B_{\text{NSF}} \qquad (7.32)$$

$$\left(\frac{d\sigma}{d\Omega}\right)_y^{\pm\mp} = \tfrac{2}{3}S_{\text{inc}}^{\text{sp}} + S_{zz}^{\text{M}} + B_{\text{SF}}, \qquad (7.33)$$

(iii) $\mathbf{P} \perp \mathbf{Q}$ with $\mathbf{P} \parallel z$

$$\left(\frac{d\sigma}{d\Omega}\right)_z^{\pm\pm} = S_{\text{coh}}^{\text{N}} + S_{\text{inc}}^{\text{iso}} + \tfrac{1}{3}S_{\text{inc}}^{\text{sp}} \pm S_z^{\text{NM}} + S_{zz}^{\text{M}} + B_{\text{NSF}} \qquad (7.34)$$

$$\left(\frac{d\sigma}{d\Omega}\right)_z^{\pm\mp} = \tfrac{2}{3}S_{\text{inc}}^{\text{sp}} + S_{yy}^{\text{M}} + B_{\text{SF}}. \qquad (7.35)$$

The terms $S_{\text{coh}}^{\text{N}}$, $S_{\text{inc}}^{\text{iso}}$ and $S_{\text{inc}}^{\text{sp}}$ are the coherent nuclear scattering, the nuclear isotopic incoherent scattering, and the nuclear spin incoherent scattering, respectively. As we are concerned here with diffraction from a magnetic structure, the magnetic terms are given by the specified components of (7.3):

$$S_{yy}^{\text{M}} = \left(\frac{\gamma r_0}{2\mu_{\text{B}}}\right)^2 |\langle M_y(\mathbf{Q})\rangle|^2, \quad S_{zz}^{\text{M}} = \left(\frac{\gamma r_0}{2\mu_{\text{B}}}\right)^2 |\langle M_z(\mathbf{Q})\rangle|^2, \qquad (7.36)$$

and the so-called *chiral term*

$$T_{yz}^{\text{M}} = \left(\frac{\gamma r_0}{2\mu_{\text{B}}}\right)^2 i\,\{\langle M_y^*(\mathbf{Q})\rangle\langle M_z(\mathbf{Q})\rangle - \langle M_z^*(\mathbf{Q})\rangle\langle M_y(\mathbf{Q})\rangle\}. \qquad (7.37)$$

The $S_\alpha^{\text{NM}}$ terms ($\alpha = y, z$) represent interference between the coherent nuclear scattering and the $\alpha$ component of $\mathbf{M}$. From (4.76),

$$S_\alpha^{\text{NM}} = \frac{\gamma r_0}{2\mu_{\text{B}}}\left\{\langle N^*(\mathbf{Q})\rangle\langle M_\alpha(\mathbf{Q})\rangle + \langle M_\alpha^*(\mathbf{Q})\rangle\langle N(\mathbf{Q})\rangle\right\}. \qquad (7.38)$$

We have also included terms $B_{\text{NSF}}$ and $B_{\text{SF}}$ to take into account any additional backgrounds in the NSF and SF channels. We assume these backgrounds to be independent of the direction of $\mathbf{P}$.[20]

[20]although this is not always the case.

We will now describe some applications of uniaxial polarization analysis in the investigation of magnetic structures.

## Separation of nuclear and magnetic scattering

For separation of nuclear coherent and electronic magnetic scattering it is often sufficient to measure the NSF and SF cross-sections with $\mathbf{P} \parallel \mathbf{Q}$, because from (7.30) and (7.31) all the nuclear coherent scattering is contained in the NSF cross-section and all the electronic magnetic scattering is contained in the SF cross-section. This method is suitable providing the nuclear incoherent scattering and any other spin-dependent backgrounds are featureless and can be treated as a constant intensity underneath the signal of interest. An example of the method is given in Fig. 7.13. The plot shows polarized neutron diffraction from a cobalt oxide which has complex short-range order involving both magnetic and charge ($Co^{2+}$ and $Co^{3+}$) degrees of freedom. By separating the magnetic and nuclear coherent scattering, the latter which arises from structural distortions associated with the charge order, one can study the different magnetic and charge correlations that coexist in the sample.

If the individual magnetic terms are required, rather than the combination that appears in (7.31), or if a more rigorous separation of the

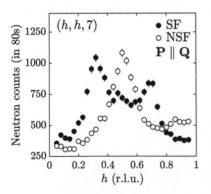

**Fig. 7.13** Polarized neutron diffraction from $La_{5/3}Sr_{1/3}CoO_4$, a complex magnetic oxide with coexisting short-range ordered phases (Babkevich *et al.*, 2016). The peaks in the SF and NSF intensities are caused by magnetic and charge correlations, respectively.

terms of interest from the backgrounds is important, then measurements of a larger set of the partial spin-state cross-sections (7.30)–(7.35) should be performed and combined as follows:[21]

$$2T_{yz}^{M} = (+-)_x - (-+)_x \tag{7.39}$$

$$2S_y^{NM} = (++)_y - (--)_y \tag{7.40}$$

$$2S_z^{NM} = (++)_z - (--)_z \tag{7.41}$$

$$S_{yy}^{M} = (+-)_x - (+-)_y - T_{yz}^{M} = (++)_y - (++)_x - S_y^{NM} \tag{7.42}$$

$$S_{zz}^{M} = (+-)_x - (+-)_z - T_{yz}^{M} = (++)_z - (++)_x - S_z^{NM} \tag{7.43}$$

Note that $S_{yy}^{M}$ and $S_{zz}^{M}$ can be obtained from either entirely SF, or entirely NSF measurements. Further, in many situations the terms $T_{yz}^{M}$, $S_y^{NM}$ and $S_z^{NM}$ are either rigorously zero or assumed to be negligible, in which case (7.39)–(7.41) are not required.[22]

### Determination of moment directions

The individual uniaxial polarization analysis cross-sections (7.30)–(7.35) probe different components of $\mathbf{M(Q)}$, and one can use this to constrain, or in favourable circumstances to establish uniquely, the direction of $\mathbf{M(Q)}$. A good application is to determine the ordered moment direction in a collinear spin structure with a single type of magnetic atom. We assume that for an individual atom $\langle \mathbf{M(Q)} \rangle = \boldsymbol{\mu} f(Q)$, where $\boldsymbol{\mu}$ is the thermal average of the atomic magnetic dipole moment and $f(Q)$ the magnetic form factor (see Section 7.2).

As above, we employ the Blume–Maleev coordinate system to describe the neutron polarization (Fig. 7.12), and take $\mathbf{Q}$ to lie in a plane defined by the $X$ and $Y$ axes of another Cartesian system, as shown in Fig. 7.14, so that $\mu_X$, $\mu_Y$, and $\mu_Z$ are the components of $\boldsymbol{\mu}$ along $X$, $Y$, and $Z$.

Table 7.2 lists the components of $\boldsymbol{\mu}^2$ that are measured in each uniaxial polarization analysis channel according to (7.30)–(7.35). There is no $T_{yz}^{M}$ term for a collinear magnetic structure, and we assume that the coherent

[21] We introduce the obvious shorthand

$$(+-)_x \equiv \left( \frac{d\sigma}{d\Omega} \right)_x^{+-}, \text{ etc.}$$

An equivalent set of equations to (7.39)–(7.43) is obtained if we reverse the signs of all the spin-state labels.

[22] Unfortunately, it is not possible to isolate $S_{coh}^{N}$ because it always comes in the combination $S_{coh}^{N} + S_{inc}^{iso} + B_{NSF}$.

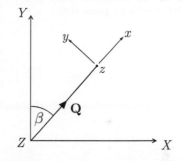

**Fig. 7.14** Definitions of the axes used for magnetic moment and neutron polarization, shown in projection onto the scattering plane.

**Table 7.2** Components of $\boldsymbol{\mu}^2$ that appear in each uniaxial polarization channel in magnetic diffraction from a collinear spin structure with a single type of magnetic atom. The $X$, $Y$, $Z$ axes and $\beta$ angle are defined in Fig. 7.14.

| Polarization | | Components of $\boldsymbol{\mu}^2$ |
|---|---|---|
| $\mathbf{P} \parallel x$ | SF | $\mu_X^2 \cos^2 \beta + \mu_Y^2 \sin^2 \beta + \mu_Z^2$ |
| | NSF | $0$ |
| $\mathbf{P} \parallel y$ | SF | $\mu_Z^2$ |
| | NSF | $\mu_X^2 \cos^2 \beta + \mu_Y^2 \sin^2 \beta$ |
| $\mathbf{P} \parallel z$ | SF | $\mu_X^2 \cos^2 \beta + \mu_Y^2 \sin^2 \beta$ |
| | NSF | $\mu_Z^2$ |

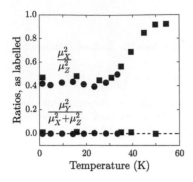

**Fig. 7.15** Results from uniaxial polarization analysis of diffraction from $La_{3/2}Sr_{1/2}CoO_4$ (Helme *et al.*, 2009). The plot shows the temperature dependence of the ratios $\mu_X^2/\mu_Z^2$ and $\mu_Y^2/(\mu_X^2 + \mu_Z^2)$, where $X$, $Y$ and $Z$ are parallel to the tetragonal [110], [001] and [1$\bar{1}$0] directions, respectively.

magnetic and nuclear scattering occurs at distinct positions in reciprocal space so that nuclear–magnetic interference terms are also zero.

One can see in Table 7.2 that the $\mathbf{P} \parallel y$ and $\mathbf{P} \parallel z$ channels of a magnetic Bragg reflection determine the ratio $(\mu_X^2 \cos^2 \beta + \mu_Y^2 \sin^2 \beta)/\mu_Z^2$. However, if two Bragg reflections with different $\beta$ angles are measured, then providing $\mu_Z \neq 0$ one can determine $\mu_X^2/\mu_Z^2$ and $\mu_Y^2/\mu_Z^2$ separately and hence obtain the direction of the moment to within the sign of $\mu_X/\mu_Z$ and $\mu_Y/\mu_Z$. Because these measurements involve ratios of intensities at each $\mathbf{Q}$ the results do not depend on the magnetic form factor and are usually more accurate than the moment direction refined from integrated intensity measurements with unpolarized neutrons. If $\mu_Z = 0$, then one can determine $\mu_X^2/\mu_Y^2$ either by using a different scattering plane ($X$–$Z$ or $Y$–$Z$), or by measuring the ratio of intensities in the $\mathbf{P} \parallel z$ SF channel at two magnetic Bragg reflections. The latter method, however, requires the magnetic form factor at the two $\mathbf{Q}$ values.

Figure 7.15 presents results obtained by this method for the ordered moment in tetragonal $La_{3/2}Sr_{1/2}CoO_4$, which exhibits antiferromagnetic correlations below $\sim 60$ K. Polarization analysis was performed at two magnetic diffraction peaks in the $(hhl)$ scattering plane (see Exercise 7.6). The data show that the moments lie in the basal plane at all temperatures, and that changes take place in the in-plane components on cooling. The latter could be due either to a canting of the moment or to a change in population of different magnetic structures with the same propagation vector.

## Determination of vector spin chirality — method 1

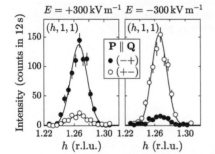

**Fig. 7.16** Polarized magnetic diffraction from the cycloidal magnetic structure in $Ni_3V_2O_8$, measured with an electric field applied parallel (left) and antiparallel (right) to the $b$ axis. The reversal in the asymmetry between the $(-+)$ and $(+-)$ spin-flip channels with electric field is caused by a reversal in the sense of rotation of the cycloid. (After Cabrera *et al.*, 2009.)

The $T_{yz}^{\mathrm{M}}$ term, which can be obtained from (7.39), is often called the chiral term. Its name comes from the fact that it can be present in diffraction from a single domain of a chiral spin structure, such as the helix shown in Fig. 7.2(b). However, the condition for non-zero $T_{yz}^{\mathrm{M}}$ is not restricted to chiral spin structures, and applies more generally to structures that have a non-zero *vector spin chirality* $\boldsymbol{\kappa}_{jk} = \mathbf{S}_j \times \mathbf{S}_k$, where $j$ and $k$ are adjacent sites. For example, the cycloid shown in Fig. 7.2(c) is not chiral, but has a non-zero component of $\boldsymbol{\kappa}_{jk}$ perpendicular to the plane of rotation of the spins.

Cabrera *et al.* (2009) exploited the chiral scattering term to show that the sense of rotation of the cycloidal magnetic structure in multiferroic $Ni_3V_2O_8$ can be controlled with an electric field. They measured a magnetic Bragg peak after cooling the sample in an electric field applied parallel to the crystallographic $b$ axis and found a large asymmetry between the $(+-)$ and $(-+)$ spin-flip scattering which reversed with the electric field (Fig. 7.16). The sign of $T_{yz}^{\mathrm{M}}$ depends on the sense of rotation of the cycloid, and Cabrera *et al.* were able to deduce that reversal of the electric field direction from $+\mathbf{b}$ to $-\mathbf{b}$ switched the sense of rotation of the cycloid from anticlockwise to clockwise.

## 7.3.2  Half-polarized neutron diffraction

A half-polarized arrangement is one in which, either (i) the incident neutron beam is polarized and the scattering intensity measured without polarization analysis, or (ii) the incident beam is unpolarized and there is polarization analysis of the scattered neutrons. The main advantage of the half-polarized setup is that only one polarizer/analyser is required, which results in significantly higher neutron intensity compared with the full uniaxial polarization analysis setup. The disadvantage is that it provides less capability to isolate the various nuclear and magnetic terms in the cross-section.

In half-polarized diffraction there are only two partial cross-sections for a given axis of polarization at the sample. In the case of an unpolarized incident beam with polarization analysis of the scattered beam the half-polarized cross-sections $\circ+$ and $\circ-$ are related to the four full-polarized cross-sections $\pm\pm$ and $\pm\mp$ by

$$\left(\frac{d\sigma}{d\Omega}\right)^{\circ+} = \frac{1}{2}\left\{\left(\frac{d\sigma}{d\Omega}\right)^{++} + \left(\frac{d\sigma}{d\Omega}\right)^{-+}\right\} \tag{7.44}$$

$$\left(\frac{d\sigma}{d\Omega}\right)^{\circ-} = \frac{1}{2}\left\{\left(\frac{d\sigma}{d\Omega}\right)^{--} + \left(\frac{d\sigma}{d\Omega}\right)^{+-}\right\}. \tag{7.45}$$

With $\mathbf{P} \parallel \mathbf{Q}$, the half-polarized setup can be used to study magnetic structure types which have a non-vanishing vector chiral term $T_{yz}^{\mathrm{M}}$, and with $\mathbf{P} \perp \mathbf{Q}$ it can be used to measure the nuclear–magnetic interference terms $S_y^{\mathrm{NM}}$ and $S_z^{\mathrm{NM}}$.

### Determination of vector spin chirality — method 2

In Section (7.3.1) we saw how the vector spin chiral term $T_{yz}^{\mathrm{M}}$ can be isolated by uniaxial polarization analysis. The $T_{yz}^{\mathrm{M}}$ term can also be measured by half-polarized diffraction, since from (7.30)–(7.31) it is given by the difference $(\circ-)_x - (\circ+)_x$, or alternatively $(+\circ)_x - (-\circ)_x$. As an example, we consider half-polarized diffraction from a transverse helix, which has left- and right-handed forms as shown in Fig. 7.3(d). Suppose the helix has a propagation vector $\mathbf{q}$, so that the magnetic Bragg peaks are located at $\mathbf{Q} = \mathbf{G} \pm \mathbf{q}$, where $\mathbf{G}$ is a reciprocal lattice vector (see Fig. 7.2). For simplicity we assume one spin per lattice point. In the Blume–Maleev coordinates the helix is described by

$$\boldsymbol{\mu}_l = \begin{pmatrix} -\mu\cos(\mathbf{q}\cdot\mathbf{l}+\phi)\sin\beta \\ \mu\cos(\mathbf{q}\cdot\mathbf{l}+\phi)\cos\beta \\ \pm\mu\sin(\mathbf{q}\cdot\mathbf{l}+\phi) \end{pmatrix}, \tag{7.46}$$

where $\beta$ is the angle between $\mathbf{Q}$ and $\mathbf{q}$ (see Fig. 7.17). From (7.46) and (7.6) one can calculate $\mathbf{M}(\mathbf{Q})$, and the partial structure factors $S_{yy}^{M}$, $S_{zz}^{M}$, and $T_{yz}^{M}$ follow from (7.36)–(7.37). The resulting half-polarized

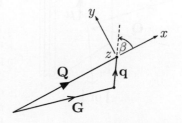

**Fig.  7.17** Relation between the Blume–Maleev axes and the propagation vector $\mathbf{q}$ for a helical magnetic structure.

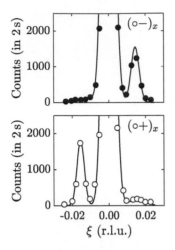

cross-sections (magnetic terms only) may be shown to be

$$\left(\frac{d\sigma}{d\Omega}\right)_x^{\circ+} = K \sum_{\mathbf{G}} (1 \pm \cos\beta)^2 \delta(\mathbf{Q} - \mathbf{q} - \mathbf{G}) + (1 \mp \cos\beta)^2 \delta(\mathbf{Q} + \mathbf{q} - \mathbf{G}),$$

$$\left(\frac{d\sigma}{d\Omega}\right)_x^{\circ-} = K \sum_{\mathbf{G}} (1 \mp \cos\beta)^2 \delta(\mathbf{Q} - \mathbf{q} - \mathbf{G}) + (1 \pm \cos\beta)^2 \delta(\mathbf{Q} + \mathbf{q} - \mathbf{G}),$$

$$(7.47)$$

where

$$K(Q) = N \frac{(2\pi)^3}{v_0} p^2 \mu^2 \frac{1}{8} f^2(Q) \exp(-2W).$$

The upper signs in (7.46) and (7.47) correspond to a right-handed (clockwise) helix, and the lower signs to a left-handed (anticlockwise) helix. If $\beta = 0$ then only one of the two magnetic satellites is non-zero, and the non-zero satellite switches sides if the final polarization is reversed.

Figure 7.18 shows half-polarized diffraction data from cubic MnSi, which has a transverse helical magnetic structure with propagation vector $\mathbf{q} = (0.017, 0.017, 0.017)$ r.l.u. The scans go through the (011) reciprocal lattice point in a direction parallel to (111), so that $\cos\beta = \sqrt{2/3}$. This gives the ratio of the factors $(1 \pm \cos\beta)^2$ to be about 100, which is sufficiently large that a magnetic satellite only appears on one side of (011) for each polarization state. We see that the data in Fig. 7.18 are consistent with the lower signs in eqns (7.47), which means that in this sample of MnSi the helix was left-handed.

**Fig. 7.18** Magnetic satellites from helimagnetic order in MnSi. The measurement used an unpolarized incident beam and polarization analysis of the scattered neutrons with $\mathbf{P} \parallel \mathbf{Q}$. The scan is along $(\xi, 1 + \xi, 1 + \xi)$, which is parallel to the propagation vector. (After Shirane *et al.*, 1983.)

## Spatial distribution of magnetization

An important application of the half-polarized setup with a polarized incident beam is to obtain the spatial distribution of magnetization in a crystal whose magnetic order has a net ferromagnetic component, either aligned or induced by an applied magnetic field. The method is described in detail in Section 7.6, and exploits the nuclear–magnetic interference scattering in the $\mathbf{P} \perp \mathbf{Q}$ NSF channels (7.32) and (7.34) to amplify the magnetic structure factors and, crucially, to determine their phases.

## Ferromagnetic single crystal polarizers

The same nuclear–magnetic interference terms that make it possible to measure magnetization distributions are also exploited in single crystal polarizers. The method uses the fact that nuclear–magnetic interference causes neutrons of one spin state to be diffracted more strongly than the other, so an initially unpolarized beam will become partially polarized after diffraction. Such devices employ ferromagnetically ordered crystals because in a ferromagnet the magnetic and chemical unit cells are the same, and so the nuclear and magnetic Bragg peaks coincide.

The practical implementation is illustrated in Fig. 7.19. A magnetic field $\mathbf{B}$ is applied parallel to a set of crystal planes which are oriented so as to satisfy the Bragg condition $\mathbf{Q} = \mathbf{G}$ for the desired wavelength. The

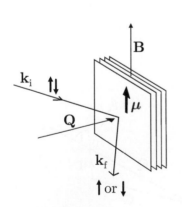

**Fig. 7.19** Schematic of a single crystal polarizer. The ferromagnetic crystal diffracts neutrons of predominantly one spin state from the unpolarized incident beam.

field aligns the magnetic moments parallel to **B**. The diffraction plane, which contains **Q**, is perpendicular to **B**, so the relevant expressions which describe the spin-dependent Bragg diffraction are eqns (7.34) and (7.35), i.e. those with $\mathbf{P} \perp \mathbf{Q}$ and $\mathbf{P} \parallel z$. We can write these (omitting the incoherent scattering and background) as[23]

[23]Note that $S_{yy}^{\mathrm{M}} = 0$ since $\boldsymbol{\mu} \parallel z$.

$$\left(\frac{\mathrm{d}\sigma}{\mathrm{d}\Omega}\right)_z^{\circ\pm} = \frac{1}{2}(S_{\mathrm{coh}}^{\mathrm{N}} + S_{zz}^{\mathrm{M}} \pm S_z^{\mathrm{NM}})$$

$$= N\frac{(2\pi)^3}{v_0}\frac{1}{2}\sum_{\mathbf{G}}\left|F_{\mathrm{N}} \pm F_{\mathrm{M}}\right|^2\delta(\mathbf{Q} - \mathbf{G}) \qquad (7.48)$$

where,

$$F_{\mathrm{N}}(\mathbf{Q}) = \sum_d \bar{b}_d \exp(-W_d)\exp(\mathrm{i}\mathbf{Q}\cdot\mathbf{d}) \qquad (7.49)$$

$$F_{\mathrm{M}}(\mathbf{Q}) = p\sum_d \mu_d f_d(Q)\exp(-W_d)\exp(\mathrm{i}\mathbf{Q}\cdot\mathbf{d}) \qquad (7.50)$$

are the nuclear and magnetic unit cell structure factors, see eqns (5.48) and (7.9). In (7.50) we have written $F_{\mathrm{M}}$ and $\mu_d$ as scalars because the moments are fully aligned.

Equation (7.48) shows that if $|F_{\mathrm{N}}| \simeq |F_{\mathrm{M}}|$ for a particular Bragg reflection then the crystal will reflect neutrons of predominantly one spin state. An additional advantage of this type of polarizer is that it also acts as monochromator.

The 111 reflection of the ferromagnetic Heusler alloy $Cu_2MnAl$ is widely used as a single crystal polarizer. This alloy has a Curie temperature of around 620 K, and at room temperature the saturated magnetic moment is about $3.2\,\mu_{\mathrm{B}}$. The nuclear and magnetic structure factors for the 111 reflection are $F_{\mathrm{N}}(111) \simeq 7.2\,\mathrm{fm}$ and $F_{\mathrm{M}}(111) \simeq -6.5\,\mathrm{fm}$, and from (7.48) it is found that the $(\circ-)$ cross-section is more than 300 times larger than the $(\circ+)$ cross-section. Therefore, the 111 reflection of $Cu_2MnAl$ produces a highly polarized beam of spin-down neutrons.

## 7.3.3 Spherical neutron polarimetry

The method of uniaxial neutron polarization analysis does not allow measurement of the cross-sections for neutrons whose final polarization is perpendicular to the initial polarization. On the other hand, the transverse terms can contain essential information for a complete magnetic structure determination. The scattering cross-section for neutrons with incident polarization in the $\alpha$ direction and scattered polarization in the $\beta$ direction is given in eqn (4.53). For pure magnetic scattering, the elastic cross-section may be written

$$\left(\frac{\mathrm{d}\sigma}{\mathrm{d}\Omega}\right)^{\alpha\beta} = 4\left(\frac{\gamma r_0}{2\mu_{\mathrm{B}}}\right)^2 |\langle\sigma_\beta|\mathbf{s}_{\mathrm{n}}|\sigma_\alpha\rangle \cdot \langle\mathbf{M}_\perp(\mathbf{Q})\rangle|^2. \qquad (7.51)$$

The neutron spin matrix elements are given in Table C.2.

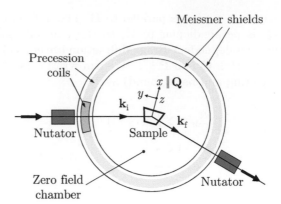

**Fig. 7.20** Diagram of the main components of the CryoPAD device used for spherical neutron polarimetry. The large arrows indicate axial guide fields on the incoming and outgoing neutron paths.

[24]With the addition of an extra precession coil and $\pi/2$ flippers to the setup for uniaxial polarization analysis it is also possible to carry out SNP with the sample in a field. The method, which involves the analysis of precessing polarization components of the scattered beam, can in principle be used on time-of-flight instruments with multidetectors (Schweika, 2003).

In practice, when full polarization analysis is required it is more common to employ the technique of *spherical neutron polarimetry* (SNP), in which the final polarization is determined for a given initial polarization, rather than to measure all the spin-dependent scattering cross-sections. We shall give a brief description of SNP as applied to magnetic diffraction. The expressions required to interpret SNP data were obtained in Section 4.8.

The best established method for realizing SNP requires the sample to be placed in a zero-field chamber[24] as implemented in the CryoPAD (Tasset, 1989; Tasset *et al.*, 1999) and MuPAD (Janoschek *et al.*, 2007) devices. CryoPAD is shown schematically in Fig. 7.20. The sample is surrounded by two concentric cylindrical shields made of superconducting niobium. These shields exploit the Meissner–Ochsenfeld effect to screen the sample from magnetic fields and to allow an abrupt change in field either side of the shields. The incident polarization and the direction along which the scattered polarization is measured are set independently by precession coils located between the Meissner shields and nutators placed in the incident and scattered beams. The systems of coils and nutators makes it possible to align the incident polarization $\mathbf{P}_i$ along any direction of space, and to analyse any component of the scattered beam polarization $\mathbf{P}_f$.

The results of a SNP measurement are usually expressed in terms of the *polarization matrix* P (Section 4.8.1) whose components are

$$\mathsf{P}_{\alpha\beta} = \frac{n_{\alpha\beta} - n_{\alpha\bar{\beta}}}{n_{\alpha\beta} + n_{\alpha\bar{\beta}}}, \qquad (7.52)$$

where $\alpha$ and $\beta$ indicate the directions of polarization of the incident and scattered beams, respectively, and $n_{\alpha\beta}$ and $n_{\alpha\bar{\beta}}$ are the numbers of scattered neutrons (corrected for background) with spins parallel and antiparallel to the $\beta$ direction.

The general expression for the polarization matrix, referred to the standard polarization axes (Fig. 7.12), is given in eqn (4.104). For the case of elastic scattering in which there is no nuclear component, i.e. pure magnetic diffraction, the matrix simplifies to

$$
\mathsf{P} = \begin{pmatrix} -S_{yy}^{\mathrm{M}} - S_{zz}^{\mathrm{M}} & 0 & 0 \\ 0 & S_{yy}^{\mathrm{M}} - S_{zz}^{\mathrm{M}} & S_{yz}^{\mathrm{M}} \\ 0 & S_{yz}^{\mathrm{M}} & -S_{yy}^{\mathrm{M}} + S_{zz}^{\mathrm{M}} \end{pmatrix} \begin{matrix} \times P_{\mathrm{i},x}/I^x \\ \times P_{\mathrm{i},y}/I^y \\ \times P_{\mathrm{i},z}/I^z \end{matrix}
$$

$$
+ \begin{pmatrix} -T_{yz}^{\mathrm{M}} & 0 & 0 \\ -T_{yz}^{\mathrm{M}} & 0 & 0 \\ -T_{yz}^{\mathrm{M}} & 0 & 0 \end{pmatrix} \begin{matrix} \times 1/I^x \\ \times 1/I^y \\ \times 1/I^z \end{matrix} , \tag{7.53}
$$

where $S_{yy}^{\mathrm{M}}$, $S_{zz}^{\mathrm{M}}$ and $T_{yz}^{\mathrm{M}}$ are the same as in (7.36)–(7.37), and the other terms are

$$
S_{yz}^{\mathrm{M}} = \left(\frac{\gamma r_0}{2\mu_{\mathrm{B}}}\right)^2 \left\{\langle M_y^*(\mathbf{Q})\rangle\langle M_z(\mathbf{Q})\rangle + \langle M_z^*(\mathbf{Q})\rangle\langle M_y(\mathbf{Q})\rangle\right\}, \tag{7.54}
$$

and

$$
I^x = S_{yy}^{\mathrm{M}} + S_{zz}^{\mathrm{M}} + T_{yz}^{\mathrm{M}} P_{\mathrm{i},x}
$$
$$
I^y = I^z = S_{yy}^{\mathrm{M}} + S_{zz}^{\mathrm{M}}. \tag{7.55}
$$

The expressions to the right of the matrices in (7.53) are multiplicative factors for the rows.

As defined in (7.52), the polarization matrix is obtained experimentally from the background-corrected count rates $n_{\alpha\beta}$ and $n_{\alpha\bar{\beta}}$. These are measured at or close to the maximum of a magnetic Bragg peak. As a consistency check, it is a good idea to measure the matrix $\mathsf{P}_{\bar{\alpha}\beta}$ as well as $\mathsf{P}_{\alpha\beta}$ (18 components in total). The measurements are repeated for several different magnetic Bragg reflections and the results compared with calculations to refine the magnetic structure.

There are several advantages of SNP over the traditional method for magnetic structure determination, which involves measurement of diffraction intensities either with unpolarized neutrons or with uniaxial polarization analysis:

- SNP can measure transverse (off-diagonal) terms in the polarization matrix. This makes it possible to distinguish a rotation of the polarization from depolarization effects due to domains, and hence makes it easier to distinguish non-collinear multi-**q** magnetic structures from single-**q** magnetic structures with a set of equivalent magnetic domains.

- Polarization can often be measured with higher accuracy than can the integrated intensity of a magnetic Bragg peak. This is because polarization is obtained from ratios of intensities without the need to move the sample ($n_{\alpha\beta}$ is simply the neutron counts at the peak centre). Integrated intensities can be difficult to measure when the peaks have intrinsic broadening or a non-trivial mosaic.

- Polarization does not depend on a magnetic form factor or on sample absorption, again because it is obtained from a ratio of counts made at the same wavevector.

While appreciating the strengths of SNP, one should also bear in mind the following:

- It is necessary to determine the magnetic propagation vector first, before performing SNP.

- Equation (7.53) for the polarization matrix only applies if the scattered beam derives from a single magnetic domain. If more than one magnetic domain scatters into a Bragg peak then the intensity terms in the polarization matrix need to be domain-averaged.[25]

- SNP cannot determine the size of the ordered magnetic moments unless there is nuclear–magnetic interference scattering in the magnetic Bragg reflections (see below).

- In its standard implementation, SNP works only in zero field.

- Magnetic structures with a ferromagnetic component cannot usually be measured because the strong internal magnetic field in a ferromagnet will depolarize the neutron beam. For this reason, SNP is mainly used to study antiferromagnets or more complex magnetic structures that have no net magnetization. The presence of a small amount of ferromagnetic impurity in a non-ferromagnetic sample can also cause depolarization. Conversely, the sudden observation of strong neutron depolarization is usually evidence that the material has entered a ferro- or ferrimagnetic phase.

We have mentioned that depolarization can be caused by magnetic domains or internal magnetic fields in the sample. One way to check for depolarization is to note that in an ideal measurement with perfect neutron beam polarization the polarization matrix for pure magnetic scattering from a single magnetic domain simplifies to

$$\mathsf{P} = \begin{pmatrix} -1 & 0 & 0 \\ B & A & C \\ B & C & -A \end{pmatrix}, \tag{7.56}$$

where,

$$A = \frac{S_{yy}^{\mathrm{M}} - S_{zz}^{\mathrm{M}}}{S_{yy}^{\mathrm{M}} + S_{zz}^{\mathrm{M}}}, \quad B = \frac{-T_{yz}^{\mathrm{M}}}{S_{yy}^{\mathrm{M}} + S_{zz}^{\mathrm{M}}}, \quad C = \frac{S_{yz}^{\mathrm{M}}}{S_{yy}^{\mathrm{M}} + S_{zz}^{\mathrm{M}}}, \tag{7.57}$$

which satisfies

$$\sum_{\beta} \mathsf{P}_{\alpha\beta}^2 = 1 \qquad \text{for all } \alpha. \tag{7.58}$$

This sum rule represents the fact that if there is no depolarization then the scattered beam is fully polarized if the incident beam is fully polarized (i.e. $|\mathbf{P}_{\mathrm{f}}| = |\mathbf{P}_{\mathrm{i}}| = 1$).

We now give two examples of the use of neutron polarimetry, the first showing how SNP can be used to discriminate between magnetic structures which are indistinguishable in intensity measurements, and the second demonstrating how the sense of rotation of the spins in a helimagnetic or cycloidal magnetic structure may be determined from the created polarization.

[25]The terms in the numerator and denominator of the matrix elements need to be separately domain-averaged. This is not the same as domain-averaging the polarization.

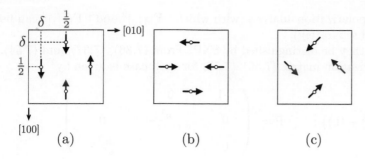

**Fig. 7.21** Magnetic structures with $\mathbf{q} = (0, 0, \frac{1}{2})$ for a tetragonal crystal. (a) and (b) are two equivalent orientation domains of a collinear antiferromagnet. (c) Non-collinear antiferromagnet. All moments are reversed in the next layer in the [001] direction.

## Distinguishing collinear and non-collinear magnetic structures

Following Brown (see Chatterji, 2006), Fig. 7.21 shows three possible magnetic structures with propagation vector $\mathbf{q} = (0, 0, \frac{1}{2})$ for a tetragonal crystal. The structures in Figs. 7.21(a) and (b) are the two equivalent orientation domains of a collinear antiferromagnet with moments pointing along the $\pm[100]$ and $\pm[010]$ directions, respectively. The structure in Fig. 7.21(c) is non-collinear, with moments pointing in the four equivalent $\langle 110 \rangle$ directions. Structure (c) is the coherent superposition of structures (a) and (b). The four magnetic atoms in the unit cell have fractional coordinates $\frac{1}{2}, \pm\delta, 0$ and $\pm\delta, \frac{1}{2}, 0$, and each carries a moment of magnitude $\mu$. The moments are stacked antiferromagnetically along the $c$ axis in all three structures.

Since $\mathbf{q} = (0, 0, \frac{1}{2})$, the magnetic reflections have the general form $(h, k, l + \frac{1}{2})$. Let us suppose the crystal has been oriented with the $a$ axis vertical, so as to access reflections in the $(0kl)$ horizontal scattering plane (Fig. 7.22). The structure factors $\mathbf{F}_M(\mathbf{Q})$ for magnetic reflections in this plane may be calculated from (7.9). Table 7.3 gives the components of $\mathbf{F}_M(\mathbf{Q})$ for each of the three magnetic structures.[26]

It is clear from Table 7.3 that the scattering intensity from structure (c) is indistinguishable from a 50:50 mixture of domains (a) and (b). This is true both for unpolarized neutrons (intensity $\propto |\mathbf{F}_{M\perp}|^2$) and for

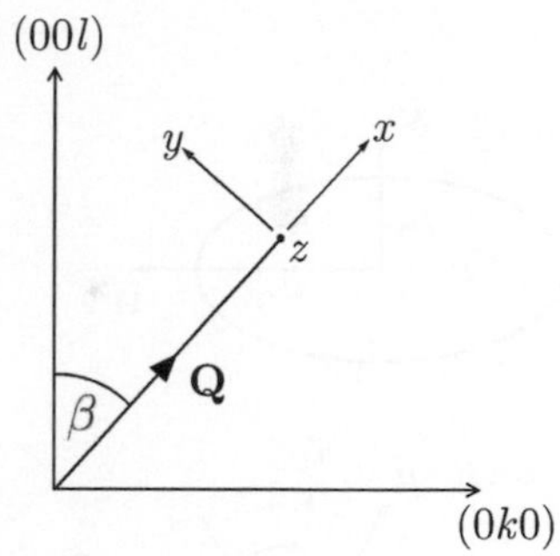

**Fig. 7.22** Polarization axes for diffraction in the $(0kl)$ scattering plane from the antiferromagnetic structures in Fig. 7.21.

[26]Remember that when referred to the polarization axes, $(\mathbf{F}_{M\perp})_y = (\mathbf{F}_M)_y$ and $(\mathbf{F}_{M\perp})_z = (\mathbf{F}_M)_z$. Hence, $|\mathbf{F}_{M\perp}|^2 = |(\mathbf{F}_M)_y|^2 + |(\mathbf{F}_M)_z|^2$.

**Table 7.3** Expressions for the magnetic structure factors $\mathbf{F}_M(\mathbf{Q})$ with $\mathbf{Q} = (0, k, l + \frac{1}{2})$ for the three magnetic structures shown in Fig. 7.21. The components are given with respect to the standard polarization axes shown in Fig. 7.22.

|  | (a) | (b) | (c) |
|---|---|---|---|
| $(\mathbf{F}_M)_x$ | 0 | $2iA \sin\beta$ | $\sqrt{2}\,iA \sin\beta$ |
| $(\mathbf{F}_M)_y$ | 0 | $-2iA \cos\beta$ | $-\sqrt{2}\,iA \cos\beta$ |
| $(\mathbf{F}_M)_z$ | $2iA$ | 0 | $\sqrt{2}\,iA$ |
| $|\mathbf{F}_{M\perp}|^2$ | $4A^2$ | $4A^2 \cos^2\beta$ | $2A^2(1 + \cos^2\beta)$ |

$$A = p\mu f(G) \exp(-W) \sin(2\pi k\delta)$$

uniaxial polarization analysis (with which $|(\mathbf{F}_M)_y|^2$ and $|(\mathbf{F}_M)_z|^2$ can be measured separately).

Could they be distinguished by SNP? From (7.36), (7.37), and (7.54), the polarization matrix (7.56)–(7.57) for each case is given by[27]

$$\frac{1}{2}\{(a)+(b)\}: \quad \mathsf{P} = \begin{pmatrix} -1 & 0 & 0 \\ 0 & \frac{-\sin^2\beta}{\cos^2\beta+1} & 0 \\ 0 & 0 & \frac{\sin^2\beta}{\cos^2\beta+1} \end{pmatrix}$$

$$(c): \quad \mathsf{P} = \begin{pmatrix} -1 & 0 & 0 \\ 0 & \frac{-\sin^2\beta}{\cos^2\beta+1} & \frac{-2\cos\beta}{\cos^2\beta+1} \\ 0 & \frac{-2\cos\beta}{\cos^2\beta+1} & \frac{\sin^2\beta}{\cos^2\beta+1} \end{pmatrix}. \quad (7.59)$$

The diagonal elements of both matrices are identical, which confirms that uniaxial polarization analysis cannot distinguish structure (c) from a 50:50 mixture of domains (a) and (b). However, the two possibilities can be distinguished by the presence or absence of the off-diagonal elements $\mathsf{P}_{zy}$ and $\mathsf{P}_{yz}$. Note also that (c) satisfies the sum rule (7.58).

### Determination of vector spin chirality — method 3

For the second example, we return to the problem of determining vector spin chirality. Consider a generalized elliptical helimagnetic structure described by

$$\boldsymbol{\mu}_\mathbf{l} = \mu_u \hat{\mathbf{u}}\cos(\mathbf{q}\cdot\mathbf{l}+\phi) \pm \mu_v \hat{\mathbf{v}}\sin(\mathbf{q}\cdot\mathbf{l}+\phi), \quad (7.60)$$

in which the spins rotate in the plane defined by the orthonormal vectors $\hat{\mathbf{u}}$ and $\hat{\mathbf{v}}$ and form an elliptical envelope with principal semi-axes $\mu_u$ and $\mu_v$, as shown in Fig. 7.23. For simplicity, we restrict the analysis to a system with one magnetic moment per unit cell, and we assume that $\hat{\mathbf{u}}$ lies in the plane containing $\mathbf{Q}$ and $\mathbf{q}$, and is at an angle of $\psi$ to $\mathbf{q}$ (Fig. 7.23). The $\pm$ sign in (7.60) determines whether the moments rotate clockwise ($+$) or anticlockwise ($-$) when looking along the propagation vector $\mathbf{q}$, corresponding to right-handed and left-handed structures, respectively.[28]

The Fourier expansion (7.1) of this structure has $+\mathbf{q}$ and $-\mathbf{q}$ components, and the Fourier coefficients are

$$\mathbf{m}_\mathbf{q} = \frac{1}{2}e^{-i\phi}(\mu_u\hat{\mathbf{u}}\pm i\mu_v\hat{\mathbf{v}}) \quad \text{and} \quad \mathbf{m}_{-\mathbf{q}} = \frac{1}{2}e^{i\phi}(\mu_u\hat{\mathbf{u}}\mp i\mu_v\hat{\mathbf{v}}). \quad (7.61)$$

Each reciprocal lattice vector $\mathbf{G}$ is flanked by $+\mathbf{q}$ and $-\mathbf{q}$ magnetic satellites, and $\langle\mathbf{M}(\mathbf{G}\pm\mathbf{q})\rangle$ is proportional to $\mathbf{m}_{\pm\mathbf{q}}$.[29] If $\beta$ is the angle between $\mathbf{Q}$ and $\mathbf{q}$ (Fig. 7.23), then the conversion to Blume–Maleev coordinates is

$$M_y = M_u\sin(\beta+\psi) \quad \text{and} \quad M_z = M_v. \quad (7.62)$$

(a)

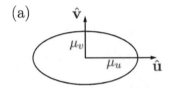

(b)

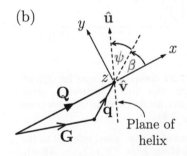

**Fig. 7.23** (a) Elliptical envelope of the helimagnetic structure defined in eqn (7.60). (b) Definition of angles and axes in the scattering plane for the calculation of the polarization matrix of the helimagnetic structure. The axis of the helix is along $\mathbf{q}$, which makes an angle $\beta$ with $\mathbf{Q}$ and $\psi$ with $\hat{\mathbf{u}}$. The $\hat{\mathbf{v}}$ direction is parallel to the Blume–Maleev $z$ axis.

[28]The special angles $\psi = 0$ and $90°$ correspond to cycloidal and helical structures, respectively. The cycloid is not chiral (i.e. it does not come in right- and left-handed forms).

[29]In the dipole approximation, $\langle\mathbf{M}(\mathbf{G}\pm\mathbf{q})\rangle = (2\pi)^3/v_0\, f(Q)\exp(-W)$ $\times \mathbf{m}_{\pm\mathbf{q}}\,\delta(\mathbf{Q}\mp\mathbf{q}-\mathbf{G})$.

From (7.61) and (7.62) we can see that in the case $\mathbf{Q} = \mathbf{G} + \mathbf{q}$,

$$\langle M_y(\mathbf{G}+\mathbf{q})\rangle \propto \mu_u e^{-i\phi}\sin(\beta+\psi)$$

$$\langle M_z(\mathbf{G}+\mathbf{q})\rangle \propto \pm i\mu_v e^{-i\phi}, \qquad (7.63)$$

where the upper and lower signs correspond to the right-handed and left-handed structures, respectively.[30] The terms in the polarization matrix are, from (7.36), (7.37), (7.54), and (7.57),

$$A = \frac{\mu_u^2\sin^2(\beta+\psi)-\mu_v^2}{\mu_u^2\sin^2(\beta+\psi)+\mu_v^2}, \quad B = \frac{\pm 2\mu_u\mu_v\sin(\beta+\psi)}{\mu_u^2\sin^2(\beta+\psi)+\mu_v^2}, \quad C = 0. \quad (7.64)$$

The $\pm$ sign in the $B$ term (the created polarization) is directly related to the direction of $\langle \mathbf{M}_\perp^\dagger(\mathbf{Q})\rangle \times \langle \mathbf{M}_\perp(\mathbf{Q})\rangle$, see eqns (4.100) and (7.37). In the case of a helix, the sign gives its handedness.

The above analysis has been used in the study of certain multiferroic materials in which a ferroelectric polarization is coupled to a helical or cycloidal magnetic structure. An example is shown in Fig. 7.24. By reversing the direction of the electric field applied parallel to the ferroelectric polarization one can convert a right-handed magnetic domain into a left-handed domain. This process is monitored by measuring the $\mathsf{P}_{yx} = \mathsf{P}_{zx} = B$ polarization matrix elements as the applied electric field is switched.

### Nuclear–magnetic interference

Finally, let us mention the possibility that magnetic and nuclear scattering occurs in the same Bragg reflection. Examples of this are $\mathbf{q} = 0$ magnetic structures, and structures with $\mathbf{q} \neq 0$ in which magnetic order is coupled to a structural distortion with the same $\mathbf{q}$. In such cases, nuclear–magnetic interference can occur, and SNP can determine the relationship between the amplitudes and phases of the magnetic and nuclear scattering. If a model for the nuclear structure factors is available then it is possible to determine the size of the ordered moments by SNP, without the need for additional intensity measurements.

[30] When $\mathbf{Q} = \mathbf{G} - \mathbf{q}$, the $\pm$ signs in (7.63) and in front of the $B$ term in (7.64) change to $\mp$.

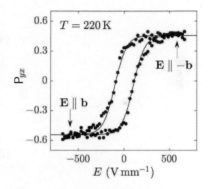

**Fig. 7.24** SNP data in the ferroelectric phase of cupric oxide (CuO) with an electric field $E$ applied along the polar $b$ axis (Babkevich *et al.*, 2012). The plot shows the $\mathsf{P}_{yx}$ polarization matrix element at $\mathbf{Q} = \mathbf{q} = (0.506, 0, -0.483)$ during a sweep of the electric field, and demonstrates the switching of the magnetic chirality as the field is reversed. Here, $\beta = 0$, $\psi = 73°$ and $\mu_u \simeq \mu_v$, so according to (7.64) $\mathsf{P}_{yx}$ should vary between $\pm 1$, not $\pm 0.5$ as observed. The discrepancy implies incomplete saturation of the ferroelectric domains.

## 7.4 Strategy to solve magnetic structures

When faced with a magnetic structure to solve, a typical approach would be as follows:

(1) First, find the propagation wavevector(s). This could be done by neutron powder or Laue (single crystal) diffraction, since these techniques probe a large range of wavevectors.[31] A good way to reveal the magnetic Bragg peaks is to compare measurements performed at temperatures above and below the magnetic ordering temperature. The difference is the magnetic intensity.

[31] See Section 10.2.1.

(2) Characterize the magnetic diffraction quantitatively, either (i) by measurement of a large set of Bragg peak integrated intensities by powder or single crystal diffraction, or (ii) by measurement of polarization matrices at several Bragg peaks by spherical neutron polarimetry. The former method can be supplemented by uniaxial polarization analysis if required, as discussed in Section 7.3.1.

(3) Refine a model against the data. Group theory can be very useful to enumerate the possible magnetic structures allowed by symmetry, and hence to restrict the number of model fitting parameters.[32] In particular, if the transition from paramagnetic to magnetically ordered phase is of second order then the magnetic structure will belong to one of the irreducible representations of the paramagnetic symmetry group of the crystal.

(4) If the intensities can be described equally well by either a multi-**q** structure or a set of single-**q** magnetic domains then one needs a single crystal sample, and either (i) perform spherical neutron polarimetry, or (ii) apply an external perturbation, such as a magnetic field or pressure, and hope that if there are magnetic domains the perturbation causes a change in the domain populations and a proportionate change in the magnetic Bragg peak intensities.

(5) Determine the size of the moments in $\mu_B$. For this one needs to place the magnetic peak intensities on an absolute scale, which can be done by calibrating them against the nuclear Bragg peaks.

## 7.5   Diffuse magnetic scattering

Materials with periodic magnetic order extending over very large distances exhibit a series of sharp Bragg peaks in the coherent magnetic scattering. Conversely, in an ideal paramagnet the magnetic moments on different sites are completely uncorrelated, so the scattering is incoherent and exhibits no structure other than a smooth decay with $Q$ due to the (single-ion) magnetic form factor. Between these extremes lies a diversity of systems which have short-range magnetic correlations: frustrated magnets, low-dimensional magnets and quantum spin liquids, to name but a few.[33] Similarly, if a relatively small concentration of defects such as domain walls are introduced into an otherwise perfectly ordered magnetic structure then the Bragg peaks show broadening, and their widths are inversely related to the correlation length.[34]

Coherent scattering from systems with short-range order is extended (i.e. not sharp) in momentum space and is known as *diffuse scattering*. The method of XYZ polarization analysis[35] is designed for the separation of magnetic diffuse scattering from nuclear scattering. By modelling magnetic diffuse scattering data one can obtain information on the nature and origin of the short-range spin correlations.

In the following sections we illustrate some of the characteristic features of magnetic diffuse scattering with reference to some typical problems. We begin with a summary of the relevant formulae.

[32]See Section 7.1.5.

[33]The term *cooperative paramagnet* is often used to describe a system which has no long-range order but which has significant magnetic correlations.

[34]See Sections 2.3.1–2.3.2.

[35]See Section 4.6.

## 7.5.1 Magnetic structure factor

Magnetic neutron scattering measures the partial response functions $S_{\alpha\beta}(\mathbf{Q},\omega)$ (Section 6.1). In diffraction measurements of diffuse scattering we are mainly concerned with systems having slow dynamics, which means that $S_{\alpha\beta}(\mathbf{Q},\omega)$ is sharply peaked around $\omega = 0$. If the width of this peak is much less than the energy resolution then the measurement in effect performs an energy integral. The measured quantity is then the partial *structure factor* $S_{\alpha\beta}(\mathbf{Q})$, which is given by[36] (see Section 3.8.3)

$$S_{\alpha\beta}(\mathbf{Q}) = \int_{-\infty}^{\infty} S_{\alpha\beta}(\mathbf{Q},\omega) \, \mathrm{d}(\hbar\omega)$$

$$= \langle M_\alpha(-\mathbf{Q},0) M_\beta(\mathbf{Q},0) \rangle. \tag{7.65}$$

The energy integral is performed at fixed $\mathbf{Q}$. The assumption that an experiment directly measures $S_{\alpha\beta}(\mathbf{Q})$ is known as the *static approximation*, and the conditions for its validity were discussed in Section 3.8.4. The second line in (7.65) shows that $S_{\alpha\beta}(\mathbf{Q})$ measures the *equal-time* or *instantaneous* correlation function, which represents a snapshot of the state of the system at $t = 0$ and is a quantity that can be calculated by standard methods from a theoretical model of the system.[37]

Following the approach in Section 3.6, we can split $S_{\alpha\beta}(\mathbf{Q})$ into separate components representing static and dynamic correlations [see also eqn (6.13)]:

$$S_{\alpha\beta}(\mathbf{Q}) = \langle M_\alpha(-\mathbf{Q}) \rangle \langle M_\beta(\mathbf{Q}) \rangle + \widetilde{S}_{\alpha\beta}(\mathbf{Q}). \tag{7.66}$$

The first term describes (elastic) Bragg scattering from static long-range order, and the second term is the diffuse scattering. When the static approximation is valid, $\widetilde{S}_{\alpha\beta}(\mathbf{Q})$ is sometimes called *elastic* diffuse scattering, but strictly speaking it is quasielastic with a width that is smaller than the energy resolution.

Diffuse scattering is related to the generalized susceptibility $\chi_{\alpha\beta}(\mathbf{Q},\omega)$ through the Fluctuation–Dissipation theorem and the Kramers–Kronig transformation (Sections D.4 and D.5). The static (zero frequency) component of the susceptibility[38] is given by eqn (D.43),

$$\chi_{\alpha\beta}(\mathbf{Q},0) = \int_{-\infty}^{\infty} \frac{\widetilde{S}_{\beta\alpha}(\mathbf{Q},\omega)}{\omega} \left\{ 1 - \exp(-\beta\hbar\omega) \right\} \mathrm{d}\omega, \tag{7.67}$$

where $\beta = 1/k_\mathrm{B}T$. When the energy width of the diffuse scattering is much less than $k_\mathrm{B}T$,

$$\chi_{\alpha\beta}(\mathbf{Q},0) \simeq \frac{1}{k_\mathrm{B}T} \int_{-\infty}^{\infty} \widetilde{S}_{\beta\alpha}(\mathbf{Q},\omega) \, \mathrm{d}(\hbar\omega)$$

$$= \frac{1}{k_\mathrm{B}T} \widetilde{S}_{\beta\alpha}(\mathbf{Q}). \tag{7.68}$$

Hence, by calibrating the intensity and extrapolating it to $\mathbf{Q} = 0$ one can relate the diffuse scattering to the magnetic susceptibility $\chi_\mathrm{M}$ measured in a laboratory magnetometer.[39]

[36] To convert into a differential cross-section in absolute units (mb sr$^{-1}$), multiply $S_{\alpha\beta}(\mathbf{Q})$ by $(\gamma r_0/2\mu_\mathrm{B})^2$.

[37] In reality, an experiment averages a series of snapshots from finite volumes of the system at time intervals long compared with the characteristic fluctuation time. If the system is ergodic then this average is equivalent to an ensemble average.

[38] At zero frequency the susceptibility is purely reactive, $\chi(\mathbf{Q},0) = \chi'(\mathbf{Q},0)$.

[39] For the conversion between the units of generalized susceptibility and magnetic susceptibility, see eqns (6.82)–(6.83).

## 7.5.2   Spherical averaging of diffuse scattering

The preferred method to study short-range magnetic correlations in a crystalline material is via magnetic diffuse scattering intensity maps in momentum space recorded from a single crystal sample. Unfortunately, single crystal samples of sufficient size are not always available, and measurements must then be made on polycrystalline materials.

Assuming the dipole approximation applies, so that $\mathbf{M}(\mathbf{Q})$ can be written in the form given in eqn (7.6), we shall obtain an expression for the spherical average of the magnetic structure factor

$$S(\mathbf{Q}) = \langle \mathbf{M}_\perp(-\mathbf{Q}) \cdot \mathbf{M}_\perp(\mathbf{Q}) \rangle$$

$$= \sum_{jk} f_j(Q) f_k(Q) \exp(-W_j - W_k) \langle \boldsymbol{\mu}_{\perp j} \cdot \boldsymbol{\mu}_{\perp k} \rangle \exp(\mathrm{i}\mathbf{Q} \cdot \mathbf{R}_{jk}),$$

$$(7.69)$$

where $\mathbf{R}_{jk} = \mathbf{R}_k - \mathbf{R}_j$ is the vector joining sites $j$ and $k$. We also assume in (7.69) that magnetic correlations are not coupled to the atomic positions and that structural correlations are static, see Sections 6.1.2 and 6.1.3. Equation (7.69) gives the $S(\mathbf{Q})$ that would be measured by unpolarized neutron scattering, but it is more usual in practice to obtain it by XYZ polarization analysis with a multidetector because this technique makes it possible to separate the magnetic scattering from any nuclear scattering, see Section 4.6.

In order to perform the spherical average we write (see Section 4.9)

$$\langle \boldsymbol{\mu}_{\perp j} \cdot \boldsymbol{\mu}_{\perp k} \rangle = \sum_{\alpha\beta} (\delta_{\alpha\beta} - \hat{Q}_\alpha \hat{Q}_\beta) \langle \mu_{\alpha j} \mu_{\beta k} \rangle, \qquad (7.70)$$

where $\alpha, \beta$ symbolize a set of local Cartesian coordinates for sites $j$ and $k$, with the local $z$ axis parallel to the vector $\mathbf{R}_{jk}$ as shown in Fig. 7.25. The direction of $\mathbf{Q}$ is defined relative to the local axes by the spherical angles $\theta, \phi$, so that

$$\hat{\mathbf{Q}} = \begin{pmatrix} \sin\theta\cos\phi \\ \sin\theta\sin\phi \\ \cos\theta \end{pmatrix}, \qquad (7.71)$$

and

$$\mathbf{Q} \cdot \mathbf{R}_{jk} = Q R_{jk} \cos\theta. \qquad (7.72)$$

A scattering measurement on a powder is equivalent to an average over all orientations of $\mathbf{Q}$ relative to the local coordinates. The fraction of all directions that are contained in the angular intervals $\theta \to \theta + \mathrm{d}\theta$ and $\phi \to \phi + \mathrm{d}\phi$ is $\sin\theta\,\mathrm{d}\theta\,\mathrm{d}\phi/4\pi$, and so the spherical average of $S(\mathbf{Q})$ is

$$S(Q) = \frac{1}{4\pi} \int_{\phi=0}^{2\pi} \int_{\theta=0}^{\pi} S(\mathbf{Q}) \sin\theta\,\mathrm{d}\theta\,\mathrm{d}\phi. \qquad (7.73)$$

Substituting (7.69)–(7.72) into (7.73) and carrying out the angular integration, we obtain

$$S(Q) = \sum_{jk} C_j C_k \left[ A_{jk} \frac{\sin Q R_{jk}}{Q R_{jk}} + B_{jk} \left\{ \frac{\sin Q R_{jk}}{(Q R_{jk})^3} - \frac{\cos Q R_{jk}}{(Q R_{jk})^2} \right\} \right],$$

$$(7.74)$$

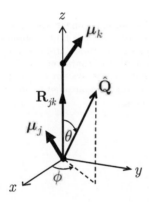

**Fig. 7.25** Local coordinate system for sites $j$ and $k$, and spherical angles for calculation of powder average.

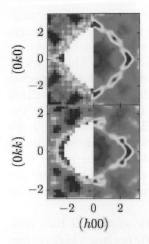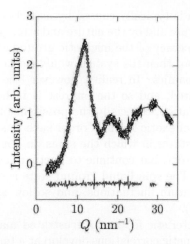

**Fig. 7.26** Magnetic diffuse scattering from $Gd_3Ga_5O_{12}$, after Paddison *et al.* (2015) and Petrenko *et al.* (1998). The right panel shows a reverse Monte Carlo (RMC) fit (red line) to powder magnetic diffuse scattering data. The blue line is the difference (fit − data). On the left are single crystal diffuse scattering maps in two reciprocal space planes. In each map, the left half is experimental data and the right half is the RMC refinement of the powder data, obtained with the SPINVERT program (http://spinvert.chem.ox.ac.uk/). (Data courtesy of Joe Paddison.)

where we have introduced the shorthand $C_j = f_j(Q)\exp(-W_j)$, and

$$A_{jk} = \langle \boldsymbol{\mu}_j \cdot \boldsymbol{\mu}_k \rangle - \langle \mu_{zj}\mu_{zk} \rangle$$
$$B_{jk} = 3\langle \mu_{zj}\mu_{zk} \rangle - \langle \boldsymbol{\mu}_j \cdot \boldsymbol{\mu}_k \rangle. \tag{7.75}$$

Terms with $j \neq k$ in the summation in eqn (7.74) arise from pair correlations between spins on different sites. If inter-site correlations are completely absent then the system is an ideal paramagnet, and only terms with $j = k$ contribute. For these terms, the expression in the square brackets reduces to[40] $(2/3)\langle \boldsymbol{\mu}_j^2 \rangle$, in agreement with the total moment sum rule, eqn (6.92).

The message to take away from this section is that magnetic diffuse scattering from powder samples contains information on three-dimensional magnetic correlations. There are two ways to extract this information. One can assume a form for the magnetic Hamiltonian for the system, use it to simulate a configuration of magnetic moments in a finite-sized system e.g. by a direct Monte Carlo method, and compare the diffuse scattering calculated from the simulation with experimental data. If required, the Hamiltonian can then be adjusted to obtain best agreement. Alternatively, one can employ a reverse Monte Carlo (RMC) method to refine the orientations of the individual magnetic moments in the configuration until the calculated diffuse scattering fits the data, without reference to any Hamiltonian. An example of the latter method is shown in Fig. 7.26. The figure compares the diffuse scattering intensity from single crystals measurements with that obtained by the RMC method from powder data. The RMC refinements are able to reproduce the main features in the single crystal data.

### 7.5.3  Frustrated magnetism

In magnetism, a *frustrated* system is one in which the magnetic interactions cannot all be satisfied simultaneously. In other words, there is no unique arrangement of magnetic moments that minimizes the energy.

[40]Write the trigonometric functions as series expansions in $x = QR_{jk}$, and take the limit as $x \to 0$. The limit corresponds to $R_{jk} = 0$.

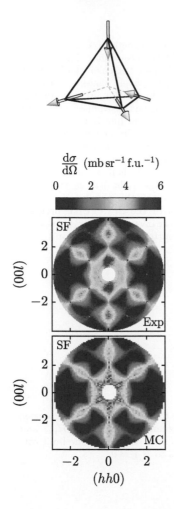

$$\frac{d\sigma}{d\Omega}\ (\mathrm{mb\,sr^{-1}\,f.u.^{-1}})$$

Fig. 7.27 Diffuse magnetic scattering from $Ho_2Ti_2O_7$ (Fennell *et al.*, 2009). At the top, one of the possible two-in-two-out configurations of Ising spins on a tetrahedron is shown. The upper colour figure is experimental data recorded with polarized neutrons, and the lower colour figure is a Monte Carlo simulation for spin ice. In Blume–Maleev coordinates (Fig. 7.12), when $\mathbf{P}\parallel z$ the SF and NSF channels measure $S_{yy}(\mathbf{Q})$ and $S_{zz}(\mathbf{Q})$, respectively (the latter is not shown). The pinch points characteristic of spin-ice correlations are only observed in the SF channel. (Data courtesy of Tom Fennell.)

Frustration is typically caused by aspects of the geometry of the magnetic structure and by the nature and strength of the magnetic couplings. If the degeneracy of the magnetic ground state were to persist down to absolute zero then the system would be in breach of the Third Law of Thermodynamics. In reality, however, the Third Law is not expected to be violated, and so the interest is in how different systems relieve their frustration, especially the possibility that they may do so via the emergence of exotic forms of order such as spin liquid phases. These are correlated states in which the spins do not show conventional forms of magnetic order, but continue to fluctuate even down to the lowest temperatures. The spin liquid ground state is only possible when there are strong quantum fluctuations, and so low spin values are required (see Section 8.7).

A characteristic feature of frustrated magnetic systems is that non-trivial magnetic correlations develop at a temperature much higher than the eventual ordering temperature $T_m$. For local moment systems, the Weiss constant $\theta$ indicates the strength of the dominant magnetic interaction (See Section C.4.1). Hence, the ratio $\theta/T_m$ provides a simple measure of the degree of frustration and is known as the *frustration index*. At low temperatures, the relaxation time (the time to reach equilibrium) can become very long, especially when the crystal has some degree of disorder. In such situations spin glass phases can develop.

Measurements of magnetic diffuse scattering from frustrated magnets can be used alongside simulations to develop models that provide an understanding of the frustration at a microscopic level. A good example is the extended family of cubic pyrochlore oxides $A_2B_2O_7$, which hosts a range of complex ground states. Many of these states have strong correlations without conventional long-range order, and are driven by geometrical frustration associated with the $A$ and $B$ sites which form three-dimensional networks of corner-sharing tetrahedra. In the classical spin ice $Ho_2Ti_2O_7$, the strong axial crystal field constrains the Ho spins to point either towards or away from the centre of a tetrahedron. The magnetic energy is minimized when two spins point in and two point out, analogous to the configuration of proton disorder in water ice. There are six different two-in-two-out configurations on a tetrahedron with the same minimum energy, so the system is frustrated. Figure 7.27 shows the magnetic diffuse scattering from $Ho_2Ti_2O_7$ in the $(hhl)$ plane, measured by XYZ polarization analysis. The characteristic spin-ice correlations are revealed by a set of 'pinch-point' singularities, where the diffuse scattering forms a narrow ridge.

### 7.5.4 Critical scattering

Critical diffuse scattering is a feature of systems that are in the vicinity of continuous phase transitions to long-range order. At temperatures below the transition $(T < T_c)$, $S(\mathbf{Q})$ has a sharp (Bragg) component and a diffuse component. The sharp component is associated with an *order parameter* which changes from zero in the disordered phase to non-zero in the ordered phase. The order parameter is usually associated with a symmetry which is broken at the phase transition, and is a function of one or more system variables which become static on crossing the phase transition point.[41] The diffuse component persists above the transition $(T > T_c)$, but decreases as $T$ moves further away from $T_c$. The critical diffuse scattering above and below $T_c$ is caused by cooperative fluctuations of the order parameter. Similar behaviour occurs if another control parameter (e.g. magnetic field or pressure) is used to drive the system across the phase transition at fixed $T = T_c$.[42]

It is found both theoretically and experimentally that certain system variables follow *scaling laws* in the vicinity of phase transitions, i.e. they vary as a control parameter raised to some power $x$. The exponent $x$ is known as a *critical exponent*. For magnetic phase transitions the critical exponents accessible to neutron scattering are defined by the following scaling laws:[43]

$$\chi_M \propto |t|^{-\gamma} \tag{7.76}$$

$$\xi \propto |t|^{-\nu} \tag{7.77}$$

$$M(T) \propto |t|^{\beta} \qquad (T < T_c) \tag{7.78}$$

$$M(H) \propto |H|^{1/\delta} \qquad (T = T_c) \tag{7.79}$$

$$G(\mathbf{r}) \propto |\mathbf{r}|^{2-d-\eta} \qquad (T = T_c) \tag{7.80}$$

where $t = (T - T_c)/T_c$ is the *reduced temperature*, and $|t| \ll 1$ in the critical regime where these laws apply. In (7.78) and (7.79), $M$ is the magnetization and $H$ is the applied magnetic field. More generally, $M(T)$ is the magnetic order parameter, and is proportional to the square root of the magnetic Bragg peak intensity. In (7.80), $d$ is the space dimensionality and $G(\mathbf{r})$ is the spin–spin correlation function[44]

$$G(\mathbf{r}) = \frac{1}{N} \int \langle \mathbf{S}(\mathbf{r}') \cdot \mathbf{S}(\mathbf{r}' + \mathbf{r}) \rangle \, d^3\mathbf{r}', \tag{7.81}$$

which is the magnetic analogue of the density–density correlation function (2.57). The magnetic structure factor $S(\mathbf{Q})$ is the Fourier transform of $G(\mathbf{r})$, see (2.59), and an approximation to $S(\mathbf{Q})$ in the critical region is

$$S(\mathbf{Q}) \propto \frac{1}{(\xi^{-2} + Q^2)^{1-\eta/2}}, \tag{7.82}$$

which contains the *correlation length* $\xi$ that appears in (7.77). The magnetic susceptibility $\chi_M$ can be obtained from the $\mathbf{Q} = 0$ limit of the scattering intensity (see Section 7.5.1), which gives access to the scaling relation (7.76).

According to the hypothesis of *universality* for continuous phase transitions, all systems belonging to the same *universality class* have the same critical exponents. For crystalline materials, the universality class depends only on the dimensionality of the lattice $d$, the dimensionality of the order parameter $n$, and the range of the interactions.

Table 7.4 lists critical exponents calculated for the Ising, $XY$, and Heisenberg models, which describe systems with nearest-neighbour interactions between spins of dimensionality $n = 1, 2$ and $3$, respectively.[45]

[41] In an antiferromagnet, for example, the order parameter is the sublattice magnetization.

[42] A special case is a *quantum phase transition*, which is a phase transition at $T = 0$ driven by a control parameter other than temperature.

[43] Magnetic systems have one other principal critical exponent, defined by $C_H \propto |t|^{-\alpha}$, where $C_H$ is the heat capacity at fixed magnetic field.

[44] Equations (7.81) and (7.82) are for an isotropic system.

[45] According to scaling theory, the critical exponents are related by the following scaling laws:

$$\gamma = \nu(2 - \eta)$$
$$\alpha + 2\beta + \gamma = 2$$
$$\gamma = \beta(\delta - 1)$$
$$\nu d = 2 - \alpha.$$

From these, one can show that

$$\delta = \frac{d + 2 - \eta}{d - 2 + \eta}.$$

**Table 7.4** Critical exponents for some common universality classes, taken from Pelissetto and Vicari (2002). The numbers in parentheses are estimated uncertainties.

| Critical exponent | Mean field | 2d Ising $(n=1)$ | 3d Ising $(n=1)$ | 2d $XY$ $(n=2)$ | 3d $XY$ $(n=2)$ | 3d Heisenberg $(n=3)$ |
|---|---|---|---|---|---|---|
| $\gamma$ | 1 | 7/4 | 1.2372(5) | — | 1.3177(5) | 1.3960(9) |
| $\nu$ | 1/2 | 1 | 0.6301(4) | — | 0.6716(3) | 0.7112(5) |
| $\beta$ | 1/2 | 1/8 | 0.3265(3) | $\simeq 0.23^*$ | 0.3485(2) | 0.3689(3) |
| $\delta$ | 3 | 15 | 4.789(2) | 15 | 4.780(2) | 4.783(3) |
| $\eta$ | 0 | 1/4 | 0.0364(5) | 1/4 | 0.0380(4) | 0.0375(5) |

$^*$The $\beta$ exponent for the 2d $XY$ model is an effective exponent for finite systems with long-range order (Bramwell and Holdsworth, 1993).

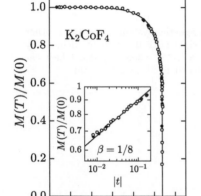

(a)

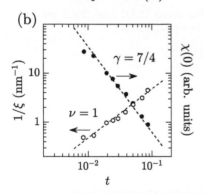

(b)

Power-law exponents found in mean-field theory, which neglects critical fluctuations, are also included in the table for comparison.

According to the Mermin–Wagner theorem, continuous symmetries cannot be spontaneously broken at $T > 0$ in $d \le 2$ for sufficiently short-range interactions. This means that conventional long-range order does not occur in 1d or 2d for the $XY$ or Heisenberg models (the theorem does not apply to the Ising model because Ising symmetry is discrete). In the 2d $XY$ model, arbitrarily large but finite systems do in fact exhibit long-range order, and the magnetization follows power-law behaviour over a restricted temperature range in the vicinity of the transition, with an effective exponent $\beta \simeq 0.23$. This exponent, which is displayed by many real systems, may be taken as an experimental signature of 2d $XY$ behaviour.

Figure 7.28 illustrates scaling behaviour in the vicinity of the magnetic ordering transition of $K_2CoF_4$, which has well separated $CoF_2$ layers and behaves as a 2d Ising antiferromagnet with $T_N = 107.85\,\mathrm{K}$. The sublattice magnetization, susceptibility and inverse correlation length data were obtained from measurements of $S(\mathbf{Q})$ as a function of temperature. The critical exponents determined from the data agree very well with the exact values for the 2d Ising model given in Table 7.4.

**Fig. 7.28** Critical scaling of $K_2CoF_4$. (a) Reduced sublattice magnetization as a function of temperature. The line is the exact form for the 2d Ising model. The insert is a log–log plot showing that the data are described quite well by the scaling law $M(T)/M(0) \propto |t|^{1/8}$. (b) Inverse correlation length and susceptibility compared with scaling laws $\xi^{-1} \propto t$ and $\chi \propto t^{-7/4}$, respectively. (After Ikeda and Hirakawa, 1974.)

## 7.6 Magnetization distributions

Diffraction with a polarized incident beam but without polarization analysis provides a practical method to determine $\mathbf{M}(\mathbf{r})$, the spatial distribution of magnetization. The method is sensitive to the wave functions of the unpaired electrons and can be used to detect the effects of covalency or ligand field interactions. It is also a direct way to measure magnetic form factors, as well as the local susceptibility at individual atomic sites.

The principle of the method is to use polarized neutron diffraction to obtain the magnetic scattering amplitude via the nuclear–magnetic

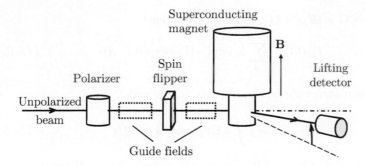

**Fig. 7.29** Schematic of a neutron diffractometer with a polarized incident beam but without polarization analysis.

interference term in the cross-section, rather than from the pure magnetic scattering term. Nuclear–magnetic interference is possible when magnetic and nuclear scattering occur at the same positions in reciprocal space. This is an intrinsic property of crystals with ferro- or ferrimagnetic order, but the more common practice is to use an external magnetic field to induce a ferromagnetic component on a sample that has no net magnetization in zero field, such as a paramagnet or a ordered magnet with a non-zero propagation wavevector. For small ferromagnetic moments the sensitivity to the magnetic amplitude is considerably enhanced through amplification by the nuclear amplitude. Another advantage of the method over pure magnetic diffraction is that the phase as well as the amplitude of the magnetic scattering is determined.

A typical experimental arrangement is shown in Fig. 7.29. The sample, which is normally a single crystal, is mounted in a superconducting magnet capable of applying a vertical field of typically several Tesla. The incident neutrons are polarized parallel or antiparallel to the direction of the magnetic field. The direction of the incident polarization can be reversed by a spin flipper placed in the incident beam.

The procedure is to record the intensities of a set of Bragg reflections, first with the incident neutron polarization parallel to the applied magnetic field and then with it reversed. The instrument shown in Fig. 7.29 has a lifting detector which makes it possible to access reflections out of the horizontal plane as well as those in the horizontal plane.

The scattering intensity for a polarized incident beam with polarization $\mathbf{P}$ is given by (4.97). In the presence of coherent nuclear and magnetic scattering the differential cross-section may be written

$$\frac{\mathrm{d}\sigma}{\mathrm{d}\Omega} = |\langle N \rangle|^2 + p^2|\langle \mathbf{M}_\perp \rangle|^2 + p\,\mathbf{P} \cdot \langle N^*\mathbf{M}_\perp + \mathbf{M}_\perp^* N \rangle + \mathrm{i}p^2\mathbf{P} \cdot \langle \mathbf{M}_\perp^* \times \mathbf{M}_\perp \rangle,$$
$$(7.83)$$

where $N(\mathbf{Q})$ and $\mathbf{M}_\perp(\mathbf{Q})$ are the nuclear and magnetic scattering amplitudes, and $p = \gamma r_0/2\mu_{\mathrm{B}}$.

We are concerned with crystalline materials, so the elastic scattering consists of sharp peaks centred on $\mathbf{Q} = \mathbf{G}$, where $\mathbf{G}$ is a reciprocal lattice vector of the chemical lattice. When there is a net ferromagnetic moment (i.e. $\mathbf{q} = 0$) in the chemical unit cell the Bragg peaks contain both coherent nuclear and magnetic scattering, and the thermally-averaged amplitudes $\langle N(\mathbf{G}) \rangle$ and $\langle \mathbf{M}_\perp(\mathbf{G}) \rangle$ are proportional to $F_{\mathrm{N}}(\mathbf{G})$

and $\mathbf{F}_{M\perp} = \hat{\mathbf{G}} \times (\mathbf{F}_M \times \hat{\mathbf{G}})$, respectively, where

$$F_N(\mathbf{G}) = \sum_d \bar{b}_d \exp(-W_d) \exp(i\mathbf{G} \cdot \mathbf{d}) \tag{7.84}$$

is the nuclear structure factor (Section 5.2.1), and

$$\mathbf{F}_M(\mathbf{G}) = p \int \mathbf{M}(\mathbf{r}) \exp(i\mathbf{G} \cdot \mathbf{r}) \, d^3\mathbf{r} \tag{7.85}$$

[46]When the dipole approximation applies, (7.85) reduces to the form given in (7.9).

is the most general form of the magnetic structure factor,[46] $\mathbf{M}(\mathbf{r})$ being the thermally averaged magnetization. The summation in (7.84) is over the atoms in the chemical unit cell, and the integration volume in (7.85) is that of the chemical unit cell.

Rather than the integrated intensities of the Bragg peaks, the experimental quantity is actually the ratio of the background-corrected count rate[47] measured with the incident beam fully polarized 'up' ($P = +1$)

[47]By 'background-corrected' we mean the count rate measured at (or close to) the maximum of the Bragg peak minus the count rate measured with the crystal turned by a few degrees so that the signal contains no Bragg scattered neutrons.

to that with it 'down' ($P = -1$). This ratio is known as the *flipping ratio*, and from (7.83) it is given by[48]

[48]In practice, the flipping ratios need to be corrected for the imperfect neutron polarization and flipper efficiency, as well as for attenuation, extinction and multiple scattering in the sample — see Sections 10.1.1, 10.1.7, and 10.2.6.

$$R = \frac{|F_N|^2 + |\mathbf{F}_{M\perp}|^2 + (F_N^* \mathbf{F}_{M\perp} + \mathbf{F}_{M\perp}^* F_N)_z + i(\mathbf{F}_{M\perp}^* \times \mathbf{F}_{M\perp})_z}{|F_N|^2 + |\mathbf{F}_{M\perp}|^2 - (F_N^* \mathbf{F}_{M\perp} + \mathbf{F}_{M\perp}^* F_N)_z - i(\mathbf{F}_{M\perp}^* \times \mathbf{F}_{M\perp})_z}, \tag{7.86}$$

where $z$ indicates the axis along which the incident neutrons are polarized. The values of the nuclear structure factors $F_N$ that appear in (7.86) are calculated from a model for the crystal structure which is usually refined from unpolarized neutron diffraction measurements made on the same crystal and at the same temperature as the polarized beam measurement. As well as providing the atomic positions and thermal parameters needed to calculate the $F_N$, the single-crystal structure refinement also determines the parameters needed to correct the polarized beam intensities for the effects of attenuation and extinction.

Equation (7.86) is completely general, but for many crystal structures it can be simplified. First of all, the chiral term (the last term in the numerator and denominator) vanishes if either $\mathbf{F}_{M\perp}^*$ is parallel to $\mathbf{F}_{M\perp}$, or if $\mathbf{P} \perp \mathbf{G}$. If the chiral term is not zero but $|\mathbf{F}_{M\perp}| \ll |F_N|$ for all $\mathbf{G}$ then it may well be a good approximation to neglect the chiral term anyway. Second, if the crystal structure has a centre of inversion symmetry then $F_N^* = F_N$ and $\mathbf{F}_{M\perp}^* = \mathbf{F}_{M\perp}$, i.e. the nuclear and magnetic structure factors are real and the chiral term is zero. Third, if an external magnetic field aligns the magnetization in the vertical direction so that $\mathbf{F}_M$ is parallel to $z$ for all $\mathbf{G}$ then $(\mathbf{F}_{M\perp})_z = F_M \sin^2 \alpha$, where $F_M = |\mathbf{F}_M|$ and $\alpha$ is the angle between the $z$ axis and $\mathbf{G}$. This relation is shown in Fig. 7.30, where it can also be seen that $F_{M\perp} = F_M \sin \alpha$.

When all three conditions are fulfilled, as well as ideal polarization, eqn (7.86) simplifies to

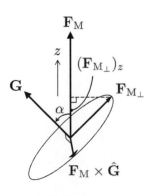

**Fig. 7.30** Geometric relation between $\mathbf{F}_M$, $\mathbf{F}_{M\perp}$ and $(\mathbf{F}_{M\perp})_z$ for the case when $\mathbf{F}_M$ is aligned along the $z$ axis.

$$R = \frac{1 + (\gamma^2 + 2\gamma) \sin^2 \alpha}{1 + (\gamma^2 - 2\gamma) \sin^2 \alpha}, \tag{7.87}$$

where $\gamma = (F_M/F_N)$. Equation (7.87) is plotted in Fig. 7.31 for two values of $\alpha$. The equation is quadratic in $\gamma$, and for a certain range of $R > 0$ there are two real roots. Identification of the correct root is not a problem when the nuclear structure factor is larger than the magnetic structure factor, as it often is, because in this case the root with $|\gamma| < 1$ is the correct one. However, if there is any doubt then the correct solution can be identified by changing the temperature or magnetic field so as to vary $F_M$. For example, let us suppose that $F_N(\mathbf{G}) > 0$ for a particular $\mathbf{G}$ and a measurement with $\mathbf{P} \perp \mathbf{G}$ gives $R(\mathbf{G}) = 3$. The two solutions to eqn (7.87) with $\alpha = 90°$ are $\gamma = 0.268$ and $3.732$. Now suppose we apply a larger magnetic field so as to increase $F_M$. If $R$ *increases* then $\gamma < 1$, whereas if $R$ *decreases* then $\gamma > 1$ (see Fig. 7.31).

When the magnetic amplitude is much smaller than the nuclear amplitude, the formula for $R$ in eqn (7.87) has the approximate form $R \simeq 1 + 4\gamma \sin^2 \alpha$. Therefore, in a polarized beam measurement the measured quantity $(R)$ depends approximately linearly on $\gamma$, in contrast to an unpolarized measurement of a Bragg peak in which the magnetic fraction of the total intensity goes as the square of $\gamma$. This explains the sensitivity of the polarized-beam method. In favourable cases it is possible to measure $R$ with a precision of one part in $10^4$, which means $F_M/F_N$ ratios as small $10^{-4}$ can be obtained.

There are two further advantages of the polarized-beam method. First, $F_M(\mathbf{G})$ is determined completely from (7.87), including the phase. Second, the flipping ratios can be measured with higher precision and more efficiently than the integrated intensity of a Bragg peak. This is because the signal is measured at the maximum of the peak and there is no need to move the sample during the measurement.

Having obtained a set of magnetic structure factors one can proceed to construct the spatial distribution of magnetization in the unit cell. There are three distinct approaches that can be used.

## 1. Direct Fourier inversion

The magnetization $\mathbf{M}(\mathbf{r})$ is a function that has the periodicity of the crystal lattice, and the magnetic structure factors $\mathbf{F}_M(\mathbf{G})$ are the Fourier coefficients of $\mathbf{M}(\mathbf{r})$ — see eqn (7.85). For simplicity we take $\mathbf{F}_M(\mathbf{G})$ to be parallel to the neutron polarization direction for all $\mathbf{G}$ so that $\mathbf{F}_M(\mathbf{G})$ and $\mathbf{M}(\mathbf{r})$ can be represented by scalar functions which are related by direct Fourier inversion,

$$pM(\mathbf{r}) = \frac{1}{v_0} \sum_{\mathbf{G}} F_M(\mathbf{G}) \exp(-i\mathbf{G} \cdot \mathbf{r}), \tag{7.88}$$

where $p = \gamma r_0/2\mu_B$, $v_0$ is the volume of the unit cell and the summation is over the complete (infinite) set of reciprocal lattice vectors $\mathbf{G}$. If the summation only includes reciprocal lattice vectors $\mathbf{G}_{2d}$ which lie in the plane perpendicular to the neutron polarization direction then the Fourier summation yields the magnetization projected down the polarization axis,[49]

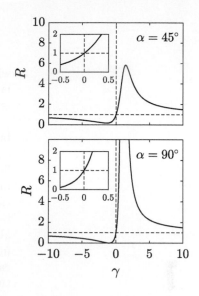

**Fig. 7.31** Relation between $R$ and $\gamma$ given by eqn (7.87) for $\alpha = 45°$ and $\alpha = 90°$. The inserts show the region near the origin.

[49] Note that in (7.89), $M$ is the magnetic moment per unit area.

$$pM(\mathbf{r}_{2\mathrm{d}}) = \frac{1}{s_0} \sum_{\mathbf{G}_{2\mathrm{d}}} F_{\mathrm{M}}(\mathbf{G}_{2\mathrm{d}}) \exp(-\mathrm{i}\mathbf{G}_{2\mathrm{d}} \cdot \mathbf{r}_{2\mathrm{d}}), \qquad (7.89)$$

where $\mathbf{r}_{2\mathrm{d}}$ is a position vector in the two-dimensional (2d) real-space lattice conjugate to the 2d reciprocal lattice defined by $\{\mathbf{G}_{2\mathrm{d}}\}$, and $s_0$ is the unit cell area of the 2d real-space lattice.

Direct Fourier transform provides a model-independent map of $M(\mathbf{r})$, but the method suffers from a number of shortcomings. Firstly, the experimental data are incomplete, being limited by the set of Bragg reflections that are both experimentally accessible and strong enough to measure. Any missing Fourier coefficients in (7.88) will cause distortions in the calculated $M(\mathbf{r})$. Second, there is no straightforward way to take into account the experimental uncertainties in the data. Instead, all $F_{\mathrm{M}}(\mathbf{G})$ are equally weighted in the Fourier transform. Finally, the method can be applied only if both the amplitude and phase of $F_{\mathrm{M}}(\mathbf{G})$ can be determined unambiguously. This rules out, for example, structures which lack a centre of symmetry.

## 2. Modelling M(r) in real space

The shortcomings of the Fourier transform method can be avoided if one has a model for $\mathbf{M}(\mathbf{r})$ and refines the parameters of the model through a comparison between the observed and calculated magnetic structure factors or flipping ratios. For example, one can base a model on the expressions developed in Section 6.3.2 for the anisotropic magnetic form factor derived from the electronic wave functions, but introduce adjustable coefficients to vary the total moment on each site, occupancy of different orbitals, radial distribution of the magnetization, etc.

Another useful way to parameterize $\mathbf{M}(\mathbf{r})$ is via a multipole expansion of the magnetization on each site. For a scalar magnetization, the expansion takes the form

$$M(\mathbf{r}) = \sum_l f_l(r)\Big[a_{l0} Z_{l0}(\hat{\mathbf{r}}) + \sum_{m=1}^{l} \big\{ a_{lm}^{\mathrm{c}} Z_{lm}^{\mathrm{c}}(\hat{\mathbf{r}}) + a_{lm}^{\mathrm{s}} Z_{lm}^{\mathrm{s}}(\hat{\mathbf{r}}) \big\} \Big], \quad (7.90)$$

where $f_l(r)$ are functions that describe the radial distribution of the magnetization,[50] $Z_{lm}^{(\mathrm{c,s})}$ are real combinations of spherical harmonics, also known as *tesseral harmonics*,[51] which describe the angular distribution of magnetization as defined by the direction of $\mathbf{r}$, and $a_{lm}$ are adjustable parameters. The local symmetry determines which $a_{lm}$ are non-zero. Although the form of $f_l(r)$ is different for spin and orbital magnetization it will typically fall off as $\exp(-\xi_l r)$, and $\xi_l$ can be varied to allow for an expansion (or contraction) relative to the free ion. For a single magnetic ion the magnetic structure factor, eqn (7.85), is given by

$$F_{\mathrm{M}}(\mathbf{G}) = 4\pi p \sum_l \mathrm{i}^l \langle j_l \rangle \Big[ a_{l0} Z_{l0}(\hat{\mathbf{G}}) + \sum_{m=1}^{l} \big\{ a_{lm}^{\mathrm{c}} Z_{lm}^{\mathrm{c}}(\hat{\mathbf{G}}) + a_{lm}^{\mathrm{s}} Z_{lm}^{\mathrm{s}}(\hat{\mathbf{G}}) \big\} \Big],$$
$$(7.91)$$

[50] For localized electrons, the spin part of $f_l(r)$ would be the square of the radial wave function, and the orbital part would be the square of the function that describes the radial variation of the orbital current — see Section 6.3.2.

[51] Tesseral harmonics are defined by

$$Z_{l0} = Y_{l,0}$$
$$Z_{lm}^{\mathrm{c}} = \frac{1}{\sqrt{2}} \{ Y_{l,-m} + (-1)^m Y_{l,m} \}$$
$$Z_{lm}^{\mathrm{s}} = \frac{\mathrm{i}}{\sqrt{2}} \{ Y_{l,-m} - (-1)^m Y_{l,m} \},$$

where $m > 0$ and $Y_{l,m}$ are the spherical harmonics, eqns (C.125) and (C.126). $Z_{lm}^{\mathrm{c}}$ and $Z_{lm}^{\mathrm{s}}$ contain the angular factors $\cos(m\phi)$ and $\sin(m\phi)$, respectively.

where $\langle j_l(G) \rangle$ is the radial integral defined in (6.27) but with $f_l(r)$ in place of $R^2(r)$, and $Z(\hat{\mathbf{G}})$ indicates that the tesseral harmonic is a function of the spherical angles that specify the direction of $\mathbf{G}$. When there is more than one magnetic ion in the unit cell the magnetic structure factor is a sum over the single-ion structure factors taking into account the phase difference associated with the positions of each ion,

$$F_{\mathrm{M}}(\mathbf{G}) = \sum_j (F_{\mathrm{M}})_j \exp(-W_j) \exp(\mathrm{i}\mathbf{G} \cdot \mathbf{r}_j). \qquad (7.92)$$

The Debye–Waller factor $\exp(-W_j)$ allows for any thermal smearing of the magnetization distribution. The analysis procedure is to fit the magnetic structure factors calculated from (7.92) to the observed magnetic structure factors, with the $a_{lm}$ and $\xi_l$ as adjustable parameters. The resulting magnetization map is then constructed from (7.90).

### 3. Bayesian reconstruction of M(r) by the MaxEnt method

The maximum entropy (MaxEnt) method is a model-free algorithm for calculating the most probable magnetization map among all the possible maps that are compatible with an incomplete data set and its experimental uncertainties — see Papoular and Gillon (1990), and Markvardsen (2000). In this method, the probability of the map is not only governed by the level of agreement of the observed and calculated magnetic structure factors, but also depends on the so-called *prior probability* of the map, i.e. the probability of the map without reference to the data. A uniform (or flat) prior map is usually employed to bias against the creation of any particular magnetization distribution.

Unlike direct Fourier inversion, MaxEnt makes no assumptions about the values of any missing data points. It follows that features appear in the maximum entropy reconstruction only if there is statistically significant evidence for them in the data. Procedures exist which allow estimation of the experimental uncertainties associated with regions in MaxEnt reconstructed maps. This allows the experimentalist to quantify the credibility of a magnetization feature.

Figure 7.32 provides an example of how the flipping ratio method can be used to study $\mathbf{M}(\mathbf{r})$. The heavy fermion superconductor $\mathrm{URu_2Si_2}$ undergoes a transition at $T_0 = 17.5\,\mathrm{K}$ to a phase with 'hidden order', so called because the order parameter is not detected in conventional diffraction measurements. In their study, Ressouche *et al.* (2012) measured a single crystal of $\mathrm{URu_2Si_2}$ in a magnetic field applied along the $c$ axis. They obtained a set of magnetic structure factors, shown in Fig. 7.32, from which they recovered $\mathbf{M}(\mathbf{r})$ by the MaxEnt method. The magnetization around the U sites is seen to be elongated along the [110] and [1$\bar{1}$0] directions. This anisotropy was not observed at temperatures above $T_0$, and calculations with an atomic model showed that the anisotropy that appears in the hidden order phase is associated with a freezing of rank-5 (triakontadipole) multipoles.

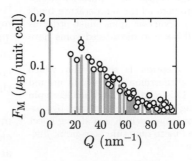

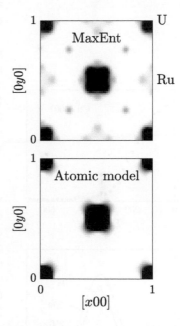

**Fig. 7.32** Magnetization distribution in $\mathrm{URu_2Si_2}$ at $T = 2\,\mathrm{K}$ measured by half-polarized neutron diffraction with a magnetic field of $9.6\,\mathrm{T}$ applied along the tetragonal $c$ axis (Ressouche *et al.*, 2012). The upper panel shows the measured magnetic structure factors (circles) together with calculated ones (vertical bars) from a MaxEnt reconstruction of $\mathbf{M}(\mathbf{r})$. The middle panel is the MaxEnt reconstruction, projected down the $c$ axis, and the lower panel shows $\mathbf{M}(\mathbf{r})$ calculated from an atomic model of the $\mathrm{U^{4+}}$ wave function. (Data courtesy of Eric Ressouche.)

# 7.7  Nuclear magnetism

[52]There are several equivalent statements of the third law. In the present context, it requires that the entropy of a perfect crystal of a pure substance tends to zero as the temperature tends to zero. Nuclear spins must order as $T \to 0$ because disordered spins have entropy.

[53]An exception is solid $^3$He, which orders antiferromagnetically at around 1 mK. This relatively high nuclear spin ordering temperature is caused by particle exchange, which is a strong effect in $^3$He due to quantum zero-point fluctuations.

According to the Third Law of Thermodynamics,[52] a system of nuclear spins must inevitably undergo a transition into an ordered phase with zero entropy. This ordering, which typically occurs at sub-$\mu$K temperatures,[53] can be detected by a number of techniques, including nuclear magnetic resonance, bulk susceptibility and neutron diffraction. The advantage of neutrons over other probes is that they can determine the spatial arrangement of the nuclear spins. The experiments are extremely challenging, and require a specially designed cryostat to access the extremely low temperatures where ordering takes place.

The principle of neutron diffraction from nuclear magnetic order is much the same as that from electronic magnetic order. When the nuclear spins undergo long-range ordering the nuclear interaction potential acquires an additional periodicity from which neutrons diffract, and the nuclear spin incoherent scattering, which is essentially $Q$-independent in the disordered phase, condenses into a series of Bragg peaks.

The nuclear interaction potential actually has two spin-dependent parts, one from the magnetic dipole interaction with the nuclear magnetic moment, and the other due to the spin-dependent part of the strong nuclear force (see Sections 1.8.2 and 4.1.2). The former is very small ($\sim 10^{-3}\mu_{\mathrm{B}}$) and hence difficult to detect by neutron diffraction, but the latter is generally a sizable fraction of the nuclear scattering length. The scattering amplitude is given by the spin-dependent scattering length, which from (4.11) may be written

$$b = A + B\mathbf{s}_{\mathrm{n}} \cdot \mathbf{I}, \qquad (7.93)$$

where $\mathbf{s}_{\mathrm{n}}$ is the neutron spin, and $\mathbf{I}$ is the nuclear spin. The coefficient $B$ is proportional to $b_+ - b_-$, the difference between the scattering amplitude when the neutron spin is parallel and antiparallel to the nuclear spin. The magnitude and sign of $B$ can be measured by nuclear diffraction, so providing a quantitative test of nuclear models.

General expressions for nuclear and magnetic scattering are given in Chapter 4 (in particular, see Table 4.1). The differential cross-section for unpolarized neutron scattering from nuclear spin order is given by

$$\frac{\mathrm{d}\sigma}{\mathrm{d}\Omega} = \frac{1}{4}B^2 \left| \sum_j \langle \mathbf{I}_j \rangle \, \exp(\mathrm{i}\mathbf{Q} \cdot \mathbf{R}_j) \, \exp(-W_j) \right|^2, \qquad (7.94)$$

Unlike electronic magnetic scattering, the neutron couples to all components of the nuclear spin, not just those perpendicular to $\mathbf{Q}$.

The first neutron diffraction studies of nuclear spin order where performed on LiH (Roinel *et al.*, 1978) and $^3$He (Benoit *et al.*, 1985). Diffraction data on nuclear ordering in Cu are shown in Fig. 7.33 (Jyrkkiö *et al.*, 1988). The Cu nucleus has a spin $I = \frac{3}{2}$ which undergoes antiferromagnetic ordering at a temperature $T_{\mathrm{N}} \simeq 60\,\mathrm{nK}$. The experiments established that the nuclear spins order in ferromagnetic planes normal to the cubic $a$ axis, with the spin direction reversed on neighbouring planes (a so-called type-I antiferromagnet).

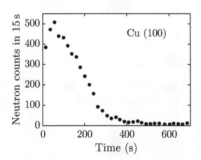

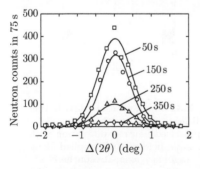

**Fig. 7.33** Neutron diffraction from nuclear spin order in Cu. To reach temperatures below $T_{\mathrm{N}}$ the nuclei were cooled by adiabatic nuclear demagnetization, and diffraction from the (100) antiferromagnetic Bragg peak was measured as a function of time and scattering angle $2\theta$ as the nuclei warmed up. (Adapted from Jyrkkiö *et al.*, 1988.)

# Chapter summary

- The basic properties of magnetic order are reviewed, including magnetic unit cells, propagation vectors, and magnetic domains.

- Common types of magnetic ordering include ferromagnets, ferrimagnets, antiferromagnets, and various non-collinear structures.

- Magnetic diffraction from a magnetically ordered crystal results in Bragg peaks at $\mathbf{Q} = \mathbf{G} \pm \mathbf{q}$, where $\mathbf{G}$ is a reciprocal lattice vector of the chemical lattice and $\mathbf{q}$ is a propagation vector.

- For unpolarized neutrons, the Bragg peak intensities are proportional to $|\mathbf{F}_{M\perp}|^2$, where $\mathbf{F}_{M\perp} = \hat{\mathbf{Q}} \times (\mathbf{F}_M \times \hat{\mathbf{Q}})$ and $\mathbf{F}_M$ is the magnetic structure factor. In the dipole approximation,

$$\mathbf{F}_M(\mathbf{Q}) = p \sum_d \mathbf{m}_{\mathbf{q},d} f_d(Q) \exp(-W_d) \exp(i\mathbf{Q} \cdot \mathbf{d}), \quad (\mathbf{Q} = \mathbf{G} \pm \mathbf{q}),$$

where $p = \gamma r_0 / 2\mu_B$, and the Fourier amplitude $\mathbf{m}_{\mathbf{q},d}$ represents the amplitude and phase of the ordered moment on site $d$ in the chemical unit cell for propagation vector $\mathbf{q}$.

- If a magnetic structure can be described by a magnetic unit cell, then magnetic Bragg peaks occur at $\mathbf{Q} = \mathbf{G}_m$, where $\mathbf{G}_m$ is a reciprocal lattice vector of the magnetic lattice, and

$$\mathbf{F}_M(\mathbf{Q}) = p \sum_d \boldsymbol{\mu}_d f_d(Q) \exp(-W_d) \exp(i\mathbf{Q} \cdot \mathbf{d}), \quad (\mathbf{Q} = \mathbf{G}_m).$$

The summation is now over sites in the magnetic unit cell, and $\boldsymbol{\mu}_d$ is the magnetic moment on site $d$.

- Diffuse magnetic scattering is an important feature of frustrated magnets, systems close to magnetic phase transitions, and systems in which long-range magnetic order is interrupted by defects.

- In the static approximation, the energy-integrated diffuse magnetic scattering gives the magnetic structure factor $S(\mathbf{Q})$, which is proportional to the equal-time, spin–spin correlation function.

- Polarized neutrons have many uses in magnetic structure determination, including (i) separating nuclear and magnetic scattering, (ii) measuring spin directions, (iii) probing vector spin chirality, (iv) measuring magnetization distributions, (v) distinguishing multi-$\mathbf{q}$ structures from a set of single-$\mathbf{q}$ magnetic domains.

- The principal polarized-neutron methods for diffraction studies are (i) uniaxial polarization analysis, (ii) the half-polarized set-up, (iii) spherical neutron polarimetry, (iv) XYZ polarization analysis, which is an implementation of the uniaxial method useful for separating magnetic diffuse scattering from nuclear scattering.

- Neutron diffraction from nuclear spin ordering is described.

# Further reading

General accounts of magnetic structures and magnetic neutron diffraction can be found in Chatterji (2006), Ressouche (2014*b*), Garlea and Chakoumakos (2015), and Rossat-Mignod (1987), as well as in the classic textbooks by Goodenough (1963), Izyumov and Ozerov (1970), and Izyumov *et al.* (1991). The key reference work for rare earth magnetism is Jensen and Mackintosh (1991). The Bilbao Crystallographic Server (http://www.cryst.ehu.es/) and the ISOTROPY software suite (http://iso.byu.edu) are deep resources of software for crystallographic and magnetic symmetry, including neutron diffraction. Applications of polar-

ized neutron diffraction in magnetism are reviewed by Ressouche (2014*a*). Experiments to detect nuclear magnetism are reviewed in Glättli and Goldman (1987), and Steiner and Siemensmeyer (2015). The book by Chatterji (2006) contains excellent chapters on spherical neutron polarimetry and on the determination of magnetization distributions with half-polarized beam diffraction, by P.J. Brown and J. Schweizer, respectively. For a summary of current research in frustrated magnetism, see Lacroix *et al.* (2011), and for a short account of magnetic critical phenomena, see Collins (1989).

# Exercises

(7.1) Verify that the triangular 120° magnetic structure shown in Fig. 7.1(d) has propagation vector $\mathbf{q} = (\frac{1}{3}, \frac{1}{3}, 0)$. Explain why it is a single-$\mathbf{q}$ structure, and identify a primitive magnetic unit cell.

(7.2) Some magnetic structures are chiral, i.e. have a definite handedness. Which of the incommensurate structures shown in Fig. 7.2 are chiral?

(7.3) $MnF_2$ has a body-centred tetragonal lattice with Mn atoms located at the corners and in the centre of the tetragonal unit cell. Below 70 K the Mn spins order antiferromagnetically, with their axes parallel to the $c$ axis and spins at the centre of the unit cell aligned opposite to those at the corners. Why is the magnetic propagation vector $\mathbf{q} = 0$? Calculate the magnetic structure factors $\mathbf{F}_M(\mathbf{G})$ for $\mathbf{G} = h\mathbf{a}^* + k\mathbf{b}^* + l\mathbf{c}^*$ and $h, k, l$ integers, and hence show that magnetic reflections are allowed when $h + k + l$ is an odd integer. Will all the magnetic reflections which satisfy this selection rule have a non-zero intensity?

(7.4) Derive the magnetic structure factor for $RMnO_3$, given in (7.21).

(7.5) Show that the $+\mathbf{q}$ and $-\mathbf{q}$ Fourier coefficients of the transverse helical magnetic structure, depicted in Fig. 7.2(b) and described by eqn (7.28), are

$$\mathbf{m}_{\pm\mathbf{q}} = \tfrac{1}{2}\mu(e^{\mp i\phi}, e^{\mp i(\phi - \pi/2)}, 0).$$

Hence, prove that the diffraction intensities for the transverse helix are the same as for the longitudinal spin density wave, eqn (7.27), except with orientation factor $(1 + \hat{Q}_z^2)$ instead of $(1 - \hat{Q}_z^2)$. Repeat the exercise to show that the orientation factor for the cycloidal structure, Fig. 7.2(c) and eqn (7.29), is $(1 + \hat{Q}_y^2)$.

(7.6) Uniaxial polarization analysis was performed at two magnetic Bragg peaks which have scattering vectors $\mathbf{Q}_1$ and $\mathbf{Q}_2$ and lie in the $X$–$Y$ plane, as shown in Fig. 7.14. Use the expressions in Table 7.2 to derive the following results:

$$\frac{\mu_X^2}{\mu_Z^2} = \frac{1}{T}(R_1 \sin^2 \beta_2 - R_2 \sin^2 \beta_1)$$

$$\frac{\mu_Y^2}{\mu_Z^2} = \frac{1}{T}(R_2 \cos^2 \beta_1 - R_1 \cos^2 \beta_2),$$

where $R_1$ and $R_2$ are the intensity ratios $I_z^{SF}/I_y^{SF}$ at $\mathbf{Q}_1$ and $\mathbf{Q}_2$, respectively, and $T = \cos^2 \beta_1 \sin^2 \beta_2 - \sin^2 \beta_1 \cos^2 \beta_2$. Measurements on $La_{3/2}Sr_{1/2}CoO_4$ at $T = 2\,K$ were made at reflections $\mathbf{Q}_1$ and $\mathbf{Q}_2$ with $\beta_1 = 9.4°$ and $\beta_2 = 80.2°$. The counts in the $I_y^{SF}$ and $I_z^{SF}$ channels were 23,461 and 9,394 at $\mathbf{Q}_1$, and 24,156 and 1,102 at $\mathbf{Q}_2$. Check that these results are consistent with Fig. 7.14.

# Magnetic Excitations

8

We now turn our attention to magnetic excitations and their measurement by neutron spectroscopy. We proceed in two stages. First, we consider the excitations of uncoupled paramagnetic ions, i.e. ions which have magnetic moments in their ground and/or excited states but which are magnetically isolated from one another. The magnetic spectrum of such ions is governed by the electronic structure of the ion, and by interactions with the ligand (crystal) field and any external magnetic field that might be present. Studies of single-ion magnetic excitations by neutron spectroscopy can provide information on magnetic anisotropy due to the local environment, and can be used to determine the wave functions of the magnetic ground state and excited states.

In the second stage we consider interactions, which introduce spatial correlations and give rise to cooperative magnetic excitations. The excitations of coupled magnetic ions are dispersive, i.e. the frequency of the magnetic modes varies with their wavevector. Inelastic neutron scattering measurements can be used to investigate the many diverse types of cooperative modes that occur in magnetic materials, and to determine the magnetic interactions which stabilize the ground state. Information can also be gleaned on the interplay between magnetic and other electronic or structural degrees of freedom.

The cross-section for magnetic neutron scattering was developed in Section 6.1. The double-differential cross-section measured in an inelastic scattering experiment is related to the response function $S(\mathbf{Q}, \omega)$ by

$$\frac{\mathrm{d}^2\sigma}{\mathrm{d}\Omega \mathrm{d}E_\mathrm{f}} = \frac{k_\mathrm{f}}{k_\mathrm{i}} \left( \frac{\gamma r_0}{2\mu_\mathrm{B}} \right)^2 S(\mathbf{Q}, \omega), \tag{8.1}$$

where $(\gamma r_0/2)^2 = 72.65\,\mathrm{mb}$. In this chapter, we shall obtain expressions for $S(\mathbf{Q}, \omega)$ for a variety of different magnetic systems of interest.

## 8.1 Single-ion magnetic excitations

### 8.1.1 Crystal field excitations

The electrons in a magnetic ion experience a potential from the surroundings made up of (i) an electrostatic (Coulomb) field from the neighbouring ions, and (ii) an effective field associated with covalent bonding. Together, these are known as the *ligand field*, although they are often colloquially called the *crystalline electric field*, or *crystal field* for short.[1]

---

[1] Section C.4.2 describes in more detail the effect of the crystal field on magnetic ions, with particular emphasis on ions containing unfilled $d$ and $f$ shells.

The crystal field potential is spatially varying, $V_{\mathrm{CF}} = V_{\mathrm{CF}}(x, y, z)$, and so it acts on the orbital part of the electronic wave functions. Operators in $V_{\mathrm{CF}}$ do not necessarily commute with the orbital angular momentum operator $\mathbf{L}$, so the eigenstates of $\mathbf{L}$ are not necessarily simultaneous eigenstates of $V_{\mathrm{CF}}$. Therefore, the crystal field will in general split the orbital degeneracy of the electrons.

We assume the magnetism of the ion derives from $n$ equivalent electrons (i.e. $n$ electrons in the same shell and with the same one-electron orbital angular momentum quantum number $l$). The crystal field Hamiltonian for the magnetic ion is obtained by integration of the crystal field potential over the distribution of electrons. The Hamiltonian can be stated in Cartesian coordinates,[2] but for most applications it is usually simpler to express it in terms of spherical operators,

$$\mathcal{H}_{\mathrm{CF}} = \sum_{k=0}^{2l} \sum_{q=-k}^{+k} B_q^k C_q^{(k)}, \qquad (8.2)$$

[2]See Hutchings (1964).

where the $B_q^k$ are *crystal field parameters*, and $k$ and $q$ are integers. The radial part of the potential is contained in the crystal field parameters, and the angular dependence is described by the $C_q^{(k)}$, which are known as *Wybourne tensor operators* and given by

$$C_q^{(k)}(\theta, \phi) = \sqrt{\frac{4\pi}{2k+1}}\, Y_{k,q}(\theta, \phi), \qquad (8.3)$$

[3]See Section C.2.1.

where $Y_{k,q}(\theta, \phi)$ are the spherical harmonic functions.[3] The local point symmetry at the site of the magnetic ion determines which Wybourne operators are required in $\mathcal{H}_{\mathrm{CF}}$, i.e. which $k$ and $q$ need to be included in the sum in eqn (8.2). Only even $k$ terms are required because the matrix elements of odd $k$ terms vanish for a single $l^n$ configuration. For practical implementations it is convenient to rewrite (8.2) as[4]

[4]Comparing (8.2) with (8.4)–(8.5), we see that for $q > 0$,

$$B_{-q}^k = B_q^k(\mathrm{c}) + \mathrm{i}B_q^k(\mathrm{s})$$
$$B_q^k = (-1)^q \{ B_q^k(\mathrm{c}) - \mathrm{i}B_q^k(\mathrm{s}) \}.$$

$$\mathcal{H}_{\mathrm{CF}} = \sum_{k=0}^{2l} B_0^k C_0^{(k)} + \sum_{k=2}^{2l} \sum_{q=1}^{k} B_q^k(\mathrm{c}) C_q^{(k)}(\mathrm{c}) + B_q^k(\mathrm{s}) C_q^{(k)}(\mathrm{s}), \qquad (8.4)$$

where

$$C_q^{(k)}(\mathrm{c}) = C_{-q}^{(k)} + (-1)^q C_q^{(k)}$$
$$C_q^{(k)}(\mathrm{s}) = \mathrm{i}\{ C_{-q}^{(k)} - (-1)^q C_q^{(k)} \} \qquad (q > 0). \qquad (8.5)$$

From the definition of the spherical harmonics, (C.125) and (C.126), it can be seen that the $q \neq 0$ tensor operators are grouped here into real-valued combinations proportional to $\cos(q\phi)$ and $\sin(q\phi)$, respectively. The associated parameters $B_q^k(\mathrm{c})$ and $B_q^k(\mathrm{s})$ are also real.

We will label the eigenstates of $\mathcal{H}_{\mathrm{CF}}$ by $\Gamma_i$. This notation is chosen because the eigenstates of $\mathcal{H}_{\mathrm{CF}}$ transform according to the irreducible representations of the crystal field point group.[5] From (6.11), the single-ion response function for unpolarized neutrons is given by

[5]The irreducible representations of a point symmetry group are often denoted by $\Gamma_1, \Gamma_2, \ldots$. The transformation properties of the irreducible representation can be found in standard tables, e.g. Koster *et al.* (1963).

$$S(\mathbf{Q}, \omega) = \exp(-2W) \sum_i p_i \sum_j |\langle \Gamma_j | \mathbf{M}_\perp(\mathbf{Q}) | \Gamma_i \rangle|^2 \delta(E_j - E_i - \hbar\omega), \qquad (8.6)$$

where the first summation is over the initial states $\Gamma_i$ with thermal population $p_i$, and the second summation is over the final states $\Gamma_j$. We have included the Debye–Waller factor $\exp(-2W)$ to take into account the thermal motion of the magnetic ion[6] (see Section 5.1.4).

The magnetic spectrum described by (8.6) consists of a series of sharp peaks centred on the transition energies $E_j - E_i$. The peak intensities depend on the thermal population of the initial state and on the matrix elements $\langle \Gamma_j | \mathbf{M}_\perp(\mathbf{Q}) | \Gamma_i \rangle$. Before considering the general case, we start with two useful approximations which apply when the initial and final states are contained in the same $l^n$ configuration and belong to a single $LS$-coupling term (see Section C.4).

## Weak crystal field

In the weak crystal field limit [case (a) described in Section C.4.2] the spin–orbit interaction couples $\mathbf{L}$ and $\mathbf{S}$ to form $\mathbf{J}$, and this coupling is not broken by the crystal field. This is the picture often used to describe ions with unfilled $f$ shells (lanthanides and actinides). When this approximation applies, the crystal field Hamiltonian can be written in a form which is simpler to diagonalize. The simplification is to replace the true multi-electron ground state of the ion by a single $J$ level from a pure $LS$ term, and then to define an effective Hamiltonian,

$$\mathcal{H}_{\mathrm{CF}} = \sum_{k=0}^{2l} B_k^0 O_k^0 + \sum_{k=2}^{2l} \sum_{q=1}^{k} B_k^q(\mathrm{c}) O_k^q(\mathrm{c}) + B_k^q(\mathrm{s}) O_k^q(\mathrm{s}), \qquad (8.7)$$

which acts on the $2J+1$ basis states $|J, M_J\rangle$ of the $J$ manifold to produce the same eigenstates as the true crystal field Hamiltonian acting on the same $J$ manifold. The $O_k^q$ in (8.7) are known as *Stevens operator equivalents*, or *Stevens operators* for short. Stevens operators transform in the same way as Wybourne operators but are written in terms of angular momentum operators.[7] In the weak-field limit, the Stevens operators are built from the components of $\mathbf{J}$. For example,[8]

$$O_2^0 = 3J_z^2 - J(J+1)$$
$$O_2^2(\mathrm{c}) = \frac{1}{2}(J_+^2 + J_-^2) = J_x^2 - J_y^2 \qquad (8.8)$$
$$O_2^2(\mathrm{s}) = \frac{-\mathrm{i}}{2}(J_+^2 - J_-^2) = J_x J_y + J_y J_x.$$

Note that the crystal field parameters that appear in (8.7) are not the same as those in (8.4).[9] The two sets of parameters are related by the multiplicative factors given in Table 8.1.

If we add a Zeeman interaction with an external magnetic flux density $\mathbf{B}$ to the crystal field Hamiltonian (8.7) then we obtain the complete single-ion Hamiltonian in the weak-field limit,

$$\mathcal{H} = \mathcal{H}_{\mathrm{CF}} + g_J \mu_{\mathrm{B}} \mathbf{J} \cdot \mathbf{B}. \qquad (8.9)$$

[6] At low temperatures, where measurements are usually made, the thermal smearing of the ionic position is much smaller than the size of the ion, so the $Q$ dependence of the Debye–Waller factor can be neglected compared with the $Q$ dependence of the magnetic form factor. Therefore, $\exp(-2W) \simeq 1$ at the $Q$ values where measurements are typically made.

[7] Some commonly occurring Stevens operators are given in Table VIII of Hutchings (1964). For a complete listing, see the McPhase Users Manual at www.mcphase.de

[8] In the literature, the cosine operators $O_k^q(\mathrm{c})$ are usually written $O_k^q$

[9] It is unfortunate that $B$ is used for both the Wybourne and Stevens parameters in much of the literature. Note, however, that the positions of the $k$ and $q$ indices are reversed, which helps to distinguish them.

**Table 8.1** Multiplicative factors $\lambda_{k0}$ and $\lambda_{kq}$ in the relations $B_k^0 = \lambda_{k0}\theta_k B_0^k$ and $B_k^q(\mathrm{c,s}) = \lambda_{kq}\theta_k B_q^k(\mathrm{c,s})$ between the Stevens crystal field parameters in (8.7) and the corresponding Wybourne parameters in (8.4). The $\theta_k$ are numerical coefficients which have been tabulated for the ground states of the trivalent $4f$ ions and given in closed form for the $3d$ ions — see Tables VI & VII of Hutchings (1964).

| | | $q$ | | | | | | |
|---|---|---|---|---|---|---|---|---|
| | | 0 | 1 | 2 | 3 | 4 | 5 | 6 |
| | 2 | $\frac{1}{2}$ | $\sqrt{6}$ | $\frac{1}{2}\sqrt{6}$ | | | | |
| $k$ | 4 | $\frac{1}{8}$ | $\frac{1}{2}\sqrt{5}$ | $\frac{1}{4}\sqrt{10}$ | $\frac{1}{2}\sqrt{35}$ | $\frac{1}{8}\sqrt{70}$ | | |
| | 6 | $\frac{1}{16}$ | $\frac{1}{8}\sqrt{42}$ | $\frac{1}{16}\sqrt{105}$ | $\frac{1}{8}\sqrt{105}$ | $\frac{3}{16}\sqrt{14}$ | $\frac{3}{8}\sqrt{77}$ | $\frac{1}{16}\sqrt{231}$ |

Diagonalization of (8.9) in the $SLJM_J$ basis splits the $(2J+1)$-fold degeneracy of the $J$ manifold. The eigenfunctions may be written

$$|\Gamma\rangle = \sum_{M_J=-J}^{+J} a_{M_J}|M_J\rangle. \tag{8.10}$$

We have not displayed the $SLJ$ indices as they are the same for all terms in the expansion.

The neutron scattering intensities of transitions within a single $J$ manifold are usually described well by the dipole approximation, which, as discussed in Section 6.2.2, allows us to replace $\mathbf{M}_\perp(\mathbf{Q})$ in (8.6) by $-g_J\mu_\mathrm{B}f(Q)\mathbf{J}_\perp$, where $g_J$ is the Landé $g$-factor and $f(Q)$ is the dipole magnetic form factor, eqn (6.30). Hence,

$$S(\mathbf{Q},\omega) = g_J^2\mu_\mathrm{B}^2 f^2(Q)\,\mathrm{e}^{-2W}\sum_i p_i \sum_j |\langle\Gamma_j|\,\mathbf{J}_\perp\,|\Gamma_i\rangle|^2\delta(E_j - E_i - \hbar\omega). \tag{8.11}$$

The operator $\mathbf{J}_\perp$ selects the components of $\mathbf{J}$ perpendicular to $\mathbf{Q}$, so if a measurement is made on a single crystal then by varying the direction of $\mathbf{Q}$ one can learn about the anisotropy of the transition matrix elements. If, further, longitudinal polarization analysis is employed (see Section 4.5), then the transition matrix elements of the three independent components $J_x$, $J_y$ and $J_z$ can be determined separately. Initial measurements are often made with unpolarized neutrons on a powder sample, in which case the powder average makes

$$\langle|\langle\Gamma_j|\,\mathbf{J}_\perp\,|\Gamma_i\rangle|^2\rangle_\mathrm{av} = \frac{2}{3}\sum_\alpha |\langle\Gamma_j|\,J_\alpha\,|\Gamma_i\rangle|^2, \qquad (\alpha = x,y,z). \tag{8.12}$$

The transition matrix elements satisfy the sum rule

$$\begin{aligned}\sum_j |\langle\Gamma_j|\,\mathbf{J}_\perp\,|\Gamma_i\rangle|^2 &= \sum_j \langle\Gamma_i|\,\mathbf{J}_\perp\,|\Gamma_j\rangle\langle\Gamma_j|\,\mathbf{J}_\perp\,|\Gamma_i\rangle \\ &= \langle\Gamma_i|\,\mathbf{J}_\perp^2\,|\Gamma_i\rangle.\end{aligned} \tag{8.13}$$

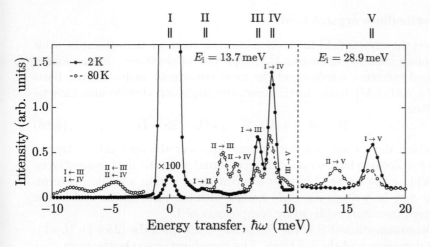

**Fig. 8.1** Spectrum of crystal field excitations of $Nd^{3+}$ in $NdPd_5Al_2$, (after Zubáč *et al.*, 2018). The spectrum was measured with a polycrystalline sample, and integrated over a wide range of scattering angle on a multidetector. Data from two different incident energies $E_i$ are shown, in order to resolve the complete spectrum. The energies of the five doublet levels are shown at the top, and each peak is labelled with the corresponding transition. (Data courtesy of J. Zubáč.)

To go from the first to the second line we have used the closure relation, eqn (C.19), which applies because the $|\Gamma_j\rangle$ form a complete set of basis states. In the case of a powder sample,

$$\langle\Gamma_i|\,\mathbf{J}_\perp^2\,|\Gamma_i\rangle = \frac{2}{3}\langle\Gamma_i|\,\mathbf{J}^2\,|\Gamma_i\rangle$$
$$= \frac{2}{3}J(J+1), \qquad (8.14)$$

and together with (8.13) this means that the total intensity of all the crystal field peaks (including any elastic scattering, if present) has a well defined value that depends only on $J$ and $Q$. At $Q = 0$, the form factor and Debye–Waller factor both equal 1, and substituting (8.14) and (8.13) into (8.11), and integrating over energy, we obtain

$$\int_{-\infty}^{\infty} S(0,\omega)\,\mathrm{d}(\hbar\omega) = \frac{2}{3}J(J+1)g_J^2\mu_B^2, \qquad \text{(powder)}. \qquad (8.15)$$

This sum rule[10], which applies for any temperature, is a useful check that all transitions have been accounted for in the measurement.

Figure 8.1 shows an example of the crystal field excitation spectrum of a lanthanide ion measured by neutron scattering. The tetragonal crystal field in $NdPd_5Al_2$ splits the $J = 9/2$ ground-state manifold of $Nd^{3+}$ into five doubly degenerate energy levels, as shown at the top of the figure. The neutron spectrum measured at $T = 2\,\mathrm{K}$ contains peaks due to transitions from the ground state doublet to the four excited doublets. These peaks appear in neutron energy-loss ($\hbar\omega > 0$). At 80 K, there is a significant thermal population of the first few excited levels, and the spectrum now contains peaks due to transitions starting from excited states as well as those from the ground state. The I → II transition is barely visible at 2 K, but the position of level II is firmly established from the 80 K spectrum which contains peaks due to transitions connecting level II to levels III, IV and V. Peaks observed in neutron energy-gain ($\hbar\omega < 0$) are due to transitions from higher to lower levels.

[10]Equation (8.15) is an example of the total moment sum rule, see (6.92).

## Intermediate crystal field

When the crystal field interaction is much stronger than the spin–orbit coupling [case (b) in Section C.4, typically $3d$ electrons], $\mathbf{J}$ is no longer a good quantum number and the most convenient multi-electron basis is the $SM_SLM_L$ basis. In this case, the single-ion Hamiltonian may be written

$$\mathcal{H} = \mathcal{H}_{\mathrm{CF}} + \lambda \mathbf{L} \cdot \mathbf{S} + \mu_{\mathrm{B}}(\mathbf{L} + 2\mathbf{S}) \cdot \mathbf{B}, \qquad (8.16)$$

in which the second and third terms represent the spin–orbit coupling and the Zeeman interaction with an external magnetic field, respectively. Providing the states of interest are contained within a single $LS$ term, $\mathcal{H}_{\mathrm{CF}}$ may be written like (8.7) except with the components of $\mathbf{J}$ in the Stevens operators replaced by components of $\mathbf{L}$.

Diagonalization of $\mathcal{H}$ in the $SM_SLM_L$ basis splits the $(2S+1)(2L+1)$-fold degeneracy of the $LS$ term. The eigenfunctions take the form

$$|\Gamma\rangle = \sum_{M_S=-S}^{+S} \sum_{M_L=-L}^{+L} a_{M_S M_L} |M_S M_L\rangle, \qquad (8.17)$$

and in the dipole approximation, the operator $\mathbf{M}_\perp(\mathbf{Q})$ in (8.6) is replaced either by $-g\mu_{\mathrm{B}} f(Q)\mathbf{S}_\perp$ or by $-2\mu_{\mathrm{B}}[\langle j_0\rangle \mathbf{S}_\perp + \frac{1}{2}(\langle j_0\rangle + \langle j_2\rangle)\mathbf{L}_\perp]$, depending on whether the orbital angular momentum is quenched or not in the absence of spin–orbit coupling (see Section 6.2.2). In the former case, the magnetic form factor $f(Q)$ is given by (6.32).

Section C.4.2 includes an example to illustrate how the electronic levels of a $3d$ ion are split by a crystal field of intermediate strength.

## General treatment

We now generalize to systems in which the initial and final states belong to the same $l^n$ configuration but contain components from different $LSJ$ manifolds. There are two situations where this is encountered. One is when the objective of the experiment is to measure transitions between states belonging to different $LSJ$ manifolds, often called[11] *intermultiplet transitions*. This topic is discussed in the next section. The other is when the spin–orbit coupling is sufficiently strong that it causes significant mixing of states from different $LS$ terms in the free ion. This is known as *intermediate coupling* (see Section C.2.2). A strong crystal field interaction may cause a further mixing of states from different $LS$ terms or from different $J$ levels within the same $LS$ term.

In both situations just described, the response function eqn (8.6) will contain transition matrix elements of the form

$$\langle SLJM_J | \mathbf{M}_\perp(\mathbf{Q}) | S'L'J'M_J' \rangle \qquad (8.18)$$

This matrix element can be calculated from the expressions given in Section 6.2. When the crystal field cannot be represented by Stevens operators in the weak or intermediate field approximations, the crystal field eigenstates must be obtained by diagonalization of $\mathcal{H}_{\mathrm{CF}}$ as given

[11] At least, in the neutron scattering literature.

in (8.2) or (8.4), including any magnetic fields, in the complete basis of $LS$-coupling or intermediate coupling single-ion states.[12]

The theory of neutron scattering from atomic electrons in the $jj$ coupling scheme has been formulated by Balcar and Lovesey (1991).

## 8.1.2 Intermultiplet transitions

Measurements of intermultiplet transitions, which have been performed for many years by optical spectroscopy, provide information on modifications to the intra-atomic electronic correlations caused by interactions with the environment in a solid. Neutron spectroscopy can provide complementary information to optical spectroscopy, but is particularly useful for studying metals since conducting samples cannot easily be probed with electromagnetic radiation. Neutron scattering has a number of other advantages: (i) absolute cross-sections can be measured, (ii) neutron scattering does not require corrections due to final-state effects, and (iii) neutrons are not surface sensitive.

At the same time, neutron intermultiplet spectroscopy is challenging because the transitions are often at high energies, usually above 100 meV, and the cross-sections are often very small, especially given the large $Q$ values enforced by kinematic constraints. High neutron incident energies, good energy resolution, and low backgrounds are needed. These conditions are achievable on spallation sources, where intermultiplet transitions at energies up to almost 2 eV have been measured (Osborn *et al.* 1991).

Intermultiplet transitions may be divided into two kinds: (1) *spin–orbit* or *inter-level* transitions, which are between different $J$ levels belonging to the same $LS$ term split by the spin–orbit interaction, and (2) *Coulomb* or *inter-term* transitions, which are between different $LS$ terms split by the Coulomb potential acting within the $l^n$ configuration. In both cases the states can be further split by the crystal field.

Measurements of spin–orbit transitions can provide additional data with which to determine the crystal field Hamiltonian. This is particularly useful when the number of observable intra-level transitions is small. An example is the spectrum of $PrO_2$ shown in Fig. 8.2 (the corresponding energy level diagram is given in Fig. C.2). The ground state of $Pr^{4+}$ has $J = 5/2$, which splits under the cubic crystal field into two eigenstates, labelled $\Gamma_8^-$ and $\Gamma_7^-$. The crystal field Hamiltonian requires two parameters, $B_0^4$ and $B_0^6$, so a measurement of the $\Gamma_8^-$ to $\Gamma_7^-$ transition alone is not sufficient. The next spin–orbit level has $J = 7/2$ and splits into three crystal field eigenstates: $\Gamma_6^-$, $\Gamma_7^-$ and $\Gamma_8^-$. Measurements of the transitions from the ground state to each of $\Gamma_6^-$, $\Gamma_7^-$ and $\Gamma_8^-$ made it possible to determine the two crystal field parameters uniquely.

The scattering cross-section for an intermultiplet transition between crystal-field-split levels, eqns (8.1) and (8.6), involves summation over many products of matrix elements of the form (8.18). The calculation is generally rather lengthy. However, the formulae simplify considerably in the limit where the crystal field splitting is negligible (i.e. cannot

[12]The SPECTRE program can do this for $f$ electrons, see https://xray.physics.ox.ac.uk/software.htm.

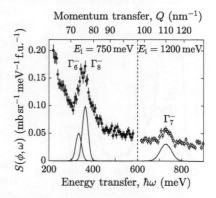

**Fig. 8.2** Intermultiplet transitions between the crystal-field-split $J = 5/2$ and $J = 7/2$ levels of $Pr^{4+}$ in $PrO_2$ (after Boothroyd *et al.*, 2001). The energy spectrum was obtained from time-of-flight data recorded at an average scattering angle $\langle \phi \rangle = 19°$ with incident energies $E_i = 750$ and $1200$ meV. The corresponding momentum transfer is shown at the top. Solid lines are cross-sections calculated from (8.6) with eigenstates obtained from the crystal field Hamiltonian (8.4) with $B_0^4 = -776$ meV and $B_0^6 = 207$ meV.

be resolved experimentally) and is much smaller than the spin–orbit splitting. In this limit, which applies to free ions and rare-earth ions in very weak crystal fields, we can sum over all states within the $SLJ$ and $S'L'J'$ manifolds, which means the double sum in (8.6) representing the average over initial states and sum over final states becomes,

$$\sum_i p_i \sum_j \cdots \quad \rightarrow \quad \sum_{M_J} \frac{1}{2J+1} \sum_{M'_J} \cdots, \qquad (8.19)$$

where $M_J$ and $M'_J$ are magnetic quantum numbers for the initial and final states, respectively. After summation over $M_J$ and $M'_J$ there is no longer any angular dependence in the cross-section. The result is (we omit the Debye–Waller factor for brevity),

$$S(\mathbf{Q}, \omega) = (2\mu_B)^2 \sum_K \frac{3}{K+2} \left\{ A(K, K+1) + B(K, K+1) \right\}^2$$
$$+ (2\mu_B)^2 \sum_K \frac{3}{2K+1} \{ B(K, K) \}^2. \qquad (8.20)$$

The $A$ and $B$ functions correspond to orbital and spin scattering, respectively. They have already been introduced in Sections 6.3.1 and 6.2, and general expressions for them are given in Section C.6. They depend on $l$ and the $SLJ$ and $S'L'J'$ quantum numbers of the initial and final states, and are linear combinations of the $\langle j_n(Q) \rangle$ radial integrals defined in eqn (6.27). Terms in the first line of (8.20) describe magnetic multipole scattering, and for these the summation index $K$ runs over even integers from 0 to $2l-2$ for $A$, and 0 to $2l$ for $B$. The second line is the electric multipole contribution, for which $K$ runs over even integers from 2 to $2l$.

The $A$ and $B$ functions satisfy certain selection rules, including

$$
\begin{aligned}
A(K, K') &= 0 \quad \text{unless} \quad \Delta S = 0, \\
B(K, K') &= 0 \quad \text{unless} \quad \Delta S = 0, \pm 1 \\
A(K, K') = B(K, K') &= 0 \quad \text{unless} \quad |\Delta J| \le K' \le J + J'
\end{aligned} \qquad (8.21)
$$

The electric multipole term $B(K, K)$ has additional constraints. It vanishes if $S = S'$, $L = L'$, and $J = J'$, and so does not contribute to the elastic scattering. It also vanishes at $Q = 0$ because there is no $\langle j_0 \rangle$ contribution (since $K \ge 2$).

The predictions of (8.20) are compared with experimental results for elemental praseodymium (Pr) in Fig. 8.3. Cross-sections for inter- and intra-multiplet transitions differ both in their magnitude and $Q$ dependence. Inter-level dipole transitions, for which $\Delta J = \pm 1$, are at least an order of magnitude weaker at $Q = 0$ than intra-level transitions, and fall off more rapidly with $Q$. Non-dipolar transitions have zero cross-section at $Q = 0$, and reach a maximum at intermediate $Q$ values typically in the range $50$–$150\,\mathrm{nm}^{-1}$. Figure 8.3 shows that the measured cross-section of the $^3H_4 \rightarrow {}^3H_5$ inter-level dipole transition of Pr agrees very well with the calculations. The inter-term transitions from $^3H$ to $^3F$ are systematically lower than predicted.

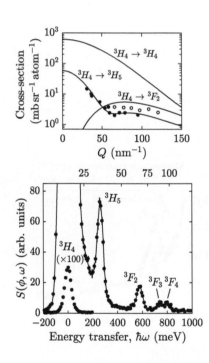

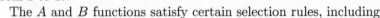

**Fig. 8.3** Intermultiplet transitions in Pr (after Taylor *et al.*, 1988). The lower figure is the energy spectrum obtained from time-of-flight data recorded at an average scattering angle $\langle \phi \rangle = 5°$ with incident energy $E_i = 1300\,\mathrm{meV}$. Peaks are labelled with the final state of the corresponding transition. The upper figure shows the calculated and measured $Q$ dependence of the cross-sections for the $^3H_4 \rightarrow {}^3H_5$ (filled circles) and $^3H_4 \rightarrow {}^3F_2$ (open circles) transitions. The calculated intra-level $^3H_4 \rightarrow {}^3H_4$ cross-section is also shown.

# 8.2   Magnetic clusters

We now turn to systems containing isolated clusters of a finite number of coupled magnetic moments. The excitations of an isolated magnetic cluster are localized, in the sense that they do not propagate, and so the spectrum consists of non-dispersive modes at discrete energies. At the same time, because the excitations are spread out over all the atoms in the cluster there are interference effects which produce a $\mathbf{Q}$ dependence characteristic of the cluster geometry.

Magnetic clusters are of fundamental interest as model systems for quantum effects, such as quantum tunneling of magnetization, quantum disordered ground states, and Bose–Einstein condensation of magnetic quasiparticles. An important class of materials in which magnetic clusters are realized is single-molecule magnets, which have potential applications in high-density information storage, magnetic refrigeration, and quantum computers.

## 8.2.1   Magnetic dimers

Magnetic dimers contain two coupled spins, and as such are the simplest class of magnetic cluster. We take the atoms carrying the spins to be centred at positions $-\mathbf{d}$ and $+\mathbf{d}$ (Fig. 8.4). To work out the magnetic spectrum we need $\mathbf{M}(\mathbf{Q})$, the Fourier transform of the magnetization. From eqns (6.17)–(6.18), we may write

$$\mathbf{M}(\mathbf{Q}) = \mathbf{M}_1(\mathbf{Q})\exp(-i\mathbf{Q}\cdot\mathbf{d}) + \mathbf{M}_2(\mathbf{Q})\exp(i\mathbf{Q}\cdot\mathbf{d}). \qquad (8.22)$$

For simplicity we will employ the dipole approximation and take the two atoms to be identical. Accordingly, we use (6.31) to replace $\mathbf{M}_j(\mathbf{Q})$ by $-g\mu_{\mathrm{B}}f(Q)\mathbf{S}_j$, where $f(Q)$ is the dipole form factor[13] and $\mathbf{S}_j$ is the spin operator for site $j$. Hence, we write (8.22) as

$$\mathbf{M}(\mathbf{Q}) = -g\mu_{\mathrm{B}}f(Q)\left\{\mathbf{S}_1\exp(-i\mathbf{Q}\cdot\mathbf{d}) + \mathbf{S}_2\exp(i\mathbf{Q}\cdot\mathbf{d})\right\}. \qquad (8.23)$$

If we assume a Heisenberg coupling between the spins and include an external magnetic field of flux density $B$ applied along the $z$ direction then the effective spin Hamiltonian for the dimer may be written

$$\mathcal{H} = J\mathbf{S}_1\cdot\mathbf{S}_2 + g\mu_{\mathrm{B}}B(S_1^z + S_2^z). \qquad (8.24)$$

The eigenstates of $\mathcal{H}$ are simultaneous eigenstates of $\mathbf{S}^2$ and $S^z$, where $\mathbf{S} = \mathbf{S}_1 + \mathbf{S}_2$ and $S^z = S_1^z + S_2^z$, and the eigenvalues are

$$E_{S,M_S} = J\left\{\tfrac{1}{2}S(S+1) - S_j(S_j+1)\right\} + g\mu_{\mathrm{B}}BM_S, \qquad (8.25)$$

with $-S \le M_S \le S$ and $0 \le S \le 2S_j$. The scattering operator for transitions between energy levels is given by (8.23), so to calculate the cross-sections we need the matrix elements $\langle S', M_S'|\mathbf{S}_j|S, M_S\rangle$, $j = 1, 2$.

As an example, we consider a spin-$\tfrac{1}{2}$ dimer with antiferromagnetic coupling, i.e. $S_1 = S_2 = \tfrac{1}{2}$ and $J > 0$. The eigenvalues and eigenfunctions of (8.24) are given in Table 8.2. The ground state is a non-magnetic

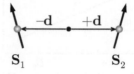

Fig. 8.4 A magnetic dimer.

[13]Other forms of the dipole approximation are given in Section 6.2.2, and more general expressions for $\mathbf{M}_j(\mathbf{Q})$ are given in Section 6.3.

**Table 8.2** Eigenvalues and eigenfunctions for the spin-$\frac{1}{2}$ antiferromagnetic dimer. The eigenstates are expressed in terms of product eigenstates of $S_1^z$ and $S_2^z$. For example, $|\uparrow\uparrow\rangle$ stands for $|M_{1S}\rangle|M_{2S}\rangle = |+\frac{1}{2}\rangle|+\frac{1}{2}\rangle$.

| $S$ | $M_S$ | $E_{S,M_S}$ | $|S, M_S\rangle$ |
|---|---|---|---|
| 0 | 0 | $-\frac{3}{4}J$ | $\frac{1}{\sqrt{2}}|\uparrow\downarrow - \downarrow\uparrow\rangle$ |
| 1 | $-1$ | $\frac{1}{4}J - g\mu_B B$ | $|\downarrow\downarrow\rangle$ |
| 1 | 0 | $\frac{1}{4}J$ | $\frac{1}{\sqrt{2}}|\uparrow\downarrow + \downarrow\uparrow\rangle$ |
| 1 | $+1$ | $\frac{1}{4}J + g\mu_B B$ | $|\uparrow\uparrow\rangle$ |

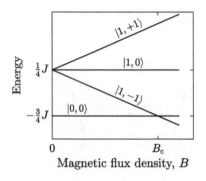

**Fig. 8.5** Zeeman splitting of the dimer energy levels with magnetic field. The levels are labelled with the corresponding eigenfunctions $|S, M_S\rangle$.

[14]The matrix elements connecting singlet to itself are all zero.

singlet with $S = 0$, and the excited states form a triplet with $S = 1$. The levels of the triplet are split by the magnetic field, as plotted in Fig. 8.5. When $B = B_c = J/(g\mu_B)$, the $M_S = -1$ component of the triplet crosses the singlet and the ground state develops a magnetic moment.

The matrix elements of $\mathbf{S}_1$ and $\mathbf{S}_2$ required to obtain the scattering intensity may be calculated from the eigenfunctions given in Table 8.2 with the aid of the single spin-$\frac{1}{2}$ matrix elements given in Table C.2. The matrix elements connecting the singlet ground state found when $B < B_c$ to each component of the triplet are[14]

$$\langle 1,0|\mathbf{S}_1|0,0\rangle = \begin{pmatrix} 0 \\ 0 \\ \frac{1}{2} \end{pmatrix}, \quad \langle 1,\pm1|\mathbf{S}_1|0,0\rangle = \frac{1}{\sqrt{2}}\begin{pmatrix} \mp\frac{1}{2} \\ \frac{1}{2}i \\ 0 \end{pmatrix}. \quad (8.26)$$

The matrix elements of $\mathbf{S}_2$ have the opposite sign to those of $\mathbf{S}_1$.

Combining (8.26) with (8.6) and (8.23), we obtain the response function for unpolarized neutron scattering from the singlet to the components of the triplet,

$$S(\mathbf{Q},\omega) = \exp(-2W)f^2(\mathbf{Q})g^2\mu_B^2\, p_s \sin^2(\mathbf{Q}\cdot\mathbf{d})$$
$$\times \Big\{ (1 - \hat{Q}_z^2)\delta(E_{1,0} - E_{0,0} - \hbar\omega)$$
$$+ \tfrac{1}{2}(1 + \hat{Q}_z^2)\delta(E_{1,\pm1} - E_{0,0} - \hbar\omega) \Big\}, \quad (8.27)$$

where $p_s$ is the thermal occupancy of the singlet. The dimer structure factor $\sin^2(\mathbf{Q}\cdot\mathbf{d})$ is the result of interference between neutrons scattered from the two antiferromagnetically coupled spins displaced by $2\mathbf{d}$.

We see from (8.26) that the $M_S = 0$ component of the triplet is connected to the singlet via the $S^z$ matrix element, whereas the $M_S = \pm1$ components are connected via the $S^{x,y}$ matrix elements. Hence, one can separate the $M_S = 0$ and $M_S = \pm1$ components by uniaxial polarization analysis. For example, with the sample in a vertical magnetic field and $\mathbf{Q}$ horizontal the transitions to the $M_S = 0$ and $M_S = \pm1$ components appear in the non-spin-flip and spin-flip channels, respectively, as can be seen from eqns (4.72) and (4.73). Moreover, the $M_S = +1$ and $-1$

levels can be distinguished from one another by measurement of the chiral scattering (see Exercise 8.1).

The spectrum of a spin-$\frac{1}{2}$ magnetic dimer in a magnetic field is illustrated by the measurements on TlCuCl$_3$ shown in Fig. 8.6. In zero field the singlet–triplet transition gives rise to a single peak, which splits into three components in applied fields up to the critical field $B_c = 5.7\,\text{T}$. Equation (8.27) shows that for $\mathbf{Q} \perp \mathbf{B}$ (i.e. $\hat{Q}_z = 0$), the scattering intensity for the transition to the $M_S = 0$ component of the triplet is twice that for transitions to the $M_S = \pm 1$ components. This prediction is borne out by the data in Fig. 8.6(a). In fields that exceed $B_c$ long-range magnetic order is induced, Fig. 8.6(b). The size of the ordered moment increases with $B > B_c$, and this behaviour is formally analogous to a Bose–Einstein condensation. Here, the condensing particles are quantized excitations of the triplet state, or *triplons*, and $B$ plays the role of the chemical potential.

### 8.2.2 Molecular nanomagnets

*Molecular nanomagnets* (MNMs) have arrays of identical molecular units each containing between two and a hundred or so paramagnetic ions. The molecular units are sufficiently well isolated from one another that inter-molecular coupling is negligible. This means there is no long-range magnetic order, and the magnetic properties are those of an ensemble of finite-size spin clusters whose intra-molecular interactions are generally well described by the Heisenberg model supplemented by anisotropic terms and/or higher-order spin couplings. Because the size of the molecular units is on the scale of nanometers, MNMs can show behaviour that is intermediate between quantum and classical. Fig. 8.7 shows the molecular units of two prototypical MNMs.

Many MNMs of interest contain $3d$ transition-metal ions with quenched orbital degrees of freedom supported on a framework of organic ligands. A suitable microscopic Hamiltonian for such systems is

$$\mathcal{H} = \sum_{\langle ij \rangle} J_{ij}\mathbf{S}_i \cdot \mathbf{S}_j + \sum_i \mathbf{S}_i \cdot \mathbf{D}_i \cdot \mathbf{S}_i + \mu_B \sum_i \mathbf{S}_i \cdot \mathbf{g}_i \cdot \mathbf{B}. \qquad (8.28)$$

The first term is a Heisenberg interaction between pairs of spins on sites $i$ and $j$, with each pair counted once. The second and third terms are, respectively, quadratic single-ion anisotropy and Zeeman interactions, with D a matrix of anisotropy coefficients, g the anisotropic $g$-tensor (see Section C.4), and $\mathbf{B}$ an external magnetic flux density.

The spin interactions and anisotropies in the Hamiltonian split the spin cluster into a series of discrete energy levels. In principle, the eigenstates can be obtained exactly in the product basis of single-spin states $|\prod_i (SM_S)_i\rangle$. This is feasible for small clusters of up to about six spins, but the dimension of the Hilbert space increases rapidly with the number of spins in the cluster and makes exact diagonalization impractical for large clusters and high spin quantum numbers. For example, a cluster of $N$ identical spins $S_i$ has dimension $(2S_i + 1)^N$. Larger spin

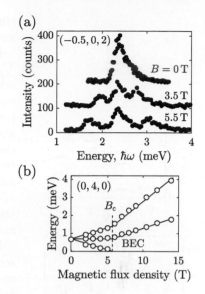

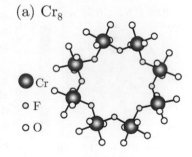

**Fig. 8.6** (a) Neutron scattering from the spin-$\frac{1}{2}$ magnetic dimer compound TlCuCl$_3$ at $T = 1.5\,\text{K}$ (after Furrer and Rüegg, 2006). The magnetic field was applied along the $b$ axis, so that $\mathbf{Q} \perp \mathbf{B}$. (b) Bose–Einstein condensation (BEC) (after Rüegg *et al.*, 2003).

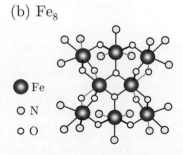

**Fig. 8.7** Molecular units of the molecular nanomagnets Cr$_8$ and Fe$_8$. Only the transition-metal ions and their ligands are shown.

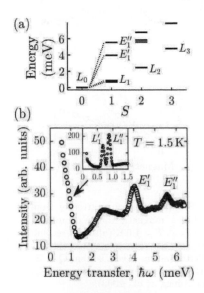

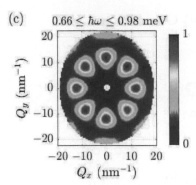

**Fig. 8.8** Neutron spectrum of the molecular wheel $Cr_8$ (after Baker *et al.*, 2012). (a) Diagram showing low-lying energy levels, labelled as $L_S$ or $E_S$, where $S$ is the total spin. (b) Low temperature powder spectrum. The transitions are indicated in (a). Inset: higher resolution data show a splitting of the $L_1$ levels. (c) **Q**-dependence of the scattering from the $L_1$ transition measured on a single crystal in the $Q_x$–$Q_y$ plane, averaged over $Q_z$, where $x$ and $y$ lie in the plane of the wheel. (Data courtesy of P. Santini.)

clusters can be analysed by numerical diagonalization procedures or by mapping to approximate effective spin models.

In the absence of single-ion anisotropy, the total spin operator $\mathbf{S} = \sum_i \mathbf{S}_i$ commutes with $\mathcal{H}$. In such cases, $S$ and $M_S$ are good quantum numbers and the spectrum of energy levels can be organized into multiplets of the allowed $S$ values. In real systems single-ion anisotropy is never zero, but in $3d$ transition metal ions it is often small compared with exchange. Systems in this regime (the strong exchange limit) can be described by total-spin basis functions $|SM_S\rangle$ with single-ion anisotropy treated as a perturbation which causes a zero-field splitting of the $S$ multiplets. The problem is then amenable to simplification through use of effective spin Hamiltonians which act on a Hilbert space built from the combined spin of a subset of the individual spins that are strongly coupled together.

The excitation spectra of MNMs can be probed by various spectroscopic techniques including optical spectroscopy, electron spin resonance, and neutron scattering, and the terms in the Hamiltonian can be obtained from analysis of the observed transition energies and intensities. In the case of neutron spectroscopy, the cross-section for transitions between energy levels may be calculated by extension of the method used for the spin-$\frac{1}{2}$ dimer in Section 8.2.1, with inclusion of spin–orbit coupling when necessary. In the strong exchange limit the calculation simplifies considerably since the cross-section is then governed by matrix elements of the form $\langle SM_S| M_q(\mathbf{Q}) |S'M_S'\rangle$, $q = 0, \pm 1$, which can be evaluated from eqn (6.73). The selection rules for allowed transitions are $\Delta S = 0, \pm 1$ and $\Delta M_S = -q$. In systems with significant magnetic anisotropy $S$ mixing can be important, leading to states that contain admixtures of several different $|SM_S\rangle$.

A well-known example of a MNM in the strong-exchange limit is the antiferromagnetic molecular wheel $Cr_8F_8[O_2CC(CH_3)_3]_{16}$, or $Cr_8$ for short, which has eight $Cr^{3+}$ ions ($S_i = \frac{3}{2}$) forming a near-perfect octahedron, Fig. 8.7(a). The magnetic spectrum is described well by eqn (8.28) with nearest-neighbour Heisenberg exchange $J = 1.46$ meV and axial single-ion anisotropy $DS_z^2$, $D = -0.038$ meV, where $z$ is perpendicular to the plane of the molecule (Waldmann *et al.*, 2003). The ground state has total spin $S = 0$, and the low-lying excited states form a series of bands which approximately follow the Landé rule $E(S) \propto S(S+1)$. These are called *rotational bands* by analogy with the rotational levels of a quantum rigid rotor. The rotational band concept is found to apply to many strong-exchange antiferromagnetic ring molecules. The lowest two rotational bands are called the $L$ and $E$ bands, and higher lying states form a quasi-continuum. Physically, excitations in the $L$ band correspond to rotations of the Néel vector, and the $E$ band is related to antiferromagnetic spin waves (see Section 8.3.2). Transitions from the $L$ band to the quasi-continuum are forbidden by the selection rules, so at low temperatures only transitions within the $L$ band or from the $L$ band to the $E$ band are observed in the neutron spectrum. The low-energy part of the spectrum of $Cr_8$, which contains transitions with $\Delta S = 1$, is shown in Fig. 8.8(a) together with the corresponding energy levels. The momentum dependence of the scattering, Fig. 8.8(b), reflects the quantum spin correlations in the wave functions involved in the transition.

*Single-molecule magnets* (SMMs) are a sub-class of molecular nanomagnets with three special characteristics: (i) they possess a large ground state total spin $S$, (ii) they exhibit very slow relaxation of the magnetization below a certain temperature known as the *blocking temperature*, $T_B$, and (iii) they exhibit quantum tunneling of the magnetization through an energy barrier between $+M_S$ and $-M_S$ states. The first important SMM to be investigated was $Mn_{12}O_{12}(CH_3COO)_{16}(H_2O)_4$, or $Mn_{12}$-acetate. Another good example is $[Fe_8O_2(OH)_{12}(tacn)_6]^{8+}$, where tacn stands for triazacyclononane. This molecule (Fe$_8$, see Fig. 8.7) has a cluster of eight $Fe^{3+}$ ions each with spin $S_i = \frac{5}{2}$ which couple to form a ground state with $S = 10$ and strong axial anisotropy. The blocking temperature is $T_B = 2.6\,K$. The ground state spin multiplet is well separated from the higher-lying multiplets, so an effective spin Hamiltonian can be used to describe the energy levels within the ground state. The spectrum of transitions within the ground state spin multiplet was found to be described very well by

$$\mathcal{H} = DS_z^2 + E(S_x^2 - S_y^2) + B_4^0 O_4^0(\mathbf{S}) + B_4^4 O_4^4(\mathbf{S}) + \mu_B \mathbf{S} \cdot \mathbf{g} \cdot \mathbf{B}, \quad (8.29)$$

with parameters $D = -25.2\,\mu eV$, $E = -4.0\,\mu eV$, $B_4^0 = 0.9 \times 10^{-4}\,\mu eV$ and $B_4^4 = 7.4 \times 10^{-4}\,\mu eV$. Here, $O_4^0(\mathbf{S})$ and $O_4^4(\mathbf{S})$ are Stevens operator equivalents built from components of $\mathbf{S}$, see eqn (8.8), The dominant anisotropy is the second-order axial term $DS_z^2$ which in zero field splits the ground state spin multiplet into an approximately quadratic band of energies $DM_S^2$, as shown in Fig. 8.9(a). The measured neutron spectrum of Fe$_8$ is reproduced in Fig. 8.9(b). The smaller terms in (8.29) are required for an accurate description of the transitions at energies below about 0.3 meV. The transverse anisotropy terms and any component of magnetic field perpendicular to $z$ do not commute with $S_z$ and allow tunneling between $+M_S$ and $-M_S$ states.

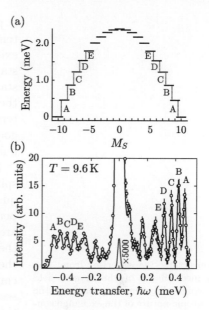

**Fig. 8.9** Neutron spectrum of the SMM Fe$_8$ (after Caciuffo *et al.*, 1998). (a) The low-lying energy levels of Fe$_8$ approximately follows the quadratic spectrum described by the Hamiltonian $\mathcal{H} = DS_z^2$. Transitions are allowed between levels which satisfy $\Delta M_S = \pm 1$. (b) Many transitions are observed at $T = 9.6\,K$ due to thermal population of excited states. The labelled peaks correspond to the transitions indicated in (a). (Data courtesy of R. Caciuffo.)

## 8.3 Spin waves

The fundamental excitations of systems with long-range magnetic order are quantized oscillations of the magnetic moments about their average directions. These oscillation modes are called *spin waves*, and the energy quanta are *magnons*. Spin waves are the magnetic equivalent of the normal modes of vibration of atoms in solids, and magnons are magnetic quasiparticles analogous to phonons.

*Spin-wave theory* is the method used to calculate the excitation spectrum in an ordered magnet. The starting point is a model for the magnetic system and its interactions, together with a stable (static) equilibrium configuration of moments. The approach is to derive the small oscillations of the magnetization about its equilibrium value on each site from some reasonable equation of motion. In ferromagnetic systems it is possible to show that the excitations are weakly interacting (at low temperature) and obey Bose statistics. It is assumed that similar states exist for more complicated magnetic systems, even though the ground states of such systems are often not known precisely. When the ground state is unknown one usually assumes a pseudo- ground state which is the lowest energy state of a system of classical spins with the same interactions as the true spins. A three-dimensional antiferromagnet is a good example of a system whose cooperative excitations can be treated quite adequately without knowledge of the exact ground state.

The lowest order theory is called *linear spin-wave theory*. Linear spin-wave theory seeks to reduce the true magnetic Hamiltonian to a quadratic form in Bose operators. The method works best at low temperatures where the deviations of the spins from their average directions are smallest. The strengths of linear spin-wave theory are (i) it generates the correct number of normal modes and their symmetries, (ii) it predicts the formation of gaps due to anisotropies,[15] (iii) closed-form expressions can always be obtained for the spin-wave dispersion and scattering cross-section,[16] (iv) the values of the exchange and anisotropy parameters in the model may be obtained by comparison with experiment, (v) the theory can be extended to calculate interactions with other excitations in the system, such as phonons, or with other spin systems such as the nuclear spin.

However, there are also limitations. Linear spin-wave theory is a semi-classical theory and is not sufficient if one wishes to extract accurate values for parameters in the Hamiltonian. It becomes increasingly inaccurate as the dimensionality of the magnetic system and the value of the spin quantum number are reduced. In such cases, which fall within the realm of *quantum magnetism*,[17] a significant proportion of the magnetic spectral weight lies outside the spin-wave modes. Also, one needs to go beyond the linear approximation to describe the temperature variation of the magnon spectrum, and this becomes rapidly more difficult as the temperature increases.

The method proceeds in two steps. First, spin operators in the Hamiltonian are replaced by expressions involving Bose operators which satisfy the spin commutation rules, and second, the quadratic part of the resulting Hamiltonian is isolated and diagonalized. There is more than one way to carry out these steps, but in the linear approximation each leads to the same result. The methods differ in the higher-order corrections.

### 8.3.1 Heisenberg ferromagnet

We start by calculating the spin-wave dispersion relation and scattering cross-section for a ferromagnetic crystal with one magnetic ion per unit cell. We consider a system of localized magnetic moments whose interactions may be described by the Heisenberg spin Hamiltonian[18]

$$\mathcal{H} = \sum_{\langle ij \rangle} J_{ij} \mathbf{S}_i \cdot \mathbf{S}_j \,, \tag{8.30}$$

[15]Anisotropy gaps arise from the combined effects of the crystal field and spin–orbit coupling, or from the anisotropic nature of the exchange coupling between neighbouring ions.

[16]Although these become rapidly more complicated as the number of magnetic subsystems increases.

[17]See Section 8.7.

[18]Note that, in reality, the Heisenberg Hamiltonian (8.30) is only the leading-order isotropic term of an effective spin Hamiltonian. There will also be a series of higher-order products of spin operators which here we neglect.

where $J_{ij} \equiv J(\mathbf{r}_i - \mathbf{r}_j)$ is the exchange parameter describing the interaction between spins at sites $\mathbf{r}_i$ and $\mathbf{r}_j$. A negative $J_{ij}$ favours ferromagnetic alignment of spins $\mathbf{S}_i$ and $\mathbf{S}_j$. The summation is over all pairs of spins on different sites (i.e. $i \neq j$) with each pair counted only once, so that $J_{ij}$ is the exchange energy per spin pair.[19] We put no constraints on the $J_{ij}$ other than that the ground state of $\mathcal{H}$ is a ferromagnet.

We use the definition of the spin raising and lowering operators with $z$ as the axis of quantization (see Section C.1.14)

$$S^+ = S^x + iS^y, \qquad S^- = S^x - iS^y, \qquad (8.31)$$

to write $\mathcal{H}$ as

$$\mathcal{H} = \sum_{\langle ij \rangle} J_{ij} \left\{ S_i^z S_j^z + \tfrac{1}{2}(S_i^+ S_j^- + S_i^- S_j^+) \right\}. \qquad (8.32)$$

The $S^+$ and $S^-$ operators connect adjacent eigenstates $|M_S\rangle$ of $S^z$,

$$S^+|M_S\rangle = (S - M_S)^{\frac{1}{2}}(S + M_S + 1)^{\frac{1}{2}}|M_S + 1\rangle, \qquad (8.33)$$

$$S^-|M_S\rangle = (S + M_S)^{\frac{1}{2}}(S - M_S + 1)^{\frac{1}{2}}|M_S - 1\rangle, \qquad (8.34)$$

where $S$ is the spin quantum number, and $M_S$ runs from $-S$ to $+S$ in integer steps. In the ferromagnetic ground state all spins have $M_S = S$ since they are fully aligned parallel to the $z$ axis. We observe that the ferromagnetic ground state is an eigenstate of $\mathcal{H}$ since it is an eigenstate of $S_i^z S_j^z$ (with eigenvalue $S^2$) as well as $S_i^+ S_j^-$ and $S_i^- S_j^+$ (both with eigenvalue 0 since $S^+|S\rangle = 0$).

Defining $n = S - M_S$ to be the spin deviation from full alignment, we can write (8.33) and (8.34) as

$$S^+|n\rangle = (2S)^{\frac{1}{2}} n^{\frac{1}{2}} \left(1 - \frac{n-1}{2S}\right)^{\frac{1}{2}} |n-1\rangle, \qquad (8.35)$$

$$S^-|n\rangle = (2S)^{\frac{1}{2}} (n+1)^{\frac{1}{2}} \left(1 - \frac{n}{2S}\right)^{\frac{1}{2}} |n+1\rangle. \qquad (8.36)$$

Looking back at the analysis of the harmonic oscillator in Section 5.1, we see that there is a close correspondence between $S^+$ and $S^-$ and the boson annihilation and creation operators $a$ and $a^\dagger$, which have the property [(5.12) and (5.13)],

$$a|n\rangle = n^{\frac{1}{2}} |n-1\rangle, \qquad (8.37)$$

$$a^\dagger|n\rangle = (n+1)^{\frac{1}{2}} |n+1\rangle, \qquad (8.38)$$

$$a^\dagger a|n\rangle = n|n\rangle, \qquad (8.39)$$

where $|n\rangle$ is now an eigenstate of the harmonic oscillator. The idea that the spin system behaves rather like a set of independent harmonic oscillators was formalized by Holstein and Primakoff (1940), who made the associations

$$S^+ = (2S)^{\frac{1}{2}} f(S) a, \qquad (8.40)$$

$$S^- = (2S)^{\frac{1}{2}} a^\dagger f(S), \qquad (8.41)$$

$$S^z = S - a^\dagger a, \qquad (8.42)$$

[19]In the literature, $J_{ij}$ is sometimes defined to be the energy *per spin* rather than *per pair of spins*. Unfortunately the choice of convention is not always stated.

where $f(S) = (1 - a^\dagger a/2S)^{\frac{1}{2}}$. These equivalences follow directly from (8.35)–(8.39). At low temperatures, the average spin deviation will be very small and so $f(S) \approx 1$. In this limit the first two transformations become linear,

$$S^+ = (2S)^{\frac{1}{2}} a, \tag{8.43}$$

$$S^- = (2S)^{\frac{1}{2}} a^\dagger. \tag{8.44}$$

[20]Note that the $a_i^\dagger a_i$ and $a_j^\dagger a_j$ terms are equal after summation.

By substituting (8.42)–(8.44) into (8.30) and neglecting terms containing products of more than two Bose operators we may write[20]

$$\mathcal{H} = \sum_{\langle ij \rangle} S J_{ij}(S - 2a_i^\dagger a_i + a_i a_j^\dagger + a_i^\dagger a_j). \tag{8.45}$$

The sum over pairs $\langle ij \rangle$ can be regarded as a sum over $\mathbf{r}_i$ followed by a sum over $\boldsymbol{\delta} = \mathbf{r}_j - \mathbf{r}_i$, with $\boldsymbol{\delta} \neq 0$. In a simple ferromagnet there is one spin per magnetic unit cell, and the magnetic lattice vectors can be chosen to coincide with the $\mathbf{r}_i$. To make this explicit we will denote the lattice vector associated with the $m$th unit cell by $\mathbf{m}$, and perform the sum over $\mathbf{m}$ rather than $\mathbf{r}_i$. Hence,

$$\sum_{\langle ij \rangle} \rightarrow \frac{1}{2} \sum_{\mathbf{m}} \sum_{\boldsymbol{\delta} \neq 0}. \tag{8.46}$$

The factor of $\frac{1}{2}$ is to avoid double counting the spin pairs. For localized spins, $J_{ij} \equiv J(\boldsymbol{\delta})$ decreases rapidly with increasing $\boldsymbol{\delta}$, so only a few terms in the sum over $\boldsymbol{\delta}$ need be retained (often just the nearest neighbours). Making the change of variables (8.46) in (8.45), we obtain

$$\mathcal{H} = \frac{1}{2} N S^2 \sum_{\boldsymbol{\delta} \neq 0} J(\boldsymbol{\delta}) - \frac{1}{2} \sum_{\mathbf{m}} \sum_{\boldsymbol{\delta} \neq 0} S J(\boldsymbol{\delta}) \big(2 a_\mathbf{m}^\dagger a_\mathbf{m} - a_\mathbf{m} a_{\mathbf{m}+\boldsymbol{\delta}}^\dagger - a_\mathbf{m}^\dagger a_{\mathbf{m}+\boldsymbol{\delta}}\big), \tag{8.47}$$

where $N$ is the total number of spins.

We are looking for wave-like solutions so it is natural to introduce Fourier transform operators

$$a_i = \frac{1}{\sqrt{N}} \sum_{\mathbf{q}} \exp(i\mathbf{q} \cdot \mathbf{r}_i) a_\mathbf{q} \tag{8.48}$$

$$a_i^\dagger = \frac{1}{\sqrt{N}} \sum_{\mathbf{q}} \exp(-i\mathbf{q} \cdot \mathbf{r}_i) a_\mathbf{q}^\dagger, \tag{8.49}$$

where the sum is over all $\mathbf{q}$ values inside the first Brillouin zone. The inverse Fourier transforms are

$$a_\mathbf{q} = \frac{1}{\sqrt{N}} \sum_{\mathbf{r}_i} \exp(-i\mathbf{q} \cdot \mathbf{r}_i) a_i \tag{8.50}$$

$$a_\mathbf{q}^\dagger = \frac{1}{\sqrt{N}} \sum_{\mathbf{r}_i} \exp(i\mathbf{q} \cdot \mathbf{r}_i) a_i^\dagger. \tag{8.51}$$

The Bose operators and their Fourier transforms obey the commutation relations[21]

$$[a_i, a_j^\dagger] = \delta_{ij}, \qquad [a_{\mathbf{q}}, a_{\mathbf{q}'}^\dagger] = \delta_{\mathbf{qq}'}. \tag{8.52}$$

It will also be convenient to define

$$J(\mathbf{q}) = \sum_{\boldsymbol{\delta} \neq 0} J(\boldsymbol{\delta}) \exp(i\mathbf{q} \cdot \boldsymbol{\delta}). \tag{8.53}$$

In replacing the Holstein–Primakoff operators by their Fourier inverses via (8.48) and (8.49) we need to evaluate sums over products like $a_{\mathbf{m}} a_{\mathbf{m}+\boldsymbol{\delta}}^\dagger$. This can be done as follows:

$$\sum_{\mathbf{m}} a_{\mathbf{m}} a_{\mathbf{m}+\boldsymbol{\delta}}^\dagger = \frac{1}{N} \sum_{\mathbf{m}} \sum_{\mathbf{q},\mathbf{q}'} \exp[i(\mathbf{q} - \mathbf{q}') \cdot \mathbf{m}] \exp(-i\mathbf{q}' \cdot \boldsymbol{\delta}) a_{\mathbf{q}} a_{\mathbf{q}'}^\dagger$$

$$= \frac{1}{N} \sum_{\mathbf{q},\mathbf{q}'} N\delta_{\mathbf{qq}'} \exp(-i\mathbf{q}' \cdot \boldsymbol{\delta}) a_{\mathbf{q}} a_{\mathbf{q}'}^\dagger$$

$$= \sum_{\mathbf{q}} \exp(-i\mathbf{q} \cdot \boldsymbol{\delta}) a_{\mathbf{q}} a_{\mathbf{q}}^\dagger. \tag{8.54}$$

Here we have used the lattice sum $\sum_{\mathbf{m}} \exp[i(\mathbf{q} - \mathbf{q}') \cdot \mathbf{m}] = N\delta_{\mathbf{qq}'}$, eqn (B.50). The other products in (8.47) are treated similarly, giving

$$\mathcal{H} = \frac{1}{2} N S^2 J(0) - \frac{S}{2} \sum_{\mathbf{q}} \left\{ 2J(0) a_{\mathbf{q}}^\dagger a_{\mathbf{q}} - J(-\mathbf{q}) a_{\mathbf{q}} a_{\mathbf{q}}^\dagger - J(\mathbf{q}) a_{\mathbf{q}}^\dagger a_{\mathbf{q}} \right\}$$

$$= \frac{1}{2} N S^2 J(0) + \sum_{\mathbf{q}} \hbar \omega_{\mathbf{q}} a_{\mathbf{q}}^\dagger a_{\mathbf{q}}, \tag{8.55}$$

where

$$\hbar \omega_{\mathbf{q}} = S\{J(\mathbf{q}) - J(0)\}. \tag{8.56}$$

To arrive at the second line in (8.55) we have used (i) the fact that $J(-\mathbf{q}) = J(\mathbf{q})$ when the spins are connected by lattice vectors, (ii) the Bose commutation relation $a_{\mathbf{q}} a_{\mathbf{q}}^\dagger - a_{\mathbf{q}}^\dagger a_{\mathbf{q}} = 1$ from eqn (8.52), and (iii) the result[22] $\sum_{\mathbf{q}} J(\mathbf{q}) = 0$.

The procedure we have just carried out, of transforming the Hamiltonian (8.30) into products of boson raising and lowering operators (8.55), is known as *second quantization*. The first term in (8.55) is the energy of a classical ferromagnet, i.e. one whose spin operators are treated as classical vectors. The second term represents $N$ independent harmonic oscillators, one for each allowed $\mathbf{q}$ in the Brillouin zone.[23] These are the spin-wave normal modes. The spin-wave dispersion relation (8.56) describes how the frequency of the spin oscillations varies with $\mathbf{q}$.[24]

Examples of simple ferromagnets which are described well by the Heisenberg model are relatively rare. Figure 8.10 shows neutron scattering measurements of the spin-wave dispersion relation in $K_2CuF_4$, which is a good approximation to a $S = 1/2$ square-lattice ferromagnet below the Curie temperature of $T_C = 6.25\,\mathrm{K}$. The data are well described by

[21] See eqn (5.76).

[22] $J(\mathbf{q})$ is an oscillatory function of $\mathbf{q}$ for any $\boldsymbol{\delta} \neq 0$, so the sum over all $\mathbf{q}$ vanishes.

[23] There is no zero-point energy because the ferromagnetic ground state is an eigenstate of the Heisenberg Hamiltonian.

[24] Note that although $\mathbf{q}$ was defined to be a wavevector in the first Brillouin zone, the periodicity of the crystal lattice means that the dispersion relation is periodic throughout reciprocal space.

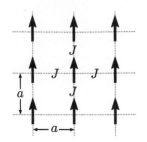

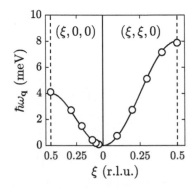

**Fig. 8.10** Spin-wave dispersion at $T \leq 2\,\mathrm{K}$ in the quasi-2D ferromagnet $K_2CuF_4$ (after Funahashi *et al.*, 1976, and Hirakawa *et al.*, 1983). The experimental points are from neutron scattering, and the line is the dispersion relation $\hbar\omega_\mathbf{q} = 2|J|S(2 - \cos q_x a - \cos q_y a)$ from (8.56) with $\mathbf{q} = (\xi, \eta, 0) \times 2\pi/a$ and nearest-neighbour exchange constant $J = -2.0\,\mathrm{meV}$. All other exchange constants are zero. The square lattice of spins is shown above.

[25]For $f$ electrons, $g\mathbf{S} \rightarrow g_J\mathbf{J}$.

the dispersion relation (8.56) with nearest-neighbour exchange interactions only (Exercise 8.2).

We know from the properties of the quantum harmonic oscillator that the amount of energy in a normal mode $\mathbf{q}$ is quantized in units of $\hbar\omega_\mathbf{q}$. The quanta in spin-wave modes are called *magnons*. The $a_\mathbf{q}^\dagger$ and $a_\mathbf{q}$ operators respectively create and destroy magnons in mode $\mathbf{q}$ according to (8.37) and (8.38). Repeated application of $a_\mathbf{q}^\dagger$ on the vacuum state $|0\rangle$ generates an infinite set of basis states $|n_\mathbf{q}\rangle$. These represent excited states of the spin system having $n_\mathbf{q}$ magnons in mode $\mathbf{q}$. As we now show, the magnetic spectrum measured by neutron scattering consists of resonances associated with the creation and destruction of these magnons.

### One-magnon cross-section

The general expression for the double differential cross-section for unpolarized neutrons was developed in Section 6.1. We shall obtain expressions for the magnetic response function $S(\mathbf{Q}, \omega)$, which is related to the cross-section by eqn (8.1). In the dipole approximation the response function for localized moments is given by eqns (6.33) and (6.34):

$$S(\mathbf{Q}, \omega) = f^2(Q)\exp(-2W)\sum_{\alpha\beta}(\delta_{\alpha\beta} - \hat{Q}_\alpha\hat{Q}_\beta)S_{\alpha\beta}(\mathbf{Q}, \omega), \qquad (8.57)$$

where the reduced partial response functions are

$$S_{\alpha\beta}(\mathbf{Q}, \omega) = g^2\mu_B^2\sum_{\lambda_i,\lambda_f} p_{\lambda_i}\langle\lambda_i| S^{\alpha\dagger}(\mathbf{Q}) |\lambda_f\rangle\langle\lambda_f| S^{\beta}(\mathbf{Q}) |\lambda_i\rangle\delta(E_{\lambda_f} - E_{\lambda_i} - \hbar\omega),$$
$$(8.58)$$

and

$$\mathbf{S}(\mathbf{Q}) = \sum_i \exp(i\mathbf{Q}\cdot\mathbf{r}_i)\mathbf{S}_i. \qquad (8.59)$$

For present purposes, $\mathbf{S}$ represents either a true spin or an effective spin (see Section C.4), and $g$ is the associated $g$-factor.[25]

In practice, we do not need to calculate all nine partial response functions $S_{\alpha\beta}(\mathbf{Q}, \omega)$. From (8.42)–(8.44) we see that $S^z$ connects initial and final states having the same number of magnons, whereas $S^x$ and $S^y$ connect initial and final states differing by one magnon. This means that the $xz$ and $yz$ response functions vanish and the $zz$ response appears as elastic scattering. When $\omega \neq 0$, therefore, only the partial response functions $S_{xx}(\mathbf{Q}, \omega)$, $S_{yy}(\mathbf{Q}, \omega)$, $S_{xy}(\mathbf{Q}, \omega)$ and $S_{yx}(\mathbf{Q}, \omega)$ are non-zero for a ferromagnet with spins aligned along $z$.

We now detail how to calculate these functions using $S_{xx}(\mathbf{Q}, \omega)$ to illustrate the method. The matrix elements in (8.58) contain the operator $S^x(\mathbf{Q})$, which from (8.43), (8.44), and (8.59) may be written

$$S^x(\mathbf{Q}) = \left(\frac{S}{2}\right)^{\frac{1}{2}} \sum_\mathbf{m} \exp(i\mathbf{Q}\cdot\mathbf{m})(a_\mathbf{m} + a_\mathbf{m}^\dagger). \qquad (8.60)$$

We write the $a_{\mathbf{m}}$ and $a_{\mathbf{m}}^\dagger$ operators in terms of the normal mode operators via (8.48) and (8.49). For a particular initial state $|\lambda_i\rangle = |n_{\mathbf{q}}\rangle$, only the operators $a_{\mathbf{q}}$ and $a_{\mathbf{q}}^\dagger$ need to be included in the Fourier sum as any Fourier components with $\mathbf{q}' \neq \mathbf{q}$ do not change $|n_{\mathbf{q}}\rangle$. The possible final states $|\lambda_f\rangle$ are $|n_{\mathbf{q}} + 1\rangle$ and $|n_{\mathbf{q}} - 1\rangle$. Considering the former of these,

$$\langle n_{\mathbf{q}} + 1 | S^x(\mathbf{Q}) | n_{\mathbf{q}} \rangle = \left( \frac{S}{2N} \right)^{\frac{1}{2}} \sum_{\mathbf{m}} \exp\{i(\mathbf{Q} - \mathbf{q}) \cdot \mathbf{m}\} \langle n_{\mathbf{q}} + 1 | a_{\mathbf{q}}^\dagger | n_{\mathbf{q}} \rangle$$

$$= \left( \frac{S}{2N} \right)^{\frac{1}{2}} (n_{\mathbf{q}} + 1)^{\frac{1}{2}} \sum_{\mathbf{m}} \exp\{i(\mathbf{Q} - \mathbf{q}) \cdot \mathbf{m}\}. \quad (8.61)$$

We used (8.38) to evaluate the matrix element of $a_{\mathbf{q}}^\dagger$. Hence,

$$|\langle n_{\mathbf{q}} + 1 | S^x(\mathbf{Q}) | n_{\mathbf{q}} \rangle|^2 = \frac{S}{2N} (n_{\mathbf{q}} + 1) \sum_{\mathbf{m}, \mathbf{m}'} \exp\{i(\mathbf{Q} - \mathbf{q}) \cdot (\mathbf{m} - \mathbf{m}')\}$$

$$= \frac{S}{2} (n_{\mathbf{q}} + 1) \frac{(2\pi)^3}{v_0} \sum_{\mathbf{G}} \delta(\mathbf{Q} - \mathbf{q} - \mathbf{G}), \quad (8.62)$$

where the identity (B.46) was used to reach the second line. Similarly,

$$|\langle n_{\mathbf{q}} - 1 | S^x(\mathbf{Q}) | n_{\mathbf{q}} \rangle|^2 = \frac{S}{2} n_{\mathbf{q}} \frac{(2\pi)^3}{v_0} \sum_{\mathbf{G}} \delta(\mathbf{Q} + \mathbf{q} - \mathbf{G}). \quad (8.63)$$

To obtain $S_{xx}(\mathbf{Q}, \omega)$ all that remains is to add together (8.62) and (8.63), average over all initial states, and append the energy $\delta$-function. The averaging entails a sum over $\mathbf{q}$ and the replacement of $n_{\mathbf{q}}$ by $\langle n_{\mathbf{q}} \rangle = n(\omega_{\mathbf{q}})$, where $n(\omega_{\mathbf{q}})$ is the boson occupation factor, i.e. the Planck distribution function, eqn (3.55). With these manipulations, we obtain

$$n(\omega_{\mathbf{q}}) = \frac{1}{\exp(\hbar\omega_{\mathbf{q}}/k_{\mathrm{B}}T) - 1}$$

$$S_{xx}(\mathbf{Q}, \omega) = g^2 \mu_{\mathrm{B}}^2 \frac{S}{2} \frac{(2\pi)^3}{v_0} \sum_{\mathbf{G}} \sum_{\mathbf{q}} \left[ n(\omega_{\mathbf{q}}) \delta(\mathbf{Q} + \mathbf{q} - \mathbf{G}) \delta(\hbar\omega + \hbar\omega_{\mathbf{q}}) \right.$$

$$\left. + \{n(\omega_{\mathbf{q}}) + 1\} \delta(\mathbf{Q} - \mathbf{q} - \mathbf{G}) \delta(\hbar\omega - \hbar\omega_{\mathbf{q}}) \right]. \quad (8.64)$$

The end result and its derivation closely parallel the treatment of coherent one-phonon scattering, and much of the subsequent discussion in Section 5.2.4 applies here too. Moreover, we can eliminate the $\delta$-functions in wavevector by the method described in Section 5.2.5, i.e. by replacing the summation in (8.64) by an integral. The result is

$$S_{xx}(\mathbf{Q}, \omega) = g^2 \mu_{\mathrm{B}}^2 \frac{NS}{2} \left[ n(\omega_{-\mathbf{Q}}) \delta(\hbar\omega + \hbar\omega_{-\mathbf{Q}}) \right.$$

$$\left. + \{n(\omega_{\mathbf{Q}}) + 1\} \delta(\hbar\omega - \hbar\omega_{\mathbf{Q}}) \right], \quad (8.65)$$

where $\hbar\omega_{\mathbf{Q}} = \hbar\omega_{-\mathbf{Q}}$ is the magnon dispersion relation extended periodically throughout reciprocal space. In other words, the response function for a particular mode $\mathbf{q}$ can be measured in a Brillouin zone centred on any reciprocal lattice vector $\mathbf{G}$, as long as $\mathbf{Q} = \mathbf{G} + \mathbf{q}$.

The other partial response functions can be obtained by the same method. We find

$$S_{yy}(\mathbf{Q},\omega) = S_{xx}(\mathbf{Q},\omega) \tag{8.66}$$

and

$$S_{xy}(\mathbf{Q},\omega) = -S_{yx}(\mathbf{Q},\omega) \tag{8.67}$$

with[26]

$$S_{xy}(\mathbf{Q},\omega) = \mathrm{i}\,g^2\mu_\mathrm{B}^2\frac{NS}{2}\big[n(\omega_{-\mathbf{Q}})\delta(\hbar\omega + \hbar\omega_{-\mathbf{Q}})$$
$$- \{n(\omega_\mathbf{Q}) + 1\}\delta(\hbar\omega - \hbar\omega_\mathbf{Q})\big]. \tag{8.68}$$

In the unpolarized cross-section (8.57) the $S_{xy}(\mathbf{Q},\omega)$ and $S_{yx}(\mathbf{Q},\omega)$ terms are added together, and so from (8.67) they cancel. Hence, the one-magnon response function for unpolarized neutrons is[27]

$$S(\mathbf{Q},\omega) = f^2(Q)\exp(-2W)(1 + \hat{Q}_z^2)S_{xx}(\mathbf{Q},\omega), \tag{8.69}$$

with $S_{xx}(\mathbf{Q},\omega)$ given by (8.65).

### One-magnon cross-section — alternative method

If all we need is the response function in the form (8.65), i.e. after integrating out the $\delta$-functions in $\mathbf{Q}$, then we can obtain it more directly by the following method. We first write the Fourier transforms of the single-site lowering and raising operators in the form[28]

$$a_\mathbf{Q} = \frac{1}{\sqrt{N}}\sum_{\mathbf{r}_i}\exp(-\mathrm{i}\mathbf{Q}\cdot\mathbf{r}_i)a_i \tag{8.70}$$

$$a_\mathbf{Q}^\dagger = \frac{1}{\sqrt{N}}\sum_{\mathbf{r}_i}\exp(\mathrm{i}\mathbf{Q}\cdot\mathbf{r}_i)a_i^\dagger, \tag{8.71}$$

where $\mathbf{Q}$ extends over all reciprocal space. The operators $a_\mathbf{Q}$ and $a_\mathbf{Q}^\dagger$ act on states $|n_\mathbf{Q}\rangle$ having energy $\hbar\omega_\mathbf{Q}$, all of which are periodic in the reciprocal lattice. In other words, we work from the outset in an extended-zone scheme in which $\mathbf{Q}$ is related to the spin-wave wavevector $\mathbf{q}$ by $\mathbf{Q} = \mathbf{G} + \mathbf{q}$, where $\mathbf{q}$ is restricted to the first Brillouin zone.

The matrix elements of $\mathbf{S}(\mathbf{Q})$ are easy to evaluate in this basis. From (8.31), (8.43)–(8.44), (8.59), and (8.70)–(8.71),

$$S^x(\mathbf{Q}) = \left(\frac{NS}{2}\right)^{\frac{1}{2}}\left(a_{-\mathbf{Q}} + a_\mathbf{Q}^\dagger\right) \tag{8.72}$$

$$S^y(\mathbf{Q}) = -\mathrm{i}\left(\frac{NS}{2}\right)^{\frac{1}{2}}\left(a_{-\mathbf{Q}} - a_\mathbf{Q}^\dagger\right), \tag{8.73}$$

and so,[29]

---

[26] Note that (8.67) follows from (8.68) and the general symmetry property $S_{\alpha\beta}(\mathbf{Q},\omega) = \{S_{\beta\alpha}(\mathbf{Q},\omega)\}^*$ — see (D.71).

[27] Note that since $\hat{\mathbf{Q}}$ is a unit vector, $1 - \hat{Q}_x^2 + 1 - \hat{Q}_y^2 = 1 + \hat{Q}_z^2$.

[28] Compare with (8.50) and (8.51).

[29] Note that from (8.37)–(8.38),

$$a_\mathbf{Q}^\dagger|n_\mathbf{Q}\rangle = (n_\mathbf{Q} + 1)^{\frac{1}{2}}|n_\mathbf{Q} + 1\rangle$$
$$a_\mathbf{Q}|n_\mathbf{Q}\rangle = (n_\mathbf{Q})^{\frac{1}{2}}|n_\mathbf{Q} - 1\rangle.$$

$$|\langle n_{\mathbf{Q}} + 1|S^x(\mathbf{Q})|n_{\mathbf{Q}}\rangle|^2 = \frac{NS}{2}\left|\langle n_{\mathbf{Q}} + 1|\, a_{-\mathbf{Q}} + a_{\mathbf{Q}}^{\dagger}\,|n_{\mathbf{Q}}\rangle\right|^2$$

$$= \frac{NS}{2}(n_{\mathbf{Q}} + 1) \tag{8.74}$$

$$|\langle n_{-\mathbf{Q}} - 1|S^x(\mathbf{Q})|n_{-\mathbf{Q}}\rangle|^2 = \frac{NS}{2} n_{-\mathbf{Q}}. \tag{8.75}$$

The sum over final states in (8.58) involves (8.74) and (8.75), and after averaging over initial states we obtain

$$S_{xx}(\mathbf{Q}, \omega) = g^2 \mu_{\mathrm{B}}^2 \frac{NS}{2}\big[n(\omega_{-\mathbf{Q}})\delta(\hbar\omega + \hbar\omega_{-\mathbf{Q}})$$
$$+ \{n(\omega_{\mathbf{Q}}) + 1\}\delta(\hbar\omega - \hbar\omega_{\mathbf{Q}})\big], \tag{8.76}$$

which is the same as (8.65).

## Polarized neutron scattering from a ferromagnet

In polarized neutron experiments on ferromagnets one must saturate the magnetization of the sample in order to counteract the tendency for domains to form. Domains are small ferromagnetic regions whose magnetization direction can vary from domain to domain (see Section 7.2.1), which invariably causes severe depolarization of the neutron beam.

The requirement of saturation constrains the neutron polarization to be parallel or antiparallel to the sample magnetization. The cross-sections of most practical interest are then given by eqns (4.68)–(4.73). Considering first the case $\mathbf{P} \parallel \mathbf{Q}$, we see from (4.68)–(4.69) and (4.74)–(4.75) that the two non-spin-flip (NSF) channels contain only nuclear scattering and are the same, whereas the two spin-flip (SF) signals contain both nuclear and magnetic scattering, with the latter given by

$$S_{\mathrm{SF}}(\mathbf{Q}, \omega) = S_{xx}(\mathbf{Q}, \omega) + S_{yy}(\mathbf{Q}, \omega) \pm \mathrm{i}\{S_{xy}(\mathbf{Q}, \omega) - S_{yx}(\mathbf{Q}, \omega)\}. \tag{8.77}$$

The $z$ axis here is both the magnetization direction and the neutron polarization axis, and the $+$ and $-$ signs refer to the $\uparrow\downarrow$ (spin-up to spin-down) and $\downarrow\uparrow$ (spin-down to spin-up) SF channels, respectively. From (8.65)–(8.68), the individual SF channels are given by

$$S_{\uparrow\downarrow}(\mathbf{Q}, \omega) = 2g^2\mu_{\mathrm{B}}^2 NS\{n(\omega_{\mathbf{Q}}) + 1\}\delta(\hbar\omega - \hbar\omega_{\mathbf{Q}})$$
$$S_{\downarrow\uparrow}(\mathbf{Q}, \omega) = 2g^2\mu_{\mathrm{B}}^2 NS n(\omega_{-\mathbf{Q}})\delta(\hbar\omega + \hbar\omega_{-\mathbf{Q}}). \tag{8.78}$$

We see that the $\uparrow\downarrow$ channel measures only magnon creation, while the $\downarrow\uparrow$ channel measures only magnon annihilation. The physical reason for this is as follows. The ground state of the ferromagnet has a total spin $-NS$ in the $z$ direction.[30] When a magnon is created, the spin of the system changes by $\Delta S_z = +1$, so to conserve angular momentum the spin of the neutron must change by $\Delta s_z = -1$ corresponding to the up-to-down process. The reverse applies for magnon annihilation. Figure 8.11 shows measurements which demonstrate that scattering by magnon creation is only found in the $\uparrow\downarrow$ SF channel.

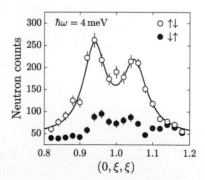

**Fig. 8.11** Spin-wave scattering from ferromagnetic $La_{0.82}Sr_{0.18}CoO_3$ with uniaxial polarization analysis. (After Ewings *et al.*, 2010.)

[30] Recall that the spin points in the opposite direction to the magnetic moment, which is parallel to $+z$ here.

The above-mentioned properties of the $\mathbf{P} \parallel \mathbf{Q}$ arrangement mean that magnon scattering can be isolated with a half-polarized set-up. One can obtain the pure magnon creation signal, for example, by subtracting successive measurements in which the incident beam is first polarized up and then down. The half-polarized method has the advantage of a higher count rate compared with full uniaxial polarization analysis, but also has a higher nuclear scattering background.

For the case $\mathbf{P} \perp \mathbf{Q}$, the NSF channel measures $S_{zz}(\mathbf{Q}, \omega)$, which contains elastic magnetic scattering, while the SF channel measures $S_{xx}(\mathbf{Q}, \omega)$ or $S_{yy}(\mathbf{Q}, \omega)$ depending on whether $\mathbf{Q}$ is parallel to $y$ or $x$. From (8.65), we see that the SF channel contains scattering by both magnon creation and annihilation, but the signals are only one-quarter the strength of the corresponding signals in (8.78).

### What is a spin wave?

The quantity that represents the motion of the spins in space and time is the spin–spin correlation function, which from eqn (6.37) is given by[31]

$$\langle \hat{S}_0^\alpha \hat{S}_1^\beta(t) \rangle \propto \sum_{\mathbf{Q}} \int_{-\infty}^{\infty} S_{\alpha\beta}(\mathbf{Q}, \omega) \exp\{-\mathrm{i}(\mathbf{Q} \cdot \mathbf{l} - \omega t)\} \mathrm{d}\omega. \qquad (8.79)$$

We can obtain a physical picture of a spin wave by considering the classical limit, which is approached when the number of excited magnons is large, i.e. $n(\omega_{\mathbf{Q}}) \gg 1$. Substituting (8.76) into (8.79) we obtain the following expression for the classical $xx$ spin correlations:

$$\langle \hat{S}_0^x \hat{S}_1^x(t) \rangle \propto \sum_{\mathbf{Q}} \exp(-\mathrm{i}\mathbf{Q} \cdot \mathbf{l}) \left[ n(\omega_{-\mathbf{Q}}) \exp(-\mathrm{i}\omega_{-\mathbf{Q}} t) + n(\omega_{\mathbf{Q}}) \exp(\mathrm{i}\omega_{\mathbf{Q}} t) \right]$$

$$\propto \sum_{\mathbf{Q}} n(\omega_{\mathbf{Q}}) \left[ \exp\{\mathrm{i}(\mathbf{Q} \cdot \mathbf{l} - \omega_{\mathbf{Q}} t)\} + \exp\{-\mathrm{i}(\mathbf{Q} \cdot \mathbf{l} - \omega_{\mathbf{Q}} t)\} \right]$$

$$\propto \sum_{\mathbf{q}} n(\omega_{\mathbf{q}}) \cos(\mathbf{q} \cdot \mathbf{l} - \omega_{\mathbf{q}} t). \qquad (8.80)$$

In the last line we used the fact that $\exp(\mathrm{i}\mathbf{G} \cdot \mathbf{l}) = 1$ for any lattice vector $\mathbf{l}$ to replace $\mathbf{Q}$ by $\mathbf{q}$. Similarly, from (8.68) the classical $xy$ correlations are given by

$$\langle \hat{S}_0^x \hat{S}_1^y(t) \rangle \propto \sum_{\mathbf{q}} n(\omega_{\mathbf{q}}) \sin(\mathbf{q} \cdot \mathbf{l} - \omega_{\mathbf{q}} t). \qquad (8.81)$$

If we just excite a single mode $\mathbf{q}$ then the $xx$ and $xy$ spin correlations are proportional to $\cos(\mathbf{q} \cdot \mathbf{l} - \omega_{\mathbf{q}} t)$ and $\sin(\mathbf{q} \cdot \mathbf{l} - \omega_{\mathbf{q}} t)$, respectively. The spin dynamics described by these correlation functions is simply a travelling wave in the $x$ and $y$ components of the spins, with the $y$ component $\pi/2$ out of phase with the $x$ component. This motion is illustrated in Fig. 8.12, which represents a snapshot of a spin wave taken at a particular time on a linear chain of spins separated by $a$. The spins precess in a circle about the average spin direction, and the precession propagates as a wave in the direction of $\mathbf{q}$ with a phase angle $qa$ between adjacent spins.

[31] Here, we write the integral over $\mathbf{Q}$ as a sum over discrete spin-wave modes.

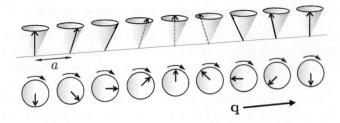

**Fig. 8.12** Illustration of a semi-classical spin wave. The spins, represented by arrows, precess in a circle about the ordered spin direction, which is vertical. The lower part of the figure is the view looking from above.

### 8.3.2 Heisenberg antiferromagnet

Linear spin-wave theory can be used to calculate the spin excitation spectrum for more general spin structures. In this section we shall extend the theory to treat a collinear, two-sublattice, Heisenberg antiferromagnet. We first consider a simple system having only isotropic interactions between nearest-neighbour spins on a square lattice, and then generalize the treatment to include anisotropic interactions, single-ions anisotropy, and interactions beyond nearest neighbours on an arbitrary lattice.

We employ the same Heisenberg Hamiltonian as before,[32]

$$
\begin{aligned}
\mathcal{H} &= \sum_{\langle ij \rangle} J_{ij} \mathbf{S}_i \cdot \mathbf{S}_j \\
&= \sum_{\langle ij \rangle} J_{ij} \left\{ S_i^z S_j^z + \tfrac{1}{2}(S_i^+ S_j^- + S_i^- S_j^+) \right\},
\end{aligned}
\tag{8.82}
$$

[32]See (8.30) and (8.32).

except that now the dominant exchange interaction is positive so that the classical ground state[33] is a collinear antiferromagnet (Section 7.1.1).

Before proceeding, we note that collinear antiferromagnetic order is not actually an eigenstate of (8.82), because the action of $\mathcal{H}$ does not leave the antiferromagnetic ground state invariant. This can be seen from the second line of (8.82). For example, if the spins on sites $i$ and $j$ start in state $|i\rangle|j\rangle = |-S\rangle|S\rangle$ then the operator $S_i^+ S_j^-$ would convert them into $|-S+1\rangle|S-1\rangle$, which is different from the initial state. Because collinear antiferromagnetism is only an approximation to the true quantum ground state, the magnetic spectrum for the Heisenberg antiferromagnet calculated by linear spin-wave theory is only approximate. Modifications to the spin-wave spectrum due to quantum effects will be discussed in Section 8.7.

[33]i.e. the lowest energy configuration when the spins $\mathbf{S}_i$ are treated as classical vectors rather than quantum-mechanical operators.

Leaving aside quantum effects for now, we can regard a collinear antiferromagnet as two sublattices of spins, which we label A and B. The spins on the A sublattice point 'up', and those on the B sublattice point 'down'. We define a set of Holstein–Primakoff annihilation and creation operators $a$ and $a^\dagger$ for sites on the A sublattice, and a second set $b$ and $b^\dagger$ for the B sublattice. The transformation between spin operators and Holstein–Primakoff operators on the two sublattices is, to first order,[34]

[34]See eqns (8.42)–(8.44).

$$
\begin{aligned}
S_A^+ &= (2S)^{\frac{1}{2}} a, & S_B^+ &= (2S)^{\frac{1}{2}} b^\dagger, \\
S_A^- &= (2S)^{\frac{1}{2}} a^\dagger, & S_B^- &= (2S)^{\frac{1}{2}} b, \\
S_A^z &= S - a^\dagger a, & S_B^z &= -S + b^\dagger b.
\end{aligned}
\tag{8.83}
$$

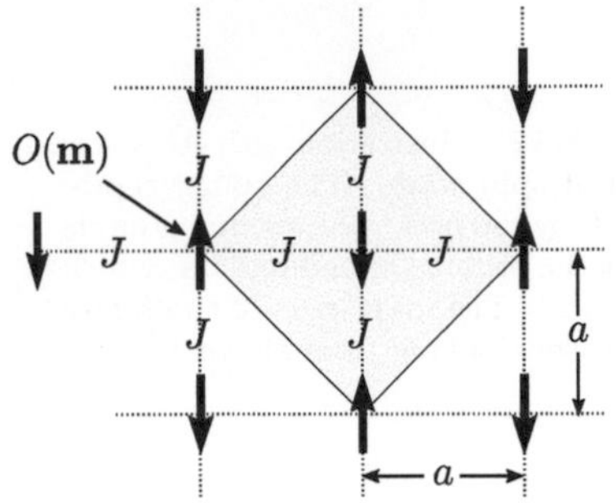

**Fig. 8.13** Square lattice collinear antiferromagnet, with nearest-neighbour interactions $J$. The shaded area is the $m$th magnetic unit cell, whose origin is at $O(\mathbf{m})$. The A sublattice spins are 'up', and the B sublattice spins are 'down'.

## Nearest-neighbour antiferromagnet on a square lattice

The classical ground state of this model antiferromagnet is shown in Fig. 8.13. We assume isotropic exchange and include interactions between nearest-neighbour spins only. In eqns (8.83), the $z$ direction is the spin quantization axis, but with isotropic interactions the choice of $z$ relative to the lattice is arbitrary as there is no preferred direction for the spins. The magnetic unit cell is defined to have its origin on an A sublattice spin, and $\mathbf{m}$ denotes the magnetic lattice vector associated with the origin of the $m$th magnetic unit cell. We perform the summation over pairs of spins in (8.82) using (8.46), replacing the double sum over $\langle ij \rangle$ by a sum over the magnetic lattice vectors and a sum over the displacements $\boldsymbol{\delta}$ of the neighbouring spins from any given spin. Hence,

$$\mathcal{H} = \frac{1}{2}\sum_{\mathbf{m}}\sum_{\boldsymbol{\delta}} J\left\{ S_{\mathbf{m}}^z S_{\mathbf{m}+\boldsymbol{\delta}}^z + \tfrac{1}{2}(S_{\mathbf{m}}^+ S_{\mathbf{m}+\boldsymbol{\delta}}^- + S_{\mathbf{m}}^- S_{\mathbf{m}+\boldsymbol{\delta}}^+) \right.$$
$$\left. + S_{\mathbf{m}+\mathbf{a}}^z S_{\mathbf{m}+\mathbf{a}+\boldsymbol{\delta}}^z + \tfrac{1}{2}(S_{\mathbf{m}+\mathbf{a}}^+ S_{\mathbf{m}+\mathbf{a}+\boldsymbol{\delta}}^- + S_{\mathbf{m}+\mathbf{a}}^- S_{\mathbf{m}+\mathbf{a}+\boldsymbol{\delta}}^+) \right\}. \quad (8.84)$$

With the substitutions of (8.83), and the neglect of terms higher than bilinear in the spin deviation operators, the Hamiltonian becomes

$$\mathcal{H} = -4NJS^2 + \frac{1}{2}\sum_{\mathbf{m}}\sum_{\boldsymbol{\delta}} SJ\left( a_{\mathbf{m}}^\dagger a_{\mathbf{m}} + b_{\mathbf{m}+\boldsymbol{\delta}}^\dagger b_{\mathbf{m}+\boldsymbol{\delta}} \right.$$
$$+ a_{\mathbf{m}} b_{\mathbf{m}+\boldsymbol{\delta}} + a_{\mathbf{m}}^\dagger b_{\mathbf{m}+\boldsymbol{\delta}}^\dagger + b_{\mathbf{m}+\mathbf{a}}^\dagger b_{\mathbf{m}+\mathbf{a}} + a_{\mathbf{m}+\mathbf{a}+\boldsymbol{\delta}}^\dagger a_{\mathbf{m}+\mathbf{a}+\boldsymbol{\delta}}$$
$$\left. + b_{\mathbf{m}+\mathbf{a}}^\dagger a_{\mathbf{m}+\mathbf{a}+\boldsymbol{\delta}}^\dagger + b_{\mathbf{m}+\mathbf{a}} a_{\mathbf{m}+\mathbf{a}+\boldsymbol{\delta}} \right), \quad (8.85)$$

where $N$ is the number of spins per sublattice. We now introduce the Fourier transform operators,

$$a_i = \frac{1}{\sqrt{N}}\sum_{\mathbf{q}} \exp(i\mathbf{q}\cdot\mathbf{r}_i)a_{\mathbf{q}} \qquad b_j = \frac{1}{\sqrt{N}}\sum_{\mathbf{q}} \exp(i\mathbf{q}\cdot\mathbf{r}_j)b_{\mathbf{q}}$$
$$(8.86)$$
$$a_i^\dagger = \frac{1}{\sqrt{N}}\sum_{\mathbf{q}} \exp(-i\mathbf{q}\cdot\mathbf{r}_i)a_{\mathbf{q}}^\dagger \qquad b_j^\dagger = \frac{1}{\sqrt{N}}\sum_{\mathbf{q}} \exp(-i\mathbf{q}\cdot\mathbf{r}_j)b_{\mathbf{q}}^\dagger,$$

where $\mathbf{q}$ is restricted to the first magnetic Brillouin zone. Substituting these operators into (8.85), and following steps analogous to those in eqn (8.54), we obtain[35]

$$\mathcal{H} = -NJ(0)S^2 + \sum_{\mathbf{q}} S\left\{ J(0)(a_{\mathbf{q}}^\dagger a_{\mathbf{q}} + b_{\mathbf{q}}^\dagger b_{\mathbf{q}}) + J(\mathbf{q})(a_{\mathbf{q}} b_{-\mathbf{q}} + a_{\mathbf{q}}^\dagger b_{-\mathbf{q}}^\dagger) \right\},$$
$$(8.87)$$

where $J(\mathbf{q})$ is defined by eqn (8.53). In the present case,

$$J(\mathbf{q}) = 2J\{\cos(\mathbf{q}\cdot\mathbf{a}) + \cos(\mathbf{q}\cdot\mathbf{b})\}$$
$$J(0) = 4J. \quad (8.88)$$

[35]Note that since the sum extends over positive and negative $\mathbf{q}$,

$$\sum_{\mathbf{q}} a_{\mathbf{q}}^\dagger a_{\mathbf{q}} = \sum_{\mathbf{q}} a_{-\mathbf{q}}^\dagger a_{-\mathbf{q}}$$
$$\sum_{\mathbf{q}} a_{\mathbf{q}} b_{-\mathbf{q}} = \sum_{\mathbf{q}} a_{-\mathbf{q}} b_{\mathbf{q}}$$
etc.

From eqn (8.87) we see that spin deviations on the A and B sites are coupled and so the Hamiltonian is not diagonal in the basis of $a_\mathbf{q}$ and $b_\mathbf{q}$ operators and their adjoints. The Hamiltonian may be brought into diagonal form by the linear transformation (Bogoliubov transformation[36])

$$a_\mathbf{q} = u_\mathbf{q}\alpha_\mathbf{q} + v_\mathbf{q}\beta^\dagger_{-\mathbf{q}}$$
$$b^\dagger_{-\mathbf{q}} = -v_\mathbf{q}\alpha_\mathbf{q} - u_\mathbf{q}\beta^\dagger_{-\mathbf{q}},$$

(8.89)

[36] Nikolay Bogoliubov, Russian mathematician (1909–1992).

giving[37]

$$\mathcal{H} = \mathcal{H}_0 + \sum_\mathbf{q}\{(\alpha^\dagger_\mathbf{q}\alpha_\mathbf{q} + \tfrac{1}{2})\hbar\omega_\mathbf{q} + (\beta^\dagger_\mathbf{q}\beta_\mathbf{q} + \tfrac{1}{2})\hbar\omega_\mathbf{q}\},$$

(8.90)

[37] The $\alpha_\mathbf{q}$ and $\beta_\mathbf{q}$ operators satisfy the Bose commutation relations (8.52). See Exercise 8.4.

where $\mathcal{H}_0 = -NJ(0)S(S+1)$ and

$$\hbar\omega_\mathbf{q} = S\sqrt{J^2(0) - J^2(\mathbf{q})}.$$

(8.91)

The coefficients in (8.89) are given by

$$u_\mathbf{q} = \sqrt{\frac{SJ(0) + \hbar\omega_\mathbf{q}}{2\hbar\omega_\mathbf{q}}}, \qquad v_\mathbf{q} = s\sqrt{\frac{SJ(0) - \hbar\omega_\mathbf{q}}{2\hbar\omega_\mathbf{q}}},$$

(8.92)

where $s$ denotes the sign of $J(\mathbf{q})$. The explicit inclusion of $s$ means that we always take the positive roots in (8.92). The Bogoliubov transformation used here to diagonalize the Hamiltonian can be verified by direct substitution, but in Section C.1.13, we describe a systematic method to obtain the Bogoliubov transformation for a general Hamiltonian which is bilinear in Bose operators.

Apart from the constant $\mathcal{H}_0$, (8.90) describes a set of independent normal modes of oscillation. These are the spin-wave excitations of the antiferromagnet. The operators $\alpha_\mathbf{q}$ and $\beta_\mathbf{q}$ and their Hermitian adjoints are the annihilation and creation operators for the modes, and the magnon dispersion is given by (8.91). For each $\mathbf{q}$ there are two degenerate modes, associated with $\alpha_\mathbf{q}$ and $\beta_\mathbf{q}$, respectively. The physical reason for this can be seen from Fig. 8.14. All spins in a normal mode precess in the same direction about the quantization axis so that the phase difference between neighbouring spins remains fixed in time. There are two degenerate modes because the spins can precess either clockwise (mode 1) or anticlockwise (mode 2). In a given mode, the up and down spins precess on different sized circles because the torque due to the exchange field acts in the opposite direction on the up spins compared with the down spins.

Figure 8.15 shows the measured spin-wave dispersion of $Rb_2MnF_4$, which is a nearly ideal realization of a nearest-neighbour Heisenberg antiferromagnet on a square-lattice. The $Mn^{2+}$ ions carry a spin $S = 5/2$, which is sufficiently large that linear spin-wave theory provides a very good description for the magnon spectrum at temperature well below the magnetic ordering temperature of $T_N = 38.4\,\mathrm{K}$.

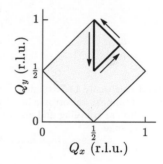

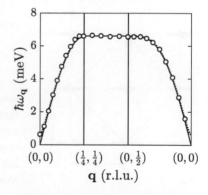

**Fig. 8.15** Spin-wave dispersion of the square-lattice Heisenberg antiferromagnet $Rb_2MnF_4$ measured at $T = 4.2\,\mathrm{K}$ (after Cowley *et al.*, 1977). The solid line is the linear spin-wave dispersion eqn (8.91) with $J = 0.66\,\mathrm{meV}$. The broken line is also from linear spin-wave theory, but includes a small exchange anisotropy which favours spin alignment along the $c$ axis. The diagram above shows the path in the magnetic Brillouin zone along which the dispersion was measured, where $\mathbf{Q} = \mathbf{G}_m + \mathbf{q}$ and $\mathbf{G}_m = (\tfrac{1}{2}, \tfrac{1}{2})$.

**Fig. 8.14** Semi-classical representation of a spin wave normal mode in an anti-ferromagnet. The spins on both sublattices precess in the same direction about the vertical axis, but the up spins precess in a larger circle than the down spins because the exchange forces act in opposite directions on them. The lower part of the figure is the view looking from above.

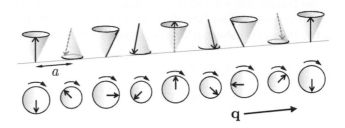

### Anisotropic antiferromagnet

We consider two types of anisotropy, *exchange anisotropy* and *single-ion anisotropy*. Exchange anisotropy means that the strength of the exchange interaction is not the same for each spin component. In general, the exchange interaction is described by a tensor whose elements $J^{\alpha\beta}$ couple the spin components $S^\alpha$ and $S^\beta$. Here we consider the simplest anisotropic exchange interaction, a diagonal tensor with three non-zero elements $J^\alpha$, $\alpha = x, y, z$, one for each of the spin components.[38] Single-ion anisotropy arises from the interaction of the spin with the crystalline environment (the ligand field) via spin–orbit coupling. It can be described phenomenologically by simple polynomials in the spin components whose form depends on the local symmetry of the ligand field.

[38]The method can easily be extended to other types of bilinear anisotropic interaction, e.g. dipolar coupling and antisymmetric exchange.

We take as an example a crystal with orthogonal axes. For such a crystal the anisotropic Hamiltonian might typically be

$$\mathcal{H} = \sum_{\langle ij \rangle} \sum_\alpha J_{ij}^\alpha S_i^\alpha S_j^\alpha + \sum_i \sum_\alpha K_\alpha (S_i^\alpha)^2, \tag{8.93}$$

where the $K_\alpha$ are constants which quantify the spin anisotropy. We further assume that the spins align parallel and antiparallel to the $x$ axis (Fig. 8.16). This means that if the anisotropy is dominated by nearest-neighbour exchange interactions then $J^x > J^y, J^z$, and if single-ion anisotropy dominates then $K_x < K_y, K_z$. With $x$ as the quantization axis the spin raising and lowering operators are defined by

$$S^+ = S^y + \mathrm{i}S^z, \qquad S^- = S^y - \mathrm{i}S^z, \tag{8.94}$$

and if the spins point along $+x$ on the A sublattice and along $-x$ on the B sublattice then [c.f. eqn. (8.83)]

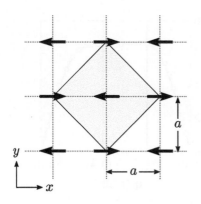

**Fig. 8.16** Square-lattice collinear antiferromagnet with $x$ as the easy axis. The shaded area is the magnetic unit cell. The A sublattice spins point along $+x$, and the B sublattice spins point along $-x$.

$$\begin{aligned}
S_\mathrm{A}^+ &= (2S)^{\frac{1}{2}} a, & S_\mathrm{B}^+ &= (2S)^{\frac{1}{2}} b^\dagger, \\
S_\mathrm{A}^- &= (2S)^{\frac{1}{2}} a^\dagger, & S_\mathrm{B}^- &= (2S)^{\frac{1}{2}} b, \\
S_\mathrm{A}^x &= S - a^\dagger a, & S_\mathrm{B}^x &= -S + b^\dagger b.
\end{aligned} \tag{8.95}$$

As a further generalization, we allow interactions with spins beyond the nearest neighbours. We must then distinguish between exchange interactions between spins on the same sublattice $J(\boldsymbol{\delta}_\mathrm{F})$ and on different sublattices $J(\boldsymbol{\delta}_{\mathrm{AF}})$. Here, $\boldsymbol{\delta}_{\mathrm{F}}$ represents the displacement of two spins on the same sublattice, and $\boldsymbol{\delta}_{\mathrm{AF}}$ is the displacement of a spin on sublattice B relative to a spin on sublattice A. We will also require the Fourier transforms[39]

[39] Note that $J_{\mathrm{F}}(\mathbf{q})$ is real because a sublattice has inversion symmetry, whereas $J_{\mathrm{AF}}(\mathbf{q})$ is complex unless sublattice A is at a centre of symmetry with respect to sublattice B.

$$J_{\mathrm{F}}(\mathbf{q}) = \sum_{\boldsymbol{\delta}_{\mathrm{F}} \neq 0} J(\boldsymbol{\delta}_{\mathrm{F}}) \exp(\mathrm{i}\mathbf{q} \cdot \boldsymbol{\delta}_{\mathrm{F}}) \qquad (8.96)$$

$$J_{\mathrm{AF}}(\mathbf{q}) = \sum_{\boldsymbol{\delta}_{\mathrm{AF}} \neq 0} J(\boldsymbol{\delta}_{\mathrm{AF}}) \exp(\mathrm{i}\mathbf{q} \cdot \boldsymbol{\delta}_{\mathrm{AF}}). \qquad (8.97)$$

We now repeat the procedure in the previous section, substituting for the spin operators in the Hamiltonian (8.93) using (8.94–8.95), and converting to Fourier transform operators using (8.86) and (8.96–8.97). After some algebra, we obtain the Hamiltonian in the same bilinear form as (8.87),

$$\mathcal{H} = \mathcal{H}_0 + \frac{1}{2} \sum_{\mathbf{q}} \mathsf{X}_{\mathbf{q}}^{\dagger} \mathsf{H}_{\mathbf{q}} \mathsf{X}_{\mathbf{q}}, \qquad (8.98)$$

where

$$\mathcal{H}_0 = -NS(S+1)\left\{ J_{\mathrm{AF}}^x(0) - J_{\mathrm{F}}^x(0) - 2K_x \right\}, \qquad (8.99)$$

and

$$\mathsf{X}_{\mathbf{q}}^{\dagger} \mathsf{H}_{\mathbf{q}} \mathsf{X}_{\mathbf{q}} = \left(a_{\mathbf{q}}^{\dagger}, b_{\mathbf{q}}^{\dagger}, a_{-\mathbf{q}}, b_{-\mathbf{q}}\right) \begin{pmatrix} A_{\mathbf{q}} & B_{\mathbf{q}} & C_{\mathbf{q}} & D_{\mathbf{q}} \\ B_{\mathbf{q}}^* & A_{\mathbf{q}} & D_{\mathbf{q}}^* & C_{\mathbf{q}} \\ C_{\mathbf{q}} & D_{\mathbf{q}} & A_{\mathbf{q}} & B_{\mathbf{q}} \\ D_{\mathbf{q}}^* & C_{\mathbf{q}} & B_{\mathbf{q}}^* & A_{\mathbf{q}} \end{pmatrix} \begin{pmatrix} a_{\mathbf{q}} \\ b_{\mathbf{q}} \\ a_{-\mathbf{q}}^{\dagger} \\ b_{-\mathbf{q}}^{\dagger} \end{pmatrix},$$

$$(8.100)$$

with

$$A_{\mathbf{q}} = S\left\{ J_{\mathrm{AF}}^x(0) - J_{\mathrm{F}}^x(0) + \tfrac{1}{2}J_{\mathrm{F}}^y(\mathbf{q}) + \tfrac{1}{2}J_{\mathrm{F}}^z(\mathbf{q}) - 2K_x + K_y + K_z \right\}$$

$$B_{\mathbf{q}} = S\left\{ \tfrac{1}{2}J_{\mathrm{AF}}^y(\mathbf{q}) - \tfrac{1}{2}J_{\mathrm{AF}}^z(\mathbf{q}) \right\}$$

$$C_{\mathbf{q}} = S\left\{ \tfrac{1}{2}J_{\mathrm{F}}^y(\mathbf{q}) - \tfrac{1}{2}J_{\mathrm{F}}^z(\mathbf{q}) + K_y - K_z \right\} \qquad (8.101)$$

$$D_{\mathbf{q}} = S\left\{ \tfrac{1}{2}J_{\mathrm{AF}}^y(\mathbf{q}) + \tfrac{1}{2}J_{\mathrm{AF}}^z(\mathbf{q}) \right\}.$$

A general procedure to diagonalize $\mathsf{H}_{\mathbf{q}}$ is given in Section C.1.13. The resulting diagonal form of $\mathcal{H}$ is

$$\mathcal{H} = \mathcal{H}_0 + \sum_{\mathbf{q}} \left\{ (\alpha_{\mathbf{q}}^{\dagger} \alpha_{\mathbf{q}} + \tfrac{1}{2})\hbar\omega_{\alpha} + (\beta_{\mathbf{q}}^{\dagger} \beta_{\mathbf{q}} + \tfrac{1}{2})\hbar\omega_{\beta} \right\}. \qquad (8.102)$$

Expressions for the eigenvalues $\hbar\omega_{\alpha}$ and $\hbar\omega_{\beta}$ that apply in the general case are rather lengthy, but for the particular case when all the $A_{\mathbf{q}}$, $B_{\mathbf{q}}$, $C_{\mathbf{q}}$, and $D_{\mathbf{q}}$ coefficients are real, the eigenvalues are given by (C.98),

$$\hbar\omega_{\alpha} = \sqrt{(A_{\mathbf{q}} + B_{\mathbf{q}})^2 - (C_{\mathbf{q}} + D_{\mathbf{q}})^2}$$

$$\hbar\omega_{\beta} = \sqrt{(A_{\mathbf{q}} - B_{\mathbf{q}})^2 - (C_{\mathbf{q}} - D_{\mathbf{q}})^2}. \qquad (8.103)$$

Notice that if we include nearest-neighbour interactions only, and set $J^x = J^y = J^z$ and $K_x = K_y = K_z$, then we recover the isotropic nearest-neighbour antiferromagnet, and (8.103) reduces to (8.91).

### One-magnon cross-section

We follow the second ('alternative') method described in Section 8.3.1. With the spins aligned along $x$, the $S_{xx}(\mathbf{Q}, \omega)$ response function contains the elastic scattering and, as explained after eqns (8.57)–(8.59), the $xy$ and $xz$ response functions vanish. Therefore, to calculate the non-zero response functions we only need the matrix elements of $S^y(\mathbf{Q})$ and $S^z(\mathbf{Q})$. From (8.59), (8.86), (8.94), and (8.95),

$$S^y(\mathbf{Q}) = \left(\frac{NS}{2}\right)^{\frac{1}{2}} \left(a_{-\mathbf{Q}} + a_{\mathbf{Q}}^\dagger + b_{\mathbf{Q}}^\dagger + b_{-\mathbf{Q}}\right) \tag{8.104}$$

$$S^z(\mathbf{Q}) = -\mathrm{i}\left(\frac{NS}{2}\right)^{\frac{1}{2}} \left(a_{-\mathbf{Q}} - a_{\mathbf{Q}}^\dagger + b_{\mathbf{Q}}^\dagger - b_{-\mathbf{Q}}\right), \tag{8.105}$$

where the Fourier transform operators are defined with $\mathbf{Q}$ extended to all reciprocal space. The Bogoliubov transformation that relates the spin deviation operators to the magnon operators is (see Section C.1.13)

$$\begin{aligned}
a_{\mathbf{Q}} &= u_\alpha^+ \alpha_{\mathbf{Q}} + u_\beta^- \beta_{\mathbf{Q}} - v_\alpha^+ \alpha_{-\mathbf{Q}}^\dagger + v_\beta^- \beta_{-\mathbf{Q}}^\dagger \\
b_{\mathbf{Q}} &= u_\alpha^+ \alpha_{\mathbf{Q}} - u_\beta^- \beta_{\mathbf{Q}} - v_\alpha^+ \alpha_{-\mathbf{Q}}^\dagger - v_\beta^- \beta_{-\mathbf{Q}}^\dagger \\
a_{-\mathbf{Q}}^\dagger &= -v_\alpha^+ \alpha_{\mathbf{Q}} + v_\beta^- \beta_{\mathbf{Q}} + u_\alpha^+ \alpha_{-\mathbf{Q}}^\dagger + u_\beta^- \beta_{-\mathbf{Q}}^\dagger \\
b_{-\mathbf{Q}}^\dagger &= -v_\alpha^+ \alpha_{\mathbf{Q}} - v_\beta^- \beta_{\mathbf{Q}} + u_\alpha^+ \alpha_{-\mathbf{Q}}^\dagger - u_\beta^- \beta_{-\mathbf{Q}}^\dagger.
\end{aligned} \tag{8.106}$$

where

$$u_\alpha^+ = \frac{1}{2}\sqrt{\frac{A_{\mathbf{Q}} + B_{\mathbf{Q}} + \hbar\omega_\alpha}{\hbar\omega_\alpha}} \qquad u_\beta^- = \frac{1}{2}\sqrt{\frac{A_{\mathbf{Q}} - B_{\mathbf{Q}} + \hbar\omega_\beta}{\hbar\omega_\beta}},$$

$$v_\alpha^+ = s\frac{1}{2}\sqrt{\frac{A_{\mathbf{Q}} + B_{\mathbf{Q}} - \hbar\omega_\alpha}{\hbar\omega_\alpha}} \qquad v_\beta^- = t\frac{1}{2}\sqrt{\frac{A_{\mathbf{Q}} - B_{\mathbf{Q}} - \hbar\omega_\beta}{\hbar\omega_\beta}},$$

$$\tag{8.107}$$

with

$$s = (C_{\mathbf{Q}} + D_{\mathbf{Q}})/|C_{\mathbf{Q}} + D_{\mathbf{Q}}|, \qquad t = (D_{\mathbf{Q}} - C_{\mathbf{Q}})/|D_{\mathbf{Q}} - C_{\mathbf{Q}}|. \tag{8.108}$$

The response functions $S_{yy}(\mathbf{Q}, \omega)$ and $S_{zz}(\mathbf{Q}, \omega)$ depend on the matrix elements $\langle n_{\alpha,\beta} \pm 1 | S^y(\mathbf{Q}) | n_{\alpha,\beta} \rangle$ and $\langle n_{\alpha,\beta} \pm 1 | S^z(\mathbf{Q}) | n_{\alpha,\beta} \rangle$, where the $\pm$ signs correspond to magnon creation or annihilation. Consider the process of magnon creation in the $\alpha$ mode. To calculate this we need to isolate the $\alpha_{\mathbf{Q}}^\dagger$ terms in (8.104) and (8.105). Substituting (8.106) into (8.104), and using (8.38) for the matrix element,[40] we obtain,

[40] $\langle n_\alpha + 1 | \alpha_{\mathbf{Q}}^\dagger | n_\alpha \rangle = (n_\alpha + 1)^{\frac{1}{2}}.$

$$|\langle n_\alpha + 1|S^y(\mathbf{Q})|n_\alpha\rangle|^2 = \frac{NS}{2}\left|\langle n_\alpha + 1|(2u_\alpha^+ - 2v_\alpha^+)a_\mathbf{Q}^\dagger|n_\alpha\rangle\right|^2$$

$$= 2NS(n_\alpha + 1)(u_\alpha^+ - v_\alpha^+)^2$$

$$= NS(n_\alpha + 1)\frac{A_\mathbf{Q} + B_\mathbf{Q} - s\sqrt{(C_\mathbf{Q} + D_\mathbf{Q})^2}}{\omega_\alpha}$$

$$= NS(n_\alpha + 1)\frac{A_\mathbf{Q} + B_\mathbf{Q} - C_\mathbf{Q} - D_\mathbf{Q}}{\omega_\alpha}. \quad (8.109)$$

Between the second and third lines we used (8.107) and the expression for $\hbar\omega_\alpha$ in (8.103), and in the third line we take the positive root and take the sign from (8.108). The expression for magnon annihilation is the same as (8.109) except the factor $(n_\alpha + 1)$ is replaced by $n_\alpha$. When we repeat these steps for the $\beta$ mode we find that the $S^y(\mathbf{Q})$ matrix element vanishes, so the $S_{yy}(\mathbf{Q},\omega)$ response function is non-zero only for the $\alpha$ mode.

The response functions are given by (8.58), and after summing over the final states $|n_\alpha \pm 1\rangle$ and averaging over all initial states, we find

$$S_{yy}(\mathbf{Q},\omega) = g^2\mu_B^2 NS\frac{A_\mathbf{Q} + B_\mathbf{Q} - C_\mathbf{Q} - D_\mathbf{Q}}{\omega_\alpha}$$
$$\times [n(\omega_\alpha)\delta(\hbar\omega + \hbar\omega_\alpha) + \{n(\omega_\alpha) + 1\}\delta(\hbar\omega - \hbar\omega_\alpha)]. \quad (8.110)$$

The corresponding expression for $S_{zz}(\mathbf{Q},\omega)$, which is non-zero only for the $\beta$ mode, is

$$S_{zz}(\mathbf{Q},\omega) = g^2\mu_B^2 NS\frac{A_\mathbf{Q} - B_\mathbf{Q} + C_\mathbf{Q} - D_\mathbf{Q}}{\omega_\beta}$$
$$\times [n(\omega_\beta)\delta(\hbar\omega + \hbar\omega_\beta) + \{n(\omega_\beta) + 1\}\delta(\hbar\omega - \hbar\omega_\beta)]. \quad (8.111)$$

In these expressions, $N$ is the number of spins on one sublattice, so division by $2N$ would give the response functions per spin. The expressions show that the $\alpha$ mode is polarized in the $y$ direction throughout the Brillouin zone (because it only appears in $S_{yy}$), while the $\beta$ mode is polarized in the $z$ direction. It follows that $S_{yz} = S_{zy} = 0$.

Figure 8.17 illustrates the magnon dispersion and response functions for a square-lattice anisotropic antiferromagnet calculated from (8.103) and (8.110)–(8.111). The examples show the effects of different types of single-ion anisotropy in the Hamiltonian (8.93), for the case when the ordered moments point along the $\pm x$ direction. Figure 8.17(a) is the isotropic case, whose spectrum has two degenerate branches as discussed previously. Including an out-of-plane anisotropy, panel (b), creates a gap in the branch with $z$ polarization (the $\beta$ mode) at the antiferromagnetic Brillouin zone centre $(\frac{1}{2}, \frac{1}{2})$, with the $y$-polarized branch (the $\alpha$ mode) ungapped. This is because when $K_z > K_x$ and $K_y = K_x$ it costs energy to tilt the spins out of the $xy$ plane, but it costs no energy to make a rigid rotation of the magnetic structure in the $xy$ plane. With easy-axis

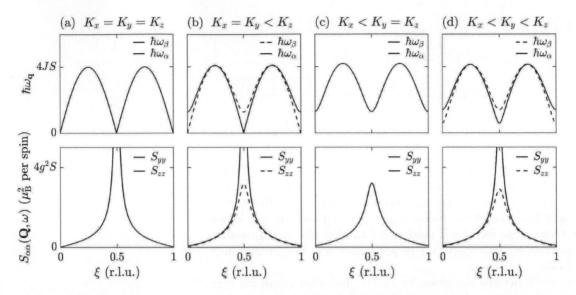

**Fig. 8.17** Magnon dispersion and response functions for a square-lattice antiferromagnet described by the Hamiltonian (8.93) with single-ion anisotropy. The horizontal axes follow the path $\mathbf{Q} = (\xi, \xi)2\pi/a$. Response functions in the lower panels are calculated for neutron energy-loss scattering at $T = 0$. Four types of anisotropy are illustrated: (a) isotropic Heisenberg model; (b) easy-plane anisotropy ($XY$-like); (c) easy-axis anisotropy (Ising-like); (d) both in-plane and out-plane anisotropy.

anisotropy, panel (c), it costs energy to tilt the spins in any direction away from the $x$ direction, so both modes are gapped and degenerate. If $K_y \neq K_z$, panel (d), then both modes are gapped but the gaps are different. Predictions for anisotropic exchange are qualitatively the same as those for single-ion anisotropy shown here.

## 8.4  Generalized susceptibility

In the previous sections, we calculated the magnetic inelastic scattering function using expressions derived from the Born approximation and Fermi's Golden Rule (Section 3.2). An alternative approach is to make use of the Fluctuation–Dissipation theorem, which relates the dynamical part of the scattering function $\widetilde{S}_{\alpha\beta}(\mathbf{Q}, \omega)$ to the absorptive part of the generalized susceptibility $\chi''_{\beta\alpha}(\mathbf{Q}, \omega)$. The relationship, which is derived in Sections D.3 and D.4, is

$$\widetilde{S}_{\alpha\beta}(\mathbf{Q}, \omega) = \{1 + n(\omega)\}\frac{1}{\pi}\chi''_{\beta\alpha}(\mathbf{Q}, \omega). \qquad (8.112)$$

The particular combination of partial scattering functions $\widetilde{S}_{\alpha\beta}(\mathbf{Q}, \omega)$ contained in the magnetic cross-section will depend on the direction of $\mathbf{Q}$ and on whether polarization analysis is employed. For example, for unpolarized neutrons,[41]

[41]See eqn (6.12).

$$\widetilde{S}(\mathbf{Q}, \omega) = \sum_{\alpha\beta}(\delta_{\alpha\beta} - \hat{Q}_\alpha\hat{Q}_\beta)\widetilde{S}_{\alpha\beta}(\mathbf{Q}, \omega). \qquad (8.113)$$

The generalized susceptibility relates the linear response of the magnetization **M** to a space- and time-varying perturbing field **H**. If the field is oscillatory, and has a single Fourier component with wavevector **Q** and angular frequency $\omega$, then[42]

$$\mathbf{M}(\mathbf{Q}, \omega) = \frac{\mu_0}{V} \boldsymbol{\chi}(\mathbf{Q}, \omega) \mathbf{H}(\mathbf{Q}, \omega). \tag{8.114}$$

The susceptibility is a complex function, which we write[43]

$$\boldsymbol{\chi}(\mathbf{Q}, \omega) = \boldsymbol{\chi}'(\mathbf{Q}, \omega) - \mathrm{i}\boldsymbol{\chi}''(\mathbf{Q}, \omega), \tag{8.115}$$

with $\boldsymbol{\chi}'(\mathbf{Q}, \omega)$ the reactive part and $\boldsymbol{\chi}''(\mathbf{Q}, \omega)$ the absorptive part (see Section D.2). Since it couples two vectors, $\boldsymbol{\chi}(\mathbf{Q}, \omega)$ is a second-rank tensor (i.e. a $3 \times 3$ matrix) with components $\chi_{\alpha\beta}(\mathbf{Q}, \omega)$, $\alpha, \beta = x, y, z$, which describe the response in the $\alpha$ direction to a probe field in the $\beta$ direction. For magnetic systems, the diagonal components $\chi'_{\alpha\alpha}(\mathbf{Q}, \omega)$ and $\chi''_{\alpha\alpha}(\mathbf{Q}, \omega)$ are real functions, and are the real and (minus the) imaginary parts of $\chi_{\alpha\alpha}(\mathbf{Q}, \omega)$, respectively. Terms with $\alpha \neq \beta$ are not in general real, but they only appear in symmetric or antisymmetric combinations, $\chi''_{\alpha\beta}(\mathbf{Q}, \omega) + \chi''_{\beta\alpha}(\mathbf{Q}, \omega)$ or $\mathrm{i}\{\chi''_{\alpha\beta}(\mathbf{Q}, \omega) - \chi''_{\beta\alpha}(\mathbf{Q}, \omega)\}$, both of which are real (see Section D.7).

Well-established theoretical methods exist to calculate the generalized susceptibility. For example, if the eigenstates of the magnetic system are known then one can use the following expression, which is derived in Section D.4.1:

$$\chi_{\alpha\beta}(\mathbf{Q}, \omega) = \lim_{\epsilon \to 0^+} \sum_{\lambda, \lambda'} (p_\lambda - p_{\lambda'}) \frac{\langle \lambda | M_\beta^\dagger(\mathbf{Q}) | \lambda' \rangle \langle \lambda' | M_\alpha(\mathbf{Q}) | \lambda \rangle}{E_{\lambda'} - E_\lambda - \hbar\omega + \mathrm{i}\hbar\epsilon}. \tag{8.116}$$

Here, $|\lambda\rangle$ and $|\lambda'\rangle$ are eigenfunctions of the magnetic Hamiltonian, and $E_\lambda$, $E_{\lambda'}$ and $p_\lambda$, $p_{\lambda'}$ are the corresponding eigenvalues and thermal occupancies.

In Section 8.5, we show how to calculate the scattering from cooperative magnetic excitations using the generalized susceptibility in the Random Phase Approximation (RPA), and in Section 8.6 we obtain $\chi_{\alpha\beta}(\mathbf{Q}, \omega)$ for itinerant magnetic systems in terms of the electron band structure.

Usually it is reasonable to assume the approximations which permit us to define a magnetic form factor and Debye–Waller factor, and to take these factors out of the generalized susceptibility (see Sections 6.1.3, 6.2.2 and 6.3.2). What remains is then the *reduced* susceptibility. We shall be sure to mention when this is done in the following sections.

## 8.5   Random Phase Approximation

Cooperative magnetic phenomena can often be described adequately by two-ion interactions, for example the Heisenberg Hamiltonian, eqn (8.30), but it is not usually possible to obtain an exact solution for the ground

[42]The units of $\chi$ are discussed in Section D.9.

[43]The minus sign is a consequence of our choice of time variation $\exp(\mathrm{i}\omega t)$. It means that if $\boldsymbol{\chi}'$ and $\boldsymbol{\chi}''$ are real and have the same sign (i.e. both positive or both negative) then the response of the system lags behind the perturbing field, as must necessarily be the case when energy is dissipated.

state and excitations of such systems. Linear spin-wave theory, described earlier in this chapter, is one scheme that can be used to obtain an approximate solution. Another is the *Random Phase Approximation* (RPA) which we discuss now.

The essence of the RPA is to replace the spin operators on sites surrounding any particular ion by their thermal averages. Each individual spin can then be regarded as interacting with an effective field, or *mean field*,[44] which depends on the average states of the surrounding ions as well as any external magnetic field. The net effect of this scheme is to replace two-ion interactions with effective single-ion terms.

The RPA is equivalent to the neglect of correlated fluctuations of the spins from their mean values. It works best for (i) magnetically ordered systems at temperatures sufficiently below the ordering temperature that the ordered moments are close to their saturated values, and (ii) paramagnetic systems in the high temperature limit where spin fluctuations on different sites are uncorrelated. The approximation breaks down in the vicinity of phase transitions, where fluctuations tend to be large and highly correlated.

The RPA method can be used as an alternative to linear spin-wave theory for calculating the dispersive excitations of localized magnetic systems. A particular strength of the RPA, however, is that it is straightforward to include single-ion anisotropy effects if the single-ion states are known. Another is that the method can be used to treat correlations in itinerant magnetic systems, see Section 8.6. Indeed, the RPA has been applied to a wide range of problems in which systems are coupled via pairwise interactions.

We first introduce the mean-field approximation, then provide a simple derivation of the generalized susceptibility $\chi(\mathbf{Q}, \omega)$ in the RPA. Finally, we illustrate the method by calculating the magnetic spectrum of the lowest exciton mode in the $4f$ element praseodymium.

### 8.5.1   The effective field

We consider a Hamiltonian of the form

$$\mathcal{H} = \sum_i \mathcal{H}_{\mathrm{CF}}(i) + \sum_i \mathcal{H}_{\mathrm{Z}}(i) - \frac{1}{2} \sum_i \sum_{j \neq i} J_{ij} \mathbf{S}_i \cdot \mathbf{S}_j. \tag{8.117}$$

The first two terms describe the crystal field and Zeeman interactions, respectively, eqns (8.2) and (C.134). The third term describes isotropic Heisenberg exchange interactions between pairs of spins.[45]

To obtain a simple approximation to the two-ion term we write

$$\mathbf{S}_i = \langle \mathbf{S}_i \rangle + \mathbf{S}_i - \langle \mathbf{S}_i \rangle$$

$$= \langle \mathbf{S}_i \rangle + \Delta \mathbf{S}_i, \tag{8.118}$$

where $\langle \mathbf{S}_i \rangle$ represents the thermal average, and substitute (8.118) into the Heisenberg term of (8.117). We neglect the product terms $\Delta \mathbf{S}_i \cdot \Delta \mathbf{S}_j$,

[44] Also known as a *molecular field*.

[45] Note that we have changed the sign of the Heisenberg interaction compared with elsewhere in this chapter, e.g. eqn (8.30). This is for consistency with other descriptions of the RPA.

Later in this section we extend the exchange term to allow a general bilinear product of spin operators.

which represent small correlated deviations from the average, so that

$$\mathcal{H} \approx \sum_i \mathcal{H}_{\mathrm{CF}}(i) + \sum_i \mathcal{H}_{\mathrm{Z}}(i) - \sum_i \mathbf{S}_i \cdot \sum_{j \neq i} J_{ij} \langle \mathbf{S}_j \rangle + \frac{1}{2} \sum_i \sum_{j \neq i} J_{ij} \langle \mathbf{S}_i \rangle \cdot \langle \mathbf{S}_j \rangle.$$
(8.119)

The last term on the right-hand side does not affect the eigenvalues of $\mathcal{H}$ relative to the ground state, and can be treated as a constant. Equation (8.119) is the *mean-field approximation* to (8.117). The effect of this is to decouple the spins, resulting in an effective single-ion Hamiltonian. The third term can be represented by a Zeeman interaction eqn (C.134) between the magnetic moments[46] $\boldsymbol{\mu}_i = -g\mu_{\mathrm{B}}\mathbf{S}_i$ and an effective magnetic field

$$\mathbf{H}_i^{\mathrm{eff}} = \frac{-1}{g\mu_{\mathrm{B}}\mu_0} \sum_{j \neq i} J_{ij} \langle \mathbf{S}_j \rangle.$$
(8.120)

The mean-field decoupling of $\mathcal{H}$ can be used to obtain the single-ion eigenstates, and hence to determine the values of $\langle \mathbf{S}_i \rangle$ self-consistently by iteration.

[46]For atoms in which the total angular momentum $\mathbf{J}$ is a good quantum number, replace $g\mu_{\mathrm{B}}\mathbf{S}_i$ with $g_J\mu_{\mathrm{B}}\mathbf{J}_i$, where $g_J$ is the Landé $g$-factor – see Section C.3.

## 8.5.2   RPA susceptibility

The RPA expression for the generalized susceptibility can be obtained by a similar approach. It will be convenient to introduce the Fourier transform operator $\mathbf{S}(\mathbf{q})$, defined by

$$\mathbf{S}(\mathbf{q}) = \sum_i \mathbf{S}_i \exp(i\mathbf{q} \cdot \mathbf{r}_i); \quad \mathbf{S}_i = \frac{1}{N} \sum_{\mathbf{q}} \mathbf{S}(\mathbf{q}) \exp(-i\mathbf{q} \cdot \mathbf{r}_i), \quad (8.121)$$

where the second sum is over all $\mathbf{q}$ in the first Brillouin zone, and $N$ is the number of magnetic unit cells. By direct substitution, and using (B.50), one can show that the Heisenberg Hamiltonian becomes

$$-\frac{1}{2} \sum_i \sum_{j \neq i} J_{ij} \mathbf{S}_i \cdot \mathbf{S}_j = -\frac{1}{2N} \sum_{\mathbf{q}} J(-\mathbf{q}) \mathbf{S}(-\mathbf{q}) \cdot \mathbf{S}(\mathbf{q}), \quad (8.122)$$

where

$$J(\mathbf{q}) = \frac{1}{N} \sum_i \sum_{j \neq i} J_{ij} \exp\left\{ i\mathbf{q} \cdot (\mathbf{r}_j - \mathbf{r}_i) \right\}. \quad (8.123)$$

We now decouple the Fourier components in (8.122) by following the steps in (8.118)–(8.119) but with $\mathbf{S}(\mathbf{q})$ instead of $\mathbf{S}_i$. Neglecting the small term $\Delta\mathbf{S}(-\mathbf{q}) \cdot \Delta\mathbf{S}(\mathbf{q})$ and omitting the constant term $\langle \mathbf{S}(-\mathbf{q}) \rangle \cdot \langle \mathbf{S}(\mathbf{q}) \rangle$, we obtain the RPA two-ion Hamiltonian

$$\mathcal{H}^{\mathrm{RPA}} = -\frac{1}{2N} \sum_{\mathbf{q}} J(-\mathbf{q}) \left\{ \mathbf{S}(-\mathbf{q}) \cdot \langle \mathbf{S}(\mathbf{q}) \rangle + \langle \mathbf{S}(-\mathbf{q}) \rangle \cdot \mathbf{S}(\mathbf{q}) \right\}$$

$$= -\frac{1}{2N} \sum_{\mathbf{q}} \{ J(-\mathbf{q}) + J(\mathbf{q}) \} \mathbf{S}(-\mathbf{q}) \cdot \langle \mathbf{S}(\mathbf{q}) \rangle. \quad (8.124)$$

In general, $J(-\mathbf{q}) = J^*(\mathbf{q})$, but for now we will assume the system to be centrosymmetric, in which case $J(-\mathbf{q}) = J(\mathbf{q})$.

[47]$\mathbf{Q}$ is not restricted to the first Brillouin zone.

To obtain $\chi(\mathbf{Q}, \omega)$ we will need to find the response of the system to a small space- and time-varying probe field $\mathbf{H}(\mathbf{r}, t)$ (see Section D.2). Without any loss of generality, we select one single Fourier component[47]

$$\mathbf{H}(\mathbf{r}, t) = \mathbf{h}(\mathbf{Q}, t) \exp(-\mathrm{i}\mathbf{Q} \cdot \mathbf{r}) \qquad (8.125)$$

so that the perturbation is described by the Hamiltonian

$$g\mu_{\mathrm{B}}\mu_0 \sum_i \mathbf{S}_i \cdot \mathbf{H}(\mathbf{r}_i, t) = g\mu_{\mathrm{B}}\mu_0 \sum_i \mathbf{S}_i \exp(-\mathrm{i}\mathbf{Q} \cdot \mathbf{r}_i) \cdot \mathbf{h}(\mathbf{Q}, t)$$

$$= g\mu_{\mathrm{B}}\mu_0 \, \mathbf{S}(-\mathbf{Q}) \cdot \mathbf{h}(\mathbf{Q}, t). \qquad (8.126)$$

The second line comes from the Fourier transform (8.121). We assume that the system responds linearly to the probe field, so that only the $\mathbf{Q}$ component has a non-zero average response $\langle \Delta \mathbf{S}(\mathbf{Q}, t) \rangle$ corresponding to a space- and time-varying mean field. With the addition of (8.126), the RPA Hamiltonian (8.124) becomes

$$\mathcal{H}^{\mathrm{RPA}} = -\frac{1}{N} J(\mathbf{Q}) \mathbf{S}(-\mathbf{Q}) \cdot \langle \Delta \mathbf{S}(\mathbf{Q}, t) \rangle + g\mu_{\mathrm{B}}\mu_0 \mathbf{S}(-\mathbf{Q}) \cdot \mathbf{h}(\mathbf{Q}, t)$$

$$= g\mu_{\mathrm{B}}\mu_0 \mathbf{S}(-\mathbf{Q}) \cdot \mathbf{H}^{\mathrm{eff}}(\mathbf{Q}, t), \qquad (8.127)$$

where

$$\mathbf{H}^{\mathrm{eff}}(\mathbf{Q}, t) = \mathbf{h}(\mathbf{Q}, t) - \frac{1}{Ng\mu_{\mathrm{B}}\mu_0} J(\mathbf{Q}) \langle \Delta \mathbf{S}(\mathbf{Q}, t) \rangle. \qquad (8.128)$$

[48]The local susceptibility does not depend on $\mathbf{Q}$ if the magnetic form factor $f^2(\mathbf{Q})$ has been factored out of it. If not, then $\chi^0(\omega)$ will have a $\mathbf{Q}$ dependence corresponding to the intra-atomic distribution of magnetization, as also will $\chi(\mathbf{Q}, \omega)$.

We may consider one angular frequency component $\omega$, and using (8.114), introduce the local and generalized susceptibilities[48] via the matrix equations,

$$-\frac{g\mu_{\mathrm{B}}}{\mu_0} \langle \Delta \mathbf{S}(\mathbf{Q}, \omega) \rangle = \chi^0(\omega) \mathbf{H}^{\mathrm{eff}}(\mathbf{Q}, \omega) \qquad \text{(local)}, \qquad (8.129)$$

$$= \chi(\mathbf{Q}, \omega) \mathbf{h}(\mathbf{Q}, \omega) \qquad \text{(generalized)}. \qquad (8.130)$$

Equating (8.129) and (8.130), and replacing $\mathbf{H}^{\mathrm{eff}}$ by (8.128), we obtain

$$\chi^0(\omega) \left\{ \frac{1}{Ng^2\mu_{\mathrm{B}}^2} J(\mathbf{Q})\chi(\mathbf{Q}, \omega)\mathbf{h}(\mathbf{Q}, \omega) + \mathbf{h}(\mathbf{Q}, \omega) \right\} = \chi(\mathbf{Q}, \omega)\mathbf{h}(\mathbf{Q}, \omega), \qquad (8.131)$$

so that

$$\chi(\mathbf{Q}, \omega) = \chi^0(\omega) \left\{ \frac{1}{Ng^2\mu_{\mathrm{B}}^2} J(\mathbf{Q})\chi(\mathbf{Q}, \omega) + \mathbf{1} \right\}, \qquad (8.132)$$

where $\mathbf{1}$ is the identity matrix. Rearranging,

$$\chi(\mathbf{Q}, \omega) = \left\{ \mathbf{1} - \frac{1}{Ng^2\mu_{\mathrm{B}}^2} \chi^0(\omega) J(\mathbf{Q}) \right\}^{-1} \chi^0(\omega), \qquad (8.133)$$

or

$$\mathbf{1} = \left\{ [\chi^0(\omega)]^{-1} - \frac{1}{Ng^2\mu_{\mathrm{B}}^2} J(\mathbf{Q}) \right\} \chi(\mathbf{Q}, \omega). \qquad (8.134)$$

Results (8.133) and (8.134) are matrix equations that can be solved for $\chi(\mathbf{Q}, \omega)$. The single-ion susceptibility $\chi^0(\omega)$ can be calculated self-consistently from the eigenstates of the mean-field (MF) Hamiltonian (8.119) via (8.116), see also eqns (D.38)–(D.39). This method is known as the MF–RPA.

At low temperatures, the description of the magnetic spectrum provided by the MF-RPA susceptibility $\chi''(\mathbf{Q}, \omega)$ is similar to that given by linear spin-wave theory, and in some simple cases it is identical at $T = 0$ (see Exercise 8.5 for a proof for the Heisenberg ferromagnet). The two methods differ at $T > 0$ and, more importantly, the MF-RPA method is better suited to describe systems with single-ion anisotropy, and is easier to generalize for more complex systems, as we now describe.

## Generalizations

We consider two extensions of the above RPA treatment. First, we replace the isotropic Heisenberg exchange interaction in (8.117) with a general bilinear interaction, so that

$$J_{ij}\mathbf{S}_i \cdot \mathbf{S}_j \to \mathbf{S}_i^{\mathrm{T}} \mathsf{J}_{ij} \mathbf{S}_j = \sum_{\alpha, \beta} J_{ij}^{\alpha\beta} S_i^{\alpha} S_j^{\beta}, \qquad (8.135)$$

where $\alpha, \beta$ are Cartesian components of the spin, and $J_{ij}^{\alpha\beta}$ is a second-rank tensor that describes the most general anisotropic two-spin interaction. The $J_{ij}^{\alpha\beta}$ are the components of the $3 \times 3$ *exchange matrix* $\mathsf{J}_{ij}$, whose Fourier transform $\mathsf{J}(\mathbf{Q})$ replaces the scalar $J(\mathbf{Q})$ in the RPA matrix equations (8.132)–(8.134).

The second extension is to systems that contain more than one inequivalent sublattice. Generalizing (8.121), we define the Fourier transform spin operators for sublattice $s$ by

$$\mathbf{S}_s(\mathbf{q}) = \sum_{i \in s} \mathbf{S}_i \exp(i\mathbf{q} \cdot \mathbf{r}_i); \quad \mathbf{S}_i = \frac{1}{N} \sum_{\mathbf{q}} \mathbf{S}_s(\mathbf{q}) \exp(-i\mathbf{q} \cdot \mathbf{r}_i), \quad (8.136)$$

where, as before, $N$ is the number of magnetic unit cells. The RPA Hamiltonian (8.127)–(8.128) becomes

$$\mathcal{H}^{\mathrm{RPA}} = g\mu_{\mathrm{B}}\mu_0 \sum_s \mathbf{S}_s(-\mathbf{Q}) \cdot \mathbf{H}_s^{\mathrm{eff}}(\mathbf{Q}, t), \qquad (8.137)$$

where,

$$\mathbf{H}_s^{\mathrm{eff}}(\mathbf{Q}, t) = \mathbf{h}_s(\mathbf{Q}, t) - \frac{1}{Ng\mu_{\mathrm{B}}\mu_0} \sum_{s'} \mathsf{J}_{ss'}(-\mathbf{Q})\langle \Delta\mathbf{S}_{s'}(\mathbf{Q}, t)\rangle, \quad (8.138)$$

and $\mathsf{J}_{ss'}(-\mathbf{Q})$ is defined by (8.123) with $i \in s$ and $j \in s'$. We generalize (8.129) and (8.130) to

$$-\frac{g\mu_{\mathrm{B}}}{\mu_0}\langle \Delta\mathbf{S}_s(\mathbf{Q}, \omega)\rangle = \chi^s(\omega)\, \mathbf{H}_s^{\mathrm{eff}}(\mathbf{Q}, \omega) \qquad (8.139)$$

$$= \chi^{ss'}(\mathbf{Q}, \omega)\, \mathbf{h}_{s'}(\mathbf{Q}, \omega), \qquad (8.140)$$

where $\chi^s(\omega)$ is the local susceptibility of a single ion on sublattice $s$, and $\chi^{ss'}(\mathbf{Q}, \omega)$ is the generalized two-ion susceptibility which describes the response of the spins on sublattice $s$ to a probing field on sublattice $s'$. Retracing the steps in (8.131)–(8.134), we obtain

$$\delta_{ss'} = \sum_t \left\{ \delta_{st} \left[\chi^s(\omega)\right]^{-1} - \frac{1}{Ng^2\mu_B^2} J_{st}(\mathbf{Q}) \right\} \chi^{ts'}(\mathbf{Q}, \omega). \quad (8.141)$$

This represents a set of simultaneous matrix equations which must be solved for each $\mathbf{Q}$ and $\omega$ to obtain the two-ion susceptibilities $\chi^{ss'}(\mathbf{Q}, \omega)$. Usually this must be done numerically,[49] although analytic solutions are possible in simple cases. Finally, the total susceptibility for all sublattices is

$$\chi(\mathbf{Q}, \omega) = \sum_{s,s'} \chi^{ss'}(\mathbf{Q}, \omega). \quad (8.142)$$

[49] An efficient algorithm has been described by Rotter *et al.*, (2012), which transforms (8.141) into a generalized eigenvalue problem at each $\mathbf{Q}$.

### Example: Crystal field excitons in praseodymium

We shall describe how the RPA approach was used to calculate the low-energy magnetic spectrum of praseodymium (Houmann *et al.*, 1975; Houmann *et al.*, 1979). Elemental praseodymium (Pr) is a paramagnetic metal which crystallizes in the double hexagonal-close-packed (dhcp) structure. Two-thirds of the Pr atoms occupy sites with hexagonal symmetry and form an hexagonal-close-packed (hcp) sublattice, while the remaining Pr sites have an approximately cubic environment. The magnetic properties of Pr are governed by localized states of the $4f^2$ configuration, whose ground state manifold ($L = 5, S = 1, J = 4$) is split by the crystal field into a series of discrete energy levels. Two-ion magnetic coupling is not strong enough to induce magnetic ordering, but it is sufficient to cause significant dispersion of the crystal field excitations.

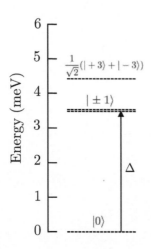

**Fig. 8.18** Lowest crystal-field levels for the hexagonal sites of dhcp Pr. The eigenfunctions are expressed in terms of $|M_J\rangle$ components with the quantization direction along the $c$ axis.

[50] Cross-coupling between excitations on the hexagonal and cubic sites is weak and can be neglected.

[51] Propagating modes of magnetically-coupled, crystal-field-split, paramagnetic ions, such as those in Pr, are distinct from the spin wave excitations found in magnetically ordered systems.

[52] We assume the dipole approximation, and factor the dipole magnetic form factor $f^2(Q)$ out of the susceptibility.

We focus here on the hexagonal sites. There are two sites per hcp unit cell, and these have identical local environments apart from an inversion. The crystal field ground state is a singlet with eigenfunction $|M_J\rangle = |0\rangle$, and the first excited state is a doublet $|\pm 1\rangle$ at $\Delta = 3.5$ meV above the ground state, Fig. 8.18. The neutron spectrum is dominated by transitions between these two levels. Magnetic coupling between pairs of Pr within the hcp sublattice[50] allows the crystal-field excitations to propagate and form a band in momentum space. These modes are called *excitons*.[51]

The basis of two Pr atoms per hcp unit cell forms two identical subsystems which we label 1 and 2. The single-ion susceptibility tensors for the two basis atoms are the same ($\chi^1_{\alpha\beta} = \chi^2_{\alpha\beta} \equiv \chi^0_{\alpha\beta}$) and may be evaluated from (8.116) with[52] $M_\alpha(\mathbf{Q}) = g_J\mu_B J^\alpha$. The only non-zero matrix elements between the singlet and the doublet are $\langle +1|J^+|0\rangle$, $\langle -1|J^-|0\rangle$, $\langle 0|J^+|-1\rangle$ and $\langle 0|J^-|+1\rangle$, all of which are equal to $\sqrt{J(J+1)} = \sqrt{20}$. Including only these two transitions, we obtain the non-zero tensor elements

$$\chi^0_{xx}(\omega) = \chi^0_{yy}(\omega) = \lim_{\epsilon \to 0^+} \left( \frac{1}{\Delta - \hbar z} + \frac{1}{\Delta + \hbar z} \right) M^2(p_0 - p_1) \quad (8.143)$$

per atom, where $z = \omega - i\epsilon$, $M = g_J \mu_B \sqrt{J(J+1)/2}$, and $p_\lambda$ is the thermal population of level $\lambda$.[53]

All the exchange interactions are contained in four matrices $\mathsf{J}_{ss'}(\mathbf{Q})$ ($s, s' = 1, 2$). The two subsystems are equivalent and possess a centre of symmetry, but a Pr site in subsystem 1 is not a centre of symmetry for subsystem 2. Hence,

$$\mathsf{J}_{11}(\mathbf{Q}) = \mathsf{J}_{11}(-\mathbf{Q}) = \mathsf{J}_{22}(\mathbf{Q})$$
$$\mathsf{J}_{12}(\mathbf{Q}) = \mathsf{J}_{21}(-\mathbf{Q}) = \mathsf{J}_{21}^*(\mathbf{Q}). \tag{8.144}$$

The RPA equations (8.141) for $s' = 1$ and $s = 1, 2$ give

$$\chi^{11}(\mathbf{Q}, \omega) = \chi^0(\omega) \left\{ 1 + \eta \mathsf{J}_{11}(\mathbf{Q})\chi^{11}(\mathbf{Q}, \omega) + \eta \mathsf{J}_{12}(\mathbf{Q})\chi^{21}(\mathbf{Q}, \omega) \right\}$$
$$\chi^{21}(\mathbf{Q}, \omega) = \eta \chi^0(\omega) \left\{ \mathsf{J}_{21}(\mathbf{Q})\chi^{11}(\mathbf{Q}, \omega) + \mathsf{J}_{22}(\mathbf{Q})\chi^{21}(\mathbf{Q}, \omega) \right\}, \tag{8.145}$$

per hcp unit cell ($N = 1$), where $\eta = 1/(g_J^2 \mu_B^2)$. The equations for $s' = 2$, $s = 1, 2$ are similar. These matrix equations can be solved analytically, but the expressions for the general case are lengthy. Instead, we will consider a special case in which the matrices $\mathsf{J}_{ss'}(\mathbf{Q})$ are diagonal,[54] so that the $x$ and $y$ components of the RPA equations decouple to give separate equations for $\chi_{xx}^{ss'}$ and $\chi_{yy}^{ss'}$. Rearranging eqns (8.145), we obtain

$$\chi_{\alpha\alpha}^{11}(\mathbf{Q}, \omega) = \frac{\chi^0(\omega) \left\{ 1 - \eta \chi^0(\omega) J_{11}^{\alpha\alpha}(\mathbf{Q}) \right\}}{D_{\alpha\alpha}(\mathbf{Q}, \omega)} \tag{8.146}$$

$$\chi_{\alpha\alpha}^{21}(\mathbf{Q}, \omega) = \frac{\eta \{\chi^0(\omega)\}^2 J_{21}^{\alpha\alpha}(\mathbf{Q})}{D_{\alpha\alpha}(\mathbf{Q}, \omega)}, \tag{8.147}$$

where $\alpha = x, y$, and

$$D_{\alpha\alpha}(\mathbf{Q}, \omega) = \left\{ 1 - \eta \chi^0(\omega) J_{11}^{\alpha\alpha}(\mathbf{Q}) - \eta \chi^0(\omega) |J_{12}^{\alpha\alpha}(\mathbf{Q})| \right\}$$
$$\times \left\{ 1 - \eta \chi^0(\omega) J_{11}^{\alpha\alpha}(\mathbf{Q}) + \eta \chi^0(\omega) |J_{12}^{\alpha\alpha}(\mathbf{Q})| \right\}. \tag{8.148}$$

The other partial susceptibilities are given by $\chi_{\alpha\alpha}^{22}(\mathbf{Q}, \omega) = \chi_{\alpha\alpha}^{11}(\mathbf{Q}, \omega)$ and $\chi_{\alpha\alpha}^{12}(\mathbf{Q}, \omega) = \chi_{\alpha\alpha}^{21}(-\mathbf{Q}, \omega)$. From (8.142), the total susceptibility is

$$\chi_{\alpha\alpha}(\mathbf{Q}, \omega) = \frac{\chi^0(\omega) \left[ 2 - \eta \chi^0(\omega) \left\{ 2 J_{11}^{\alpha\alpha}(\mathbf{Q}) - J_{21}^{\alpha\alpha}(\mathbf{Q}) - J_{12}^{\alpha\alpha}(\mathbf{Q}) \right\} \right]}{D_{\alpha\alpha}(\mathbf{Q}, \omega)}. \tag{8.149}$$

Since $J_{11}^{\alpha\alpha}(\mathbf{Q})$ and $|J_{12}^{\alpha\alpha}(\mathbf{Q})|$ are both real, $\chi_{\alpha\alpha}(\mathbf{Q}, \omega)$ has poles, i.e. singularities, at values of $\omega$ where the real part of $D(\mathbf{Q}, \omega)$ vanishes. The real part of $\chi^0(\omega)$ is, from (8.143),

$$\{\chi^0(\omega)\}' = \frac{2M^2 \Delta (p_0 - p_1)}{\Delta^2 - (\hbar\omega)^2}, \tag{8.150}$$

and the zeroes of $D(\mathbf{Q}, \omega)$ are found by substitution of (8.150) into (8.148). These occur when $\hbar\omega = E_\alpha^+$ and $E_\alpha^-$, $\alpha = x, y$, where

$$E_\alpha^\pm(\mathbf{Q}) = \left[ \Delta^2 - J(J+1)\Delta(p_0 - p_1) \left\{ J_{11}^{\alpha\alpha}(\mathbf{Q}) \pm |J_{12}^{\alpha\alpha}(\mathbf{Q})| \right\} \right]^{1/2}. \tag{8.151}$$

[53]Here, $p_1$ is the population of one component of the doublet. When $k_B T \ll \Delta$, $p_0 \approx 1$ and $p_1 \approx 0$.

[54]This applies when $\mathbf{q}$ is parallel to the $y$ axis, for reasons of symmetry. In general, however, the off-diagonal exchange terms cannot be neglected in praseodymium.

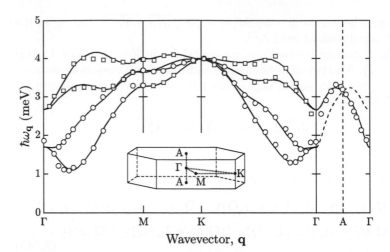

**Fig. 8.19** Dispersion relations for the singlet–doublet exciton propagating on the hexagonal sites in praseodymium. The points are the experimental data taken at 6 K, and the lines are the result of RPA calculations. The wavevector path in the hexagonal Brillouin zone is shown. Acoustic and optic modes are denoted by squares and circles, respectively, and are split in most directions into two different polarizations. (After Houmann *et al.*, 1979.)

[55] This is not in fact the case for Pr, according to experiment (see Fig. 8.19).

[56] In the acoustic mode, the spin fluctuations are in-phase within the unit cell in limit of $\mathbf{Q} \to 0$. In the optic mode, they are not in-phase.

[57] Equal to minus the imaginary part.

There are, therefore, a total of four modes for each $\mathbf{Q}$, one pair with pure $x$ polarization and the other with pure $y$ polarization. If the exchange interactions were isotropic, i.e. $J^{xx}_{ss'} = J^{yy}_{ss'}$, then the $x$- and $y$- modes would be degenerate.[55] For a given polarization, the modes are identified as *acoustic* and *optic*,[56] depending on the sign in (8.151).

In the vicinity of the poles, which occur at $\hbar\omega = E^{\pm}_{\alpha}(\mathbf{Q})$, the generalized susceptibility (8.149) has the approximate form (omitting the $\mathbf{Q}$ arguments on the right-hand side)

$$\chi_{\alpha\alpha}(\mathbf{Q},\omega) \approx \lim_{\epsilon \to 0^+} \frac{C^{\pm}_{\alpha\alpha}}{(E^{\pm}_{\alpha})^2 - (\hbar z)^2}$$

$$= \lim_{\epsilon \to 0^+} \left( \frac{1}{E^{\pm}_{\alpha} - \hbar z} + \frac{1}{E^{\pm}_{\alpha} + \hbar z} \right) \frac{C^{\pm}_{\alpha\alpha}}{2E^{\pm}_{\alpha}}, \qquad (8.152)$$

where $z = \omega - i\epsilon$, and

$$C^{\pm}_{\alpha\alpha}(\mathbf{Q}) = 2M^2\Delta(p_0 - p_1)\left[1 \pm \mathrm{Re}\{J^{\alpha\alpha}_{12}(\mathbf{Q})\}/|J^{\alpha\alpha}_{12}(\mathbf{Q})|\right]. \qquad (8.153)$$

The neutron inelastic scattering cross-section, eqn (8.1), is proportional to $S(\mathbf{Q},\omega)$, which for unpolarized neutrons is given in the dipole approximation by (8.57). The dynamical part of $S_{\alpha\alpha}(\mathbf{Q},\omega)$ is related to the absorptive part of the generalized susceptibility by (8.112). From Dirac's formula (B.20), the absorptive part[57] of (8.152) may be written

$$\chi''_{\alpha\alpha}(\mathbf{Q},\omega) = \pi \left\{ \delta(\hbar\omega - E^{\pm}_{\alpha}) - \delta(\hbar\omega + E^{\pm}_{\alpha}) \right\} \frac{C^{\pm}_{\alpha\alpha}}{2E^{\pm}_{\alpha}}. \qquad (8.154)$$

Neutron scattering measurements of the singlet–doublet excitation spectrum for the hexagonal sites in Pr were carried out in the 1970s (Houmann *et al.*, 1975 and 1979). Figure 8.19 shows measurements of the dispersion relations along some of the high-symmetry directions in the Brillouin zone. The RPA method was found to be very successful in describing the dispersion relations and temperature dependence of some of the modes. The acoustic and optic modes and their splitting into different polarizations (where allowed by symmetry) could be identified unambiguously from intensity measurements. Along $\Gamma M$ the excitation spectrum is described by the analytic formulae given in (8.146)–(8.154).

# 8.6   Itinerant magnetism

When atoms come together to form a solid, orbital overlap causes the discrete electronic levels of the isolated atoms to broaden into bands of states. The solid is metallic if there is a non-vanishing density of electron states at the Fermi energy.[58] In the simplest picture, metals contain *valence electrons* and *conduction electrons*. Valence electrons are bound to the atoms, whereas conduction electrons are *itinerant*, i.e. free to roam around the lattice and carry electrical current.[59] Magnetism arising from unpaired valence electrons can usually be described by local-moment models, such as those described earlier in this chapter. The conduction electrons, on the other hand, have wave functions that are typically coherent over many atoms. This means that the operators representing orbital and spin moments cannot be assigned to individual atomic sites, and the magnetic behaviour is best treated within a band model.

The energy eigenfunctions of electrons in a periodic potential take the form of *Bloch functions*, which may be written

$$\psi_{\mathbf{k}n}(\mathbf{r}) = \exp(i\mathbf{k} \cdot \mathbf{r})u_{\mathbf{k}n}(\mathbf{r}), \qquad (8.155)$$

where $\mathbf{k}$ is a discrete wavevector and $n$ is a band index. The function $u_{\mathbf{k}n}(\mathbf{r}) = u_{\mathbf{k}n}(\mathbf{r}+\mathbf{l})$ is periodic in the lattice ($\mathbf{l}$ is any lattice vector). For a given $n$, the eigenvalues form a band of states whose energies $E_{\mathbf{k}} = E_{\mathbf{k}+\mathbf{G}}$ are periodic in the reciprocal lattice ($\mathbf{G}$ is any reciprocal lattice vector). Different bands derive from different atomic orbitals, or combinations of atomic orbitals if there are two or more atoms per primitive unit cell. Each band is split into two sub-bands for the two spin states of the electrons. In a paramagnetic metal with inversion symmetry these sub-bands are degenerate, but they can be split by an external magnetic field, by magnetic order, or by loss of inversion symmetry.

## 8.6.1   Non-interacting magnetic susceptibility

The magnetic response of a system of itinerant electrons is described by the generalized susceptibility, which we may obtain from (8.116). For simplicity, we assume a single band and drop the band index $n$. We denote the Bloch states by

$$|\lambda\rangle \equiv |\mathbf{k}\sigma\rangle,$$

where $\sigma$ represents electron spin ($\sigma = \uparrow$ or $\downarrow$). As electrons are fermions, the occupation probabilities $p_\lambda$ are given by the Fermi–Dirac distribution

$$p_{\mathbf{k}\sigma} = f_{\mathbf{k}\sigma} = \frac{1}{\exp\{\beta(E_{\mathbf{k}\sigma} - \mu)\} + 1}, \qquad (8.156)$$

where $\beta = 1/k_{\mathrm{B}}T$ and $\mu$ is the chemical potential.

[58] More generally, the *chemical potential*. The Fermi energy is defined as the chemical potential at zero temperature.

[59] The real world is not so simple. Electrons usually have a character which is somewhere between the extreme localized and itinerant limits. This is one of the reasons why magnetic phenomena are so varied and challenging.

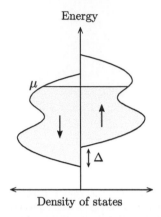

Energy

Density of states

**Fig. 8.20** Splitting of spin-polarized sub-bands in a metal. The densities of states for the spin-up and spin-down bands are shifted in energy by $\Delta$. The chemical potential $\mu$ lies at the boundary between the filled and empty states.

[60]To prove these, write the spin density

$$\mathbf{S}(\mathbf{r}) = \sum_j \mathbf{S}_j \delta(\mathbf{r} - \mathbf{r}_j),$$

so that

$$\mathbf{S}(\mathbf{q}) = \sum_j \mathbf{S}_j \exp(\mathrm{i}\mathbf{q} \cdot \mathbf{r}_j).$$

The spin raising operator on site $j$ is

$$S_j^+ = c_{j\uparrow}^\dagger c_{j\downarrow},$$

and the Fourier transform relations for the fermion operators are

$$c_j = \frac{1}{\sqrt{N}} \sum_{\mathbf{k}} c_{\mathbf{k}} \exp(\mathrm{i}\mathbf{k} \cdot \mathbf{r}_j)$$

$$c_j^\dagger = \frac{1}{\sqrt{N}} \sum_{\mathbf{k}} c_{\mathbf{k}}^\dagger \exp(-\mathrm{i}\mathbf{k} \cdot \mathbf{r}_j),$$

where $N$ is the number of unit cells in the crystal. Hence,

$$S^+(\mathbf{q}) = \sum_j c_{j\uparrow}^\dagger c_{j\downarrow} \exp(\mathrm{i}\mathbf{q} \cdot \mathbf{r}_j)$$

$$= \sum_{\mathbf{k},\mathbf{k}'} c_{\mathbf{k}\uparrow}^\dagger c_{\mathbf{k}'\downarrow} \frac{1}{N} \sum_j e^{\mathrm{i}(\mathbf{q}-\mathbf{k}+\mathbf{k}')\cdot \mathbf{r}_j}$$

$$= \sum_{\mathbf{k},\mathbf{k}'} c_{\mathbf{k}\uparrow}^\dagger c_{\mathbf{k}'\downarrow} \delta_{\mathbf{q}-\mathbf{k}+\mathbf{k}'}$$

$$= \sum_{\mathbf{k}} c_{\mathbf{k}+\mathbf{q}\uparrow}^\dagger c_{\mathbf{k}\downarrow}.$$

Identity (B.50) has been used to go from the second to the third line. The expressions for $S^-(\mathbf{q})$ and $S^z(\mathbf{q})$ in (8.160) are obtained similarly.

In this section we shall allow the possibility that the spin-polarized sub-bands are rigidly split by an amount $\Delta$, Fig. 8.20, so that

$$E_{\mathbf{k}\uparrow} = E_{\mathbf{k}}^0 + \Delta/2$$
$$E_{\mathbf{k}\downarrow} = E_{\mathbf{k}}^0 - \Delta/2, \tag{8.157}$$

where $E_{\mathbf{k}}^0$ represents the unsplit bands ($\Delta = 0$). The splitting could be induced by an applied magnetic field or, as we shall see in Section 8.6.2, due to magnetic order. Any other effects of electronic correlations will be neglected for now.

We shall assume that the magnetic interaction operator can be written in the form

$$\mathbf{M}(\mathbf{Q}) = -g\mu_{\mathrm{B}}\mathbf{S}(\mathbf{Q}), \tag{8.158}$$

which applies when the magnetization is spin-only (see Section 6.3.2) or is adequately represented by the dipole approximation (Section 6.2.2). Despite being itinerant, conduction electrons tend to retain a strong local character through the functions $u_{\mathbf{k}n}(\mathbf{r})$ in (8.155) which usually resemble localized atomic-like orbitals or combinations of orbitals (Wannier functions). We may take into account the local nature of the associated spin distribution by assuming that $\mathbf{S}(\mathbf{Q})$ factorizes into a product $f(\mathbf{Q})\mathbf{S}(\mathbf{q})$, where $f(\mathbf{Q})$ is an appropriate magnetic form factor and $\mathbf{q} = \mathbf{Q} - \mathbf{G}$ is a wavevector in the first Brillouin zone [$\mathbf{S}(\mathbf{Q})/f(\mathbf{Q})$ is periodic in $\mathbf{Q}$ because of translational symmetry]. Accordingly, the generalized susceptibility may be written

$$\chi(\mathbf{Q},\omega) = f^2(\mathbf{Q})\chi(\mathbf{q},\omega), \tag{8.159}$$

where $\chi(\mathbf{q},\omega)$ is a reduced susceptibility which does not depend on any spatial variations in spin density.

It is convenient to express $\mathbf{S}(\mathbf{q})$ in the formalism of second quantization (White, 2007). In brief, we introduce fermion creation and annihilation operators, $c_\lambda^\dagger$ and $c_\lambda$. The operator $c_\lambda^\dagger$ creates an electron in the state $|\lambda\rangle$ providing this state is not already occupied, and $c_\lambda$ destroys an electron in $|\lambda\rangle$. These operators satisfy the anticommutation relations $c_\lambda c_{\lambda'}^\dagger + c_{\lambda'}^\dagger c_\lambda = \delta_{\lambda\lambda'}$ and $c_\lambda c_{\lambda'} + c_{\lambda'} c_\lambda = c_\lambda^\dagger c_{\lambda'}^\dagger + c_{\lambda'}^\dagger c_\lambda^\dagger = 0$. The occupation number operator $c_\lambda^\dagger c_\lambda$ has eigenvalue 1 if the state $|\lambda\rangle$ is occupied, and 0 if not.

The components of $\mathbf{S}(\mathbf{q})$ are given by[60]

$$S^+(\mathbf{q}) = \sum_{\mathbf{k}} c_{\mathbf{k}+\mathbf{q}\uparrow}^\dagger c_{\mathbf{k}\downarrow}$$

$$S^-(\mathbf{q}) = \sum_{\mathbf{k}} c_{\mathbf{k}+\mathbf{q}\downarrow}^\dagger c_{\mathbf{k}\uparrow} \tag{8.160}$$

$$S^z(\mathbf{q}) = \frac{1}{2} \sum_{\mathbf{k}} (c_{\mathbf{k}+\mathbf{q}\uparrow}^\dagger c_{\mathbf{k}\uparrow} - c_{\mathbf{k}+\mathbf{q}\downarrow}^\dagger c_{\mathbf{k}\downarrow}),$$

with, as usual, $S^+ = S^x + \mathrm{i}S^y$ and $S^- = S^x - \mathrm{i}S^y$. The up and down arrows denote the two spin states relative to the quantization axis.

Because the electrons exist in two sub-bands there are two types of spin excitation, either spin-flip or non-spin-flip. These processes are described by the *transverse* and *longitudinal* susceptibilities, respectively. As we are neglecting electronic correlations and calculating only the single-particle response, the susceptibilities are termed *non-interacting* or *bare*.[61] The transverse non-interacting susceptibility is given by

$$\chi^0_{xx} + \chi^0_{yy} = \frac{1}{2}(\chi^0_{+-} + \chi^0_{-+}), \qquad (8.161)$$

where, from (8.116) and (8.158),

$$\chi^0_{+-}(\mathbf{q},\omega) = g^2\mu_B^2 \lim_{\epsilon\to 0^+} \sum_{\lambda,\lambda'} (f_\lambda - f_{\lambda'}) \frac{\langle\lambda|S^-(-\mathbf{q})|\lambda'\rangle\langle\lambda'|S^+(\mathbf{q})|\lambda\rangle}{E_{\lambda'} - E_\lambda - \hbar\omega + i\hbar\epsilon}. \qquad (8.162)$$

From (8.160), the product $\langle\lambda|S^-(-\mathbf{q})|\lambda'\rangle\langle\lambda'|S^+(\mathbf{q})|\lambda\rangle$ equals 1 if $|\lambda\rangle = |\mathbf{k}\downarrow\rangle$ and $\langle\lambda'| = \langle\mathbf{k}+\mathbf{q}\uparrow|$, or zero otherwise. Therefore,

$$\begin{aligned}\chi^0_{+-}(\mathbf{q},\omega) &= g^2\mu_B^2 \lim_{\epsilon\to 0^+} \sum_{\mathbf{k}} \frac{f_{\mathbf{k}\downarrow} - f_{\mathbf{k}+\mathbf{q}\uparrow}}{E_{\mathbf{k}+\mathbf{q}\uparrow} - E_{\mathbf{k}\downarrow} - \hbar\omega + i\hbar\epsilon}\\ &= g^2\mu_B^2 \lim_{\epsilon\to 0^+} \sum_{\mathbf{k}} \frac{f_{\mathbf{k}\downarrow} - f_{\mathbf{k}+\mathbf{q}\uparrow}}{E^0_{\mathbf{k}+\mathbf{q}} - E^0_{\mathbf{k}} + \Delta - \hbar\omega + i\hbar\epsilon}.\end{aligned} \qquad (8.163)$$

The second form is obtained from (8.157). The expression for $\chi^0_{-+}$ is the same but with $\uparrow$ and $\downarrow$ interchanged and $\Delta \to -\Delta$ in the second form. Similarly, the longitudinal non-interacting susceptibility is

$$\chi^0_{zz}(\mathbf{q},\omega) = \chi^0_\uparrow(\mathbf{q},\omega) + \chi^0_\downarrow(\mathbf{q},\omega), \qquad (8.164)$$

where

$$\chi^0_\sigma(\mathbf{q},\omega) = \frac{g^2\mu_B^2}{4} \lim_{\epsilon\to 0^+} \sum_{\mathbf{k}} \frac{f_{\mathbf{k}\sigma} - f_{\mathbf{k}+\mathbf{q}\sigma}}{E_{\mathbf{k}+\mathbf{q}\sigma} - E_{\mathbf{k}\sigma} - \hbar\omega + i\hbar\epsilon}. \qquad (8.165)$$

All other susceptibility components $\chi^0_{\alpha\beta}$ are zero, as may be seen if other pairs of spin operators from (8.160) are used in (8.116).

The scattering intensity is proportional to the absorptive part of the generalized susceptibility, eqn (8.112). Applying Dirac's formula (B.20) to (8.163) and (8.164) we obtain for the absorptive part of the transverse non-interacting susceptibility[62]

$$-\mathrm{Im}\,\chi^0_{+-}(\mathbf{q},\omega) = \pi g^2\mu_B^2 \sum_{\mathbf{k}} (f_{\mathbf{k}\downarrow} - f_{\mathbf{k}+\mathbf{q}\uparrow})\delta(E_{\mathbf{k}+\mathbf{q}\uparrow} - E_{\mathbf{k}\downarrow} - \hbar\omega), \qquad (8.166)$$

and for the longitudinal part

$$-\mathrm{Im}\,\chi^0_{zz}(\mathbf{q},\omega) = \pi \frac{g^2\mu_B^2}{4} \sum_{\mathbf{k},\sigma} (f_{\mathbf{k}\sigma} - f_{\mathbf{k}+\mathbf{q}\sigma})\delta(E_{\mathbf{k}+\mathbf{q}\sigma} - E_{\mathbf{k}\sigma} - \hbar\omega). \qquad (8.167)$$

If the system is a paramagnet in zero field then there is no splitting of the sub-bands, $E_{\mathbf{k}\uparrow} = E_{\mathbf{k}\downarrow} = E^0_{\mathbf{k}}$. If, additionally, the $g$-factor is

[61]We denote this by the zero superscript on $\chi^0$.

[62]The absorptive part is minus the imaginary part.

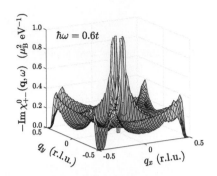

**Fig. 8.21** Absorptive part of the transverse non-interacting susceptibility, eqn (8.166), for a single-band tight-binding model on a square lattice, eqn (8.168). The parameters used in the calculation are $t = 1\,\mathrm{eV}$, $E_0 = 4t$ and $\mu = E_0 + 0.8t$, and the energy is $\hbar\omega = 0.6t$. The $g$-factor is $g = 2$.

[63] The simplicity of the model is deceptive. It is not at all simple to solve exactly, except in certain limiting cases.

[64] Named after John Hubbard, British theoretical physicist (1931–1980).

[65] According to the Pauli Exclusion Principle, two electrons of the same spin can never occupy the same orbital. Coulomb interactions between electrons on different sites, or different orbitals on the same site, can also be included, in which case the model is known as the *extended Hubbard model*.

[66] Generalization to more than one orbital per site involves the inclusion of inter- as well as intra-orbital hopping integrals.

isotropic then $\chi^0_{+-}(\mathbf{q},\omega) = \chi^0_{-+}(\mathbf{q},\omega) = 2\chi^0_{zz}(\mathbf{q},\omega)$, i.e. the generalized susceptibility is isotropic. The non-interacting magnetic susceptibility for a paramagnetic metal is analogous to the *Lindhard function*, which describes the charge susceptibility (i.e. the dielectric response) of a metal.

As an example, Fig. 8.21 presents the zero-temperature Lindhard susceptibility calculated from a single band, nearest-neighbour, tight-binding model for a two-dimensional square lattice. The electronic dispersion is given by [see (8.171)]

$$E_{\mathbf{k}} = E_0 - 2t(\cos k_x a + \cos k_y a). \qquad (8.168)$$

### 8.6.2   Interacting susceptibility

The susceptibilities given in (8.163) and (8.164) correspond to the spectrum of single-particle or *Stoner* excitations. These excitations are, by definition, uncorrelated, meaning that the occupation of an up-spin Bloch state $|\mathbf{k}\uparrow\rangle$ is independent of the occupation of the corresponding down-spin state $|\mathbf{k}\downarrow\rangle$. In order to describe magnetic ordering phenomena we must find a way to include the exchange correlations which have so far been neglected. A simple model[63] that does this is the *Hubbard model*.

The Hubbard model[64] describes interacting electrons hopping between sites on a lattice, and is defined by the Hamiltonian

$$\mathcal{H} = -\sum_{\langle ij \rangle, \sigma} t_{ij}(c^\dagger_{i\sigma}c_{j\sigma} + c^\dagger_{j\sigma}c_{i\sigma}) + U\sum_i n_{i\uparrow}n_{i\downarrow}. \qquad (8.169)$$

The first term is the kinetic energy associated with electrons hopping between atomic-like orbitals on different sites, and the second term describes the on-site Coulomb potential energy. In the Coulomb term, $U$ is the repulsive energy of two electrons of opposite spin on the same site,[65] and $n_{i\sigma} = c^\dagger_{i\sigma}c_{i\sigma}$ is the electron number operator. When $U = 0$ the Hamiltonian reduces to the tight-binding model, which may be written in second-quantized form

$$\mathcal{H} = \sum_{\mathbf{k},\sigma} E_{\mathbf{k}} c^\dagger_{\mathbf{k}\sigma}c_{\mathbf{k}\sigma} \qquad (U = 0). \qquad (8.170)$$

The tight-binding dispersion is

$$E_{\mathbf{k}} = E_0 - \sum_{\boldsymbol{\delta} \neq \mathbf{0}} t_{\boldsymbol{\delta}} \exp(i\mathbf{k}\cdot\boldsymbol{\delta}), \qquad (8.171)$$

where $\boldsymbol{\delta} = \mathbf{r}_j - \mathbf{r}_i$ is a vector connecting nearest-neighbour sites, and $E_0$ is a constant. This model describes a band of non-interacting electron states with a sharp spin-independent dispersion in $\mathbf{k}$-space whose band width is determined by $t_{\boldsymbol{\delta}}$, the hopping or transfer integrals.[66]

## Mean-field approximation

We can obtain an approximate solution to the Hubbard model (8.169) by applying the mean-field approximation, Section 8.5.1, to simplify the two-body interaction:

$$n_{i\uparrow}n_{i\downarrow} \approx n_{i\uparrow}\langle n_{\downarrow}\rangle + n_{i\downarrow}\langle n_{\uparrow}\rangle - \langle n_{\uparrow}\rangle\langle n_{\downarrow}\rangle, \qquad (8.172)$$

where $\langle n_{\sigma}\rangle$ is the thermal average of $n_{\sigma}$. Using the identity[67] $\sum_i n_{i\sigma} = \sum_i c_{i\sigma}^{\dagger}c_{i\sigma} = \sum_{\mathbf{k}} c_{\mathbf{k}\sigma}^{\dagger}c_{\mathbf{k}\sigma}$, together with (8.170), and neglecting the constant term $\langle n_{\uparrow}\rangle\langle n_{\downarrow}\rangle$, we may write the mean-field Hubbard Hamiltonian

$$\mathcal{H} = \sum_{\mathbf{k},\sigma}(E_{\mathbf{k}} + U\langle n_{-\sigma}\rangle)c_{\mathbf{k}\sigma}^{\dagger}c_{\mathbf{k}\sigma}. \qquad (8.173)$$

[67]See, for example, (8.54).

This corresponds to the tight-binding Hamiltonian (8.170) but with spin-dependent sub-bands $E_{\mathbf{k}\sigma} = E_{\mathbf{k}} + U\langle n_{-\sigma}\rangle$, the up-spin band being higher in energy than the down-spin band by

$$\Delta = U(\langle n_{\downarrow}\rangle - \langle n_{\uparrow}\rangle). \qquad (8.174)$$

If $U$ exceeds a critical value, which we shall obtain later, then $\langle n_{\downarrow}\rangle \neq \langle n_{\uparrow}\rangle$ and this model describes a ferromagnet. Irrespective of whether the system is a ferromagnet or a paramagnet, the mean-field eigenstates are the same as those of the non-interacting system and the magnetic response is described by the expressions derived in the previous section, eqns (8.163) and (8.164), which represent the spectrum of single-particle (i.e. electron–hole) excitations.

## Random Phase Approximation

The presence of exchange correlations in the Hubbard model gives rise to cooperative excitations which are not described by the mean-field approximation. To a first approximation we may introduce correlations into the dynamic magnetic response by employing the Random Phase Approximation. The method[68] involves calculating the average space- and time-dependent magnetization to lowest order in the perturbing field and is analogous to that described in Section 8.5.2. It leads to expressions analogous to eqn (8.133) for the transverse and longitudinal parts of the interacting susceptibility:

[68]For the details, see Moriya (1985).

$$\chi_{+-}(\mathbf{q},\omega) = \frac{\chi_{+-}^0(\mathbf{q},\omega)}{1 - \alpha\chi_{+-}^0(\mathbf{q},\omega)}, \qquad (8.175)$$

and

$$\chi_{zz}(\mathbf{q},\omega) = \frac{\chi_{\uparrow}^0 + \chi_{\downarrow}^0 + 8\alpha\chi_{\uparrow}^0\chi_{\downarrow}^0}{1 - 16\alpha^2\chi_{\uparrow}^0\chi_{\downarrow}^0}, \qquad (8.176)$$

where $\alpha = U/(Ng^2\mu_{\mathrm{B}}^2)$, $N$ is the number of lattice points, and the spin-dependent non-interacting susceptibilities $\chi_{\uparrow}^0$ and $\chi_{\downarrow}^0$ are defined in (8.165). If the metal is in the paramagnetic state then $\chi_{\uparrow}^0 = \chi_{\downarrow}^0 = \chi_{zz}^0/2$ and $\chi_{+-}^0 = 2\chi_{zz}^0$. It then follows from (8.175) and (8.176) that $\chi_{+-}(\mathbf{q},\omega) = 2\chi_{zz}(\mathbf{q},\omega)$, as must be the case since all directions are equivalent in an isotropic paramagnet.

### 8.6.3  Magnetic order and spin waves in metals

According to the itinerant-electron theory, magnetic order is associated with a divergence in the static ($\omega = 0$) susceptibility. This is because in the limit when $\chi(\mathbf{q}, 0) \to \infty$ it requires a vanishingly small perturbing field to develop a non-zero magnetic response. To find the condition for magnetic order we consider the RPA interacting susceptibilities in the paramagnetic state. Setting $\chi^0_{+-} = 2\chi^0$ in (8.175) and $\chi^0_\uparrow = \chi^0_\downarrow = \chi^0/2$ in (8.176), and introducing the shorthand $\chi(\mathbf{q})$ for $\chi(\mathbf{q}, 0)$, we see that both the transverse and longitudinal forms of the static susceptibility for a paramagnet may be written in the form

$$\chi(\mathbf{q}) = \frac{\chi^0(\mathbf{q})}{1 - 2\alpha\chi^0(\mathbf{q})}, \tag{8.177}$$

with $\alpha = U/(Ng^2\mu_B^2)$ as before, and

$$\chi^0(\mathbf{q}) = \frac{g^2\mu_B^2}{2} \sum_{\mathbf{k}} \frac{f_{\mathbf{k}} - f_{\mathbf{k}+\mathbf{q}}}{E^0_{\mathbf{k}+\mathbf{q}} - E^0_{\mathbf{k}}}. \tag{8.178}$$

If the divergence is at $\mathbf{q} = 0$, the system becomes a ferromagnet. This is the simplest type of itinerant magnetic order, and we shall consider its static and dynamic properties in more detail below. If instead, the divergence is at $\mathbf{q} \neq 0$ then the order corresponds to a *spin density wave* (SDW), a periodic modulation of the spin density in a metal which can be either transverse or longitudinal depending on the magnetic anisotropy in the paramagnetic state. A transverse SDW is one in which the spins order in a direction perpendicular to the propagation vector $\mathbf{q}$, whereas in a longitudinal SDW the spins order in a direction parallel to $\mathbf{q}$.

A divergence in $\chi(\mathbf{q})$ requires a peak in $\chi^0(\mathbf{q})$ in order that the condition $1 - 2\alpha\chi^0(\mathbf{q}) = 0$ [see (8.177)] occurs at a particular value of some control parameter e.g. temperature. The $\mathbf{q}$ dependence of $\chi^0(\mathbf{q})$ is determined by the shape of the Fermi surface. In particular, if there are parallel sheets of the Fermi surface separated by a wavevector $\mathbf{q}_n$, a situation known as *Fermi surface nesting*, then many $\mathbf{k}$ vectors contribute to the sum in (8.178) and there will usually be an enhancement in $\chi^0(\mathbf{q})$ at $\mathbf{q} = \mathbf{q}_n$. Figure 8.22 illustrates Fermi surface nesting for a square Brillouin zone containing two approximately circular Fermi surface pockets. Such a metal is susceptible to SDW formation with one or both of the nesting wavevectors $\mathbf{q}_1$ and $\mathbf{q}_2$ as the propagation vector(s). The Fermi surface depicted in Fig. 8.22 is similar to that found in the iron-based superconductors, many of which have a SDW phase with propagation vector $(0.5, 0)$ on the approximately square grid of Fe atoms.

If $\chi(\mathbf{q})$ peaks at a particular $\mathbf{q}$, but does not actually diverge, then the effect of exchange enhancement is to produce strongly correlated spin fluctuations in the paramagnetic state, known as *paramagnons*. These can be observed as a diffuse peak in the magnetic neutron scattering centred on $\mathbf{q}$. An example is the nearly-ferromagnetic metal palladium, see Doubble *et al.* (2010).

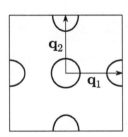

**Fig. 8.22** Illustration of Fermi surface nesting. The diagram shows a square Brillouin zone containing two Fermi surface pockets, of approximately similar shape and size, connected by nesting wavevectors $\mathbf{q}_n = \mathbf{q}_1$ and $\mathbf{q}_2$.

Let us now consider the particular case of itinerant ferromagnetism, in which the leading instability occurs at $\mathbf{q} = 0$. The non-interacting static susceptibility $\chi^0(0,0)$ is then just the uniform susceptibility. The summation in (8.178) may be replaced by an integral containing the density of states function[69] $G(E)$. The integral can then be evaluated by the introduction of an infinitesimal splitting $\delta E$ between initial and final states:

[69] $G(E)\,dE$ is the number of electron states between $E$ and $E+dE$, including the spin degeneracy.

$$\chi^0(0,0) = \frac{g^2\mu_B^2}{2} \lim_{\delta E \to 0} \int_0^\infty \frac{G(E)}{2} \left[ \frac{f(E) - f(E + \delta E)}{(E + \delta E) - E} \right] dE$$

$$= \frac{g^2\mu_B^2}{4} \int_0^\infty G(E) \left[ -\frac{\partial f}{\partial E} \right] dE$$

$$= \frac{g^2\mu_B^2 G(E_F)}{4}. \tag{8.179}$$

Here we have used the identity $-\partial f/\partial E = \delta(E - E_F)$, which is proved in Section B.2.4. The factor of 2 appears in the denominator because $G(E)$ is the total number of electron states per unit energy and there are two electron spin states per $\mathbf{k}$. If we multiply $\chi^0(\mathbf{q} = 0)$ by $\mu_0/V$ [see eqn (D.88)] then we obtain the well-known Pauli susceptibility (in SI units) for a paramagnetic metal without exchange enhancement,

$$\chi_{\text{Pauli}} = \frac{\mu_0 g^2 \mu_B^2 g(E_F)}{4}, \tag{8.180}$$

where $g(E) = G(E)/V$ is the density of electron states per unit volume. Substitution of (8.179) in the denominator of (8.177) gives the condition for ferromagnetic instability in a metal with exchange enhancement,

$$\frac{UG(E_F)}{2N} > 1. \tag{8.181}$$

This is the well-known *Stoner criterion* for ferromagnetism.[70]

[70] Edmund C. Stoner, British theoretical physicist (1899–1968) who developed the theory of ferromagnetism in metals while at the University of Leeds.

## 8.6.4   Magnetic excitations in ferromagnetic metals

First, we note that when a metal is in its paramagnetic phase ($T > T_C$), or the exchange coupling is small, the spectrum is dominated by single-particle (Stoner) spin-flip excitations which are described by the non-interacting susceptibility (8.163). The spectrum is constrained by the band dispersion and any band splitting $\Delta$ to a finite region of $\omega$–$\mathbf{q}$ space called the *Stoner continuum*. The Stoner continuum is shown in Figs. 8.23(a) and (b), as calculated from eqn (8.166) for a free electron gas with $\Delta = 0$ and $\Delta \neq 0$, respectively.

When exchange-coupling is important the transverse spectrum can be calculated in the RPA from the imaginary part of (8.175):

$$\text{Im}\,\chi_{+-}(\mathbf{q}, \omega) = \frac{\text{Im}\,\chi_{+-}^0(\mathbf{q}, \omega)}{[1 - \alpha \,\text{Re}\,\chi_{+-}^0(\mathbf{q}, \omega)]^2 + \alpha^2 [\text{Im}\,\chi_{+-}^0(\mathbf{q}, \omega)]^2}. \tag{8.182}$$

There are now regions within the Stoner continuum where the effect of

**Fig. 8.23** Spectrum of spin-flip excitations for an electron gas with and without exchange correlations, see Moriya (1985). (a),(b) $-\mathrm{Im}\,\chi^0_{+-}(\mathbf{q},\omega)$ calculated from eqn (8.166) with (a) $\Delta = 0$, and (b) $\Delta = 1.3E^0_{\mathrm{F}}$. (c) Spectrum in the ferromagnetic state calculated in the RPA from $-\mathrm{Im}\,\chi_{+-}(\mathbf{q},\omega)$, eqn (8.175), with $\alpha = 0.37$. The spin-wave (SW) mode is indicated. The axes are normalized by the Fermi energy ($E^0_{\mathrm{F}}$) and wavevector ($k^0_{\mathrm{F}}$) of the non-interacting gas, and the intensities are normalized to $\chi^0(0,0)$, eqn (8.179).

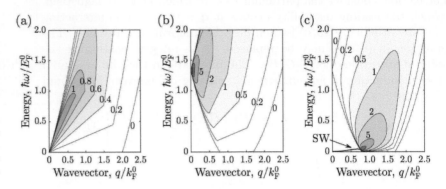

exchange enhancement is large and the spectrum deviates strongly from the single-particle spectrum. This can be seen in Fig. 8.23(c), which shows the RPA spectrum for a ferromagnetic electron gas having the same band splitting as the non-interacting electron gas in Fig. 8.23(b).

Outside the Stoner continuum $\mathrm{Im}\,\chi^0_{+-}(\mathbf{q},\omega) = 0$, and so the intensity is expected to be zero. However, if

$$1 - \alpha\,\mathrm{Re}\,\chi^0_{+-}(\mathbf{q},\omega) = 0 \tag{8.183}$$

then (8.182) becomes $0/0$. This apparently ill-defined limit actually corresponds to a collective mode whose frequency $\omega_{\mathbf{q}}$ disperses from $(\mathbf{q},\omega) = (0,0)$ with an initial form $\omega_{\mathbf{q}} = Dq^2$, the same as for spin waves in a local-moment ferromagnet (see Exercises 8.2 and 8.6). The spin-wave dispersion is found from the solution to (8.183), and the spectrum may be obtained from (8.182) with the term in the denominator

$$1 - \alpha\,\mathrm{Re}\,\chi^0_{+-}(\mathbf{q},\omega) = \alpha\,\mathrm{Re}\,\chi^0_{+-}(\mathbf{q},\omega_{\mathbf{q}}) - \alpha\,\mathrm{Re}\,\chi^0_{+-}(\mathbf{q},\omega)$$

$$= \frac{U}{N}\sum_{\mathbf{k}}(f_{\mathbf{k}\downarrow} - f_{\mathbf{k}+\mathbf{q}\uparrow})$$

$$\times\left(\frac{1}{E^0_{\mathbf{k}+\mathbf{q}} - E^0_{\mathbf{k}} + \Delta - \hbar\omega_{\mathbf{q}}} - \frac{1}{E^0_{\mathbf{k}+\mathbf{q}} - E^0_{\mathbf{k}} + \Delta - \hbar\omega}\right). \tag{8.184}$$

At small $\mathbf{q}$, we have (i) $\sum_{\mathbf{k}}(f_{\mathbf{k}\downarrow} - f_{\mathbf{k}+\mathbf{q}\uparrow}) \approx N(\langle n_{\downarrow}\rangle - \langle n_{\uparrow}\rangle) = N\Delta/U$ from (8.174), (ii) $E^0_{\mathbf{k}+\mathbf{q}} \approx E^0_{\mathbf{k}}$, and (iii) $\hbar\omega_{\mathbf{q}} \ll \Delta$. With these approximations, (8.184) may be written

$$1 - \alpha\,\mathrm{Re}\,\chi^0_{+-}(\mathbf{q},\omega) \approx \frac{\hbar}{\Delta}(\omega_{\mathbf{q}} - \omega),$$

so that (8.182) becomes

$$-\mathrm{Im}\,\chi_{+-}(\mathbf{q},\omega) = \frac{\pi\Delta}{\alpha}\frac{\Gamma/\pi}{(\hbar\omega_{\mathbf{q}} - \hbar\omega)^2 + \Gamma^2}, \tag{8.185}$$

where $\Gamma = -\alpha\Delta\,\mathrm{Im}\,\chi^0_{+-}(\mathbf{q},\omega)$. In the limit $\Gamma \to 0$, the Lorentzian function in (8.185) becomes a $\delta$-function (see Sections B.1 and B.2), and so at small $\mathbf{q}$,

$$-\mathrm{Im}\,\chi_{+-}(\mathbf{q},\omega) = \frac{\pi N \Delta g^2 \mu_{\mathrm{B}}^2}{U}\delta(\hbar\omega_{\mathbf{q}} - \hbar\omega). \qquad (8.186)$$

This shows that outside the Stoner continuum, where $\mathrm{Im}\,\chi^0_{+-}(\mathbf{q},\omega) = 0$, the spin-wave mode is sharp with an intensity proportional to $\Delta/U$. When the spin-wave mode enters the Stoner continuum it becomes broadened by decay into electron–hole pairs. This behaviour can be seen in the spectrum of the ferromagnetic electron gas shown in Fig. 8.23(c).

Finally, we comment briefly on the spectrum of longitudinal excitations, which can be calculated in the RPA from eqn (8.176). Unlike with the transverse susceptibility, there is no obvious condition for the imaginary part of the longitudinal susceptibility to diverge, so we do not expect to observe spin-wave-like propagating longitudinal modes. Instead, the longitudinal spectrum will typically be broad in energy and wavevector, and largest in the critical regime where there exists strong exchange-enhanced spin fluctuations.

## 8.7   Quantum magnetism

Many magnetic phenomena, such as those found in paramagnets and long-range ordered phases, can be described adequately without regard for the underlying quantum nature of the magnetism. *Quantum magnetism* is a term reserved for magnetic states that contain a high degree of *quantum entanglement*, in other words states in which one spin is entangled with many others. Entanglement increases fluctuations and tends to destabilize long-range order. An extreme example is the *quantum spin liquid*, which has short-range spin correlations but no long-range magnetic order at any temperature.[71] The tendency towards macroscopic quantum ground states instead of conventional long-range order is favoured by low spin values, continuous spin symmetries,[72] and interactions which are antiferromagnetic, short-range, low-dimensional,[73] and frustrated.

The spin-$\frac{1}{2}$ dimer with antiferromagnetic exchange encountered in Section 8.2.1 is a simple example of an entangled spin system. The spins are entangled to form a singlet ground state with energy $-\frac{3}{4}J$ and a triplet excited state — see Table 8.2. In the equivalent semiclassical (non-entangled) dimer both the ground and excited states are doublets[74] and the ground state energy is $-\frac{1}{4}J$. This illustrates how quantum entanglement can lower the energy of a spin system.

Quantum zero-point fluctuations and the inherent non-linearity of the spin commutation relations have important consequences for the spin dynamics of quantum magnets. As classical long-range order is suppressed, the spectral weight associated with sharp linear spin-wave modes decreases and is typically transferred into regions of broad continua. The spectrum might also develop features that signal the presence of non-classical excited states. One example is $\omega/T$ scaling,[75] which can be observed in the vicinity of a *quantum phase transition*.[76] Other examples

[71] Quantum spin liquids are instead characterized by a form of topological order.

[72] The nearest-neighbour Heisenberg model, eqn (8.30), has continuous symmetry because the Hamiltonian is invariant under continuous rotations in three-dimensional spin space. The $XY$ model (the Heisenberg model without the $z$ components of spin) is another example, since it is invariant under continuous rotations about the $z$ axis.

[73] One- and two-dimensional interactions, and spin clusters.

[74] The eigenfunctions are

$$|g\rangle = (|\uparrow\downarrow\rangle, |\downarrow\uparrow\rangle)$$
$$|e\rangle = (|\uparrow\uparrow\rangle, |\downarrow\downarrow\rangle).$$

[75] This means that a series of spectra measured at different temperatures can be collapsed onto a single curve if the energy is divided by temperature and the response scaled by a power of temperature, i.e. for a given $\mathbf{Q}$, $\chi''(\omega) \sim T^{-\alpha}f(\omega/T)$, where $f$ is a universal function and $\alpha$ is a scaling exponent.

[76] A quantum phase transition is a transition between two different phases at zero temperature.

are quantum bound states, longitudinal (Higgs) modes, and fractional spin excitations. Being a direct probe of the space- and time-dependent magnetic response, neutron spectroscopy is a valuable tool for identifying such emergent phenomena and for testing theoretical models.

### 8.7.1   The $S = \frac{1}{2}$ Heisenberg antiferromagnetic chain

There are many interesting model quantum magnets. We do not have the space to review them all, and we shall instead focus on the one-dimensional Heisenberg antiferromagnet (HAF), particularly the spin-$\frac{1}{2}$ chain, a model which highlights many of the important emergent properties of quantum magnets.[77] The Hamiltonian is

$$\mathcal{H} = J \sum_n \mathbf{S}_n \cdot \mathbf{S}_{n+1}, \tag{8.187}$$

where $\mathbf{S}_n$ is the operator for a spin on site $n$, and $J > 0$. We shall consider a chain of $N$ spins with separation $a$ between neighbouring spins.

The classical ground state of (8.187) is the Néel phase, in which each spin is aligned in the opposite direction to its neighbour, and the classical excited states are spin-wave modes whose dispersion relation and transverse response function per spin (for $\omega > 0$) are[78]

$$\hbar\omega_q = 2SJ|\sin qa|$$

$$S_{\alpha\alpha}(q,\omega) = g^2\mu_B^2 \frac{S}{2}|\tan qa|\{n(\omega_q)+1\}\delta(\hbar\omega - \hbar\omega_q), \tag{8.188}$$

where $q$ is the wavevector component along the chain and $\alpha$ is a spin component perpendicular to the ordered moment.

The true ground state of (8.187) for $S = \frac{1}{2}$, however, is very different. It is a macroscopic singlet entangling all spins in the chain. Like the Néel phase it has zero total spin, but unlike Néel order the individual spin projections along any axis are all zero, $\langle S_n^\alpha \rangle = 0, \alpha = x, y, z$. The ground state energy[79] is $E_0 = (1/4 - \ln 2)NJ = -0.443NJ$, compared with $-NJ/4$ for the Néel phase. The difference is due to entanglement, which allows the system to gain energy from transverse spin fluctuations. Although quantum disordered, the ground state contains strong antiferromagnetic correlations which decay inversely with distance (with weak logarithmic corrections). These long-range correlations strongly influence the spin excitation spectrum, especially at low energies.

To visualize the elementary excitations of the spin-$\frac{1}{2}$ HAF chain it is helpful to consider first the Ising spin chain, which has the Hamiltonian

$$\mathcal{H} = J \sum_n S_n^z S_{n+1}^z. \tag{8.189}$$

The spins in a spin-$\frac{1}{2}$ Ising magnet point up or down, i.e. $\langle S_n^z \rangle = \pm\frac{1}{2}$, and the Néel phase is the exact ground state. The lowest excitations are states in which the direction of one spin is reversed — see Fig. 8.24. This causes two antiferromagnetic nearest-neighbour bonds to become

[77]The spin-$\frac{1}{2}$ HAF chain is an important model because it is one of the few quantum many-body systems that can be solved exactly. In 1931, Hans Bethe (1906–2005) presented a method for obtaining the eigenvalues and eigenfunctions. The method, now known as the Bethe ansatz, has been extended and applied to obtain solutions for many other quantum many-body systems.

[78]These can be obtained from eqns (8.97), (8.103), and (8.110)–(8.111). For a one-dimensional antiferromagnet with nearest-neighbour interactions only, $J(q) = 2J\cos qa$.

[79]First obtained by L. Hulthén (1938).

ferromagnetic, which increases the total spin of the chain by $\Delta S = 1$ and costs energy $2 \times \frac{1}{2}J = J$. Single spin-flip excitations are eigenstates of (8.189), so once created they do not move along the chain. However, if a small transverse term of the form $\delta\mathcal{H} = \lambda J \sum_n (S_n^x S_{n+1}^x + S_n^y S_{n+1}^y)$, $\lambda \ll 1$, is added to (8.189) then the bilinear products of spin raising and lowering operators contained in $\delta\mathcal{H}$ can move a ferromagnetic bond along the chain. For example, the term $S_{n+1}^- S_{n+2}^+$ acting on the excited state shown in Fig. 8.24(b) flips the spins on sites $n+1$ and $n+2$ which moves the ferromagnetic bond joining sites $n$ and $n+1$ two places to the right. This hopping process creates two localized domain walls, each of which separates regions of $\pi$ phase-shifted Néel order. Localized domain walls that can propagate without changing shape are a type of excitation known as a *soliton*. Therefore, the excited states of a spin-$\frac{1}{2}$ Ising AFM chain with weak transverse coupling are pairs of solitons which, once formed, can move independently of one another and which each carry a spin of $\frac{1}{2}$ (because the excited state has a total spin of 1). The two domain walls have individual momenta $\hbar q_1$ and $\hbar q_2$, and since there are many possible combinations of $q_1$ and $q_2$ that give the same total wavevector $q = q_1 + q_2$ the spectrum contains a continuous band of states for each $q$, the band width being of order $\lambda J$.

The notion of domain wall excitations remains useful for the spin-$\frac{1}{2}$ HAF chain. Inclusion of transverse coupling allows the domain walls to spread out until in the Heisenberg limit they delocalize over the whole chain. In the HAF chain the analogues of the domain wall excitations are called *spinons*. The process by which an excitation with spin 1 splits into two deconfined spinons each carrying spin $\frac{1}{2}$ is termed *fractionalization*.

The momentum-dependent excitation spectrum of the spin-$\frac{1}{2}$ HAF chain is dominated by a continuum of two-spinon excitations which extends from a lower boundary $\hbar\omega_L(q)$ to an upper boundary $\hbar\omega_U(q)$, where (des Cloizeaux and Pearson, 1962; Yamada, 1969)

$$\hbar\omega_L(q) = \frac{\pi}{2}J|\sin qa|, \qquad \hbar\omega_U(q) = \pi J|\sin\frac{1}{2}qa|. \quad (8.190)$$

Apart from a factor of $\pi/2$ difference in the energy scale, the dispersion of the lower threshold $\hbar\omega_L(q)$ is the same as the semi-classical spin-wave dispersion for the spin-$\frac{1}{2}$ HAF chain, eqn (8.188), despite the fact that the quantum disordered chain has translational invariance with period $a$, whereas the Néel-ordered classical chain has period $2a$. This reflects the strong AFM correlations still present in the quantum chain. A fundamental difference, however, is that the classical chain has a magnetic Bragg peak at $q = \pi/a$ whereas the quantum chain does not (it only has a divergence). Moreover, the spinon spectrum as a whole has period $2\pi/a$, consistent with translational invariance.

Exact calculations of the response function $S_{\alpha\alpha}(q,\omega)$ have only recently been performed (Caux and Hagemans, 2006), but a useful analytic approximation (Müller *et al.*, 1979; Müller *et al.*, 1981) is[80]

$$S_{\alpha\alpha}(q,\omega) \approx g^2\mu_B^2 \frac{A}{2\pi} \frac{\Theta(\omega - \omega_L(q))\Theta(\omega_U(q) - \omega)}{\sqrt{\hbar^2\omega^2 - \hbar^2\omega_L(q)^2}}, \quad (8.191)$$

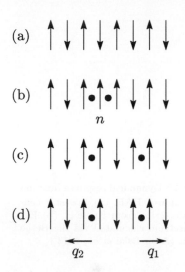

**Fig. 8.24** The $S = \frac{1}{2}$ Ising antiferromagnetic chain. (a) Néel ground state. (b) Single spin-flip excitation on site $n$. (c) A term $S_{n+1}^- S_{n+2}^+$ shifts the position of the ferromagnetic domain wall two places to the right. (d) Motion of domain walls in an Ising chain with small transverse coupling.

[80] $S_{\alpha\alpha}(q,\omega)$ for the spin-$\frac{1}{2}$ HAF chain is isotropic, $S_{xx} = S_{yy} = S_{zz}$, because the Hamiltonian (8.187) has full rotational symmetry and there is no conventional long-range order.

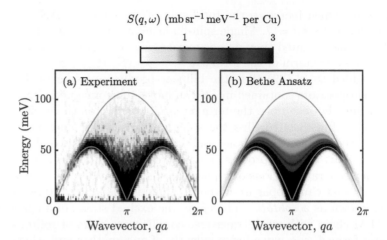

**Fig. 8.25** Dynamical response function of the spin-$\frac{1}{2}$ HAF. (a) Inelastic neutron scattering from KCuF$_3$. (b) Exact spectrum calculated via the Bethe ansatz. (After Lake *et al.*, 2013.)

per spin, where $\Theta(x)$ is the Heaviside step function and $A = 1$–$1.5$ (we have omitted the magnetic form factor and Debye–Waller factor). This formula, known as the *Müller ansatz*, is constructed to satisfy the two-spinon boundaries (8.190) together with various sum rules (which determine $A$) and the exact form near the lower cut-off. It shows that the spectrum is largest at $\hbar\omega_{\mathrm{L}}(q)$ where there is a divergence, but there is also a tail which extends in energy up to $\hbar\omega_{\mathrm{U}}(q)$ where there is a cut-off. In reality, the exact $S_{\alpha\alpha}(q,\omega)$ contains spectral weight above $\hbar\omega_{\mathrm{U}}(q)$ due to 4-, 6-, ..., spinon excitations. Figure 8.25 shows calibrated neutron scattering measurements of $S_{\alpha\alpha}(q,\omega)$ made on KCuF$_3$, a near-ideal realization of the spin-$\frac{1}{2}$ HAF chain. A simulation of the exact spectrum is shown beside the data. The agreement is impressive, especially as there is no adjustable parameter to scale the intensity.

Coupling between neighbouring HAF chains, always present in real systems, tends to suppress quantum fluctuations and induces long-range magnetic order. The fundamental excitations of the ordered component of the spins are spin waves. However, when the coupling is weak the ordered moment is smaller than the fully aligned moment, and a gapped longitudinal mode corresponding to collective fluctuations in the amplitude of the spins appears in addition to the gapless transverse spin-wave modes. This longitudinal mode, which is analogous to the *Higgs mode* in the standard model of particle physics, has been observed in the spectrum of KCuF$_3$ by Lake *et al.* (2000).

We conclude this section with a few remarks about HAF chains with $S > \frac{1}{2}$. All HAF chains are quantum disordered with zero total spin, but as first conjectured by Haldane (1983*a* and 1983*b*), chains of integer and half-integer spins have fundamentally different ground states.[81] Locally, the $S = \frac{1}{2}$ chain looks like the classical Néel state, but the presence of spin-$\frac{1}{2}$ domain wall defects makes the antiferromagnetic correlations decay algebraically with distance. For $S = 1$, the spins at each site can be viewed as composites of two sub-spins each with $S = \frac{1}{2}$ strongly bound in a spin-1 triplet, while at the same time each sub-spin forms a

[81]Work for which Haldane was rewarded with a share of the 2016 Nobel Prize in Physics.

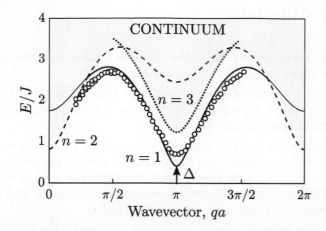

**Fig. 8.26** Schematic magnetic spectrum of the $S = 1$ HAF chain. The Haldane gap $\Delta$ is indicated. The low-lying branch is identified with one-particle $(n = 1)$ and two-particle $(n = 2)$ excitations near $qa = \pi$ and 0, respectively. The dashed and dotted lines indicate the lower boundaries of the two- and three-particle continua. The symbols are measurements of the low-lying excitations at $T = 6.2\,\text{K}$ in CsNiCl$_3$, which is an approximation to an ideal $S = 1$ HAF chain (Kenzelmann *et al.*, 2002). The disagreement between the data and theoretical curve near $qa = \pi$ is due to weak inter-chain interactions.

singlet, or *valence bond*, with a sub-spin on the neighbouring site. The spin correlations in the Haldane ground state decay exponentially, but it is not a simple disordered phase. It has a form of hidden order known as *string order*, which arises because a spin defect in the Haldane state has a predictable effect on the spin sequence.

A low-lying excitation of the Haldane chain is generated by a spin flip of one of the sub-spins in a $S = 1$ composite spin, replacing one of the singlet valence bonds with a triplet. This costs a discrete amount of energy and results in a gap $\Delta \approx 0.41J$ in the single-particle spectrum at $q = \pi/a$. This *Haldane gap* appears even though the Hamiltonian is rotationally invariant, and in this sense is analogous to the singlet–triplet excitation of the spin-$\frac{1}{2}$ HAF dimer. Neutron scattering experiments on various model compounds have confirmed the existence and size of the Haldane gap and mapped out the full spectrum for $S = 1$, which contains a continuum of 2- and 3-particle excitations with a well-defined lower boundary in addition to the isolated triplet branch (see Fig. 8.26). A Haldane gap is predicted for all integer-$S$ HAF chains, but it decreases in size exponentially with $S$. With increasing $S > 1$ the spectrum of both integer and half-integer HAF chains tends rapidly towards the semi-classical spin-wave spectrum.

## Chapter summary

- The response function $S(\mathbf{Q}, \omega)$ for magnetic neutron spectroscopy has been given for a variety of systems of interest.

- The neutron spectra of isolated magnetic systems have been described, including crystal-field transitions, intermultiplet transitions, and magnetic clusters.

- Spin waves are propagating modes in systems with long-range magnetic order. Linear spin-wave theory can be used to calculate the spin-wave spectrum. The theory treats magnetic moments as semi-classical effective spins coupled by exchange interactions, and includes single-ion anisotropy phenomenologically.

- The Random Phase Approximation (RPA) is a method to calculate the generalized susceptibility $\chi(\mathbf{Q}, \omega)$ approximately. It replaces two-ion interactions with effective single-ion terms. The RPA susceptibility is given by

$$\boldsymbol{\chi}(\mathbf{Q}, \omega) = \left\{ 1 - \frac{1}{Ng^2\mu_{\mathrm{B}}^2} \boldsymbol{\chi}^0(\omega) J(\mathbf{Q}) \right\}^{-1} \boldsymbol{\chi}^0(\omega),$$

where $\boldsymbol{\chi}^0(\omega)$ is the non-interacting susceptibility and $J(\mathbf{Q})$ is the Fourier transform of the two-ion interactions. Generalizations to include anisotropic interactions and an arbitrary number of inequivalent subsystems are described.

- The RPA is an alternative to spin-wave theory for ordered magnets, but can also be used for excitonic-like propagating modes in paramagnets and other systems containing pairwise interactions.

- In itinerant (metallic) magnets there are single-particle (Stoner) spin excitations, which are diffuse in character, as well as sharp spin-wave-like propagating modes.

- The generalized susceptibility for itinerant magnets can be calculated by the RPA from a model that describes a set of non-interacting electron states with exchange correlations, such as the Hubbard model.

- Quantum magnets are systems with a high degree of quantum entanglement in which conventional magnetic order is suppressed. The dynamical response of quantum magnets often contains fractionalized excitations and regions of multiparticle continua.

# Further reading

For a comprehensive review of neutron intermultiplet spectroscopy, see Osborn *et al.* (1991). Molecular nanomagnets and cluster excitations are reviewed by Gatteschi *et al.* (2006) and Furrer and Waldmann (2013). Short reviews of neutron scattering from magnetic excitations are given by Chatterji and Regnault in Chatterji (2006), and by Raymond (2014). Fazekas (1999) contains a very clear account of quantum magnetism and model magnetic systems.

# Exercises

(8.1) Explain how one could use measurements of the chiral scattering by uniaxial polarization analysis (see Section 4.5.2) to identify the $M_S = +1$ and $M_S = -1$ singlet–triplet transitions of a spin-$\frac{1}{2}$ antiferromagnetic dimer.

(8.2) Derive the spin-wave dispersion relation for the spin-$S$ nearest-neighbour Heisenberg ferromagnet on a square lattice (see Fig. 8.10), $\hbar\omega_\mathbf{q} = 2|J|S(2 - \cos q_x a - \cos q_y a)$. Show that when $q_x a \ll 1$ and $q_y a \ll 1$ the dispersion takes the form $\hbar\omega_\mathbf{q} \simeq Dq^2$, where $D = |J|Sa^2$.

(8.3) Show that the spin-wave dispersion relation of a ferromagnet in a magnetic field applied parallel to the magnetization direction is the same as that in zero field but shifted up in energy by $g\mu_B B$, where $B$ is the magnetic flux density in the medium.

(8.4) Starting from the Bogoliubov transformation (8.89), prove that the operators $\alpha_\mathbf{q}$ and $\beta_\mathbf{q}$ satisfy the Bose commutation relations (8.52).

(8.5) In this exercise you will prove for the special case of a spin-$S$, isotropic, Heisenberg ferromagnet with one spin per unit cell, that the RPA-MF response functions at $T = 0$ are the same as those obtained from linear spin-wave theory.

(a) Show for spin $S$ that the separation between energy levels in the molecular field (8.120) is $\Delta = SJ(0)$, where $J(\mathbf{q})$ is given by (8.123).

(b) Use (8.116) to derive the following mean-field single-ion susceptibility components:

$$\chi^0_{+-}(\omega) = \chi^0_{-+}(-\omega) = \frac{2NSg^2\mu_B^2}{\Delta - \hbar\omega}, \qquad \chi^0_{zz}(\omega) = 0,$$

where $N$ is the number of spins, and $\chi(\omega)$ is understood to mean $\lim_{\epsilon \to 0^+} \chi(\omega - i\epsilon)$. Here we have excluded the magnetic form factor and Debye–Waller factor from $\chi^0_{\alpha\beta}(\omega)$.

(c) Hence, show that the single-ion susceptibility tensor may be written

$$\boldsymbol{\chi}^0(\omega) = \frac{NSg^2\mu_B^2}{\Delta^2 - (\hbar\omega)^2} \begin{pmatrix} \Delta & i\hbar\omega & 0 \\ -i\hbar\omega & \Delta & 0 \\ 0 & 0 & 0 \end{pmatrix}.$$

(d) From (8.133), derive the following generalized susceptibility components:

$$\chi_{xx}(\mathbf{Q},\omega) = \chi_{yy}(\mathbf{Q},\omega) = g^2\mu_B^2 \frac{NS}{2}$$
$$\times \lim_{\epsilon \to 0^+} \left\{ \frac{1}{\hbar\omega_\mathbf{Q} - \hbar(\omega - i\epsilon)} + \frac{1}{\hbar\omega_\mathbf{Q} + \hbar(\omega - i\epsilon)} \right\}$$

$$\chi_{xy}(\mathbf{Q},\omega) = -\chi_{yx}(\mathbf{Q},\omega) = g^2\mu_B^2 \frac{NS}{2}$$
$$\times \lim_{\epsilon \to 0^+} i\left\{ \frac{1}{\hbar\omega_\mathbf{Q} - \hbar(\omega - i\epsilon)} - \frac{1}{\hbar\omega_\mathbf{Q} + \hbar(\omega - i\epsilon)} \right\},$$

where $\hbar\omega_\mathbf{Q} = \hbar\omega_{-\mathbf{Q}} = S\{J(0) - J(\mathbf{Q})\}$.

(e) Use (D.31) to obtain $\chi''_{xx}(\mathbf{Q},\omega)$ and $\chi''_{xy}(\mathbf{Q},\omega)$, and hence show that the MF-RPA response functions $S_{xx}(\mathbf{Q},\omega)$ and $S_{xy}(\mathbf{Q},\omega)$ are identical to those found by linear spin-wave theory, eqns (8.65) and (8.68).

(8.6)  Show that eqn (8.183) may be written

$$1 = \frac{U}{N} \sum_{\mathbf{k}} \left( \frac{f_{\mathbf{k}\downarrow}}{E^0_{\mathbf{k}+\mathbf{q}} - E^0_{\mathbf{k}} + \Delta - \hbar\omega} \right.$$
$$\left. - \frac{f_{\mathbf{k}\uparrow}}{E^0_{\mathbf{k}} - E^0_{\mathbf{k}-\mathbf{q}} + \Delta - \hbar\omega} \right).$$

By expanding this expression up to order $q^2$ show that the spin-wave dispersion relation for an itinerant ferromagnet takes the form $\hbar\omega_{\mathbf{q}} \simeq Dq^2$ at small $q$. For simplicity, you could assume the electronic band dispersion has the form $E^0_{\mathbf{k}} = \hbar^2 k^2/2m^*$.

# Neutron Optics

**9**

Optical phenomena, such as refraction, reflection and dispersion, arise when materials behave as continuous media. Although most closely associated with light, optical phenomena occur for all waves propagating in media, including neutrons.

On the whole, neutron optical phenomena do not significantly influence the kind of scattering experiment done to probe the atomic-scale structure and dynamics of matter. They can, however, be observed under the right conditions, and are exploited in a number of important devices and techniques. Reflection, for example, is used to polarize neutrons and to transport neutrons along guides, and neutron reflectometry is a powerful method to study the surface and interfacial structure in a wide variety of solid and liquid systems. Neutron imaging uses variations in the neutron refractive index to make images of macroscopic objects, often exposing features that are hidden to electromagnetic radiation. On a more fundamental level, neutron optics are exploited in various ways to measure basic quantities such as the lifetime of the neutron.

We begin with an outline of the dynamical theory of neutron diffraction, introducing the basic principles behind optical phenomena in media that can be considered continuous and homogeneous. We go on to describe two important applications: neutron reflectivity and neutron imaging. Finally, we show how dynamical diffraction theory can explain effects such as the fine structure observed in Bragg diffraction from perfect or near-perfect crystals and the extinction of Bragg peak intensities.

## 9.1 Dynamical theory

The kinematical theory employed up to this point is based on the assumption that the interaction of neutrons with matter is very weak, which justifies a number of approximations: (i) the wave amplitude is the same at each scattering centre; (ii) neutrons are scattered only once (or not at all); (iii) the scattered waves are spherical, and can be approximated by plane waves at the large distances where diffraction is observed; (iv) the total scattered amplitude is the sum of the amplitudes from each scattering centre taking into account the phase differences due to geometry and due to the interaction potential. In this approach, which corresponds to the Born approximation, the scattering amplitude is found to be the Fourier transform of the scattering potential.[1]

The Born approximation is valid for sufficiently small samples or weak scattering. If the probability of multiple scattering is small but not

[1]See Section 3.2.

negligible then corrections can be calculated within the framework of the kinematical theory (see Section 10.1.2). Neutron optical phenomena, on the other hand, are inherently the result of multiple scattering and so a very different approach is needed.

### 9.1.1   Coherent wave

The approach builds on the fact that optical phenomena arise from interference of the scattered waves with each other and with the incident wave. This requires that the incident and scattered waves must all be coherent,[2] i.e. have a constant phase difference with respect to one another, which means that the wavelength must not change on scattering. If these conditions are to be satisfied then the scattering must not result in any change in the microscopic state of the system. The only types of scattering involved in neutron optics, therefore, are those that involve static correlations,[3] i.e. coherent elastic scattering from solids[4] as well as forward ($\mathbf{Q} = 0$) scattering from any phase of matter.[5]

Since all waves involved in a neutron optical process are coherent and static we can describe the scattering state of the neutron by a single wave function $\Psi(\mathbf{r})$, known as the *coherent wave*. The coherent wave is the solution of a one-body Schrödinger equation

$$\left\{ -\frac{\hbar^2}{2m_{\mathrm{n}}}\nabla^2 + V(\mathbf{r})\right\}\Psi(\mathbf{r}) = E\Psi(\mathbf{r}), \tag{9.1}$$

where $E$ is the incident energy, and $V(\mathbf{r})$ is an effective potential known as the *optical potential* which represents the interaction of the neutron with the medium.

Of course, coherent scattering of the type which gives rise to neutron optical phenomena is not the only type of collision process that can occur. Neutrons can be absorbed in the nucleus, and there are other scattering processes which in optics are referred to collectively as *diffuse scattering*. For neutrons these processes are (i) incoherent scattering, and (ii) coherent scattering not derived from static correlations. They are called *diffuse* because they give rise to scattering that is not sharply peaked in $\mathbf{Q}$. Absorption and diffuse scattering, when present, attenuate $\Psi(\mathbf{r})$ and can be represented by a complex optical potential.

### 9.1.2   Optical potential

For simplicity, let us consider a non-magnetic medium and unpolarized neutrons.[6] To a good approximation, the optical potential $V(\mathbf{r})$ may then be represented by the time-averaged total Fermi pseudopotential,[7] eqn (4.6),

$$V(\mathbf{r}) = \frac{2\pi\hbar^2}{m_{\mathrm{n}}} \sum_j \langle b_j\, \delta(\mathbf{r} - \mathbf{r}_j)\rangle, \tag{9.2}$$

The time average, denoted by the angular brackets, ensures that the scattering is elastic. The optical potential given by (9.2) is a smooth function of position because atomic vibrations smear out the sharp $\delta$-function

[2]This is the crucial feature. In contrast, the kinematical theory is restricted to interference effects in the scattered radiation (and then only in the absence of multiple scattering), so only the scattered waves have to be coherent.

[3]Those in which the mean position of each atom is independent of time — see Section 3.6.

[4]For example, Bragg scattering from crystalline solids. The atoms in gases and liquids are mobile and so their positions are uncorrelated over long times.

[5]Forward scattering is always elastic. To see this, consider the correlation function

$I(\mathbf{Q},t) = \langle \exp(-i\mathbf{Q}\cdot\hat{\mathbf{r}}_j) \exp\{i\mathbf{Q}\cdot\hat{\mathbf{r}}_k(t)\}\rangle.$

When $\mathbf{Q} = 0$, $I(\mathbf{Q},t) = 1$, and so from (5.1–5.2) and (B.16)

$S(\mathbf{Q},\omega) \propto \int_{-\infty}^{\infty} I(\mathbf{Q},t)\exp(-i\omega t)\,\mathrm{d}t$
$\propto \delta(\hbar\omega).$

Hence, scattering with zero momentum transfer is truly elastic.

[6]Magnetic effects and polarized neutrons are covered in Section 9.2.3.

[7]Use of the pseudopotential neglects so-called *local field* corrections which take into account the attenuation of the coherent wave due to diffuse scattering.

pseudopotential associated with each nucleus. To reflect this, we can write $V(\mathbf{r})$ in terms of the scattering length density (1.25)

$$V(\mathbf{r}) = \frac{2\pi\hbar^2}{m_\text{n}} \langle n_b(\mathbf{r}) \rangle. \tag{9.3}$$

If the medium is homogeneous, like a liquid or amorphous solid, then $\langle n_b(\mathbf{r}) \rangle$ and hence $V(\mathbf{r})$ is everywhere constant. We may then write

$$\left. \begin{aligned} V(\mathbf{r}) &= V \\ \langle n_b(\mathbf{r}) \rangle &= n_b \\ &= \sum_j n_j \bar{b}_j \end{aligned} \right\} \text{homogeneous medium}, \tag{9.4}$$

where $n_j$ is the number density of species $j$. For a crystal, $\langle n_b(\mathbf{r}) \rangle$ is lattice periodic, but the periodicity only matters when the neutron wavevector satisfies the Bragg condition. This case will be considered in Section 9.4. Away from the Bragg condition $n_b$ can be used for $\langle n_b(\mathbf{r}) \rangle$.

We now introduce a complex optical potential and show that the imaginary part represents attenuation of the coherent wave. We also obtain an expression relating the imaginary part of $V$ and the neutron absorption cross-section. We define[8]

$$V = V' - iV'', \tag{9.5}$$

which from (9.2) and (9.3) implies a complex scattering length and scattering length density

$$b = b' - ib'' \quad \text{and} \quad n_b = n_b' - in_b''. \tag{9.6}$$

In a homogeneous medium the neutron flux $\Phi = v|\Psi|^2$ will decay exponentially with distance. If the beam is travelling in the $z$ direction and enters the medium at $z = 0$ then we may write

$$\Phi(z) = \Phi(0) \exp(-\mu z), \tag{9.7}$$

where $\mu$ is the *attenuation coefficient*. The coherent wave function that describes the beam takes the form

$$\Psi(z) = \Psi(0) \exp(ikz), \tag{9.8}$$

where $k$ is a complex wavevector $k = k' + ik''$. Comparing $|\Psi(z)|^2$ from (9.8) with (9.7) we see that

$$k'' = \mu/2. \tag{9.9}$$

On substituting (9.8) into the Schrödinger equation (9.1) we find

$$-\frac{\hbar^2}{2m_\text{n}} \left\{ k_0^2 + (ik' - k'')^2 \right\} + V' - iV'' = 0, \tag{9.10}$$

where $k_0$ is the wavevector for $V = 0$, i.e. $E = \hbar^2 k_0^2 / 2m_\text{n}$.

[8]The negative sign for the imaginary parts in (9.5) and (9.6) ensures that a plane wave of the form $\exp(ikz)$ with positive $V''$ decays exponentially as it propagates in the $+z$ direction, see eqns (9.8) and (9.12).

Consider first the solution for $E > V'$. Equating the real and imaginary parts of (9.10) we find

$$k' = \left\{ \tfrac{1}{2}(k_0^2 - 4\pi n_b') + \tfrac{1}{2}\sqrt{(k_0^2 - 4\pi n_b')^2 + (4\pi n_b'')^2} \right\}^{\frac{1}{2}}, \qquad (9.11)$$

$$k'' = \frac{m_n}{\hbar^2 k'}V'' = \frac{2\pi}{k'}n_b''. \qquad (9.12)$$

Equations (9.8) and (9.11) show that without attenuation ($k'' = 0$) neutrons propagate in the medium as a wave whose wavevector $k'$ is different from that in vacuum ($k_0$) because some kinetic energy is converted into potential energy on entering the medium. When $k'' \neq 0$, eqn (9.12) shows that the attenuation is proportional to the imaginary part of the optical potential. Suppose that the medium consists of only one type of nucleus, and that it has absorption cross-section $\sigma_a$ and number density $n$. If the attenuation is purely due to absorption then $\mu = n\sigma_a$ (see Section 1.7.4), and so from (9.9) and (9.12),

$$n\sigma_a = \frac{2m_n}{\hbar^2 k'}V''. \qquad (9.13)$$

For a single type of nucleus $n_b = nb$, and so from (9.3) and (9.13)

$$\sigma_a = \frac{4\pi}{k'}b''. \qquad (9.14)$$

Hence, if $b''$ is independent of energy then the absorption cross-section varies inversely with the neutron wavevector and hence its speed. This so-called '1/v law' applies when the neutron incident energy is far from a nuclear resonance.

As mentioned above, the coherent wave $\Psi(\mathbf{r})$ is attenuated by diffuse scattering as well as by absorption.[9] If we denote the total cross-section per atom for diffuse scattering by $\sigma_d$, then the attenuation coefficient becomes

$$\mu = n(\sigma_a + \sigma_d). \qquad (9.15)$$

Relations (9.13) and (9.14) can be generalized accordingly. For a material containing different nuclei, the same formulae apply but with $n(\sigma_a + \sigma_d)$ replaced by $\sum_j n_j(\sigma_a + \sigma_d)_j$. The diffuse scattering cross-section is discussed further in Section 9.1.3.

The solution of (9.10) for $E \leq V'$ is

$$k'' = \left\{ \tfrac{1}{2}(4\pi n_b' - k_0^2) + \tfrac{1}{2}\sqrt{(4\pi n_b' - k_0^2)^2 + (4\pi n_b'')^2} \right\}^{\frac{1}{2}}, \qquad (9.16)$$

$$k' = 2\pi n_b''/k''. \qquad (9.17)$$

In the absence of absorption ($n_b'' = 0$) the wavevector is purely imaginary and the wave function (9.8) decays exponentially in the medium. So despite the fact that neutrons do not have sufficient kinetic energy to surmount the barrier they can still penetrate some distance into it. The part of the wave function that penetrates into the medium is known as the *evanescent wave*.[10]

[9]This is a result that can be obtained from the rigorous theory of dynamical neutron diffraction — see Sears (1989).

[10]The word *evanescent* comes from the Latin *evanescere*, which means to vanish, or fade out of sight.

### 9.1.3 Diffuse scattering

Diffuse scattering is relatively weak and does not usually give rise to strong multiple scattering under normal experimental conditions, so the kinematical theory can be used to calculate it. As mentioned in Section 9.1.1, diffuse scattering in optics is all the scattering that is not perfectly elastic. This comprises the coherent and incoherent scattering contained in $\widetilde{S}(\mathbf{Q}, \omega)$, the part of the response function that derives from dynamical correlations (Section 3.6).

To calculate the attenuation we must calculate the total cross-section for diffuse scattering, i.e. we must integrate the double differential cross-section over all directions and over all neutron final energies:[11]

$$\sum_j n_j (\sigma_d)_j = \int \int \frac{\mathrm{d}^2\sigma}{\mathrm{d}\Omega \mathrm{d}E_f} \,\mathrm{d}\Omega \,\mathrm{d}E_f. \tag{9.18}$$

The outcome depends on the neutron incident energy, the microscopic details of the material under investigation and on temperature.

For example, if the scattering extends over only a small range of energy transfer relative to the incident neutron energy then we can use the static approximation (see Section 3.8.4) to simplify the energy integral. In the static approximation the incoherent scattering is elastic and isotropic, and the integral satisfies the sum rule (5.137)

$$\int \int \left( \frac{\mathrm{d}^2\sigma}{\mathrm{d}\Omega \mathrm{d}E_f} \right)_{\mathrm{inc}} \mathrm{d}\Omega \,\mathrm{d}E_f = \sum_j n_j (\sigma_{\mathrm{inc}})_j. \tag{9.19}$$

For the coherent scattering, the static approximation is expressed by

$$\int \int \left( \frac{\mathrm{d}^2\sigma}{\mathrm{d}\Omega \mathrm{d}E_f} \right)_{\mathrm{coh}} \mathrm{d}\Omega \,\mathrm{d}E_f = \int \widetilde{S}_c(\mathbf{Q}) \,\mathrm{d}\Omega, \tag{9.20}$$

where $\widetilde{S}_c(\mathbf{Q})$ is the energy integral of $\widetilde{S}(\mathbf{Q}, \omega)$, eqn (3.59). The solid-angle integral in (9.20) extends up to a maximum $\mathbf{Q}$ corresponding to back-scattering. This will control how much coherent diffuse scattering is included. If the neutron incident energy is very large then the maximum $\mathbf{Q}$ is also very large and the value of the integral tends towards the sum rule limit for coherent nuclear scattering, $4\pi \sum_j n_j |\bar{b}_j|^2$, see (5.136). This gives an upper bound to (9.20). The lower bound is close to zero, because when the incident energy is small $\mathbf{Q}$ is small, and[12] $\widetilde{S}_c(0) \simeq 0$.

To summarize, the attenuation coefficient (9.15) in the static approximation is given by

$$\mu = \sum_j n_j (\sigma_a + \sigma_{\mathrm{inc}})_j + \int \widetilde{S}_c(\mathbf{Q}) \,\mathrm{d}\Omega, \tag{9.21}$$

where the integral varies from 0 to $4\pi \sum_j n_j |\bar{b}_j|^2$ with increasing neutron incident energy. The energy dependence of $\mu$ is shown schematically in Fig. 9.1. Equation (9.21) is often sufficient because in many cases of interest the bulk of the coherent and incoherent spectral weight in $\widetilde{S}(\mathbf{Q}, \omega)$ is found close to $\hbar\omega = 0$, and so the static approximation holds well for the calculation of total cross-sections.

[11]The cross-section here is per unit volume. Any perfectly elastic scattering is excluded from the integral on the right-hand side of (9.18).

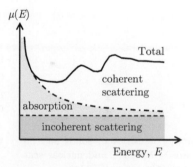

**Fig. 9.1** Sketch of the attenuation coefficient $\mu$ as a function of neutron energy $E$ for a non-crystalline substance.

[12]$\widetilde{S}_c(0)$ is not strictly zero if there are density fluctuations, see Section 2.4.4, but this need not concern us here.

### 9.1.4  Refractive index

[13]Not to be confused with number density, which is also denoted by $n$.

In optics, the definition of refractive index[13] $n$ is $v = c/n$, where $v$ and $c$ are the magnitudes of the phase velocities of light in the medium and in vacuum. Since $v = \omega/k$, and the angular frequency $\omega$ is continuous across a boundary between two media, we can write

$$n = \frac{k}{k_0},\qquad(9.22)$$

where $k$ and $k_0$ are the wavevectors in the medium and in vacuum. Using (9.8) we may write the coherent wave in the medium

$$\Psi(z) = \Psi(0)\exp(ink_0z).\qquad(9.23)$$

For an attenuating medium, $k$ is complex and so $n$ is complex,

$$n' + in'' = \frac{1}{k_0}(k' + ik'').\qquad(9.24)$$

Comparing (9.23) with (9.8) and (9.9) we see that $\mu = 2k_0n''$, and on substituting (9.23) into the Schrödinger equation (9.1) with incident energy $E = \hbar^2 k_0^2/2m_n$ we obtain

$$n^2 = 1 - \frac{2m_n}{\hbar^2 k_0^2}(V' - iV'')\qquad(9.25)$$

$$= 1 - n'_b\frac{4\pi}{k_0^2} + i\mu\frac{k'}{k_0^2}.\qquad(9.26)$$

On the second line, $n'_b$ is the real part of the scattering length density, and we have used (9.3) and (9.12) to substitute for the complex optical potential. Equation (9.26) shows that the neutron refractive index is dispersive, i.e. it depends on the neutron wavelength.

[14]Number densities in solids and liquids are $\sim 10^{28}\,\mathrm{m^3}$, and nuclear scattering lengths are $\sim 10^{-14}$ m.

In the absence of attenuation ($\mu = 0$) $n$ is real, and with typical values[14] $n_b \sim 10^{14}\,\mathrm{m^{-2}}$ and $k_0 \sim 10^{10}\,\mathrm{m^{-1}}$, eqn (9.26) gives

$$1 - n^2 \sim 10^{-5}.\qquad(9.27)$$

Therefore, $n$ is very close to 1, and for the usual case of a positive scattering length density it is very slightly less than 1. If we now assume a typical attenuation coefficient $\mu = 10^2\,\mathrm{m^{-1}}$ and take $k' = k_0$ then the imaginary term in (9.26) is $\sim10^{-8}$. In fact, the imaginary part of the refractive index is only important for $\mu \gtrsim 10^4\,\mathrm{m^{-1}}$, which is only found in materials containing one of the strongly absorbing nuclei with $\sigma_a > 10^3$ b, such as $^{10}$B, $^{113}$Cd or $^{155,157}$Gd. The attenuation due to diffuse scattering is much less than that due to strong absorption, so the effect of diffuse scattering on the refractive index can be neglected.

These estimates show that (9.26) can almost always be written as $n^2 = 1 + x$ with $|x| \ll 1$, so we can use the binomial theorem to write

$$n \approx 1 - \lambda^2\frac{n'_b}{2\pi} + i\lambda\frac{\mu}{4\pi},\qquad(9.28)$$

[15]The imaginary term in (9.28) is exact, and comes directly from eqns (9.9) and (9.24).

where $\lambda = 2\pi/k_0$ is the wavelength of the incident neutrons.[15] The neutron refractive index for magnetic media and polarized neutrons is given in Section 9.2.3.

## 9.1.5 Specular reflection from a plane boundary

Consider the reflection of neutrons from a smooth boundary lying in the $z = 0$ plane between two media which have refractive indices $n_1$ and $n_2$, as shown in Fig. 9.2. Conservation of momentum parallel to the boundary requires that the reflection is *specular*, i.e. that the angle of incidence equals the angle of reflection.[16]

Since the neutron is a free particle in the $xy$ plane we may write the coherent wave $\Psi(\mathbf{r}) = \psi(z)\exp(\mathrm{i}k_x x)\exp(\mathrm{i}k_y y)$. Substituting this into the Schrödinger equation (9.1) we find that $\psi(z)$ satisfies

$$\left\{ -\frac{\hbar^2}{2m_\mathrm{n}}\frac{\mathrm{d}^2}{\mathrm{d}z^2} + V(z) \right\}\psi(z) = E_z\psi(z), \qquad (9.29)$$

where $E_z = E - \hbar^2(k_x^2 + k_y^2)/2m_\mathrm{n}$ is the total energy associated with motion in the $z$ direction, and $V(z)$ steps up from $V_1$ to $V_2$ at $z = 0$ (Fig. 9.3). Following the standard procedure, we introduce trial solutions for the wave functions in each medium,

$$\psi_1(z) = \exp(\mathrm{i}k_1 z \sin\theta_1) + r\exp(-\mathrm{i}k_1 z \sin\theta_1)$$
$$\psi_2(z) = t\exp(\mathrm{i}k_2 z \sin\theta_2), \qquad (9.30)$$

where $r$ and $t$ are the complex amplitudes of the reflected and transmitted waves, and equate the wave functions and their first derivatives at $z = 0$. Solving the boundary equations we obtain

$$r = \frac{n_1 \sin\theta_1 - n_2 \sin\theta_2}{n_1 \sin\theta_1 + n_2 \sin\theta_2}, \qquad t = \frac{2n_1 \sin\theta_1}{n_1 \sin\theta_1 + n_2 \sin\theta_2}. \qquad (9.31)$$

These are known as the *Fresnel coefficients*. The intensity of the scattered beam relative to the incident beam is known as the *reflectance* or *reflectivity*, and is given by

$$R = |r|^2 = \left| \frac{n_1 \sin\theta_1 - n_2 \sin\theta_2}{n_1 \sin\theta_1 + n_2 \sin\theta_2} \right|^2. \qquad (9.32)$$

The relation between $\sin\theta_1$ and $\sin\theta_2$ can be obtained by substitution of (9.30) into the one-dimensional Schrödinger equation (9.29). By conservation of energy, $E_z$ is the same in both media, and so we find that

$$k_2^2 \sin^2\theta_2 = k_1^2 \sin^2\theta_1 - \frac{2m_\mathrm{n}}{\hbar^2}\Delta V, \qquad (9.33)$$

where $\Delta V = V_2 - V_1$. Using (9.22) and (9.25) to simplify (9.33) we obtain *Snell's law*,

$$n_1 \cos\theta_1 = n_2 \cos\theta_2. \qquad (9.34)$$

Now, if we assume that $V_1$ and $V_2$ are real and take $V_2 > V_1$, i.e. $n_2 < n_1$, then it follows from (9.34) that $\theta_2 < \theta_1$. Hence, the beam bends away from the normal as it crosses the boundary, see Fig. 9.2, and there exists a critical incident angle $\theta_1 = \theta_\mathrm{c}$ at which $\theta_2 = 0$. For $\theta_1 < \theta_\mathrm{c}$ the right-hand side of eqn (9.33) is negative and so $k_2 \sin\theta_2$ is imaginary.

[16]In optics, the angles of incidence and reflection are usually defined with respect to the normal of the boundary. For neutrons, these angles are close to 90° and it is more convenient to measure angles with respect to the boundary plane.

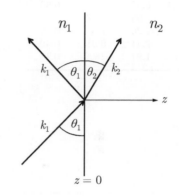

**Fig. 9.2** Reflection and transmission of neutrons from a plane boundary. The boundary is defined by the plane $z = 0$.

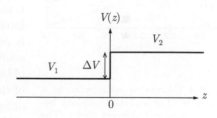

**Fig. 9.3** Effective one-dimensional potential barrier to represent specular reflection from a plane boundary.

**Table 9.1** Refractive index parameters in (9.26) and critical angles for selected materials at $\lambda = 0.4\,\mathrm{nm}$.

|  | $10^4 n_b'$ ($\mathrm{nm}^{-2}$) | $\mu$ ($\mathrm{cm}^{-1}$) | $\theta_c$ (deg) |
|---|---|---|---|
| Si | 2.07 | 0.01 | 0.19 |
| $H_2O$ | −0.56 | 2.7 | — |
| $D_2O$ | 6.37 | 0.07 | 0.33 |
| Ti | −1.95 | 0.77 | — |
| Co | 2.26 | 7.6 | 0.19 |
| Ni | 9.41 | 0.91 | 0.40 |
| Gd | 1.97 | 2800 | 0.18 |

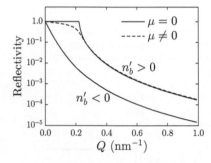

**Fig. 9.4** Calculated reflectivity as a function of $Q$ at a plane boundary between vacuum and different media. The full lines are for media with positive and negative scattering densities and no absorption, and the broken line shows the effect of strong absorption on the reflectivity of the medium with positive scattering length density. The values used in the simulations are $n_b' = 10 \times 10^{-4}\,\mathrm{nm}^{-2}$, $n_b' = -3 \times 10^{-4}\,\mathrm{nm}^{-2}$, $\mu = 150\,\mathrm{cm}^{-1}$.

This gives $R = 1$ in (9.32), and results in a transmitted wave (9.30) which decays exponentially into the barrier as an evanescent wave. Any attenuation in the barrier will then result in $R < 1$.

The critical angle for neutrons incident from vacuum ($n_1 = 1$) onto a material without strong attenuation ($\mu \ll 10^4\,\mathrm{m}^{-1}$) whose scattering length density has a positive real part $n_b'$ is [from (9.33) and (9.3)]

$$\sin \theta_c = \lambda \left( \frac{n_b'}{\pi} \right)^{1/2}. \tag{9.35}$$

Table 9.1 lists values of $n_b'$, $\mu$ and $\theta_c$ for a few materials. The critical angles at a wavelength of $0.4\,\mathrm{nm}$ are all less that $0.5°$.

Neutron reflectivity is usually measured either (i) as a function of incident angle $\theta_1$ for a fixed incident wavelength $\lambda$, or (ii) as a function of $\lambda$ for a fixed $\theta_1$. It is convenient to plot $R$ as a function of the elastic scattering vector $Q = (4\pi/\lambda) \sin \theta_1$ because the function $R(Q)$ is independent of which method was used to obtain the data. Figure 9.4 shows curves of $R$ as a function of $Q$ for neutrons incident from vacuum on materials with and without strong absorption. The main effects of absorption are to reduce the reflectivity below the critical edge and to round off the critical edge.

### 9.1.6   Neutron guides

The existence of a critical angle, eqn (9.35), below which there is total reflection is exploited in neutron guides. These are devices which can transport neutrons without the usual inverse-square law reduction in intensity that governs the radiation from a point source. Guides make it possible to locate an instrument at the optimum distance from the source as determined by considerations such as space constraints, levels of background radiation, and the flight path that best suits the operation of the instrument. Neutron losses below 0.2% per metre are achievable, allowing transport over distances of up to several hundred metres.

Neutron guides consist of an evacuated channel with a rectangular cross-section and smooth walls. The walls are typically made from glass with an inner surface that is coated with a metal film. They work in much the same way as optical fibres do for light. Neutrons which enter the guide with angles less than $\theta_c$ to the guide axis are totally reflected off the walls and therefore continue along the guide without leaking out. At the same time, neutrons entering the guide at angles greater than $\theta_c$ are transmitted through the wall and lost. Because $\theta_c$ increases with wavelength, shorter wavelength neutrons are more likely to escape from the guide, and so the distribution of neutrons at the exit of the guide is shifted towards longer wavelengths compared with that at the entrance. Nickel is a popular choice of guide coating as it has a high scattering length density and hence a relatively large $\theta_c$ (see Table 9.1).

Fig. 9.5 shows different types of guide that are in use. A disadvantage of the straight guide, Fig. 9.5(a), is that the exit of the guide views the source directly, so if the source emits undesirable high-energy neutrons or $\gamma$-rays then some of these will pass through the guide and cause extra background. This can be avoided by curving the guide so that there is no direct line-of-sight from exit to entrance, Fig. 9.5(b).

Guides are often designed with the objective of maximizing the flux on the sample. Two examples of how this can be achieved are shown in Fig. 9.5(c) and (d). In both cases, the increase in flux is achieved at the expense of an increase in the divergence of the transmitted beam.[17] With the elliptical guide the walls are not parallel and so the acceptance angle is wider, whereas the tapered guide concentrates the intensity on a smaller area by analogy with the way a converging lens focuses light.

Although nickel-coated guides are still widely used, the best performance nowadays is achieved with *supermirror guides*. These are able to reflect neutrons at angles which greatly exceed the critical angle of Ni-coated guides, and so they transmit shorter wavelengths more efficiently and have substantially higher acceptance angles. Supermirrors are also very efficient polarizers for cold neutrons (see Section 9.2.3).

Supermirror coatings consist of alternating layers of two different materials repeated thousands of times. Nickel and titanium are very suitable bilayer materials as they create a large scattering contrast between adjacent layers (see Table 9.1). The thickness of the bilayer slowly decreases with depth, resulting in a superlattice with a continuously varying period that acts like a diffraction grating with a depth-dependent Bragg condition, see Fig. 9.6. Neutrons with long wavelengths are Bragg-reflected from the near surface region where the superlattice period is relatively large. Shorter wavelength neutrons are transmitted through the uppermost layers but are diffracted further beneath the surface where the period is smaller. The net effect is a continuum of overlapping Bragg peaks which ensures that neutrons are reflected over a continuous range of $Q$ up to a limit set by the shortest bilayer period in the supermirror.

Supermirrors are characterized by their $m$ value, which is the ratio of the critical wavevector transfer $Q_c$ of the supermirror to that of a

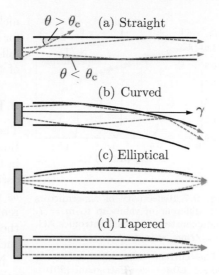

**Fig. 9.5** Different types of neutron guide.

[17] This is an example of Liouville's theorem, which states that the density of particles in phase space remains constant. Phase space contains points that represent all the coordinates and momenta of the particles in the beam, so if the particles move closer together in real space then according to Liouville's theorem they must move further apart in momentum space, i.e. the beam becomes more divergent.

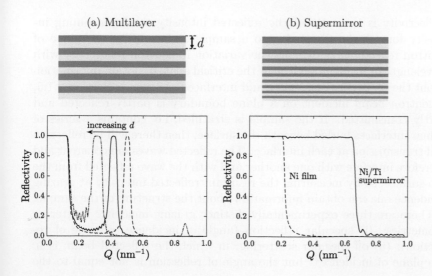

**Fig. 9.6** Schematic showing how a supermirror works. (a) Reflectivity of a multilayer. The diffraction peak moves to smaller $Q$ as the period $d$ increases. (b) Reflectivity of a supermirror. The period decreases with depth, so diffraction from deeper layers contributes at larger $Q$. The net effect is a continuum of overlapping Bragg peaks. The simulation is for an $m = 3$ Ni/Ti supermirror with 250 bilayers on a glass substrate. The reflectivity of a simple Ni film is shown for comparison.

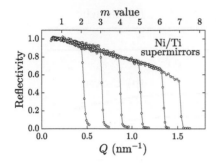

**Fig. 9.7** Reflectivity of supermirrors with different $m$ value up to $m = 7$. (Data courtesy of SwissNeutronics AG, http://www.swissneutronics.ch.)

[18]The current best value is $\tau_n = 880.2 \pm 1.0\,\text{s}$ — see Patrignani (2016). This value is an average of several measurements, some of which, however, differ by up to seven standard deviations.

[19]The electric dipole moment is a measure of the distribution of positive and negative charge in the neutron. So far, no electric dipole moment has been found. The current upper limit is $1.8 \times 10^{-26}\,e\,\text{cm}$ (Abel, 2020).

[20]This is equivalent to the condition that the critical angle for total reflection is $90°$, see eqn (9.35).

nickel-coated guide. Figure 9.7 shows the measured reflectivity curves for a set of Ni/Ti supermirrors with $m$ values up to 7. The gradual fall-off in reflectivity with $Q$ is mainly due to the increasing attenuation of the neutron beam with increasing depth.

### 9.1.7   Storage of ultracold neutrons

The phenomenon of total reflection of neutrons finds an extreme application in neutron storage traps, or *neutron bottles*. Neutrons can be stored in bottles if their kinetic energies are below a few hundred neV, which corresponds to temperatures below a few mK. Such neutrons are known as *ultracold neutrons* (UCN). Storage times of a few minutes can currently be achieved, which is sufficiently long to allow measurements of the neutron lifetime[18] and other fundamental properties of the neutron, such as its electric dipole moment,[19] and tests of Newtonian gravity.

Neutrons will be trapped if they are reflected by the walls of a container for any angle of incidence. For this to apply, the neutron's kinetic energy $E = \frac{1}{2}m_n v^2$ must be less than the optical potential $V = 2\pi\hbar^2 n_b/m_n$, eqn (9.3). Put another way, the velocity normal to the surface must be less than a critical value given by[20] $E = V$,

$$v_c = \frac{\hbar}{m_n}\left(4\pi n_b\right)^{1/2}. \tag{9.36}$$

One of the best wall materials for neutron bottles is diamond-like carbon (DLC), which combines a high potential barrier with relatively low losses. The best samples of DLC have $v_c \simeq 7.2\,\text{ms}^{-1}$, which corresponds to $E \simeq 270\,\text{neV}$ or an effective temperature $T = E/k_B = 3.1\,\text{mK}$. Losses in the walls are caused by absorption and by energy-gain inelastic scattering due to thermal motion of the atoms in the walls.

## 9.2   Neutron reflectometry

Reflectivity is the ratio of the reflected intensity to the incoming intensity for a beam directed onto a sample surface. The technique of neutron reflectometry exploits the variation in neutron reflectivity with wavelength or incident angle near the critical angle to obtain information about the properties of surfaces and interfaces. As seen in Section 9.1.5, a neutron beam incident on a plane boundary is partly reflected and partly transmitted. If the sample is stratified, i.e. contains a series of planar interfaces buried beneath the surface, then there will be reflection and transmission at each interface. The reflected waves are coherent and therefore interfere with one another and with the wave reflected from the top surface, so by measuring the resultant reflected intensity at grazing incidence one can obtain information about the structure of the sample.

There are three experimentally distinct grazing-incidence scattering geometries: (i) specular reflection (angle of incidence = angle of reflection); (ii) off-specular scattering, in which the reflected beam is in the plane of incidence[21] but the angle of reflection is not equal to the

[21]The plane of incidence is the plane containing the incident wavevector and the normal to the surface. Here, this is the plane containing the $x$ and $z$ axes.

angle of incidence; (iii) grazing incidence small-angle neutron scattering (GISANS), in which the scattering is specular with a component perpendicular to the plane of incidence. These are illustrated in Fig. 9.8. Specular reflectivity depends on the refractive index profile normal to the surface on length scales up to ~100 nm. Off-specular scattering carries information about the in-plane structure of the surface and interfaces on length scales up to ~10 μm, and GISANS probes lateral structure on length scales ~100 nm.

Neutron reflectometry has a number of strengths. Because neutrons penetrate large distances in most materials they can probe interfaces beneath the surface. Different features in a sample can be highlighted by the use of contrast variation (Section 2.6.5). This is achieved by isotopic substitution within a part of the sample, which changes the refractive index of the selected feature without affecting its chemical or structural properties. Substitution of hydrogen by deuterium is a particularly powerful method of contrast variation to study interfaces within aqueous and organic materials. The spin dependence of the neutron refractive index makes it possible to study layered magnetic media with polarized neutron reflectometry (Section 9.2.3).

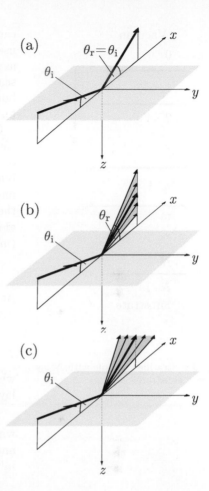

**Fig. 9.8** Different grazing incidence scattering geometries: (a) specular ($\theta_r = \theta_i$), (b) off-specular scattering in the plane of incidence ($\theta_r \neq \theta_i$), (c) grazing incidence small-angle scattering (GISANS).

## 9.2.1    Specular reflectivity

The principal aim of a specular reflectivity experiment is to obtain information about the variation of the scattering length density normal to the surface, $n_b(z)$. Because of the phase problem there is no direct way to invert $R(Q)$ to obtain $n_b(z)$. The challenge, therefore, is to find the most probable form for $n_b(z)$ given the data. For this purpose we need to be able to calculate $R(Q)$ from a model for $n_b(z)$.

For specular reflection we need only consider the direction normal to the surface. It will be convenient to express the neutron wave function in terms of $k_z = nk_0 \sin\theta$, the wavevector component normal to the surface, and to express the reflectivity as a function of $Q = Q_z$, the wavevector transfer normal to the surface. For example, from (9.31) the Fresnel coefficients for a plane boundary are,

$$r = \frac{k_{z1} - k_{z2}}{k_{z1} + k_{z2}} \qquad \text{and} \qquad t = \frac{2k_{z1}}{k_{z1} + k_{z2}} \qquad (9.37)$$

and from (9.32)–(9.35) the corresponding reflectivity may be written

$$R(Q) = \left| \frac{Q - (Q^2 - Q_c^2)^{1/2}}{Q + (Q^2 - Q_c^2)^{1/2}} \right|^2, \qquad (9.38)$$

where in this case $Q = 2k_{z1}$, and in the absence of strong attenuation

$$Q_c = 4\{\pi(n'_{b2} - n'_{b1})\}^{1/2}. \qquad (9.39)$$

**Transfer matrix method**

In general, specular reflectivity profiles can be calculated from the solution of the one-dimensional Schrödinger equation (9.29). Exact solutions

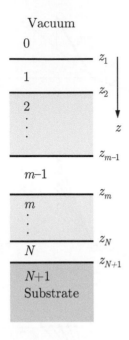

Vacuum

0

$z_1$

1

$z_2$

2
.
.
.

$z_{m-1}$

m–1

$z_m$

m
.
.
.

$z_N$

N

$z_{N+1}$

N+1
Substrate

**Fig. 9.9** Stratified sample with $N$ layers on a substrate.

[22] $\mathsf{C}_1$ has $\phi_0 = 0$ as a special case, because in layer 0 the relevant complex amplitudes are those at $z = 0$.

can be obtained in closed form for a few simple cases, but the most widely used approach is the *transfer matrix method*, which treats the medium as a series of discrete layers each with its own individual thickness and scattering length density. For naturally stratified samples with sharp boundaries the method is exact, but good approximations are available for dealing with more diffuse boundaries.

Consider the system shown in Fig. 9.9 comprising $N$ uniform layers. Neutrons incident from the vacuum are reflected from the top surface of the first layer, and can exit but cannot enter the sample through the bottom surface of the $N^{\text{th}}$ layer. In each intervening layer there are multiple reflections, but as the waves are all coherent we can combine them into two waves, one travelling along $+z$ and the other travelling along $-z$. We may write the solution of the Schrödinger equation in the $(m-1)^{\text{th}}$ layer (remembering that $k_z = k_z' + ik_z''$ is complex) as

$$\psi_{m-1}(z) = A_{m-1}e^{ik_{z,m-1}z} + B_{m-1}e^{-ik_{z,m-1}z}. \quad (9.40)$$

At the top of the $(m-1)^{\text{th}}$ layer $(z = z_{m-1})$,

$$\psi_{m-1}(z_{m-1}) = A_{m-1}e^{ik_{z,m-1}z_{m-1}} + B_{m-1}e^{-ik_{z,m-1}z_{m-1}}$$

$$= \psi_{m-1}^+ + \psi_{m-1}^-, \quad (9.41)$$

where $\psi_{m-1}^\pm$ are complex amplitudes. At the bottom of the $(m-1)^{\text{th}}$ layer

$$\psi_{m-1}(z_m) = \psi_{m-1}^+ e^{i\phi_{m-1}} + \psi_{m-1}^- e^{-i\phi_{m-1}}, \quad (9.42)$$

where $\phi_{m-1} = k_{z,m-1}(z_m - z_{m-1})$. Applying continuity of the wave amplitudes and derivatives at the boundary between the $(m-1)^{\text{th}}$ and $m^{\text{th}}$ layers we find

$$\psi_{m-1}^+ e^{i\phi_{m-1}} + \psi_{m-1}^- e^{-i\phi_{m-1}} = \psi_m^+ + \psi_m^- \quad (9.43)$$

$$k_{z,m-1}\left(\psi_{m-1}^+ e^{i\phi_{m-1}} - \psi_{m-1}^- e^{-i\phi_{m-1}}\right) = k_{zm}\left(\psi_m^+ - \psi_m^-\right). \quad (9.44)$$

Solving (9.43) and (9.44) simultaneously we obtain the following relation between the amplitudes at the top of the $(m-1)^{\text{th}}$ and $m^{\text{th}}$ layers,

$$\begin{pmatrix} \psi_{m-1}^+ \\ \psi_{m-1}^- \end{pmatrix} = \frac{1}{t_m}\begin{pmatrix} e^{-i\phi_{m-1}} & r_m e^{-i\phi_{m-1}} \\ r_m e^{i\phi_{m-1}} & e^{i\phi_{m-1}} \end{pmatrix}\begin{pmatrix} \psi_m^+ \\ \psi_m^- \end{pmatrix}, \quad (9.45)$$

where $r_m$ and $t_m$ are Fresnel coefficients for the $(m-1)$ to $m$ boundary,

$$r_m = \frac{k_{z,m-1} - k_{zm}}{k_{z,m-1} + k_{zm}}, \qquad t_m = \frac{2k_{z,m-1}}{k_{z,m-1} + k_{zm}}. \quad (9.46)$$

The $2 \times 2$ matrix in eqn (9.45) may be written

$$\mathsf{C}_m = \frac{1}{t_m}\begin{pmatrix} e^{-i\phi_{m-1}} & 0 \\ 0 & e^{i\phi_{m-1}} \end{pmatrix}\begin{pmatrix} 1 & r_m \\ r_m & 1 \end{pmatrix}, \quad (9.47)$$

and is the transfer matrix which propagates the wave amplitude from the top of the $m^{\text{th}}$ layer to the top of the $(m-1)^{\text{th}}$ layer.[22] In layer 0 (the vacuum) and layer $N + 1$

$$\psi_0(z) = e^{ik_{z0}z} + re^{-ik_{z0}z}, \qquad \psi_{N+1}(z) = te^{ik_{z,N+1}(z-z_{N+1})}. \quad (9.48)$$

Hence $\psi_{N+1}^{+} = t$ and $\psi_{N+1}^{-} = 0$, and multiplication of the characteristic matrices for the entire sample gives us

$$\begin{pmatrix} 1 \\ r \end{pmatrix} = \mathsf{C}_1 \mathsf{C}_2 \ldots \mathsf{C}_{N+1} \begin{pmatrix} t \\ 0 \end{pmatrix}$$

$$= \begin{pmatrix} \mathsf{M}_{11} & \mathsf{M}_{12} \\ \mathsf{M}_{21} & \mathsf{M}_{22} \end{pmatrix} \begin{pmatrix} t \\ 0 \end{pmatrix}. \quad (9.49)$$

Finally, the reflectivity is given by

$$R = |r^2| = \frac{|\mathsf{M}_{21}|^2}{|\mathsf{M}_{11}|^2}. \quad (9.50)$$

The transfer matrix method for the calculation of $R$ is very simple to implement on a computer, requiring only the multiplication of the $N+1$ characteristic transfer matrices defined in (9.47). The matrices depend on the thickness of each layer and on the real and imaginary parts of $k_z$ for each layer, which are obtained either from (9.11) and (9.12), or from (9.16) and (9.17), with $k_z$ in place of $k$.

As an example, Fig. 9.10 shows neutron reflectivity data for a film of nickel on glass. The oscillations, which are known as *Kiessig fringes*, are due to interference between waves scattered from the surface and the nickel–glass interface. The thickness $d$ of the film can be determined from the oscillation period, which is $\Delta Q = 2\pi/d$ (see Exercise 9.2).

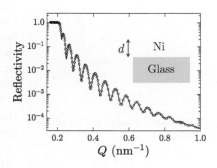

**Fig. 9.10** Reflectivity of a film of nickel (Ni) on a quartz glass substrate (Cubitt *et al.*, 2015). The symbols are data, and the line is calculated from a model with a film thickness of $d = 96.1$ nm and a Gaussian roughness of $\sigma = 3.6$ nm (see Section 9.2.2). The model also includes a surface oxide layer (NiO) of thickness 2.2 nm and roughness 0.6 nm, and a correction for the angular divergence of the incident beam. (Data courtesy of R. Cubitt.)

### Kinematic (Born) approximation

An approximate solution for the reflectivity profile may be obtained via the kinematical theory of diffraction which is based on the first Born approximation. The differential cross-section for coherent elastic scattering in the Born approximation is given in eqn (1.45). For a continuous medium the sum over plane waves from point nuclei becomes the Fourier transform of the scattering length density (Section 1.6.3),

$$\left( \frac{d\sigma}{d\Omega} \right)_{\text{coh}} = \left| \int n_b(\mathbf{r}) \exp(i\mathbf{Q} \cdot \mathbf{r}) \, d^3\mathbf{r} \right|^2. \quad (9.51)$$

Here we are concerned with a layered sample, so $n_b(\mathbf{r}) = n_b(z)$. We assume that the beam illuminates a rectangular area on the sample's surface of dimensions $2L_x \times 2L_y$. The Fourier transform in the $x$ and $y$ direction is the same as that for a single slit of width $2L$, which is a sinc function.[23] For large $L \gg 1/Q$, the sinc function tends to a $\delta$-function,[24] so

[23] $\int_{-L}^{+L} e^{iQx} \, dx = 2\sin(QL)/Q$

[24] See Section B.2.2

$$\left( \frac{d\sigma}{d\Omega} \right)_{\text{coh}} = 16\pi^2 L_x L_y \delta(Q_x)\delta(Q_y) \left| \int_{-\infty}^{\infty} n_b(z) \exp(iQ_z z) \, dz \right|^2. \quad (9.52)$$

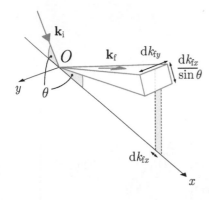

**Fig. 9.11** Calculation of $d\Omega$ for specular reflection.

This tells us that the coherent scattering is sharply peaked about the specular condition, $Q_x = Q_y = 0$. The differential cross-section is related to the reflectivity by

$$R = \frac{1}{4L_x L_y \sin\theta} \int \left(\frac{d\sigma}{d\Omega}\right)_{coh} d\Omega. \tag{9.53}$$

The solid angle integral appears because the reflectivity includes all the neutrons scattered close to the specular condition, whereas the differential cross-section describes the scattering per unit solid angle, see eqn (1.31). The factor $4L_x L_y \sin\theta$ is the projection of the surface area illuminated by the incident beam onto the plane perpendicular to the beam. The product of this factor with the incident flux, contained in the denominator of the differential cross-section, gives the total intensity incident on the sample, which is the required denominator for the reflectivity.

The solid angular element $d\Omega$ can be related to the wavevector with the help of Fig. 9.11. We define $d\Omega$ by a rectangular distribution of $\mathbf{k}_f$ directions within intervals $dk_{fx}$ and $dk_{fy}$ in the $x$ and $y$ directions. The end points of the $\mathbf{k}_f$ vector map out a rectangular area of dimensions $(dk_{fx}/\sin\theta) \times dk_{fy}$. The solid angle subtended by this area at $O$ is therefore $dk_{fx}dk_{fy}/k_f^2 \sin\theta$. As $\mathbf{k}_i$ is fixed and $\mathbf{Q} = \mathbf{k}_i - \mathbf{k}_f$ it follows that $d\mathbf{k}_f = -d\mathbf{Q}$. Hence, we may write

$$d\Omega = \frac{dQ_x dQ_y}{k_0^2 \sin\theta}, \tag{9.54}$$

where $k_0 = k_i = k_f$. Combining (9.52)–(9.54), and writing $Q = Q_z = 2k_0 \sin\theta$, we obtain the reflectivity profile

$$R(Q) = \frac{16\pi^2}{Q^2} \left| \int_{-\infty}^{\infty} n_b(z) \exp(iQz)\, dz \right|^2. \tag{9.55}$$

To illustrate the kinematical approximation for $R(Q)$ we consider again the case of a plane slab with scattering length density $n_b$ and surface at $z = 0$. Performing the integral in (9.55) we obtain[25]

$$R(Q) = \frac{16\pi^2}{Q^4} n_b^2. \tag{9.56}$$

[25]The upper limit of integration requires some attention. In practice, resolution effects limit the range of coherence of the neutron wave, so the part of the integrand at very large $z$ does not contribute. We can model this by including a factor $\exp(-\epsilon z)$ in the integral for $z > 0$ and taking the limit as $\epsilon \to 0$ at the end.

The result is compared with the exact form in Fig. 9.12. The agreement is very good over much of the plotted $Q$ range, but deviates near the critical edge where $R \to 1$.

The Born approximation is valid when the neutron wave function in the medium is not significantly different from its free-space form. This clearly breaks down near the critical edge, but nevertheless, the kinematical approximation for $R(Q)$ can be useful for obtaining a qualitative understanding of reflectivity profiles.

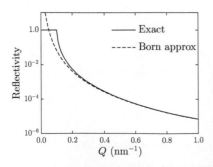

**Fig. 9.12** Reflectivity of a plane slab of silicon calculated from (9.38) and from the Born approximation (9.56).

## 9.2.2 Surface imperfections

The analysis of specular reflection presented in Sections 9.1.5 and 9.2.1 applies to surfaces and interfaces which are smooth and sharp. In reality, layer boundaries can be broadened, e.g. due to inter-diffusion, and there can be roughness within the plane of the surface. Surfaces can also have macroscopic imperfections, such as undulations, or smooth variations in the thickness or curvature of a solid film. Figure 9.13 illustrates these different imperfections.

We consider first the case of microscopic roughness. Because the neutron beam has a small angular divergence (typically 0.02°) and impinges on the surface at grazing incidence the coherence of the neutron beam can be substantial, typically extending ∼100 nm beneath the surface and up to several tens of $\mu$m in the plane of the surface. Surface imperfections on this scale or smaller will cause diffuse scattering, spraying neutrons out of the specularly reflected beam and reducing its intensity.[26]

This reduction in the specular reflectivity due to microscopic interfacial roughness cannot be distinguished from any rounding-off of the interfaces, and in practice both are typically modelled by a Gaussian smoothing of the boundary. A simple way to see the effect of this is to work in the kinematical approximation. Integration of (9.55) by parts leads to an alternative expression for the reflectivity in terms of the derivative of $n_b$,

$$R(Q) = \frac{16\pi^2}{Q^4} \left| \int_{-\infty}^{\infty} \frac{\mathrm{d}n_b}{\mathrm{d}z} \exp(\mathrm{i}Qz)\,\mathrm{d}z \right|^2 . \tag{9.57}$$

For a sharp boundary, $n_b(z)$ is a step function, and the derivative of a step function is a $\delta$-function (see Section B.2.4). For simplicity, consider a boundary at $z = 0$ between the vacuum and a medium with scattering length density $n_b$. In this case

$$\frac{\mathrm{d}n_b}{\mathrm{d}z} = n_b \delta(z). \tag{9.58}$$

To model the broadened interface we replace the $\delta$-function by a normalized Gaussian, eqn (B.1):

$$\frac{\mathrm{d}n_b}{\mathrm{d}z} \rightarrow \frac{n_b}{(2\pi\sigma^2)^{\frac{1}{2}}} \exp\left(-\frac{z^2}{2\sigma^2}\right). \tag{9.59}$$

Here, $\sigma = \sqrt{\langle \delta z_s^2 \rangle}$ can be interpreted as the standard deviation of the surface height fluctuations $(z_s)$, and is called the *Gaussian roughness*. Substitution of (9.59) into (9.57) leads to

$$R(Q) = \frac{16\pi^2 n_b^2}{Q^4} \exp\left(-\sigma^2 Q^2\right). \tag{9.60}$$

The ideal reflectivity for a sharp boundary, eqn (9.56), is multiplied by a Gaussian function of $Q$. We see, therefore, that the effect of a roughened or diffuse boundary is to reduce the reflectivity from the ideal case described by eqn (9.56). Physically, this is analogous to the effect of the Debye–Waller factor which describes the reduction in scattering intensity due to thermal smearing of the atomic positions (Section 5.1.4). The effect is significant for $Q \geq 1/\sigma$, and is illustrated in Fig. 9.14.

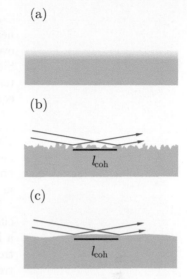

(a)

(b)

(c)

**Fig. 9.13** Types of surface imperfection. (a) Diffuse boundary. (b) Surface with microscopic roughness on a scale less than the coherence length $l_{\mathrm{coh}}$ of the neutron beam. (c) Macroscopic undulations on a scale greater than $l_{\mathrm{coh}}$.

[26] Note that this is distinct from the diffuse scattering that is included in the imaginary part of the refractive index. Here we are talking about scattering from imperfections localized at the boundaries.

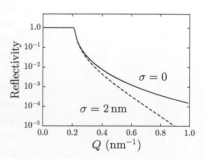

**Fig. 9.14** Reflectivity of a Ni film calculated with and without Gaussian roughness. The roughness is included via the modified Fresnel coefficient (9.61).

As already mentioned, the kinematical approximation breaks down near the critical edge. A rigorous treatment of Gaussian roughness valid over the entire $Q$ range has been given by Névot and Croce (1980) and Sinha *et al.* (1988). They showed that Gaussian roughening at an interface modifies the Fresnel reflection coefficient for a sharp interface, eqn (9.37), by

$$r \to r \exp(-2k_{z1}k_{z2}\sigma^2). \tag{9.61}$$

This is very convenient because it enables us to adapt the transfer matrix method to calculate reflectivity profiles for samples with diffuse or rough boundaries. All that is required is replace the Fresnel coefficient (9.46) in the characteristic matrix (9.47) by the modified form given in (9.61).

Having dealt with microscopic roughness, we now consider briefly the effect of large-scale imperfections. Macroscopic surface undulations on a length scale greater than the in-plane coherence length of the neutron beam, Fig. 9.13(c), will not cause interference effects and can be treated as a spread of local incident angles. The effect can be modelled in a similar way to that of resolution, by a convolution of the reflectivity profile for a macroscopically flat sample with a broadening function such as Gaussian whose width corresponds to the spread of local incident angles. The result is a smearing of the ideal reflectivity profile. If the reflected beam is sufficiently broad that not all of it arrives at the detector then there can also be a reduction in intensity.

### 9.2.3   Polarized neutron reflectometry

Polarized neutron reflectometry (PNR) has been used extensively to study the magnetic properties of thin films and multilayers. The use of polarized neutrons in a reflectivity measurement offers the ability to determine separately the chemical and magnetization profile beneath the surface of a layered sample. In contrast to conventional magnetometers, which can only measure the total magnetization of a sample, PNR is able to determine the size and direction of the magnetization of the individual layers in a multilayered sample. This has been exploited to study the properties of samples containing two or more magnetic layers which are antiparallel to one another (antiferromagnetic) or even in non-collinear orientations. PNR has also been used to measure the magnetization of ultrathin magnetic layers. Bulk magnetometry can be very difficult for such systems because the diamagnetic signal from the substrate can swamp the signal from the magnetic layer. The advantages of PNR are that the substrate does not contribute significantly to the magnetic signal and that information on the depth profile of the magnetic film can be obtained at the same time.

**Optical potential including magnetic interaction**

As discussed in Chapter 4, the sensitivity to magnetism comes from the Zeeman interaction between the neutron's magnetic moment $\boldsymbol{\mu}_\mathrm{n}$ and the magnetization in the material. In the present context we can simply extend the form of the optical potential given in (9.3) to include the magnetic potential. For a homogeneous medium

$$V = \frac{2\pi\hbar^2}{m_\mathrm{n}}n_b - \boldsymbol{\mu}_\mathrm{n}\cdot\mathbf{B},\tag{9.62}$$

where $\mathbf{B}$ is the magnetic induction. Reflectivity is caused by changes in $\mathbf{B}$. If $\mathbf{B}_0$ is the magnetic induction outside the sample due to an applied field then Maxwell's equations show that $\mathbf{B} - \mathbf{B}_0 = \mu_0\mathbf{M}_\mathrm{t}$, where $\mathbf{M}_\mathrm{t}$ is the component of the magnetization tangential to the surface.[27] The magnetic moment of the neutron is related to its spin by $\boldsymbol{\mu}_\mathrm{n} = -2\gamma\mu_\mathrm{N}\mathbf{s}_\mathrm{n}$, see eqn (1.10), so the optical potential relative to vacuum may be written

$$V = \frac{2\pi\hbar^2}{m_\mathrm{n}}n_b + 2\gamma\mu_0\mu_\mathrm{N}\mathbf{s}_\mathrm{n}\cdot\mathbf{M}_\mathrm{t}.\tag{9.63}$$

Since the eigenvalues of the component of $\mathbf{s}_\mathrm{n}$ along the quantization direction, which may be taken to be parallel to $\mathbf{M}_\mathrm{t}$, are $\pm 1/2$ we may write

$$V^\pm = \frac{2\pi\hbar^2}{m_\mathrm{n}}(n_b \pm n_M),\tag{9.64}$$

where $n_M$ is a magnetic scattering length density, given by

$$n_M = \frac{m_\mathrm{n}}{2\pi\hbar^2}\gamma\mu_0\mu_\mathrm{N}\sum_j n_j\,\mu_{tj}$$

$$= \sum_j n_j\,b_{Mj}.\tag{9.65}$$

The summation is over all magnetic species $j$, with $n_j$ the corresponding number density, $\mu_{tj}$ the tangential component of the magnetic moment, and $b_{Mj}$ a magnetic scattering length.[28] Finally, by replacing $n_b$ with $(n_b \pm n_M)$ in (9.26) we can obtain expressions for the refractive index of neutrons with spins parallel or antiparallel to the in-plane magnetization,

$$(n^\pm)^2 = 1 - \frac{4\pi}{k_0^2}(n_b' \pm n_M) + \mathrm{i}\mu\frac{k'}{k_0^2},\tag{9.66}$$

where $n_b'$ and $k'$ are the real parts of $n_b$ and $k$, respectively, and $\mu$ is the attenuation coefficient. For a material without strong attenuation, the corresponding critical angles for total reflection, eqn (9.35), are given by

$$\sin\theta_\mathrm{c}^\pm = \lambda\left(\frac{n_b' \pm n_M}{\pi}\right)^{1/2}.\tag{9.67}$$

### Transfer matrix method for PNR

We shall now obtain general formulae for the spin-resolved specular reflectivity of a layered magnetic sample using an extension of the transfer matrix method described in Section 9.2.1 (Blundell and Bland, 1992). As in the unpolarized case, the specular reflectivity can be calculated from the solution to the one-dimensional Schrödinger equation (9.29)

$$\left\{-\frac{\hbar^2}{2m_\mathrm{n}}\frac{\mathrm{d}^2}{\mathrm{d}z^2} + V(z)\right\}\psi(z) = E_z\Psi(z).\tag{9.68}$$

[27] In magnetic media $\mathbf{B} = \mu_0(\mathbf{H}+\mathbf{M})$, where $\mathbf{H}$ is the auxiliary magnetic field and $\mathbf{M}$ is the magnetization in the medium ($\mathbf{M}=0$ outside the medium). Let us resolve the fields normal (n) and tangential (t) to a boundary. Then,

$$\mathbf{B}-\mathbf{B}_0 = \mathbf{B}_\mathrm{n}-\mathbf{B}_{0\mathrm{n}}+\mathbf{B}_\mathrm{t}-\mathbf{B}_{0\mathrm{t}}$$
$$= \mathbf{B}_\mathrm{t}-\mathbf{B}_{0\mathrm{t}}$$
$$= \mu_0(\mathbf{H}_\mathrm{t}-\mathbf{H}_{0\mathrm{t}}+\mathbf{M}_\mathrm{t})$$
$$= \mu_0\mathbf{M}_\mathrm{t}.$$

We have used the boundary conditions $\mathbf{B}_\mathrm{n}=\mathbf{B}_{0\mathrm{n}}$ and $\mathbf{H}_\mathrm{t}=\mathbf{H}_{0\mathrm{t}}$, which come from Gauss' law $\nabla\cdot\mathbf{B}=0$ and Ampère's law for stationary fields $\nabla\times\mathbf{H}=0$ (in the absence of free currents).

[28] If $b_M$ is expressed in femtometres ($1\,\mathrm{fm}=10^{-15}$ m) and $\mu_\mathrm{t}$ in Bohr magnetons ($1\,\mu_\mathrm{B}=9.274\times10^{-24}$ JT$^{-1}$) then

$$b_M = 2.699\mu_\mathrm{t}.$$

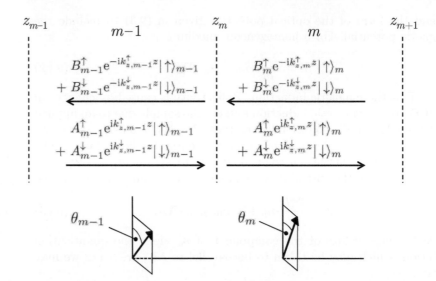

**Fig. 9.15** Spin-dependent incident and reflected waves at the boundary between layers $m-1$ and $m$. The directions of the in-plane component of the magnetization in each layer is shown below.

This time, however, we use the form of the potential given in (9.63) so as to include the neutron spin. Equation (9.68) has two eigenfunctions corresponding to the neutron spin parallel ($\uparrow$) and antiparallel ($\downarrow$) to the quantization direction.[29] These eigenfunctions may be written

$$\psi^\sigma(z) = e^{ik_z^\sigma z}|\sigma\rangle, \qquad \sigma = \uparrow \text{ or } \downarrow, \tag{9.69}$$

where $|\sigma\rangle$ is the spin wave function and $k_z^\sigma$ is given by the previous expressions (9.11)–(9.12) or (9.16)–(9.17) with $n_b$ generalized to $n_b \pm n_M$ in accord with (9.64).

For a general direction of $\mathbf{s_n}$ the solution of (9.68) is a coherent superposition of $\psi^\uparrow(z)$ and $\psi^\downarrow(z)$ which we shall write

$$\psi(z) = A^\uparrow \psi^\uparrow(z) + A^\downarrow \psi^\downarrow(z). \tag{9.70}$$

As before, we consider a system with $N$ layers such that the $(N+1)^{\text{th}}$ layer is the substrate, as shown in Fig. 9.9. Within each layer both $n_b$ and $n_M$ are constant. At the boundary between layer $m-1$ and layer $m$ there are incident and reflected waves from the left and right, as indicated in Fig. 9.15. Continuity of the amplitudes and derivatives of the wave functions at the boundary applies separately to the spin-up and spin-down components when the spin wave functions either side of the boundary are expressed with respect to the same quantization direction. Let us suppose that $\mathbf{M}_t$, which defines the quantization direction, rotates through an angle $\beta_m = \theta_{m-1} - \theta_m$ around the $z$ axis on crossing from layer $m$ to layer $m-1$, as shown in Fig. 9.15. Under this rotation the spin basis functions in the layers are related by (see Section C.1.16)

$$\begin{pmatrix} \uparrow \\ \downarrow \end{pmatrix}_m = \begin{pmatrix} \cos\frac{\beta_m}{2} & -\sin\frac{\beta_m}{2} \\ \sin\frac{\beta_m}{2} & \cos\frac{\beta_m}{2} \end{pmatrix} \begin{pmatrix} \uparrow \\ \downarrow \end{pmatrix}_{m-1}. \tag{9.71}$$

Applying this transformation, and imposing continuity of the amplitudes and derivatives of the wave functions at $z_m$, we obtain four equations analogous to (9.43) and (9.44), two for each spin state $|\uparrow\rangle_{m-1}$ and $|\downarrow\rangle_{m-1}$. After rearrangement we obtain the following relationships between the amplitudes at the top of the $(m-1)^{\text{th}}$ and $m^{\text{th}}$ layers,[30]

$$\begin{pmatrix} \psi_{m-1}^{\uparrow+} \\ \psi_{m-1}^{\uparrow-} \\ \psi_{m-1}^{\downarrow+} \\ \psi_{m-1}^{\downarrow-} \end{pmatrix} = \mathsf{C}_m \begin{pmatrix} \psi_{m}^{\uparrow+} \\ \psi_{m}^{\uparrow-} \\ \psi_{m}^{\downarrow+} \\ \psi_{m}^{\downarrow-} \end{pmatrix}, \tag{9.72}$$

where

$$\mathsf{C}_m = \begin{pmatrix} \mathsf{U}_m^{\uparrow\uparrow} \cos \frac{\beta_m}{2} & \mathsf{U}_m^{\downarrow\uparrow} \sin \frac{\beta_m}{2} \\ -\mathsf{U}_m^{\uparrow\downarrow} \sin \frac{\beta_m}{2} & \mathsf{U}_m^{\downarrow\downarrow} \cos \frac{\beta_m}{2} \end{pmatrix} \tag{9.73}$$

with

$$\mathsf{U}_m^{\sigma_1 \sigma_2} = \frac{1}{2} \begin{pmatrix} e^{-i\phi_{m-1}^{\sigma_2}} & 0 \\ 0 & e^{i\phi_{m-1}^{\sigma_2}} \end{pmatrix} \begin{pmatrix} 1 + \frac{k_m^{\sigma_1}}{k_{m-1}^{\sigma_2}} & 1 - \frac{k_m^{\sigma_1}}{k_{m-1}^{\sigma_2}} \\ 1 - \frac{k_m^{\sigma_1}}{k_{m-1}^{\sigma_2}} & 1 + \frac{k_m^{\sigma_1}}{k_{m-1}^{\sigma_2}} \end{pmatrix}, \tag{9.74}$$

and $\phi_{m-1}^{\sigma} = k_{z,m-1}^{\sigma}(z_m - z_{m-1})$.

We can use (9.72)–(9.74) to obtain the spin-dependent Fresnel coefficients for a single boundary between two magnetic layers 1 and 2. Putting $\phi_1^{\sigma} = 0$ to relate the amplitudes on the boundary we obtain

$$r^{\uparrow\uparrow} = \frac{(k_{z1}^{\uparrow} - k_{z2}^{\uparrow})(k_{z1}^{\downarrow} + k_{z2}^{\downarrow}) \cos^2 \frac{\beta}{2} + (k_{z1}^{\uparrow} - k_{z2}^{\downarrow})(k_{z1}^{\downarrow} + k_{z2}^{\uparrow}) \sin^2 \frac{\beta}{2}}{(k_{z1}^{\uparrow} + k_{z2}^{\uparrow})(k_{z1}^{\downarrow} + k_{z2}^{\downarrow}) \cos^2 \frac{\beta}{2} + (k_{z1}^{\uparrow} + k_{z2}^{\downarrow})(k_{z1}^{\downarrow} + k_{z2}^{\uparrow}) \sin^2 \frac{\beta}{2}}$$

$$r^{\uparrow\downarrow} = \frac{2k_{z1}^{\uparrow}(k_{z2}^{\uparrow} - k_{z2}^{\downarrow}) \sin \frac{\beta}{2} \cos \frac{\beta}{2}}{(k_{z1}^{\uparrow} + k_{z2}^{\uparrow})(k_{z1}^{\downarrow} + k_{z2}^{\downarrow}) \cos^2 \frac{\beta}{2} + (k_{z1}^{\uparrow} + k_{z2}^{\downarrow})(k_{z1}^{\downarrow} + k_{z2}^{\uparrow}) \sin^2 \frac{\beta}{2}}$$

$$t^{\uparrow\uparrow} = \frac{2k_{z1}^{\uparrow}(k_{z1}^{\downarrow} + k_{z2}^{\downarrow}) \cos \frac{\beta}{2}}{(k_{z1}^{\uparrow} + k_{z2}^{\uparrow})(k_{z1}^{\downarrow} + k_{z2}^{\downarrow}) \cos^2 \frac{\beta}{2} + (k_{z1}^{\uparrow} + k_{z2}^{\downarrow})(k_{z1}^{\downarrow} + k_{z2}^{\uparrow}) \sin^2 \frac{\beta}{2}}$$

$$t^{\uparrow\downarrow} = \frac{2k_{z1}^{\uparrow}(k_{z1}^{\downarrow} + k_{z2}^{\uparrow}) \sin \frac{\beta}{2}}{(k_{z1}^{\uparrow} + k_{z2}^{\uparrow})(k_{z1}^{\downarrow} + k_{z2}^{\downarrow}) \cos^2 \frac{\beta}{2} + (k_{z1}^{\uparrow} + k_{z2}^{\downarrow})(k_{z1}^{\downarrow} + k_{z2}^{\uparrow}) \sin^2 \frac{\beta}{2}}.$$

$$\tag{9.75}$$

For $\beta = 0$ these reduce to the expressions given in (9.37) for unpolarized neutrons. The formulae for incident spin down are the same as those in (9.75) but with the arrows reversed.

Multiplication of the characteristic matrices $\mathsf{C}_m$ for a multilayered sample yields the relation between the amplitudes of the waves on the incident and exit surfaces of the sample. For example, if the incident neutrons are in the $\uparrow$ state then we obtain

$$
\begin{pmatrix} 1 \\ r^{\uparrow\uparrow} \\ 0 \\ r^{\uparrow\downarrow} \end{pmatrix} = \begin{pmatrix} \mathsf{M}_{11} & \mathsf{M}_{12} & \mathsf{M}_{13} & \mathsf{M}_{14} \\ \mathsf{M}_{21} & \mathsf{M}_{22} & \mathsf{M}_{23} & \mathsf{M}_{24} \\ \mathsf{M}_{31} & \mathsf{M}_{32} & \mathsf{M}_{33} & \mathsf{M}_{34} \\ \mathsf{M}_{41} & \mathsf{M}_{42} & \mathsf{M}_{43} & \mathsf{M}_{44} \end{pmatrix} \begin{pmatrix} t^{\uparrow\uparrow} \\ 0 \\ t^{\uparrow\downarrow} \\ 0 \end{pmatrix}, \tag{9.76}
$$

where the matrix $\mathsf{M} = \mathsf{C}_1 \mathsf{C}_2 \ldots \mathsf{C}_{N+1}$. For spin $\downarrow$ incident neutrons all that changes is that the vectors become $(0, r^{\downarrow\uparrow}, 1, r^{\downarrow\downarrow})$ and $(t^{\downarrow\uparrow}, 0, t^{\downarrow\downarrow}, 0)$ reading downwards. The four equations summarized in (9.76) can easily be solved to obtain the spin-resolved reflectivities $R^{\sigma_1 \sigma_2} = |r^{\sigma_1 \sigma_2}|^2$:

$$
R^{\uparrow\uparrow} = \left| \frac{\mathsf{M}_{21}\mathsf{M}_{33} - \mathsf{M}_{23}\mathsf{M}_{31}}{\mathsf{M}_{11}\mathsf{M}_{33} - \mathsf{M}_{13}\mathsf{M}_{31}} \right|^2
$$

$$
R^{\uparrow\downarrow} = \left| \frac{\mathsf{M}_{41}\mathsf{M}_{33} - \mathsf{M}_{43}\mathsf{M}_{31}}{\mathsf{M}_{11}\mathsf{M}_{33} - \mathsf{M}_{13}\mathsf{M}_{31}} \right|^2
$$

$$
R^{\downarrow\uparrow} = \left| \frac{\mathsf{M}_{23}\mathsf{M}_{11} - \mathsf{M}_{21}\mathsf{M}_{13}}{\mathsf{M}_{11}\mathsf{M}_{33} - \mathsf{M}_{13}\mathsf{M}_{31}} \right|^2
$$

$$
R^{\downarrow\downarrow} = \left| \frac{\mathsf{M}_{43}\mathsf{M}_{11} - \mathsf{M}_{41}\mathsf{M}_{13}}{\mathsf{M}_{11}\mathsf{M}_{33} - \mathsf{M}_{13}\mathsf{M}_{31}} \right|^2. \tag{9.77}
$$

When the external field and in-plane magnetization in each layer are parallel or antiparallel to a single direction, i.e. $\beta_m = 0$ or $\pi$ for all $m$, the characteristic matrix splits into two $2 \times 2$ blocks which decouples the two spin states. In this case $R^{\uparrow\downarrow} = R^{\downarrow\uparrow} = 0$. Conversely, if non-zero spin-flip reflectivities $R^{\uparrow\downarrow}$ and $R^{\downarrow\uparrow}$ are observed then there exists in-plane magnetization in the sample which is not parallel or antiparallel to the incident polarization.

### Optical method for PNR

The transfer matrix method for PNR is a straightforward extension of that for unpolarized neutrons, but one disadvantage is that it is not generally possible to include a Debye–Waller factor to describe the effect of interfacial roughness as is possible for non-magnetic samples (Section 9.2.2). The reason is that, with the exception of collinear magnetic samples ($\beta_m = 0$ or $\pi$), the elements of the characteristic matrix are not simple functions of the spin-dependent Fresnel coefficients (9.75). One option is to simulate the effect of inter-diffusion by dividing the interfaces into many thin layers to smooth the change in scattering length density from one layer to the next, but this is inefficient and does not have the same effect as microscopic roughness. An alternative formulation for PNR, described next, expresses the reflectivity in terms of a sum of multiple reflections and transmissions at each interface (Blundell and Bland, 1993). The advantage of this optical method is that roughness can be modelled with a Debye–Waller factor analogous to (9.61).

Consider the three-layer system shown in Fig. 9.16. For a unit incident wave there are multiple reflections and transmissions at each boundary. The total amplitude of the reflected waves is given by the matrix equation

$$r_1 = r_{12} + t_{12}P_2r_{23}P_2\left\{I + (r_{21}P_2r_{23}P_2) + (r_{21}P_2r_{23}P_2)^2 + \ldots\right\}t_{21}$$

$$= r_{12} + t_{12}P_2r_{23}P_2\left\{I - r_{21}P_2r_{23}P_2\right\}^{-1}t_{21}, \qquad (9.78)$$

where

$$r_{ij} = \begin{pmatrix} r_{ij}^{\uparrow\uparrow} & r_{ij}^{\uparrow\downarrow} \\ r_{ij}^{\downarrow\uparrow} & r_{ij}^{\downarrow\downarrow} \end{pmatrix}, \quad t_{ij} = \begin{pmatrix} t_{ij}^{\uparrow\uparrow} & t_{ij}^{\uparrow\downarrow} \\ t_{ij}^{\downarrow\uparrow} & t_{ij}^{\downarrow\downarrow} \end{pmatrix}, \quad P_m = \begin{pmatrix} e^{i\phi_m^{\uparrow}} & 0 \\ 0 & e^{i\phi_m^{\downarrow}} \end{pmatrix},$$

(9.79)

and $I$ is the $2 \times 2$ identity matrix. The $r_{ij}$ and $t_{ij}$ matrices contain the spin-dependent Fresnel coefficients given in eqn (9.75), and $P_m$ propagates the wave across layer $m$.[31]

The reflectivity for the complete system of $N$ layers + substrate, Fig. 9.9, can be obtained recursively from the effective Fresnel coefficient $r_{m-1}$ for the boundary between layers $m-1$ and $m$,

$$r_{m-1} = r_{m-1,m} + t_{m-1,m}P_mr_mP_m\left\{I - r_{m,m-1}P_mr_mP_m\right\}^{-1}t_{m,m-1},$$

which includes the multiple reflections in all the layers below the boundary. The first term to be evaluated is $r_{N-1}$, with $r_N$ the Fresnel coefficient for the boundary between layer $N$ and the substrate. Successive terms $r_{N-2}$, $r_{N-3}$ ... depend on the previous one. The last term is $r_0$, which gives the total reflectivity of the whole sample. The roughness of each interface can be included by modification of the Fresnel reflection coefficients (9.75) at each interface with a Debye–Waller factor

$$r_{ij}^{\sigma_1\sigma_2} \rightarrow r_{ij}^{\sigma_1\sigma_2} \exp(-2k_{zi}^{\sigma_1}k_{zj}^{\sigma_2}\sigma_{ij}^2), \qquad (9.80)$$

where $\sigma_{ij}$ is the Gaussian roughness, see (9.59). The chemical and magnetic roughness are constrained to be the same.

## Examples of PNR

Figure 9.17 shows specular reflectivity measurements on a single ferromagnetic layer of FeGe. A field of $\mu_0H = 0.667\,\text{T}$ applied parallel to the surface of the film served to saturate the magnetization of the film in the direction of the field. With this arrangement, only the non-spin-flip reflectivities $R^{\uparrow\uparrow}$ and $R^{\downarrow\downarrow}$ are non-zero. A half-polarized measurement without spin analysis of the reflected neutrons could therefore be performed, since $R^{\uparrow} \equiv R^{\uparrow\uparrow}$ and $R^{\downarrow} \equiv R^{\downarrow\downarrow}$. The $R^{\uparrow}$ and $R^{\downarrow}$ data in Fig. 9.17 have different critical edges and exhibit a clear separation over the entire $Q$ range, as expected from the spin dependence of the refractive index, eqn (9.66). The oscillations in the reflectivities are Kiessig fringes related to the thickness of the film.

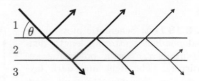

**Fig. 9.16** Multiple interference in a three-layer magnetic system. The incident angle $\theta$ has been greatly exaggerated.

[31] $\phi_m^{\sigma} = k_{z,m}^{\sigma}(z_{m+1} - z_m)$ is the phase difference across layer $m$ in the spin channel $\sigma$.

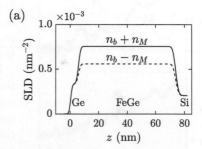

(a)

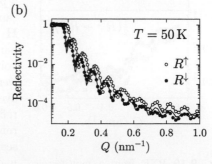

(b)

**Fig. 9.17** PNR from a single ferromagnetic layer of FeGe, after Porter *et al.* (2015). A nominally 70 nm thick layer of FeGe was deposited on a Si wafer, and capped with 5 nm of Ge. (a) Scattering length density (SLD) for ↑ and ↓ spin neutrons. (b) $R^{\uparrow}$ and $R^{\downarrow}$ reflectivities. (Data courtesy of Christopher Marrows.)

**Fig. 9.18** Left: Schematic plots of the spin-dependent scattering length densities $n_b \pm n_M$ for a polarizing supermirror. The light and dark layers are the non-magnetic and magnetic layers, respectively. Right: Spin-dependent neutron reflectivities $R^\uparrow$ and $R^\downarrow$ and polarization $(P)$ for an Fe/Si polarizing supermirror. (Data courtesy of SwissNeutronics AG, http://www.swissneutronics.ch.)

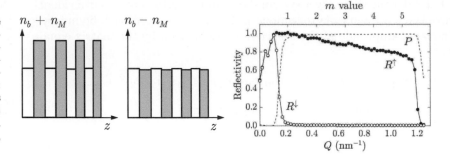

Very efficient neutron polarizers can be produced with supermirrors that contain alternating ferromagnetic and non-magnetic layers. The non-magnetic spacer layer is chosen so that its scattering length density $n_b$ is closely matched to $n_b - n_M$ of the magnetic layer (Fig. 9.18). That way, the contrast and hence reflectivity for the $\downarrow$ neutrons is very small, and the reflected beam will comprise predominantly $\uparrow$ neutrons over a range of $Q$ from $Q_c^\downarrow$ to $Q_c^\uparrow$. Typical supermirror material pairs are Fe/Si, Co/Ti or FeCoV/TiN$_x$. Spin-dependent reflectivity data for an Fe/Si polarizing supermirror are shown in Fig. 9.18. This supermirror would achieve a beam polarization above 98% over a neutron wavelength bandwidth from 0.2 to 1.1 nm for an incident angle of 1°.

Figure 9.19 gives an example of PNR in which important information is contained in the spin-flip reflectivity channels. The sample is a multilayered ferromagnet grown on a piezoelectric substrate. What is special about this particular sample is that the direction of the magnetization $\mathbf{M}$ in the plane of the multilayer can be varied by application of a potential difference across the sample. In the PNR measurement, the sample was first magnetically saturated by an in-plane magnetic field $\mathbf{H}$, after which the field was reduced to 1 mT and the four neutron-spin partial reflectivities $R^{\uparrow\uparrow}$, $R^{\downarrow\downarrow}$, $R^{\uparrow\downarrow}$ and $R^{\downarrow\uparrow}$ were measured as the applied voltage was varied. The two spin-flip (SF) reflectivities $(R^{\uparrow\downarrow}$ and $R^{\downarrow\uparrow})$ were found to be the same to within experimental error and so were averaged to improve the statistics, whereas the two non-spin-flip (NSF) reflectivities $(R^{\uparrow\uparrow}$ and $R^{\downarrow\downarrow})$ are significantly different due to the large remanent magnetization of the sample. To understand the behaviour seen in Fig. 9.19, recall that the NSF reflectivities are sensitive to the nuclear scattering length densities and to the in-plane component of $\mathbf{M}$ parallel to $\mathbf{H}$, whereas the SF reflectivities probe the in-plane component of $\mathbf{M}$ perpendicular to $\mathbf{H}$. The large change observed in the SF relative to the NSF reflectivities with applied voltage implies that the in-plane component of $\mathbf{M}$ perpendicular to $\mathbf{H}$ changes much more than the component parallel to $\mathbf{H}$. In fact, the data are very well described by a simple model in which the direction of $\mathbf{M}$ rotates from $\phi \simeq 15°$ to $\phi \simeq 40°$ as the voltage is increased from 0 to 400 V, where $\phi$ is the angle between $\mathbf{M}$ and $\mathbf{H}$.

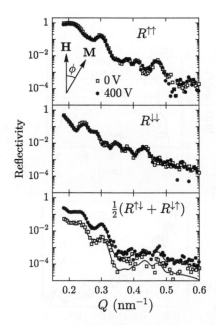

**Fig. 9.19** PNR from a multiferroic multilayer comprising alternating layers of Fe$_{86}$Ga$_{14}$ and Ni$_{80}$Fe$_{20}$ deposited on a piezoelectric PbMg$_{1/3}$Nb$_{2/3}$O$_3$–PbTiO$_3$ substrate, after Jamer *et al.* (2018). The angle $\phi$ between the in-plane magnetization $\mathbf{M}$ and the applied magnetic field $\mathbf{H}$ depends on the voltage applied across the sample. (Data courtesy of Michelle Jamer, Julie Borchers, and Brian Kirby.)

## 9.2.4   Off-specular scattering and GISANS

Pure specular reflection is only observed for surfaces and/or buried interfaces which are laterally smooth. Imperfections such as interfacial roughness, domain boundaries or lateral compositional variations cause off-specular diffuse scattering. Observations of off-specular scattering are often restricted to the plane of incidence (the $xz$ plane) so that $\mathbf{Q} = (Q_x, 0, Q_z)$. However, if one tilts the wavevector out of the plane of incidence while maintaining the specular condition then $\mathbf{Q} = (0, Q_y, Q_z)$. This is the technique of GISANS (grazing incidence small-angle neutron scattering). The two distinct geometries are shown in Fig. 9.8. Reflectometers that are equipped with a two-dimensional multidetector can measure both types of off-specular scattering. The experimentally accessible range for off-specular scattering in the plane of incidence is typically $10^{-4} \leq Q_x \leq 10^{-2}\,\mathrm{nm}^{-1}$, which gives access to lateral correlations on a $\sim\mu$m scale. For GISANS, the typical range is $10^{-2} \leq Q_y \leq 1\,\mathrm{nm}^{-1}$, corresponding to tens of nm.

The Born approximation can be used to calculate off-specular diffuse scattering for $Q \gg Q_c$. Near to the critical edge the Born approximation breaks down, and the approach that has often been used is the distorted-wave Born approximation (DWBA). The DWBA uses the wave functions calculated for ideal smooth surfaces as basis functions for perturbation theory. Details of the method can be found elsewhere (Sinha *et al.*, 1988). Recent advances in the theory, combined with the development of high-flux, high-resolution reflectometers, have made it possible to study a range of different thin film phenomena, such as Yoneda fringes/wings, lateral self-assembly, fractal surface roughness, liquid surface waves, etc.

## 9.3   Neutron imaging

At first glance, neutron imaging appears to have little in common with neutron scattering. Imaging uses transmission, the most basic of all optical properties, to make real-space maps of the internal structure of an object, whereas scattering measures structure in reciprocal space. On the other hand, the information obtained by the two techniques is complementary and, moreover, imaging stations are increasingly being installed at neutron scattering facilities and involve some of the same operational considerations as scattering instruments.

In its simplest implementation, neutron imaging involves measurement of the transmission of a beam of neutrons through the object, the transmitted intensity being recorded on a position-sensitive area detector to determine its spatial variation (Anderson *et al.*, 2009). This version of the technique is called *neutron radiography* (Fig. 9.20), and is directly analogous to the method used to obtain a conventional medical X-ray image. An extension involves measurement of several hundred transmission images as the sample is turned by small angular increments through either 180° or 360°. The set of measurements is used to reconstruct a

**Fig. 9.20** X-ray (top) and neutron (bottom) radiographic images of a hard disk drive. (Courtesy of the Neutron Imaging & Activation Group, Paul Scherrer Institut, Switzerland.)

**Fig. 9.21** Neutron tomographic images of a Swiss army knife showing how internal structure can be revealed without physically dismantling the specimen. The lower image has been processed to reveal the oil between the blades (in yellow). (Courtesy of the Neutron Imaging & Activation Group, Paul Scherrer Institut, Switzerland.)

three-dimensional image through a mathematical procedure known as an *inverse Radon transform*. This technique is called *computed tomography* (CT). An example is shown in Fig. 9.21.

Neutrons have a number of advantages for imaging compared with X-rays, including the higher penetration especially for metals, the sensitivity to light elements such as hydrogen and lithium, and the ability to detect magnetic materials. The main disadvantages are the longer measurement times that are often required and the possibility that the object might become radioactive during exposure. The complementary power of neutron and X-ray imaging is illustrated in Fig. 9.20.

The technique has found employment in engineering, materials science, geology, archeology, and other disciplines. Examples of applications include non-destructive testing of machine components, the composition of rocks, the distribution of lithium in Li-ion batteries, the study of hydrogenous materials such as water in fuel cells, hydrogen uptake in concretes and steels, visualization of magnetic domains, the composition of paleontological and heritage artefacts, and many others.

Neutron imaging has largely developed separately from neutron diffraction, although both techniques measure internal structure and are complementary. Imaging makes real-space maps with a spatial resolution of order $\mu$m and above, whereas diffraction can probe structural correlations on length scales from atomic dimensions up to about 100 nm but averages over the whole volume of sample in the neutron beam.

One distinction between imaging and diffraction is that neutron images reveal spatial variations in the *attenuation* of the object whereas diffraction is sensitive to variations in the *scattering amplitude*. Attenuation is caused by nuclear absorption as well as incoherent and coherent (nuclear and magnetic) scattering of neutrons out of the beam (Sections 9.1.2–9.1.3). For crystalline materials there can be significant attenuation due to Bragg diffraction. The overall attenuation of the beam is described by the Beer–Lambert law (Section 1.7.4) integrated over the path taken by the neutrons through the sample. Specifically, the transmitted intensity may be written

$$I = I_0 \exp\left\{ -\int \mu(z)\mathrm{d}z \right\}, \qquad (9.81)$$

where $I_0$ is the incident intensity and $\mu(z)$ is the attenuation coefficient eqn (9.21) which in general varies with position $z$ along the path.

The sample's internal structure causes the transmitted intensity $I$ to vary with position across the sample. A highly collimated neutron beam with a large area (up to $30 \times 30\,\mathrm{cm}^2$) is used to record this spatial variation. The beam is obtained from a pinhole of diameter $D$ at a distance $L$ from the object. The collimation can be varied, but the ratio $L/D$ is typically a few hundred. The transmitted intensity is recorded on a pixellated area detector, for example a scintillator coupled to a photomultiplier, with a spatial resolution down to about 10 $\mu$m in the best cases. Conversely, by relaxing the resolution one can collect images in a matter of seconds and follow changes in the object in real time. For example, through the use of stroboscopic measurements on a material undergoing a cycle it has been possible to achieve $\sim 10\,\mu s$ temporal resolution after averaging over many cycles.

The standard neutron imaging technique uses the full polychromatic spectrum of incident neutrons to obtain the maximum intensity in the image, but the option to use a monochromatic beam to perform energy-selective imaging exists on some beam lines. This achieves sensitivity to particular materials of interest in the object via the contrast between images taken above and below Bragg diffraction edges or in the vicinity of nuclear absorption resonances, where there are strong changes in the attenuation. An example is the use of fast neutrons ($E > 1\,\mathrm{MeV}$) in radiography to map ratios of the elements of C, N and O via their characteristic fast neutron-induced nuclear resonances to detect drugs and explosives in luggage at border crossings. A similar principle is employed to gain magnetic contrast through the use of polarized neutrons.

Lately, both neutron imaging and diffraction have been performed at the same time, and simultaneous neutron and X-ray imaging has been achieved through the installation of an X-ray source in the neutron cabin. Another development has been to put diffraction gratings in the incident beam and perform interferometric experiments which are sensitive to phase contrast. These advances suggest that the synergy between imaging and scattering will strengthen in the future as more imaging instruments are added to the portfolio of techniques at neutron facilities.

## 9.4   Perfect crystal diffraction

We now return to the dynamical theory of neutron diffraction, and show how it describes a number of interesting optical phenomena that occur in diffraction from perfect crystals.

According to Bloch's theorem, the general solution $\Psi(\mathbf{r})$ to the Schrödinger equation (9.1) for waves in a periodic potential is a superposition of Bloch functions, each of which can be expanded as a Fourier series (Singleton, 2001)

$$\psi_{\mathbf{k}}(\mathbf{r}) = \sum_{\mathbf{G}} C_{\mathbf{G}} e^{i(\mathbf{k}-\mathbf{G})\cdot\mathbf{r}}, \qquad (9.82)$$

where $\mathbf{G}$ is a reciprocal lattice vector. The periodic potential can also be expanded as a Fourier series which we shall write as[32]

$$V(\mathbf{r}) = \frac{\hbar^2}{2m_{\mathrm{n}}} \sum_{\mathbf{G}} V_{\mathbf{G}} e^{-i\mathbf{G}\cdot\mathbf{r}}. \qquad (9.83)$$

The coefficient $V_{\mathbf{G}}$ has dimensions of wavevector-squared, and is related in a simple way to the nuclear unit-cell structure factor $F_{\mathrm{N}}(\mathbf{G})$ defined in eqn (5.48). The relation is (see Exercise 9.5)

$$V_{\mathbf{G}} = \frac{4\pi}{v_0} F_{\mathrm{N}}(\mathbf{G}). \qquad (9.84)$$

[32]This form is periodic in the lattice, $V(\mathbf{r}+\mathbf{l}) = V(\mathbf{r})$, because reciprocal lattice vectors satisfy $e^{i\mathbf{G}\cdot\mathbf{l}} = 1$ for all $\mathbf{l}$.

Substitution of (9.82) and (9.83) into the Schrödinger equation (9.1) with energy $E = \hbar^2 k_0^2 / 2 m_\mathrm{n}$ results in

$$\sum_{\mathbf{G}} C_{\mathbf{G}} \left\{ k_0^2 - (\mathbf{k} - \mathbf{G})^2 \right\} e^{i(\mathbf{k} - \mathbf{G}) \cdot \mathbf{r}} = \sum_{\mathbf{G}'} V_{\mathbf{G}'} \, e^{-i\mathbf{G}' \cdot \mathbf{r}} \sum_{\mathbf{G}''} C_{\mathbf{G}''} \, e^{i(\mathbf{k} - \mathbf{G}'') \cdot \mathbf{r}}, \tag{9.85}$$

and equating the coefficients of $e^{i(\mathbf{k} - \mathbf{G}) \cdot \mathbf{r}}$ we find

$$C_{\mathbf{G}} \left\{ k_0^2 - (\mathbf{k} - \mathbf{G})^2 \right\} = \sum_{\mathbf{G}'} V_{\mathbf{G}'} C_{\mathbf{G} - \mathbf{G}'}. \tag{9.86}$$

Because the potential is weak, $|V_{\mathbf{G}}| \ll k_0^2$ for all $\mathbf{G}$. Hence, we can restrict attention to transmitted waves that are close to the incident wave, i.e. $\mathbf{k} \simeq \mathbf{k}_0$. For the same reason, the $C_{\mathbf{G}}$ coefficients in (9.86) are negligible except when $k_0^2 - (\mathbf{k} - \mathbf{G})^2 \simeq 0$. When $\mathbf{G} \neq 0$ this condition corresponds to Bragg scattering of the incident wave.[33]

These constraints guide us to consider two distinct cases, depending on whether $\mathbf{k}_0$ does or does not satisfy the Bragg (Laue) condition.

(i) $\mathbf{k}_0$ *away from any Bragg condition.* If $k_0^2 - (\mathbf{k} - \mathbf{G})^2 \simeq 0$ only holds for $\mathbf{G} = 0$ then (9.86) shows that all the $C_{\mathbf{G}}$ coefficients are negligible except $C_0$. Retention of only the $C_0$ term reduces (9.86) to

$$C_0 (k_0^2 - k^2) = V_0 C_0, \tag{9.87}$$

where $V_0$ is the the uniform component of the potential, which from (9.83) is given by[34] $V_0 = 2 m_\mathrm{n} V / \hbar^2$. Since the refractive index is defined by $n = k / k_0$, eqn (9.87) gives

$$n^2 = \frac{k^2}{k_0^2} = 1 - \frac{2 m_\mathrm{n}}{\hbar^2 k_0^2} V, \tag{9.88}$$

which reproduces the expression for $n^2$ derived earlier, eqn (9.25). This means that as long as the Bragg condition is not satisfied the coherent wave inside a crystal is given by (9.8), which is the $\mathbf{G} = 0$ term in (9.82). All the results derived previously for optical phenomena in homogeneous media then apply equally to a crystal.

(ii) $\mathbf{k}_0$ *near a Bragg condition.* When $k_0^2 - (\mathbf{k} - \mathbf{G})^2 \simeq 0$ for one or more $\mathbf{G} \neq 0$ the corresponding $C_{\mathbf{G}}$ join $C_0$ in being non-negligible. We shall assume without loss of generality that the Bragg condition is satisfied for only one $\mathbf{G}$, in which case (9.86) reduces to two simultaneous equations. These can be represented in matrix form as

$$\begin{pmatrix} k_0^2 - k^2 - V_0 & -V_{-\mathbf{G}} \\ -V_{\mathbf{G}} & k_0^2 - (\mathbf{k} - \mathbf{G})^2 - V_0 \end{pmatrix} \begin{pmatrix} C_0 \\ C_{\mathbf{G}} \end{pmatrix} = 0. \tag{9.89}$$

A necessary condition for a non-trivial solution is that the matrix determinant vanishes. This gives

$$(k_0^2 - k^2 - V_0) \left\{ k_0^2 - (\mathbf{k} - \mathbf{G})^2 - V_0 \right\} - V_{\mathbf{G}} V_{-\mathbf{G}} = 0. \tag{9.90}$$

[33] Recall that Bragg scattering occurs when the Laue condition $\mathbf{k} - \mathbf{k}' = \mathbf{G}$ is satisfied for some $\mathbf{G} \neq 0$ and $|\mathbf{k}'| = |\mathbf{k}|$. With $|\mathbf{k}'| \simeq |\mathbf{k}_0|$ the Laue condition reduces to $k_0^2 - (\mathbf{k} - \mathbf{G})^2 \simeq 0$.

[34] Here, $V = \langle V(\mathbf{r}) \rangle$ is the average potential in the crystal. Also, $V_0 = 4\pi n_b$ from (9.3), (9.4) and (9.83).

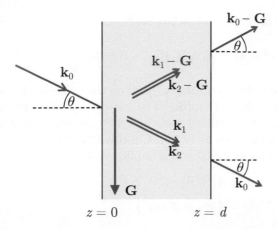

Fig. 9.22 Parallel-sided crystal showing arrangement for Bragg diffraction in transmission with **G** in the plane of the large face. The wave reflected from incident surface has been omitted because its amplitude is negligible unless $\theta \simeq 90°$.

As $|V_\mathbf{G}| \ll k_0^2$, the solutions of eqn (9.90) are found when one or other of the two factors enclosed in parentheses is very close to zero. In addition, a boundary condition at the crystal surface dictates that the momentum parallel to the surface is conserved,[35] and so the parallel components of $\mathbf{k}_0$ and $\mathbf{k}$ are equal. With this constraint eqn (9.90) can be transformed into a quadratic equation in $k^2$, and for a given $\mathbf{k}_0$ and $\mathbf{G}$ there are two solutions $\mathbf{k}_1$ and $\mathbf{k}_2$ both of which are very close to $\mathbf{k}_0$. The coherent wave in the crystal is then a superposition of four Bloch waves (9.82), two of which are primary waves and two the Bragg-diffracted waves,

$$\Psi(\mathbf{r}) = A_1 \left[ e^{i\mathbf{k}_1 \cdot \mathbf{r}} + \alpha_1 e^{i(\mathbf{k}_1 - \mathbf{G}) \cdot \mathbf{r}} \right] + A_2 \left[ e^{i\mathbf{k}_2 \cdot \mathbf{r}} + \alpha_2 e^{i(\mathbf{k}_2 - \mathbf{G}) \cdot \mathbf{r}} \right], \quad (9.91)$$

where $\alpha_1$ and $\alpha_2$ are the amplitude ratios $C_\mathbf{G}/C_0$ for $\mathbf{k}_1$ and $\mathbf{k}_2$, which can be obtained from (9.89).

[35]Because there is no change in optical potential in the plane of the surface, and hence no force parallel to the surface.

### 9.4.1 Pendellösung interference

The various internal and external waves for the two-beam case are represented in Fig. 9.22 for the special case of a parallel-sided crystal prepared so that the reciprocal lattice vector **G** for a particular set of lattice planes lies in the plane of the large face of the crystal. The two primary internal waves are in phase at the incident surface, but because they have very slightly different wavelengths they interfere to produce beats as they propagate through the crystal. When they exit the crystal the two waves combine into a single wave whose intensity varies sinusoidally according to the phase of the beat pattern at the exit surface. The same applies to the Bragg-diffracted intensity, except that due to conservation of neutron flux the beats are in antiphase with those in the transmitted beam. This behaviour is known as *Pendellösung*, or 'pendulum solution'.

To illustrate the dynamical diffraction phenomena that are observable near the Bragg condition we shall obtain the Pendellösung for the parallel-sided crystal shown in Fig. 9.22. We shall assume $V(\mathbf{r})$ to be real, so that from (9.83) $V_{-\mathbf{G}} = V_\mathbf{G}^*$. In this symmetric geometry, the boundary condition at the surface means that $k^2 = (\mathbf{k} - \mathbf{G})^2$, and so

[36]The upper and lower signs correspond to the $\mathbf{k}_1$ and $\mathbf{k}_2$ solutions, respectively.

the two solutions of (9.89) and (9.90) are[36]

$$k^2 = k_0^2 - V_0 \pm |V_{\mathbf{G}}|, \qquad \alpha = \mp \frac{V_{\mathbf{G}}}{|V_{\mathbf{G}}|}. \tag{9.92}$$

The incident wave is $\Psi_0 = \exp(i\mathbf{k}_0 \cdot \mathbf{r})$, and on the exit side the coherent wave takes the form $\Psi_0' = t \exp(i\mathbf{k}_0 \cdot \mathbf{r}) + r \exp\{i(\mathbf{k}_0 - \mathbf{G}) \cdot \mathbf{r}\}$, which is a superposition of a transmitted wave and a Bragg-reflected wave. The wave in the medium is given by (9.91). Continuity of the wave amplitude across the boundaries at $z = 0$ and $d$ implies that[37]

[37]We are neglecting the wave reflected from the incident surface. This is a good approximation except near grazing incidence ($\theta \simeq 90°$).

$$\left. \begin{array}{l} 1 = A_1 + A_2 \\[2mm] 0 = A_1 \alpha_1 + A_2 \alpha_2 \end{array} \right\} \quad z = 0, \tag{9.93}$$

$$\left. \begin{array}{l} rcl A_1 Y_1 + A_2 Y_2 = t Y_0 \\[2mm] \alpha_1 A_1 Y_1 + \alpha_2 A_2 Y_2 = r Y_0 \end{array} \right\} \quad z = d, \tag{9.94}$$

where $Y_j = e^{ik_j d / \cos\theta}$ and $Y_0 = e^{ik_0 d / \cos\theta}$. Solving (9.93) we find $A_1 = A_2 = 1/2$, and putting these into (9.94) we find that the intensities of the diffracted and transmitted waves are

$$\begin{aligned} |r|^2 &= \frac{1}{4} |Y_1 - Y_2|^2 \\[2mm] |t|^2 &= \frac{1}{4} |Y_1 + Y_2|^2. \end{aligned} \tag{9.95}$$

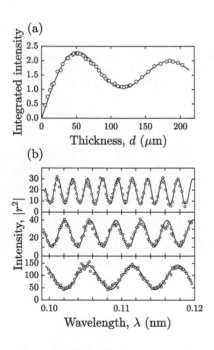

(a)

Integrated intensity

Thickness, $d$ ($\mu$m)

(b)

Intensity, $|r|^2$

Wavelength, $\lambda$ (nm)

Since $k_0^2 \gg |V_0|, |V_{\mathbf{G}}|$, the expression for $k$ in (9.92) can be simplified to $k \simeq k_0 - (V_0 \mp |V_{\mathbf{G}}|)/2k_0$. In this approximation, eqns (9.95) become

$$\begin{aligned} |r|^2 &= \sin^2 \left( \frac{d\lambda |F_{\mathrm{N}}(\mathbf{G})|}{v_0 \cos\theta} \right) \\[2mm] |t|^2 &= \cos^2 \left( \frac{d\lambda |F_{\mathrm{N}}(\mathbf{G})|}{v_0 \cos\theta} \right), \end{aligned} \tag{9.96}$$

where $\lambda = 2\pi/k_0$ is the incident wavelength, and eqn (9.84) has been used to express the result in terms of the nuclear structure factor $F_{\mathrm{N}}(\mathbf{G})$ and unit cell volume $v_0$.

The functions in (9.96) display Pendellösung interference, i.e. the oscillatory transfer of neutron intensity between the transmitted and diffracted beams. Examples of some impressive early measurements of Pendellösung fringes are shown in Fig. 9.23. In one case the fringes were observed as $d$ was reduced by etching the crystal. In the other, $\lambda$ and $\theta$ were varied simultaneously so that the Bragg condition was satisfied at each point. The lines show that the data are in agreement with the dynamical theory (see Exercise 9.6).

Finally, we wish to point out that the quantity $|r|^2$ in eqns (9.95) and (9.96) is the intensity (or reflectivity) strictly at the Bragg condition. Because the $\mathbf{k}_1$ and $\mathbf{k}_2$ waves propagate in slightly different directions, the Bragg-diffracted beam has in reality a non-zero width and exhibits fine structure within the peak envelope. The intrinsic width is typically

**Fig. 9.23** Pendellösung interference effects observed in Bragg diffraction peaks from perfect crystals of silicon. (a) Oscillations as a function of thickness. (After Sippel *et al.*, 1965.) (b) Oscillations at the centre of the 111 Bragg reflection as the incident wavelength and Bragg angle are varied synchronously to maintain the Bragg condition. The three measurements are for samples with thickness (top to bottom) 10.0, 5.94 and 3.32 mm. (After Shull, 1968.) The lines are calculated from the dynamical theory. The formula for $|r^2|$ in (b) is given in eqn (9.96).

a few arcseconds in a *rocking curve*, i.e. an angle scan in which the crystal is rotated through the Bragg condition about an axis perpendicular to **G**. Needless to say, it requires a highly collimated beam to reveal the intrinsic broadening and fine structure. In diffraction studies the experimental quantity of interest is usually not $|r|^2$ but the *integrated intensity*, which is the area of the peak measured in a scan through the Bragg condition such as a rocking curve. The data and theoretical curves shown in Fig. 9.23(a) are integrated intensities, whereas those in Fig. 9.23(b) are reflectivities at the centre of the Bragg reflection.

## 9.4.2 Extinction

The observation of Pendellösung interference and other dynamical diffraction effects of the kind shown in Fig. 9.23 requires highly monochromatic and collimated neutron beams and near-perfect single crystals. However, even routine single crystal diffraction measurements can be influenced strongly by dynamical effects, through a process known as *extinction*.

Figure 9.23(a) provides an illustration of what happens as the neutron path length through a perfect crystal is increased. For very thin crystals the Bragg peak intensity is proportional to thickness because only a tiny fraction of the incident neutrons are diffracted out of the primary beam. This is the limit in which the kinematical theory applies. As the thickness increases, the data start to deviate towards lower intensity from the initial linear behaviour. This reduction in intensity relative to the kinematical approximation for a perfect crystal is due to Pendellösung-type multiple beam interference and is called *primary extinction* (Fig. 9.24).

In reality, typical bulk crystals contain defects (dislocations, stacking faults, vacancies, etc) which tend to divide the crystal volume into a mosaic of smaller blocks (grains or domains), each of which is a perfect crystal in itself but with an orientation that varies from block to block (Fig. 9.25). The diffracted waves from different blocks are not coherent, and so the Bragg peak is the sum of the beams Bragg-reflected from the individual blocks. Hence, the angular width, or *mosaic spread*, of a Bragg peak from a mosaic crystal is determined by the angular distribution of the block orientations.

Primary extinction is much smaller in a mosaic crystal than in a perfect bulk crystal, but the relative orientations of different blocks are such that a particular beam diffracted from one block may be re-diffracted by other blocks, as seen in Fig. 9.25. The attenuation caused by this type of process is known as *secondary extinction*. Although primary and secondary extinction are both caused by multiple diffraction they are fundamentally difference processes. In primary extinction, the reduction in intensity occurs within one perfect crystal block where there is complete coherence between the primary and Bragg-reflected waves, so the resultant effect is obtained by summing amplitudes. Secondary extinction, on the other hand, is caused by multiple diffraction from different blocks, and as the diffracted waves from different blocks are incoherent the resultant effect is obtained by summing intensities.

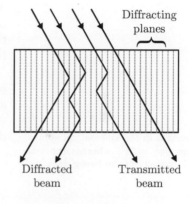

**Fig. 9.24** Multiple beam interference in a perfect crystal causes primary extinction of Bragg peaks.

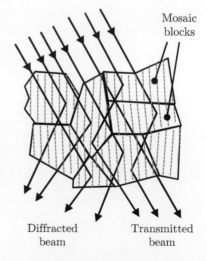

**Fig. 9.25** Multiple Bragg diffraction in a mosaic crystal causes secondary extinction.

The results derived in Section 9.4.1 for symmetric Bragg diffraction from a slab crystal provide some useful indicators as to the importance of primary extinction. First, we can introduce a characteristic *extinction length* (or distance) $\xi$ defined by

$$\xi = \frac{\pi v_0 \cos\theta}{\lambda |F_{\mathrm{N}}(\mathbf{G})|}. \tag{9.97}$$

This is one period of the Pendellösung interference pattern (9.96). By way of example, the extinction length for the 111 reflection of Si at $\lambda = 0.1\,\mathrm{nm}$ is $\xi = 0.21\,\mathrm{mm}$ (see Exercise 9.7). Extinction lengths for neutron diffraction are typically about an order of magnitude larger than those for X-ray diffraction, and at least two orders of magnitude larger than for electron diffraction.

Second, eqn (9.96) allows us to define a primary extinction correction $y$ which relates the Bragg peak intensities calculated in the dynamical and kinematical theories. For the slab crystal the diffracted intensity in (9.96) may be written $\sin^2(\pi d/\xi)$. The kinematical theory applies when $d$ is very small, in which limit $\sin^2(\pi d/\xi) \simeq (\pi d/\xi)^2$. Therefore,[38]

$$y = \frac{\sin^2(\pi d/\xi)}{(\pi d/\xi)^2} = 1 - \frac{1}{3}\left(\frac{\pi d}{\xi}\right)^2 + \dots. \tag{9.98}$$

[38] Note that in the kinematical approximation $|r|^2 \propto d^2$, whereas the integrated intensity of the Bragg peak $\propto d$.

When $d = \xi/2\pi$ the correction is $y = 0.92$, so as a rough guide one can say that primary extinction influences strong Bragg peak intensities at the $\sim 10\%$ level when the linear dimension of the mosaic blocks is approximately $\xi/2\pi$. For Si 111 at $\lambda = 0.1\,\mathrm{nm}$ this is about $30\,\mu\mathrm{m}$.

Third, it follows from (9.97) and (9.98) that primary extinction is most severe for Bragg reflections with large structure factors, and that it can be reduced by a decrease in either the neutron wavelength or the mosaic block size. In Section 10.2.6 we shall employ a more useful definition of the extinction correction in terms of integrated intensity, and give an approximate form that extends over a wider range of extinction. Both primary and secondary extinction will be considered.

# Chapter summary

- Neutron optical phenomena occur in continuous media and are described by dynamical scattering theory. This involves the solution of the one-body Schrödinger equation

$$\left\{ -\frac{\hbar^2}{2m_\mathrm{n}}\nabla^2 + V(\mathbf{r}) \right\}\Psi(\mathbf{r}) = E\Psi(\mathbf{r}),$$

  where $V(\mathbf{r})$ is the optical potential.

- The neutron refractive index is very close to 1, and is given to a very good approximation by

$$n = 1 - \lambda^2 \frac{n_b'}{2\pi} + \mathrm{i}\lambda\frac{\mu}{4\pi},$$

  where $n_b'$ is the real part of the scattering length density, $\mu$ is the attenuation coefficient and $\lambda$ is the wavelength of neutrons incident on the medium from vacuum.

- Total reflection occurs when $n_b' > 0$. For $\mu \lesssim 10^4\,\mathrm{m}^{-1}$ there is a critical angle for total reflection given by

$$\sin\theta_\mathrm{c} = \lambda\left(\frac{n_b'}{\pi}\right)^{1/2}.$$

- For polarized neutrons and magnetic media, replace $n_b'$ by $n_b' \pm n_M$, where $n_M$ is the magnetic scattering length density.

- Reflectivity is the ratio of the reflected intensity to the incoming intensity. Neutron reflectometry is a powerful technique used to probe the structural properties of surfaces and interfaces in stratified samples.

- We have described two ways to calculate the specular reflectivity profile $R(Q)$: the transfer matrix method and the optical method.

- Away from the critical edge, $R(Q)$ is given to a good approximation by the kinematical (Born) approximation,

$$R(Q) = \frac{16\pi^2}{Q^2}\left|\int_{-\infty}^{\infty} n_b(z)\exp(\mathrm{i}Qz)\,\mathrm{d}z\right|^2.$$

- The technique of neutron imaging uses transmission measurements to make real-space maps of the internal structure of an object with $\mu$m spatial resolution.

- We have described several optical phenomena that occur in Bragg diffraction from near-perfect crystals, including Pendellösung oscillations, and primary and secondary extinction.

# Further reading

Dynamical theory of neutron diffraction is covered in more detail in Sears (1989). The technique of neutron reflectometry has been reviewed by Penfold and Thomas (1990), Majkrzak *et al.* in Chatterji (2006), Zabel *et al.* in Kronmüller and Parkin (2007), Ott (2014), and Toperverg (2015). More information about developments in neutron imaging can be found in *Neutron News* **26:2**, 6–43 (2015).

# Exercises

(9.1) A time-of-flight reflectometer with a wavelength range that extends from 0.05 to 0.65 nm is used to study the surface of a dilute solution of molecules in deuterated water ($D_2O$). Calculate the critical wavevector transfer $Q_c$ of $D_2O$, and hence suggest a suitable incident angle to use.

(9.2) Show that in the Born approximation the neutron reflectivity from a film of thickness $d$ and of scattering length density $n_b$ on a substrate of scattering length density $n_b^s$ is

$$R(Q) = \frac{16\pi^2}{Q^4}\left\{n_b^2 + 2n_b\Delta n_b \cos Qd + (\Delta n_b)^2\right\},$$

where $\Delta n_b = n_b^s - n_b$. Confirm that the data in Fig. 9.10 correspond to a film thickness of 96.1 nm.

(9.3) Show that the $Q$ resolution $\Delta Q$ of a neutron reflectometer is given by

$$\left(\frac{\Delta Q}{Q}\right)^2 = \left(\frac{\Delta\lambda}{\lambda}\right)^2 + \left(\frac{\Delta\theta}{\theta}\right)^2,$$

where $\lambda$ and $\theta$ are the incident wavelength and angle. Assume that $\lambda$ and $\theta$ are independent random variables with normal distributions, and that $\theta \ll 1$ radian.

(9.4) Calculate the scattering lengths $b^\pm = b \pm b_M$ for neutron spins parallel and antiparallel to the magnetization in ferromagnetic nickel (Ni), and hence show that the spin-dependent critical wavevectors are $Q_c^+ = 0.234\,\text{nm}^{-1}$ and $Q_c^- = 0.199\,\text{nm}^{-1}$.

[For Ni: density = $8908\,\text{kg m}^{-3}$, atomic mass = $58.69\,\text{g mol}^{-1}$, magnetic moment = $0.62\,\mu_B$, $b = 10.3\,\text{fm}$.]

(9.5) [Harder] Derive the relation given in eqn (9.84),

$$V_\mathbf{G} = \frac{4\pi}{v_0}F_N(\mathbf{G}),$$

where $v_0$ is the volume of the unit cell. Hint: put $\mathbf{r}_j = \mathbf{l} + \mathbf{d} + \mathbf{u}_{ld}$ in (9.2) and use (B.15) to write the $\delta$-function as an integral.

(9.6) Use the data in Fig. 9.23(b) to calculate the coherent scattering length of silicon (Si), given that Si has a face-centred cubic lattice with lattice parameter $a = 0.5431\,\text{nm}$ and a basis comprising Si at 0,0,0 and $\frac{1}{4},\frac{1}{4},\frac{1}{4}$.

(9.7) Verify that the extinction length for the 111 reflection of silicon at $\lambda = 0.1\,\text{nm}$ is $\xi = 0.21\,\text{mm}$.

# Practical Aspects of Neutron Scattering

# 10

Much of this book is preoccupied with the derivation of formulae for neutron scattering cross-sections. What is measured in an experiment, however, is not a formula but a neutron count rate. Depending on the particular type of experiment or instrument employed it may be necessary to carry out one or more intermediate steps, e.g. transformations, corrections, normalization, etc, to convert the data into a cross-section or, alternatively, to convert the expression for a cross-section into a measured intensity. These tasks can be quite trivial, like dividing by a reference intensity, or they can involve rather more detailed theoretical calculation or modelling, as might be needed to convolve a theoretical scattering cross-section with the instrumental resolution function.

In this chapter we aim to provide the reader with a working understanding of some of the most important of these experimental issues. For ease of reference, we discuss the particular procedures used in diffraction and spectroscopy separately. Before that, though, we shall introduce some topics and concepts of more general applicability.

## 10.1 General practicalities

The goal of most neutron scattering experiments from bulk samples is to measure either the static or the dynamical structure factor, $S(\mathbf{Q})$ or $S(\mathbf{Q}, \omega)$. Usually there are choices to be made about how the experiment is to be performed, and with careful planning and optimization of the experimental parameters one can ensure the best possible data quality and avoid unnecessary complications in its analysis and interpretation.

Experiment planning usually focuses on the instrument and the sample. Since neutron scattering is inherently a weak probe, the choice of instrument parameters often comes down to a trade-off between high neutron flux and good resolution. The level of unwanted background from the sample environment and other sources may also be a factor.

As far as the sample is concerned, if it is small enough then apart from corrections for background and resolution the measured scattering is directly proportional to $S(\mathbf{Q})$ or $S(\mathbf{Q}, \omega)$. For reasons of intensity, however, there is inevitably a certain minimum sample size below which a measurement is simply not feasible in the time available. In practice, experiments are generally performed on samples of order 1 to 10 mm in linear dimension. Assuming the experimentalist has some control over the size and shape of the sample they will need to consider a number of factors. Foremost amongst these is the desire to maximize the count rate subject to an acceptable amount of attenuation and multiple scattering.

Finally, as well as measuring the sample itself, the experimentalist might also need to perform supplementary measurements that enable them to correct for detector efficiency, normalize the data, determine backgrounds, etc. Time spent performing these measurements must be weighed against the time spent measuring the sample itself.

### 10.1.1   Attenuation: absorption and self-shielding

The Beer–Lambert law, eqn (1.40), describes the transmission of neutrons through a sample of uniform thickness. The transmission is the factor by which the count rate is attenuated, and here we shall call this the attenuation factor $A$:

$$A = \frac{I(t)}{I_0} = \exp(-\mu t),  \tag{10.1}$$

where $I_0$ is the incident intensity, $I(t)$ is the intensity for thickness $t$ and $\mu$ is the linear attenuation coefficient. For simplicity, we shall assume that the sample is homogeneous so that $\mu$ is independent of position.[1] Attenuation is caused by scattering (coherent and incoherent) and nuclear absorption, and to distinguish these processes we shall write

$$\mu = n(\sigma_s + \sigma_a),  \tag{10.2}$$

where $n$ is the number density of some convenient grouping of atoms, such as a chemical formula unit (f.u.) or a crystallographic unit cell, and $\sigma_s$ and $\sigma_a$ are the total scattering and absorption cross-sections, respectively, for the grouping.

The attenuation of a neutron beam due to scattering is also called *self-shielding*. It can be difficult to calculate self-shielding accurately because $\sigma_s$ tends to be a complicated function of neutron energy. There are three aspects to the energy dependence. First, if the kinetic energy of the neutrons is greater than the binding energy of the atoms, then the nuclei will recoil under impact from the neutrons. The scattering cross-sections are then reduced from the values tabulated for bound atoms, and in the limit of high incident energy they are replaced by the free-atom cross-sections (see Section 1.8.6). Second, due to the motion of the atoms there is an inelastic contribution to $\sigma_s$ which at low energies ($E_i \ll k_B T$) becomes proportional to $1/k_i$.[2] Third, the removal of neutrons out of the beam by coherent scattering is limited by the fact that the maximum possible scattering angle is $180°$, which imposes a cut-off on $Q$.

As an example of the third effect, consider diffraction from a polycrystalline material. Fig. 10.1 plots the calculated total coherent scattering cross-section as a function of energy and wavelength for polycrystalline aluminium. The condition for coherent scattering is given by Bragg's law, $\lambda = 2d\sin\theta_B$ — see Section 2.2.1. As $\sin\theta_B \leq 1$, Bragg's law can

[1]This contrasts with the situation encountered in imaging (Section 9.3) where the sample has internal structure which is probed through spatial variations in $\mu$.

[2]This may be seen from

$$\sigma_s = \int\int \frac{d^2\sigma}{d\Omega dE_f}\, d\Omega dE_f$$
$$= \int\int \frac{k_f}{k_i} S(\mathbf{Q},\omega)\, d\Omega dE_f.$$

As $k_i \to 0$, $k_f \to Q$ and $E_f \to \hbar\omega$, so the integration variables become independent of $k_i$. The $1/k_i$ factor can then be taken outside the integral, giving $\sigma_s \propto 1/k_i$.

only be satisfied for a given set of crystal planes if the wavelength $\lambda$ is smaller than $2d$, where $d$ is the inter-planar spacing. The maximum wavelength $\lambda_c = 2d_{max}$ is known as the *Bragg cut-off*. If $\lambda$ exceeds $\lambda_c$ then no Bragg scattering is possible. As $\lambda$ decreases it passes the threshold condition $\lambda = 2d$ for the onset of diffraction for crystal planes with successively smaller $d$-spacings. At each Bragg onset there is a sharp increase in the coherent scattering cross-section. As $\lambda \to 0$, the jump at each Bragg edge diminishes until the cross-section tends towards the limiting bound-atom value of $4\pi|\bar{b}|^2$. For a single crystal sample, Bragg diffraction requires not only energies above the Bragg cut-off but also that the crystal is in the correct alignment to satisfy the Bragg condition for one or more sets of planes. Therefore, the energy dependence of the total coherent scattering is highly dependent on the crystal orientation.

The absorption cross-section was discussed in Section 1.7.2. Away from any nuclear resonances the absorption cross-section usually follows a $1/v$ law, which means the energy dependence can be written

$$\sigma_a(E) = \sigma_a(25.3\,\text{meV}) \times \left(\frac{25.3}{E}\right)^{\frac{1}{2}}, \qquad (10.3)$$

where $E$ is in meV, and $25.3\,\text{meV}$ is the reference energy for which absorption cross-sections are usually tabulated (see Appendix A).

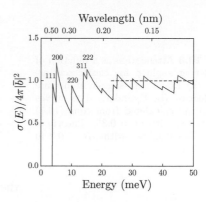

**Fig. 10.1** Energy dependence of the total coherent scattering cross-section $\sigma(E)$ of polycrystalline aluminium, which has a face-centred cubic lattice with cell parameter $a = 0.405\,\text{nm}$. The first five Bragg edges are labelled. $\sigma(E)$ is given by $I_{\text{cone}}(G)/\Phi$ from (10.94), and has been normalized by the total single-atom coherent scattering cross-section $\sigma_{\text{coh}} = 4\pi|\bar{b}|^2$. As $E$ increases, $\sigma(E)/4\pi|\bar{b}|^2 \to 1$ (broken line).

## Correction for attenuation

To correct for attenuation we must first average the attenuation factor (10.1) over the whole sample taking into account (i) the sample shape and size, (ii) the direction of the incident and scattered beams, and (iii) the incident and final neutron energies, and then divide the measured count rate by the averaged attenuation factor $A$.

Suppose that scattering takes place in a volume $dV$ such that the path lengths in the sample for incident and scattered neutrons are $L_i$ and $L_f$, respectively, as depicted in Fig. 10.2. The contribution from this volume element to the attenuation of the sample as a whole is

$$dA = \frac{1}{V}\exp(-\mu_i L_i - \mu_f L_f)\,dV, \qquad (10.4)$$

where $\mu_i$ and $\mu_f$ stand for $\mu(E_i)$ and $\mu(E_f)$, the attenuation coefficients for the incident and final energies, respectively. Therefore,

$$A = \frac{1}{V}\int_V \exp(-\mu_i L_i - \mu_f L_f)\,dV. \qquad (10.5)$$

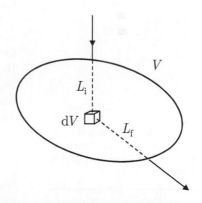

**Fig. 10.2** Neutron path through the sample for calculation of attenuation.

In general, this integral must be evaluated numerically, but an analytic expression can be obtained for a plane slab which is a common sample geometry, Fig. 10.3(a). Suppose that the slab has a thickness $t$ and that the size of the large face is much greater than $t$. For simplicity we treat the special case in which the slab is normal to the incident beam. As the attenuation does not depend on position in the plane of the large face

**Fig. 10.3** Attenuation as a function of scattering angle $\phi$ for elastic scattering. (a) Plane slab sample at normal incidence. (b) Cylinder. The curves in (a) are calculated from eqns (10.7) and (10.8) with $\mu t = 0.35$. The curve in (b) is calculated with $\mu R = 0.2$ in eqn (10.9).

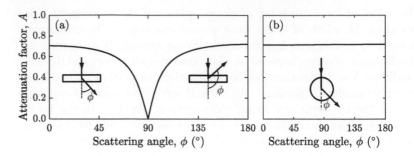

the volume integral in (10.5) reduces to the one-dimensional integral

$$A = \frac{1}{t} \int_0^t \exp\left\{-\mu_i z - \mu_f (t - z) \sec \phi\right\} \, dz, \qquad (10.6)$$

where $\phi$ is the scattering angle, which in (10.6) is taken to be in the range $0 \le \phi \le 90°$. Performing the integration, we find

$$A = \frac{\exp(-\mu_f t \sec \phi) - \exp(-\mu_i t)}{(\mu_i - \mu_f \sec \phi)t}. \qquad (10.7)$$

The corresponding result for $90° \le \phi \le 180°$ is

$$A = \frac{1 - \exp\left\{(-\mu_i + \mu_f \sec \phi)t\right\}}{(\mu_i - \mu_f \sec \phi)t}. \qquad (10.8)$$

Illustrative calculations are shown in Fig. 10.3(a) for elastic scattering ($\mu_i = \mu_f = \mu$).

Other common sample shapes include spheres, cylinders, and annular cylinders. The integral in (10.5) cannot be performed analytically for these shapes, so the attenuation must be calculated numerically. Results for elastic scattering have been tabulated (Rouse *et al.*, 1970; Schmitt and Ouladdiaf, 1998). For spheres and cylinders with radius $R$ such that $\mu R \le 1$, the attenuation is given to better than 0.5% by the formula

$$A = \exp\left\{-(a_1 + b_1 \sin^2 \frac{\phi}{2})\mu R - (a_2 + b_2 \sin^2 \frac{\phi}{2})(\mu R)^2\right\}, \qquad (10.9)$$

where the $a$ and $b$ coefficients are listed in Table 10.1. In cylindrical sample geometry eqn (10.9) applies when the cylinder axis is perpendicular to the scattering plane. The attenuation of cylinders and spheres varies only weakly with scattering angle, as illustrated in Fig. 10.3(b).

**Table 10.1** Coefficients in eqn (10.9) for the attenuation factor of a cylinder and a sphere.

|       | Cylinder | Sphere  |
|-------|----------|---------|
| $a_1$ | 1.7133   | 1.5108  |
| $b_1$ | −0.0368  | −0.0315 |
| $a_2$ | −0.0927  | −0.0951 |
| $b_2$ | −0.0375  | −0.2898 |

## Worked example

To illustrate the points in the previous two sections we shall estimate the transmission of a powder sample of LiFeAs as used in a neutron inelastic scattering experiment (Taylor *et al.*, 2011). In the actual experiment the sample was in the form of a cylindrical annulus, but for simplicity we shall assume a plane slab geometry here.

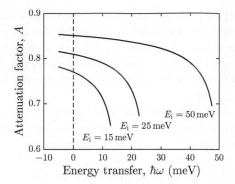

**Fig. 10.4** Attenuation as a function of energy transfer of a powder sample of LiFeAs in slab geometry. Results for three different incident energies are shown.

The unit cell of LiFeAs is primitive tetragonal with lattice parameters $a = b = 0.378$ nm and $c = 6.36$ nm. There are two LiFeAs formula units per unit cell, so the number density is $n = 2/(a^2c) = 2.20 \times 10^{28}$ m$^{-3}$.

The total scattering and absorption cross-sections for Li, Fe, As, and for the LiFeAs f.u. are listed in Table 10.2. The attenuation is dominated by the nuclear absorption of Li, which can be assumed to follow the $1/v$ law as described by eqn (10.3). The lowest incident energy used in the experiment was $E_i = 15$ meV which corresponds to a wavelength $\lambda = 0.233$ nm. Crystal planes with $d > \lambda/2$ can diffract neutrons out of the incident beam, and for this energy there are about fifteen such planes in LiFeAs with non-negligible structure factors. It is reasonable to assume, therefore, that the total coherent scattering cross-section is close to its asymptotic bound-atom value of 17.11 b per f.u. The highest $E_i$ used in the experiment was 50 meV, which is not high enough to cause the target nuclei to recoil, so bound atom cross-sections are used.

Based on these considerations we find from (10.2) that the attenuation coefficient of LiFeAs may be written

$$\mu(E) = n(\sigma_{\text{coh}} + \sigma_{\text{inc}} + \sigma_{\text{a}}) = 40.7 + \frac{859}{\sqrt{E}} \quad \text{m}^{-1}, \tag{10.10}$$

where $E$ is in meV.

For the present exercise we shall assume that the effective thickness of the sample at full density (100% packing fraction) was $t = 1$ mm, and that the scattering angle $\phi$ of the signal of interest (which was magnetic) was sufficiently small that $\sec\phi \simeq 1$. With these assumptions, the attenuation factor $A$ can be calculated as a function of energy transfer $\hbar\omega = E_i - E_f$ for a given incident energy from eqns (10.7) and (10.10). Results are shown in Fig. 10.4 for the three incident energies used in the experiment, $E_i = 15, 25$ and 50 meV.[3]

## 10.1.2 Multiple scattering

Samples used in neutron scattering experiments tend to be large, and so the probability that a neutron scattered once is scattered a second time (or more times) may not be negligible. In contrast to self-shielding, which removes neutrons from the beam, *multiple scattering* adds intensity onto the single-scattered signal. If at least one incoherent scattering

**Table 10.2** Neutron bound-atom cross-sections for LiFeAs. The absorption cross-sections are for an energy of 25.3 meV.

|  | $\sigma_{\text{coh}}$ (b) | $\sigma_{\text{inc}}$ (b) | $\sigma_{\text{a}}$ (b) |
|---|---|---|---|
| Li | 0.45 | 0.92 | 70.5 |
| Fe | 11.22 | 0.40 | 2.6 |
| As | 5.44 | 0.06 | 4.5 |
| LiFeAs | 17.11 | 1.38 | 77.6 |

Units: 1 barn (b) $= 10^{-28}$ m$^2$.

[3]When the scattered neutron energy $E_f$ becomes smaller than about 5 meV only a few planes satisfy the Bragg condition and so the coherent scattering cross-section for the scattered beam could be less than the asymptotic value assumed here. However, this will not be significant for the overall attenuation as the total cross-section of LiFeAs is dominated by absorption.

process is involved then multiple scattering contributes only a diffuse background, and corrections for multiple scattering are generally unnecessary as long as the signal of interest in $S(\mathbf{Q}, \omega)$ is relatively sharply peaked. Multiple scattering is a more serious problem if $S(\mathbf{Q}, \omega)$ is a smooth function of $\mathbf{Q}$, especially if the multiple scattering intensity is a significant fraction of $S(\mathbf{Q}, \omega)$.

Multiple coherent scattering tends to broaden features in $S(\mathbf{Q}, \omega)$, but a special case is multiple Bragg scattering in crystalline materials which can reduce the Bragg intensity for the desired reflection (extinction) or cause a Bragg peak to appear at a forbidden position. Multiple Bragg scattering is discussed in more detail in Section 10.2.6.

Methods to calculate multiple scattering generally fall into one of two types, either analytic calculations based on neutron transport theory, or Monte Carlo methods. The former approach was the first to be developed (Vineyard, 1954) and was reviewed and extended by Sears (1975). Analytic calculations can be efficient if the multiple scattering is not too severe, and provide useful physical insights into the effects. Relatively simple formulae have been obtained for the ratio of double to single scattering in the quasi-isotropic approximation (see below), in which $S(\mathbf{Q}, \omega) \propto \delta(\hbar\omega)$, but results for arbitrary $S(\mathbf{Q}, \omega)$ require numerical computation. Monte carlo methods are an alternative to numerical integration and allow simulation of the actual experiment for a realistic $S(\mathbf{Q}, \omega)$. For example, the effects of both multiple scattering and resolution can be evaluated simultaneously.[4]

[4]See, for example, Lin *et al.* (2016).

### Analytic expression for multiple scattering intensity

In the Born approximation, the double-differential cross-section is related to the intrinsic response function of the scattering system by

$$\frac{\mathrm{d}^2\sigma}{\mathrm{d}\Omega \mathrm{d}E_\mathrm{f}} = \frac{k_\mathrm{f}}{k_\mathrm{i}} S(\mathbf{Q}, \omega). \qquad (10.11)$$

This relation applies to samples which are sufficiently small compared with the neutron mean free path that neutrons either pass straight through or are scattered once before leaving the sample. Suppose now that multiple scattering and absorption are not negligible. The measured cross-section determined from the neutrons that emerge from the sample is no longer proportional to $S(\mathbf{Q}, \omega)$ and instead may be written

$$\frac{\mathrm{d}^2\sigma}{\mathrm{d}\Omega \mathrm{d}E_\mathrm{f}} = \frac{k_\mathrm{f}}{k_\mathrm{i}} s(\mathbf{k}_\mathrm{i}, \mathbf{k}_\mathrm{f}), \qquad (10.12)$$

where $s(\mathbf{k}_\mathrm{i}, \mathbf{k}_\mathrm{f})$ is an effective scattering function which includes both attenuation and multiple scattering, and which depends on the shape and size of the sample. This function takes the form

$$s(\mathbf{k}_\mathrm{i}, \mathbf{k}_\mathrm{f}) = \sum_{j=1}^{\infty} s_j(\mathbf{k}_\mathrm{i}, \mathbf{k}_\mathrm{f}), \qquad (10.13)$$

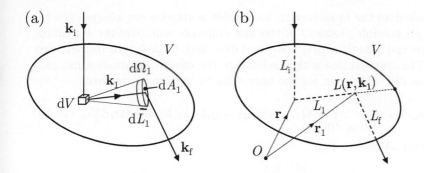

Fig. 10.5 Neutron paths for calculating double scattering in a sample of volume $V$. (a) Wavevectors and scattering volumes. (b) Position vectors and distances.

where $s_j(\mathbf{k}_i, \mathbf{k}_f)$ is the contribution from neutrons which have been scattered $j$ times.

The effective scattering function for single scattering is given by

$$s_1(\mathbf{k}_i, \mathbf{k}_f) = S(\mathbf{Q}, \omega) A_1(\mathbf{k}_i, \mathbf{k}_f), \qquad (10.14)$$

where $A_1(\mathbf{k}_i, \mathbf{k}_f)$ is the sample attenuation factor $A$ given in (10.5).

To calculate $s_2(\mathbf{k}_i, \mathbf{k}_f)$ we must average over all double-scattering processes in which a neutron enters the sample with initial wavevector $\mathbf{k}_i$ and emerges from the sample with wavevector $\mathbf{k}_f$. These processes involve an intermediate state with wavevector $\mathbf{k}_1$, as shown in Fig. 10.5(a). The corresponding wavevectors and energy transfers are

$$\begin{aligned} \mathbf{Q}_1 &= \mathbf{k}_i - \mathbf{k}_1, \quad \hbar\omega_1 = E_i - E_1, \\ \mathbf{Q}_2 &= \mathbf{k}_1 - \mathbf{k}_f, \quad \hbar\omega_2 = E_1 - E_f. \end{aligned} \qquad (10.15)$$

Consider the path represented in Fig. 10.5, in which the first collision takes place in a volume $dV$ centred on the point denoted by $\mathbf{r}$, and the second takes place in $dV_1$ centred on $\mathbf{r}_1$. For reasons that will become apparent later, $dV_1$ is chosen to be slab-shaped with an area $dA_1$ perpendicular to $\mathbf{k}_1$ and a thickness $dL_1$, as shown in Fig. 10.5(a). From (10.11) and the definition of the double-differential cross-section, eqn (1.34), the number of neutrons scattered per second from $dV$ into solid angle $d\Omega_1$ about $\mathbf{k}_1$ with energies between $E_1$ and $E_1 + dE_1$ is given by

$$\frac{k_1}{k_0} \Phi_0 S(\mathbf{Q}_1, \omega_1) \exp(-\mu_i L_i) \, dV \, d\Omega_1 dE_1, \qquad (10.16)$$

where $\Phi_0$ is the incident flux, and $S(\mathbf{Q}, \omega)$ is normalized to unit volume. The factor $\exp(-\mu_i L_i)$ is the attenuation of the incident beam from the point of entry to the point $\mathbf{r}$, with $\mu_i \equiv \mu(E_i)$ as before. Similarly, the number of neutrons scattered per second per unit solid angle per unit energy from $dV_1 = dA_1 dL_1$ and emerging from the sample is

$$\frac{k_f}{k_1} \Phi_1 S(\mathbf{Q}_2, \omega_2) \exp(-\mu_f L_f) \, dA_1 dL_1. \qquad (10.17)$$

The flux $\Phi_1$ at $\mathbf{r}_1$ is obtained from (10.16) by division by the area $dA_1$ and multiplication by the attenuation over the distance $L_1$ between collisions:

$$\Phi_1 = \frac{k_1}{k_0} \Phi_0 S(\mathbf{Q}_1, \omega_1) \exp(-\mu_i L_i - \mu_1 L_1) \frac{dV \, d\Omega_1 dE_1}{dA_1}. \qquad (10.18)$$

To calculate the cross-section for double scattering we average (10.17) over all possible positions of the first collision, sum over the remaining differential quantities $d\Omega_1$, $dE_1$, and $dL_1$, and divide by the incident flux $\Phi_0$. The cross-section is then related to the effective scattering function by (10.12). The result for the latter can be written in the form

$$s_2(\mathbf{k}_i, \mathbf{k}_f) = \int_{-\infty}^{\infty} \int_{\Omega_1} S(\mathbf{Q}_1, \omega_1) S(\mathbf{Q}_2, \omega_2) A_2(\mathbf{k}_i, \mathbf{k}_1, \mathbf{k}_f) \, d\Omega_1 dE_1, \quad (10.19)$$

per unit volume, where

$$A_2(\mathbf{k}_i, \mathbf{k}_1, \mathbf{k}_f) = \frac{1}{V} \int_V \int_0^{L(\mathbf{r}, \mathbf{k}_1)} \exp(-\mu_i L_i - \mu_1 L_1 - \mu_f L_f) \, dL_1 dV \quad (10.20)$$

describes the average attenuation for double scattering. The volume integral in (10.20) extends over all points $\mathbf{r}$ in $V$, and the $L_1$ integral is along the line joining $\mathbf{r}$ to the surface of the crystal in the direction of $\mathbf{k}_1$, see Fig. 10.5(b).

Higher orders of multiple scattering can be calculated by straightforward extension of this method. On the whole, as long as the attenuation is not too great the ratio of third- to second-order scattering is roughly the same as the ratio of second- to first-order scattering, and the same applies to higher orders (Vineyard, 1954). Therefore, to assess the effect of multiple scattering it is usually sufficient to estimate the first- and second-order contributions.

## Quasi-isotropic approximation

Anyone who wishes to evaluate multiple scattering in detail has no choice but to perform the multi-dimensional integrals contained in (10.19) and (10.20), either by numerical or Monte Carlo methods. The results will of course be system- and sample-dependent. However, a number of theoretical investigations have been performed within the so-called *quasi-isotropic approximation*, which assumes purely isotropic elastic scattering. This model would be realistic for samples dominated by incoherent scattering, such as vanadium or hydrogen-rich materials, or samples having diffuse elastic scattering with little $\mathbf{Q}$ dependence. The quasi-isotropic approximation also provides some more general insights into how multiple scattering is influenced by the size and shape of the sample, and it has been used to estimate multiple scattering in experiments with neutron polarization analysis (Harders *et al.*, 1985).

In the quasi-isotropic approximation, the scattering function (per unit volume) is taken to be

$$S(\mathbf{Q}, \omega) = n \frac{\sigma}{4\pi} \delta(\hbar\omega), \quad (10.21)$$

where $n$ is the number density of some appropriate scattering unit and $\sigma$ is the corresponding scattering cross-section. For example, if the sample were vanadium then $\sigma \simeq \sigma_{\text{inc}} = 5.08$ b and $n = 7.23 \times 10^{28}$ m$^{-3}$.

Sears (1975) has given formulae for multiple scattering in the quasi-isotropic approximation for various sample geometries. Figure 10.6 shows

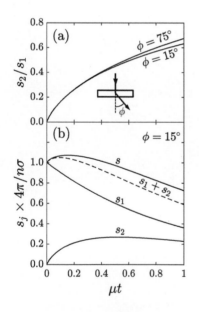

**Fig. 10.6** Multiple scattering in the quasi-isotropic approximation for a slab sample at normal incidence as a function of $\mu t$. (a) Ratio of double to single scattering. (b) Effective scattering functions for single ($s_1$), double ($s_2$), single+double ($s_1 + s_2$), and sum of all orders ($s = \sum_j s_j$). The curves are calculated from eqns (7.2.1), (7.2.4) and (6.2.9) in Sears (1975).

results calculated for an infinite slab sample of thickness $t$ and attenuation coefficient $\mu$. The presented curves have been calculated for zero absorption ($\sigma_a = 0$). In Fig. 10.6(a) the ratio of double to single scattering is plotted for two different scattering angles. The ratio increases with $\mu t$ but is relatively insensitive to the scattering angle until $\phi$ approaches 90°. Figure 10.6(b) shows results for the effective scattering function $s$ together with the single and double scattering contributions, $s_1$ and $s_2$. It can be seen that $s$ does not change much with $\mu t$ when $\mu t \lesssim 0.5$.

Figure 10.7 shows results calculated in the quasi-isotropic approximation for cylinders. The curves are for different ratios of radius ($r$) to height ($h$) for the case when $\mathbf{k}_i$ and $\mathbf{k}_f$ are both perpendicular to the cylinder axis. For this geometry the ratio $s_2/s_1$ of double to single scattering is virtually independent of scattering angle, but the ratio is seen to increase with $\mu r$, and for a given $\mu r$ is largest for smallest $r/h$. A more detailed analysis shows that for a given sample volume $V = \pi r^2 h$, $s_2/s_1$ is largest when $r/h \simeq 0.5$. Given a fixed sample volume, therefore, a cylinder with $r/h \ll 0.5$ or $r/h \gg 0.5$ would be preferable to reduce multiple scattering effects.

As the foregoing has illustrated, multiple scattering effects can sometimes be reduced through careful choice of the size and shape of a sample. Another consideration, however, is absorption. Because multiple scattering involves extra internal paths any absorption will reduce higher-order scattering compared to first-order scattering. In fact, one can see by comparing eqn (10.14) with (10.19)–(10.21) that in the quasi-isotropic approximation the ratio $s_2/s_1$ is proportional to $\mu_s/\mu$,[5] where $\mu = \mu_s + \mu_a$ is the total attenuation due to self-shielding ($\mu_s = n\sigma_s$) and absorption ($\mu_a = n\sigma_a$). One way to exploit this for practical advantage is to mix an appropriate amount of neutron-absorbing material in with the sample, which makes the proportion of multiple scattering smaller at the expense of the overall scattering intensity. Another is to introduce absorbing spacers into the sample to divide it up into smaller blocks which have correspondingly reduced multiple scattering.

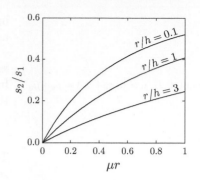

**Fig. 10.7** Ratio of double to single scattering in the quasi-isotropic approximation for cylindrical samples with different ratios of radius ($r$) to height ($h$). The curves are calculated from eqn (9.3.13) in Sears (1975) scaled to match the numerical results of Blech and Averbach (1965).

[5] For $n^{\text{th}}$ order scattering, the relation is $s_n/s_1 \propto (\mu_s/\mu)^{n-1}$.

## 10.1.3 Optimizing sample size and shape

If the scattering is to be measured over a wide range of angle then a cylinder or sphere is ideal, to take advantage of the fact that attenuation and multiple scattering are nearly isotropic for these shapes. A cylinder is also a good choice if the sample needs to be rotated about an axis perpendicular to the incident beam. If the sample is to be fixed relative to the beam and measurements restricted to scattering angles below about 60° then slab geometry works well because it minimizes the path through the sample for a given beam area.

As far as size is concerned, the ideal sample is the one that gives the highest count-rate with an acceptable level of multiple scattering. To illustrate the factors that need to be balanced, consider a slab sample at normal incidence such that the cross-sectional area of the incident beam is smaller than the area of the slab, and suppose that the signal of interest is at small scattering angles such that $\cos\phi \simeq 1$. For simplicity, assume also that the sample attenuation is independent of energy.

The fraction of the neutron beam that scatters in an interval $dx$ in the sample and subsequently escapes from the far side of the sample without further collision is $\mu_s dx \, \exp(-\mu t)$.[6] Therefore, the fraction of neutrons that scatter once in the sample and escape is

$$f_1 = \mu_s t \exp(-\mu t). \tag{10.22}$$

To maximize $f_1$ we differentiate (10.22) with respect to $t$ and equate to zero. This gives

$$\frac{df_1}{dt} = \mu_s \exp(-\mu t) - \mu t \mu_s \exp(-\mu t) = 0,$$

so

$$\mu t = 1. \tag{10.23}$$

A slab with this thickness would have an attenuation factor (i.e. transmission) of $A = 1/e = 0.368$. This would be a good choice for a strongly absorbing material, but if $\mu$ is dominated by scattering, i.e. $\mu \simeq \mu_s$, then a transmission of $1/e$ would result in a significant amount of multiple scattering. This is because a non-absorbing sample with transmission $A$ scatters a fraction $1 - A$ of the incident neutrons, and a fraction of order $1 - A$ of these scattered neutrons will in turn scatter a second time, and so on. If $A = 1/e$ then $1 - A = 0.632$, and so around 63% of the deflected neutrons emerging from the sample will have been scattered two or more times. For many experiments this would be unacceptably high, and a smaller sample with less multiple scattering would be preferable.

There is a rule-of-thumb in neutron scattering which states that a good choice of sample is one that scatters ∼10% of the incident neutrons. With much more than 10%, corrections for attenuation, multiple scattering, and perhaps also resolution (due to sample size) could be a problem. With much less, the scattered intensity is lower than necessary. For a non-absorbing sample, the fraction of neutrons scattered out of the incident beam is given by $1 - \exp(-\mu_s t)$. Equating this to 0.1 we obtain

$$\mu_s t = 0.105 \qquad \text{(10\% scatterer, no absorption).} \tag{10.24}$$

What about samples that absorb neutrons as well as scatter them? As discussed at the end of Section 10.1.2, absorption reduces the proportion of multiple scattering relative to single scattering, which suggests that the sample could be made larger than prescribed by (10.24) so as to benefit from a higher count rate.

A simple solution is to apply the 10% rule to the neutrons that scatter once followed either by additional scattering or escape from the sample, but not absorption. These neutrons are a fraction $f$ of the total, where $f$ depends on three factors: (i) $\exp(-\mu x)$, the probability of a neutron surviving to $x$; (ii) $\mu_s dx$, the probability of a neutron scattering in the interval $x$ to $x + dx$; (iii) $\exp\{-\mu_a(t - x)\}$, the probability of a neutron

[6]From (10.2), $\mu = n(\sigma_s + \sigma_a) = \mu_s + \mu_a$.

travelling from $x$ to $t$ without being absorbed. Multiplying these three probabilities together and integrating over $x$, we obtain

$$f = \exp(-\mu_a t)\{1 - \exp(-\mu_s t)\}. \qquad (10.25)$$

The numerical solution of (10.25) for $f = 0.1$ is plotted in Fig. 10.8 as a function of $\mu_a/\mu_s$. The optimum thickness at 10% scattering increases with $\mu_a/\mu_s$, and reaches $\mu t = 1$ when $\mu_a/\mu_s = 3.16$. The $f = 0.1$ curve continues on to higher $\mu t$, but as we have already shown from eqn (10.22), a thickness corresponding to $\mu t = 1$ gives the maximum intensity of neutrons that scatter once then escape from the sample. Therefore, at $\mu_a/\mu_s = 3.16$ the optimum sample switches from the 10% scatterer to the $1/e$ attenuator. For zero absorption, the $f = 0.1$ solution reduces to that given in (10.24).

Another impression of the optimum thickness can be obtained from the plot of $\mu_s t$ vs $\mu_a/\mu_s$, shown by the dashed line in the lower part of Fig. 10.8. For a fixed $\mu_s$ this curve would describe how the optimum thickness varies with $\mu_a$. With increasing $\mu_a/\mu_s$ the thickness for 10% scattering increases, reaching a maximum at $\mu_a/\mu_s = 3.16$ which is 2.5 times larger than at $\mu_a = 0$ for the same $\mu_s$. When $\mu_a/\mu_s > 3.16$ the optimum $t$ decreases to avoid excessive absorption.

Similar considerations will apply to other sample geometries. It is important to realize, though, that these size criteria are only a guide. One area to which they do not necessarily apply is in single-crystal diffraction. For crystals without significant absorption the main concern is usually to ensure that extinction of the Bragg peak intensities is not too severe. The maximum crystal size permitted by this requirement depends on the mosaic spread and mosaic block size (see Sections 9.4.2 and 10.2.6). In practice, crystals used in diffraction studies are typically spherical in shape and ~1 mm in diameter.

Another exception is small-angle scattering from hydrogenous materials. The multiple scattering signal from such samples can be significant because of the large nuclear spin incoherent cross-section of hydrogen, but at small $Q$ it is virtually independent of $Q$ and so just contributes a flat background. Samples that are thicker than dictated by the 10% rule are often preferred, therefore, in order to benefit from a stronger coherent scattering signal.

## Worked examples

First, we shall calculate the thickness $t$ of water at room temperature which makes it a 10% total scatterer at a wavelength of 0.6 nm. We assume that the sample cell is parallel-sided and that neutrons are incident normally. As the absorption cross-section of water is much smaller than the scattering cross-section we can calculate $t$ from eqn (10.24). The density of water is $1000\,\text{kg m}^{-3}$, and the corresponding number density of water molecules ($H_2O$) is $n = 3.35 \times 10^{28}\,\text{m}^{-3}$. The total scattering cross-section of water is wavelength-dependent due to inelastic effects. At 0.6 nm $\sigma_s \simeq 200\,\text{b}$ per $H_2O$ molecule (Heinloth, 1961), so $\mu_s = n\sigma_s = 0.67\,\text{mm}^{-1}$. Putting this in (10.24) we obtain $t = 0.16\,\text{mm}$.

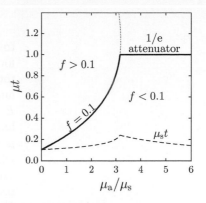

**Fig. 10.8** Optimum thickness of a slab sample as the ratio of absorption to attenuation is varied (solid line). The thickness $t$ is scaled by the attenuation coefficient $\mu = \mu_s + \mu_a$. The solution to eqn (10.25) with $f = 0.1$ is labelled and corresponds to a 10% scattering sample. When $\mu t = 1$ the attenuation is $1/e$, the value required to maximize the intensity of single-scattered neutrons escaping from the sample. The dashed line near the bottom axis is a plot of $\mu_s t$ on the same scale.

As a second example, consider the optimum radius of a cylindrical can to contain a powder sample of LiFeAs (a moderately absorbing material) for a diffraction experiment at an energy of 50 meV ($\lambda = 0.128$ nm). The total attenuation coefficient of LiFeAs is determined from eqn (10.10), and the relevant bound-atom cross-sections are given in Table 10.2. For $E = 50$ meV we find $\mu = 0.162$ mm$^{-1}$, but if we assume our powder has a packing fraction of 60% then the attenuation coefficient drops to $\mu = 0.6 \times 0.162 = 0.097$ mm$^{-1}$. The cross-sections give $\mu_a/\mu_s = 2.97$, which from Fig. 10.8 would suggest $\mu t \simeq 0.8$ for 10% scattering if the sample were a slab. We shall assume this result also applies to a cylinder providing the radius is adjusted to give the same attenuation as the slab. Comparing eqns (10.7) and (10.9) one can see that at small scattering angles the dimensions of a cylinder and a slab having the same attenuation are related by $R \simeq t/1.7$.[7] Therefore, our optimum cylinder has radius $R = 0.8/(1.7 \times 0.097) \simeq 5$ mm.

[7]This approximation holds for small $\mu t$.

### 10.1.4  Data normalization

There are several neutron scattering data normalization procedures which can be carried out in order to fulfil different needs.

#### Incident flux and counting time

In the Born approximation the scattering rate is proportional to the incident flux and the counting time, and so the measured intensity must be divided by (i.e. normalized to) the time-integrated flux of neutrons incident on the sample in order to obtain the intrinsic scattering probability — see Section 1.7.1. One way to measure the incident flux is to place a monitor in the incident beam.[8] The monitor gives a signal which for a given energy is proportional to the incident flux and the counting time. If a measurement involves different incident energies, for example as on a white-beam time-of-flight instrument, then after correction for the energy dependence of the monitor efficiency the monitor measures the spectrum of incident neutrons.

[8]A beam monitor is a low efficiency detector which is placed in the beam and records a count rate that is proportional to the intensity in the beam. The monitor is designed to have a very high transmission, typically close to 99.9%. For practical details, see Willis and Carlile (2009).

#### Detector solid angle and efficiency

Neutron scattering instruments are often equipped with detectors which comprise multiple detecting elements (multidetectors). If they are unequal in size, or located at different distances from the sample, then the individual detector cells that make up a multidetector can subtend different solid angles at the sample position. Even if the cells are nominally identical, small differences in their setups can result in slightly different detection efficiencies. The best way to quantify such differences is to measure a sample which scatters isotropically, since the signal recorded in each detector cell is then proportional to the detector solid angle and efficiency. Normalization of the intensity measured from the sample under investigation by that from the isotropic scatterer cancels out any cell-by-cell variations.

**Table 10.3** Scattering lengths and cross-sections of the nuclides of vanadium. The incoherent cross-section of the natural isotopic mixture is almost entirely due to spin incoherence of the dominant $^{51}$V nuclide.

| nuclide | nuclear spin | abundance (%) | $b_{coh}$ (fm) | $\sigma_{coh}$ (b) | $\sigma_{inc}$ (b) | $\sigma_a$ (b) |
|---|---|---|---|---|---|---|
| $^{nat}$V |  |  | $-0.382$ | 0.0184 | 5.08 | 5.1 |
| $^{50}$V | 6 | 0.25 | 7.6 | 7.3 | 0.5 | $\sim$60 |
| $^{51}$V | 7/2 | 99.75 | $-0.402$ | 0.0203 | 5.07 | 4.9 |

The primary standard used for detector normalization is elemental vanadium, because vanadium has a moderately large incoherent scattering cross-section and a very small coherent cross-section[9] (see Table 10.3). The latter means that the intensities of the Bragg peaks are very small compared with the incoherent diffuse scattering. Vanadium is a very good approximation to an ideal isotropic scatterer,[10] but the time required to perform a vanadium run with sufficient counts for use in detector normalization can make it inefficient to measure vanadium frequently. An alternative option is to exploit the strong incoherent scattering of hydrogen and use a hydrogen-rich material as a secondary standard, such water or polythene. However, although the scattering from such materials is mainly incoherent, it tends not to be precisely isotropic due to inelastic and multiple scattering effects. If this is a concern then a wavelength- and angle-dependent correction must be determined and applied.

**Absolute intensity calibration**

A great strength of neutron scattering is the ability to calibrate the measured intensity and hence convert the data into a cross-section in absolute units. This is possible because the nuclear scattering lengths and amplitudes are known very accurately,[11] and because corrections for absorption, multiple scattering, etc, are usually small and therefore straightforward to calculate.

There are various calibration methods one can use. One is to measure a standard sample with a known cross-section. For accurate work the standard should be measured in the same scattering geometry as the sample, and should occupy the same area of the main beam as did the sample to ensure that the intensity of neutrons incident on the standard is the same as that on the sample. The primary standard used for this calibration method is vanadium.

Suppose the incident beam delivers a steady flux of neutrons $\Phi_0$, and the scattering is measured in a detector element which subtends a small solid angle $\Delta\Omega$ at the sample position. By definition, the count rate recorded in the detector element is given by

$$R_s = \left(\frac{d\sigma}{d\Omega}\right)_s \Delta\Omega\Phi_0. \tag{10.26}$$

[9] Due to the fortuitous near-cancellation of the scattering lengths for neutron spins parallel and antiparallel to the spin of the nucleus.

[10] A vanadium–niobium alloy containing about 5% niobium to suppress the coherent scattering completely is even closer to an isotropic scatterer than pure vanadium, and is regularly used in practice.

[11] To better than 1% for most elements in the periodic table, see Appendix A.

Similarly, for the vanadium standard,

$$R_V = \left(\frac{d\sigma}{d\Omega}\right)_V \Delta\Omega\Phi_0. \qquad (10.27)$$

In practice, we do not usually measure count rates. Instead, we record the number of counts in the detector when a preset number of counts in the incident beam monitor has been reached. The ratio of the detector counts to the monitor counts, hereafter denoted by $C$, is proportional to the count rate $R$. Hence, dividing (10.26) and (10.27) we may write

$$\left(\frac{d\sigma}{d\Omega}\right)_s = \left(\frac{d\sigma}{d\Omega}\right)_V \frac{C_s}{C_V}. \qquad (10.28)$$

If attenuation, multiple scattering, and the tiny coherent scattering signal are neglected (or corrected for) then the differential cross-section for a sample of vanadium containing $N_V$ atoms is given by

$$\left(\frac{d\sigma}{d\Omega}\right)_V = N_V \frac{\sigma_{inc}(V)}{4\pi}, \qquad (10.29)$$

where $\sigma_{inc}(V) = 5.08\,\mathrm{b}$ is the bound-atom incoherent scattering cross-section of vanadium. We can express $N_V$ in terms of mass via

$$N_V = \frac{m_V}{A_r(V)m_u}, \qquad (10.30)$$

where $m_V$ is the mass of the vanadium sample, $A_r(V) = 50.94$ is the relative atomic mass of vanadium, and $m_u$ is the unified atomic mass unit.[12] For the sample, we usually want to express the cross-section relative to some appropriate formula unit (f.u.), rather than for the sample as a whole. If the formula unit has relative atomic mass $A_r(\text{f.u.})$ and there are $N_{\text{f.u.}}$ of them in a sample of total mass $m_s$, then

$$\left(\frac{d\sigma}{d\Omega}\right)_s = N_{\text{f.u.}} \left(\frac{d\sigma}{d\Omega}\right)_{\text{f.u.}}$$

$$= \frac{m_s}{A_r(\text{f.u.})m_u} \left(\frac{d\sigma}{d\Omega}\right)_{\text{f.u.}}. \qquad (10.31)$$

[12] $m_u$ is equal to 1/12 the mass of an atom of carbon-12 ($^{12}$C).

Combining (10.28)–(10.31), we obtain

$$\left(\frac{d\sigma}{d\Omega}\right)_{\text{f.u.}} = \frac{\sigma_{inc}(V)}{4\pi} \frac{m_V}{m_s} \frac{A_r(\text{f.u.})}{A_r(V)} \frac{C_s}{C_V}. \qquad (10.32)$$

Equation (10.32) displays the essence of the vanadium calibration method, but there are several caveats. First, (10.32) is written in terms of the differential cross-section, which is what is measured on an instrument without energy analysis. On an instrument with energy analysis, $C_s$ is a function of energy transfer and represents the counts/monitor *per unit energy*, and $d\sigma/d\Omega$ is replaced on the left-hand side by $(k_i/k_f) \times d^2\sigma/d\Omega dE_f$. By contrast, the quantity $C_V$ in (10.32) must be the *integrated intensity* (integral over energy) of the scattering from the vanadium sample. On instruments without energy analysis the detector performs the energy integral directly, but on instruments with energy analysis a separate procedure to integrate the measured vanadium spectrum must be performed to obtain $C_V$. Second, there will be some attenuation and multiple scattering in the vanadium sample, and as a result the factor $\sigma_{inc}(V)/4\pi$ must be multiplied by the dimensionless factor $s \times 4\pi/n\sigma_{inc}(V)$, where $n = 7.23 \times 10^{28}\,\mathrm{m}^{-3}$ is the number density of vanadium and $s$ is the effective scattering function per unit volume defined in Section 10.1.2. An example of this correction is shown in Fig. 10.6(b) for a slab sample, but in general it needs to be calculated for the particular size and geometry of the vanadium sample by the methods described in Section 10.1.2. Finally, there will also be attenuation in the sample itself, and to correct for this the measured $C_s$ must be divided by the sample-averaged attenuation factor (i.e. transmission), as described in Section 10.1.1.

Absolute calibration can also be performed with a secondary standard, i.e. a reference sample whose scattering has previously been calibrated against vanadium and whose **Q** variation is known in detail, or from a scattering feature of the sample itself whose cross-section is known. For example, with a crystalline material one can use the integrated intensities of a set of Bragg peaks whose structure factors are known or, if energy analysis is available, the scattering from an acoustic phonon as described in Section 5.2.6.

## 10.1.5   Counting statistics

### Poisson distribution

Neutron scattering measurements usually involve the counting of events[13] in a finite interval, e.g. a given interval of time, a preset monitor value, or a spatially distinct detector element. The data are discrete, since counts are restricted to integer values, and noisy[14] due to the stochastic nature of the scattering process. Moreover, providing the detector is not saturating, the events occur independently, i.e. the recording of one event does not influence the next one.

Data with these characteristics are described by the *Poisson distribution*,

$$P(N) = \frac{\mu^N e^{-\mu}}{N!}, \tag{10.33}$$

where $N$ is a discrete random variable that represents the number of counts recorded in a given interval. The mean $\langle N \rangle$ and standard deviation $\sigma_N$ of the Poisson distribution are given by[15]

$$\langle N \rangle = \mu \tag{10.34}$$

$$\sigma_N = \sqrt{\langle N^2 \rangle - \langle N \rangle^2}$$

$$= \sqrt{\mu}. \tag{10.35}$$

[13] Here, an 'event' is the process whereby a neutron is captured in the detector and, after several intermediate steps, creates an electrical pulse which increments the counter by one unit.

[14] Meaning that for a constant scattering probability, the number of counts varies when repeated measurements are made over the same interval.

[15] These follow from (10.33) and the series expansion of the exponential function

$$e^x = \sum_{n=0}^{\infty} \frac{x^n}{n!}.$$

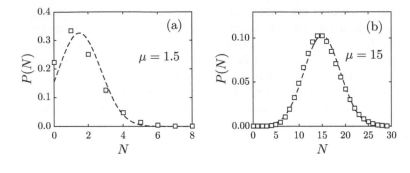

**Fig. 10.9** The Poisson distribution (square symbols) calculated from eqn (10.33) with (a) $\mu = 1.5$, and (b) $\mu = 15$. The dashed curves are Gaussian distributions with the same mean and standard deviation as the Poisson distributions.

Figure 10.9 plots the Poisson probabilities for two different mean values $\mu$. In both cases, the Poisson distribution is seen to be skewed, with the maximum of the distribution falling slightly below the mean. A useful approximation to the Poisson distribution is that of a normal distribution (or normalized Gaussian function — see Section B.1) with the same mean and standard deviation as the Poisson distribution. The level of approximation improves as $\mu$ increases, as shown on Fig. 10.9, such that the two distributions differ by only a few per cent at $\mu = 15$.

Figure 10.10 shows an example of simulated data onto which Poisson noise has been added. The underlying signal is a peak on a flat background; the effect of Poisson noise is to scatter the data points about the true signal.

### Statistical errors

Statistical scatter is often represented by *error bars*. The error bar is a line that extends one standard deviation above and below the data point to indicate the range in which the true signal is most likely to lie. For Poisson statistics, the error bar would ideally be $\pm\sqrt{\mu}$ centred on $\mu$, since from eqn (10.35) the standard deviation is $\sigma_N = \sqrt{\mu}$ and the mean is $\mu$. In an experiment, however, we do not know the value of $\mu$; we only know the number of counts $N$ recorded in each channel. The best we can do, therefore, is to assume that the number of counts is close to the mean of the Poisson distribution for that channel, i.e. $N \simeq \mu$. It then follows that $\sigma_N \simeq \sqrt{N}$. Accordingly, the error bars on Fig. 10.10 are plotted as $\pm\sqrt{N}$.

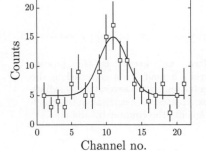

**Fig. 10.10** Data with Poisson noise. The line shows the true signal, and the points are simulated counts in an array of detector channels.

The data in Fig. 10.10 are relatively noisy because the counts per channel are relatively low. Nevertheless, the amplitude of the peak observed near the middle of the scan is several times larger than the error bars, which indicates that the peak is a real feature and not simply an artefact generated by statistical fluctuations. To strengthen our confidence in the veracity of the peak one could increase the measuring time so as to increase the number of counts recorded in each channel. The counts $N$ increase in proportion to time, and the Poisson error bar increases as $\sqrt{N}$, so the error as a fraction of the signal (the fractional error) goes as $\sqrt{N}/N = 1/\sqrt{N}$. Hence, to reduce the fractional error by a factor of 2 one must count four times longer.

As an aside, we note that the symmetric error bar $\pm\sqrt{N}$ defined above is not a particularly good way of expressing uncertainty in data with low statistics. This is because the asymmetry of the Poisson distribution is particularly marked for small $N$, see Fig 10.9. To take an extreme case, if zero counts are recorded in a particular channel then the uncertainty can only extend in the positive direction because a negative count is impossible. Moreover, $\sqrt{0} = 0$ suggests zero error, which is usually not the case. One should be cautious, therefore, with the use of $\pm\sqrt{N}$ errors when the counts are very small, e.g. low background signals.

## Propagation of errors

We next describe how to propagate statistical errors when measurements are combined arithmetically. Let us start with some well-known results.

Consider two independent random variables $X$ and $Y$. Under addition or subtraction, the mean values of the distributions that describe $X$ and $Y$ satisfy[16]

$$\langle Z \rangle = \langle X \rangle \pm \langle Y \rangle. \tag{10.36}$$

[16]This is rigorously true for any type of probability distribution.

By considering small deviations $\delta Z = Z - \langle Z \rangle$ etc, about the mean, then squaring and averaging both sides one quickly finds that

$$\sigma_Z^2 = \sigma_X^2 + \sigma_Y^2, \tag{10.37}$$

where $\sigma_Z^2 = \langle \delta Z^2 \rangle$ is the variance of the distribution. The standard deviation, $\sigma_Z$, is a measure of the spread of values in the distribution and can be used to quantify the error. Hence, eqn (10.37) shows that for addition and subtraction of independent measurements the errors add in quadrature.

Using a similar approach, one can show that for multiplication and division of two independent random variables

$$\langle Z \rangle = \langle X \rangle \langle Y \rangle \qquad \text{(multiplication)}, \tag{10.38}$$

$$\langle Z \rangle \simeq \frac{\langle X \rangle}{\langle Y \rangle} \qquad \text{(division)}, \tag{10.39}$$

$$\frac{\sigma_Z^2}{\langle Z \rangle^2} \simeq \frac{\sigma_X^2}{\langle X \rangle^2} + \frac{\sigma_Y^2}{\langle Y \rangle^2} \qquad \text{(multiplication and division)}. \tag{10.40}$$

Equation (10.38) holds regardless of the probability distributions for $X$ and $Y$, whereas the approximate relations (10.39) and (10.40) are dependent on the detailed form of the distributions. In general, the approximations hold best for narrow distributions, i.e. $\sigma_X \ll \langle X \rangle$ and $\sigma_Y \ll \langle Y \rangle$, and in the case of the ratio distribution ($Z = X/Y$), when the outcome $Y = 0$ has negligible probability. Equation (10.40) shows that the fractional errors add in quadrature for multiplication and division.

Let us now consider variables which are Poisson distributed. It can be shown that the sum of two independent Poisson random variables $X$ and $Y$ with corresponding means $\mu_X$ and $\mu_Y$ has a Poisson distribution with mean $\mu_Z = \mu_X + \mu_Y$ and standard deviation $\sigma_Z = \sqrt{\mu_Z} = \sqrt{\sigma_X^2 + \sigma_Y^2}$,

[17]The Skellam distribution is

$$P(N) = e^{-(\mu_X + \mu_Y)} \left( \frac{\mu_X}{\mu_Y} \right)^{N/2}$$
$$\times I_N(2\sqrt{\mu_X \mu_Y}),$$

where $I_N(x)$ is a modified Bessel function of the first kind.

in accord with (10.36) and (10.37), and that the difference $Z = X - Y$ follows a *Skellam distribution*[17] with mean $\mu_Z = \mu_X - \mu_Y$ and standard deviation $\sigma_Z = \sqrt{\sigma_X^2 + \sigma_Y^2}$, again consistent with (10.36) and (10.37).

As an example, suppose we have made two separate measurements and recorded counts $N_1$ and $N_2$ in a given detector channel. As Poisson variables, the estimated errors are $\sigma_{1,2} \simeq \sqrt{N_{1,2}}$. Hence, the sum and difference and their corresponding errors are

$$N = N_1 \pm N_2$$
$$\sigma_N = \sqrt{N_1 + N_2}. \tag{10.41}$$

How do we normalize the combined data and find the corresponding error? Suppose that the monitor counts for runs 1 and 2 were $M_1$ and $M_2$. In the case of addition, we combine the runs according to (10.41) and then divide both $N$ and $\sigma_N$ by the total monitor $M_1 + M_2$. Subtraction of neutron data, e.g. to correct a sample run for a measured background, is usually only performed after scaling each run to the same monitor count. Therefore, eqns (10.41) would need to be modified to $N = N_1 - \alpha N_2$ and $\sigma_N = \sqrt{N_1 + \alpha^2 N_2}$, where $\alpha = M_1/M_2$ is the factor needed to scale the background counts and error to the same monitor as the sample run,[18] and $N$ and $\sigma_N$ would by divided by $M_1$ to normalize to monitor.

[18]The estimated error on the background count is $\sqrt{N_2}$, so the error in the scaled background count is $\alpha\sqrt{N_2}$. Hence, for the propagation of the errors by (10.37) we have $\sigma_X^2 = N_1$ and $\sigma_Y^2 = \alpha^2 N_2$.

Another common procedure is division of a sample run by a vanadium run. If the sample and vanadium counts in a detector channel were $N_1$ and $N_2$ then from (10.39) and (10.40) the best estimates for the ratio and its uncertainty would be

$$R \simeq \frac{N_1}{N_2}$$
$$\sigma_R \simeq R\sqrt{\frac{\sigma_1^2}{N_1^2} + \frac{\sigma_2^2}{N_2^2}} \simeq \frac{N_1}{N_2}\sqrt{\frac{1}{N_1} + \frac{1}{N_2}}. \tag{10.42}$$

### 10.1.6    Resolution

Counting statistics are one important source of error in neutron scattering experiments. Another is beam divergence and wavelength spread due to non-idealities in the neutron optics, which cause neutrons to have a range of momentum and energy transfers distributed around the nominal values $\mathbf{Q}_0$ and $\omega_0$ set by the experimentalist. As a result, scattering features which are intrinsically sharp will tend to appear blurred. The ability to resolve sharp features can often be controlled through the choice of instrument configuration, but any improvement in resolving power will inevitably be accompanied by a reduction in intensity. When limited flux is available, the trade-off between resolution and intensity can be an important factor in the success of an experiment.

Figure 10.11 illustrates a signal containing two identical peaks which can just be resolved. The peaks have Gaussian line shapes to model the effect of instrumental broadening, and their centres are separated

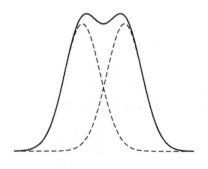

**Fig. 10.11** Two Gaussian peaks which are just resolved.

by exactly one full width at half maximum (FWHM) of the Gaussian function. With the peak positions chosen this way there is a small but unmistakable dip at the centre of the signal, which makes it possible to detect the presence of two peaks without any prior assumptions. If, however, the peaks were moved closer together then the dip would soon vanish and one would need to perform a line shape analysis of the signal using an accurate model of the broadening function in order to establish that the signal contained two peaks. This simple example shows that the FWHM of the broadening function is approximately the limit below which details in a measurement are not directly resolvable.[19]

Returning to neutron scattering, instrumental effects may be taken into account through the four-dimensional convolution

$$I(\mathbf{Q}_0, \omega_0) = \int \int R(\mathbf{Q}_0 - \mathbf{Q}, \omega_0 - \omega) S(\mathbf{Q}, \omega) \, \mathrm{d}^3\mathbf{Q} \, \mathrm{d}(\hbar\omega), \quad (10.43)$$

where $I(\mathbf{Q}_0, \omega_0)$ is proportional to the measured intensity, $S(\mathbf{Q}, \omega)$ is the intrinsic scattering function of the sample, and the instrumental *resolution function* $R(\mathbf{Q}_0 - \mathbf{Q}, \omega_0 - \omega)$ represents the probability of detecting a neutron with $\mathbf{Q}$ and $\omega$ when the instrument is set to measure $\mathbf{Q}_0$ and $\omega_0$. The resolution function is the sum of the probabilities of all possible neutron paths that have $\mathbf{Q}$ and $\omega$.

Exact calculations of $R(\mathbf{Q}_0 - \mathbf{Q}, \omega_0 - \omega)$ are usually difficult, but advanced Monte-Carlo ray-tracing algorithms incorporating realistic performance curves for the instrument components have been developed (e.g. Willendrup *et al.*, 2014). For some purposes, however, it is more convenient to use an approximate analytic expression than a numerical simulation, and in these cases the Gaussian approximation is usually employed. This approach assumes that the resolution function is close to a Gaussian, as expected from the Central Limit Theorem,[20] and further assumes that the probability distribution for each individual instrument component is Gaussian too. The combined resolution function is then a four-dimensional Gaussian of the form (see Section B.1)

$$R(\mathbf{Q}_0 - \mathbf{Q}, \omega_0 - \omega) = \frac{|\mathsf{M}|^{1/2}}{(2\pi)^2} \exp(-\tfrac{1}{2}\mathsf{X}^{\mathrm{T}}\mathsf{M}\mathsf{X}), \quad (10.44)$$

where $\mathsf{X}$ is the four-component column vector $(\mathbf{Q}_0 - \mathbf{Q}, \omega_0 - \omega)$, $\mathsf{X}^{\mathrm{T}}$ is the row matrix transpose of $\mathsf{X}$, and $\mathsf{M}$ is a $4 \times 4$ matrix known as the *resolution matrix*. The pre-factor $|\mathsf{M}|^{1/2}/(2\pi)^2$, where $|\mathsf{M}|$ is the determinant of $\mathsf{M}$, normalizes $R$ to unity.

The resolution function decreases in amplitude with distance from the point $(\mathbf{Q}_0, \omega_0)$, where it is a maximum. The contours surrounding $(\mathbf{Q}_0, \omega_0)$ along which $R(\mathbf{Q}_0 - \mathbf{Q}, \omega_0 - \omega)$ has a constant value form ellipsoidal surfaces in $(\mathbf{Q}, \omega)$-space, and the particular surface corresponding to the half-maximum contour is known as the *resolution ellipsoid*. In the Gaussian approximation (10.44) the equation for the resolution ellipsoid is

$$\mathsf{X}^{\mathrm{T}}\mathsf{M}\mathsf{X} = 2\ln 2. \quad (10.45)$$

[19]This result is very similar to the Rayleigh criterion in optics, which states that two point sources are just resolved when the principal maximum of the diffraction image of one source coincides with the first minimum of the other. The difference here is that we are assuming Gaussian line shapes, whereas the Rayleigh criterion applies to simple optical apertures whose diffraction images have strong principle maxima surrounded by weaker oscillatory fringes.

[20]In simple terms, the Central Limit Theorem states that the distribution of the sum of many independent random variables will be normal (i.e. Gaussian), irrespective of the shape of the individual distributions.

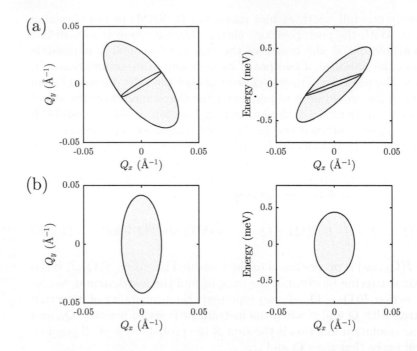

**Fig. 10.12** Resolution ellipsoids viewed on different planes when the principal axes of the ellipsoid are (a) not aligned and (b) aligned with the $(\mathbf{Q}, \omega)$ axes. The shaded areas are projections of the resolution ellipsoid onto the plotting axes, and the smaller enclosed ellipses in (a) are sections through the centre of the resolution ellipsoid.

The extent to which features in the scattering are affected by resolution broadening will depend on the shape, size and orientation of the resolution ellipsoid relative to the signal of interest. Therefore, in planning an experiment it is often worthwhile to calculate and plot out the relevant sections of the resolution ellipsoid, as illustrated in Fig. 10.12.

In general, the principal axes of the resolution ellipsoid will not be aligned with the axes of the $(\mathbf{Q}, \omega)$ coordinate system, Fig. 10.12(a), but if the resolution in one or more $(\mathbf{Q}, \omega)$ components is not coupled to the other components then the independent components will be principal axes of the resolution ellipsoid, as in Fig. 10.12(b). For example, on some time-of-flight spectrometers the energy resolution is decoupled from the $\mathbf{Q}$ resolution. In such a case one could calculate, say, the energy resolution, i.e. the width of the resolution ellipsoid along the energy direction, in the Gaussian approximation by adding in quadrature the variances of the independent contributions, as in (10.37), and using the relation FWHM = $2.355\sigma$ for a Gaussian[21] to obtain the final result.

[21] Section B.1.

### 10.1.7 Polarization analysis corrections

Polarized neutron techniques make use of various neutron optical components, including polarizers (devices that transmit one neutron spin state preferentially over the other), spin flippers (devices that reverse the neutron spin), guide fields (to maintain the neutron polarization between beam line components), and rotators (to rotate the plane of polarization). For practical details, readers are referred to Williams (1988), Willis and Carlile (2009), and Andersen *et al.* (2006).

Needless to say, the efficiencies of polarized-neutron devices are never perfect, and procedures are needed to correct for shortcomings in their performance. These corrections are particularly important when the efficiencies are significantly less than 100%, or when the cross-sections are very strongly polarization-dependent.

Instrument performance is often expressed in terms of a *flipping ratio*, $R$, defined as the ratio of the intensity of the 'right' spin state to that of the 'wrong' spin state. The larger the value of $R$, the better the performance. In order to determine $R$ one needs a device that can reverse the direction of polarization, such as a spin flipper. A common procedure is to use this device to measure the non-spin-flip (NSF) and spin-flip (SF) intensities from a pure nuclear coherent scatterer,[22] whereupon

$$R = \frac{I^{\text{NSF}}}{I^{\text{SF}}}. \tag{10.46}$$

[22] Nuclear coherent scattering preserves the spin of the neutron, i.e. $\sigma^{\text{SF}} = 0$. The incoherent background must be subtracted from the NSF and SF signals first.

The scattering could be provided by the sample itself or by a separate standard sample. Structural Bragg peaks are often used because they can be very intense, but glassy materials can be better for multidetector instruments as they produce nuclear diffuse scattering over an extended angular range which can be used to calibrate many detectors simultaneously. The flipping ratio can also be measured on the main beam if the detector can withstand the intensity.

In general, it is best to measure flipping ratios and calibrate an instrument under conditions as close as possible to those of the experiment itself because the beam polarization and device efficiencies depend on experimental variables such as wavelength, angular divergence, position in the beam, sample environment, etc. In addition, the type of coherent scattering used to measure $R$ should ideally be the same as what will be studied in the experiment itself. This is because different types of scattering, e.g. single-crystal diffraction, powder diffraction, diffuse scattering, inelastic scattering, etc, tend to give different values of $R$.

### General treatment

In this section we shall describe a general correction procedure for neutron polarization analysis.[23] Figure 10.13 shows a typical instrument configuration. The polarizer produces a polarized beam whose direction of polarization can be reversed by the first spin flipper. The analyser and second flipper make it possible to measure the polarization of the neutrons scattered from the sample. The detector records all the neutrons that are transmitted by the analyser. At any point after the polarizer, the neutron polarization is either parallel ($+$) or antiparallel ($-$) to a magnetic guide field which preserves the polarization along the instrument. If the guide field slowly changes orientation then the polarization rotates adiabatically with the guide field.

[23] The approach outlined here follows that of Wildes (2006).

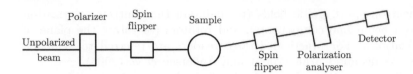

**Fig. 10.13** Schematic of an instrument for neutron polarization analysis.

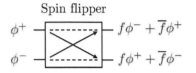

**Fig. 10.14** Neutron spin reversal in an imperfect spin flipper. $\phi^+$ and $\phi^-$ are the fractions of (+) and (−) spin neutrons before the flipper ($\phi^+ + \phi^- = 1$). $f$ is the flipper efficiency ($0 \leq f \leq 1$), and $\bar{f} = 1 - f$.

[24]Note that an alternative arrangement in uniaxial polarization analysis is to have polarizer and analyser 'crossed', i.e. set to transmit opposite spin states.

[25]The superscripts on the measured intensity denote the two flipper states: $0 \equiv$ 'off', $1 \equiv$ 'on'.

The neutron beam is initially unpolarized, and after the polarizer the fractions of neutrons in the 'right' and 'wrong' polarization states are $p$ and $1 - p$, respectively ($0 \leq p \leq 1$). Hence, the polarizer can be considered as a device whose transmission is proportional to $p$ for one spin state and $1 - p$ for the other. A perfect polarizer would have $p = 1$. The actual value of $p$ is the result of several factors, including incomplete polarization produced by the polarizing device, depolarization in the guide field, and any fields used to rotate the polarization before the sample, and depolarization in the sample itself before scattering. An efficiency $a$ is defined similarly for the analyser side of the instrument. The flipper efficiency $f$ is defined as the fraction of the neutrons in one spin state which undergo a spin reversal when the flipper is 'on', as illustrated in Fig. 10.14. An ideal flipper has $f = 1$. If the flipper is 'off' then any effect of the flipper on the polarization is absorbed into $p$ or $a$.

We consider an instrument for uniaxial polarization analysis like that shown in Fig. 10.13, and take as an example the case where both the polarizer and analyser are set to transmit neutrons of (+) polarization state.[24] First, suppose that both flippers are 'off'. The spin-dependent intensities along the instrument are shown in Fig. 10.15, and the intensity recorded in the detector is[25]

$$I^{00} = I_0 \left\{ pa\sigma^{++} + p\bar{a}\sigma^{+-} + \bar{p}a\sigma^{-+} + \bar{p}\bar{a}\sigma^{--} \right\}, \qquad (10.47)$$

where $\sigma^{\alpha\beta}$ are the spin-dependent cross-sections, $I_0$ is the constant that converts the cross-sections into scattered intensities, and $\bar{x}$ is shorthand for $1 - x$. Next, suppose that flipper $F_P$ on the polarizer side is turned 'on'. The effect of $F_P$ is to perform the transformation shown in Fig. 10.14, which is a spin reversal combined with a degree of mixing. The spin-dependent intensities along the instrument are now those shown in Fig. 10.16, and the intensity measured in the detector will be

**Fig. 10.15** Schematic showing the spin-dependent intensities at different points along an instrument configured to measure $I^{00}$. The instrument components are, from left to right: polarizer (P), flipper ($F_P$), sample (S), flipper ($F_A$) and analyser (A). The arrows denote the spin direction of the neutrons, with spin-'up' corresponding to the (+) polarization state. Both P and A are set to transmit 'up' spins.

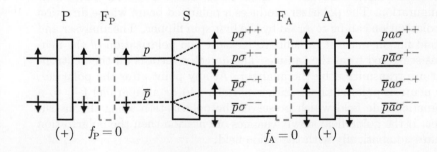

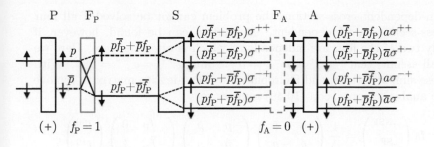

$$I^{10} = I_0 \left\{ (p\overline{f}_{\rm P} + \overline{p}f_{\rm P})a\sigma^{++} + (p\overline{f}_{\rm P} + \overline{p}f_{\rm P})\overline{a}\sigma^{+-} \right.$$
$$\left. + (pf_{\rm P} + \overline{p}\overline{f}_{\rm P})a\sigma^{-+} + (pf_{\rm P} + \overline{p}\overline{f}_{\rm P})\overline{a}\sigma^{--} \right\}. \quad (10.48)$$

Equations (10.47) and (10.48), together with the equivalent expressions for $I^{01}$ and $I^{11}$ which are found in a similar way, give a set of four simultaneous linear equations which can be solved to find the corrected intensities $I_0\sigma^{++}$, etc. The results are concisely expressed in matrix form,

$$I_0 \begin{pmatrix} \sigma^{++} \\ \sigma^{+-} \\ \sigma^{-+} \\ \sigma^{--} \end{pmatrix} = {\rm AF_A PF_P} \begin{pmatrix} I^{00} \\ I^{01} \\ I^{10} \\ I^{11} \end{pmatrix}, \quad (10.49)$$

where A, $F_A$, P and $F_P$ are characteristic matrices for the efficiencies of the analyser, analyser flipper, polarizer, and polarizer flipper, and are given by

$$A = \frac{1}{\overline{a} - a} \begin{pmatrix} -a & \overline{a} & 0 & 0 \\ \overline{a} & -a & 0 & 0 \\ 0 & 0 & -a & \overline{a} \\ 0 & 0 & \overline{a} & -a \end{pmatrix}, \quad F_A = \frac{1}{f_A} \begin{pmatrix} f_A & 0 & 0 & 0 \\ -\overline{f}_A & 1 & 0 & 0 \\ 0 & 0 & f_A & 0 \\ 0 & 0 & -\overline{f}_A & 1 \end{pmatrix},$$

$$P = \frac{1}{\overline{p} - p} \begin{pmatrix} -p & 0 & \overline{p} & 0 \\ 0 & -p & 0 & \overline{p} \\ \overline{p} & 0 & -p & 0 \\ 0 & \overline{p} & 0 & -p \end{pmatrix}, \quad F_P = \frac{1}{f_P} \begin{pmatrix} f_P & 0 & 0 & 0 \\ 0 & f_P & 0 & 0 \\ -\overline{f}_P & 0 & 1 & 0 \\ 0 & -\overline{f}_P & 0 & 1 \end{pmatrix}.$$

$$(10.50)$$

The same formalism can be applied to other instrument configurations and devices, including polarizers that can switch the direction of polarization and so have no need for spin flippers, and polarizers whose total transmission depends on whether they are set to transmit $(+)$ or $(-)$ spin neutrons. These are described in the review by Wildes (2006), which also includes the calibration procedures needed to determine the individual efficiencies of the different optical devices.

We shall now illustrate the method by calculating the polarization corrections for a configuration with one flipper, located before the sample. Such an instrument can measure $I^{00}$ and $I^{10}$, but as there are four

**Fig. 10.16** The same as Fig. 10.15 but for an instrument configured to measure $I^{10}$.

[26]The other occasions when $\sigma^{++}$ and $\sigma^{-+}$ are the only independent terms are: (i) $\sigma^{+-} = \sigma^{--} = 0$; (ii) $\sigma^{++} = \sigma^{--}$ and $\sigma^{+-} = 0$; (iii) $\sigma^{+-} = \sigma^{-+}$ and $\sigma^{--} = 0$.

[27]Note that in eqn (5) of Wildes (2006) there is an error in the flipper matrix which was corrected in an erratum (Wildes, 2006).

spin-dependent cross-sections the problem cannot be solved if all four cross-sections are independent. A solution can be found, however, if only two of the four are independent, which is often the case. Here we shall assume that[26] $\sigma^{++} = \sigma^{--} = \sigma^{\mathrm{NSF}}$ and $\sigma^{+-} = \sigma^{-+} = \sigma^{\mathrm{SF}}$. With these simplifications, and $\mathsf{F_A}$ replaced by the identity matrix to remove the analyser flipper, eqn (10.49) reduces to[27]

$$I_0 \begin{pmatrix} \sigma^{\mathrm{NSF}} \\ \sigma^{\mathrm{SF}} \end{pmatrix} = \frac{1}{\Delta} \begin{pmatrix} -a & \bar{a} \\ \bar{a} & -a \end{pmatrix} \begin{pmatrix} -p & \bar{p} \\ \bar{p} & -p \end{pmatrix} \begin{pmatrix} f_{\mathrm{P}} & 0 \\ -\bar{f}_{\mathrm{P}} & 1 \end{pmatrix} \begin{pmatrix} I^{00} \\ I^{10} \end{pmatrix},$$
(10.51)

where $\Delta = f_{\mathrm{P}}(p - \bar{p})(a - \bar{a})$. The two equations represented by (10.51) can be rearranged to give the flipping ratio (10.46), which is found to be

$$R = \frac{I^{00}}{I^{10}} = \frac{pa + \bar{p}\bar{a}}{(pa + \bar{p}\bar{a})\bar{f}_{\mathrm{P}} + (\bar{p}a + p\bar{a})f_{\mathrm{P}}}.$$
(10.52)

If it can be assumed that the flipper is ideal then (10.52) reduces to

$$R = \frac{pa + \bar{p}\bar{a}}{\bar{p}a + p\bar{a}} \qquad (f_{\mathrm{P}} = 1),$$
(10.53)

and eqns (10.51) may be written

$$I_0 \begin{pmatrix} \sigma^{\mathrm{NSF}} \\ \sigma^{\mathrm{SF}} \end{pmatrix} = \frac{1}{R-1} \begin{pmatrix} R & -1 \\ -1 & R \end{pmatrix} \begin{pmatrix} I^{00} \\ I^{10} \end{pmatrix}.$$
(10.54)

In this particular case, therefore, all that is required to correct the data is an experimental measurement of $R$.

Finally, let us note that one would ideally like to characterize instrument performance in terms of the beam polarization $\mathbf{P}$, as defined in eqn (1.9), which varies along the instrument. At the sample position, the polarization[28] with the flipper off is (Fig. 10.15)

[28]More precisely, the polarization along the quantization direction.

$$P_{\mathrm{P}} = \frac{p - \bar{p}}{p + \bar{p}} = p - \bar{p},$$
(10.55)

and with the flipper on (Fig. 10.16),

$$P_{\mathrm{P}} = (p - \bar{p})(f_{\mathrm{P}} - \bar{f}_{\mathrm{P}}).$$
(10.56)

If we treat the analyser like a polarizer then analogous formulae can be used to define $P_{\mathrm{A}}$, the effective polarization of the analyser side of the instrument. Hence, the overall instrument performance can be represented by an effective instrument polarization $P_{\mathrm{eff}} = P_{\mathrm{P}}P_{\mathrm{A}}$. When both flippers are off or both flippers are ideal ($f_{\mathrm{P}} = f_{\mathrm{A}} = 1$),

$$\begin{aligned} P_{\mathrm{eff}} &= (p - \bar{p})(a - \bar{a}) \\ &= \frac{R-1}{R+1}. \end{aligned}$$
(10.57)

The second line follows from (10.53) and applies only for ideal flippers.

## 10.1.8    Spurions

The term *spurion* is a colloquialism for any type of parasitic process which results in a signal that can be mistaken for a real signal from the sample under investigation. When confronted with a new or unexpected feature in the data it is always prudent to investigate and eliminate all other possible explanations before claiming a discovery.[29] The processes that cause spurions depend on the source, instrument, sample and sample environment. Here we give a short account of some of the more common types of spurion.

### Sample environment spurions

Sample mounts and sample environment equipment, such as the walls of cryostats, furnaces, magnets, etc, are often unavoidably in the neutron beam and can produce spurious signals, especially Bragg peaks. Typical materials used are aluminium and copper. Sheets of these metals are generally polycrystalline but tend to be textured, so the Bragg peak intensities can be stronger in certain directions than in others. Parasitic scattering due to multiple scattering processes involving both the sample and sample environment also occurs. Sample environment spurions are usually more of a problem on instruments that employ focusing or double-focusing monochromators and analysers to increase signal. Focusing devices generally require more open beam tubes, and collimation tends to be dispensed with.

Several steps can be taken to reduce the incidence of sample environment spurions. Sample mounts should be as small as possible, and as much as possible of the sample mount and addenda should be masked with absorbing materials such as cadmium or $B_4C$. Conventional glues and other hydrogenous materials should be avoided because of the strong incoherent scattering from natural hydrogen (H vibrations in glues have sometimes been mistaken for excitation modes in the sample). Solid samples can be attached to the mount with aluminium wire, PTFE tape,[30] or a fluoropolymer-based glue with a low hydrogen content, such as CYTOP$^{TM}$. If one has the option to use a neutron wavelength that is below the Bragg cut-off of any materials present in the vicinity of the sample then this has the advantage of eliminating Bragg peaks not originating from the sample.

### Crystal monochromator/analyser spurions

When crystal monochromators and analysers are used for wavelength selection a variety of different spurious processes can occur.[31] Often these result from the fact that monochromator and analyser crystals reflect not only the fundamental wavelength $\lambda$, but also the harmonics of $\lambda$, i.e. $\lambda/2$, $\lambda/3$, etc, as described by Bragg's law, eqn (2.9), $n\lambda = 2d\sin\theta_B$. For an instrument with both a monochromator an analyser, this means that if we intend to measure scattering that involves a change in neutron wavevector from $k_i$ to $k_f$ then what we actually observe are $nk_i$ to

[29] *'How often have I said to you that when you have eliminated the impossible, whatever remains, however improbable, must be the truth?'* Sherlock Holmes in *The Sign of the Four*, by Sir Arthur Conan Doyle (1890).

[30] PTFE stands for polytetrafluoroethylene.

[31] A more detailed account can be found in Shirane *et al.* (2002).

[32]See Section 10.3.2.

$mk_f$ processes, where $n$ and $m$ are positive integers. When $n, m > 1$ such scattering can give rise to spurions, particularly when the condition $nk_i = mk_f$ is met and there is strong elastic scattering from the sample. A well-known example is the $2k_i \rightarrow 3k_f$ spurion that occurs at 18.3 meV on a triple-axis spectrometer[32] when $k_f$ is set to $26.62 \, \text{nm}^{-1}$ ($E_f = 14.7 \, \text{meV}$), as is often the case. The energy equivalent of $3k_f$ is 132.1 meV, so if there are neutrons in the incident beam with this energy then they can be elastically scattered by the sample onto the analyser and reach the detector. Neutrons with this energy can be present in the incident beam as second harmonic contamination when the monochromator is set to $k_i = 3k_f/2 = 39.93 \, \text{nm}^{-1}$, which corresponds to $E_i = 33.0 \, \text{meV}$. Therefore, any $2k_i \rightarrow 3k_f$ elastic scattering will appear in the spectrum at an energy transfer of $\hbar\omega = 33.0 - 14.7 = 18.3 \, \text{meV}$.

Monochromator and analyser crystals are usually chosen to have good Bragg reflectivity combined with small absorption and incoherent cross-sections. However, weak elastic incoherent and inelastic scattering cannot be avoided and can be a source of spurions. For example, if a strong Bragg peak from the sample hits the analyser then there can be sufficient non-Bragg scattering from the analyser to register a significant signal in the detector. A similar process would be elastic incoherent scattering from the monochromator followed by Bragg scattering from the sample and analyser. These types of process result in sharp peaks which are elastic in origin but appear in inelastic parts of the spectrum, and are known as *accidental Bragg scattering*. It is relatively easy to identify accidental Bragg scattering because a small change in incident energy will make the spurious peak move, and in the case of a single crystal sample the spurious peak has the width of a Bragg peak when the sample is rotated.

One way to avoid $\lambda/2$ contamination is to select a monochromator or analyser reflection which is forbidden in second order. The 111 reflections of silicon (Si) and germanium (Ge) have this property (see Exercise 10.7). In fact, all even orders of Si and Ge 111 are forbidden, but all odd orders are allowed so $\lambda/3$ is still present.

If there is any doubt about a scattering feature then there are several checks that can be applied: (i) measure the same $(\mathbf{Q}, \omega)$ but with a different incident energy, or at a different but equivalent $\mathbf{Q}$ (if one exists); (ii) measure the temperature dependence — if the signal is from the sample then it will usually vary with temperature; (iii) turn the analyser away from the Bragg condition for the chosen final neutron energy — if the signal remains it is an artefact; (iv) install a beam monitor between sample and analyser to detect the presence of Bragg peaks from the sample in the scattered beam.

### Neutron beam filters

Filters are often employed to reduce higher-order contamination in a neutron beam. Commonly used filter materials are beryllium (Be), beryllium oxide (BeO), and pyrolytic graphite (PG). Filters containing

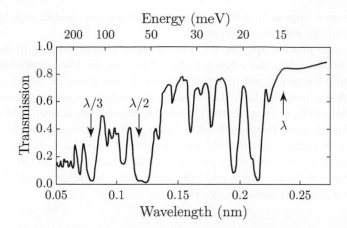

**Fig. 10.17** The transmission of a graphite filter of thickness 41.3 mm measured on the Harwell LINAC in 1974. The graphite filter is often used with a wavelength of $\lambda = 0.236$ nm (wavevector $k = 26.62$ nm$^{-1}$, energy $E = 14.7$ meV), for which both $\lambda/2$ and $\lambda/3$ coincide with dips in the filter transmission, as indicated.

these crystalline materials have a high transmission for wavelengths greater than the Bragg cut-off (see Section 10.1.1) but scatter shorter wavelengths out of the beam. A filter containing polycrystalline Be will strongly suppress neutrons with wavelengths below 0.4 nm (energies above 5 meV). For BeO, suppression occurs when $\lambda < 0.46$ nm ($E > 3.8$ meV). The performance of Be and BeO filters is significantly improved when they are cooled to 77 K (Fig. 10.18).

In the case of PG, its operation as a filter can be understood from its wavelength-dependent transmission shown in Fig. 10.17. Pyrolytic graphite is crystalline in the direction perpendicular to the graphene sheets but has a high degree of orientational disorder within the sheets. The transmission for neutrons perpendicular to the sheets is large for wavelengths above about 0.23 nm, but at lower wavelengths there are a series of transmission minima which are caused by Bragg diffraction. These minima can be used to suppress higher orders. The graphite filter is often used at a wavelength of $\lambda = 0.236$ nm because at this wavelength the second order ($\lambda/2$) and third order ($\lambda/3$) both coincide with dips in the transmission and are strongly suppressed.

Single crystal sapphire (Al$_2$O$_3$) is an effective filter to suppress fast neutrons ($\lambda < 0.1$ nm). This is largely due to an increase in the nuclear inelastic scattering cross-section with increasing energy. Other types of filters for higher energies employ neutron resonances. For example, if $\lambda$ is chosen so that $\lambda/2$ coincides with a resonance of the filter material then the second-order neutrons will be suppressed.

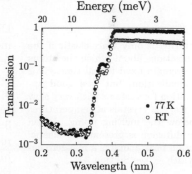

**Fig. 10.18** Transmission of a Be filter as a function of wavelength in the vicinity of the Bragg cut-off (Groitl *et al.*, 2016). Measurements at 77 K and room temperature (RT) are shown. The transmission above the cut-off is significantly improved with cooling. (Data courtesy of F. Groitl.)

### Time-of-flight spurions

It is sometimes said that time-of-flight instruments are 'spurion-free'. This is not entirely true, but the use of choppers and time-of-flight to select the incident energy and separate the final energies does substantially reduce the spurions associated with monochromators and analysers. In addition, time-of-flight instruments on pulsed sources have an intrinsically low background because the fast neutrons and gamma rays from the source appear in a very narrow time window which can be excluded from the measured spectrum.

One spurious process specific to time-of-flight instruments is frame overlap. It occurs when the intensity of neutrons in the low-energy tail from one pulse frame has not dropped to background levels by the start of the next frame (see Fig. 1.8). The effect is to produce a spurious signal at short times-of-flight. Frame overlap can be avoided by insertion of a frame-overlap chopper to remove every other pulse. Another process is multiple elastic scattering from the sample and its surroundings which makes the neutrons take a longer path to the detectors and hence arrive at a later time-of-flight. On a spectrometer, these neutrons would be recorded as a spurious inelastic signal.

## 10.2   Practical aspects of diffraction

[33]Scattering is only truly elastic if the mean positions about which the atoms vibrate remain fixed for all time. This is an idealization, but is a good approximation for systems with very slow dynamics, e.g. crystals at temperatures well below the melting point. When atomic diffusion is present the 'elastic' diffraction peak acquires a non-zero energy width — see Section 3.6.

Diffractometers are designed to provide information on structure through measurements of coherent elastic or total scattering.[33] Most diffractometers do not have energy analysis and so integrate over a range of energies. If this integral includes all the relevant scattering then the static approximation applies and the quantity obtained is $S(\mathbf{Q})$, which contains information on instantaneous correlations (see Sections 3.8.3 and 3.8.4). When an analyser is used, its purpose is usually to transmit a narrow range of energies centred on $\hbar\omega = 0$ and filter out any unwanted inelastic scattering. This can be useful for reducing background when the scattering of interest is essentially elastic, e.g. Bragg peaks.

The following sections are concerned mainly with diffraction from crystalline materials, although some of the instrument considerations will apply to studies of non-crystalline materials, too.[34]

[34]Small-angle scattering, which is used to investigate large-scale structure, is discussed separately in Section 2.6.

Crystal structure investigations can be performed on single crystals or polycrystalline powders. With a single crystal, the Bragg reflections from distinct sets of crystal planes can be resolved because they reflect in different directions. Moreover, the intensity in a Bragg reflection is concentrated in a single sharp peak giving the best possible ratio of signal to background. By contrast, with a powder sample the Bragg peak intensities are spread out into rings, and Bragg reflections from all planes with the same $d$-spacing merge into the same ring. For these reasons, single-crystal diffraction would normally be the technique of choice. However, powder diffraction is more straightforward to perform (the shape and size of the sample is well defined by the container, and there is no need to align the sample), and some of the complications in single crystal data analysis (e.g. attenuation, extinction, domain averaging, multiple scattering, etc) are either less severe or absent with powders. Another strength of powder diffraction is that it is usually possible to obtain more accurate lattice parameters than with single crystal diffraction. Finally, for certain complex materials it can be very difficult or even impossible to grow single crystals of sufficient size for neutron diffraction, in which case powder diffraction is the only option.

## 10.2.1 Types of neutron diffractometer

### Single-crystal diffractometers

The traditional type of instrument for single crystal diffraction at continuous sources is the four-circle diffractometer, shown schematically in Fig. 10.19. Neutrons of fixed wavelength are used, obtained by Bragg reflection from a monochromator crystal. The sample is mounted on a device called an Eulerian cradle, which has three independent rotations $\omega$, $\chi$ and $\phi$. The detector makes an angle $2\theta$ with the incident beam direction,[35] and the $2\theta$ rotation axis is coincident with the $\omega$ axis, both being vertical for a horizontal scattering plane. For a given neutron wavelength, any set of crystal planes (within practical limits) can be brought into the Bragg condition through an appropriate choice of the Eulerian angles and $2\theta$. The Bragg reflection is measured by scanning $\omega$ or $2\theta$ (or a combination of the two) in small steps through the Bragg condition. The area under the resulting peak is called the *integrated intensity*. Use of a position-sensitive multidetector can be an advantage for characterizing the intensity distribution as well as the background in the neighbourhood of the Bragg peak. Ultimately, the aim is to measure the integrated intensities of a large number of Bragg reflections, and then to refine a structural model until the best possible agreement is obtained between the data and the intensities calculated from the model.

The four-circle goniometer is not always suitable for experiments that require large or heavy sample-environment equipment. For such experiments there are diffractometers available which replace the Eulerian cradle with a large sample table capable of withstanding the weight of cryostats, cryomagnets, etc (Fig. 10.20). The table can rotate about a vertical axis, and may also include tilt goniometers which permit a small range of rotation about two orthogonal horizontal axes. The detector rotates in the horizontal plane about the $2\theta$ axis, but can sometimes be mounted on a lifting arm to access reflections out of the horizontal scattering plane. Other options are an analyser and a multidetector.

A third type of continuous-source single-crystal diffractometer is the Laue diffractometer. The Laue method employs a broad spectrum of neutrons, and has a large position-sensitive detector to record the scattering from a crystal in a fixed orientation (Fig. 10.21). A Bragg-reflected beam will be produced from a set of planes if the incident spectrum contains neutrons whose wavelength satisfies Bragg's law, eqn (2.9). The reflected beam will be monochromatic, and will appear at an angle $2\theta_B$ to the incident beam in the plane of incidence,[36] where $\theta_B$ is the Bragg angle (Section 2.2.1). As the incident spectrum is continuous, neutrons diffract off many different planes simultaneously, resulting in an array of Bragg spots on the detector. Traditionally, the Laue method has been used for aligning single crystals, since the spot pattern exhibits all the symmetries of the crystal as viewed along the incident beam. More recently, however, detector technology has advanced to the point where crystal structures can now be solved from Laue diffraction data, and diffuse scattering in between the Bragg peaks can be detected.

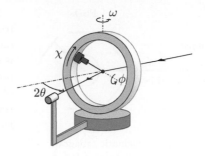

**Fig. 10.19** Schematic diagram of a four-circle diffractometer.

[35]Elsewhere we use $\phi$ for the scattering angle. In Bragg diffraction, however, the scattering angle is twice the Bragg angle $\theta_B$, so $2\theta$ is a convenient choice of notation. This choice also avoids confusion with the Eulerian angle $\phi$.

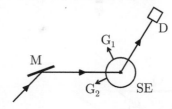

**Fig. 10.20** Plan view of a single-crystal diffractometer for bulky sample environment (SE) kit. $G_1$ and $G_2$ are the orthogonal rotation axes of two tilt goniometers mounted on the horizontal sample table. The monochromator (M) and detector (D) are also shown.

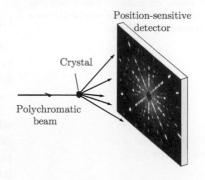

**Fig. 10.21** Schematic diagram of a Laue diffractometer.

[36]The plane of incidence is the plane containing the incident beam and the normal to the Bragg planes.

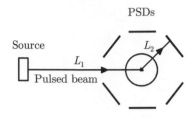

**Fig. 10.22** Schematic diagram of a single-crystal diffractometer on a pulsed source. Six banks of position-sensitive detectors (PSDs) are shown. The PSDs are two-dimensional.

[37]The method is sometimes called time-sorted Laue diffraction.

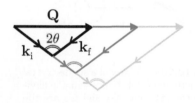

**Fig. 10.23** Diagram showing the radial path taken by $\mathbf{Q}$ on a time-of-flight diffractometer for a fixed $2\theta$ and incident beam direction. The scattering triangles for each $\mathbf{k}_i$ are similar because the scattering is elastic.

[38]$\mathbf{G} = h\mathbf{a}^* + k\mathbf{b}^* + l\mathbf{c}^*$
— see Section 2.2.2.

Single-crystal diffractometers on pulsed sources resemble Laue diffractometers in that they have large position-sensitive detectors and record an array of Bragg peaks produced when a polychromatic neutron beam is diffracted by the crystal (Fig. 10.22). The pulsed nature of the beam, however, means that diffraction images can be measured as a function of neutron time-of-flight.[37] From the de Broglie equation (1.1) and non-relativistic momentum $p = m_\mathrm{n} v$ the flight time $t$ for neutrons of wavelength $\lambda$ is

$$t = \frac{m_\mathrm{n}}{h} L\lambda, \tag{10.58}$$

where $L = L_1 + L_2$ is the total flight path from source to detector.

The time-of-flight method offers two important advantages. First, on a continuous source all orders of reflection from a set of Bragg planes are superimposed at the same point on the detector and cannot be separated; on a pulsed source different orders also hit the same point on the detector, but because the diffracted wavelengths are different they can be separated by their flight times. Second, on a pulsed source the background around a Bragg peak only comes from the incident wavelength that produced the Bragg peak, whereas on a continuous source the background is from all wavelengths in the incident spectrum, so is much higher.

A notable property of the time-of-flight diffractometer is that the scattering vector $\mathbf{Q}$ follows a radial path in reciprocal space for a fixed scattering angle $2\theta$, as illustrated in Fig. 10.23. Therefore, if $2\theta = 2\theta_\mathrm{B}$ for some Bragg reflection then the time-of-flight scan recorded in a detector at $2\theta$ corresponds to a radial scan in reciprocal space through the reciprocal lattice vector at which the reflection is observed. Combining (1.18) and (10.58) we find that the variation of $Q$ with time-of-flight is given by

$$Q = \frac{2m_\mathrm{n}}{\hbar} \frac{L\sin\theta}{t}. \tag{10.59}$$

### Crystal orientation for single crystal diffraction

The values of the diffractometer setting angles for observation of a particular Bragg reflection $hkl$ are calculated from a matrix equation of the form

$$\mathbf{G}^{\mathrm{lab}} = \mathsf{MUBG}, \tag{10.60}$$

where $\mathbf{G}$ is the reciprocal lattice vector for the $hkl$ reflection in reciprocal lattice coordinates[38] and $\mathbf{G}^{\mathrm{lab}}$ is the Bragg scattering vector referred to a set of Cartesian (i.e. orthonormal) axes fixed with respect to the laboratory. $\mathsf{B}$ is the reciprocal lattice metric tensor which relates the reciprocal lattice to a set of Cartesian axes fixed with respect to the reciprocal lattice. If all the setting angles are zero then $\mathsf{M}$ is the $3 \times 3$ identity matrix and $\mathsf{U}$ is a $3 \times 3$ rotation matrix which relates the reciprocal lattice Cartesian axes to the laboratory axes. $\mathsf{U}$ is called the orientation matrix since it depends on how the crystal was mounted. If the setting angles are not zero then $\mathsf{M}$ is a combination of rotation matrices corresponding to the setting angles. For example, on a four-circle diffractometer $M = \Omega X \Phi$, the product of three rotation matrices corresponding to the Eulerian angles $\omega$, $\chi$, and $\phi$, respectively.

The product UB is known as the *UB matrix* (pronounced 'you be'). It depends on the crystal structural parameters (lattice parameters and inter-axial angles) and on how the crystal is orientated on the goniometer, and is determined by refinement against measurements of a set of Bragg reflections. Once the UB matrix is known, the crystal can be driven to the Bragg condition for any desired reflection within practical limits.

## Powder diffractometers

Powder diffraction is one of the most widely used neutron scattering techniques. The sample need not actually be a powder (although often it is). It could be a sintered pellet, ceramic, or any other type of polycrystalline solid. The key requirement is that it contains millions of tiny, randomly oriented, crystalline grains,[39] so that every possible orientation is represented with equal probability.[40] This means that the Bragg reflections from a given set of planes form a cone of semi-angle $2\theta_B$ about the incident beam direction, known as a Debye–Scherrer cone (Fig. 10.24).

Powder diffractometers on continuous sources are *angle-dispersive*. This means that they employ a single wavelength and achieve a wide $Q$ coverage through a multidetector that forms an arc around the sample extending over as large a range of $2\theta$ as possible, see Fig. 10.25(a). Bragg's law, $\lambda = 2d \sin\theta_B$, tells us that for fixed $\lambda$, planes with smaller $d$-spacing diffract at larger $2\theta$.

Pulsed source powder diffractometers, on the other hand, are *wavelength-dispersive*. They receive a broad bandwidth of wavelengths from the moderator and separate them by time-of-flight. For a fixed $2\theta$, longer time-of-flight corresponds to larger $d$-spacing. Detectors on wavelength-dispersive instruments vary from single banks at a few specific angles, Fig. 10.25(b), to large curved multidetectors. Detector pixels at different $2\theta$ measure the same Bragg peak but with different wavelengths. After a number of $\lambda$- and $\theta$-dependent corrections have been made, the time-of-flight spectra recorded in each pixel within a bank are usually combined into a single diffraction pattern, or *diffractogram*.

Powder diffraction measurements on a continuous source can be plotted as a function of $2\theta$, $d$-spacing or $Q$. The advantage of $d$-spacing and $Q$ is that the resulting diffractogram is independent of the wavelength

[39] The size of the crystallites should ideally be $\sim 1\,\mu m$. Too large, and there may be too few particles to give an adequate statistical average. Too small, and peak broadening due to finite-size effects may become significant.

[40] It is sometimes necessary to rotate the sample to remove the effects of texturing and achieve true randomness.

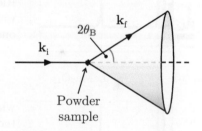

**Fig. 10.24** Debye–Scherrer cone of diffraction from a powder sample. $\theta_B$ is the Bragg angle.

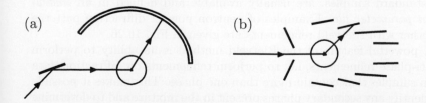

**Fig. 10.25** Schematic diagrams of a powder diffractometer at (a) a continuous source, and (b) a pulsed source. The detector banks in (b) are tilted relative to $k_f$ so as to achieve a constant resolution across the bank.

used to collect the data. Similarly, although time-of-flight, $t$, is a natural variable for pulsed-source diffraction, if $t$ is used to present data then information about the diffractometer (flight path and detector angle) is needed to interpret the data in terms of crystallography. In general, therefore, it is preferable to use $d$-spacing (which is proportional to $t$) or $Q$ as the variable on a diffractogram.

### Rietveld profile refinement of powder diffraction data

A major challenge for the analysis of powder diffraction data is the presence of overlapping peaks. Powder-averaging has the effect of contracting three-dimensional $\mathbf{Q}$ space down to just the radial component $Q$, which means that all Bragg reflections with the same $d$-spacing are superimposed into a single peak on a powder diffractogram. Moreover, reflections with nearly the same $d$-spacing may not be resolved due to resolution-broadening, even though the reflecting planes may be oriented in very different directions in the crystal.

[41]Hugo Rietveld, Dutch crystallographer (1932–2016).

In contrast to routine single-crystal diffraction, therefore, it is often not feasible to extract the integrated intensities of individual Bragg reflections from powder data for comparison to the structure factor amplitudes calculated from a structural model. Instead, a full pattern profile analysis must generally be performed. The method, first described by Rietveld[41] (1969), uses a least-squares fitting procedure to minimize the weighted squared difference between the measured diffraction profile and one calculated from a structural model,

$$\sum_i w_i \{y_i - y_i^c(p)\}^2,$$

where $y_i$ is the intensity in channel $i$ of the measured diffraction pattern, $y_i^c(p)$ is the calculated intensity in $i$, which depends on the variable parameters $p$ of the model. The statistical weight is $w_i = 1/\sigma_i^2$, the reciprocal of the variance of $y_i$. If the experimental uncertainty is dominated by counting statistics then $\sigma_i^2 = y_i$ is an appropriate choice (see Section 10.1.5).

In addition to the crystallographic parameters, the model includes refinable parameters to describe the shape and widths of the Bragg peaks, the background, and any texturing (preferred orientation) of the crystallites in the samples, if present. The form of the peak shape function will depend on the particular diffractometer and experiment configuration employed, as well as on the sample size and shape. Appropriate functions for a particular instrument, often determined from measurements on standard samples, are usually available and defined in an *instrument parameter file*. Examples of neutron powder diffraction patterns together with Rietveld refinements are given in Fig. 10.26.

A powerful feature of the Rietveld method is its ability to perform multi-phase refinements, i.e. to perform refinements on diffraction data from samples that contain more than one phase. This makes it possible to identify any secondary phases present in the mixture and to determine

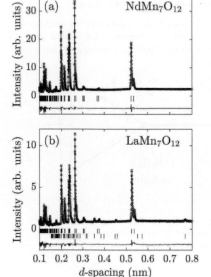

**Fig. 10.26** Neutron powder diffractograms and Rietveld structure refinements, after Johnson *et al.* (2018). (a) Single-phase refinement of $NdMn_7O_{12}$. The ticks indicate the peak positions, and the line under the ticks is the difference between the data and the calculated pattern. (b) Multi-phase refinement of $LaMn_7O_{12}$. Several small impurity peaks can be seen, and the second set of ticks corresponds to the peak positions for $LaMnO_3$, which makes up about 1% of the sample by weight. (Data courtesy of R. D. Johnson.)

the relative concentration of each phase. This capability is illustrated in Fig. 10.26. A number of searchable databases of crystal structures and related information exist to assist with phase identification. The International Union of Crystallography (IUCr) provides links to these, and much more besides.[42]

[42]http://www.iucr.org.

## 10.2.2 Debye–Waller factor and ADPs

When performing a crystal structure refinement, whether from single-crystal or powder data, there is a set of parameters to refine which describe the Debye–Waller factor (Sections 5.1.4 and 5.2). The Debye–Waller factor takes into account the effect of the random thermal motion (or indeed any source of random positional disorder) of the atoms on the intensities of the Bragg reflections, and has the form

$$\exp(-W) = \exp\{-\tfrac{1}{2}\langle(\mathbf{Q}\cdot\hat{\mathbf{u}})^2\rangle\}, \tag{10.61}$$

where $\hat{\mathbf{u}}$ is the operator for the displacement of the atom from its equilibrium position. The factor in the exponent can be written in matrix form as

$$\langle(\mathbf{Q}\cdot\hat{\mathbf{u}})^2\rangle = \mathbf{Q}^{\mathrm{T}}\mathsf{U}\mathbf{Q}, \tag{10.62}$$

where $\mathsf{U}$ is a $3 \times 3$ symmetric matrix whose components are $U_{11} = \langle\hat{u}_x^2\rangle$, $U_{12} = U_{21} = \langle\hat{u}_x\hat{u}_y\rangle$, etc. There are up to six independent components of $\mathsf{U}$. These are known as the *anisotropic displacement parameters*,[43] or ADPs for short, and can be treated as refinable parameters in the Rietveld method.

Equations (10.61) and (10.62) describe a general three-dimensional Gaussian (see Section B.1) in reciprocal space. The Fourier transform of this function is proportional to the probability distribution for the atomic displacements, which after normalization is given by[44]

$$P(\mathbf{r}) = \frac{|\mathsf{U}^{-1}|^{1/2}}{(2\pi)^{3/2}} \exp(-\tfrac{1}{2}\mathbf{r}^{\mathrm{T}}\mathsf{U}^{-1}\mathbf{r}). \tag{10.63}$$

This is a trivariate Gaussian whose shape can be visualized by a surface of constant probability. The 50% probability contour is often used for this purpose, and is the solution to

$$\mathbf{r}^{\mathrm{T}}\mathsf{U}^{-1}\mathbf{r} = 2\ln 2. \tag{10.64}$$

The surface defined by (10.64) is an ellipsoid, and is called the *displacement ellipsoid* or *thermal ellipsoid*. The shape and orientation of the ellipsoid will be determined by the local symmetry and bonding around the atom.

The ADPs introduced in (10.62) are referred to a Cartesian basis whose axes are parallel to the axes of $\mathbf{Q}$. These are not the ADPs found in standard structural refinement packages in crystallography.[45] Instead, the ADPs are usually defined with respect to the crystal axes, and are given in one of three ways, denoted by $\beta_{ij}$, $U_{ij}$ or $B_{ij}$. We shall give a brief explanation of the difference between these.

[43]Also known as *atomic displacement parameters*. Fortunately, the acronym is the same.

[44]To show this, write $\mathsf{U} = \mathsf{A}^{\mathrm{T}}\mathsf{D}\mathsf{A}$, where $\mathsf{D}$ is a diagonal matrix and $\mathsf{A}$ is the orthogonal matrix which diagonalizes $\mathsf{U}$ (n.b. for an orthogonal matrix $\mathsf{A}^{\mathrm{T}} = \mathsf{A}^{-1}$). Hence, $\mathbf{Q}^{\mathrm{T}}\mathsf{U}\mathbf{Q} = (\mathsf{A}\mathbf{Q})^{\mathrm{T}}\mathsf{D}\mathsf{A}\mathbf{Q} = \mathbf{P}^{\mathrm{T}}\mathsf{D}\mathbf{P}$, so that the components of $\mathbf{P}$ are the principal axes of $\mathsf{D}$ in the basis of $\mathbf{Q}$. The Fourier transform of the trivariate Gaussian $\exp(-\tfrac{1}{2}\mathbf{P}^{\mathrm{T}}\mathsf{D}\mathbf{P})$ factorizes into the product of three one-dimensional Gaussian Fourier transforms, one for each principal axis. By performing these Fourier transforms one arrives at (10.63).

[45]For a detailed discussion of ADPs, see Trueblood (1996).

[46] As elsewhere in this book, we use the convention $\mathbf{a} \cdot \mathbf{a}^* = 2\pi$ for the relation between the direct and reciprocal bases. In crystallography, $\mathbf{a} \cdot \mathbf{a}^* = 1$ is sometimes used instead.

In the crystal basis, we may write $\mathbf{u} = \Delta x \mathbf{a} + \Delta y \mathbf{b} + \Delta z \mathbf{c}$ and $\mathbf{Q} = h\mathbf{a}^* + k\mathbf{b}^* + l\mathbf{c}^*$, where $(\mathbf{a}, \mathbf{b}, \mathbf{c})$ and $(\mathbf{a}^*, \mathbf{b}^*, \mathbf{c}^*)$ are basis vectors in direct and reciprocal space,[46] respectively, for the cell on which the Bragg reflections are indexed. It then follows from (10.61) that

$$\exp(-W) = \exp(-\mathbf{h}^{\mathrm{T}} \boldsymbol{\beta} \mathbf{h}), \tag{10.65}$$

where $\mathbf{h}^{\mathrm{T}} = (h, k, l)$ and $\boldsymbol{\beta}$ is a second-rank symmetric tensor with components

$$\beta_{11} = 2\pi^2 \langle \Delta x^2 \rangle, \qquad \beta_{12} = 2\pi^2 \langle \Delta x \Delta y \rangle, \qquad \text{etc.} \tag{10.66}$$

This is the definition of the dimensionless ADPs denoted by $\beta_{ij}$.

Alternatively, one may write $\mathbf{u} = (\Delta\xi a^* \mathbf{a} + \Delta\eta b^* \mathbf{b} + \Delta\zeta c^* \mathbf{c})/2\pi$ and define the ADP tensor by

$$U_{11} = \langle \Delta\xi^2 \rangle, \qquad U_{12} = \langle \Delta\xi \Delta\eta \rangle \qquad \text{etc.} \tag{10.67}$$

This definition has the advantage that for a crystal lattice with orthogonal axes, the displacements $(\Delta\xi, \Delta\eta, \Delta\zeta)$ are just the components of the Cartesian displacement vector $\mathbf{u}$ along crystal axes. Hence, the $U_{ij}$ parameters, which have dimensions of $(\text{length})^2$, are the same as (for orthogonal axes), or closely related to, the mean squared displacement amplitudes $\langle \hat{u}_x^2 \rangle$, etc, defined with $x, y, z$ axes parallel to the crystal axes. The $U_{ij}$ and $\beta_{ij}$ parameters are related by

$$\beta_{11} = \tfrac{1}{2}(a^*)^2 U_{11}, \qquad \beta_{12} = \tfrac{1}{2}a^* b^* U_{12}, \qquad \text{etc.} \tag{10.68}$$

The third common form of ADPs is

$$B_{ij} = 8\pi^2 U_{ij}, \tag{10.69}$$

which is a simple scaling of the $U_{ij}$ parameters.

Finally, we mention that it is not always possible to refine the full set of ADPs allowed by symmetry if, for example, the anisotropy is small or the data are not of sufficient quality. In such cases one can assume an isotropic Debye–Waller factor, that is take $\langle \hat{u}_x^2 \rangle = \langle \hat{u}_y^2 \rangle = \langle \hat{u}_z^2 \rangle = \langle \hat{u}^2 \rangle / 3$. This allows us to define a single isotropic equivalent ADP

$$U_{\text{eq}} = \tfrac{1}{3}\langle \hat{u}^2 \rangle = \tfrac{1}{3}(U_{11} + U_{22} + U_{33}), \tag{10.70}$$

and from (10.69), $B_{\text{eq}} = 8\pi^2 U_{\text{eq}}$. The right-hand side of (10.70) depends only on the trace of $\mathsf{U}$, which is rotationally invariant. Note, though, that (10.70) holds only for a Cartesian basis as it will not give the correct isotropic average if the axes are scaled by different amounts. The approximate isotropic Debye–Waller factor is

$$\exp(-W) = \exp(-\tfrac{1}{2}U_{\text{eq}} Q^2). \tag{10.71}$$

### 10.2.3 Thermal diffuse scattering

The scattering from a crystalline material will normally be dominated by sharp Bragg peaks, but diffraction measurements may also exhibit an additional broad component called *thermal diffuse scattering* (TDS) which arises from inelastically scattered neutrons. As with absorption and extinction, TDS can limit the accuracy with which Bragg peak intensities are determined. The TDS occurs because the recorded intensity is an integral of all the scattering (elastic + inelastic) weighted by the probability function that describes the energy resolution of the instrument. The form of the TDS depends on many factors, but it will be particularly significant on instruments without energy transfer analysis, for example diffractometers with a monochromatic incident neutron beam but without an analyser in the scattered beam to reject neutrons whose energies change in the sample.[47]

The strongest TDS usually originates from low-energy acoustic phonons with wavevectors close to the reciprocal lattice vectors.[48] In this case, the TDS increases with temperature and takes the form either of a diffuse peak under the Bragg peaks or of enhanced scattering surrounding the Bragg peaks. Which of these two possibilities is actually observed depends on the ratio of the speed of sound in the crystal to the speed of the neutron (Willis and Pryor, 1975). In a few cases, the speed of sound has actually been determined from measurements of TDS (see Willis *et al.*, 1986, for example). Most studies of TDS, however, have been made in order to develop corrections that enable more accurate Bragg peak intensities to be obtained. That said, the neglect of TDS in structure refinements tends to result in errors in the atomic displacement parameters but has little influence on the refined atomic coordinates.

### 10.2.4 Ewald sphere construction

The *Ewald sphere construction* is a convenient way to understand the occurrence of Bragg diffraction in terms of the reciprocal lattice. The Bragg condition is described mathematically by the Laue equation (see Section 2.2.2),

$$\mathbf{Q} = \mathbf{k}_i - \mathbf{k}_f = \mathbf{G}, \tag{10.72}$$

which tells us that the reciprocal lattice vectors $\mathbf{G}$ are the $\mathbf{Q}$ vectors which satisfy the Bragg condition.[49]

The first task, therefore, is to construct the reciprocal lattice. The reciprocal lattice can be generated from $\mathbf{G} = h\mathbf{a}^* + k\mathbf{b}^* + l\mathbf{c}^*$, together with any selection rules on $h$, $k$, $l$. The reciprocal basis vectors $\mathbf{a}^*$, $\mathbf{b}^*$, $\mathbf{c}^*$ are related to the direct basis vectors $\mathbf{a}$, $\mathbf{b}$, $\mathbf{c}$ through eqns (2.20). Figure 10.27(a) shows a section through such a reciprocal lattice. In this particular example the section contains the vectors $\mathbf{a}^*$ and $\mathbf{b}^*$.

Suppose the incident wavevector is constrained to lie in the $\mathbf{a}^*$–$\mathbf{b}^*$ plane and you want to measure the 210 reflection. Choose any reciprocal lattice point as the origin $O$, and draw the reciprocal lattice vector $\mathbf{G}$ which connects $O$ to the point 210. On the same scale, construct the

[47] More details on the different designs of neutron diffractometers are given in Section 10.2.1.

[48] The scattering cross-section for phonons is discussed in Chapter 5.

[49] Crystallographers often define the scattering vector as $\mathbf{Q} = \mathbf{k}_f - \mathbf{k}_i$, i.e. the negative of that used here. This reverses the arrows on some of the vectors but does not make any changes of substance. We have not adopted the crystallography convention here because there are considerable advantages in preserving a consistent definition of $\mathbf{Q}$ throughout the book. We are certain that our readers have the mental agility to switch the sign of the vectors as and when necessary.

(a)   Reciprocal lattice

(b)   Direct lattice

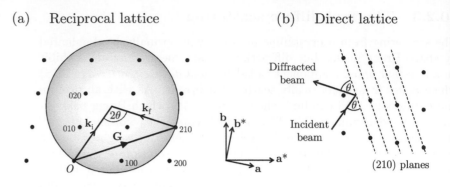

**Fig. 10.27** (a) Ewald sphere construction for measurement of the 210 reflection. (b) Crystal lattice showing diffraction from the (210) planes. The relation between the basis vectors of the direct and reciprocal lattices is shown in the centre.

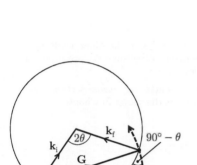

**Fig. 10.28** $\omega$-scan in which a reciprocal lattice point crosses the Ewald sphere.

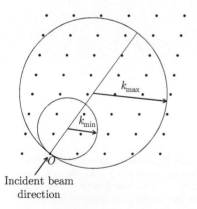

**Fig. 10.29** Ewald sphere construction for a time-of-flight scan with a quasi-continuous beam. The two circles are the sections through the Ewald spheres for the minimum and maximum wavevectors. There will be diffracted beams for each of the reciprocal lattice points contained in the shaded region.

scattering triangle corresponding to eqn (10.72). The triangle is isosceles since $k_i = k_f = 2\pi/\lambda$. The scattering angle $2\theta$ and the direction of $\mathbf{k}_i$ are now fully determined by geometry. In real space, the same diffraction process corresponds to specular reflection from the (210) planes of the crystal lattice, as shown in Fig. 10.27(b).

Now suppose we keep the same incident beam direction relative to the crystal but change the scattering angle. As $2\theta$ varies, the end point of the vector $\mathbf{Q}$ marks out a circle of radius $2\pi/\lambda$ centred on the end point of $\mathbf{k}_i$, as shown in Fig. 10.27(a). This circle represents the path in reciprocal space on which $\mathbf{Q}$ must lie. If we allow $\mathbf{k}_f$ to point out of the plane then the circle becomes a sphere, which is called the *Ewald sphere*. If a reciprocal lattice point lies on the surface of the Ewald sphere then diffraction will occur.

The measurement of Bragg reflections always involves some form of scan, as detailed in Section 10.2.1. On a fixed wavelength instrument, such as a four-circle diffractometer, we might perform an $\omega$-scan, i.e. rotate the crystal about a vertical axis so that it passes through the Bragg condition (also known as a *rocking curve*). In this type of scan the reciprocal lattice rotates around $O$ such that the reciprocal lattice point at which diffraction is to be measured moves along a circular path, as shown in Fig. 10.28. A peak in intensity is observed as the path crosses the surface of the sphere. The angle between the scan direction and the surface of the Ewald sphere in the scattering plane is $90° - \theta$.

On a time-of-flight instrument with a quasi-continuous beam, the crystal remains fixed while the neutron wavevector varies between some lower and upper limits $k_{\min}$ and $k_{\max}$. Figure 10.29 shows the intersection of the Ewald spheres for $k_{\min}$ and $k_{\max}$ with a plane of reciprocal space. As a function of time-of-flight, the radius of the Ewald sphere shrinks and the surface of the sphere passes through all the reciprocal lattice points contained in the volume between the limiting spheres. Therefore, the points in the shaded region correspond to allowed diffraction peaks. As illustrated in Fig. 10.23, reflections corresponding to lines of reciprocal lattice points that pass through $O$ are all observed at the same scattering angle.

## 10.2.5  Integrated intensities and the Lorentz factor

In this section we shall establish the connection between the differential cross-section for Bragg scattering (see Section 5.2.1),

$$\left(\frac{d\sigma}{d\Omega}\right) = N\frac{(2\pi)^3}{v_0}\sum_{\mathbf{G}}|F(\mathbf{G})|^2\delta(\mathbf{Q}-\mathbf{G}),\qquad(10.73)$$

where $N$ is the number of unit cells in the crystal, $v_0$ is the volume of the unit cell and $F(\mathbf{G})$ is the structure factor,[50] and the quantity that is actually measured in an experiment, which is usually the integrated intensity $I(\mathbf{G})$ of a peak in some type of scan. Specifically, we shall show how to obtain $|F(\mathbf{G})|^2$ from $I(\mathbf{G})$.

One way to do this is to make intensity measurements on a three-dimensional grid in $\mathbf{Q}$-space surrounding the Bragg peak. As the volume of the delta function in (10.73) is 1, the $\mathbf{Q}$-integrated cross-section is directly proportional to $|F(\mathbf{G})|^2$. This method, however, is inefficient, and it is usually preferable to measure Bragg peaks in some kind of scan. We shall consider here the most common such methods.

The diffracted beam described by eqn (10.73) is infinitely narrow, but in reality the beam is broadened by the effects of instrumental resolution and sample mosaic. If the diffractometer is set to measure a scattering vector $\mathbf{Q}_0$ in the neighbourhood of the reciprocal lattice vector $\mathbf{G}$ then the effective differential cross-section after folding in the instrumental resolution is given by [see eqn (10.43)]

$$\left(\frac{d\sigma}{d\Omega}\right)_{\text{eff}} = N\frac{(2\pi)^3}{v_0}\sum_{\mathbf{G}}|F(\mathbf{G})|^2\int\delta(\mathbf{Q}-\mathbf{G})R(\mathbf{Q}_0-\mathbf{Q})\,d^3\mathbf{Q}$$

$$= N\frac{(2\pi)^3}{v_0}\sum_{\mathbf{G}}|F(\mathbf{G})|^2R(\mathbf{Q}_0-\mathbf{G}).\qquad(10.74)$$

### Open detector

With an open detector, i.e. no collimation or analyser after the sample, all the scattered neutrons reach the detector and are counted. In this arrangement the detector integrates over a solid angle, and the measured count rate is

$$\Sigma = \Phi\int\left(\frac{d\sigma}{d\Omega}\right)_{\text{eff}}d\Omega.\qquad(10.75)$$

[50]The results derived here apply equally to nuclear and magnetic scattering. For nuclear scattering, $N$, $v_0$ and $F(\mathbf{G})$ refer to the chemical unit cell and $F(\mathbf{G})$ contains the nuclear scattering amplitudes, and for magnetic scattering they are replaced by the corresponding magnetic quantities, see Sections 7.2 and 7.3.

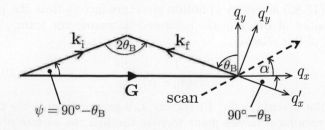

**Fig. 10.30** Reciprocal space coordinates for calculation of the integrated intensity of a peak in a wavevector scan.

This follows from the definition of the differential cross-section, eqn (1.31), $\Phi$ being the incident flux. We define

$$\mathbf{q} = \mathbf{Q}_0 - \mathbf{G} = \mathbf{k}_i - \mathbf{k}_f - \mathbf{G}, \qquad (10.76)$$

and introduce the Cartesian axes shown in Fig. 10.30. When $\mathbf{Q}_0 = \mathbf{G}$ the $q'_x$ axis is along $-\mathbf{k}_f$ and $q'_y$ and $q'_z$ are tangential to the Ewald sphere. If the dimensions of the resolution ellipsoid are sufficiently small we may neglect the curvature of the Ewald sphere and write

$$d\Omega = \frac{dk'_{fy}\,dk'_{fz}}{k_f^2} = \frac{dq'_y\,dq'_z}{k^2}, \qquad (10.77)$$

so that from (10.74)–(10.77) we have

$$\Sigma(q'_x) = \Phi N \frac{(2\pi)^3}{v_0} |F(\mathbf{G})|^2 \frac{1}{k^2} \int\!\!\int R(\mathbf{q}')\,dq'_y\,dq'_z, \qquad (10.78)$$

where $R(\mathbf{q}')$ is the resolution function in the $(q'_x, q'_y, q'_z)$ coordinates.

Now consider a scan along a path in the scattering plane inclined at an angle $\alpha$ to the $q_x$ axis (which is parallel to $\mathbf{G}$), as shown in Fig. 10.30. If $s$ is the distance from the reciprocal lattice point along this path then

$$\begin{aligned} q'_x &= s\cos(90° - \theta_B + \alpha) \\ &= s\sin(\theta_B - \alpha), \end{aligned} \qquad (10.79)$$

where $\theta_B$ is the Bragg angle. From (10.78) and (10.79), the integrated intensity of the peak in $\Sigma(q'_x)$ along the scan path is

$$\begin{aligned} I_s(\mathbf{G}) &= \int \Sigma(q'_x)\,ds \\ &= \frac{1}{|\sin(\theta_B - \alpha)|} \int \Sigma(q'_x)\,dq'_x \\ &= \Phi N \frac{(2\pi)^3}{v_0} |F(\mathbf{G})|^2 \frac{1}{k^2} \frac{1}{|\sin(\theta_B - \alpha)|} \int R(\mathbf{q}')\,d^3\mathbf{q}' \\ &= \Phi \frac{2\pi V}{v_0^2} \lambda^2 |F(\mathbf{G})|^2 \frac{1}{|\sin(\theta_B - \alpha)|}, \end{aligned} \qquad (10.80)$$

where $V = Nv_0$ is the volume of the crystal, and $k = 2\pi/\lambda$. We have used the fact that $R(\mathbf{q}')$ is normalized to unity, see eqn (10.44). The modulus in the denominator ensures that the integrated intensity is positive.

Equation (10.80) allows us to obtain structure factors from the integrated intensities of Bragg peaks measured in wavevector scans. The angular part of this relation,

$$L = \frac{1}{|\sin(\theta_B - \alpha)|}, \qquad (10.81)$$

is known as the *Lorentz factor*. Physically, $L$ is proportional to the time it takes the reciprocal lattice point to pass through the surface of the

Ewald sphere,[51] which is where diffraction occurs. Hence, a scan that is tangential to the Ewald sphere ($\alpha = \theta_B$) has a very large integrated intensity, whereas a scan along the radius ($\theta_B - \alpha = 90°$) has a relatively small integrated intensity. The inverse Lorentz factor $L^{-1}$ corrects the measured integrated intensities for this effect.

We shall now derive formulae equivalent to (10.80) and (10.81) for several other standard types of scan.

[51] The surface of the Ewald sphere has some thickness due to the effects of resolution (i.e. the non-zero wavelength spread and divergence of the incident beam, and the sample mosaic).

*(1) $\omega$ and $\omega$–$2\theta$ scans*

In an $\omega$-scan, the sample rotates about an axis perpendicular to the scattering plane so that the path in reciprocal space makes an angle $\alpha = \pm 90°$ to $\mathbf{G}$ (Fig. 10.31). A small step along the path is given by

$$ds = G d\omega$$
$$= 2k \sin\theta_B \, d\omega, \tag{10.82}$$

and using eqn (10.79) with $|\sin(\theta_B \pm 90°)| = \cos\theta_B$ we find

$$d\omega = \frac{dq_x'}{2k \sin\theta_B \cos\theta_B} = \frac{dq_x'}{k \sin 2\theta_B}. \tag{10.83}$$

By following the steps in (10.80) we obtain the integrated intensity

$$I_\omega(\mathbf{G}) = \int \Sigma(q_x') \, d\omega$$
$$= \Phi \frac{V}{v_0^2} \lambda^3 |F(\mathbf{G})|^2 \frac{1}{\sin 2\theta_B} \qquad (\omega\text{-scan}). \tag{10.84}$$

In an $\omega$–$2\theta$ scan, the incident beam direction is fixed and the detector rotates at twice the angular speed of the sample, so the angles $\omega$ and $\theta$ are the same apart from a zero offset. This means that the specular reflection condition for the Bragg planes is maintained throughout the scan, and so the scan follows a longitudinal path in reciprocal space, i.e. $\alpha = 0°$ or $180°$ (Fig. 10.31). The scan variable $s$ is proportional to $Q_0 = 2k \sin\theta$, and so when the scan is close to the Bragg condition

$$ds = dQ_0 = 2k \cos\theta_B \, d\omega. \tag{10.85}$$

This, together with $dq_x' = ds \sin\theta_B$ from (10.79), gives

$$d\omega = \frac{dq_x'}{k \sin 2\theta_B}, \tag{10.86}$$

and hence,

$$I_{\omega-2\theta}(\mathbf{G}) = \Phi \frac{V}{v_0^2} \lambda^3 |F(\mathbf{G})|^2 \frac{1}{\sin 2\theta_B} \qquad (\omega\text{–}2\theta \text{ scan}), \tag{10.87}$$

the same as for the $\omega$-scan, eqn (10.84). In both types of angle scan considered here, therefore, the Lorentz factor is seen to be[52]

$$L = \frac{1}{\sin 2\theta_B}. \tag{10.88}$$

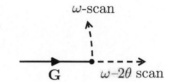

$\omega$-scan

$\mathbf{G}$     $\omega$–$2\theta$ scan

**Fig. 10.31** Diagram showing the direction in reciprocal space of an $\omega$-scan and an $\omega$–$2\theta$ scan.

[52] Lorentz factors for more general angular scans have been given by McIntyre and Stansfield (1988).

### (2) Laue method

In the time-of-flight Laue method (Section 10.2.1), the sample is fixed relative to $\mathbf{k_i}$ and all that changes is $\lambda$. At a fixed scattering angle $2\theta$ the scattering vector $\mathbf{Q_0}$ makes a radial scan in reciprocal space ($\alpha = 0°$ or $180°$), as shown above in Fig. 10.23. Hence,

$$ds = dQ_0 = -\frac{4\pi}{\lambda^2}\sin\theta_B\, d\lambda, \tag{10.89}$$

and so with $dq'_x = ds\sin\theta_B$ from (10.79) we have

$$d\lambda = -\frac{\lambda^2\, dq'_x}{4\pi\sin^2\theta_B}. \tag{10.90}$$

Let $\phi(\lambda)d\lambda$ be the flux of incident neutrons with wavelengths between $\lambda$ and $\lambda + d\lambda$. Replacing $\Phi$ with $\phi(\lambda)d\lambda$ in (10.78) and converting the integral over $\lambda$ into an integral over $dq'_x$ with (10.90), we obtain

$$I_\lambda(\mathbf{G}) = \int \Sigma(q'_x)\, d\lambda$$

$$= \phi(\lambda)\frac{V}{v_0^2}\lambda^4|F(\mathbf{G})|^2\frac{1}{2\sin^2\theta_B}. \tag{10.91}$$

The minus sign has been dropped to ensure that the integrated intensity is positive. The value of $\lambda$ to be used in (10.91) is that which satisfies the Bragg condition $G = (4\pi/\lambda)\sin\theta_B$.

### (3) Powder method

Consider first a constant-wavelength powder diffractometer. We do not strictly perform a scan, but the random distribution of grain orientations has the same effect as a crystal rotation scan ($\omega$-scan). The main difference is that only a fraction of the crystal grains contribute.

A grain diffracts into a Debye–Scherrer cone if the angle between $\mathbf{G}$ and $\mathbf{k_i}$ is $\psi = 90° - \theta_B$ (Fig. 10.30). The fraction of grains whose orientation lies in the range $\psi$ to $\psi + d\psi$ is $\frac{1}{2}\sin\psi\, d\psi = \frac{1}{2}\cos\theta_B\, d\theta$ (Fig. 10.32). Hence, the total intensity in the Debye-Scherrer cone for a particular $\mathbf{G}$ is

$$I_{\text{cone}}(\mathbf{G}) = \int \Sigma(q'_x)\tfrac{1}{2}\cos\theta_B\, d\theta, \tag{10.92}$$

where $\Sigma(q'_x)$ is given by eqn (10.78). Variation of $\psi$ (or $\theta$) has the same effect as an $\omega$-scan, so $d\theta$ is related to $dq'_x$ by eqn (10.83), giving

$$I_{\text{cone}}(\mathbf{G}) = \frac{1}{4k\sin\theta_B}\int \Sigma(q'_x)\, dq'_x. \tag{10.93}$$

All planes with the same $d$-spacing diffract into the same Debye-Scherrer cone, so the total intensity is given by

$$I_{\text{cone}}(G) = \sum_{|\mathbf{G}|=G} I_{\text{cone}}(\mathbf{G})$$
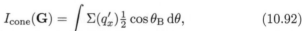
$$= \Phi\frac{V}{v_0^2}\frac{\lambda^3}{4\sin\theta_B}\sum_{|\mathbf{G}|=G}|F(\mathbf{G})|^2, \tag{10.94}$$

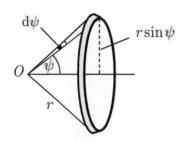

**Fig. 10.32** Calculation of the fraction of directions between polar angles $\psi$ and $\psi + d\psi$. The shaded strip resides on the surface of a sphere of radius $r$ and total surface area $4\pi r^2$. The area of the strip is $2\pi r^2\sin\psi\, d\psi$, which is a fraction $\frac{1}{2}\sin\psi\, d\psi$ of the total surface area of the sphere.

where the summation includes all reciprocal lattice vectors $\mathbf{G}$ having length $G = 2k \sin\theta_B$.[53]

Constant-wavelength powder diffractometers usually have a bank of detectors of constant height $h$ at a fixed distance $r$ from the sample. A detector at a scattering angle $2\theta_B$ will intercept only a fraction $h/(2\pi r \sin 2\theta_B)$ of the Debye–Scherrer cone, so will record an intensity

$$I_{\text{bank}}(G) = \Phi \frac{V}{v_0^2} \frac{h}{2\pi r} \frac{\lambda^3}{4 \sin 2\theta_B \sin \theta_B} \sum_{|\mathbf{G}|=G} |F(\mathbf{G})|^2. \quad (10.95)$$

The calculation for a time-of-flight powder diffractometer is the same except $\Phi$ is replaced by $\phi(\lambda)\mathrm{d}\lambda$ and the integration is over $\lambda$. Differentiating Bragg's law in the vicinity of the Bragg condition we obtain

$$\mathrm{d}\lambda = 2d\cos\theta_B\, \mathrm{d}\theta$$
$$= \lambda \frac{\cos\theta_B}{\sin\theta_B}\, \mathrm{d}\theta$$
$$= \frac{\lambda^2}{4\pi} \frac{\mathrm{d}q_x'}{\sin^2\theta_B}, \quad (10.96)$$

and continuing as above,

$$I_{\text{cone}}(G) = \phi(\lambda)\frac{V}{v_0^2} \frac{\lambda^4 \cos\theta_B}{4\sin^2\theta_B} \sum_{|\mathbf{G}|=G} |F(\mathbf{G})|^2 \quad \text{(TOF)}, \quad (10.97)$$

and

$$I_{\text{bank}}(G) = \phi(\lambda)\frac{V}{v_0^2} \frac{h}{2\pi r} \frac{\lambda^4}{8\sin^3\theta_B} \sum_{|\mathbf{G}|=G} |F(\mathbf{G})|^2 \quad \text{(TOF)}. \quad (10.98)$$

**Three-axis diffractometer**

A diffractometer with an open detector is not always the best instrument for elastic scattering measurements. If the scattering is weak then a better signal-to-background ratio can be achieved on a triple-axis spectrometer (Section 10.3.2). The use of an analyser to select the elastic scattering, sometimes in conjunction with collimators before and/or after the analyser, can suppress unwanted inelastic scattering and improve wavevector resolution, albeit at the cost of intensity.

With an analyser in place, not all the elastically scattered neutrons reach the detector, and so the detector does not integrate the scattering over solid angle as does an open detector. The intensity in a single measurement is proportional to the resolution-folded differential cross-section given in eqn (10.74). The resolution function of a triple-axis spectrometer may be written in the form of eqn (10.44). For simplicity, we shall assume that $\mathbf{k}_i$, $\mathbf{k}_f$, and the scan direction all lie in a plane, in which case a two-dimensional resolution function can be used

$$R(\mathbf{Q}_0 - \mathbf{Q}) = \frac{|\mathsf{M}|^{1/2}}{2\pi} \exp(-\tfrac{1}{2}M_{11}q_x^2 - M_{12}q_xq_y - \tfrac{1}{2}M_{22}q_y^2), \quad (10.99)$$

[53]The sum is often replaced by $m(\mathbf{G})|F(\mathbf{G})|^2$, where $m(\mathbf{G})$ is the multiplicity of the Bragg reflection described by $\mathbf{G}$. When there are two or more sets of reflections with the same $G$, each set unrelated to the others by symmetry, then you must sum $m_j(\mathbf{G})|F_j(\mathbf{G})|^2$ over all the distinct sets $j$.

[54]Explicit expressions can be found in Shirane *et al.* (2002), and Popovici (1975).

where $|\mathsf{M}| = M_{11}M_{22} - M_{12}^2$. The axes are those shown in Fig. 10.30. In general, the $M_{ij}$ coefficients are functions of $\theta$.[54] From (10.74), the integrated intensity of a peak in a linear scan in reciprocal space passing through **G** may be written

$$I_s(\mathbf{G}) = C|F(\mathbf{G})|^2 \int R(\mathbf{Q}_0 - \mathbf{G})\,\mathrm{d}s, \qquad (10.100)$$

where $C$ is a constant. If the scan makes an angle $\alpha$ with the $q_x$ axis, as shown in Fig. 10.30, then along the scan

$$q_x = s\cos\alpha, \qquad q_y = s\sin\alpha. \qquad (10.101)$$

Combining (10.99)–(10.101), and doing the Gaussian integral, we find

$$I_s(\mathbf{G}) = \frac{C|F(\mathbf{G})|^2}{(2\pi)^{1/2}} \frac{|\mathsf{M}|^{1/2}}{(M_{11}\cos^2\alpha + 2M_{12}\cos\alpha\sin\alpha + M_{22}\sin^2\alpha)^{1/2}}. \qquad (10.102)$$

Hence, the Lorentz factor is

$$L = \frac{|\mathsf{M}|^{1/2}}{(M_{11}\cos^2\alpha + 2M_{12}\cos\alpha\sin\alpha + M_{22}\sin^2\alpha)^{1/2}}. \qquad (10.103)$$

## 10.2.6 Extinction and multiple scattering

### Extinction

The formulae given in Section 10.2.5 are based on the kinematical approximation, which assumes that there is no reduction in the incident beam intensity inside the sample due to scattering, and no reduction in the diffracted intensity from re-scattering. These assumptions generally hold well in diffraction from powders and very small single crystals, but Bragg peak intensities measured from larger single crystals are often systematically smaller than predicted in the kinematical approximation. This effect is known as *extinction*, and is illustrated in Fig. 10.33.

The causes of extinction were outlined in Section 9.4.2 within the frame of the mosaic crystal model, which views a single crystal as a collection of perfect-crystal blocks in slightly different orientations. In this model there are two extreme types of extinction: (i) *Primary extinction*, due to coherent multiple-beam interference that takes place within a single mosaic block and is explained by the dynamical theory of diffraction (Section 9.4); (ii) *Secondary extinction*, which occurs when the diffracted beam from one mosaic block is re-diffracted by other mosaic blocks.

Primary extinction is only important if the individual perfect-crystal blocks are larger than a certain *extinction length*, given by eqn (9.97). Secondary extinction is most significant for large crystals with small *mosaic spread*[55] and negligible absorption. Crystals with small mosaic blocks and sufficient mosaic spread to make re-diffraction negligible are said to be *ideally imperfect*, meaning good enough for diffraction studies but not so good as to suffer from significant extinction.

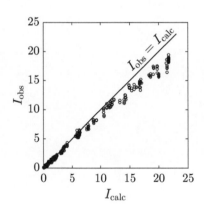

**Fig. 10.33** Observed vs calculated diffraction intensities from a single-crystal structure refinement of $\mathrm{ErNi_2^{11}B_2C}$. Extinction corrections have not been applied, which accounts for the deviation from the line. (Data courtesy of J. P. Barratt and G. J. McIntyre.)

[55]The mosaic spread is the width of the distribution of mosaic block orientations.

Various methods have been proposed to calculate extinction. The most widely used approach is based on the mosaic block model and involves the solution of a set of intensity transfer equations. A formulation of this method presented by Becker and Coppens (1974) forms the basis for the extinction corrections included in many crystal structure refinement procedures. We shall give a brief account of its main results.

The overall effect of extinction is described by a coefficient $y$ which relates the integrated intensity of a Bragg reflection calculated with extinction, $I(\mathbf{G})$, to that calculated in the kinematic approximation, $I_k(\mathbf{G})$:

$$I(\mathbf{G}) = y I_k(\mathbf{G}), \tag{10.104}$$

where $\mathbf{G}$ is the reciprocal lattice vector corresponding to the reflection.[56] Expressions for $I_k(\mathbf{G})$ for various types of diffraction scan are given in Section 10.2.5. The theory assumes that primary and secondary extinction coexist and are independent, in which case

$$y = y_p y_s, \tag{10.105}$$

where $y_p$ and $y_s$ are the coefficients for primary and secondary extinction, respectively. In the case of a spherical crystal containing idealized spherical crystal blocks, both coefficients can be satisfactorily described by the analytic form

$$y_i = \left\{ 1 + 2x_i + \frac{A_i(\theta_B)x_i^2}{1 + B_i(\theta_B)x_i} \right\}^{-1/2}, \qquad i = (\text{p, s}). \tag{10.106}$$

Here, $\theta_B$ is the Bragg angle, and approximate analytic expressions for $A_i(\theta)$ and $B_i(\theta)$ are given by Becker and Coppens (1974). The dimensionless variable $x_p$ is given by[57]

$$x_p = \frac{3\lambda^2 r^2}{2v_0^2} |F(\mathbf{G})|^2, \tag{10.107}$$

where $r$ is the radius of the spherical crystallites. For a spherical crystal the $x_s$ variable may be written

$$x_s = \frac{\lambda^3 \overline{T}}{v_0^2 \sin 2\theta_B} f(r,g)|F(\mathbf{G})|^2, \tag{10.108}$$

where $\overline{T}$ is the absorption-weighted mean path through the crystal, and $f(r,g)$ is a function which depends on the form of the distribution of misalignment angles of the mosaic blocks. Models often use a Gaussian or Lorentzian distribution. For a Lorentzian,

$$f(r,g) = \frac{gr \sin 2\theta_B}{g\lambda + r \sin 2\theta_B}, \tag{10.109}$$

where $g$ is essentially the inverse width of the mosaic distribution.[58] The expression for $\overline{T}$ is

$$\overline{T} = \frac{1}{AV} \int_V (L_i + L_f) \exp\{-\mu(L_i + L_f)\}\, dV, \tag{10.110}$$

[56]In other words, to apply the extinction correction to diffraction data one multiplies the measured integrated intensities by $1/y$.

[57]When $x_p \ll 1$ one can expand the function in (10.106) to give $y_p \simeq 1 - x_p$ which, after substituting for $x_p$ from (10.107), has the same form (apart from a numerical prefactor) as the result obtained from dynamical diffraction theory for a perfect slab crystal, eqns (9.97)–(9.98). To see this, one must replace the factor $d/\cos\theta_B$, which is the path length in the slab, by $3r/2$, the mean path length through a sphere.

[58]Small $g$ means large mosaic spread.

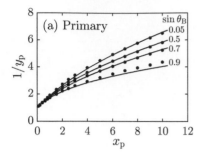

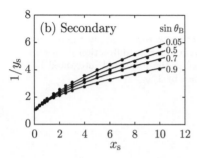

**Fig. 10.34** Correction factors for (a) primary and (b) secondary extinction calculated by the method of Becker and Coppens (1974). The curves are for four different values of $\sin\theta_B$. The points are 'exact' values from numerical integration, and the lines are from the approximate analytic formula (10.106).

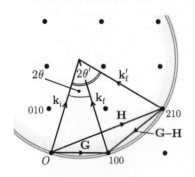

**Fig. 10.35** Reciprocal space diagram to illustrate the condition for multiple diffraction. The reciprocal lattice points corresponding to the 100 and 210 reflections both lie on the Ewald sphere, so both reflections satisfy the diffraction condition simultaneously.

which is similar to eqn (10.5) for the mean attenuation factor $A$. In fact, the denominator of $\overline{T}$ contains $A$, as well as the crystal volume $V$.

Figure 10.34 shows the behaviour of the functions (10.106) for primary and secondary extinction. The inverse extinction coefficient $1/y_p$ is seen to increase monotonically with $x_p$, Fig. 10.34(a), and the behaviour of $1/y_s$ is qualitatively similar, Fig. 10.34(b).

Equations (10.107) and (10.108) for $x_p$ and $x_s$ suggest some practical steps that can be taken in order to control the effects of extinction. Ideally, one should aim to measure a crystal which is roughly spherical in shape because, strictly speaking, the formulae apply only to mosaic crystals which are macroscopically spherical and which have monodisperse spherical crystallites. In practice, however, the spherical approximation can be adequate for crystals of other shapes too providing the extinction is not too large. Secondary extinction decreases with decreasing mean path length, so small crystals are best as long as they give sufficient diffraction intensity (1–3 mm is usually ideal). When fitting the data to a structural model one can optimize the extinction model by allowing the two parameters $r$ and $g$ to vary during the refinement. Inclusion of data recorded at several wavelengths, to exploit the strong wavelength dependence of $x_p$ and $x_s$, can also help, as can the omission from the refinement of reflections that suffer from particularly large extinction.

For very accurate work the extinction model should be extended to take into account the crystal shape, the anisotropy of the mosaic blocks, and the true angular and size distributions of the mosaic blocks. Such calculations, however, could be very laborious.

## Multiple Bragg diffraction

Multiple diffraction (also known as the *Renninger effect*) occurs when two or more sets of crystal planes satisfy the Bragg condition for the same incident wavevector. The process is illustrated in Fig. 10.35. Suppose the diffractometer is set up to measure the 100 reflection, which has scattering angle $2\theta$ and is associated with the reciprocal lattice vector $\mathbf{G}$. Now suppose that by coincidence another reciprocal lattice point, say the 210 (scattering angle $2\theta'$, reciprocal lattice vector $\mathbf{H}$), also lies on the surface of the Ewald sphere. In other words, both the 100 and 210 reflections satisfy the Bragg condition simultaneously. This double diffraction will affect the observed intensity of the desired 100 reflection, making it either lower or higher than expected. It could be lower because intensity will be removed from the incident beam by both reflections instead of just the 100, and it could be higher because the beam diffracted along $\mathbf{k}_f'$ from the 210 reflection becomes the incident beam for the $\bar{1}\bar{1}0$ reflection (reciprocal lattice vector $\mathbf{G} - \mathbf{H}$), which produces a doubly diffracted beam along $\mathbf{k}_f$. If the 100 is a systematic absence in the space group then the latter effect can be particularly confusing as it causes a peak to be measured at a position where one should not exist.

The condition for two or more reflections to satisfy the Bragg condition simultaneously is not as unlikely as one might think. The secondary refections can lie anywhere on the surface of the Ewald sphere, not just in the scattering plane of the primary reflection, and moreover, reciprocal lattice points do not have to be exactly at the surface for diffraction to occur. They just need to be within $\Delta k$ of the surface, where $\Delta k$ corresponds to the instrumental resolution. A further consideration is that multiple diffraction is most serious when there is a very large number of reciprocal lattice points inside the Ewald sphere, i.e. when the neutron wavelength is much smaller than the lattice parameters of the crystal.

One way to identify and avoid the effects of multiple diffraction is to measure the Bragg peak intensity as the crystal is rotated about the normal to the Bragg plane. This is called an *azimuthal scan*. In reciprocal space the azimuthal scan corresponds to a rotation of the reciprocal lattice about the scattering vector. For the setting shown in Fig. 10.35 this would be a rotation around $\mathbf{G}$, which would take the 210 reciprocal lattice point off the surface of the Ewald sphere while keeping the 100 point on it. If it is impossible to perform an azimuthal scan then another option is to change $\lambda$ and $2\theta$ together so as to maintain the Laue condition for $\mathbf{G}$ but move off the multiple diffraction condition.

## 10.2.7   Inelasticity corrections for $S(Q)$

Neutron diffraction experiments on liquids and non-crystalline solids aim to measure $S(Q)$, which is defined by

$$S(Q) = \int_{-\infty}^{\infty} S(Q, \omega) \, \mathrm{d}(\hbar\omega). \tag{10.111}$$

One can take the static approximation (see Section 3.8.4) as a starting point, i.e assume that the atoms are stationary. The measured diffraction intensity is then directly proportional to $S(Q)$. In reality, however, thermal agitation and recoil effects mean that atoms move during the scattering process, causing inelastic scattering. The intensity in a diffraction measurement may then differ significantly from the static approximation, and if the diffraction instrument does not have energy analysis it might be necessary to apply *inelasticity corrections*. Such corrections can be important for an accurate separation of the interference scattering, which contains the desired structural information, from the essentially featureless single-atom (or 'self') scattering.

In a diffraction experiment without energy analysis all neutrons scattered into to the detector are recorded irrespective of their final energy. If the incident energy $E_i$ and scattering angle $\phi$ are fixed, then the measurement yields an effective structure factor $S_{\text{eff}}$ given by

$$S_{\text{eff}}(Q_0) = \int_0^{\infty} \frac{\epsilon(E_f)}{\epsilon(E_i)} \frac{\mathrm{d}^2\sigma}{\mathrm{d}\Omega\mathrm{d}E_f} \, \mathrm{d}E_f$$

$$= \int_{-\infty}^{E_i} \frac{\epsilon(E_f)}{\epsilon(E_i)} \frac{k_f}{k_i} S(Q, \omega) \, \mathrm{d}(\hbar\omega), \tag{10.112}$$

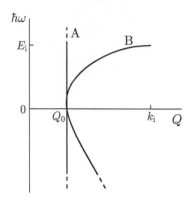

**Fig. 10.36** Integration paths in $Q$–$\omega$ space. Integration of $S(Q,\omega)$ along path A (constant $Q$) gives $S(Q)$, whereas a diffraction experiment which has fixed $E_i$ and fixed scattering angle measures the integral along path B.

[59] Because the lighter the mass, the greater the recoil.

[60] In other words, $\epsilon(E) \propto 1/E^{1/2}$. The result for a general $\epsilon(E)$ is given by Yarnell *et al.* (1973).

[61] See Exercise 3.5.

where $Q_0 = 2k_i \sin(\phi/2)$ is the nominal elastic scattering vector for $E_f = E_i$, and $\epsilon(E)$ is the (energy-dependent) detector efficiency.

In the static approximation $S(Q,\omega) = S(Q)\delta(\hbar\omega)$, in which case eqn (10.112) gives $S_{\text{eff}}(Q_0) = S(Q_0)$. In general, however, expression (10.112) differs in several important ways from (10.111). First, the integration in (10.112) is cut off at $E_i$ because this is the maximum amount of energy that neutrons can transfer to the sample. Second, the integral is weighted by the detector efficiency and the $k_f/k_i$ factor in the cross-section. Third, the energy integral in (10.111) is performed at constant $Q$ whereas that in (10.112) is performed at constant scattering angle, so the integration paths in $Q$–$\omega$ space are not the same.

The latter point is illustrated in Fig. 10.36. The two integration paths almost coincide when $|\hbar\omega| \ll E_i$, but they differ increasingly as $\hbar\omega$ becomes a significant fraction of $E_i$. The difference becomes greater with increasing scattering angle. In practical terms, therefore, high incident energies and small scattering angles are best to minimize inelasticity corrections. As far as the sample is concerned, the static approximation will work best when $\hbar\omega_{\max} \ll E_i$, where $\hbar\omega_{\max}$ is a characteristic energy above which $S(Q,\omega)$ becomes negligibly small. In general, $\hbar\omega_{\max}$ depends on temperature and on the atomic mass of the atoms,[59] so inelasticity corrections are largest for high temperature and small masses.

Inelasticity corrections in diffraction were first calculated by Placzek (1952). Placzek's method applies when $\hbar\omega_{\max} \ll E_i$, and involves an expansion of the integrand in (10.112) in powers of $x = \hbar\omega/E_i$. For the particular case of a detector whose efficiency varies inversely with neutron velocity,[60] the result is

$$S_{\text{eff}}(Q_0) = S(Q_0) - \frac{Q_0^2}{2E_i} S_1'(Q_0)$$
$$+ \frac{1}{8E_i^2}\{(2k_i^2 - Q_0^2)S_2'(Q_0) + Q_0^4 S_2''(Q_0)\} + \ldots, \qquad (10.113)$$

where $S_n(Q)$ is the $n$th energy moment of $S(Q,\omega)$, eqn (3.69), and the primes indicate differentiation with respect to $Q^2$.

The Placzek correction formula (10.113) applies to all terms that make up the structure factor, but the correction is most important for the 'self' part, i.e. the first term in eqn (2.52). This is because in the interference scattering (the 'difference' term) the impact of the neutron interaction is distributed over all the atoms in the coherence volume, whereas in the 'self' part the interaction is with individual nuclei.

To estimate the inelasticity correction for the 'self' term in $S(Q)$ we can use results from the dynamical structure factor of an ideal gas at temperature $T$. The first and second energy moments are given by[61]

$$S_1(Q) = E_r, \qquad S_2(Q) = 2E_r k_B T + E_r^2, \qquad (10.114)$$

where $E_r = \hbar^2 Q^2/2M$ is the recoil energy of an atom of mass $M$. By substituting these into (10.113) we find

$$S_{\text{eff}}(Q_0) = S(Q_0) + \frac{m_n}{M} \left\{ \frac{k_B T}{2E_i} - \frac{Q_0^2}{2k_i^2} \left( 1 + \frac{k_B T}{2E_i} \right) \right\} + O\left( \frac{m_n^2}{M^2} \right). \tag{10.115}$$

For a multicomponent liquid, sum the correction over all the atomic species with the concentrations of each species as weights.

For a time-of-flight diffractometer, the integral over the double differential cross-section in eqn (10.112) is over a path of constant time-of-flight. Both the incident and final wavevectors vary along this path, which depends on the ratio of the final to initial flight path lengths. In addition, the incident flux of neutrons depends on $E_i$. Calculation of the inelasticity correction by the Placzek moment expansion method is more laborious for time-of-flight diffraction than for the fixed $E_i$ case, but the result is available, see e.g. Howe *et al.* (1989).

When the moment expansion is unsuitable, e.g. for light atoms, an alternative approach is needed. One approach is to use an *ad hoc* fitting function to fit the single-atom scattering and thereby remove it from the interference scattering. Another is to identify a reasonably realistic model for $S(Q,\omega)$ and integrate it numerically over the relevant paths in $Q$–$\omega$ space. A model scattering function based on the exact solution for a simple harmonic oscillator has been proposed, which is found to be applicable over a wide range of $Q$ and $\omega$ and which can be used to calculate inelasticity corrections for both the 'self' and 'difference' parts of $S(Q)$, see Soper (2009).

# 10.3  Practical aspects of spectroscopy

Neutron spectroscopy (or neutron inelastic scattering) is the technique used to investigate excitations. In general terms, a neutron spectrometer is an instrument with energy analysis which measures the scattering rate from a sample as a function of neutron energy and momentum transfer. The data are directly related to the scattering function $S(\mathbf{Q},\omega)$. The dynamic range accessible to neutron spectroscopy is considerable, covering $10^{-8} \lesssim \hbar\omega \lesssim 100\,\text{eV}$ and $10^{-2} \lesssim Q \lesssim 1000\,\text{nm}^{-1}$, see Fig. 10.37. This encompasses a very wide variety of dynamic processes in condensed matter, extending from diffusive motions of large molecules, through propagating lattice and magnetic waves in crystalline solids, to bond vibrations and intra-atomic transitions.[62]

## 10.3.1  Types of spectrometer

No single design of instrument can access the entire $(\mathbf{Q},\omega)$ space spanned by neutron scattering. In this section we shall briefly describe some of the more widely used types of neutron spectrometer, specifically: (i) the triple-axis spectrometer — a continuous-source instrument which allows the incident and scattered neutron energies and the scattering angle to be selected so as to concentrate measurement time on a relatively

[62] Energy transfer and time are Fourier conjugate variables ($\omega \sim 2\pi/t$), and so are momentum transfer and distance ($Q \sim 2\pi/d$). One can therefore convert the $Q$–$\omega$ space spanned by neutron scattering into a conjugate space of length and time scales, as is done in Fig. 10.37.

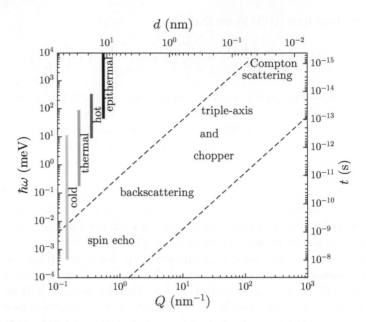

**Fig. 10.37** Indicative regions of momentum transfer ($Q$) and energy transfer ($\hbar\omega$) accessible to different types of neutron spectrometer. The energy range covered by different moderator types is shown on the left. The corresponding length ($d$) and time ($t$) scales of the correlations are also shown. The boundary of the region accessible to neutron energy-loss scattering is marked by the broken lines.

local region in $(\mathbf{Q}, \omega)$ space; (ii) time-of-flight spectrometers, which survey large volumes of $(\mathbf{Q}, \omega)$ space through use of two-dimensional multi-detectors; (iii) backscattering spectrometers, which are designed to achieve very high-energy resolution. Two other types of spectrometer were discussed in Chapter 5 in association with particular methods: neutron Compton scattering, which involves scattering of very high-energy neutrons typically in excess of $1\,\mathrm{eV}$ to measure atomic momentum distributions (Section 5.5), and neutron spin echo spectroscopy, which uses polarizing optics to measure slow dynamics in the time domain (Section 5.9). The range of $(\mathbf{Q}, \omega)$ space covered by different instruments is illustrated in Fig. 10.37.

## 10.3.2   Triple-axis spectrometers

[63]Bertram N. Brockhouse (1918–2003), Canadian physicist who shared the 1994 Nobel Prize in Physics for the development of neutron spectroscopy.

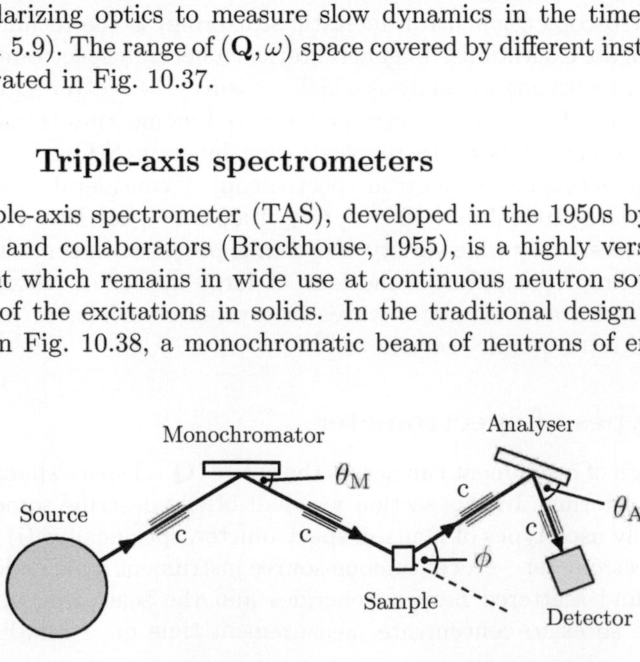

The triple-axis spectrometer (TAS), developed in the 1950s by Brockhouse[63] and collaborators (Brockhouse, 1955), is a highly versatile instrument which remains in wide use at continuous neutron sources for studies of the excitations in solids. In the traditional design of TAS, shown in Fig. 10.38, a monochromatic beam of neutrons of energy $E_\mathrm{i}$

**Fig. 10.38** Diagram showing the components of a traditional triple-axis spectrometer. Collimators (c) can be placed at the positions indicated.

strikes the sample, and the neutrons that are scattered into a particular direction and with a chosen final energy $E_f$ are recorded in the detector. The divergence of the incident and final beams can be constrained through the use of Soller collimators.[64] The incident neutrons are picked out from the broad spectrum of neutrons produced in the source by Bragg reflection from a crystal monochromator. Similarly, the final neutrons are selected by Bragg reflection from a crystal analyser. Pyrolytic graphite, copper, silicon, and Heusler alloy[65] are commonly used monochromator and analyser materials. The values of $E_i$ and $E_f$ are determined by the Bragg angles $\theta_M$ and $\theta_A$ at the monochromator and analyser, respectively.

When used for elastic scattering ($E_i = E_f$) the TAS can give a better signal-to-noise ratio than a conventional diffractometer on a continuous source because the analyser cuts out most of the background due to inelastic scattering, which can be significant. More often, however, the TAS is used for inelastic scattering, in which case $E_i \neq E_f$.

The standard procedure for inelastic scattering is to hold $E_f$ constant while $E_i$ is varied. This is called the *constant-$k_f$ mode* of operation. An alternative is to fix $E_i$ and vary $E_f$, which is the *constant-$k_i$ mode*.[66] An advantage of the constant-$k_f$ mode is that when the detector count rate is normalized to a beam monitor placed just before the sample the normalized intensity is directly proportional to $S(\mathbf{Q}, \omega)$.[67]

The three angles $\theta_M$, $\theta_A$ and $\phi$ (the scattering angle at the sample), which all have common (vertical) rotation axes, are what give the TAS its name, and the ability to set these independently endows the TAS with great flexibility. With the sample mounted on a goniometer to provide the capability for rotation around orthogonal horizontal axes in addition to the vertical axis (coincident with the $\phi$ axis), scans can be performed along any trajectory in $(\mathbf{Q}, \omega)$ space.[68]

## Constant-energy and constant-Q scans

Two particular types of scan that are widely used on a TAS are illustrated in Fig. 10.39. These are (i) the *constant-energy scan*, in which $\mathbf{Q}$ varies along a line in reciprocal space while the energy transfer remains fixed, and (ii) the *constant-Q scan*, in which the energy varies at a fixed $\mathbf{Q}$. These scan types are a natural way to investigate dispersive excitations in single crystals, as measurements can be concentrated along high-symmetry directions to aid comparison with theory.

## Focusing

The ability to perform constant-energy and constant-$\mathbf{Q}$ scans is one of the strengths of the TAS for studies of dispersion relations. Another is the facility to tailor the size and orientation of the resolution function (Section 10.1.6) to best suit the measurement. This is achieved through judicious choice of $E_i$, $E_f$, beam divergence, monochromator and analyser material, and the scattering sense at each of the three rotation axes.

[64] Soller collimators consist of an array of parallel absorbing plates made from e.g. plastic sheet coated in paint that contains $Gd_2O_3$. The neutron transmission is approximately a triangular function of angle measured horizontally with respect to the collimator axis, with FWHM usually in the range 0.2° to 2°.

[65] Heusler alloy ($Cu_2MnAl$) is used for polarized neutrons, see Section 7.3.2.

[66] $E_i$ and $E_f$ can also be varied simultaneously, but this is not often done.

[67] The monitor efficiency varies as $1/k_i$, so the monitor count rate is proportional to $\Phi_0/k_i$, where $\Phi_0$ is the incident flux. Hence, normalization of the detector count rate by the monitor gives a signal proportional to $k_i \times d^2\sigma/(d\Omega dE_f)$, which from (10.11) is proportional to $S(\mathbf{Q}, \omega)$ for fixed $k_f$.

[68] Subject to the constraints imposed by neutron kinematics and the limits of the rotation axes.

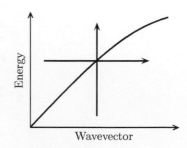

**Fig. 10.39** The constant-energy scan (horizontal line) and constant-Q scan (vertical line) can be performed on a TAS to measure dispersion relations of propagating excitations.

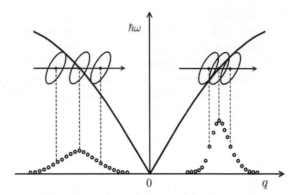

**Fig. 10.40** Focusing property of a TAS. The figure shows a constant-energy scan crossing the acoustic branches of a phonon dispersion curve $\hbar\omega(q)$. The orientation of the resolution ellipse is focused for the $+q$ branch and defocused for the $-q$ branch. The corresponding peak shapes are shown together with the positions of the resolution ellipsoid at the peak maximum and half-maxima. There is a small asymmetry in the peak shape due to the curvature of the dispersion relation.

[69]The dispersion surface is the surface in $(\mathbf{Q},\omega)$ space formed by the dispersion relation $\omega(\mathbf{Q})$.

The orientation of the resolution function matters because it affects the shape of the peak measured in a scan that crosses a dispersion surface.[69]

Consider the constant-energy scan shown schematically in Fig. 10.40. The scan crosses the $+q$ and $-q$ acoustic branches of a phonon dispersion relation which, for simplicity, we consider to be a function of only one wavevector coordinate. At any step in the scan the measured intensity is proportional to the intersection of the dispersion curve with the resolution function, here indicated by an ellipse corresponding to the half-height contour (see Section 10.1.6). The scattering intensity reaches its maximum value when the centre of the resolution ellipse sits directly on the dispersion curve, and falls to half its maximum when the ellipse just touches the dispersion curve on either side of the peak. The $+q$ peak, therefore, is sharper and higher than the $-q$ peak.

### Recent developments

Although the basic design of the TAS has remained unaltered since its inception, there have been two important recent modifications: (i) implementation of horizontally and vertically curved monochromators and analysers, constructed from arrays of co-aligned single crystals, to focus the beam in order to increase the intensity at the expense of $\mathbf{Q}$ resolution; (ii) development of multiple analyser/detector systems which can measure at a number of positions in $(\mathbf{Q},\omega)$ space simultaneously, and which can in some designs lift out of the horizontal plane to makes it possible to measure out-of-plane components of $\mathbf{Q}$.

### 10.3.3    Time-of-flight spectroscopy

The great strength of the TAS is its ability to concentrate intensity on a particular $\mathbf{Q}$ and $\omega$ of interest, and perform steps scans through this point along a specified path in $(\mathbf{Q},\omega)$ space. By contrast, time-of-flight (ToF) spectrometers are efficient when one is interested in studying a spectrum over a broad range of energies and/or wavevectors simultaneously. ToF spectrometers have a pulsed incident beam and a large pixellated area detector, and use the flight time over a known distance to separate the neutron energies. They are the natural type of instrument

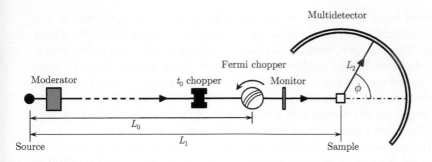

**Fig. 10.41** Schematic diagram of a typical direct-geometry chopper spectrometer at a pulsed spallation source.

for inelastic scattering at pulsed neutron sources because they can make optimal use of the inherent time structure of the source, but they are also common at continuous sources, where choppers can be used to pulse the neutron beam. The use of ToF to survey large regions of $(\mathbf{Q}, \omega)$ space provides a detailed understanding of excitation spectra and accompanying backgrounds, which is important for complex materials with multiple modes and for cases when it is not known in advance which parts of the spectrum are of most interest to study.

Time-of-flight spectrometers at pulsed sources have two additional advantages: (i) there is a high flux of epithermal neutrons (neutrons which have not reached thermal equilibrium and whose energies are therefore greater than those of thermal neutrons) which can be used to measure high-energy excitations and momentum distributions (Compton scattering); (ii) they have an intrinsically low background. This is largely because the fast neutron background produced by the source arrives at very short times and has generally decayed to zero by the time the interesting part of the spectrum is recorded. Moreover, the fast neutron burst can be suppressed by a background chopper if necessary. By contrast, at continuous sources the fast neutrons are present at all times (see Fig. 1.8).

### Direct and indirect geometry

One can measure a time-of-flight spectrum either by fixing $k_i$ and using ToF to define $k_f$, or *vice versa*. These two modes of operation are called *direct* and *indirect* (or *inverted*) geometry, respectively, and are analogues of the constant-$k_i$ and constant-$k_f$ modes of a TAS. Unlike a TAS, however, ToF spectrometers are permanently set up to work in either one mode or the other, and do not switch between the two. This is because the arrangement of ToF hardware components on a beamline designed for direct geometry is very different from that for indirect geometry.

Figure 10.41 shows the basic components of a typical direct-geometry ToF spectrometer at a spallation source. The neutron flight time is measured from the initial burst of high-energy neutrons at the target. The moderator slows these down towards thermal energies. A background (or $t_0$) chopper[70] is used to suppress $\gamma$-rays and fast neutrons that would otherwise be a source of unwanted background. The incident energy is

[70]Background choppers are made from a very thick piece of absorbing material (such as Nimonic or Inconel alloy) and are phased to close the beam aperture at the moment of impact of the proton pulse on the target.

selected from the polychromatic beam by a *Fermi chopper*, which is a cylinder containing an array of curved channels separated by absorbing slats. The chopper spins at an integer multiple of the source frequency. The incident energy is determined by the phase of the chopper relative to the source pulse, and the higher the frequency the narrower the bandwidth of $E_i$ transmitted by the chopper. The monochromatic pulse scatters from the sample into a large bank of position-sensitive detectors, usually tubes containing pressurized $^3$He gas.

On instruments designed for small $E_i$, the Fermi chopper is usually replaced by a disk chopper (a disk with a hole in it), and additional pulse-shaping and frame overlap choppers can be added. The latter reduces the effective pulse frequency by removing pulses so that the intensity of neutrons in the low-energy tail from one frame is negligible by the start of the next frame (see Fig. 1.8). On continuous sources, the pulsed monochromatic incident beam is usually created by a crystal monochromator and a chopper, but two choppers can also be used.

The essential components of an indirect-geometry ToF spectrometer at a spallation source are depicted in Fig. 10.42. There is a $t_0$ chopper in the incident beam to suppress very fast neutrons, but no monochromating chopper. Frame overlap choppers can be installed when required. The final energy $E_f$ is selected by Bragg reflection from an analyser crystal. The 002 reflection from pyrolytic graphite is often used. The analyser will also reflect higher orders (004, 006, ...) and a filter is placed after the analyser to suppress these (see Section 10.1.8). In modern instruments the analyser–detector system usually comprises multiple banks in order to increase the volume of $\mathbf{Q}$–$\omega$ space sampled.

The choice of direct or indirect geometry for a given experiment depends on several factors. One of these, illustrated in Fig. 10.43, is the kinematically accessible region of $(Q, \omega)$ space. The curves plotted in Fig. 10.43 are the trajectories of ToF scans in $(Q, \omega)$ space for a number of representative scattering angles. They are obtained from the relation between $Q$ and $\omega$ given in eqn (1.17), with either $E_i$ or $E_f$ fixed at 10 meV (chosen purely for illustration), and $\hbar\omega = E_i - E_f$. The trajectories for direct and indirect geometry are related by $\omega \to -\omega$.

An important property of indirect-geometry spectrometers, evident from Fig. 10.43, is that for a single $E_f$ they can access a wide range of energy transfer in neutron energy-loss[71] ($\hbar\omega > 0$). On direct-geometry

[71]Experiments that seek to probe excitations at energies appreciably in excess of $k_B T$ must be performed in energy-loss because of the suppression of energy-gain scattering by detailed balance (see Section 3.5.3).

**Fig. 10.42** Schematic diagram of an indirect-geometry crystal analyser spectrometer at a spallation neutron source. $L_2$ is the sample-to-detector distance via the analyser.

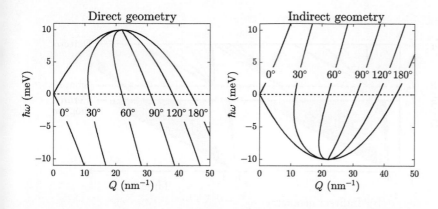

**Fig. 10.43** Energy and momentum transfer diagrams for direct-geometry ($E_i = 10\,\text{meV}$) and indirect-geometry ($E_f = 10\,\text{meV}$) neutron spectrometers. The lines are time-of-flight trajectories for the specific scattering angles shown.

spectrometers the energy-loss transfer cannot exceed $E_i$,[72] and the energy resolution of the entire spectrum increases with $E_i$, so if a large $E_i$ is chosen to access large $\hbar\omega$ then the resolution at small $\hbar\omega$ will be poor. Therefore, experiments on direct-geometry ToF spectrometers that require coverage of a wide range of energy-loss transfer must be performed with several $E_i$s, each optimized for a different range of $\hbar\omega$, and the data from each run combined after the measurement.

A drawback of indirect-geometry crystal analyser spectrometers is that the spectrum can contain spurious background features due to the analyser (see Section 10.1.8). Direct-geometry spectrometers, on the other hand, are celebrated for their very low backgrounds especially on spallation sources, and are usually preferred for studies of weak signals.

Two other advantages of direct geometry are, first, the flexibility to choose the incident energy and energy resolution to match the feature under investigation, and second, the large pixellated area detectors found on the latest instruments, which are like gigantic digital cameras. The development of these detectors has made direct-geometry ToF spectrometers very powerful for studies of excitations in single-crystal samples, as discussed below.

[72]In practice, high quality data are not usually obtainable for energies greater than 80–90% of $E_i$.

## Distance–time plots

Distance–time plots are a useful way to illustrate the propagation of neutrons of different energies in a ToF spectrometer. Figure 10.44 presents distance–time plots for direct- and indirect-geometry ToF spectrometers. In both cases, the flight times from source to sample ($t_1$) and from sample to detector ($t_2$) are given by

$$t_1 = \frac{L_1}{v_i} \qquad \text{and} \qquad t_2 = \frac{L_2}{v_f}, \qquad (10.116)$$

where $v_i = \hbar k_i / m_n$ and $v_f = \hbar k_f / m_n$ are the initial and final neutron speeds, respectively. The total flight time is $t = t_1 + t_2$.

Figure 10.44(a) shows how the phase of the Fermi chopper relative to the source pulse determines $v_i$, and hence $E_i$, on a direct-geometry spectrometer, and how the spectrum of scattered neutron speeds gives

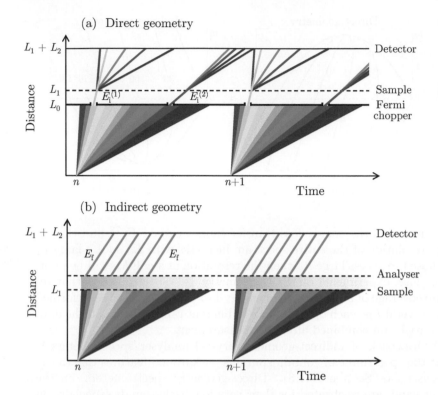

(a) Direct geometry

(b) Indirect geometry

**Fig. 10.44** Distance–time plots for direct- and indirect-geometry ToF spectrometers. The speed of the neutrons is colour-coded and is equal to the gradient of the lines.

rise to a spread of arrival times at the detector. The diagram also illustrates the process known as *repetition-rate multiplication*, in which the Fermi chopper is run at a multiple $p$ of the source frequency so that $p$ monoenergetic pulses of neutrons reach the sample per source pulse. Figure 10.44(a) depicts the case when $p = 2$, giving incident energies $E_i^{(1)}$ and $E_i^{(2)}$ ($E_i^{(2)} < E_i^{(1)}$). To avoid frame overlap the two incident energies need to be chosen such that the time-of-flight spectrum at the detector from $E_i^{(1)}$ does not overlap with that from $E_i^{(2)}$. This becomes easier to achieve as $L_1$ increases and $L_2$ decreases.

For the indirect geometry case, Fig. 10.44(b), the array of parallel lines drawn from analyser to detector represents the paths of neutrons scattered with a fixed final speed $v_f$ (energy $E_f$) which are transmitted by the analyser. These monoenergetic neutrons are recorded continuously in the detector for the duration of each frame.

### Converting spectra from ToF to energy transfer

The traditional method to obtain ToF spectra is to divide the useful time frame into a set of time channels and record the counts in each channel.[73] The detector counts are then divided by the width of the time channel and normalized to the incident flux, which is measured by a monitor located before the sample.[74] After calibration, this procedure yields the differential cross-section per unit time $\mathrm{d}^2\sigma/\mathrm{d}\Omega\mathrm{d}t$, such that $(\mathrm{d}^2\sigma/\mathrm{d}\Omega\mathrm{d}t)\mathrm{d}t$ is the cross-section for times between $t$ and $t + \mathrm{d}t$.

[73]The time channels do not have to be equally spaced.

[74]On a direct-geometry ToF spectrometer the incident flux is the same for each detector time channel because $E_i$ is constant, but on an indirect-geometry spectrometer different flight times correspond to different incident energies. In the latter case, a particular detector time channel $t \to t + \delta t$ corresponds to a range of incident energies $E_i \to E_i + \delta E_i$. The incident flux is then the integral of the monitor ToF spectrum between limits $t_M(E_i)$ and $t_M(E_i + \delta E_i)$, where $t_M$ is the flight time at the monitor.

It is usually preferable to express the spectrum as a cross-section per unit energy $d^2\sigma/d\Omega dE_f = d^2\sigma/d\Omega d\hbar\omega$, where $(d^2\sigma/d\Omega d\hbar\omega)d\hbar\omega$ is the cross-section for energy transfers between $\hbar\omega$ and $\hbar\omega + d\hbar\omega$. The energy interval $d\hbar\omega$ corresponds to the time interval $dt$, and so the time and energy cross-sections are related by

$$\frac{d^2\sigma}{d\Omega d\hbar\omega} \, d\hbar\omega = \frac{d^2\sigma}{d\Omega dt} \, dt. \tag{10.117}$$

We see that the conversion from time-of-flight to energy transfer requires the derivative $dt/d\hbar\omega$, which can be obtained from

$$\hbar\omega = E_i - E_f$$
$$= \frac{m_n}{2}\left(v_i^2 - v_f^2\right)$$
$$= \frac{m_n}{2}\left(\frac{L_1^2}{t_1^2} - \frac{L_2^2}{t_2^2}\right), \tag{10.118}$$

in which we used (10.116). The lengths $L_1$ and $L_2$ are fixed. For direct geometry $t_1$ is fixed, and so substituting $t_2 = t - t_1$ we have

$$\hbar\omega = \frac{m_n}{2}\left\{\frac{L_1^2}{t_1^2} - \frac{L_2^2}{(t-t_1)^2}\right\}. \tag{10.119}$$

After differentiating, we find[75]

$$\frac{dt}{d\hbar\omega} = \frac{(t-t_1)^3}{m_n L_2^2}, \tag{10.120}$$

so that

$$\frac{d^2\sigma}{d\Omega d\hbar\omega} = \frac{(t-t_1)^3}{m_n L_2^2}\frac{d^2\sigma}{d\Omega dt}. \tag{10.121}$$

Finally, let us express the results in terms of the more familiar scattering function $S(\mathbf{Q}, \omega)$. For present purposes we shall write the scattering function as $S(\theta, \phi, \omega)$ to indicate that the spectrum is recorded in an individual detector pixel whose direction is specified by the polar angles $(\theta, \phi)$ relative to the incident beam axis.[76] From eqns (10.11) and (10.121), we have, for direct geometry,

$$S(\theta, \phi, \omega) = \frac{k_i}{k_f}\frac{d^2\sigma}{d\Omega d\hbar\omega}$$
$$= \frac{(t-t_1)^4}{t_1 m_n L_2^2}\frac{d^2\sigma}{d\Omega dt}. \tag{10.122}$$

We made use of (10.116) and $t_2 = t - t_1$ to write $k_i/k_f = (t-t_1)/t_1$. The corresponding expression for indirect geometry is obtained from (10.118) with $t_2$ fixed and $t_1 = t - t_2$.

Figure 10.45 illustrates the correspondence between spectra measured in ToF and energy transfer for a direct-geometry ToF spectrometer. The symmetric ToF spectrum becomes a skewed function of energy transfer

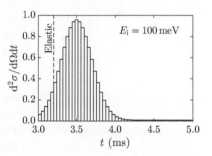

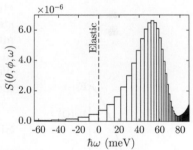

**Fig. 10.45** Conversion from time-of-flight $t$ to energy transfer $\hbar\omega$ on a direct-geometry ToF spectrometer. The time and energy bins are related by eqn (10.120). The ToF spectrum contains a Gaussian centred on $t = 3.5\,\mathrm{ms}$ together with a small constant background. The spectrum is calculated for an incident energy $E_i = 100\,\mathrm{meV}$ and primary and secondary flight paths $L_1 = 10\,\mathrm{m}$ and $L_2 = 4\,\mathrm{m}$, respectively. The sharp rise in $S(\theta, \phi, \omega)$ as $\hbar\omega \to E_i$ is a result of the compression of the background signal into increasingly narrower energy bins.

owing to the form of eqn (10.122)). The time channels are all equal, but the corresponding energy bins become narrower with increasing energy transfer in accord with (10.120). A consequence of this energy bin compression is that any long-time tail in the ToF spectrum due, for example, to a delayed neutron background will appear as a divergence in the energy spectrum as the energy transfer approaches the incident energy, as can be seen in Fig. 10.45 at energies above 80 meV.

### Event mode data recording

Rather than sorting the raw neutron counts into pre-defined time channels (*histogram mode*), the latest data acquisition systems tag each detected neutron with two pieces of information: (i) its location on the detector (pixel identifier), and (ii) a time stamp, which gives the flight time from source to detector and identifies which pulse the neutron is associated with. Information about the instrument and sample environment can also be stored alongside each neutron 'event' using the same timing clock. The complete list of neutron events for each run is stored for later processing.

This so-called *event mode* data acquisition is very flexible, and is advantageous for all types of time-of-flight instrument not just spectrometers. Neutrons are recorded at the full time resolution of the data acquisition electronics, allowing the user to choose appropriate bin sizes for analysis after the experiment based on features found in the data. Similarly, data can be recorded while some sample environment parameter such as temperature is varied continuously, rather than in discrete steps, and the systematic dependence on the parameter can be extracted afterwards. File size is also a consideration. Event mode data files scale in size linearly with the number of events, whereas histogrammed data files are always the same size. Hence, for short measurements, event mode provides more efficient data storage.

### Single-crystal ToF spectroscopy

With single crystal samples it is possible to determine the anisotropy of $S(\mathbf{Q}, \omega)$ through measurements along different crystallographic directions. Direct- and indirect-geometry ToF spectrometers each have particular strengths for single-crystal studies. Indirect-geometry spectrometers offer a wide range of energy transfer in neutron energy-loss, and if large focusing analyser crystals are used then a high count rate can be achieved at the expense of relaxed wavevector resolution. On the other hand, the array of analyser crystals tends to be confined to one dimension, which means that only certain directions in reciprocal space can be probed in one setting.

Studies requiring access to a large continuous volume of $(\mathbf{Q}, \omega)$ space are better suited to direct-geometry ToF spectrometers equipped with large solid angle position-sensitive detectors, which have the added advantage of an intrinsically low background. Most of this section will

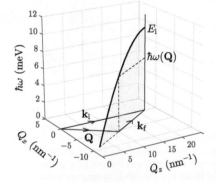

**Fig. 10.46** Time-of-flight trajectory in $(\mathbf{Q}, \hbar\omega)$ space for a single detector lying in the horizontal plane. The calculation is for direct geometry with $E_i = 10$ meV and a scattering angle of $25°$ ($\theta = 25°$, $\phi = 0$). Only the energy-loss side is shown. The components $Q_x$ and $Q_z$ are perpendicular and parallel to the incident beam, respectively.

be devoted to the methodology of single crystal measurements on such instruments.

With a fixed crystal angle there are three independent degrees of freedom on a ToF spectrometer. Two of these are associated with the direction of $\mathbf{k}_f$, as specified e.g. by the spherical angles $\theta$ and $\phi$, and the third is the time-of-flight $t$. One can transform these variables to wavevector and energy transfer via the usual scattering equations $\mathbf{Q} = \mathbf{k}_i - \mathbf{k}_f$ and $\hbar\omega = E_i - E_f$. Choosing Cartesian axes such that $z$ is parallel to $\mathbf{k}_i$ (assumed horizontal), $x$ is in the horizontal plane perpendicular to $\mathbf{k}_i$ and $y$ is vertical, we have

$$Q_x = -k_f \sin\theta \cos\phi$$
$$Q_y = -k_f \sin\theta \sin\phi$$
$$Q_z = k_i - k_f \cos\theta$$
(10.123)

$$\hbar\omega = \frac{\hbar^2}{2m_n}\left(k_i^2 - k_f^2\right).$$
(10.124)

Figure 10.46 illustrates the transformation described by (10.123)–(10.124) in the case of direct geometry, for which $k_i$ is fixed and $k_f$ varies with $t$. The figure shows the time-of-flight trajectory through $(\mathbf{Q}, \hbar\omega)$ space for a single detector. This example shows how the ToF spectrum recorded in a single detector pixel measures the spectrum along a curved path in $(\mathbf{Q}, \hbar\omega)$ space, and more generally, that energy and wavevector transfer are intrinsically coupled on ToF spectrometers.

In Fig. 10.47 we have repeated the calculation but this time for a row of detectors covering a range of scattering angles, and added a model dispersion relation $\hbar\omega(\mathbf{Q})$ representative of phonon or spin-wave excitations. The ToF trajectories now form a surface in $(\mathbf{Q}, \hbar\omega)$ space, and we expect to observe spin-wave scattering along the contour where this surface intersects the dispersion surface.

By extension, the ToF trajectories for the entire position-sensitive area detector will sweep out a volume in $(\mathbf{Q}, \hbar\omega)$ space, so that $S(\mathbf{Q}, \omega)$ is measured on a three-dimensional manifold embedded in the four-dimensional space of $\mathbf{Q}$ and $\hbar\omega$. In other words, we measure $S(\mathbf{Q}, \omega)$ as a function of only three independent $(\mathbf{Q}, \omega)$ components, and the fourth component is an implicit function of the other three. By way of example, Fig. 10.48 depicts the ToF scan for a square detector in the space spanned by $(Q_x, Q_y, Q_z)$, with $\hbar\omega$ being the implicit variable. The manifold is shown as a series of constant-$\hbar\omega$ surfaces.

The direct-geometry ToF method with position-sensitive detectors is very well suited for studying materials whose excitation spectra have reduced dimensionality, i.e. behave either one- or two-dimensionally. There is a large class of magnetic materials which have this property, including many important quantum magnets (see Section 8.7).

Quasi-one-dimensional (1d) magnetism is the result of structural factors which create chains of magnetic ions with very strong intra-chain magnetic interactions but very weak inter-chain coupling. The $S(\mathbf{Q}, \omega)$ for such materials has a strong dispersion for $Q_\parallel$, the component of $\mathbf{Q}$

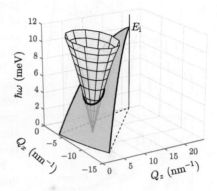

**Fig. 10.47** Time-of-flight trajectory in $(\mathbf{Q}, \hbar\omega)$ space, similar to Fig. 10.46 except calculated for a row of detectors covering the angular range from $10°$ to $35°$. A model spin-wave dispersion surface is shown emerging from a point in the $Q_x$–$Q_z$ plane. Scattering from spin waves will be observed along the contour where the ToF and dispersion surfaces intercept.

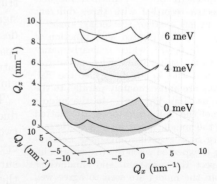

**Fig. 10.48** Constant energy-transfer surfaces in $\mathbf{Q}$ space for a direct-geometry ToF spectrometer. The simulation is for $E_i = 10$ meV and a detector of dimensions $3 \times 3$ m$^2$ located 4 m from the sample. The three surfaces are for energy transfers $\hbar\omega = 0$ meV (elastic), 4 meV, and 6 meV.

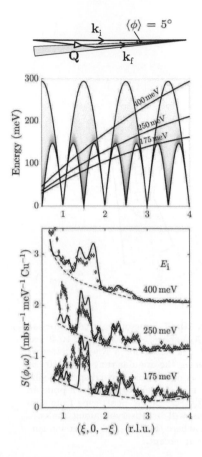

**Fig. 10.49** Magnetic spectrum of CuO, a quasi-1d $S = \frac{1}{2}$ antiferromagnet. Spectra recorded with three different $E_i$ at an average scattering angle of $\langle\phi\rangle = 5°$ are plotted as a function of $\mathbf{Q} = (\xi, 0, -\xi)$, which is nearly parallel to the spin chains. The crystal was aligned with $\mathbf{k}_i \parallel (\xi, 0, -\xi)$, so the ToF scans are approximately parallel to the chains (top diagram). In the middle panel the ToF trajectories are plotted on the spinon spectrum of a 1d Heisenberg $S = \frac{1}{2}$ antiferromagnet. Scattering is expected when the trajectories intersect the spectrum, and simulations from the model are seen to agree well with the data. Discrepancies at lower $Q$ are due to scattering from phonons and the effects of inter-chain coupling. (Adapted from Boothroyd *et al.*, 1997.)

---

[77]With event mode data acquisition the sample can be rotated continuously, avoiding the artificial discretization of the data.

parallel to the chains, but little or no variation perpendicular to the chains apart from the trivial effect of the magnetic form factor.

One way to measure a quasi-one-dimensional spectrum is to align the chains *parallel* to $\mathbf{k}_i$ and integrate over both directions perpendicular to $\mathbf{k}_i$. The ToF scan then measures $S(\mathbf{Q}, \omega)$ along a curved trajectory relative to the coordinates $\hbar\omega$ and $Q_\parallel$. An example of this method is presented in Fig. 10.49, which shows ToF data from a single crystal of cupric oxide (CuO). The spectrum contains diffuse features at high energies which are characteristic of the spinon spectrum for a spin-$\frac{1}{2}$ Heisenberg antiferromagnetic chain, Section 8.7.1.

Nowadays it is more usual to align the chains *perpendicular* to $\mathbf{k}_i$ and integrate over $Q_{\perp y}$, the component of $\mathbf{Q}$ which is perpendicular to both the chains and $\mathbf{k}_i$. The $(\hbar\omega, Q_\parallel)$ plane then contains the complete spectrum, the only caveat being that $\hbar\omega$ is an implicit function of $Q_{\perp z}$, the component of $\mathbf{Q}$ perpendicular to the chains and parallel to $\mathbf{k}_i$. This method was employed to obtain the spectrum shown in Fig. 8.25.

For two-dimensional (2d) magnets, $S(\mathbf{Q}, \omega)$ has a strong $\mathbf{Q}$ dependence parallel to the layers and a trivial dependence perpendicular to the layers. The natural way to measure $S(\mathbf{Q}, \omega)$ is then to align the layers perpendicular to $\mathbf{k}_i$, so that $Q_x$, $Q_y$ and $\hbar\omega$ are the independent variables, with $\hbar\omega$ an implicit function of the component $Q_z$ perpendicular to the layers. Figure 10.50(a) provides an example of such a measurement. The quasi-2d antiferromagnet $La_2CoO_4$ has layers of Co atoms aligned perpendicular to the $c$ axis. The spectrum is plotted to show the dispersion parallel to the 2d wavevector $\mathbf{Q}_{2D} = (\xi, \xi)$, and also illustrates how the component of $\mathbf{Q}$ along $c$ varies with energy transfer.

The methods just described are very efficient ways to measure spectra of one- and two-dimensional systems because the crystal remains in a fixed orientation and it is usually possible to obtain the entire spectrum with only two or three different values of $E_i$. When $S(\mathbf{Q}, \omega)$ has significant variation in all three directions, however, the methods just described are problematic. The curved ToF trajectories through $(\mathbf{Q}, \omega)$ space will not in general follow any simple path through the reciprocal lattice of the crystal, which makes it hard to visualize the spectrum and almost impossible to analyse the data.

The only efficient way to investigate 3D spectra by ToF is to introduce another independent measurement degree of freedom. There are two possible approaches. Either (i) perform measurements with a fixed crystal orientation and a large number of different $E_i$ values, or (ii) measure a large number of different crystal orientations with a fixed $E_i$. Of these, the second is almost always preferable because a larger volume of reciprocal space can be covered and the energy resolution (which is determined largely by $E_i$) is constant.

For practical implementation of the multi-angle method, the crystal is mounted on an axis of rotation perpendicular to $\mathbf{k}_i$ and rotated in small steps[77] of typically 0.5°–2°. ToF spectra are recorded at each step. The full range of the crystal rotation angle $\psi$ is chosen so as to give appropriate coverage of reciprocal space for the energy transfers of interest. The

(a)
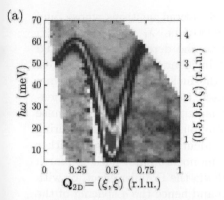

(b)
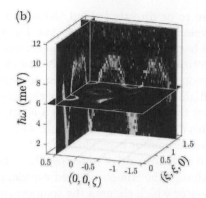

**Fig. 10.50** (a) Magnon spectrum of the quasi-2d antiferromagnet $La_2CoO_4$. The spectrum was measured with $k_i$ parallel to the $c$ axis, and the plot shows the dispersion parallel to the 2d wavevector $(\xi, \xi)$. (Adapted from Babkevich *et al.*, 2010). (b) Magnon spectrum of $RbMnF_3$ measured by the multi-angle method for single-crystal ToF spectroscopy. The data analysis was performed with the Horace software (Ewings *et al.*, 2016, courtesy of Russell Ewings).

complete dataset contains the spectrum measured on a four-dimensional manifold $(\theta, \phi, t, \psi)$, which is then transformed onto the physically relevant variables $(Q_x, Q_y, Q_z, \hbar\omega)$ for analysis. For example, with appropriate software one can take line cuts and cross-sections through the resulting four-dimensional map of $S(\mathbf{Q}, \omega)$ along high-symmetry directions and planes, as required. An example of multi-angle data from a 3D antiferromagnet is given in Fig. 10.50(b). The plot presents a set of three cross-sections through $S(\mathbf{Q}, \omega)$, each defined by two independent components of $(Q_x, Q_y, Q_z, \hbar\omega)$.

## 10.3.4   Backscattering spectrometers

Neutron spin echo is the technique which offers the best energy resolution (Fig. 10.37), but very high resolution ($\Delta E \sim 1\,\mu eV$) can also be achieved with crystal analyser spectrometers if the scattering angle at the analyser is close to $180°$, i.e. near backscattering.

This can be understood from Bragg's law, $\lambda = 2d\sin\theta$, which gives the wavelength diffracted from a set of Bragg planes with spacing $d$ when the beam is incident at an angle $\theta$ to the planes. As $d$ and $\theta$ are independent variables we can obtain the wavelength spread $\Delta\lambda$ by differentiating Bragg's law with respect to each variable and adding the two contributions in quadrature [c.f. eqn (10.37]:

$$(\Delta\lambda)^2 = (2\Delta d \sin\theta)^2 + (2d\Delta\theta \cos\theta)^2, \qquad (10.125)$$

or

$$\left(\frac{\Delta\lambda}{\lambda}\right)^2 = \left(\frac{\Delta d}{d}\right)^2 + (\Delta\theta \cot\theta)^2. \qquad (10.126)$$

As $\theta \to 90°$, $\cot\theta \to 0$ and the resolution becomes dominated by the $\Delta d$ term. The size of this term depends on the crystal perfection and the chosen reflection. For example, the 002 reflection of pyrolytic graphite close to backscattering gives an energy resolution[78] of $\Delta E \simeq 10\,\mu eV$ (see Exercise 10.10).

Backscattering spectrometers typically have a large bank of analyser crystals surrounding the sample and a set of detectors located either

[78] From the relation $E = h^2/(2m_n\lambda^2)$ the energy resolution is given by

$$\left(\frac{\Delta E}{E}\right)^2 = 4\left(\frac{\Delta\lambda}{\lambda}\right)^2.$$

close to or directly behind the sample. Scattered neutrons whose final wavelengths satisfy Bragg's law for $\theta \simeq 90°$, i.e. $\lambda_f \simeq 2d$ ($k_f \simeq \pi/d$) are backscattered from the analysers into the detectors. A balanced spectrometer will have closely matched incident and final energy resolutions and some mechanism to scan the incident energy. On time-of-flight sources long flight paths are used to allow the incident neutrons to disperse in time, and a chopper is employed to define the pulse width and incident energy band. On continuous sources a backscattering monochromator is employed, and the incident energy is scanned either with a Doppler drive which oscillates the monochromator back and forth to change the Bragg-reflected wavelength via the Doppler effect, or with a heater which changes the temperature and hence the $d$ spacing of the monochromator.[79]

[79]The same technique is used on very high-resolution inelastic X-ray scattering spectrometers to scan the incident X-ray energy.

## 10.3.5 Multiple scattering in spectroscopy

Elastic–elastic double scattering is not important in spectroscopy as it does not affect the data at non-zero energy transfers. Inelastic–inelastic processes are generally weak, but if the temperature is low they will tend to shift spectral weight to higher energies. For example, if there are two strong sharp modes at $\omega_1$ and $\omega_2$ then two energy-loss collisions will produce peaks at energies $2\omega_1$, $\omega_1 + \omega_2$, and $2\omega_2$ which can appear at sum or difference wavevectors.

Elastic–inelastic scattering produces a diffuse, energy-dependent background which can be a particular problem for magnetic scattering measurements. Typically, in such experiments a weak magnetic signal, which is strongest at small $Q$ due to the magnetic form factor and which can be broad in energy, must be isolated from the non-magnetic scattering. Data at high $Q$ give the phonon spectrum, but the scattering at very small $Q$ is often approximately independent of $Q$ and much larger than that calculated from the usual $Q^2$ dependence of the one-phonon spectrum. This extra signal at low $Q$ is mainly due to elastic + one-phonon multiple scattering. One way to estimate the multiple scattering signal is to measure a non-magnetic reference sample with a similar phonon spectrum to the material of interest. A straight subtraction of the scattering from the reference is not always satisfactory because (i) the coherent nuclear scattering lengths will not in general be the same as for the sample, and (ii) there may be differences in the phonon spectra of the two materials. A more accurate procedure is to scale the high $Q$ spectrum of the sample by the ratio of the low-$Q$ and high-$Q$ spectra of the reference material. The validity of this method for certain sample geometries has been verified by Monte Carlo simulations incorporating realistic model scattering cross-sections (Goremychkin and Osborn, 1993).

# Chapter summary

- In this chapter, aspects of the planning and optimization of a neutron scattering experiment have been covered, including attenuation, multiple scattering, data normalization, counting statistics, resolution, corrections for polarization analysis, and spurions.

- Practical aspects of diffraction experiments have been summarized, including instrumentation, Rietveld refinement, anisotropic displacement parameters, the Ewald sphere construction, Lorentz factors, extinction, and multiple scattering.

- Practical aspects of spectroscopy have been summarized, including triple-axis, time-of-flight, and backscattering spectrometers, direct and indirect geometry, and some specific points arising in time-of-flight inelastic scattering.

# Further reading

More details of neutron instrumentation and beamline components can be found in Willis and Carlile (2009), Carpenter and Loong (2015), and Furrer *et al.* (2009). Experimental techniques for neutron diffraction are described by Kisi and Howard (2012) and Nield and Keen (2000), and for neutron spectroscopy by Mitchell *et al.* (2005). Neutron techniques associated with the triple-axis spectrometer can be found in Shirane *et al.* (2002). A comprehensive account of time-of-flight neutron scattering techniques is given by Windsor (1981). A helpful introduction to probability and statistics, especially Bayesian methods for data analysis, is given by Sivia and Skilling (2006).

# Exercises

(10.1) Calculate the Bragg cut-off wavelength $\lambda_c$ for aluminium (see Fig. 10.1).

(10.2) Show that when the scattering angle is zero and $\mu_i = \mu_f = \mu$, the attenuation factor for transmission through a slab at normal incidence, eqn (10.7), reduces to the Beer–Lambert law, eqn (10.1):

$$A = \exp(-\mu t).$$

(10.3) Calculate the thickness of cadmium required to absorb 99.9% of neutrons at an energy of 15 meV.

(10.4) What is the optimum thickness of a parallel-sided container for small-angle neutron scattering from a dilute solution of macromolecules in (a) ordinary water ($H_2O$), and (b) heavy water ($D_2O$)?

(10.5) A detector records 200 counts when the monitor is preset to 1000. What is the fractional error in the counts? What monitor should be chosen to achieve a fractional error of 0.02?

(10.6) A measurement on a sample has a signal-to-background ratio of about 10:1. For how long should the background run be measured relative to the sample run in order to minimize the statistical error on the difference between the sample and background count rates?

(10.7) The crystal structure of silicon (Si) has a face-centred cubic lattice and a basis comprising Si at $0,0,0$ and $\frac{1}{4},\frac{1}{4},\frac{1}{4}$. Show that there is no second-order ($\lambda/2$) contamination in the beam from a Si 111 monochromator.

(10.8) By considering the limiting forms of the secondary extinction variable $x_s$ given by eqns (10.108) and (10.109), show that secondary extinction does not depend on mosaic when $g\lambda \gg r$, and does not depend on the crystallite size when $g\lambda \ll r$.

(10.9) The data shown in Fig. 10.33 were obtained with neutrons of wavelength $\lambda = 0.0841$ nm from a roughly spherical crystal of $ErNi_2{}^{11}B_2C$ of diam-

eter of 2 mm. The crystal structure is tetragonal with cell parameters $a = b = 0.35$ nm, $c = 1.06$ nm. Given the extinction parameters $r = 70$ nm and $g = 0.26 \times 10^4$ rad$^{-1}$, calculate the extinction coefficients $y_p$ and $y_s$ for the 400, 004 and 330 reflections. The structure factors for these reflections are $F(400) = 96.7$ fm, $F(004) = 45.3$ fm and $F(330) = 14.3$ fm. Attenuation can be neglected.

(10.10) The (002) planes of pyrolytic graphite have spacing $d = 0.335$ nm and imperfection $\Delta d/d = 2.5 \times 10^{-3}$. Show that the 002 reflection gives an energy resolution of $\Delta E = 9\,\mu eV$ in backscattering. If the angular divergence is $\Delta\theta = 1.5°$, at what value of $\theta$ does the angular contribution to the resolution match that from the crystal imperfection? Calculate $\Delta E$ in backscattering from the silicon 111 reflection, for which $d = 0.314$ nm and $\Delta d/d = 2.5 \times 10^{-4}$.

# Neutron Scattering Lengths and Cross-Sections

A

The table below lists experimental values of scattering lengths and cross-sections for the natural elements and selected isotopes. The data are from `https://www.ncnr.nist.gov/resources/n-lengths` and `http://www.ati.ac.at/~neutropt/scattering/table.html`, which are based on the listings of Koester *et al.* (1991), Sears (1992), and Rauch and Waschkowski (2000), together with recent results for $^{13}$C (Fischer *et al.*, 2008), $^{17}$O and $^{18}$O (Fischer *et al.*, 2012), $^{154}$Sm (Kohlmann *et al.*, 2016; Phelan *et al.*, 2016), $^{153}$Eu (Kohlmann *et al.*, 2016), $^{160}$Gd (Kennedy *et al.*, 2011), and Ir isotopes (Hannon *et al.*, 2018). Experimental errors are in the last digit, or thereabouts.

The scattering lengths and scattering cross-sections of most nuclides are independent of neutron wavelength in the thermal region, and the absorption cross-sections are inversely proportional to neutron speed $v$ (the $1/v$ law). By convention, the absorption cross-sections are tabulated for $v = 2,200\,\mathrm{ms^{-1}}$ ($E = 25.30\,\mathrm{meV}$, $\lambda = 0.1798\,\mathrm{nm}$, $k = 34.94\,\mathrm{nm^{-1}}$). One can obtain the absorption at other neutron energies using

$$\sigma_\mathrm{a}(E) = \sigma_\mathrm{a}(25.3\,\mathrm{meV}) \times \left[ \frac{25.3}{E(\mathrm{meV})} \right]^{1/2}$$

or

$$\sigma_\mathrm{a}(\lambda) = \sigma_\mathrm{a}(0.1798\,\mathrm{nm}) \times \frac{\lambda(\mathrm{nm})}{0.1798}.$$

Complex scattering lengths ($b = b' - ib''$) are given for strongly absorbing nuclides. The imaginary part is related to the absorption cross-section by $b'' = (k/4\pi)\sigma_\mathrm{a}$.

| Symbol | Quantity | $[1\,\text{fm} = 10^{-15}\,\text{m}; 1\,\text{barn (b)} = 10^{-28}\,\text{m}^2]$ |
|---|---|---|
| $Z$ | Atomic number | |
| $A$ | Mass number | |
| $p$ | Natural abundance (%) | |
| | (for radionuclides, the half-life in years (a) is given instead) | |
| $\bar{b}$ | Bound coherent scattering length (fm) | |
| $\sigma_{\text{coh}}$ | Bound coherent scattering cross-section (b) | |
| $\sigma_{\text{inc}}$ | Bound incoherent scattering cross-section (b) | |
| $\sigma_{\text{s}}$ | Total bound scattering cross-section (b) | |
| $\sigma_{\text{a}}$ | Absorption cross-section (b) for $v = 2,200\,\text{ms}^{-1}$ | |

| | $Z$ | $A$ | $p$ (%) | $\bar{b}$ (fm) | $\sigma_{\text{coh}}$ (b) | $\sigma_{\text{inc}}$ (b) | $\sigma_{\text{s}}$ (b) | $\sigma_{\text{a}}$ (b) |
|---|---|---|---|---|---|---|---|---|
| H | 1 | | | −3.741 | 1.757 | 80.26 | 82.02 | 0.3326 |
| | | 1 | 99.985 | −3.742 | 1.758 | 80.27 | 82.03 | 0.3326 |
| | | 2 | 0.015 | 6.671 | 5.592 | 2.05 | 7.64 | 0.0005 |
| | | 3 | (12.3 a) | 4.792 | 2.89 | 0.14 | 3.03 | 0 |
| He | 2 | | | 3.26 | 1.34 | 0.00 | 1.34 | 0.00747 |
| | | 3 | 0.0001 | 5.74 | 4.42 | 1.6 | 6.0 | 5333 |
| | | | | −1.483i | | | | |
| | | 4 | 99.9999 | 3.26 | 1.34 | 0 | 1.34 | 0 |
| Li | 3 | | | −1.90 | 0.454 | 0.92 | 1.37 | 70.5 |
| | | 6 | 7.5 | 2.0 | 0.51 | 0.46 | 0.97 | 940 |
| | | | | −0.261i | | | | |
| | | 7 | 92.5 | −2.22 | 0.619 | 0.78 | 1.40 | 0.0454 |
| Be | 4 | 9 | 100 | 7.79 | 7.63 | 0.0018 | 7.63 | 0.0076 |
| B | 5 | | | 5.30 | 3.54 | 1.70 | 5.24 | 767 |
| | | | | −0.213i | | | | |
| | | 10 | 20.0 | −0.1 | 0.144 | 3.0 | 3.1 | 3835 |
| | | | | −1.066i | | | | |
| | | 11 | 80.0 | 6.65 | 5.56 | 0.21 | 5.77 | 0.0055 |
| C | 6 | | | 6.648 | 5.551 | 0.001 | 5.552 | 0.00350 |
| | | 12 | 98.89 | 6.653 | 5.559 | 0 | 5.559 | 0.00353 |
| | | 13 | 1.11 | 6.542 | 5.38 | 0.034 | 5.41 | 0.00137 |
| N | 7 | | | 9.36 | 11.01 | 0.50 | 11.51 | 1.90 |
| | | 14 | 99.63 | 9.37 | 11.03 | 0.5 | 11.53 | 1.91 |
| | | 15 | 0.37 | 6.44 | 5.21 | 0.0001 | 5.21 | 0.00002 |
| O | 8 | | | 5.805 | 4.232 | 0.000 | 4.232 | 0.00019 |
| | | 16 | 99.75 | 5.805 | 4.232 | 0 | 4.232 | 0.0001 |
| | | 17 | 0.04 | 5.867 | 4.33 | 0.004 | 4.33 | 0.24 |
| | | 18 | 0.21 | 6.01 | 4.54 | 0.03 | 4.57 | 0.0002 |
| F | 9 | 19 | 100 | 5.654 | 4.017 | 0.0008 | 4.018 | 0.0096 |
| Ne | 10 | | | 4.566 | 2.620 | 0.008 | 2.628 | 0.039 |
| Na | 11 | 23 | 100 | 3.63 | 1.66 | 1.62 | 3.28 | 0.530 |
| Mg | 12 | | | 5.375 | 3.631 | 0.08 | 3.71 | 0.063 |
| Al | 13 | 27 | 100 | 3.449 | 1.495 | 0.0082 | 1.503 | 0.231 |
| Si | 14 | | | 4.151 | 2.163 | 0.004 | 2.167 | 0.171 |

*(continued)*

| | Z | A | $p$ (%) | $\bar{b}$ (fm) | $\sigma_{\text{coh}}$ (b) | $\sigma_{\text{inc}}$ (b) | $\sigma_{\text{s}}$ (b) | $\sigma_{\text{a}}$ (b) |
|---|---|---|---|---|---|---|---|---|
| P | 15 | 31 | 100 | 5.13 | 3.307 | 0.005 | 3.312 | 0.172 |
| S | 16 | | | 2.847 | 1.019 | 0.007 | 1.026 | 0.53 |
| Cl | 17 | | | 9.577 | 11.526 | 5.3 | 16.8 | 33.5 |
| | | 35 | 75.77 | 11.65 | 17.06 | 4.7 | 21.8 | 44.1 |
| | | 37 | 24.23 | 3.08 | 1.19 | 0.001 | 1.19 | 0.433 |
| Ar | 18 | | | 1.909 | 0.458 | 0.225 | 0.683 | 0.675 |
| K | 19 | | | 3.67 | 1.69 | 0.27 | 1.96 | 2.1 |
| Ca | 20 | | | 4.70 | 2.78 | 0.05 | 2.83 | 0.43 |
| Sc | 21 | 45 | 100 | 12.29 | 19.0 | 4.5 | 23.5 | 27.5 |
| Ti | 22 | | | −3.438 | 1.485 | 2.87 | 4.35 | 6.09 |
| V | 23 | | | −0.382 | 0.0184 | 5.08 | 5.10 | 5.08 |
| Cr | 24 | | | 3.635 | 1.660 | 1.83 | 3.49 | 3.05 |
| Mn | 25 | 55 | 100 | −3.73 | 1.75 | 0.40 | 2.15 | 13.3 |
| Fe | 26 | | | 9.45 | 11.22 | 0.40 | 11.62 | 2.56 |
| Co | 27 | 59 | 100 | 2.49 | 0.779 | 4.8 | 5.6 | 37.18 |
| Ni | 28 | | | 10.3 | 13.3 | 5.2 | 18.5 | 4.49 |
| | | 58 | 68.27 | 14.4 | 26.1 | 0 | 26.1 | 4.6 |
| | | 60 | 26.10 | 2.8 | 0.99 | 0 | 0.99 | 2.9 |
| Cu | 29 | | | 7.718 | 7.485 | 0.55 | 8.03 | 3.78 |
| | | 63 | 69.17 | 6.43 | 5.2 | 0.006 | 5.2 | 4.50 |
| | | 65 | 30.83 | 10.61 | 14.1 | 0.40 | 14.5 | 2.17 |
| Zn | 30 | | | 5.680 | 4.054 | 0.077 | 4.131 | 1.11 |
| Ga | 31 | | | 7.288 | 6.675 | 0.16 | 6.83 | 2.75 |
| Ge | 32 | | | 8.185 | 8.42 | 0.18 | 8.60 | 2.20 |
| As | 33 | 75 | 100 | 6.58 | 5.44 | 0.060 | 5.50 | 4.5 |
| Se | 34 | | | 7.970 | 7.98 | 0.32 | 8.30 | 11.7 |
| Br | 35 | | | 6.795 | 5.80 | 0.10 | 5.90 | 6.9 |
| | | 79 | 50.69 | 6.80 | 5.81 | 0.15 | 5.96 | 11.0 |
| | | 81 | 49.31 | 6.79 | 5.79 | 0.05 | 5.84 | 2.7 |
| Kr | 36 | | | 7.81 | 7.67 | 0.01 | 7.68 | 25 |
| Rb | 37 | | | 7.09 | 6.32 | ∼ 0.5 | 6.8 | 0.38 |
| Sr | 38 | | | 7.02 | 6.19 | 0.06 | 6.25 | 1.28 |
| Y | 39 | 89 | 100 | 7.75 | 7.55 | 0.15 | 7.70 | 1.28 |
| Zr | 40 | | | 7.16 | 6.44 | 0.02 | 6.46 | 0.185 |
| Nb | 41 | 93 | 100 | 7.054 | 6.253 | 0.0024 | 6.255 | 1.15 |
| Mo | 42 | | | 6.715 | 5.67 | 0.04 | 5.71 | 2.48 |
| Tc | 43 | 99 | $(2.1 \times 10^5$ a) | 6.8 | 5.8 | ∼ 0.5 | 6.3 | 20 |
| Ru | 44 | | | 7.03 | 6.21 | 0.4 | 6.6 | 2.56 |
| Rh | 45 | 103 | 100 | 5.88 | 4.34 | ∼ 0.3 | 4.6 | 144.8 |
| Pd | 46 | | | 5.91 | 4.39 | 0.093 | 4.48 | 6.9 |
| Ag | 47 | | | 5.922 | 4.407 | 0.58 | 4.99 | 63.3 |
| | | 107 | 51.83 | 7.555 | 7.17 | 0.13 | 7.30 | 37.6 |
| | | 109 | 48.17 | 4.165 | 2.18 | 0.32 | 2.50 | 91.0 |
| Cd | 48 | | | 4.87 −0.70i | 3.04 | 3.46 | 6.50 | 2520 |
| | | 110 | 12.51 | 5.9 | 4.4 | 0 | 4.4 | 11 |
| | | 111 | 12.81 | 6.5 | 5.3 | ∼ 0.3 | 5.6 | 24 |
| | | 112 | 24.13 | 6.4 | 5.1 | 0 | 5.1 | 2.2 |
| | | 113 | 12.22 | −8.0 −5.73i | 12.1 | ∼ 0.3 | 12.4 | 20600 |
| | | 114 | 28.72 | 7.5 | 7.1 | 0 | 7.1 | 0.34 |
| | | 116 | 7.47 | 6.3 | 5.0 | 0 | 5.0 | 0.075 |

*(continued)*

| | $Z$ | $A$ | $p$ (%) | $\bar{b}$ (fm) | $\sigma_{\mathrm{coh}}$ (b) | $\sigma_{\mathrm{inc}}$ (b) | $\sigma_{\mathrm{s}}$ (b) | $\sigma_{\mathrm{a}}$ (b) |
|---|---|---|---|---|---|---|---|---|
| In | 49 | | | 4.065 −0.0539i | 2.08 | 0.54 | 2.62 | 193.8 |
| | | 113 | 4.3 | 5.39 | 3.65 | 0.0000 | 3.65 | 12.0 |
| | | 115 | 95.7 | 4.01 −0.0562i | 2.02 | 0.55 | 2.57 | 202 |
| Sn | 50 | | | 6.225 | 4.870 | 0.022 | 4.892 | 0.626 |
| Sb | 51 | | | 5.57 | 3.90 | 0.00 | 3.90 | 4.91 |
| Te | 52 | | | 5.80 | 4.23 | 0.09 | 4.32 | 4.7 |
| I | 53 | 127 | 100 | 5.28 | 3.50 | 0.31 | 3.81 | 6.15 |
| Xe | 54 | | | 4.92 | 3.04 | | | 23.9 |
| Cs | 55 | 133 | 100 | 5.42 | 3.69 | 0.21 | 3.90 | 29.0 |
| Ba | 56 | | | 5.07 | 3.23 | 0.15 | 3.38 | 1.1 |
| La | 57 | | | 8.24 | 8.53 | 1.13 | 9.66 | 8.97 |
| Ce | 58 | | | 4.84 | 2.94 | 0.00 | 2.94 | 0.63 |
| Pr | 59 | 141 | 100 | 4.58 | 2.64 | 0.015 | 2.66 | 11.5 |
| Nd | 60 | | | 7.69 | 7.43 | 9.2 | 16.6 | 50.5 |
| | | 142 | 27.16 | 7.7 | 7.5 | 0 | 7.5 | 18.7 |
| | | 143 | 12.18 | ∼ 14 | 25 | 55 | 80 | 334 |
| | | 144 | 23.80 | 2.8 | 1.0 | 0 | 1.0 | 3.6 |
| | | 145 | 8.29 | ∼ 14 | 25 | ∼ 5 | 30 | 42 |
| | | 146 | 17.19 | 8.7 | 9.5 | 0 | 9.5 | 1.4 |
| | | 148 | 5.75 | 5.7 | 4.1 | 0 | 4.1 | 2.5 |
| | | 150 | 5.63 | 5.3 | 3.5 | 0 | 3.5 | 1.2 |
| Pm | 61 | 147 | (2.62 a) | 12.6 | 20.0 | 1.3 | 21.3 | 168.4 |
| Sm | 62 | | | 0.80 −1.65i | 0.422 | 39 | 39 | 5922 |
| | | 144 | 3.1 | ∼ −3 | ∼ 1 | 0 | ∼ 1 | 0.7 |
| | | 147 | 15.1 | 14 | 25 | 14 | 39 | 57 |
| | | 148 | 11.3 | ∼ −3 | ∼ 1 | 0 | ∼ 1 | 2.4 |
| | | 149 | 13.9 | −19.2 −11.7i | 63.5 | 137 | 200 | 42080 |
| | | 150 | 7.4 | 14 | 25 | 0 | 25 | 104 |
| | | 152 | 26.6 | −5.0 | 3.1 | 0 | 3.1 | 206 |
| | | 154 | 22.6 | 8.9 | 10 | 0 | 10 | 8.4 |
| Eu | 63 | | | 7.22 −1.26i | 6.75 | 2.5 | 9.2 | 4530 |
| | | 151 | 47.8 | 6.13 −2.53i | 5.5 | 3.1 | 8.6 | 9100 |
| | | 153 | 52.2 | 8.85 | 9.8 | 1.3 | 11.1 | 312 |
| Gd | 64 | | | 6.5 −13.82i | 29.3 | 151 | 180 | 49700 |
| | | 152 | 0.2 | ∼ 10 | ∼ 13 | 0 | ∼ 13 | 735 |
| | | 154 | 2.1 | ∼ 10 | ∼ 13 | 0 | ∼ 13 | 85 |
| | | 155 | 14.8 | 6.0 −17.0i | 40.8 | 25 | 66 | 61100 |
| | | 156 | 20.6 | 6.3 | 5.0 | 0 | 5.0 | 1.5 |
| | | 157 | 15.7 | −1.14 −71.9i | 650 | 394 | 1044 | 259000 |
| | | 158 | 24.8 | 9 | 10 | 0 | 10 | 2.2 |
| | | 160 | 21.8 | 8.9 | 10.0 | 0 | 10.0 | 0.77 |

*(continued)*

| | Z | A | $p$ (%) | $\bar{b}$ (fm) | $\sigma_{\text{coh}}$ (b) | $\sigma_{\text{inc}}$ (b) | $\sigma_{\text{s}}$ (b) | $\sigma_{\text{a}}$ (b) |
|---|---|---|---|---|---|---|---|---|
| Tb | 65 | 159 | 100 | 7.38 | 6.84 | 0.004 | 6.84 | 23.4 |
| Dy | 66 | | | 16.9 | 35.9 | 54.4 | 90.3 | 994 |
| | | | | −0.276i | | | | |
| | | 160 | 2.3 | 6.7 | 5.6 | 0 | 5.6 | 56 |
| | | 161 | 19.0 | 10.3 | 13.3 | 3 | 16 | 600 |
| | | 162 | 25.5 | −1.4 | 0.25 | 0 | 0.25 | 194 |
| | | 163 | 24.9 | 5.0 | 3.1 | 0.21 | 3.3 | 124 |
| | | 164 | 28.1 | 49.4 | 307 | 0 | 307 | 2840 |
| | | | | −0.79i | | | | |
| Ho | 67 | 165 | 100 | 8.01 | 8.06 | 0.36 | 8.42 | 64.7 |
| Er | 68 | | | 7.79 | 7.63 | 1.1 | 8.7 | 159 |
| | | 164 | 1.6 | 8.2 | 8.4 | 0 | 8.4 | 13 |
| | | 166 | 33.4 | 10.6 | 14.1 | 0 | 14.1 | 19.6 |
| | | 167 | 22.9 | 3.0 | 1.1 | 0.13 | 1.2 | 659 |
| | | 168 | 27.1 | 7.4 | 6.9 | 0 | 6.9 | 2.74 |
| | | 170 | 14.9 | 9.6 | 11.6 | 0 | 11.6 | 5.8 |
| Tm | 69 | 169 | 100 | 7.07 | 6.28 | 0.10 | 6.38 | 100 |
| Yb | 70 | | | 12.43 | 19.42 | 4.0 | 23.4 | 34.8 |
| | | 168 | 0.14 | −4.07 | 2.13 | 0 | 2.13 | 2230 |
| | | | | −0.62i | | | | |
| | | 170 | 3.06 | 6.77 | 5.8 | 0 | 5.8 | 11.4 |
| | | 171 | 14.3 | 9.66 | 11.7 | 3.9 | 15.6 | 48.6 |
| | | 172 | 21.9 | 9.43 | 11.2 | 0 | 11.2 | 0.8 |
| | | 173 | 16.1 | 9.56 | 11.5 | 3.5 | 15.0 | 17.1 |
| | | 174 | 31.8 | 19.3 | 46.8 | 0 | 46.8 | 69.4 |
| | | 176 | 12.7 | 8.72 | 9.6 | 0 | 9.6 | 2.85 |
| Lu | 71 | | | 7.21 | 6.53 | 0.7 | 7.2 | 74 |
| | | 175 | 97.39 | 7.24 | 6.59 | 0.6 | 7.2 | 21 |
| | | 176 | 2.61 | 6.1 | 4.7 | 1.2 | 5.9 | 2065 |
| | | | | −0.57i | | | | |
| Hf | 72 | | | 7.77 | 7.6 | 2.6 | 10.2 | 104 |
| | | 176 | 5.2 | 6.61 | 5.5 | 0 | 5.5 | 23.5 |
| | | 177 | 18.6 | ∼ 0.8 | ∼ 0.1 | ∼ 0.1 | ∼ 0.2 | 373 |
| | | 178 | 27.1 | 5.9 | 4.4 | 0 | 4.4 | 84 |
| | | 179 | 13.7 | 7.46 | 7.0 | 0.14 | 7.1 | 41 |
| | | 180 | 35.2 | 13.2 | 21.9 | 0 | 21.9 | 13.04 |
| Ta | 73 | | | 6.91 | 6.00 | 0.01 | 6.01 | 20.6 |
| W | 74 | | | 4.86 | 2.97 | 1.63 | 4.60 | 18.3 |
| | | 182 | 26.3 | 6.97 | 6.10 | 0 | 6.10 | 20.7 |
| | | 183 | 14.3 | 6.53 | 5.36 | ∼ 0.3 | 5.7 | 10.1 |
| | | 184 | 30.7 | 7.48 | 7.03 | 0 | 7.03 | 1.7 |
| | | 186 | 28.6 | −0.72 | 0.065 | 0 | 0.065 | 37.9 |
| Re | 75 | | | 9.2 | 10.6 | ∼ 0.9 | 11.5 | 89.7 |
| | | 185 | 37.40 | 9.0 | 10.2 | ∼ 0.5 | 10.7 | 112 |
| | | 187 | 62.60 | 9.3 | 10.9 | ∼ 1.0 | 11.9 | 76.4 |
| Os | 76 | | | 10.7 | 14.4 | ∼ 0.3 | 14.7 | 16.0 |
| | | 186 | 1.58 | 11.6 | 17 | 0 | 17 | 80 |
| | | 187 | 1.6 | 10 | 13 | ∼ 0.3 | 13 | 320 |
| | | 188 | 13.3 | 7.6 | 7.3 | 0 | 7.3 | 4.7 |
| | | 189 | 16.1 | 10.7 | 14.4 | ∼ 0.5 | 14.9 | 25 |
| | | 190 | 26.4 | 11.0 | 15.2 | 0 | 15.2 | 13.1 |
| | | 192 | 41.0 | 11.5 | 16.6 | 0 | 16.6 | 2.0 |

*(continued)*

| | $Z$ | $A$ | $p$ (%) | $\bar{b}$ (fm) | $\sigma_{\mathrm{coh}}$ (b) | $\sigma_{\mathrm{inc}}$ (b) | $\sigma_{\mathrm{s}}$ (b) | $\sigma_{\mathrm{a}}$ (b) |
|---|---|---|---|---|---|---|---|---|
| Ir | 77 | | | 10.6 | 14.1 | $\sim 0$ | 14 | 425 |
| | | 191 | 37.3 | 12.1 | 18.4 | | | 954 |
| | | 193 | 62.7 | 9.7 | 11.8 | | | 111 |
| Pt | 78 | | | 9.60 | 11.58 | $\sim 0.1$ | 11.7 | 10.3 |
| | | 192 | 0.79 | 9.9 | 12.3 | 0 | 12.3 | 10.0 |
| | | 194 | 32.9 | 10.55 | 14.0 | 0 | 14.0 | 1.44 |
| | | 195 | 33.8 | 8.83 | 9.8 | 0.13 | 9.9 | 27.5 |
| | | 196 | 25.3 | 9.89 | 12.3 | 0 | 12.3 | 0.72 |
| | | 198 | 7.2 | 7.8 | 7.6 | 0 | 7.6 | 3.66 |
| Au | 79 | 197 | 100 | 7.63 | 7.32 | 0.43 | 7.75 | 98.65 |
| Hg | 80 | | | 12.69 | 20.24 | 6.6 | 26.8 | 372 |
| | | 196 | 0.2 | 30.3 | 115 | 0 | 115 | 3080 |
| | | 198 | 10.1 | | | 0 | | 2.0 |
| | | 199 | 17.0 | 16.9 | 36 | 30 | 66 | 2150 |
| | | 200 | 23.1 | | | 0 | | < 60 |
| | | 201 | 13.2 | | | | | 8 |
| | | 202 | 29.6 | | | 0 | | 4.89 |
| | | 204 | 6.8 | | | 0 | | 0.43 |
| Tl | 81 | | | 8.776 | 9.678 | 0.21 | 9.89 | 3.43 |
| | | 203 | 29.52 | 6.99 | 6.1 | 0.14 | 6.3 | 11.4 |
| | | 205 | 70.48 | 9.52 | 11.39 | 0.007 | 11.40 | 0.104 |
| Pb | 82 | | | 9.405 | 11.115 | 0.0030 | 11.118 | 0.171 |
| | | 204 | 1.4 | 9.90 | 12.3 | 0 | 12.3 | 0.65 |
| | | 206 | 24.1 | 9.22 | 10.68 | 0 | 10.68 | 0.030 |
| | | 207 | 22.1 | 9.28 | 10.82 | 0.002 | 10.82 | 0.699 |
| | | 208 | 52.4 | 9.50 | 11.34 | 0 | 11.34 | 0.000 |
| Bi | 83 | 209 | 100 | 8.532 | 9.148 | 0.0084 | 9.156 | 0.034 |
| Po | 84 | | | | | | | |
| At | 85 | | | | | | | |
| Rn | 86 | | | | | | | |
| Fr | 87 | | | | | | | |
| Ra | 88 | 226 | $(1.6 \times 10^3$ a) | 10.0 | 13 | 0 | 13 | 12.8 |
| Ac | 89 | | | | | | | |
| Th | 90 | 232 | 100 | 10.31 | 13.36 | 0 | 13.36 | 7.37 |
| Pa | 91 | 231 | $(3.3 \times 10^4$ a) | 9.1 | 10.4 | $\sim 0.1$ | 10.5 | 201 |
| U | 92 | | | 8.417 | 8.903 | 0.005 | 8.908 | 7.57 |
| | | 233 | $(1.6 \times 10^5$ a) | 10.1 | 12.8 | $\sim 0.1$ | 12.9 | 574.7 |
| | | 235 | 0.72 | 10.47 | 13.78 | 0.2 | 14.0 | 680.9 |
| | | 238 | 99.28 | 8.402 | 8.871 | 0 | 8.871 | 2.68 |
| Np | 93 | 237 | $(2.1 \times 10^6$ a) | 10.55 | 14.0 | $\sim 0.5$ | 14.5 | 176 |
| Pu | 94 | | | | | | | |
| | | 238 | (87.74 a) | 14.1 | 25.0 | 0 | 25.0 | 558 |
| | | 239 | $(2.4 \times 10^4$ a) | 7.7 | 7.5 | 0.2 | 7.7 | 1017 |
| | | 240 | $(6.6 \times 10^3$ a) | 3.5 | 1.54 | 0 | 1.54 | 289.6 |
| | | 242 | $(3.8 \times 10^5$ a) | 8.1 | 8.2 | 0 | 8.2 | 18.5 |
| Am | 95 | 243 | $(7.4 \times 10^3$ a) | 8.3 | 8.7 | 0.3 | 9.0 | 75.3 |
| Cm | 96 | | | | | | | |
| | | 244 | (18.1 a) | 9.5 | 11.3 | 0 | 11.3 | 16.2 |
| | | 246 | $(4.7 \times 10^3$ a) | 9.3 | 10.9 | 0 | 10.9 | 1.36 |
| | | 248 | $(3.5 \times 10^5$ a) | 7.7 | 7.5 | 0 | 7.5 | 3.0 |

# Mathematical Definitions

B

This appendix contains a number of mathematical definitions and results that are used in the main text.

## B.1 Common peak functions

### B.1.1 The Gaussian

The Gaussian function is defined by

$$f(x) = \frac{1}{\sqrt{2\pi\sigma^2}} \exp\left\{-\frac{(x-a)^2}{2\sigma^2}\right\}, \tag{B.1}$$

with the multiplicative prefactor giving unit normalization:

$$\int_{-\infty}^{\infty} f(x)\,dx = 1.$$

The Gaussian, which is known as the normal distribution in statistics, is symmetrically peaked about $x = a$, Fig. B.1. The standard deviation of the peak is $\sigma$, and the full width at half maximum (FWHM) is given by $2(2\ln 2)^{\frac{1}{2}}\sigma = 2.355\sigma$.

The most general $d$-dimensional Gaussian with unit normalization is given by

$$f(X) = \frac{|M|^{1/2}}{(2\pi)^{d/2}} \exp(-\tfrac{1}{2}X^{T}MX) \tag{B.2}$$

where X is a $d$-dimensional column vector, $X^{T}$ is the row matrix transpose of X, and M is a $d \times d$ symmetric matrix. This form of the Gaussian can be anisotropic. In three dimensions, the anisotropy can be characterized by the surface in $\mathbf{r}$ space corresponding to a fixed value of $f(\mathbf{r})$, e.g. the half-maximum. Such a surface is an ellipsoid centred at the origin. The principal axes of the ellipsoid are given by the eigenvectors of M. Referred to the principal axes,

$$f(\mathbf{R}) = \frac{1}{(2\pi)^{\frac{3}{2}}} \frac{1}{\sigma_X\sigma_Y\sigma_Z} \exp\left(-\frac{X^2}{2\sigma_X^2} - \frac{Y^2}{2\sigma_Y^2} - \frac{Z^2}{2\sigma_Z^2}\right), \tag{B.3}$$

where the coordinates $X, Y, Z$ are along the principal axes. An isotropic three-dimensional Gaussian is given by

$$f(\mathbf{r}) = \frac{1}{(2\pi\sigma^2)^{\frac{3}{2}}} \exp\left(-\frac{r^2}{2\sigma^2}\right). \tag{B.4}$$

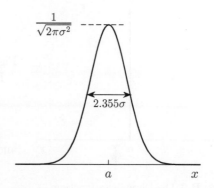

**Fig. B.1** The normalized Gaussian function.

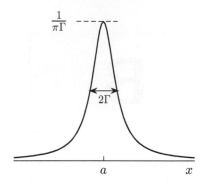

**Fig. B.2** The normalized Lorentzian function.

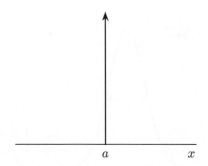

**Fig. B.3** The Dirac delta function.

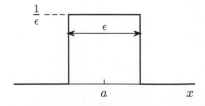

**Fig. B.4** Normalized top-hat function.

### B.1.2   The Lorentzian

The Lorentzian function normalized to unity is

$$f(x) = \frac{\Gamma/\pi}{(x-a)^2 + \Gamma^2}. \tag{B.5}$$

It is symmetrically peaked at $x = a$, with a FWHM of $2\Gamma$, Fig. B.2. The standard deviation of a Lorentzian is infinite and therefore not useful.

## B.2   The Dirac delta function

The Dirac delta function $\delta(x-a)$ is defined as the extreme limit of any sharply peaked function centred on $x = a$ that has an integral of unity:

$$\delta(x-a) = 0 \qquad x \neq a,$$
$$= \infty \qquad x = a,$$
$$\int_{-\infty}^{\infty} \delta(x-a)\,\mathrm{d}x = 1. \tag{B.6}$$

Fig. B.3 gives one way to represent a delta function.

It follows from the definition of $\delta(x-a)$ that

$$\int_{-\infty}^{\infty} f(x)\delta(x-a)\,\mathrm{d}x = f(a), \tag{B.7}$$

and that a re-scaling of the $x$ variable by a real constant $c$ leads to

$$\delta\{c(x-a)\} = \frac{1}{|c|}\,\delta(x-a). \tag{B.8}$$

The extension of these properties to two (three) dimensions is straightforward: $x$ is replaced by a two (three)-dimensional vector $\mathbf{r}$, and the integrals are performed over an area (volume).

A number of common functions behave as delta functions in certain limits, for example the zero width limits of the normalized Gaussian and Lorentzian functions, eqns (B.1) and (B.5). Here we list some other representations used elsewhere in this book.

### B.2.1   Top-hat function

$$\delta(x-a) = \lim_{\epsilon \to 0} H(x), \tag{B.9}$$

where

$$H(x) = \frac{1}{\epsilon} \qquad (a - \epsilon/2) \leq x \leq (a + \epsilon/2)$$
$$= 0 \qquad \text{otherwise.} \tag{B.10}$$

The 'top-hat' function $H(x)$ is illustrated in Fig. B.4.

## B.2.2   Sinc function

The sinc function has the form $\sin x / x$. An example of where it appears is in the Fourier transform of a slit of width $2L$, which is $2 \sin QL / Q$. For large $L$ this is a sharply peaked symmetric function centred at $Q = 0$, Fig. B.5. The sinc and sinc$^2$ functions have the property[1]

$$\int_{-\infty}^{\infty} \frac{\sin x}{x} \, \mathrm{d}x = \int_{-\infty}^{\infty} \frac{\sin^2 x}{x^2} \, \mathrm{d}x = \pi. \qquad \text{(B.11)}$$

Hence,

$$\delta(Q) = \lim_{L \to \infty} \frac{1}{\pi} \frac{\sin QL}{Q} \qquad \text{(B.12)}$$

$$= \lim_{L \to \infty} \frac{1}{\pi} \frac{\sin^2 QL}{Q^2 L}. \qquad \text{(B.13)}$$

## B.2.3   Integral representations

In one dimension

$$\delta(x) = \frac{1}{2\pi} \int_{-\infty}^{\infty} \exp(\mathrm{i}kx) \, \mathrm{d}k, \qquad \text{(B.14)}$$

and in three dimensions

$$\delta(\mathbf{r}) = \frac{1}{(2\pi)^3} \int \exp(\mathrm{i}\mathbf{Q} \cdot \mathbf{r}) \, \mathrm{d}^3\mathbf{Q}, \qquad \text{(B.15)}$$

where the integral is performed over all of $\mathbf{Q}$ space.

From (B.14) and (B.8),

$$\delta(\hbar\omega) = \frac{1}{2\pi\hbar} \int_{-\infty}^{\infty} \exp(\mathrm{i}\omega t) \, \mathrm{d}t, \qquad \text{(B.16)}$$

and similarly,

$$\delta(E_{\lambda_\mathrm{f}} - E_{\lambda_\mathrm{i}} - \hbar\omega) = \frac{1}{2\pi\hbar} \int_{-\infty}^{\infty} \exp\{\mathrm{i}(E_{\lambda_\mathrm{f}} - E_{\lambda_\mathrm{i}})t/\hbar\} \, \exp(-\mathrm{i}\omega t) \, \mathrm{d}t. \qquad \text{(B.17)}$$

## B.2.4   Derivative of Fermi–Dirac distribution

The Fermi–Dirac distribution is

$$f(E) = \frac{1}{\exp\{\beta(E - \mu)\} + 1}, \qquad \text{(B.18)}$$

where $\beta = 1/(k_\mathrm{B}T)$ and $\mu$ is the chemical potential. At $T = 0$ the chemical potential becomes the Fermi energy $E_\mathrm{F}$. As may be seen in Fig. B.6, $f(E)$ steps down abruptly from 1 to 0 at $E = \mu$. The derivative $-(\partial f / \partial E)_T$ is a sharply peaked symmetric function of $E$ centred on $\mu$ with a width (FWHM) of $3.53 k_\mathrm{B}T$. The area of $-(\partial f / \partial E)_T$ is

$$\int_0^{\infty} -\left(\frac{\partial f}{\partial E}\right)_T \, \mathrm{d}E = f(0) - f(\infty) = 1.$$

[1]The integrals can be done by contour integration.

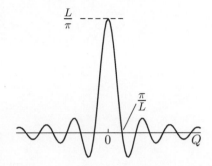

**Fig. B.5** Normalized sinc function.

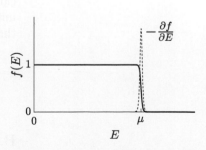

**Fig. B.6** The Fermi–Dirac distribution for $k_\mathrm{B}T \ll \mu$, and its derivative.

Hence,

$$-\left(\frac{\partial f}{\partial E}\right)_{T=0} = \delta(E - E_\mathrm{F}).\qquad\text{(B.19)}$$

## B.2.5   Dirac's formula

$$
\begin{aligned}
\lim_{\epsilon \to 0^+} \frac{1}{\omega' - \omega \mp \mathrm{i}\epsilon} &= \lim_{\epsilon \to 0^+} \frac{\omega' - \omega \pm \mathrm{i}\epsilon}{(\omega' - \omega)^2 + \epsilon^2}\\
&= \lim_{\epsilon \to 0^+} \frac{\omega' - \omega}{(\omega' - \omega)^2 + \epsilon^2} \pm \frac{\mathrm{i}\epsilon}{(\omega' - \omega)^2 + \epsilon^2}\\
&= \frac{1}{\omega' - \omega} \pm \mathrm{i}\pi\delta(\omega' - \omega).
\end{aligned}
\qquad\text{(B.20)}
$$

This is known as *Dirac's formula*. The imaginary term on the second line is a Lorentzian function which, from (B.5), has FWHM $2\epsilon$ and area $\pi$. As $\epsilon$ decreases, the Lorentzian becomes sharper while the area remains constant. It therefore behaves as a $\delta$-function in the limit when $\epsilon = 0$.

## B.3   Fourier transforms

### B.3.1   Definitions

The Fourier transform of a function $f(x)$ is defined by

$$f(k) = \int_{-\infty}^{\infty} f(x)\exp(\mathrm{i}kx)\,\mathrm{d}x.\qquad\text{(B.21)}$$

The transform can be inverted:

$$f(x) = \frac{1}{2\pi}\int_{-\infty}^{\infty} f(k)\exp(-\mathrm{i}kx)\,\mathrm{d}k.\qquad\text{(B.22)}$$

The constants 1 and $1/2\pi$ in front of the integrals in (B.21) and (B.22) can be replaced by any two constants whose product is $1/2\pi$. The definition extends in a simple way to higher dimensions. For example, in three dimensions

$$f(\mathbf{k}) = \int f(\mathbf{r})\exp(\mathrm{i}\mathbf{k}\cdot\mathbf{r})\,\mathrm{d}^3\mathbf{r}\qquad\text{(B.23)}$$

$$f(\mathbf{r}) = \frac{1}{(2\pi)^3}\int f(\mathbf{k})\exp(-\mathrm{i}\mathbf{k}\cdot\mathbf{r})\,\mathrm{d}^3\mathbf{k}.\qquad\text{(B.24)}$$

### B.3.2   Convolution theorem

The convolution of two functions $f(x)$ and $g(x)$ is defined as

$$f * g = \int_{-\infty}^{\infty} f(x')g(x - x')\,\mathrm{d}x'.\qquad\text{(B.25)}$$

The Fourier transform (FT) of a convolution is

$$FT\{f * g\} = FT\{f\}FT\{g\}. \tag{B.26}$$

This is known as the convolution theorem. To prove it, substitute (B.22) into (B.25) and make use of (B.14). Another form of the theorem is

$$FT\{fg\} = \frac{1}{2\pi}FT\{f\} * FT\{g\}. \tag{B.27}$$

## B.3.3   Fourier transform of a Gaussian

If $f(x) = \exp(-ax^2)$, then

$$f(k) = \sqrt{\frac{\pi}{a}}\,\exp(-k^2/4a), \tag{B.28}$$

which is a Gaussian function of $k$. For the isotropic three-dimensional Gaussian $f(r) = \exp(-ar^2)$,

$$f(k) = \left(\frac{\pi}{a}\right)^{\frac{3}{2}}\exp(-k^2/4a). \tag{B.29}$$

## B.3.4   Fourier transforms of other functions

(i)
$$\int \frac{\mathbf{R}}{R^3}\exp(i\mathbf{Q}\cdot\mathbf{R})\,d^3\mathbf{R} = \frac{4\pi i}{Q}\hat{\mathbf{Q}}. \tag{B.30}$$

Proof: Write $\mathbf{R}$ in spherical coordinates with the polar axis along $\mathbf{Q}$. The angle between $\mathbf{Q}$ and $\mathbf{R}$ is the polar angle $\theta$, and $d^3\mathbf{R} = R^2\sin\theta\,d\phi\,d\theta\,dR$. As the integrand is a vector pointing in the radial direction, only the component along $\mathbf{Q}$ contributes in the the $\phi$ integration. This allows us to replace $\mathbf{R}$ with $R\cos\theta\,\hat{\mathbf{Q}}$, so that

$$\int \frac{\mathbf{R}}{R^3}\exp(i\mathbf{Q}\cdot\mathbf{R})\,d^3\mathbf{R} = 2\pi\hat{\mathbf{Q}}\int_0^\infty\left\{\int_0^\pi\cos\theta\sin\theta\exp(iQR\cos\theta)\,d\theta\right\}dR$$

$$= 4\pi i\hat{\mathbf{Q}}\int_0^\infty\left(\frac{\sin QR}{Q^2R^2} - \frac{\cos QR}{QR}\right)dR$$

$$= \frac{4\pi i}{Q}\hat{\mathbf{Q}}.$$

Integration by parts has been used to evaluate the $\theta$ integral and the first term in the $R$ integral.

(ii)
$$\int \frac{1}{q^2}\exp(i\mathbf{q}\cdot\mathbf{R})\,d^3\mathbf{q} = \frac{2\pi^2}{R}. \tag{B.31}$$

Proof: Write $\mathbf{q}$ in spherical coordinates with the polar axis along $\mathbf{R}$. The angle between $\mathbf{q}$ and $\mathbf{R}$ is the polar angle $\theta$, and $d^3\mathbf{q} = 2\pi q^2\sin\theta\,d\theta\,dq$.

Hence,

$$\int \frac{1}{q^2} \exp(i\mathbf{q} \cdot \mathbf{R}) \, d^3\mathbf{q} = 2\pi \int_0^\infty \left\{ \int_0^\pi \exp(iqR\cos\theta) \sin\theta \, d\theta \right\} dq$$

$$= 2\pi \int_0^\infty \left[ -\frac{1}{iqR} \exp(iqR\cos\theta) \right]_0^\pi dq$$

$$= 4\pi \int_0^\infty \frac{\sin(qR)}{qR} \, dq$$

$$= \frac{2\pi^2}{R}.$$

The last line follows from (B.11).

(iii)    $$\nabla \times \left( \frac{\mathbf{s} \times \mathbf{R}}{R^3} \right) = \frac{1}{2\pi^2} \int \hat{\mathbf{q}} \times (\mathbf{s} \times \hat{\mathbf{q}}) \exp(i\mathbf{q} \cdot \mathbf{R}) \, d^3\mathbf{q}. \qquad \text{(B.32)}$$

Proof: It follows from the differential operator $\nabla = (\partial/\partial X, \partial/\partial Y, \partial/\partial Z)$ that

$$\nabla \left( \frac{1}{R} \right) = -\frac{\mathbf{R}}{R^3} \qquad\qquad \text{(B.33)}$$

$$\nabla \{ \exp(i\mathbf{q} \cdot \mathbf{R}) \} = i\mathbf{q} \exp(i\mathbf{q} \cdot \mathbf{R}) \qquad\qquad \text{(B.34)}$$

$$\nabla \times \{ \mathbf{a} \exp(i\mathbf{q} \cdot \mathbf{R}) \} = i\mathbf{q} \times \mathbf{a} \exp(i\mathbf{q} \cdot \mathbf{R}), \qquad\qquad \text{(B.35)}$$

where $\mathbf{R} = (X, Y, Z)$ and $\mathbf{a}$ is a constant vector. Now

$$\nabla \times \left( \frac{\mathbf{s} \times \mathbf{R}}{R^3} \right) = -\nabla \times \left\{ \mathbf{s} \times \nabla \left( \frac{1}{R} \right) \right\} \qquad \text{[from (B.33)]}$$

$$= -\frac{1}{2\pi^2} \int \frac{1}{q^2} \nabla \times \{ \mathbf{s} \times \nabla [\exp(i\mathbf{q} \cdot \mathbf{R})] \} \, d\mathbf{q}$$

$$\text{[from (B.31)]}$$

$$= \frac{1}{2\pi^2} \int \hat{\mathbf{q}} \times (\mathbf{s} \times \hat{\mathbf{q}}) \exp(i\mathbf{q} \cdot \mathbf{R}) \, d\mathbf{q},$$

as required. The last line follows from (B.34) and (B.35).

# B.4  Lattices

## B.4.1  Lattice and reciprocal lattice

A lattice is an array of points generated by

$$\mathbf{l} = n_1 \mathbf{a}_1 + n_2 \mathbf{a}_2 + n_3 \mathbf{a}_3, \qquad\qquad \text{(B.36)}$$

where $\mathbf{l}$ represents the displacement of lattice point $l$ from some arbitrary origin (also a lattice point), $\mathbf{a}_1$, $\mathbf{a}_2$, $\mathbf{a}_3$ are primitive lattice vectors and

$n_1$, $n_2$, $n_3$ are integers. A lattice may be represented mathematically by

$$L(\mathbf{r}) = \sum_{\mathbf{l}} \delta(\mathbf{r} - \mathbf{l}). \qquad (B.37)$$

The Fourier transform of $L(\mathbf{r})$ is

$$\begin{aligned} L(\mathbf{Q}) &= \int \sum_{\mathbf{l}} \delta(\mathbf{r} - \mathbf{l}) \exp(i\mathbf{Q} \cdot \mathbf{r}) \, d^3\mathbf{r} \\ &= \sum_{\mathbf{l}} \exp(i\mathbf{Q} \cdot \mathbf{l}), \qquad (B.38) \end{aligned}$$

where $\mathbf{Q}$ is the Fourier transform conjugate variable. For a general $\mathbf{Q}$, the real and imaginary parts of $\exp(i\mathbf{Q} \cdot \mathbf{l})$ will oscillate from positive to negative over the lattice, and so $L(\mathbf{Q}) \simeq 0$. However, for any $\mathbf{Q}$ that satisfies

$$\exp(i\mathbf{Q} \cdot \mathbf{l}) = 1 \text{ for all } \mathbf{l}, \qquad (B.39)$$

we have that $L(\mathbf{Q}) = N$ (the total number of lattice points). A $\mathbf{Q}$ that satisfies (B.39) will be denoted by $\mathbf{G}$. The set of all $\mathbf{G}$ vectors may be generated by

$$\mathbf{G} = m_1 \mathbf{a}_1^* + m_2 \mathbf{a}_2^* + m_3 \mathbf{a}_3^*, \qquad (B.40)$$

where $m_1$, $m_2$, $m_3$ are integers, and

$$\mathbf{a}_1^* = \frac{2\pi}{v_0} \mathbf{a}_2 \times \mathbf{a}_3, \quad \mathbf{a}_2^* = \frac{2\pi}{v_0} \mathbf{a}_3 \times \mathbf{a}_1, \quad \mathbf{a}_3^* = \frac{2\pi}{v_0} \mathbf{a}_1 \times \mathbf{a}_2. \qquad (B.41)$$

Here $v_0$ is the volume of the real-space primitive unit cell[2]

Equation (B.40) defines a lattice in $\mathbf{Q}$-space (or *reciprocal space*), called the *reciprocal lattice*. The $\mathbf{G}$ vectors are *reciprocal lattice vectors*. From (B.41), the real and reciprocal space basis vectors satisfy

$$\mathbf{a}_1^* \cdot \mathbf{a}_1 = \mathbf{a}_2^* \cdot \mathbf{a}_2 = \mathbf{a}_3^* \cdot \mathbf{a}_3 = 2\pi,$$

and

$$\mathbf{a}_1^* \cdot \mathbf{a}_2 = \mathbf{a}_1^* \cdot \mathbf{a}_3 = \mathbf{a}_2^* \cdot \mathbf{a}_1 = \ldots = 0. \qquad (B.42)$$

The volume of the unit cell in reciprocal space[3] (the Brillouin zone) is $v_0^* = (2\pi)^3/v_0$.

To verify that the reciprocal lattice vectors defined in (B.40) solve eqn (B.39) we take the scalar product of $\mathbf{G}$ with $\mathbf{l}$ as given by (B.36) and make use of (B.42):

$$\mathbf{G} \cdot \mathbf{l} = 2\pi(m_1 n_1 + m_2 n_2 + m_3 n_3). \qquad (B.43)$$

As $m_1$, $m_2$, $m_3$, and $n_1$, $n_2$, $n_3$ are all integers, the right-hand side of (B.43) is an integer multiple of $2\pi$ for all $\mathbf{l}$. Hence, the $\mathbf{G}$ vectors defined by eqns (B.40) and (B.41) are solutions to eqn (B.39).

[2] The volume of a primitive unit cell in real space is given by

$$\begin{aligned} v_0 &= \mathbf{a}_1 \cdot \mathbf{a}_2 \times \mathbf{a}_3 \\ &= \mathbf{a}_2 \cdot \mathbf{a}_3 \times \mathbf{a}_1 \\ &= \mathbf{a}_3 \cdot \mathbf{a}_1 \times \mathbf{a}_2 . \end{aligned}$$

[3] The volume of the unit cell in reciprocal space is given by

$$\begin{aligned} v_0^* &= \mathbf{a}_1^* \cdot \mathbf{a}_2^* \times \mathbf{a}_3^* \\ &= \mathbf{a}_2^* \cdot \mathbf{a}_3^* \times \mathbf{a}_1^* \\ &= \mathbf{a}_3^* \cdot \mathbf{a}_1^* \times \mathbf{a}_2^* . \end{aligned}$$

## B.4.2 Lattice sums

The arguments in the previous section imply that the lattice sum (B.38) is sharply peaked at the reciprocal lattice vectors and so may be written

$$\sum_{\mathbf{l}} \exp(i\mathbf{Q} \cdot \mathbf{l}) = c \sum_{\mathbf{G}} \delta(\mathbf{Q} - \mathbf{G}). \tag{B.44}$$

We can determine the constant $c$ by integrating both sides of (B.44) over a Brillouin zone (a unit cell of the reciprocal lattice). A Brillouin zone contains one reciprocal lattice point, so the integral of the right-hand side of (B.44) is $c$. On the left-hand side, the $\mathbf{l} = 0$ term in the sum is equal to 1, so its contribution to the integral is simply the volume of the Brillouin zone, which is $(2\pi)^3/v_0$. For the terms with $\mathbf{l} \neq 0$ we can write $\mathbf{Q} = q_1\mathbf{a}_1^* + q_2\mathbf{a}_2^* + q_3\mathbf{a}_3^*$, and use (B.36) and (B.42) to obtain

$$\mathbf{Q} \cdot \mathbf{l} = 2\pi(q_1 n_1 + q_2 n_2 + q_3 n_3). \tag{B.45}$$

The lattice sum in (B.44) then factors into the product of three geometric series of the form $\sum \exp(2\pi i q n)$, each of which integrates to zero as may be seen by the following:[4]

$$\int_{-\frac{1}{2}}^{\frac{1}{2}} \exp(2\pi i q n)\, \mathrm{d}q = \frac{1}{2\pi i n}\left[\exp(2\pi i q n)\right]_{-\frac{1}{2}}^{\frac{1}{2}} = 0.$$

Hence,

$$\sum_{\mathbf{l}} \exp(i\mathbf{Q} \cdot \mathbf{l}) = \frac{(2\pi)^3}{v_0} \sum_{\mathbf{G}} \delta(\mathbf{Q} - \mathbf{G}). \tag{B.46}$$

Note that

$$\left|\sum_{\mathbf{l}} \exp(i\mathbf{Q} \cdot \mathbf{l})\right|^2 = \sum_{\mathbf{l}}\sum_{\mathbf{l}'} \exp\{i\mathbf{Q} \cdot (\mathbf{l} - \mathbf{l}')\}$$

$$= N \sum_{\mathbf{l}} \exp(i\mathbf{Q} \cdot \mathbf{l}). \tag{B.47}$$

The last line follows because $\mathbf{l} - \mathbf{l}'$ is a lattice vector and for very large $N$ the sum over $\mathbf{l}$ is the same for any $\mathbf{l}'$. One can simply choose $\mathbf{l}' = 0$ and replace the sum over $\mathbf{l}'$ by $N$.

A relation similar to (B.46) applies to the discrete wavevectors that describe waves in crystals. Consider a finite crystal of dimensions $N_1\mathbf{a}_1 \times N_2\mathbf{a}_2 \times N_3\mathbf{a}_3$, where $N = N_1 N_2 N_3$ is the total number of primitive unit cells. The allowed $\mathbf{q}$ are obtained by imposing periodic boundary conditions, as in (5.59) but now in all three dimensions. This gives[5]

$$\mathbf{q} = \frac{m_1}{N_1}\mathbf{a}_1^* + \frac{m_2}{N_2}\mathbf{a}_2^* + \frac{m_3}{N_3}\mathbf{a}_3^*, \tag{B.48}$$

where $m_1$, $m_2$, $m_3$ are integers. Now, consider the lattice sum in (B.46) but with $\mathbf{q}$ in place of $\mathbf{Q}$. From (B.36),

$$\sum_{\mathbf{l}} \exp(i\mathbf{q} \cdot \mathbf{l}) = S_1 S_2 S_3, \tag{B.49}$$

[4]Without loss of generality, we choose to integrate over the first Brillouin zone, for which $-\frac{1}{2} \leq q_j \leq \frac{1}{2}$.

[5]Periodic boundary conditions in three dimensions mean that a translation through $N_1\mathbf{a}_1$ leaves the wave invariant, i.e.

$$\exp\{i\mathbf{q} \cdot (\mathbf{r} + N_1\mathbf{a}_1)\} = \exp(i\mathbf{q} \cdot \mathbf{r}),$$

and similarly for $N_2\mathbf{a}_2$ and $N_3\mathbf{a}_3$. Using (B.42) one can see that any $\mathbf{q}$ of the form (B.48) satisfies the periodic boundary conditions.

where

$$S_j = \sum_{n=0}^{N_j-1} e^{2\pi i m_j n/N_j} \qquad (j = 1, 2, 3)$$

$$= \sum_{n=0}^{N_j-1} x_j^n \qquad (x_j = e^{2\pi i m_j/N_j}) \qquad\qquad \sum_{n=0}^{N-1} x^n = \frac{x^N - 1}{x - 1}.$$

$$= \frac{\sin(m_j\pi)}{\sin(m_j\pi/N_j)} \, x_j^{(N_j-1)/2}$$

$$= N_j \qquad (m_j = 0)$$
$$= 0 \qquad (m_j \neq 0).$$

Hence, from (B.48) and (B.49),

$$\sum_l \exp(i\mathbf{q} \cdot \mathbf{l}) = N \qquad (\mathbf{q} = 0),$$
$$= 0 \qquad (\mathbf{q} \neq 0).$$

It follows that if $\mathbf{q}$ and $\mathbf{q}'$ are any two allowed wavevectors in the first Brillouin zone then

$$\sum_l \exp\{i(\mathbf{q} - \mathbf{q}') \cdot \mathbf{l}\} = N\delta_{\mathbf{q}\mathbf{q}'}. \qquad (B.50)$$

# Quantum Mechanics, Atomic Physics, and Magnetism

## C.1 Quantum mechanics

We give a brief overview of non-relativistic quantum theory in the framework of the so-called Copenhagen interpretation.[1]

[1]The conventional interpretation of quantum mechanics was developed in the 1920s by a group of theoretical physicists centred around Niels Bohr in Copenhagen.

### C.1.1 Quantum probability

A fundamental difference between quantum and classical physics is that in the quantum world the outcome of a measurement is governed by a probability distribution. This means that repeated physical measurements on identical copies of a quantum system do not have to give the same result. The quantum probabilities that control physical measurements take the form

$$P(c) = |c|^2, \tag{C.1}$$

where $c$ is a complex number called the *probability amplitude*. Moreover, the rule for combining probabilities is sometimes different in classical and quantum theory. Specifically, for two independent outcomes $c_1$ and $c_2$, the standard way of calculating the probability of *either $c_1$ or $c_2$* is

$$P(c_1 \text{ or } c_2) = P(c_1) + P(c_2) \qquad \text{(classical)}, \tag{C.2}$$

but for certain quantum systems the combination law is

$$\begin{aligned} P(c_1 \text{ or } c_2) &= |c_1 + c_2|^2 \qquad \text{(quantum)} \\ &= |c_1|^2 + |c_2|^2 + 2\,\mathrm{Re}(c_1^* c_2). \end{aligned} \tag{C.3}$$

The additional term[2] $2\,\mathrm{Re}(c_1^* c_2)$ present in the quantum case depends on the relative phases of the two probability amplitudes $c_1$ and $c_2$ and leads to *quantum interference* effects. Coherent neutron diffraction is an example of quantum interference, the $c_i$ being the amplitudes of neutron waves scattered from different sites.

[2]The symbol Re means 'take the real part'.

### C.1.2 State vectors

According to quantum theory, the physical state of a system is described by the set of probability amplitudes that allow you to predict the possible

[3]David Hilbert (1862–1943), German mathematician.

[4]*Linearly independent* means that it is not possible to express any one of the basis vectors as a linear combination of the others.

outcomes of measurements performed on the system. The complete set of probability amplitudes for all possible measurements on the system is collected together in the *state vector*, or *state function*, $\Psi$. The state vector inhabits a many-dimensional *system space* also known as a *Hilbert space*,[3] just as a position vector $\mathbf{r}$ inhabits a three-dimensional Euclidean space and represents a spatial displacement.

Any state vector can be written as a linear combination of linearly-independent[4] *basis vectors* $\phi_i$ which span the system space, i.e.

$$\Psi = \sum_i c_i \phi_i. \tag{C.4}$$

As a shorthand, this can be written as a vector

$$\Psi = \begin{pmatrix} c_1 \\ c_2 \\ . \\ . \end{pmatrix}. \tag{C.5}$$

If $\phi$ varies continuously then in effect there are an infinite number of basis functions and the vector becomes a continuous function $c(\phi)$.

The set of probability amplitudes $\{c_i\}$ is specific to the $\phi_i$ basis. There will be different sets of amplitudes if $\Psi$ is defined on another basis, just as the components of a vector $\mathbf{r}$ are different in a Cartesian basis with system coordinates $(x, y, z)$ than in a spherical basis with coordinates $(r, \theta, \phi)$.

### C.1.3   The adjoint space and Dirac notation

In general, the vectors that appear in quantum theory have components that are complex numbers. On the other hand, the physical results predicted by quantum theory must be real. To make this possible it is necessary to introduce a dual or *adjoint* space containing vectors whose components are the complex conjugates of the vectors in the first space.

[5]Paul A. M. Dirac (1924–1984), British mathematical physicist who shared the 1933 Nobel Prize in Physics with Erwin R. J. A. Schrödinger.

Paul Dirac[5] introduced the notation $|\Psi\rangle$, pronounced 'ket psi', to represent the complete set of probability amplitudes contained in $\Psi$. The corresponding adjoint is written $\langle\Psi|$ and pronounced 'bra psi'. The bra and ket are related by the adjoint operation,

$$\langle\Psi| = |\Psi\rangle^\dagger. \tag{C.6}$$

[6]The inner product is a generalization of the scalar or dot product

$$\mathbf{a} \cdot \mathbf{b} = \sum_i a_i b_i.$$

The usefulness of Dirac's notation becomes apparent when applied to inner products, which occur a lot in quantum mechanics. The inner product of two complex vectors $\mathbf{a}$ and $\mathbf{b}$ is defined as[6]

$$\langle \mathbf{a}, \mathbf{b} \rangle = \mathbf{a}^\dagger \cdot \mathbf{b} = \sum_i a_i^* b_i, \tag{C.7}$$

where $\dagger$ is the adjoint operation, whose explicit meaning is to take the complex conjugate and transpose. If $a(x)$ and $b(x)$ are continuous functions then the inner product is

$$\langle a, b \rangle = \int a^*(x) b(x) \, \mathrm{d}x. \tag{C.8}$$

In Dirac notation, the inner product between two state vectors $\Psi_1$ and $\Psi_2$ is written as $\langle\Psi_1|\Psi_2\rangle$, irrespective of whether they are discrete or continuous. In the case when both state vectors are expressed in the same basis then

$$\langle\Psi_1|\Psi_2\rangle = \sum_i a_i^* b_i \qquad \text{(discrete)}$$

$$= \int a^*(x)b(x)\,\mathrm{d}x \qquad \text{(continuous)}, \tag{C.9}$$

where $a_i$ and $a(x)$ are the discrete or continuous probability amplitudes in $\Psi_1$, and $b_i$ and $b(x)$ are likewise for $\Psi_2$. It follows that

$$\langle\Psi_1|\Psi_2\rangle^* = \langle\Psi_2|\Psi_1\rangle. \tag{C.10}$$

The definitions in (C.9) imply that

$$\langle\phi_i|\phi_j\rangle = \delta_{ij}, \tag{C.11}$$

with $\delta_{ij}$ the Kronecker delta. In other words, the basis is orthonormal.

From (C.4) and (C.11) it can be seen that the probability amplitude for a particular state $j$ can be projected out of the state function via the inner product

$$c_j = \langle\phi_j|\Psi\rangle. \tag{C.12}$$

Moreover,

$$\langle\Psi|\Psi\rangle = \sum_i |c_i|^2 \geq 0, \tag{C.13}$$

so that both $\langle\Psi|\Psi\rangle$ and each of the $|c_i|^2$ terms is real and positive. Normalization of $|\Psi\rangle$ so that $\langle\Psi|\Psi\rangle = 1$ then leads us to interpret

$$p_i = |c_i|^2 \tag{C.14}$$

as the probability for the system to be in state $i$.

Dirac's notation[7] greatly simplifies the formalism of quantum mechanics by avoiding the need to deal with the specifics of the Hilbert space.

[7] The rationale for the terms 'bra' and 'ket' should now be apparent.

## C.1.4 Wave functions

Suppose that $\Psi$ is the state vector for a particle, and that $|\mathbf{r}\rangle$ represents the position of particle. The projection of the state vector onto the position basis,

$$\psi(\mathbf{r}) = \langle\mathbf{r}|\Psi\rangle, \tag{C.15}$$

is traditionally called the *wave function* of the particle. The interpretation given to the wave function is that $\langle\psi(\mathbf{r})|\psi(\mathbf{r})\rangle = |\psi(\mathbf{r})|^2$ is the probability density for detecting the particle at a point $\mathbf{r}$. In other words, $|\psi(\mathbf{r})|^2\,\mathrm{d}V$ is the probability of finding the particle in a volume $\mathrm{d}V$ centred on $\mathbf{r}$.

## C.1.5  Measurement and operators

The basis functions contained in the state vector represent states of well-defined values of a particular physical observable. Consider a quantum system described by the state vector

$$|\Psi\rangle = \sum_i c_i |i\rangle, \tag{C.16}$$

where $|i\rangle$ is a basis ket for an observable $q$. According to the Copenhagen interpretation, a measurement of $q$ will give a value $q_i$ with probability $|c_i|^2$. Physically, we expect measurement to be reproducible,[8] so subsequent measurements will yield $q_i$ with probability 1. We must conclude, then, that the initial measurement changed the quantum state of the system from $|\Psi\rangle$ to $|i\rangle$. This abrupt change is called the *collapse of the wave function*, and is one of the enduring mysteries of quantum mechanics.

Physical observables in quantum mechanics are represented by linear operators.[9] Consider the following mathematical entity

$$Q = \sum_i q_i |i\rangle\langle i|. \tag{C.17}$$

This odd-looking thing has some interesting and useful properties.

First, for a state vector in the form (C.16),

$$
\begin{aligned}
\langle\Psi|Q|\Psi\rangle &= \sum_i q_i \langle\Psi|i\rangle\langle i|\Psi\rangle \\
&= \sum_i q_i |c_i|^2 \\
&= \sum_i q_i p_i \\
&= \langle q\rangle.
\end{aligned}
\tag{C.18}
$$

Here, $\langle q\rangle$ is the mean value of the observable $q$ obtained from many measurements on identical copies of a system described by the ket $|\Psi\rangle$. We call this the *expectation value* of $Q$, and it is often written as $\langle Q\rangle$. Results (C.10), (C.12), and (C.14) were used in deriving (C.18).

Second, if $q_i = 1$ for all $i$ then we have the identity operator,

$$1 = \sum_i |i\rangle\langle i|. \tag{C.19}$$

This equation is known as the *closure* or *completeness* relation.

Third, when we apply $Q$ to the basis ket $|j\rangle$ we obtain

$$
\begin{aligned}
Q|j\rangle &= \sum_i q_i |i\rangle\langle i|j\rangle \\
&= \sum_i q_i |i\rangle \delta_{ij} \\
&= q_j |j\rangle.
\end{aligned}
\tag{C.20}
$$

This has the form of an eigenvalue equation, with $q_j$ the eigenvalue and $|j\rangle$ the eigenvector.

[8] Providing nothing is done to disturb the system in between measurements.

[9] An operator $A$ is linear if it satisfies $A(a_1\Psi_1 + a_2\Psi_2) = a_1 A\Psi_1 + a_2 A\Psi_2$ for any complex numbers $a_j$ and for any functions $\Psi_j$.

Fourth, by a similar set of manipulations it can be shown that

$$\langle\Psi_1|Q|\Psi_2\rangle = \langle\Psi_2|Q|\Psi_1\rangle^*. \tag{C.21}$$

An operator which has this property is said to be *Hermitian*. Hermitian operators have real eigenvalues and mutually orthogonal eigenvectors. The former is needed if eigenvalues are to be the results of physical measurements, and the latter is a requirement of the basis functions of the state vector — see (C.11). These and other considerations justify one of the fundamental postulates of quantum mechanics, that all physical observables are represented by Hermitian operators.

One very special Hermitian operator is the energy operator $\mathcal{H}$, which is also called the *Hamiltonian*. Suppose that $|E_i\rangle$ denotes an energy eigenvector. Then by solving

$$\mathcal{H}|E_i\rangle = E_i|E_i\rangle. \tag{C.22}$$

one obtains the energy eigenvalues $E_i$. For historical reasons (C.22) is known as the time-independent Schrödinger equation.[10]

The easiest way to solve an operator eigenvalue equation is to form the matrix $Q_{ij} = \langle i|Q|j\rangle$, where $\{|j\rangle\}$ is a complete set of basis functions which in general will not be the eigenstates of $Q$.[11] The individual matrix elements satisfy (C.21) and so $Q_{ij}$ is a Hermitian matrix.

Non-Hermitian operators also crop up in quantum mechanics, though not in association with physical observables. If $A$ is such a non-Hermitian operator then there exists a dual operator $A^\dagger$, called the Hermitian adjoint, related to $A$ by

$$\langle\Psi_1|A^\dagger|\Psi_2\rangle = \langle\Psi_2|A|\Psi_1\rangle^* = \langle A\Psi_1|\Psi_2\rangle. \tag{C.23}$$

The second equality follows from (C.10). Hermitian operators are seen from (C.21) to be self-adjoint. Examples of non-Hermition adjoint operators are the spin raising and lowering operators $S^+$ and $S^-$.

Operators frequently need to be multiplied together. The product $AB$ means operate first with $B$ then apply $A$ to the result. The adjoint of a product can be found from repeated application of (C.23):

$$\begin{aligned}
\langle\Psi_1|(AB)^\dagger|\Psi_2\rangle &= \langle\Psi_2|AB|\Psi_1\rangle^* \\
&= \langle AB\Psi_1|\Psi_2\rangle \\
&= \langle B\Psi_1|A^\dagger|\Psi_2\rangle \\
&= \langle\Psi_1|B^\dagger A^\dagger|\Psi_2\rangle.
\end{aligned} \tag{C.24}$$

Hence,

$$(AB)^\dagger = B^\dagger A^\dagger. \tag{C.25}$$

## C.1.6  Commutation relations

The commutator of two operators $A$ and $B$ is defined to be

$$[A, B] \equiv AB - BA. \tag{C.26}$$

[10] Erwin Schrödinger (1887–1961) was one of the pioneers of quantum mechanics. He developed (C.22) in the form of a differential equation for the wave function of a particle. His famous cat was devised as a thought experiment to illustrate what Schrödinger saw as the absurdity that a physical system can be in a mixture of states which only becomes well-defined upon the act of measurement (the collapse of the wave function).

[11] If the basis functions are eigenstates of $Q$ then $Q_{ij}$ is trivially a diagonal matrix.

A *commutation relation* is an expression for $[A, B]$. If $[A, B] = 0$ then the two operators are said to *commute*, which means that there is a complete set of eigenfunctions common to both $A$ and $B$. If $A$ and $B$ represent physical observables and commute then after a measurement of $A$ a measurement of $B$ does not affect the value of $A$, and *vice versa*.

### C.1.7   Baker–Campbell–Hausdorff formula

We shall need the following relation,

$$\exp A \exp B = \exp(A + B) \exp(\tfrac{1}{2}[A, B]), \qquad (C.27)$$

which is a special case of the Baker–Campbell–Hausdorff formula that applies when $A$ and $B$ are operators whose commutator $[A, B] = c$ is a complex number (i.e. not an operator).

We first prove that

$$[\exp(\lambda A), B] = \lambda c \exp(\lambda A), \qquad (C.28)$$

where $\lambda$ is a c-number. Consider a general term in the series expansion of $\exp(\lambda A)$:

$$\begin{aligned}
[(\lambda A)^n, B] &= \lambda^n (A^n B - B A^n) \\
&= \lambda^n (A^{n-1} A B - B A^n) \\
&= \lambda^n (c A^{n-1} + A^{n-1} B A - B A^n),
\end{aligned}$$

The last line follows from $AB = c + BA$. Repetition $n - 1$ more times shifts $B$ to the front of the middle term on the right-hand side, giving

$$[(\lambda A)^n, B] = n \lambda c (\lambda A)^{n-1}. \qquad (C.29)$$

Relation (C.28) then follows, as (C.29) applies to each term in the series expansion of $\exp(\lambda A)$.

Now, consider the function

$$f(\lambda) = \exp(\lambda A) \exp(\lambda B). \qquad (C.30)$$

Using (C.28) we can write the derivative of $f$ as

$$\begin{aligned}
\frac{\mathrm{d}f}{\mathrm{d}\lambda} &= A \exp(\lambda A) \exp(\lambda B) + \exp(\lambda A) B \exp(\lambda B) \\
&= (A + B + \lambda c) f(\lambda),
\end{aligned}$$

and after integration we obtain

$$f(\lambda) = K \exp(\lambda A + \lambda B) \exp(\tfrac{1}{2}\lambda^2 c). \qquad (C.31)$$

By comparing (C.30) and (C.31) for the case $\lambda = 0$ we see that the integration constant is $K = 1$. Formula (C.27) is obtained when $\lambda = 1$.

## C.1.8   Time dependence

Another of the postulates of quantum mechanics states that the time dependence of a state vector is governed by an equation called the time-dependent Schrödinger equation,

$$\mathcal{H}|\Psi\rangle = i\hbar\frac{\partial}{\partial t}|\Psi\rangle. \tag{C.32}$$

This equation cannot be derived from first principles, but it is possible to show that it is reasonable on various physical grounds.

For an energy eigenstate $|E_i\rangle$ which satisfies (C.22) we obtain

$$\mathcal{H}|E_i\rangle = i\hbar\frac{\partial|E_i\rangle}{\partial t} = E_i|E_i\rangle, \tag{C.33}$$

which has solution

$$|E_i(t)\rangle = \exp(-iE_it/\hbar)|E_i(0)\rangle. \tag{C.34}$$

This can be used to obtain a simple expression for the time evolution of an arbitrary state vector. First, we express the state vector as a linear combination of energy eigenstates

$$|\Psi(t)\rangle = \sum_i c_i(t)|E_i(t)\rangle. \tag{C.35}$$

Application of (C.32) leads to

$$i\hbar\sum_i \dot{c}_i|E_i(t)\rangle = 0, \tag{C.36}$$

where the dot stands for $\partial/\partial t$. As (C.36) is true for all time we must have $\dot{c}_i = 0$ for all $i$. It then follows from (C.34) and (C.35) that

$$|\Psi(t)\rangle = \sum_i c_i \exp(-iE_it/\hbar)|E_i(0)\rangle. \tag{C.37}$$

Another way to write (C.37) is

$$|\Psi(t)\rangle = \exp(-i\mathcal{H}t/\hbar)|\Psi(0)\rangle. \tag{C.38}$$

The exponential function of $\mathcal{H}$ represents the power series expansion $e^x = 1 + x + x^2/2! + \dots$ with $x = -i\mathcal{H}t/\hbar$. To derive (C.38), write $|\Psi(0)\rangle$ in the form (C.35) and apply the first equality in (C.33) repeatedly to convert the right-hand side of (C.38) into the right-hand side of (C.37). The adjoint of (C.38) is

$$\langle\Psi(t)| = \langle\Psi(0)|\exp(i\mathcal{H}t/\hbar). \tag{C.39}$$

The time-dependent Schrödinger equation (C.32) and its Hermitian adjoint $\langle\Psi|\mathcal{H} = -i\hbar\,\partial\langle\Psi|/\partial t$ can be used to obtain the rate of change of the expectation value $\langle q \rangle$ of an operator $Q$:

$$\frac{\mathrm{d}\langle q \rangle}{\mathrm{d}t} = \frac{\mathrm{d}}{\mathrm{d}t}\langle \Psi|Q|\Psi \rangle$$

$$= \frac{1}{\mathrm{i}\hbar}\langle \Psi|Q\mathcal{H}|\Psi \rangle - \frac{1}{\mathrm{i}\hbar}\langle \Psi|\mathcal{H}Q|\Psi \rangle + \langle \Psi|\frac{\partial Q}{\partial t}|\Psi \rangle$$

$$= \frac{1}{\mathrm{i}\hbar}\langle [Q,\mathcal{H}] \rangle + \left\langle \frac{\partial Q}{\partial t} \right\rangle. \tag{C.40}$$

If $Q$ does not depend explicitly on time then the rate of change of $\langle q \rangle$ is proportional to the expectation value of the commutator $[Q,\mathcal{H}]$. It follows that if $Q$ commutes with the Hamiltonian then $\langle q \rangle$ is conserved.

## C.1.9 Fermi's Golden Rule

In scattering theory, we need the transition rate from an initial energy eigenstate into other energy eigenstates due to a perturbation.

We begin with a state vector $|\Psi(t)\rangle$ expanded as (C.37) in terms of the energy eigenstates $|E_i\rangle$ of a Hamiltonian $\mathcal{H}_0$. At $t = 0$ we switch on a perturbation $V$ so that the Hamiltonian becomes $\mathcal{H} = \mathcal{H}_0 + V$. We can obtain an expression for the change in the state vector by combining $\mathcal{H}|\Psi\rangle = \mathrm{i}\hbar\partial|\Psi\rangle/\partial t$ from (C.32) with $\mathcal{H}_0|E_i\rangle = E_i|E_i\rangle$ from (C.22),

$$\mathrm{i}\hbar\sum_i \frac{\partial c_i}{\partial t}\exp(-\mathrm{i}E_it/\hbar)|E_i\rangle = \sum_i c_i(t)\exp(-\mathrm{i}E_it/\hbar)V|E_i\rangle. \tag{C.41}$$

Applying $\langle E_j|$ to both sides we obtain

$$\mathrm{i}\hbar\frac{\partial c_j}{\partial t} = \sum_i \langle E_j|V|E_i \rangle\, c_i(t)\exp\{-\mathrm{i}(E_i - E_j)t/\hbar\}. \tag{C.42}$$

Suppose now that the perturbation has the explicit time dependence $V\exp(-\mathrm{i}\omega t)$, where $V$ is independent of time,[12] and that at $t = 0$ the system is in a pure eigenstate $|E_0\rangle$. If $V$ is very weak then at short times the system will remain close to its initial state, so we can put $c_0(t) \approx c_0(0) = 1$ and $c_{j\neq 0}(t) \approx 0$ on the right side of (C.42). Integrating between 0 and $t$ we obtain the first-order correction to the amplitude:

$$\mathrm{i}\hbar c_j(t) = 2\langle E_j|V|E_0 \rangle\exp(-\mathrm{i}\Omega t/2)\frac{\sin(\Omega t/2)}{\Omega}, \tag{C.43}$$

where $\Omega = (E_0 - E_j + \hbar\omega)/\hbar$. The transition rate into $|E_j\rangle$ is then[13]

$$R_{0\to j} = \frac{\mathrm{d}|c_j|^2}{\mathrm{d}t} = \frac{2}{\hbar^2}|\langle E_j|V|E_0 \rangle|^2\frac{\sin\Omega t}{\Omega}. \tag{C.44}$$

The function $\sin\Omega t/\Omega$ is a sharply peaked function centred on $\Omega = 0$ with amplitude $t$, so the transition rate is largest when $E_j = E_0 + \hbar\omega$, and increases linearly with time.

In practice, many systems of interest have a large number of final states with energies close to $E_j$, and we usually want to sum over all these states. We can do this by introducing a continuous final energy $E$ in place of the discrete levels $E_j$, and define the density of states function $g(E)$ such that $g(E)\mathrm{d}E$ is the number of states between $E$ and $E+\mathrm{d}E$. For sufficiently large $t$, the function $\sin\Omega t/\Omega$ can be represented by $\pi\hbar\delta(E - E_0 - \hbar\omega)$ — see eqns (B.8) and (B.11)–(B.13). Integration of (C.44) over $E$ then gives

$$R_{0\to 1} = \frac{2\pi}{\hbar}\,|\langle E_1|V|E_0\rangle|^2 g(E_1), \qquad (C.45)$$

where $E_1 = E_0 + \hbar\omega$. Expression (C.45) is *Fermi's Golden Rule*.[14]

In neutron scattering, the system usually consists of a sample with discrete energy levels and the neutron with its quasi-continuous plane-wave states $|k\rangle$ and discrete spin states $|\sigma\rangle$. In this case, the initial and final states of the system are $|\mathrm{i}\rangle \equiv |k_\mathrm{i}\sigma_\mathrm{i}E_0\rangle$ and $|\mathrm{f}\rangle \equiv |k_\mathrm{f}\sigma_\mathrm{f}E_1\rangle$, and the density of states now refers to the scattered neutron energies $E_\mathrm{f}$. Moreover, the time dependence $\exp(-\mathrm{i}\omega t)$ of the perturbation comes from the time dependence of the neutron planes waves, $\exp(-\mathrm{i}Et/\hbar)$, so that $\hbar\omega = E_\mathrm{i} - E_\mathrm{f}$. Hence, Fermi's Golden Rule takes the form

$$R_{\mathrm{i}\to\mathrm{f}} = \frac{2\pi}{\hbar}\,|\langle \mathrm{f}|V|\mathrm{i}\rangle|^2 g(E_\mathrm{f}), \qquad (C.46)$$

with energy conservation condition $E_1 + E_\mathrm{f} = E_0 + E_\mathrm{i}$.

## C.1.10   Heisenberg operators

Up to now, we have associated physical observables with operators that are constant in time, and built all the time dependence into the state vector $|\Psi(t)\rangle$ which obeys eqn (C.32). This formulation is known as the Schrödinger picture, or Schrödinger representation. An alternative approach is to contain all the time dependence in the operators and have the state vectors independent of time. This is the Heisenberg picture.[15]

The identities (C.38) and (C.39) allow us to construct a time-dependent *Heisenberg operator* $\hat{A}(t)$ from an operator $A$ as follows[16]

$$\hat{A}(t) = \exp(\mathrm{i}\mathcal{H}t/\hbar)\, A \, \exp(-\mathrm{i}\mathcal{H}t/\hbar). \qquad (C.47)$$

A simple way to demonstrate the equivalence of the Schrödinger and Heisenberg pictures is to consider the time dependence of the expectation value of an operator $A$,

$$\langle A(t)\rangle = \langle\Psi(t)|A|\Psi(t)\rangle \qquad \text{(Schrödinger)}$$
$$= \langle\Psi(0)|\exp(\mathrm{i}\mathcal{H}t/\hbar)A\exp(-\mathrm{i}\mathcal{H}t/\hbar)|\Psi(0)\rangle$$
$$= \langle\Psi(0)|\hat{A}(t)|\Psi(0)\rangle \qquad \text{(Heisenberg)}. \qquad (C.48)$$

By differentiating (C.47) with respect to time we obtain the time dependence of a Heisenberg operator:

$$\mathrm{i}\hbar\frac{\mathrm{d}\hat{A}}{\mathrm{d}t} = [\hat{A}, \mathcal{H}]. \qquad (C.49)$$

Equation (C.40) is recovered if we pre- and post-multiply both sides by $\langle\Psi(0)|$ and $|\Psi(0)\rangle$.

[14]A result very similar to (C.45) was derived earlier by Dirac (Dirac, 1927).

[15]Werner Heisenberg (1901–1976) was another pioneer of quantum mechanics. His key insight was to see the importance of measurement in the description of quantum systems, and together with Born and Jordan he developed the matrix formulation of quantum mechanics.

[16]The 'hat' over the $A$ in $\hat{A}(t)$ is used throughout this book to distinguish a Heisenberg time-dependent operator from the operator that represents the physical observable in the Schrödinger picture. Note that $\hat{A}(0) = A$.

## C.1.11   Correlation functions

A correlation function describes how the value of a physical quantity at a particular place and time relates to the value of another physical quantity at a different place and different time. Consider two properties $A(\mathbf{r},t)$ and $B(\mathbf{r},t)$ described by Heisenberg operators $\hat{A}(\mathbf{r},t)$ and $\hat{B}(\mathbf{r},t)$. The correlation function for these quantities is defined by

$$\langle \hat{A}(\mathbf{r},t)\hat{B}(\mathbf{r}',t')\rangle = \sum_{\lambda} p_\lambda \langle\lambda| \hat{A}(\mathbf{r},t)\hat{B}(\mathbf{r}',t') |\lambda\rangle, \tag{C.50}$$

where $p_\lambda$ is the probability that the system is in energy eigenstate $|\lambda\rangle$. If the system is in thermal equilibrium with its surroundings,[17] then

$$p_\lambda = \frac{1}{Z}\exp(-\beta E_\lambda), \tag{C.51}$$

where

$$Z = \sum_{\lambda}\exp(-\beta E_\lambda) \tag{C.52}$$

is the partition function, and $\beta = 1/(k_{\rm B}T)$. We shall omit the $\mathbf{r}$ and $\mathbf{r}'$ labels for the rest of this section to simplify the notation.

If the scattering system is in thermal equilibrium then we would not expect the correlation function to depend on the origin of time. To prove this, we use the closure relation (C.19) and (3.26) to write (C.50) as

$$\langle \hat{A}(t)\hat{B}(t')\rangle = \sum_{\lambda,\lambda'} p_\lambda\langle\lambda|\hat{A}(t)|\lambda'\rangle\langle\lambda'|\hat{B}(t')|\lambda\rangle$$

$$= \sum_{\lambda,\lambda'} p_\lambda\langle\lambda|A|\lambda'\rangle\langle\lambda'|B|\lambda\rangle \exp\{{\rm i}(E_\lambda - E_{\lambda'})(t-t')/\hbar\}. \tag{C.53}$$

This shows that the correlation function depends on the difference $t-t'$, and so it is independent of the origin of time, as expected. In particular, the correlation function that appears in neutron scattering satisfies,

$$\langle \hat{A}^\dagger\hat{A}(t)\rangle = \langle \hat{A}^\dagger(t_0)\hat{A}(t_0+t)\rangle. \tag{C.54}$$

By interchanging $\lambda$ and $\lambda'$ in (C.53) we can obtain the following expression for the thermal average of a commutator,

$$\langle[\hat{A}(t),\hat{B}(t')]\rangle = \langle\hat{A}(t)\hat{B}(t')\rangle - \langle\hat{B}(t')\hat{A}(t)\rangle$$

$$= \sum_{\lambda,\lambda'}(p_{\lambda'} - p_\lambda)\langle\lambda|B|\lambda'\rangle\langle\lambda'|A|\lambda\rangle$$

$$\times \exp\{{\rm i}(E_{\lambda'} - E_\lambda)(t-t')/\hbar\}. \tag{C.55}$$

Another identity we require is

$$\langle\hat{A}\hat{B}(t)\rangle = \langle\hat{B}\hat{A}(-t+{\rm i}\hbar\beta)\rangle. \tag{C.56}$$

This is proved as follows. From (C.38), (C.47), (C.50), and (C.51),

$$\langle\hat{A}\hat{B}(t)\rangle = \frac{1}{Z}\sum_{\lambda}\exp(-\beta E_\lambda)\langle\lambda| \hat{A}\hat{B}(t) |\lambda\rangle$$

[17]If the system can exchange particles as well as heat with its surroundings then the summation in eqn (C.52) must extend over the energy levels of all available states with different numbers of particles.

$$= \frac{1}{Z} \sum_\lambda \langle\lambda| \hat{A}\hat{B}(t) \exp(-\beta\hat{\mathcal{H}}) |\lambda\rangle$$

$$= \frac{1}{Z} \sum_\lambda \langle\lambda| \hat{B}(t) \exp(-\beta\hat{\mathcal{H}})\hat{A} |\lambda\rangle$$

$$= \frac{1}{Z} \sum_\lambda \exp(-\beta E_\lambda)\langle\lambda| \exp(\beta\hat{\mathcal{H}})\hat{B}(t) \exp(-\beta\hat{\mathcal{H}})\hat{A} |\lambda\rangle$$

$$= \langle\hat{B}(t - i\hbar\beta)\hat{A}\rangle$$

$$= \langle\hat{B}\hat{A}(-t + i\hbar\beta)\rangle.$$

The last step follows from property (C.54). Between the second and third lines we have used the identity $\sum_\lambda \langle\lambda| \hat{A}\hat{B} |\lambda\rangle = \sum_\lambda \langle\lambda| \hat{B}\hat{A} |\lambda\rangle$, which follows from the closure relation (C.19).[18]

The final result we need is

$$\langle\hat{A}^\dagger \hat{B}(t)\rangle^* = \langle\hat{B}^\dagger(t)\hat{A}\rangle. \tag{C.57}$$

This follows from the definition (C.23) of the Hermition adjoint operator and from the adjoint of a product of operators, eqn (C.25). Both (C.23) and (C.25) hold for Heisenberg operators, and (C.57) follows when they are applied to the definition of the correlation function (C.50).

## C.1.12   Quantum harmonic oscillator

In this section we obtain some properties of operators used to describe the one-dimensional quantum harmonic oscillator (QHO), which has energy eigenstates $|n\rangle$ with integer $n$ and eignevalues given by

$$E_n = (n + \tfrac{1}{2})\hbar\omega_0. \tag{C.58}$$

In Section 5.1.1 we introduced the operators $a$ and $a^\dagger$,

$$a = \left(\frac{1}{2M\hbar\omega_0}\right)^{1/2} (M\omega_0 x + ip_x)$$

$$a^\dagger = \left(\frac{1}{2M\hbar\omega_0}\right)^{1/2} (M\omega_0 x - ip_x), \tag{C.59}$$

and showed that the one-dimensional QHO Hamiltonian can be written

$$\mathcal{H} = (a^\dagger a + \tfrac{1}{2})\hbar\omega_0. \tag{C.60}$$

The $a$ and $a^\dagger$ operators satisfy the Bose commutation relations

$$[a, a^\dagger] = 1, \quad [a, a] = [a^\dagger, a^\dagger] = 0, \tag{C.61}$$

and convert between eigenstates that differ by one energy quantum:

$$a|n\rangle = n^{\frac{1}{2}} |n - 1\rangle \tag{C.62}$$

$$a^\dagger|n\rangle = (n + 1)^{\frac{1}{2}} |n + 1\rangle. \tag{C.63}$$

[18] Specifically,

$$\sum_\lambda \langle\lambda| \hat{A}\hat{B} |\lambda\rangle = \sum_{\lambda,\lambda'} \langle\lambda|\hat{A}|\lambda'\rangle\langle\lambda'|\hat{B}|\lambda\rangle$$

$$= \sum_{\lambda,\lambda'} \langle\lambda'|\hat{B}|\lambda\rangle\langle\lambda|\hat{A}|\lambda'\rangle$$

$$= \sum_{\lambda'} \langle\lambda'| \hat{B}\hat{A} |\lambda'\rangle.$$

432  *Quantum Mechanics, Atomic Physics, and Magnetism*

In Section 5.1.2 we found that

$$\langle a^\dagger a \rangle = \frac{1}{\exp(\beta\hbar\omega_0) - 1},\tag{C.64}$$

which gives the average number of quanta in the oscillator when it is in thermal equilibrium at temperature $T$.

The first result we need is the following matrix element for the product of $m$ annihilation and $m$ creation operators:

$$\langle n| a^m (a^\dagger)^m |n\rangle = \frac{(n+m)!}{n!}.\tag{C.65}$$

This follows immediately from repeated application of (C.62) and (C.63).

Second, we obtain the explicit time dependence of the Heisenberg operators corresponding to $a$ and $a^\dagger$. The Heisenberg operator corresponding to $a$ is[19]

$$\hat{a}(t) = \exp(i\mathcal{H}t/\hbar)\, a \, \exp(-i\mathcal{H}t/\hbar),$$

and so

$$\frac{d}{dt}\hat{a}(t) = \frac{i}{\hbar}[\mathcal{H}, \hat{a}(t)] \qquad \text{[from (C.49)]}$$

$$= \frac{i}{\hbar}[a^\dagger a, \hat{a}(t)]\,\hbar\omega_0 \qquad \text{[from (C.60)]}$$

$$= -i\omega_0\hat{a}(t).\tag{C.66}$$

Integration of (C.66) and a similar equation for $\hat{a}^\dagger(t)$ leads to

$$\hat{a}(t) = a\exp(-i\omega_0 t),\tag{C.67}$$

$$\hat{a}^\dagger(t) = a^\dagger \exp(i\omega_0 t).\tag{C.68}$$

Third, we obtain $\langle x^2 \rangle$, the mean squared displacement of the oscillator. From (C.59),

$$x = Aa + A^* a^\dagger,\tag{C.69}$$

where $A = (\hbar/2M\omega_0)^{1/2}\exp(-i\omega_0 t)$. Hence,[20]

$$\langle x^2 \rangle = \langle (Aa + A^* a^\dagger)(Aa + A^* a^\dagger)\rangle$$

$$= |A|^2\langle aa^\dagger + a^\dagger a\rangle$$

$$= 2|A|^2(\langle a^\dagger a\rangle + \tfrac{1}{2}) \qquad \text{[from(C.61)]}$$

$$= \frac{\hbar}{2M\omega_0}\coth(\tfrac{1}{2}\beta\hbar\omega_0) \qquad \text{[from(C.64)].}\tag{C.70}$$

Finally, we prove Bloch's identity

$$\langle \exp x\rangle = \exp(\tfrac{1}{2}\langle x^2\rangle).\tag{C.71}$$

We start from (C.69), with $A = (\hbar/2M\omega_0)^{1/2}\exp(-i\omega_0 t)$, and make use of the identity

$$\exp A \exp B = \exp(A + B)\exp\{\tfrac{1}{2}[A, B]\},\tag{C.72}$$

[19]Heisenberg operators were introduced in Section C.1.10, and are identified by a 'hat'.

[20]The thermal average is given by

$$\langle x^2\rangle = \sum_{n=0}^{\infty} p_n \langle n|x^2|n\rangle,$$

where $p_n$ is the thermal occupancy of state $|n\rangle$. The matrix elements of the products $aa$ and $a^\dagger a^\dagger$ which are contained in $x^2$ vanish because of properties (C.62) and (C.63).

which is proved in Section C.1.7. The Bose commutation relations (C.61) enable us to exploit (C.72) to write

$$\exp x = \exp(Aa)\exp(A^*a^\dagger)\exp(-\tfrac{1}{2}|A|^2), \tag{C.73}$$

so that the thermal average is[21]

$$\langle \exp x \rangle = \exp(-\tfrac{1}{2}|A|^2)\frac{1}{Z}\sum_{n=0}^{\infty}\langle n|\exp(Aa)\exp(A^*a^\dagger)|n\rangle \, \exp(-\beta E_n). \tag{C.74}$$

Using the series expansion for the exponential function, and identity (C.65), we obtain for the matrix element in (C.74)

$$\langle n|\exp(Aa)\exp(A^*a^\dagger)|n\rangle = \sum_{m=0}^{\infty}\frac{|A|^{2m}}{(m!)^2}\frac{(n+m)!}{n!}, \tag{C.75}$$

so that

$$\langle \exp x \rangle = \exp(-\tfrac{1}{2}|A|^2)\exp(-\tfrac{1}{2}\beta\hbar\omega_0)$$

$$\times \frac{1}{Z}\sum_{n=0}^{\infty}\sum_{m=0}^{\infty}\frac{|A|^{2m}}{(m!)^2}\frac{(n+m)!}{n!}\,\exp(-n\beta\hbar\omega_0). \tag{C.76}$$

Here we substituted for $E_n$ from (C.58). From the binomial theorem

$$\frac{1}{(1-u)^{m+1}} = \sum_{n=0}^{\infty}\frac{(n+m)!}{n!\,m!}u^n, \tag{C.77}$$

and so with $u = \exp(-\beta\hbar\omega_0)$ and $Z = u^{\frac{1}{2}}/(1-u)$ [see (5.18)] we obtain

$$\langle \exp x \rangle = \exp(-\tfrac{1}{2}|A|^2)\sum_{m=0}^{\infty}\frac{|A|^{2m}}{m!}\frac{1}{(1-u)^m}$$

$$= \exp(-\tfrac{1}{2}|A|^2)\exp\left(\frac{|A|^2}{1-u}\right)$$

$$= \exp\{|A|^2(\langle a^\dagger a\rangle + \tfrac{1}{2})\}$$

$$= \exp(\tfrac{1}{2}\langle x^2\rangle). \tag{C.78}$$

Between the second and third lines we used $\langle a^\dagger a\rangle = u/(1-u)$, see (C.64), and to reach the final result we used $\langle x^2\rangle = 2|A|^2(\langle a^\dagger a\rangle + \tfrac{1}{2})$ from the second-to-last line of (C.70).

## C.1.13   Quadratic form of Bose operators: diagonalization

The low-energy degrees of freedom in condensed matter are often described by quantized fields whose collective excitations are bosonic quasi-particles. For example, phonons are quantized lattice vibrations, and magnons are quantized excitations of coupled magnetic moments. A minimal Hamiltonian for such systems is quadratic in the mode amplitudes and so contains bilinear products of boson annihilation and creation operators. We first outline the method presented by White *et al.* (1965) for diagonalizing such a Hamiltonian, and then apply it to the case of two coupled Bose fields.

### Basic theory

A general quadratic form of Bose operators may be written

$$\mathcal{H} = \frac{1}{2}X^\dagger H X, \tag{C.79}$$

where $X$ is a column vector of annihilation and creation operators, $X^\dagger$ is the transposed Hermitian adjoint of $X$, and $H$ is a Hermitian matrix of $c$-number coefficients.[22] We seek the Bogoliubov linear transformation

$$X = SX' \tag{C.80}$$

which converts $\mathcal{H}$ into diagonal form

$$\mathcal{H} = \frac{1}{2}X'^\dagger \Omega X', \tag{C.81}$$

where $\Omega$ is the diagonal matrix of eigenvalues $\hbar\omega_i$ of $\mathcal{H}$.

The operators in $X$ satisfy the Bose commutation relations,[23] which may be expressed as

$$[X, X^\dagger] \equiv X(X^*)^T - (X^*X^T)^T = g, \tag{C.82}$$

where $X^*$ is the Hermitian adjoint of $X$, $X^T$ means the transpose of $X$, and $g$ is a diagonal matrix containing the values of the commutators. Substituting (C.80) in (C.82) we find[24]

$$[X, X^\dagger] = [SX', X'^\dagger S^\dagger]$$

$$= SX'(X'^*)^T(S^*)^T - (S^*X'^*X'^T S^T)^T$$

$$= SX'(X'^*)^T(S^*)^T - S(X'^*X'^T)^T(S^*)^T$$

$$= S[X', X'^\dagger]S^\dagger$$

$$= Sg'S^\dagger, \tag{C.83}$$

where $g'$ is the diagonal matrix of commutators for the new Bose operators $X'$. By equating (C.82) and (C.83) we see that

$$S^{-1} = g'S^\dagger g^{-1} \quad \text{and} \quad (S^\dagger)^{-1} = g^{-1}Sg'. \tag{C.84}$$

[22] That is, real or complex numbers as opposed to quantum operators.

[23] The method is actually more general than this, applying to any operators whose commutators are $c$-numbers, providing a diagonal form exists.

[24] We use the following property of the transpose of $c$-number matrices,

$$(AB\ldots C)^T = C^T\ldots B^T A^T.$$

To obtain an expression for H we first combine (C.79)–(C.81) to obtain[25]

$$H = (S^\dagger)^{-1}\Omega S^{-1},  \tag{C.85}$$

and then use the second expression in (C.84) to give

$$gHS = Sg'\Omega.  \tag{C.86}$$

Since $g$ and $g'$ are diagonal matrices (for Bose operators), eqn (C.86) is an eigenvalue equation in the usual form $Mv = \lambda v$, where $M \equiv gH$, the eigenvectors $v$ are the columns of $S$, and the eigenvalue corresponding to the $i$th column of $S$ is

$$\lambda_i = g'_{ii}\hbar\omega_i.  \tag{C.87}$$

The $\lambda_i$ are the solutions to the secular equation

$$|gH - \lambda I| = 0,  \tag{C.88}$$

where $I$ is the identity matrix with the same dimensions as H, and the eigenvectors are obtained from

$$gHS_i = \lambda_i S_i,  \tag{C.89}$$

where $S_i$ denotes the $i$th column of $S$.

## Two interacting Bose fields

We consider a Hamiltonian containing bilinear products of Bose operators. The general bilinear Hamiltonian for two coupled Bose fields is (Lindgård *et al.*, 1967)

$$\mathcal{H} = \mathcal{H}_0 + \frac{1}{2}\sum_{\mathbf{q}}\left\{ A_{\mathbf{q}}^a(a_{\mathbf{q}}^\dagger a_{\mathbf{q}} + a_{\mathbf{q}}a_{\mathbf{q}}^\dagger) + A_{\mathbf{q}}^b(b_{\mathbf{q}}^\dagger b_{\mathbf{q}} + b_{\mathbf{q}}b_{\mathbf{q}}^\dagger) \right.$$

$$+ C_{\mathbf{q}}^a a_{\mathbf{q}}^\dagger a_{-\mathbf{q}}^\dagger + C_{\mathbf{q}}^{a*} a_{-\mathbf{q}}a_{\mathbf{q}} + C_{\mathbf{q}}^b b_{\mathbf{q}}^\dagger b_{-\mathbf{q}}^\dagger + C_{\mathbf{q}}^{b*} b_{-\mathbf{q}}b_{\mathbf{q}}$$

$$\left. + 2(B_{\mathbf{q}}a_{\mathbf{q}}^\dagger b_{\mathbf{q}} + B_{\mathbf{q}}^* b_{\mathbf{q}}^\dagger a_{\mathbf{q}}) + 2(D_{\mathbf{q}}a_{\mathbf{q}}^\dagger b_{-\mathbf{q}}^\dagger + D_{\mathbf{q}}^* b_{-\mathbf{q}}a_{\mathbf{q}}) \right\}.  \tag{C.90}$$

This may be written in the matrix form (C.79)

$$\mathcal{H} = \mathcal{H}_0 + \frac{1}{2}X_{\mathbf{q}}^\dagger H_{\mathbf{q}} X_{\mathbf{q}},  \tag{C.91}$$

with

$$X_{\mathbf{q}}^\dagger H_{\mathbf{q}} X_{\mathbf{q}} = \left(a_{\mathbf{q}}^\dagger, b_{\mathbf{q}}^\dagger, a_{-\mathbf{q}}, b_{-\mathbf{q}}\right)\begin{pmatrix} A_{\mathbf{q}}^a & B_{\mathbf{q}} & C_{\mathbf{q}}^a & D_{\mathbf{q}} \\ B_{\mathbf{q}}^* & A_{\mathbf{q}}^b & D_{-\mathbf{q}} & C_{\mathbf{q}}^b \\ C_{\mathbf{q}}^{a*} & D_{-\mathbf{q}}^* & A_{-\mathbf{q}}^a & B_{-\mathbf{q}}^* \\ D_{\mathbf{q}}^* & C_{\mathbf{q}}^{b*} & B_{-\mathbf{q}} & A_{-\mathbf{q}}^b \end{pmatrix}\begin{pmatrix} a_{\mathbf{q}} \\ b_{\mathbf{q}} \\ a_{-\mathbf{q}}^\dagger \\ b_{-\mathbf{q}}^\dagger \end{pmatrix}.  \tag{C.92}$$

The elements on the leading diagonal must be real, but the off-diagonal elements are in general complex.[26]

[25] Note that $(S^{-1})^\dagger = (S^\dagger)^{-1}$.

[26] H is a Hermitian matrix, so $H^T = H^*$.

## Example

To illustrate the diagonalization procedure we consider a simplified matrix in which all the elements are real,

$$
H_{\mathbf{q}} = \begin{pmatrix} A_{\mathbf{q}} & B_{\mathbf{q}} & C_{\mathbf{q}} & D_{\mathbf{q}} \\ B_{\mathbf{q}} & A_{\mathbf{q}} & D_{\mathbf{q}} & C_{\mathbf{q}} \\ C_{\mathbf{q}} & D_{\mathbf{q}} & A_{\mathbf{q}} & B_{\mathbf{q}} \\ D_{\mathbf{q}} & C_{\mathbf{q}} & B_{\mathbf{q}} & A_{\mathbf{q}} \end{pmatrix}.
\tag{C.93}
$$

The Bogoliubov transformation (C.80) may be written

$$
\begin{pmatrix} a_{\mathbf{q}} \\ b_{\mathbf{q}} \\ a^\dagger_{-\mathbf{q}} \\ b^\dagger_{-\mathbf{q}} \end{pmatrix} = \begin{pmatrix} S_{11} & S_{12} & S_{13} & S_{14} \\ S_{21} & S_{22} & S_{23} & S_{24} \\ S_{31} & S_{32} & S_{33} & S_{34} \\ S_{41} & S_{42} & S_{43} & S_{44} \end{pmatrix} \begin{pmatrix} \alpha_{\mathbf{q}} \\ \beta_{\mathbf{q}} \\ \alpha^\dagger_{-\mathbf{q}} \\ \beta^\dagger_{-\mathbf{q}} \end{pmatrix}.
\tag{C.94}
$$

For convenience, in both $X_{\mathbf{q}}$ and $X'_{\mathbf{q}}$ we have put the annihilation operators first, followed by the corresponding creation operators.[27] The commutator matrices for this choice are

[27]The order of the operators is arbitrary, but if a different order is chosen then the diagonal elements of $\mathbf{g}$ and $\mathbf{g}'$ in (C.95) will need to be reordered too.

$$
\mathbf{g} = \mathbf{g}' = \begin{pmatrix} 1 & 0 & 0 & 0 \\ 0 & 1 & 0 & 0 \\ 0 & 0 & -1 & 0 \\ 0 & 0 & 0 & -1 \end{pmatrix}.
\tag{C.95}
$$

From (C.88), the values for $\lambda$ are obtained from

$$
\begin{vmatrix} A_{\mathbf{q}} - \lambda & B_{\mathbf{q}} & C_{\mathbf{q}} & D_{\mathbf{q}} \\ B_{\mathbf{q}} & A_{\mathbf{q}} - \lambda & D_{\mathbf{q}} & C_{\mathbf{q}} \\ -C_{\mathbf{q}} & -D_{\mathbf{q}} & -A_{\mathbf{q}} - \lambda & -B_{\mathbf{q}} \\ -D_{\mathbf{q}} & -C_{\mathbf{q}} & -B_{\mathbf{q}} & -A_{\mathbf{q}} - \lambda \end{vmatrix} = 0,
\tag{C.96}
$$

which gives

$$
\lambda^2 = (A_{\mathbf{q}} \pm B_{\mathbf{q}})^2 - (C_{\mathbf{q}} \pm D_{\mathbf{q}})^2.
\tag{C.97}
$$

[28]There are two normal modes for each $\mathbf{q}$ because there are two coupled Bose fields. One mode is associated with the $\alpha$ operator, and the other with the $\beta$ operator. The dimension of $H_{\mathbf{q}}$ is twice the number of eigenvalues of $\mathcal{H}$ because $X_{\mathbf{q}}$ contains both the field operators and their adjoints.

From (C.87) we identify the two eigenvalues[28] of $\mathcal{H}$ as

$$
\hbar\omega_\alpha \equiv \hbar\omega_1 = \sqrt{(A_{\mathbf{q}} + B_{\mathbf{q}})^2 - (C_{\mathbf{q}} + D_{\mathbf{q}})^2}
$$
$$
\hbar\omega_\beta \equiv \hbar\omega_2 = \sqrt{(A_{\mathbf{q}} - B_{\mathbf{q}})^2 - (C_{\mathbf{q}} - D_{\mathbf{q}})^2},
\tag{C.98}
$$

where the positive roots are taken, and we index the solutions for $\lambda$

$$
\lambda_1 = -\lambda_3 = \hbar\omega_\alpha, \qquad \lambda_2 = -\lambda_4 = \hbar\omega_\beta.
\tag{C.99}
$$

We now take the $\lambda_i$ one at a time and solve the four simultaneous equations contained in (C.89) to obtain the corresponding column of S. For example, for the first column ($\lambda_1 = \hbar\omega_\alpha$) we find

$$
S_1 = S_{11} \begin{pmatrix} 1 \\ 1 \\ x \\ x \end{pmatrix},
\tag{C.100}
$$

where $x = -(C_{\mathbf{q}} + D_{\mathbf{q}})/(A_{\mathbf{q}} + B_{\mathbf{q}} + \omega_\alpha)$. Now we have to find $\mathsf{S}_{11}$ and three other multiplicative constants, one for each of $\mathsf{S}_2$, $\mathsf{S}_3$ and $\mathsf{S}_4$. The magnitudes of these constants are determined from the condition[29] $\mathsf{S}^{-1}\mathsf{S} = I$, and the phase[30] is fixed by the requirement that the expressions for the operators and their adjoints in eqn (C.94) must be consistent, i.e. compare rows 1 and 3 and rows 2 and 4. The final result may be written

[29]$\mathsf{S}^{-1}$ can be obtained from (C.84).

[30]The multiplicative constants for each column are in general complex.

$$\mathsf{S} = \begin{pmatrix} u_\alpha^+ & u_\beta^- & -v_\alpha^+ & v_\beta^- \\ u_\alpha^+ & -u_\beta^- & -v_\alpha^+ & -v_\beta^- \\ -v_\alpha^+ & v_\beta^- & u_\alpha^+ & u_\beta^- \\ -v_\alpha^+ & -v_\beta^- & u_\alpha^+ & -u_\beta^- \end{pmatrix}, \qquad (C.101)$$

where

$$u_\alpha^+ = \frac{1}{2}\sqrt{\frac{A_{\mathbf{q}} + B_{\mathbf{q}} + \hbar\omega_\alpha}{\hbar\omega_\alpha}}, \qquad u_\beta^- = \frac{1}{2}\sqrt{\frac{A_{\mathbf{q}} - B_{\mathbf{q}} + \hbar\omega_\beta}{\hbar\omega_\beta}},$$

$$v_\alpha^+ = s\frac{1}{2}\sqrt{\frac{A_{\mathbf{q}} + B_{\mathbf{q}} - \hbar\omega_\alpha}{\hbar\omega_\alpha}}, \qquad v_\beta^- = t\frac{1}{2}\sqrt{\frac{A_{\mathbf{q}} - B_{\mathbf{q}} - \hbar\omega_\beta}{\hbar\omega_\beta}},$$

$$(C.102)$$

with $s = (C_{\mathbf{q}} + D_{\mathbf{q}})/|C_{\mathbf{q}} + D_{\mathbf{q}}|$ and $t = (D_{\mathbf{q}} - C_{\mathbf{q}})/|D_{\mathbf{q}} - C_{\mathbf{q}}|$. The explicit inclusion of the phase factors $s$ and $t$ means that we take the positive roots in (C.102).

## C.1.14 Angular momentum

In quantum mechanics, angular momentum is represented by a dimensionless operator $\mathbf{J}$ whose components satisfy the commutation relations[31]

[31]Including dimensions, the angular momentum is $\mathbf{J}\hbar$ and the commutation relations (C.103) are $[J_x, J_y] = \mathrm{i}\hbar J_z$, etc.

$$[J_x, J_y] = \mathrm{i}J_z, \qquad [J_y, J_z] = \mathrm{i}J_x, \qquad [J_z, J_x] = \mathrm{i}J_y. \qquad (C.103)$$

As the components do not commute with one another it is impossible to find a complete set of simultaneous eigenstates of two components of $\mathbf{J}$. We can, however, find a complete set of simultaneous eigentates of $\mathbf{J}^2 = J_x^2 + J_y^2 + J_z^2$ and one component of $\mathbf{J}$, since $\mathbf{J}^2$ commutes with any one of $J_x$, $J_y$ or $J_z$. By convention we choose this component to be $J_z$, so that $[\mathbf{J}^2, J_z] = 0$. We also define the operators

$$J_+ = J_x + \mathrm{i}J_y, \qquad J_- = J_x - \mathrm{i}J_y. \qquad (C.104)$$

The $J_\pm$ operators commute with $\mathbf{J}^2$, and from (C.103) and (C.104)

$$[J_z, J_+] = J_+, \qquad [J_z, J_-] = -J_-. \qquad (C.105)$$

From (C.103)–(C.105) it can be shown that the eigenvalues of $\mathbf{J}^2$ are given by $J(J+1)$, where $J = 0, \frac{1}{2}, 1, \frac{3}{2}, \ldots$, and that the eigenvalues $M_J$ of $J_z$ run from $-J$ to $J$ in integer steps, i.e. $M_J = -J, -J+1, \ldots, J-1, J$.

Hence, if we represent the simultaneous eigenstates of $\mathbf{J}^2$ and $J_z$ by the kets $|J, M_J\rangle$, then

$$\mathbf{J}^2|J, M_J\rangle = J(J+1)|J, M_J\rangle \tag{C.106}$$

$$J_z|J, M_J\rangle = M_J|J, M_J\rangle. \tag{C.107}$$

Furthermore, it can be shown that

$$J_\pm|J, M_J\rangle = \sqrt{(J \mp M_J)(J \pm M_J + 1)}\,|J, M_J \pm 1\rangle, \tag{C.108}$$

from which $J_+|J, J\rangle = J_-|J, -J\rangle = 0$.

In atomic physics there are three types of angular momentum operator, denoted by $\mathbf{L}$ (orbital), $\mathbf{S}$ (spin), and $\mathbf{J} = \mathbf{L} + \mathbf{S}$ (the total, or combined, angular momentum). Orbital angular momentum corresponds to physical rotations, whereas spin angular momentum has no classical analogue and is an intrinsic property of particles. The operators satisfy all the commutation relations given above, but the quantum numbers $L$ and $M_L$ are confined to integer values. This is because a rotation through $2\pi$ in real space must leave the orbital angular momentum eigenstates invariant.[32] The quantum numbers $S$ and $M_S$ can take integer or half-odd-integer values. It follows from the addition of angular momentum that $J$ and $M_J$ can also take integer or half-odd-integer values.

### C.1.15  Matrix elements for $S = \frac{1}{2}$

The simultaneous eigenkets $|S, M_S\rangle = |\frac{1}{2}, \pm\frac{1}{2}\rangle$ form a complete basis for $S = \frac{1}{2}$. From (C.107), they satisfy

$$S_z|+\tfrac{1}{2}\rangle = \tfrac{1}{2}|+\tfrac{1}{2}\rangle, \qquad S_z|-\tfrac{1}{2}\rangle = -\tfrac{1}{2}|-\tfrac{1}{2}\rangle, \tag{C.109}$$

where $|\frac{1}{2}, \pm\frac{1}{2}\rangle$ is shortened to $|\pm\frac{1}{2}\rangle$ for simplicity. Orthonormality means

$$\langle\pm\tfrac{1}{2}|\pm\tfrac{1}{2}\rangle = 1, \qquad \langle\pm\tfrac{1}{2}|\mp\tfrac{1}{2}\rangle = 0, \tag{C.110}$$

so that

$$\langle\pm\tfrac{1}{2}|S_z|\pm\tfrac{1}{2}\rangle = \pm\tfrac{1}{2}, \qquad \langle\pm\tfrac{1}{2}|S_z|\mp\tfrac{1}{2}\rangle = 0, \tag{C.111}$$

and from (C.108),

$$S_\pm|\mp\tfrac{1}{2}\rangle = |\pm\tfrac{1}{2}\rangle, \qquad S_\pm|\pm\tfrac{1}{2}\rangle = 0. \tag{C.112}$$

We now consider the $S_x$ and $S_y$ spin components. From (C.104),

$$S_x = \tfrac{1}{2}(S_+ + S_-), \qquad S_y = -\tfrac{1}{2}i(S_+ - S_-). \tag{C.113}$$

The eigenkets of $S_x$ and $S_y$ may be expressed in the $S_z$ basis as

$$|\pm x\rangle = \tfrac{1}{\sqrt{2}}\left\{|+\tfrac{1}{2}\rangle \pm |-\tfrac{1}{2}\rangle\right\}$$
$$|\pm y\rangle = \tfrac{1}{\sqrt{2}}\left\{|+\tfrac{1}{2}\rangle \pm i|-\tfrac{1}{2}\rangle\right\}, \tag{C.114}$$

where $|\pm x\rangle$ stands for the eigenkets of $S_x$ with eigenvalues $\pm\frac{1}{2}$.[33]

[32] The following is illustrative. In spherical polar coordinates, the $L_z$ operator is given by $L_z = -i\partial/\partial\phi$. Hence, the simultaneous wave functions $\psi(r, \theta, \phi)$ of $\mathbf{L}^2$ and $L_z$ satisfy

$$-i\frac{\partial\psi}{\partial\phi} = M_L\psi,$$

which integrates to

$$\psi = f(r, \theta)\exp(iM_L\phi).$$

The requirement that $\psi(\phi+2\pi) = \psi(\phi)$ then restricts $M_L$ to integer values.

[33] In this notation, $|\pm z\rangle = |\pm\frac{1}{2}\rangle$.

**Table C.1** Quantum amplitudes $\langle\alpha|\beta\rangle$ for a spin $S = \frac{1}{2}$ calculated from the eigen-states given in (C.114). The spin state $\beta$ is shown along the top of the table and $\alpha$ is down the left side. The symbols $\omega = \exp(i\pi/4)$ and $\omega^* = \exp(-i\pi/4)$ are used.

|     | $+x$ | $-x$ | $+y$ | $-y$ | $+z$ | $-z$ |
|-----|------|------|------|------|------|------|
| $+x$ | $1$ | $0$ | $\frac{1}{\sqrt{2}}\omega$ | $\frac{1}{\sqrt{2}}\omega^*$ | $\frac{1}{\sqrt{2}}$ | $\frac{1}{\sqrt{2}}$ |
| $-x$ | $0$ | $1$ | $\frac{1}{\sqrt{2}}\omega^*$ | $\frac{1}{\sqrt{2}}\omega$ | $\frac{1}{\sqrt{2}}$ | $-\frac{1}{\sqrt{2}}$ |
| $+y$ | $\frac{1}{\sqrt{2}}\omega^*$ | $\frac{1}{\sqrt{2}}\omega$ | $1$ | $0$ | $\frac{1}{\sqrt{2}}$ | $-\frac{1}{\sqrt{2}}i$ |
| $-y$ | $\frac{1}{\sqrt{2}}\omega$ | $\frac{1}{\sqrt{2}}\omega^*$ | $0$ | $1$ | $\frac{1}{\sqrt{2}}$ | $\frac{1}{\sqrt{2}}i$ |
| $+z$ | $\frac{1}{\sqrt{2}}$ | $\frac{1}{\sqrt{2}}$ | $\frac{1}{\sqrt{2}}$ | $\frac{1}{\sqrt{2}}$ | $1$ | $0$ |
| $-z$ | $\frac{1}{\sqrt{2}}$ | $-\frac{1}{\sqrt{2}}$ | $\frac{1}{\sqrt{2}}i$ | $-\frac{1}{\sqrt{2}}i$ | $0$ | $1$ |

Results (C.110)–(C.114) can be used to obtain the quantum amplitudes $\langle\alpha|\beta\rangle$ and matrix elements $\langle\alpha|\mathbf{S}|\beta\rangle$ given in Tables C.1 and C.2, respectively, where $\alpha, \beta = \pm x, \pm y, \pm z$. Two points about these matrix elements are worth noting. First, the spin eigenkets (C.114) are defined only to within an arbitrary phase factor $\exp(i\theta)$. The choice of the phase $\theta$ affects the values of the matrix elements but does not affect any physical observable derived from them, such as the neutron scattering cross-section. Second, the operators in these matrix elements are Hermitian, so interchanging $\alpha$ and $\beta$ gives the complex conjugate.

## C.1.16  Rotation matrices for angular momentum

Let $R_\xi(\theta)$ be a rotation through an angle $\theta$ about an axis $\xi$,[34] and let us consider the effect of $R_\xi(\theta)$ on the eigenstates of $J_z$, which we arrange in a row matrix $\mathsf{U} = (u_j, u_{j-1}, \ldots, u_{-j})$. The rotation operation performs a linear transformation which can be represented by a square matrix $\mathsf{D}_\xi(\theta)$ of order $2j + 1$, such that

$$R_\xi(\theta)\mathsf{U} = \mathsf{U}\mathsf{D}_\xi(\theta). \tag{C.115}$$

If $\theta$ is infinitesimally small, then we can expand the rotation operator as a Taylor series in $\theta$,

$$R_\xi(\theta) = I + i\theta J_\xi + \cdots, \tag{C.116}$$

where $I$ is the identity operator and $J_\xi = -i\frac{\partial}{\partial\theta}$ is the angular momentum operator for rotations about the $\xi$ axis. Non-infinitesimal rotations can then be considered as a succession of infinitesimal rotations,

$$R_\xi(\theta) = \lim_{n\to\infty}\left(I + i\frac{\theta}{n}J_\xi\right)^n$$
$$= \sum_{n=0}^{\infty}\frac{(i\theta J_\xi)^n}{n!}. \tag{C.117}$$

[34]Specifically, an anticlockwise rotation of the coordinates about the $\xi$ axis.

The matrix $\mathsf{D}_\xi(\theta)$ is formed from the matrix elements of $R_\xi(\theta)$ between all the states in $\mathsf{U}$.

As an example, consider the case $S = 1/2$. The basis states are $\mathsf{U} = (|+z\rangle, |-z\rangle)$, and $J_\xi$ in (C.117) is represented by a $2\times2$ matrix containing the matrix elements of $S_\xi$ between the states $|\pm z\rangle$. The matrix elements for $\xi = x, y, z$ are given in Table C.2, and the corresponding matrices, without the factor of $\frac{1}{2}$, are the Pauli spin matrices

$$\sigma_x = \begin{pmatrix} 0 & 1 \\ 1 & 0 \end{pmatrix}, \quad \sigma_y = \begin{pmatrix} 0 & -i \\ i & 0 \end{pmatrix}, \quad \sigma_z = \begin{pmatrix} 1 & 0 \\ 0 & -1 \end{pmatrix}. \quad \text{(C.118)}$$

Taking $\xi = z$ as an example, we substitute $J_z = \frac{1}{2}\sigma_z$ into (C.117) and split the sum into even and odd $n$ terms. This gives

$$\mathsf{D}_z(\theta) = \sum_{n=0}^{\infty} \frac{1}{n!} \left( \frac{i\theta\sigma_z}{2} \right)^n$$

$$= \sum_{n\,\text{even}} \frac{1}{n!} \left( \frac{i\theta}{2} \right)^n I + \sum_{n\,\text{odd}} \frac{1}{n!} \left( \frac{i\theta}{2} \right)^n \sigma_z$$

$$= \begin{pmatrix} e^{i\theta/2} & 0 \\ 0 & e^{-i\theta/2} \end{pmatrix}. \quad \text{(C.119)}$$

Here, $I$ is the $2 \times 2$ identity matrix. Similarly, we find

$$\mathsf{D}_x(\theta) = \begin{pmatrix} \cos\theta/2 & i\sin\theta/2 \\ i\sin\theta/2 & \cos\theta/2 \end{pmatrix}, \quad \mathsf{D}_y(\theta) = \begin{pmatrix} \cos\theta/2 & \sin\theta/2 \\ -\sin\theta/2 & \cos\theta/2 \end{pmatrix}. \quad \text{(C.120)}$$

The $\mathsf{D}_y(\theta)$ matrix quoted in eqn (9.71) is the transpose of that in (C.120) because in (9.71) the eigenstates of $S_z$ are written as a column vector $\mathsf{U}^\mathsf{T}$, so the matrix equation describing the rotation of the angular momentum basis states is the transpose of (C.115).

**Table C.2** Matrix elements $\langle\alpha|S_j|\beta\rangle$, $j = x, y, z$, for a spin $S = \frac{1}{2}$ calculated from (C.111)–(C.114). The spin state $\beta$ is shown along the top of the table and $\alpha$ is down the left side. The symbols $\omega = \exp(\mathrm{i}\pi/4)$ and $\omega^* = \exp(-\mathrm{i}\pi/4)$ are used.

|        | $+x$ | $-x$ | $+y$ | $-y$ | $+z$ | $-z$ |
|--------|------|------|------|------|------|------|
| **$S_x$** | | | | | | |
| $+x$ | $\frac{1}{2}$ | $0$ | $\frac{1}{2\sqrt{2}}\omega$ | $\frac{1}{2\sqrt{2}}\omega^*$ | $\frac{1}{2\sqrt{2}}$ | $\frac{1}{2\sqrt{2}}$ |
| $-x$ | $0$ | $-\frac{1}{2}$ | $-\frac{1}{2\sqrt{2}}\omega^*$ | $-\frac{1}{2\sqrt{2}}\omega$ | $-\frac{1}{2\sqrt{2}}$ | $\frac{1}{2\sqrt{2}}$ |
| $+y$ | $\frac{1}{2\sqrt{2}}\omega^*$ | $-\frac{1}{2\sqrt{2}}\omega$ | $0$ | $-\frac{1}{2}\mathrm{i}$ | $-\frac{1}{2\sqrt{2}}\mathrm{i}$ | $\frac{1}{2\sqrt{2}}$ |
| $-y$ | $\frac{1}{2\sqrt{2}}\omega$ | $-\frac{1}{2\sqrt{2}}\omega^*$ | $\frac{1}{2}\mathrm{i}$ | $0$ | $\frac{1}{2\sqrt{2}}\mathrm{i}$ | $\frac{1}{2\sqrt{2}}$ |
| $+z$ | $\frac{1}{2\sqrt{2}}$ | $-\frac{1}{2\sqrt{2}}$ | $\frac{1}{2\sqrt{2}}\mathrm{i}$ | $-\frac{1}{2\sqrt{2}}\mathrm{i}$ | $0$ | $\frac{1}{2}$ |
| $-z$ | $\frac{1}{2\sqrt{2}}$ | $\frac{1}{2\sqrt{2}}$ | $\frac{1}{2\sqrt{2}}$ | $\frac{1}{2\sqrt{2}}$ | $\frac{1}{2}$ | $0$ |
| **$S_y$** | | | | | | |
| $+x$ | $0$ | $\frac{1}{2}\mathrm{i}$ | $\frac{1}{2\sqrt{2}}\omega$ | $-\frac{1}{2\sqrt{2}}\omega^*$ | $\frac{1}{2\sqrt{2}}\mathrm{i}$ | $-\frac{1}{2\sqrt{2}}\mathrm{i}$ |
| $-x$ | $-\frac{1}{2}\mathrm{i}$ | $0$ | $\frac{1}{2\sqrt{2}}\omega^*$ | $-\frac{1}{2\sqrt{2}}\omega$ | $-\frac{1}{2\sqrt{2}}\mathrm{i}$ | $-\frac{1}{2\sqrt{2}}\mathrm{i}$ |
| $+y$ | $\frac{1}{2\sqrt{2}}\omega^*$ | $\frac{1}{2\sqrt{2}}\omega$ | $\frac{1}{2}$ | $0$ | $\frac{1}{2\sqrt{2}}$ | $-\frac{1}{2\sqrt{2}}\mathrm{i}$ |
| $-y$ | $-\frac{1}{2\sqrt{2}}\omega$ | $-\frac{1}{2\sqrt{2}}\omega^*$ | $0$ | $-\frac{1}{2}$ | $-\frac{1}{2\sqrt{2}}$ | $-\frac{1}{2\sqrt{2}}\mathrm{i}$ |
| $+z$ | $-\frac{1}{2\sqrt{2}}\mathrm{i}$ | $\frac{1}{2\sqrt{2}}\mathrm{i}$ | $\frac{1}{2\sqrt{2}}$ | $-\frac{1}{2\sqrt{2}}$ | $0$ | $-\frac{1}{2}\mathrm{i}$ |
| $-z$ | $\frac{1}{2\sqrt{2}}\mathrm{i}$ | $\frac{1}{2\sqrt{2}}\mathrm{i}$ | $\frac{1}{2\sqrt{2}}\mathrm{i}$ | $\frac{1}{2\sqrt{2}}\mathrm{i}$ | $\frac{1}{2}\mathrm{i}$ | $0$ |
| **$S_z$** | | | | | | |
| $+x$ | $0$ | $\frac{1}{2}$ | $\frac{1}{2\sqrt{2}}\omega^*$ | $\frac{1}{2\sqrt{2}}\omega$ | $\frac{1}{2\sqrt{2}}$ | $-\frac{1}{2\sqrt{2}}$ |
| $-x$ | $\frac{1}{2}$ | $0$ | $\frac{1}{2\sqrt{2}}\omega$ | $\frac{1}{2\sqrt{2}}\omega^*$ | $\frac{1}{2\sqrt{2}}$ | $\frac{1}{2\sqrt{2}}$ |
| $+y$ | $\frac{1}{2\sqrt{2}}\omega$ | $\frac{1}{2\sqrt{2}}\omega^*$ | $0$ | $\frac{1}{2}$ | $\frac{1}{2\sqrt{2}}$ | $-\frac{1}{2\sqrt{2}}\mathrm{i}$ |
| $-y$ | $\frac{1}{2\sqrt{2}}\omega^*$ | $\frac{1}{2\sqrt{2}}\omega$ | $\frac{1}{2}$ | $0$ | $\frac{1}{2\sqrt{2}}$ | $-\frac{1}{2\sqrt{2}}\mathrm{i}$ |
| $+z$ | $\frac{1}{2\sqrt{2}}$ | $\frac{1}{2\sqrt{2}}$ | $\frac{1}{2\sqrt{2}}$ | $\frac{1}{2\sqrt{2}}$ | $\frac{1}{2}$ | $0$ |
| $-z$ | $-\frac{1}{2\sqrt{2}}$ | $\frac{1}{2\sqrt{2}}$ | $-\frac{1}{2\sqrt{2}}\mathrm{i}$ | $\frac{1}{2\sqrt{2}}\mathrm{i}$ | $0$ | $-\frac{1}{2}$ |

## C.2   Atomic physics

### C.2.1   Hydrogen-like atoms

Hydrogen-like atoms have one electron of mass $m_e$, which is bound in a Coulomb potential $V(r) = -Ze^2/(4\pi\epsilon_0 r)$ to a nucleus of mass $m_N$ containing $Z$ protons. The eigenvalues and eigenvectors of the one-electron states obtained from the solution of the time-independent Schrödinger equation (C.22) are given by[35]

$$E_n = -\frac{\mu e^4}{(4\pi\epsilon_0\hbar)^2}\frac{Z^2}{n^2} \tag{C.121}$$

$$\psi(\mathbf{r}) = R_{nl}(r)Y_{l,m}(\theta,\phi), \tag{C.122}$$

[35]The Schrödinger equation is non-relativistic so the wave function (C.122) does not include spin.

where $\mu = m_e m_N/(m_e + m_N) \simeq m_e$ is the reduced mass. The wave function is the product of a radial part $R_{nl}(r)$ and an angular part $Y_{l,m}(\theta,\phi)$.

The radial wave function is given by

$$R_{nl}(r) = -\left(\frac{2Z}{na_0}\right)^{\frac{3}{2}}\left\{\frac{(n-l-1)!}{2n[(n+l)!]^3}\right\}^{\frac{1}{2}}\rho^l L_{n+l}^{2l+1}(\rho)\exp(-\rho/2), \quad \text{(C.123)}$$

where

$$a_0 = \frac{4\pi\epsilon_0\hbar^2}{\mu e^2}, \qquad \rho = \frac{2Z}{na_0}r,$$

and $L_\lambda^\mu(\rho)$ are the *associated Laguerre polynomials*:

$$L_\lambda^\mu(\rho) = (\lambda!)^2\sum_{\nu=0}^{\lambda-\mu}\frac{(-1)^{\mu+\nu}}{\nu!(\mu+\nu)!(\lambda-\mu-\nu)!}\rho^\nu. \tag{C.124}$$

The angular wave function is a *spherical harmonic* of rank $l$ and order $m$, defined by[36]

[36]We use the Condon–Shortley phase convention (Condon and Shortley, 1935).

$$Y_{l,m}(\theta,\phi) = (-1)^m\left[\frac{2l+1}{4\pi}\frac{(l-m)}{(l+m)}\right]^{\frac{1}{2}}P_l^m(\cos\theta)\exp(im\phi) \quad (m \geq 0) \tag{C.125}$$

and for negative order

$$Y_{l,-m} = (-1)^m Y_{l,m}^* \quad (m \geq 0). \tag{C.126}$$

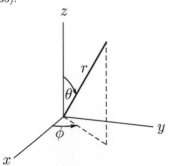

The $P_l^m(\cos\theta)$ are the *associated Legendre polynomials*. The angles $\theta$ and $\phi$ specify the direction of the vector $\mathbf{r}$ in spherical coordinates — see Fig. C.1.

The following normalization conditions apply:

$$1 = \int_0^\infty R_{nl}^2(r)r^2\,\mathrm{d}r \quad \text{and} \quad 1 = 2\pi\int_0^\pi |Y_{l,m}|^2\sin\theta\,\mathrm{d}\theta. \tag{C.127}$$

**Fig. C.1** Relation between the spherical coordinates $(r,\theta,\phi)$ and the Cartesian coordinates $(x,y,z)$.

The electron wave functions are specified by three integers $n$, $l$ and $m$, known as the *principal*, *angular momentum*, and *magnetic* quantum numbers, respectively. These satisfy the constraints $n \geq 1$, $0 \leq l \leq n-1$ and $-l \leq m \leq l$. The principal quantum number alone determines the energy levels — see eqn (C.121). Solutions with $l = 0, 1, 2, 3, \ldots$ are designated $s$, $p$, $d$, $f$, $\ldots$ states. By convention, states are labelled by prefixing the $n$ value to the letter indicating the $l$ value. For example, $n = 3, l = 2$ is a $3d$ state. The $m$ quantum number specifies the component of angular momentum along the quantization direction. For a given $l$ there are $(2l + 1)$ values of $m$.

## C.2.2 Many-electron atoms

It is not possible to solve the Schrödinger equation exactly for atoms with more than one electron. The standard way to describe the electronic state of a multi-electron atom or ion is to establish the hierarchy of interactions and consider them in order of decreasing strength.

The starting point is the *central-field approximation*, in which the states of a single electron are calculated in the Coulomb field of the nucleus screened by the spherically averaged repulsive field from the other electrons. Computational schemes such as the Hartree–Fock approximation are employed to solve the Schrödinger equation self-consistently. The many-electron wave functions are written as products of one-electron basis functions which are combined in a Slater determinant to take into account exchange effects. This procedure results in quantized electronic levels grouped into *configurations*. For example, the ground state configuration of $Cr^{3+}$ is $1s^2 2s^2 2p^6 3s^2 3p^6 3d^3$, often written [Ar]$3d^3$ or just $3d^3$. Here, [Ar] is shorthand for the closed shell electronic configuration of Ar, and the quantum numbers $n$ and $l$ are those that appear in the solution for hydrogen-like atoms, Section C.2.1. When relativistic effects are included the states have additional angular momentum quantum numbers $s$ (spin) and $j$, with $s = 1/2$ and $j = l \pm 1/2$.

The central potential means that the one-electron functions can still be separated into radial and angular parts as for hydrogenic atoms, eqn (C.122), but the radial wave function is no longer given by (C.123) and must be obtained numerically.[37] One way to do this is to express the radial function in an analytic form with adjustable parameters which are varied until a self-consistent solution of the Hartree–Fock equations is found. The standard approach uses a linear combination of functions called Slater-type orbitals:

$$R_{nl}(r) = \sum_j C_j \frac{(2\xi_j)^{n+\frac{1}{2}}}{[(2n)!]^{\frac{1}{2}}} r^{n-1} \exp(-\xi_j r) \qquad \text{(C.128)}$$

Values of the parameters $C_j$ and $\xi_j$ have been tabulated for $3d$ and $4d$ elements and ions (Clementi and Roetti, 1974), and also for trivalent $4f$ ions (Freeman and Watson, 1962).[38] The advantage of the analytic form is in the ease with which certain matrix elements in atomic physics can be calculated, such as the radial averages $\langle r^n \rangle$ and $\langle j_n(Q) \rangle$ that appear in expressions for magnetic form factors. Values of $\langle r^n \rangle$ and $\langle j_n(Q) \rangle$ from

[37] The Cowan code is a useful resource for such calculations (Cowan, 1981). See https://www.tcd.ie/Physics/research/groups/xray-spectroscopy/cowan/.

[38] Freeman and Watson absorb the normalizing prefactor $(2\xi)^{n+\frac{1}{2}}/[(2n)!]^{\frac{1}{2}}$ into the $C_j$ parameter, and incorporate the $r^2$ factor in the radial normalization, eqn (C.127), into $R_{nl}(r)$.

relativistic Dirac–Fock calculations have been tabulated for $4f$ and $5f$ ions (Freeman and Desclaux, 1979, and Desclaux and Freeman, 1978).

The next strongest interactions are (i) residual Coulomb interactions between the outermost electrons, not included in the central field approximation, and (ii) spin–orbit coupling. In most cases of interest the residual Coulomb energy exceeds that of the spin–orbit interaction and the configuration splits into *terms* with well-defined $L$ and $S$ quantum numbers, where $L$ represents the combined orbital angular momentum of the electrons and $S$ the combined spin angular momentum. This is the so-called *Russell–Saunders* or *LS-coupling* scheme. Terms have a degeneracy of $(2S+1)(2L+1)$, and are often represented in spectroscopy by the term symbol $^{2S+1}L$, in which the letters $S$, $P$, $D$, $F$, ... are used to denote $L = 0, 1, 2, 3, \ldots$.

As long as the spin–orbit interaction is much weaker than the residual electrostatic interactions its main effect is to split the $LS$-coupling terms into *levels*, each level having a well-defined value of $J$, the combined spin and orbital angular momentum quantum number. The spectroscopic notation for a level is $^{2S+1}L_J$. The ground state level in $LS$-coupling can be obtained by following Hund's rules.[39] For the example of $3d^3$, Hund's rules predict $S = \frac{3}{2}$, $L = 3$, and $J = \frac{3}{2}$ ($^4F_{3/2}$ in spectroscopic notation) for the ground state $J$ level. This level is a multiplet with a degeneracy of $2J + 1 = 4$ (a quartet).

A further effect of the spin–orbit interaction is to couple, and therefore mix, levels with different $L$ and $S$ but the same $J$. For light elements the $LS$-coupling scheme is an excellent approximation, and the admixture of higher levels in the Hund's rules ground state can usually be neglected. However, with increasing atomic number the spin–orbit interaction gains strength and it becomes necessary to include the admixture of higher levels to obtain accurate results. When this is done the scheme is called *intermediate coupling*.

At very high atomic numbers the spin–orbit interaction can eventually exceed the residual electrostatic interactions. In this limit, $L$ and $S$ are no longer good quantum numbers and a better approximation is to combine the spin and orbital angular momentum of each individual electron to form $j$, and then couple these one-electron angular momenta together to form $J$. This scheme is called *jj-coupling*.

[39]Hund's rules: (1) combine the individual electron spins so as to maximize $S$ subject to the Pauli principle; (2) given the state determined by the first rule, combine the individual orbital angular momenta to maximize $L$ subject to the Pauli principle; (3) the value of $J$ is then given by $J = |L - S|$ if the shell is less than half full, and by $J = L + S$ if the shell is greater than half full.

## C.3   Magnetic moments

Magnetic materials are characterized by the existence of a magnetic dipole moment on some or all of their constituent atoms or ions. The size of the magnetic moment depends on (i) the electronic state of the atoms in the material, (ii) the applied magnetic field (if present), (iii) temperature, and (iv) the type and strength of any magnetic coupling.

Every atom (or ion) is to some extent *diamagnetic*, i.e. on application of a magnetic field a moment is induced anti-parallel to the field. As far as magnetic phenomena are concerned, however, by far the most important atomic species are those that are also *paramagnetic*, i.e. the induced moment is parallel to the applied magnetic field.

Paramagnetic atoms and ions usually carry permanent intrinsic magnetic moments which are randomly oriented in zero applied field but which tend to align with an applied field. It is also possible for an atom to have no magnetic moment in zero field but to develop a positive induced magnetic moment on application of a field. The latter effect is called *Van Vleck paramagnetism*. If the atoms in a paramagnetic material are coupled by exchange or dipolar interactions then cooperative effects can occur, and below a certain temperature the moments can align spontaneously into an ordered magnetic structure.[40]

The paramagnetic moment of an atom is the sum of the moments associated with the orbital and spin angular momenta. In quantum mechanics, the magnetic moment is represented by the operator[41]

$$\boldsymbol{\mu} = -\mu_\mathrm{B}(\mathbf{L} + 2\mathbf{S}). \tag{C.129}$$

For perfect $LS$-coupling states with well defined $J$, only the component of $\boldsymbol{\mu}$ along $\mathbf{J}$ is a conserved quantity.[42] In this case we can use[43]

$$\boldsymbol{\mu} = -g_J \mu_\mathrm{B} \mathbf{J}, \tag{C.130}$$

(see Exercise 6.1) where $\mathbf{J} = \mathbf{L} + \mathbf{S}$ and $g_J$ is the Landé $g$-factor,

$$g_J = 1 + \frac{J(J+1) - L(L+1) + S(S+1)}{2J(J+1)}. \tag{C.131}$$

The magnetic moment (C.130) can be written in the form

$$\boldsymbol{\mu} = -(g_S + g_L)\mu_\mathrm{B}\mathbf{J}, \tag{C.132}$$

where

$$
\begin{aligned}
g_S &= \frac{J(J+1) - L(L+1) + S(S+1)}{J(J+1)} \\
g_L &= \frac{J(J+1) + L(L+1) - S(S+1)}{2J(J+1)},
\end{aligned}
\tag{C.133}
$$

and $g_J = g_S + g_L$. The terms associated with $g_S$ and $g_L$ in (C.132) are the spin and orbital components of the magnetic moment, respectively.

## C.3.1  Free atoms and ions in a magnetic field

The interaction of a magnetic moment with a magnetic field is described by the Zeeman Hamiltonian

$$\mathcal{H}_\mathrm{Z} = -\boldsymbol{\mu} \cdot \mathbf{B}. \tag{C.134}$$

The ground state of an isolated or free paramagnetic atom or ion (i.e. one not interacting with other atoms or ions) usually has a well-defined $J$, and the application of a weak magnetic field[44] splits the $J$ manifold into

---

[40] Magnetic ordering is discussed in Section 7.1.

[41] Here we have taken the electron $g$-factor $g_e$ to be 2, see eqn (4.27). The minus sign is due to the negative charge on the electron, which makes the charge current and the mass current flow in opposite directions. Hence, the magnetic moment points in the opposite direction to the angular momentum.

[42] The components of $\boldsymbol{\mu}$ perpendicular to $\mathbf{J}$ average to zero due to quantum-mechanical fluctuations, and cannot be observed.

[43] Equations (C.129) and (C.130) do not imply that the operators $\mathbf{L}+2\mathbf{S}$ and $g_J\mathbf{J}$ are identical. What they mean is that $\boldsymbol{\mu}$ can be replaced by $-g_J\mu_\mathrm{B}\mathbf{J}$ inside any matrix element taken between $LS$-coupling states within the same $J$ multiplet.

[44] Equation (C.135) is valid as long as the Zeeman energy $-\boldsymbol{\mu} \cdot \mathbf{B}$ is much less than the spin–orbit interaction.

a set of $2J + 1$ equally spaced levels (the *Zeeman effect*) with energies

$$E = g_J \mu_B B M_J, \tag{C.135}$$

where $B$ is the flux density and $M_J$ is the quantum number representing the component of total angular momentum parallel to the applied field.[45] Each Zeeman level has a well-defined component of magnetic moment along the field direction (here taken to be the $z$ axis) given by

$$\mu_z = -g_J \mu_B M_J, \tag{C.136}$$

so that the maximum possible magnetic moment induced by the field (the saturation moment) is when $M_J = -J$ and is

$$\mu_{sat} = g_J \mu_B J. \tag{C.137}$$

The magnetization of a gas of $N$ atoms per unit volume in thermal equilibrium at temperature $T$ is found to be

$$M = N g_J \mu_B J B_J(x), \tag{C.138}$$

where $x = g_J \mu_B J B / (k_B T)$ and $B_J(x)$ is the Brillouin function

$$B_J(x) = \frac{2J+1}{2J} \coth\left(\frac{2J+1}{2J}x\right) - \frac{1}{2J} \coth\left(\frac{1}{2J}x\right). \tag{C.139}$$

If $x \ll 1$, the susceptibility $\chi = M/H = \mu_0 M/B$ is given by Curie's Law

$$\chi = \frac{N \mu_0 \mu_{eff}^2}{3 k_B T} = \frac{C}{T}, \tag{C.140}$$

where $C$ is the Curie constant and $\mu_{eff}$ is the effective moment,

$$\mu_{eff} = g_J \mu_B \sqrt{J(J+1)}. \tag{C.141}$$

## C.4   Magnetic ions in solids

The formulae given above for the magnetization and susceptibility tend to break down for magnetic ions in solids, except at very high temperatures. The most important additional factors that need to be considered are (i) exchange interactions, (ii) ligand field effects and (iii) itinerant effects. For the time being we will ignore itinerant effects, the analysis of which requires the additional framework of band theory,[46] and focus on the behaviour of localized electrons.

### C.4.1   Exchange interactions

Exchange interactions introduce correlations between magnetic moments on different sites, which can drive the system into a magnetically ordered state. At temperatures above the magnetic ordering temperature the effect of exchange interactions is like that of a magnetic field acting on the spin part of the wave function. The paramagnetic susceptibility of local-moment systems is often found to obey the Curie–Weiss law,

$$\chi = \frac{C}{T - \theta}, \tag{C.142}$$

where $C$ is again the Curie constant and $\theta$ is the *Weiss constant*, which depends on the strength and nature of the exchange interaction.[47]

## C.4.2   The ligand field

In reality, values of $C$ determined experimentally[48] are often smaller than predicted by the free atom expressions (C.140) and (C.141), an effect known as *quenching*.

The discrepancy is due to the *ligand field*, which is an anisotropic interaction that acts on the orbital part of the wave function. It derives partly from the electrostatic potential due to the other ions in the material (the *crystalline electric field*, or *crystal field* for short) and partly from covalency effects due to the overlap of the electronic wave function with that on different ions. In the ionic limit the ligand field is just the Coulomb potential, and in the simplest model it is represented by a set of point charges on the neighbouring sites.[49]

The ligand field disturbs the orbital motion of the electrons and makes the orbital angular momentum fluctuate.[50] In a magnetic field these fluctuations compete against the tendency for the orbital moment to align with the field, and make the induced moment less than it would be in the absence of the ligand field. The induced magnetization will only approach the free-atom value predicted by (C.138) when the temperature is high enough for the randomizing effect of thermal fluctuations to dominate over the quenching effect of the ligand field.

Most magnetic substances contain ions with incomplete $d$ or $f$ shells. The effect of the ligand field on the total magnetic moment can be very different, and depends on the relative energy scales of the ligand field, spin–orbit coupling and intra-atomic Coulomb interactions. Magnetic systems are usually classified into three regimes of behaviour which we consider separately.

### (a) Ligand field < spin–orbit < intra-atomic Coulomb

This hierarchy of interactions applies to the lanthanide series,[51] which has partially filled $4f$ orbitals, and to lighter actinides (partially filled $5f$ orbitals) with weak covalent bonds. The ligand field is weak for these systems because the mean radius of the $f$ shell is significantly smaller than the ionic radius, and so the $f$ electrons are usually well shielded from external interactions. Consequently, the $LS$-coupling scheme is a good approximation. The main effect of the ligand field is then to lift, either partially or fully, the $(2J+1)$-fold degeneracy of the $^{2S+1}L_J$ levels. A secondary effect is to mix states from different $^{2S+1}L_J$ levels. The degree of mixing is usually small and often neglected, but with strong ligand fields it can become significant, especially for the lanthanides Ce,

[47] For simple ferromagnets and antiferromagnets $\theta$ is positive for a net ferromagnetic exchange and negative for a net antiferromagnetic exchange.

[48] e.g. from a plot of $1/\chi$ versus $T$.

[49] See M.T Hutchings, (1964).

[50] Providing the ligand field interaction is much weaker than the intra-atomic Coulomb interaction the orbital angular momentum fluctuates in direction only. If, however, the ligand field interaction is greater than the intra-atomic Coulomb interaction then $L$ is no longer well defined and the orbital angular momentum fluctuates in both magnitude and direction.

[51] Also known as the *rare earths*.

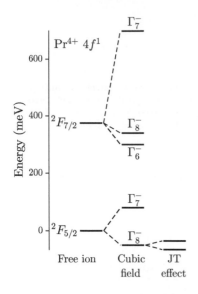

**Fig. C.2** Single-ion energy level scheme for Pr$^{4+}$ in PrO$_2$ (Boothroyd *et al.*, 2001; Gardiner *et al.*, 2004).

[52]Systems with large orbital degeneracy often exhibit the *Jahn–Teller effect*, i.e. undergo a small structural distortion that splits the orbitally degenerate ground state so as to lower the electronic energy at the expense of a small(er) increase in elastic energy.

[53]The spin–orbit interaction takes the form $\lambda \mathbf{L} \cdot \mathbf{S}$, where $\lambda > 0$ if the shell is less than half full, and $\lambda < 0$ if the shell is greater than half full.

Pr, Nd, Pm, Sm, Eu, and Tb, whose spin–orbit splittings are relatively small ($\leq 0.3\,\mathrm{eV}$). The most accurate treatments of ligand field effects in lanthanides include ligand field mixing among the complete set of intermediate coupling basis states.

The ligand field acts on $\mathbf{L}$, but being small relative to the spin–orbit interaction it cannot decouple $\mathbf{L}$ from $\mathbf{S}$. Hence, the directions of $\mathbf{L}$, $\mathbf{S}$ and $\mathbf{J}$ fluctuate under the action of the ligand field but their magnitudes remain constant. When $k_\mathrm{B}T$ is smaller than the ligand field splitting the magnetic moment is partially quenched relative to the free ion moment. Except for a few special cases in which $L = 0$, such as Gd$^{3+}$, the magnetic moment is formed from a mixture of spin and orbital states.

Figure C.2 illustrates how the ligand field partially quenches the magnetic moment of a $4f$ ion. In the absence of a ligand field, the $4f^1$ configuration has a single term $^2F$ which is split by spin–orbit coupling into two levels $^2F_{5/2}$ and $^2F_{7/2}$. A ligand field of cubic symmetry, as found in PrO$_2$, splits the levels into a series of two- and fourfold degenerate eigenstates which transform according to the odd-parity irreducible representations $\Gamma_i^-$ of the cubic point group $m3m$. The saturated moment of the $\Gamma_8^-$ ground state is $\mu_\mathrm{sat} = 1.50\,\mu_\mathrm{B}$, less than the free ion value of $\mu_\mathrm{sat} = g_J J \mu_\mathrm{B} = 2.14\,\mu_\mathrm{B}$. Below 120 K, the fourfold degenerate $\Gamma_8^-$ ground state is further split by the Jahn–Teller effect[52] into two doublets. This reduces $\mu_\mathrm{sat}$ further, and together with the dynamic Jahn–Teller effect, which is also present, leads to the observed low temperature ordered moment of about $0.6\,\mu_\mathrm{B}$.

### (b) Spin–orbit < ligand field < intra-atomic Coulomb

This hierarchy applies to $3d$ transition-metal ions in many environments. $LS$ coupling is a good approximation for these systems, but the spin and orbital angular momenta are almost completely decoupled by the ligand field and $J$ is no longer a good quantum number. The ligand field splits the $(2L+1)$-fold orbital degeneracy of the $^{2S+1}L$ terms, while retaining the full $(2S+1)$-fold spin degeneracy. Admixtures of excited $LS$ terms may be needed to give an adequate description of the ground state in cases where the ligand field is not negligible relative to the intra-atomic Coulomb interaction.

The magnetic moment in $3d$ ions derives predominantly from the spin degrees of freedom, which in the absence of exchange interactions behave as if for a free ion. If the ligand field ground state is orbitally degenerate then the orbital moment can be substantial, but more often than not the ground state is an orbital singlet or non-magnetic doublet and the orbital moment is almost fully quenched. In the latter case, there may exist a small orbital moment induced by the spin via the spin–orbit interaction. The direction of this induced orbital moment can be either parallel or antiparallel to the spin moment depending on the sign of the spin–orbit coupling parameter.[53] For these systems it is possible to represent the effects of both spin and orbital angular momentum by an *effective* (or *fictitious*) *spin operator* $\widetilde{\mathbf{S}}$ whose interaction with a magnetic field is

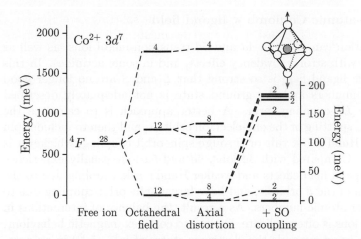

**Fig. C.3** Single-ion energy level scheme of $Co^{2+}$ for the axially distorted octahedral ligand field of $La_2CoO_4$, showing the effect of successively smaller interactions (Helme *et al.*, 2009). The numbers attached to each level give the degeneracies. On the far right, the spin–orbit (SO) splitting of the lowest orbital quasitriplet is shown on an expanded scale.

described by the *spin Hamiltonian*[54]

$$\mathcal{H}_{\text{spin}} = g\mu_B\widetilde{\mathbf{S}} \cdot \mathbf{B} + O(B^2). \tag{C.143}$$

The first term in $\mathcal{H}_{\text{spin}}$ represents the interaction of a pure spin moment with a magnetic field. It is analogous to the Zeeman Hamiltonian, eqns (C.134) and (C.130), but has a modified $g$-factor, often called the *spectroscopic splitting factor*, which can differ from the spin-only value of the Landé $g$-factor, $g_S = 2$, eqn (C.131). Any deviation of $g$ from 2 is due to an orbital component induced by the spin.[55] The second term in (C.143) comes from the second-order Van Vleck orbital moment, and is usually much smaller than the first term. When that is the case, the Zeeman splitting is proportional to magnetic field and can be written

$$\Delta E = g\mu_B B, \tag{C.144}$$

which defines the $g$-factor of the level for practical purposes.

As an example of a $3d$ ion in a ligand field, consider the states of $Co^{2+}$ ($3d^7$) surrounded by six $O^{2-}$ ions in an axially distorted octahedron (i.e. an octahedron that has been stretched along the axis joining opposite vertices) as found, for example, in $La_2CoO_4$ (Fig. C.3). The 7-fold orbital degeneracy of the Hund's rule free-ion ground state term $^4F$ ($L = 3$, $S = \frac{3}{2}$) is split by the octahedral ligand field into two orbital triplets and an orbital singlet. The triplet ground state is further split by the Jahn–Teller effect, which favours an axial distortion of the octahedron and results in an orbital singlet ground state and an orbital doublet first excited state. The ordered moment of the ground state in the absence of spin–orbit coupling would be $3\,\mu_B$, from $S = 3/2$ and $g = 2$ (isotropic). However, with inclusion of spin–orbit coupling the 4-fold spin degeneracy is lifted and the quartet splits into two doublets, each with $\widetilde{S} = \frac{1}{2}$. The $g$-factors for the doublets are anisotropic. For the ground state doublet $g_\parallel = 2.0$ and $g_\perp = 5.4$ parallel and perpendicular to the distortion axis, the latter reflecting a significant admixture of orbital angular momentum. The corresponding ordered moments are $\mu_\parallel = 1.0\,\mu_B$ and $\mu_\perp = 3.1\,\mu_B$, with $\mu_\perp^{\text{spin}} = 2.2\,\mu_B$ and $\mu_\perp^{\text{orb}} = 0.9\,\mu_B$.

[54]The spin operator $\widetilde{\mathbf{S}}$ is called fictitious because it has the properties of a spin operator but does not necessarily correspond to the true spin operator. There are two reasons for this. Firstly, when spin–orbit coupling is added to the ligand field interaction the spin and orbital angular momenta are no longer completely independent. This means the 'spin' part of the wave function now contains a small admixture of orbital states, in other words $\widetilde{\mathbf{S}}$ includes a little bit of $\mathbf{L}$ in addition to $\mathbf{S}$. Second, the spin–orbit ground state can have a lower degeneracy than the full $2S + 1$ spin degeneracy, as in the example shown in Fig. C.3 and described in the text.

[55]In general, the $g$-factor is anisotropic and the first term in eqn (C.143) should be a tensor coupling, $\mu_B\widetilde{\mathbf{S}}\cdot\mathbf{g}\cdot\mathbf{B}$, with $\mathbf{g}$ a second rank tensor (i.e. a $3 \times 3$ matrix). The anisotropy arises because the orbital moment induced by the field (via the spin–orbit coupling) depends on the direction of the field relative to the electron distribution, whose orientation is fixed in space by the ligand field.

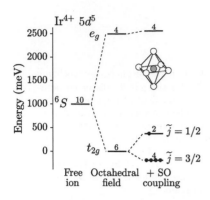

**Fig. C.4** One-electron states resulting from a splitting of the $5d^5$ configuration by a strong octahedral ligand field and by spin–orbit coupling. The $\widetilde{j} = 1/2$ state is proposed for $Ir^{4+}$ in $Sr_2IrO_4$ (Kim *et al.*, 2008). The numbers over each level give the degeneracies, and the black dots show the occupancies.

[56] In the recent literature $\widetilde{j}$ is often denoted by $j_{eff}$.

## (c) Intra-atomic Coulomb < ligand field

This situation can occur in $4d$ and $5d$ transition-metal ions, as well as in $3d$ ions with strong covalency effects, and in some actinides. In this regime, the ligand field is so strong that $L$ and $S$ are no longer good quantum numbers and the ground state is not adequately described by a pure $LS$-coupling term. A better approach is to calculate the ligand field splitting of the one-electron orbitals, and then to include spin either via Hund's first rule or through spin–orbit coupling, whichever is dominant. Compared with $3d$ ions, $4d$ and $5d$ ions usually experience stronger ligand field effects and weaker Hund's rule coupling due to the larger radii of the $d$ orbitals, and have larger spin–orbit coupling due to their higher atomic numbers. As a result, the balance of interactions in $4d$ and $5d$ ions is often delicate, leading to complex magnetic behaviour.

We take as an example the electronic states of $Ir^{4+}$ ($5d^5$) in an octahedral ligand field, as found approximately in $Sr_2IrO_4$. Application of Hund's rules gives $L = 0, S = J = 5/2$ for the free ion. However, as just mentioned, Hund's coupling does not apply in a strong ligand field, and we should consider the one-electron $d$ orbitals instead. As illustrated in Fig. C.4, the octahedral field splits the $d$ orbitals into a triplet and a doublet, labelled $t_{2g}$ and $e_g$, respectively. Each orbital has two-fold spin degeneracy at this point. Inclusion of spin–orbit coupling splits the $t_{2g}$ states into a lower quartet and a higher doublet. We could introduce fictitious spins $\widetilde{S} = 3/2$ and $\widetilde{S} = 1/2$ for these manifolds, but because they derive from an orbital triplet it is more usual to introduce a fictitious orbital quantum number $\widetilde{l} = 1$ to describe the $t_{2g}$ states, and then characterize the spin–orbit split states by $\widetilde{j} = \widetilde{l} \pm s$, where $s = 1/2$ is the true spin.[56] In the $5d^5$ configuration the $\widetilde{j} = 3/2$ quartet is completely filled and the $\widetilde{j} = 1/2$ doublet is half filled, so the magnetic behaviour is determined by the latter only. The $\widetilde{j} = 1/2$ doublet is very different from a spin-only doublet. The $\widetilde{j} = 1/2$ wave functions contain strong entanglement of spin and orbital degrees of freedom, and although the $g$-factor is isotropic its value is negative, $g = -2$, which means that in a magnetic field the lower Zeeman-split component has $\widetilde{j}_z = 1/2$, opposite to the usual case. The total field-induced magnetic moment of $\mu = 1\,\mu_B$ is dominated by the orbital contribution, with $\mu^{spin} = 1/3\,\mu_B$ and $\mu^{orbital} = 2/3\,\mu_B$. Owing to the large orbital component, interactions between $\widetilde{j} = 1/2$ moments can be highly anisotropic.

# C.5   Multipole expansion

Consider a charge density $\rho(\mathbf{r'})$ of finite spatial extent. From the principle of superposition, the electrostatic potential $V(\mathbf{r})$ at a distant point displaced by $\mathbf{r}$ from an origin inside $\rho(\mathbf{r'})$ is given by

$$V(\mathbf{r}) = \frac{1}{4\pi\epsilon_0} \int \frac{\rho(\mathbf{r'})}{|\mathbf{r} - \mathbf{r'}|} \, d^3\mathbf{r'}. \tag{C.145}$$

The integration is performed over the entire volume of $\rho(\mathbf{r}')$. The integral can be expressed as a series of successively smaller terms via the Legendre expansion,

$$\frac{1}{|\mathbf{r} - \mathbf{r}'|} = \frac{1}{r} \sum_{l=0}^{\infty} \left(\frac{r'}{r}\right)^l P_l(\cos\gamma), \tag{C.146}$$

where $\gamma$ is the angle between $\mathbf{r}$ and $\mathbf{r}'$, Fig. C.5, and the $P_l(\cos\gamma)$ are Legendre polynomials. By substituting (C.146) into (C.145) we can write

$$V(\mathbf{r}) = \frac{1}{4\pi\epsilon_0 r} \sum_{l=0}^{\infty} \frac{Q_l}{r^l}, \tag{C.147}$$

where

$$Q_l = \int (r')^l P_l(\cos\gamma)\rho(\mathbf{r}')\,\mathrm{d}^3\mathbf{r}'. \tag{C.148}$$

Equation (C.147) is the *multipole expansion* of the electrostatic potential. The terms in the series originate from successively higher frequency components of the angular variation of $\rho(\mathbf{r}')$, and correspond to the *electric multipoles*. The angular-dependent functions $Q_l$ contain the multipole moments of $\rho(\mathbf{r}')$. For example, $Q_0 = \int \rho(\mathbf{r}')\,\mathrm{d}^3\mathbf{r}' = Q$ (the total charge) is the electric monopole, and $Q_1 = \int r' \cos\gamma\, \rho(\mathbf{r}')\,\mathrm{d}^3\mathbf{r}' = \mathbf{p} \cdot \hat{\mathbf{r}}$, where $\mathbf{p} = \int \mathbf{r}'\rho(\mathbf{r}')\,\mathrm{d}^3\mathbf{r}'$ is the electric dipole moment. The electric multipole terms in $V(\mathbf{r})$ have parity $(-1)^l$, since $P_l \to (-1)^l P_l$ under the inversion operation $\mathbf{r} \to -\mathbf{r}$.

The magnetic vector potential $\mathbf{A}(\mathbf{r})$ for a steady current distribution $\mathbf{j}(\mathbf{r}')$ may be chosen as[57]

$$\mathbf{A}(\mathbf{r}) = \frac{\mu_0}{4\pi} \int \frac{\mathbf{j}(\mathbf{r}')}{|\mathbf{r} - \mathbf{r}'|}\,\mathrm{d}^3\mathbf{r}', \tag{C.149}$$

and by (C.146) it can similarly be written as a multipole expansion

$$\mathbf{A}(\mathbf{r}) = \frac{\mu_0}{4\pi r} \sum_{l=0}^{\infty} \frac{\mathbf{m}_l}{r^l}, \tag{C.150}$$

where

$$\mathbf{m}_l = \int (r')^l P_l(\cos\gamma)\,\mathbf{j}(\mathbf{r}')\,\mathrm{d}^3\mathbf{r}'. \tag{C.151}$$

The terms in the series (C.150) correspond to magnetic multipoles.[58] The magnetic scattering potential $V_\mathrm{M}$ is proportional to[59] $\mathbf{B}(\mathbf{r}) = \nabla \times \mathbf{A}$, and so the magnetic multipole terms in $V_\mathrm{M}$ have parity $(-1)^{l+1}$, since $\mathbf{A} \to (-1)^l \mathbf{A}$ and $\nabla \to -\nabla$ under $\mathbf{r} \to -\mathbf{r}$.

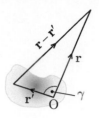

**Fig. C.5** Definition of $\mathbf{r}$, $\mathbf{r}'$ and $\gamma$.

**Table C.3** The nomenclature for multipoles of rank $l$ comes from the Greek for $2^l$, the number of poles in the angular function.

| Rank, $l$ | Name |
| --- | --- |
| 0 | monopole |
| 1 | dipole |
| 2 | quadrupole |
| 3 | octupole |
| 4 | hexadecapole |
| 5 | triakontadipole |
| 6 | tetrahexacontapole |

[57] $\mathbf{A}(\mathbf{r})$ is not defined uniquely. There is a gauge freedom which permits the transformation $\mathbf{A} \to \mathbf{A} + \nabla\psi$ without changing $\mathbf{B}(\mathbf{r})$, where $\psi(\mathbf{r})$ is any scalar function.

[58] For $l = 0$, $\mathbf{m}_0 = \int \mathbf{j}(\mathbf{r}')\,\mathrm{d}^3\mathbf{r}' = 0$, so the magnetic monopole is zero. For $l = 1$, it can be shown that $\mathbf{m}_1 = \boldsymbol{\mu} \times \hat{\mathbf{r}}$, with $\boldsymbol{\mu}$ the magnetic dipole moment.

[59] See eqn (4.23).

# C.6 Atomic scattering matrix elements

In the Racah formulation introduced by Johnston (1966), the scattering matrix elements for an atom with many equivalent electrons contain $A$ and $B$ coefficients which are associated with orbital (or more precisely, current) and spin scattering, respectively. Here we give formulae for these coefficients based on the expressions in Lovesey and Rimmer (1969) and Lovesey (1984b).

We consider transitions between states of a configuration $l^n$, formed from $n$ equivalent electrons with angular momentum quantum number $l$. We work in the $LS$-coupling scheme, in which the initial and final states are described by total angular momentum quantum numbers $\nu LSJ$ and $\nu'L'S'J'$, respectively, where $\nu$ represents any quantum numbers needed to distinguish between different terms with the same $LS$.

The $A$ and $B$ coefficients depend on two integer indices, and are non-zero only in the following cases:

$$A(K \pm 1, K) \neq 0 \quad \text{when} \quad K = 1, 3, ..., 2l - 1, \tag{C.152}$$

$$B(K \pm 1, K) \neq 0 \quad \text{when} \quad K = 1, 3, ..., 2l + 1, \tag{C.153}$$

$$B(K, K) \neq 0 \quad \text{when} \quad K = 2, 4, ..., 2l. \tag{C.154}$$

In addition, $A(K \pm 1, K) = 0$ when $S \neq S'$, and $B(K, K) = 0$ when $S = S'$, $L = L'$ and $J = J'$. The following relations also apply:

$$A(K + 1, K) = \left(\frac{K}{K+1}\right)^{1/2} A(K - 1, K), \tag{C.155}$$

$$B(K + 1, K) = \left(\frac{K}{K+1}\right)^{1/2} B(K - 1, K). \tag{C.156}$$

Subject to compliance with these conditions, we may separate the radial averages $\langle j_n \rangle$ of the spherical Bessel functions, eqn (6.27), and write

$$A(K - 1, K) = a_K\big(\langle j_{K-1}\rangle + \langle j_{K+1}\rangle\big), \tag{C.157}$$

$$B(K - 1, K) = b_K\langle j_{K-1}\rangle + b'_K\langle j_{K+1}\rangle, \tag{C.158}$$

$$B(K, K) = b''_K\langle j_K\rangle. \tag{C.159}$$

An explicit expressions for $a_K$ is

$$a_K = (-1)^{l+L+L'+S+J'} i^{K+1}(2l+1)^2 A(K,K,l) \begin{Bmatrix} K & L' & L \\ S & J & J' \end{Bmatrix}$$

$$\times \left\{ \tfrac{1}{3}(2L+1)(2L'+1)(2J'+1)(K+1)\right\}^{1/2}$$

$$\times n \sum_{\bar{\theta}} (-1)^{\bar{L}} (\theta\{|\bar{\theta}) (\bar{\theta}|\}\theta') \begin{Bmatrix} K & l & l \\ \bar{L} & L & L' \end{Bmatrix}, \tag{C.160}$$

where

$$A(K,K,l) = \frac{(-1)^{(K-1)/2}}{2(2l+1)} \left(\frac{2l+1-K}{2K+1}\right)^{1/2}$$

$$\times \frac{\{(2l+1+K)/2\}!}{\{(K+1)/2\}!\{(K-1)/2\}!\{(2l+1-K)/2\}!}$$

$$\times \left\{ \frac{(K+1)!(K-1)!(2l+1-K)!}{(2l+1+K)!}\right\}^{1/2}.$$

The coefficients $b_K$, $b'_K$ and $b''_K$ are given by

$$b_K = \mathrm{i}^{K-1}\frac{K+1}{\{3(2K+1)\}^{1/2}}\, c(K-1,K), \qquad \text{(C.161)}$$

$$b'_K = \mathrm{i}^{K+1}\left\{\frac{K(K+1)}{3(2K+1)}\right\}^{1/2} c(K+1,K), \qquad \text{(C.162)}$$

$$b''_K = (-1)^{K/2}\left(\frac{2K+1}{3}\right)^{1/2} c(K,K), \qquad \text{(C.163)}$$

where

$$c(\dot{K},K) = \mathrm{i}(-1)^{L'+S'}\left(\tfrac{3}{2}\right)^{1/2}(2l+1)\begin{pmatrix} l & \dot{K} & l \\ 0 & 0 & 0 \end{pmatrix}\begin{Bmatrix} 1 & \dot{K} & K \\ S' & L' & J' \\ S & L & J \end{Bmatrix}$$

$$\times \{(2\dot{K}+1)(2K+1)(2S+1)(2S'+1)(2L+1)(2L'+1)(2J'+1)\}^{1/2}$$

$$\times n\sum_{\bar{\theta}}(-1)^{\bar{S}+\bar{L}}\left(\theta\{|\bar{\theta}\right)\left(\bar{\theta}|\}\theta'\right)\begin{Bmatrix} S & 1 & S' \\ \tfrac{1}{2} & \bar{S} & \tfrac{1}{2} \end{Bmatrix}\begin{Bmatrix} L & \dot{K} & L' \\ l & \bar{L} & l \end{Bmatrix}. \qquad \text{(C.164)}$$

Equations (C.160) and (C.164) contain Wigner 3$j$, 6$j$ and 9$j$ symbols. These are the 6-element arrays in round and curly brackets, and the 9-element array in curly brackets, respectively.[60] The equations also include a sum over states $\bar{\theta}$ belonging to the $l^{n-1}$ configuration that are common parents to the $\theta \equiv \nu LSJ$ and $\theta' \equiv \nu'L'S'J'$ states, with $\left(\theta\{|\bar{\theta}\right)$ and $\left(\bar{\theta}|\}\theta'\right)$ the corresponding fractional parentage coefficients.[61]

[60] Wigner symbols arise in the coupling of angular momentum in quantum mechanics.

[61] Racah (1943) showed that the correctly antisymmetrized eigenfunction of a term $\nu LS$ of the configuration $l^n$ can be expressed as a linear combination of states obtained by angular-momentum coupling one additional equivalent $l$ electron to antisymmetric states of the $l^{n-1}$ configuration. The appropriate $l^{n-1}$ states are the *parents* of the $l^n$ state, and the coefficients of the linear combination are known as *coefficients of fractional parentage*. The coefficients have been tabulated by Nielson and Koster (1963).

# Linear Response Theory

# D

Experimental techniques which probe matter weakly often measure linear properties of the system. By *linear* we mean that the response of the system is directly proportional to the size of the stimulus applied during the measurement. The quantity obtained in such measurements is called a *linear response coefficient*. An example is the magnetic susceptibility, which is the ratio of the induced magnetization to the applied magnetic field. If the relation between response and stimulus involves higher powers of the stimulus then the system is *non-linear*.

We have stated previously that neutron scattering can be considered a weak probe of matter because it obeys the Born approximation (Section 3.2). We have also shown (Section 3.4) that neutron scattering measures certain correlation functions which reflect thermal equilibrium properties of the scattering system. It is now of interest to ask whether one can link the correlation functions probed by neutron scattering to the linear response coefficients measured by other techniques.

The answer is unequivocally yes! As shown by Kubo,[1] there are in fact exact expressions connecting the linear response coefficients with the correlation functions for systems in equilibrium (Kubo, 1957, Kubo, 1966). In this appendix we shall give an account of linear response theory and derive several useful results, including a relationship between the scattering function and the generalized susceptibility.[2] The latter, which provides an example of the *Fluctuation–Dissipation theorem*, is useful for two reasons. First, it enables us to make a direct comparison between the results of neutron scattering and other techniques, and second, it provides us with an alternative apparatus for calculating the neutron cross-section from a microscopic model for the system.

[1]Ryogo Kubo (1920–1995), Japanese mathematical physicist.

[2]The present approach is similar to those of Lovesey (1984a), Appendix B, and Jensen and Mackintosh (1991), Chapter 3.

## D.1 The density matrix

Consider a quantum-mechanical system described by a state function $\Psi$. The mean value of a dynamical variable represented by an operator $A$ is given by (see Section C.1.5)

$$\langle A \rangle = \langle \Psi | A | \Psi \rangle. \tag{D.1}$$

We can expand $\Psi$ in terms of any complete set of basis functions $\{\phi_i\}$,

$$\Psi = \sum_i c_i \phi_i, \tag{D.2}$$

so that

$$\langle A \rangle = \sum_j \sum_i c_i^* c_j \, A_{ij}, \tag{D.3}$$

where $A_{ij} = \langle \phi_i | A | \phi_j \rangle$.

This formalism applies to an isolated system in a pure state completely described by the state function $\Psi$. On the other hand, we usually find ourselves concerned with a small sub-system coupled to a much larger system, e.g. a single molecule in a gas of many molecules, and in such circumstances it is not realistic to have knowledge of the state function for the entire system.

Statistical mechanics deals with our incomplete knowledge as follows. We define a set of pure states $\{\Psi\}$ for the sub-system in the absence of any interactions with the surroundings. We then assume that interactions, such as exchanges of energy or particles, cause the sub-system to fluctuate between different $\Psi$. The probability of the sub-system being in a given state is then governed by a statistical distribution, and the mean properties of the sub-system must be obtained as statistical averages over this distribution.

This approach suggests that the physical properties of our sub-system might be derivable from a statistical average of (D.3) over the probability of occupation of different $\Psi$ states. The coefficient of $A_{ij}$ would then be $\overline{c_i^* c_j}$, the averaged product, which we denote by $\rho_{ji}$. We can then regard the summation in (D.3) as equivalent to finding the trace[3] of a matrix formed from the product of two square matrices $\rho_{ji}$ and $A_{ij}$. Further, just as $A_{ij}$ is the matrix representation of the operator $A$, so we can regard $\rho_{ji}$ as the matrix representation of some operator $\rho$ defined by

$$\rho_{ij} = \langle \phi_i | \rho | \phi_j \rangle. \tag{D.4}$$

Hence, eqn (D.3) for $\langle A \rangle$ can be written as

$$\langle A \rangle = \mathrm{Tr}\{\rho A\} = \mathrm{Tr}\{A\rho\}, \tag{D.5}$$

the second equality coming from the cyclic invariance of the trace. $\rho_{ij}$ is known as the *density matrix*, and $\rho$ the *density operator*.

Now suppose that the sub-system is disturbed from equilibrium so that in the Schrödinger representation the basis states vary with time. In this representation the matrix $\rho_{ij}$ does not vary with time (since it originates from fixed $c_i$ coefficients), but the operator $\rho$ does vary with time through the time dependence of the basis states. We can then obtain an equation of motion for $\rho(t)$ by differentiating eqn (D.4):

$$0 = \frac{\partial \langle \phi_i |}{\partial t} \rho |\phi_j\rangle + \langle \phi_i | \frac{\partial \rho}{\partial t} |\phi_j\rangle + \langle \phi_i | \rho \frac{\partial |\phi_j\rangle}{\partial t}$$

$$= -\frac{1}{i\hbar} \langle \phi_i | \mathcal{H}\rho |\phi_j\rangle + \langle \phi_i | \frac{\partial \rho}{\partial t} |\phi_j\rangle + \frac{1}{i\hbar} \langle \phi_i | \rho\mathcal{H} |\phi_j\rangle. \tag{D.6}$$

In the second line we have used the time-dependent Schrödinger equation $\mathcal{H}|\phi\rangle = i\hbar \partial|\phi\rangle/\partial t$ and its complex conjugate to replace the time derivatives. Hence,

$$i\hbar \frac{\partial \rho}{\partial t} = [\mathcal{H}, \rho]. \tag{D.7}$$

[3] The trace of a matrix $\mathbf{M}$ is the sum of the elements on the leading diagonal, and is denoted by Tr. Thus

$$\mathrm{Tr}\{\mathbf{M}\} = \sum_j \mathbf{M}_{jj}.$$

$\mathrm{Tr}\{ABC\} = \mathrm{Tr}\{CAB\}$, etc.

Equation (D.7) resembles the equation of motion of a Heisenberg operator, eqn (C.49), but the order of the operators in the commutator is reversed. This difference reflects that $\rho(t)$ is not a Heisenberg operator for a physical observable, but is instead a generalized state function.

## D.2   Generalized susceptibility

Suppose that the sub-system is weakly perturbed from equilibrium by a position- and time-dependent generalized 'force' $f(\mathbf{r}, t)$. For simplicity, we consider a force comprising a single Fourier component,

$$f(\mathbf{r}, t) = f(\mathbf{Q}, \omega) \exp \left\{ i(\omega t - \mathbf{Q} \cdot \mathbf{r}) \right\}. \tag{D.8}$$

As we have a linear system, the principle of superposition allows us to construct the response to an arbitrary space- and time-dependent force from the response of the system to each individual Fourier component.

Suppose further that the response of the system is observed in a dynamical variable described by an operator $A$ without explicit time dependence. The force (D.8) causes a deviation in the mean value of $A$ from its equilibrium value

$$\langle A(\mathbf{r}, t) \rangle - \langle A \rangle_0 = \langle \Delta A(\mathbf{r}, t) \rangle = a(\mathbf{Q}, \omega) \exp \left\{ i(\omega t - \mathbf{Q} \cdot \mathbf{r}) \right\}. \tag{D.9}$$

For a *linear response*, the Fourier amplitudes in (D.8) and (D.9) are directly proportional to one another. This proportionality defines the *generalized susceptibility* $\chi(\mathbf{Q}, \omega)$,

$$a(\mathbf{Q}, \omega) = \chi(\mathbf{Q}, \omega) f(\mathbf{Q}, \omega). \tag{D.10}$$

The generalized susceptibility is a complex number which we write as

$$\chi(\mathbf{Q}, \omega) = \chi'(\mathbf{Q}, \omega) - i\chi''(\mathbf{Q}, \omega), \tag{D.11}$$

where $\chi'$ and $\chi''$ are the *real* and *imaginary* parts,[4] also known as the *reactive* and *absorptive* parts of the susceptibility, and give the components of the response that are in-phase and out-of-phase with the force, respectively.[5] Work is done by the component of the force $f(t)$ that is in-phase with the velocity $v(t) = \partial \Delta A / \partial t$, so the mean rate of dissipation of energy is given by (from now on, we do not show the $\mathbf{Q}$ dependence)

$$\langle P \rangle = \langle \mathrm{Re}\{f(t)\} \mathrm{Re}\{v(t)\} \rangle$$

$$= \langle f(\omega) \cos \omega t \, f(\omega) \omega \{ \chi''(\omega) \cos \omega t - \chi'(\omega) \sin \omega t \} \rangle$$

$$= \frac{1}{2} \omega f^2(\omega) \chi''(\omega) \tag{D.12}$$

Here, the angular brackets denote a time-average, and we have used the identities $\langle \cos^2 \omega t \rangle = 1/2$ and $\langle \cos \omega t \sin \omega t \rangle = 0$. Equation (D.12) shows that $\chi''(\omega)$ determines the mean rate of dissipation.

[4]Later, we shall use a more general definition of $\chi'$ and $\chi''$ which means that they are not always real functions equal to the real and imaginary parts of $\chi$ — see eqns (D.30) and (D.31). For example, the expressions in eqns (D.38) and (D.39) can be complex if $\hat{B} \neq \hat{A}^\dagger$.

[5]The negative sign in (D.11) together with the time dependence $\exp(i\omega t)$ in (D.9) means that if $\chi'$ and $\chi''$ are real and have the same sign (both positive or both negative) then the response of the system lags behind the driving force, as must necessarily be the case when energy is dissipated.

## D.3   Response function

Suppose that for $t < 0$ the sub-system is in thermal equilibrium and described by a Hamiltonian $\mathcal{H}_0$. At $t = 0$ the Hamiltonian suddenly changes to[6]

$$\mathcal{H} = \mathcal{H}_0 + \mathcal{H}_1(t), \quad (t > 0), \tag{D.13}$$

where $\mathcal{H}_1(t)$ represents an external perturbation[7] which couples to the system through an operator $B$ without explicit time dependence. The time dependence of $\mathcal{H}_1$ is contained explicitly in the force $f(t)$ defined such that

$$\mathcal{H}_1(t) = -Bf(t). \tag{D.14}$$

It is the time variation of the perturbation that causes dissipation, so we omit the position and wavevector labels to simplify the notation.

To determine the linear response of the system we need an expression for $\langle \Delta A(t) \rangle$ accurate to first order in the perturbation $\mathcal{H}_1$. To find this, we solve the equation of motion (D.7) for the density operator to first order in $\mathcal{H}_1$, and substitute the solution in (D.5).

Let us write the solution for the density operator as a series

$$\rho(t) = \rho_0(t) + \rho_1(t) + \dots, \tag{D.15}$$

where $\rho_n(t)$ is of $n$th order in $\mathcal{H}_1$. Substituting (D.15) into the equation of motion (D.7) with $\mathcal{H} = \mathcal{H}_0 + \mathcal{H}_1$ we find for the first two terms

$$\begin{aligned}
i\hbar \frac{\partial \rho_0}{\partial t} &= [\mathcal{H}_0, \rho_0(t)], \\
i\hbar \frac{\partial \rho_1}{\partial t} &= [\mathcal{H}_0, \rho_1(t)] + [\mathcal{H}_1(t), \rho_0(t)].
\end{aligned} \tag{D.16}$$

The boundary conditions at $t = 0$, are

$$\begin{aligned}
\mathcal{H}_1(0) &= 0, \\
\rho_0(0) &= \rho(0), \\
\rho_n(0) &= 0, \qquad n \geq 1.
\end{aligned} \tag{D.17}$$

With some elementary calculus one can verify that the solutions of (D.16) with boundary conditions (D.17) are

$$\rho_0(t) = \rho(0), \tag{D.18}$$

$$\rho_1(t) = -\frac{i}{\hbar} \int_0^t \exp\{-i\mathcal{H}_0(t - t')/\hbar\} [\mathcal{H}_1(t'), \rho(0)] \exp\{i\mathcal{H}_0(t - t')/\hbar\} \, dt', \tag{D.19}$$

and on substituting these solutions into (D.5) we find that the mean value of $A$ at time $t$, to first order in $\mathcal{H}_1$, is

$$\begin{aligned}
\langle A(t) \rangle &= \mathrm{Tr}\{\rho(0)A\} + \mathrm{Tr}\{\rho_1(t)A\} \\
&= \langle A \rangle_0 - \frac{i}{\hbar} \int_0^t \mathrm{Tr}\{ T_t^\dagger T_{t'} [\mathcal{H}_1(t'), \rho(0)] T_t T_{t'}^\dagger A \} \, dt',
\end{aligned} \tag{D.20}$$

where we have used the shorthand $T_t = \exp(\mathrm{i}\mathcal{H}_0 t/\hbar)$ and $T_t^\dagger = \exp(-\mathrm{i}\mathcal{H}_0 t/\hbar)$. We now expand the commutator, and use the cyclic invariance of the trace to write

$$
\begin{aligned}
\langle \Delta A(t) \rangle = &-\frac{\mathrm{i}}{\hbar} \int_0^t \mathrm{Tr}\{\, \rho(0) T_t T_{t'}^\dagger A T_t^\dagger T_{t'} \mathcal{H}_1(t') \,\}\, \mathrm{d}t' \\
&+\frac{\mathrm{i}}{\hbar} \int_0^t \mathrm{Tr}\{\, \rho(0) \mathcal{H}_1(t') T_t T_{t'}^\dagger A T_t^\dagger T_{t'} \,\}\, \mathrm{d}t' \\
\\
= &-\frac{\mathrm{i}}{\hbar} \int_0^t \mathrm{Tr}\{\, \rho(0) T_t A T_t^\dagger T_{t'} \mathcal{H}_1(t') T_{t'}^\dagger \,\}\, \mathrm{d}t' \\
&+\frac{\mathrm{i}}{\hbar} \int_0^t \mathrm{Tr}\{\, \rho(0) T_{t'} \mathcal{H}_1(t') T_{t'}^\dagger T_t A T_t^\dagger \,\}\, \mathrm{d}t'. \quad (\text{D.21})
\end{aligned}
$$

In the second step of (D.21) we have used the fact that $\rho(0)$ and $\mathcal{H}_0$ commute, a property that derives from the equation of motion (D.7) with $\rho(0)$ a constant.

Remembering the form of $\mathcal{H}_1(t')$ specified by (D.14), we see that eqn (D.21) contains Heisenberg operators $\hat{A}(t)$ and $\hat{B}(t')$ defined in the usual way by the operator products $T_t A T_t^\dagger$ and $T_{t'} B T_{t'}^\dagger$, respectively.[8] Therefore,

$$
\langle \Delta A(t) \rangle = \int_0^t \phi_{AB}(t - t')\, f(t')\, \mathrm{d}t', \quad (\text{D.22})
$$

where,

$$
\phi_{AB}(t - t') = \frac{\mathrm{i}}{\hbar} \langle [\hat{A}(t), \hat{B}(t')] \rangle. \quad (\text{D.23})
$$

These results show that the linear response of the system depends on the *thermal equilibrium* expectation value of the products of Heisenberg operators $\hat{A}(t)$ and $\hat{B}(t')$.

Equation (D.23) is known as the *Kubo formula*. The *response function* $\phi_{AB}(t - t')$ has a number of properties worth mentioning:

(1) It represents the response of the system at time $t$ to an impulsive perturbation applied at an earlier time $t'$ ($0 < t' < t$), as becomes obvious if one substitutes $f(t') = \delta(t')$ into (D.22).

(2) It depends only on the difference $t - t'$, so that[9]

$$
\phi_{AB}(t) = \frac{\mathrm{i}}{\hbar} \langle [\hat{A}(t), \hat{B}] \rangle, \quad (\text{D.24})
$$

where $\hat{B} \equiv \hat{B}(0)$.

(3) $\phi_{AB}(t)$ is only physically meaningful when $t > 0$, since perturbations applied in the future cannot affect the present state of the system. However, $\phi_{AB}(t)$ does exist mathematically for $t \le 0$, and satisfies[10]

$$
\phi_{AB}(-t) = -\phi_{BA}(t). \quad (\text{D.25})
$$

(4) $\phi_{AB}$ is real if the perturbing function $f(t')$ is real, since eqn (D.22) applies for all $t$. If not, then from (D.23) and (C.57),

$$
\phi_{AB}^*(t) = -\phi_{B^\dagger A^\dagger}(-t). \quad (\text{D.26})
$$

[8] As elsewhere, the hat on the operator shows that it is a Heisenberg operator. See Section C.1.10.

[9] This follows from property (C.54), according to which a shift in the origin of time (in this case by subtracting $t'$) does not affect the value of a correlation function.

[10] In detail,
$$
\begin{aligned}
\phi_{AB}(-t) &= \frac{\mathrm{i}}{\hbar} \langle [\hat{A}(-t), \hat{B}] \rangle \\
&= \frac{\mathrm{i}}{\hbar} \langle [\hat{A}, \hat{B}(t)] \rangle \\
&= -\frac{\mathrm{i}}{\hbar} \langle [\hat{B}(t), \hat{A}] \rangle \\
&= -\phi_{BA}(t).
\end{aligned}
$$

If the perturbation is such that, after a sufficiently long time, the value of $\langle \Delta A(t) \rangle$ does not depend on the time at which the perturbation was switched on (i.e. all transients have decayed to zero by time $t$), then the lower limit of integration in (D.22) may be extended to $-\infty$. To ensure that this is valid, we can replace $f(t')$ by $f(t') \exp(\epsilon t')$, $\epsilon > 0$, and take the limit as $\epsilon \to 0$ after evaluating the integral. This trick makes the perturbation switch on slowly over a long period of time, rather than suddenly at $t = 0$, and so ensures that the system is in equilibrium at any time. If, in addition, we represent $f(t')$ by the Fourier component (D.8), make a substitution for $t - t'$, and compare the resulting expression with eqns (D.9) and (D.10), then we obtain the following equation for the generalized susceptibility:

$$\chi_{AB}(\omega) = \lim_{\epsilon \to 0^+} \int_0^\infty \phi_{AB}(t) \exp(-\mathrm{i}\omega t) \exp(-\epsilon t) \, \mathrm{d}t. \qquad (D.27)$$

Hence,

$$\chi_{AB}(\omega) = \lim_{\epsilon \to 0^+} \chi_{AB}(z) \qquad (D.28)$$

$$\chi_{AB}(-\omega) = \lim_{\epsilon \to 0^+} \chi_{AB}(-z^*) = \lim_{\epsilon \to 0^+} \chi^*_{A^\dagger B^\dagger}(z), \qquad (D.29)$$

where $z = w - \mathrm{i}\epsilon$. The second form of (D.29) follows from (D.25)–(D.27).

The susceptibility divides into terms which are even and odd under simultaneous sign reversal of $\omega$ and interchange of $\hat{A}$ and $\hat{B}$:

$$\chi'_{AB}(\omega) = \chi'_{BA}(-\omega) = \lim_{\epsilon \to 0^+} \frac{1}{2} \left\{ \chi_{BA}(-z^*) + \chi_{AB}(z) \right\} \qquad (D.30)$$

$$\chi''_{AB}(\omega) = -\chi''_{BA}(-\omega) = \lim_{\epsilon \to 0^+} \frac{1}{2\mathrm{i}} \left\{ \chi_{BA}(-z^*) - \chi_{AB}(z) \right\}. \qquad (D.31)$$

These relations give $\chi_{AB}(\omega) = \chi'_{AB}(\omega) - \mathrm{i}\chi''_{AB}(\omega)$, consistent with (D.11). If $\hat{A}$ and $\hat{B}$ are Hermitian conjugates $(\hat{B} = \hat{A}^\dagger)$,[11] then $\chi'_{AB}(\omega)$ and $\chi''_{AB}(\omega)$ are real functions, and are the real (reactive) and imaginary (absorptive) parts of $\chi_{AB}(\omega)$, respectively.[12]

[11] If, in addition, $\hat{A}$ and $\hat{B}$ are the same and Hermitian $(\hat{A} = \hat{B} = \hat{A}^\dagger)$ then $\phi_{AB}(t)$ is real — see (D.25) and (D.26).

[12] In general, the functions $\chi'_{AB}(\omega)$ and $\chi''_{AB}(\omega)$ defined by (D.30) and (D.31) are not the real and imaginary parts of $\chi_{AB}(\omega)$. However, these functions are convenient because when defined this way they simplify some useful expressions that will be encountered later, e.g. eqns (D.36) and (D.38)–(D.39).

# D.4 Fluctuation–Dissipation theorem

Let us now establish the relevance of these results to neutron scattering theory. In a scattering process, the neutron both acts as the perturbation as well as the probe which enables us to observe the response of the system. The scattering probability is proportional to a function that has the general form [e.g. eqn (3.36)]

$$S_{BA}(\omega) = \frac{1}{2\pi\hbar} \int_{-\infty}^\infty \langle \hat{B}\hat{A}(t) \rangle \exp(-\mathrm{i}\omega t) \, \mathrm{d}t. \qquad (D.32)$$

Continuing to suppress the $\mathbf{Q}$ dependence, we consider the time Fourier transform of the response function $\phi_{AB}(t)$ given by (D.24):

$$\frac{1}{2\pi}\int_{-\infty}^{\infty}\phi_{AB}(t)\exp(-\mathrm{i}\omega t)\,\mathrm{d}t = \frac{\mathrm{i}}{2\pi\hbar}\int_{-\infty}^{\infty}\langle\hat{A}(t)\hat{B} - \hat{B}\hat{A}(t)\rangle\exp(-\mathrm{i}\omega t)\,\mathrm{d}t$$

$$= \frac{\mathrm{i}}{2\pi\hbar}\int_{-\infty}^{\infty}\langle\hat{B}\hat{A}(t+\mathrm{i}\hbar\beta) - \hat{B}\hat{A}(t)\rangle$$
$$\times\exp(-\mathrm{i}\omega t)\,\mathrm{d}t$$

$$= \mathrm{i}\{\exp(-\beta\hbar\omega) - 1\}\widetilde{S}_{BA}(\omega), \qquad (\text{D.33})$$

with $\beta = 1/(k_{\mathrm{B}}T)$. The steps required to proceed from the first to the second lines are analogous to those used to prove identity (C.56). In the last line, $\widetilde{S}(\omega)$ is the dynamic part of the response function (D.32) as defined in eqn (3.50), and appears instead of $S(\omega)$ because the static parts $\langle\hat{A}(\pm\infty)\hat{B}\rangle$ and $\langle\hat{B}\hat{A}(\pm\infty)\rangle$ of the two correlation functions contained in $\phi_{AB}(t)$ both equal $\langle\hat{A}\rangle\langle\hat{B}\rangle$ and so cancel each other out.[13] To arrive at the last line, substitute $t' = t + \mathrm{i}\hbar\beta$ in the first term of the integral. The expression for the dynamic part of the response function is therefore

$$\widetilde{S}_{BA}(\omega) = \{1 + n(\omega)\}\frac{\mathrm{i}}{2\pi}\int_{-\infty}^{\infty}\phi_{AB}(t)\exp(-\mathrm{i}\omega t)\,\mathrm{d}t, \qquad (\text{D.34})$$

where

$$n(\omega) = \frac{1}{\exp(\beta\hbar\omega) - 1} \qquad (\text{D.35})$$

is the Planck distribution. Now we apply (D.25) and (D.27) to obtain

$$\widetilde{S}_{BA}(\omega) = \{1 + n(\omega)\}$$
$$\times\frac{\mathrm{i}}{2\pi}\lim_{\epsilon\to 0^{+}}\int_{0}^{\infty}\{\phi_{AB}(t)\exp(-\mathrm{i}zt) - \phi_{BA}(t)\exp(\mathrm{i}z^{*}t)\}\,\mathrm{d}t$$

$$= \{1 + n(\omega)\}\frac{\mathrm{i}}{2\pi}\lim_{\epsilon\to 0^{+}}\{\chi_{AB}(z) - \chi_{BA}(-z^{*})\}$$

$$= \{1 + n(\omega)\}\frac{1}{\pi}\chi''_{AB}(\omega), \qquad (\text{D.36})$$

where $\chi''_{AB}(\omega)$ is the absorptive part of $\chi_{AB}(\omega)$ — see (D.31). In the above, $z = \omega - \mathrm{i}\epsilon$ and we take the limit as $\epsilon\to 0^{+}$ after integration for the reasons given before eqn (D.27). Equation (D.36) is an expression of the *Fluctuation–Dissipation theorem*, which links the spectrum of spontaneous fluctuations $\widetilde{S}(\omega)$ to the strength of the system's response to external perturbations $\chi''(\omega)$.

[13]This means that $\widetilde{S}(\omega)$ does not contain any purely elastic scattering, e.g. Bragg scattering from static order, but $\widetilde{S}(\omega)$ may contain a signal at $\omega = 0$ arising from quasi-static correlations.

## D.4.1   Calculating the generalized susceptibility

We can calculate $\chi_{AB}(\omega)$ from the eigenvalues and eigenfunctions of $\mathcal{H}_0$, if they are known. From (D.24), (D.27) and the identity (C.55), it can be shown that

$$\chi_{AB}(\omega) = \frac{i}{\hbar} \sum_{\lambda,\lambda'} (p_{\lambda'} - p_{\lambda}) \langle \lambda | B | \lambda' \rangle \langle \lambda' | A | \lambda \rangle$$

$$\times \lim_{\epsilon \to 0^+} \int_0^\infty \exp\{i(E_{\lambda'} - E_\lambda - \hbar\omega + i\hbar\epsilon)t/\hbar\}\, dt$$

$$= \lim_{\epsilon \to 0^+} \sum_{\lambda,\lambda'} (p_\lambda - p_{\lambda'}) \frac{\langle \lambda | B | \lambda' \rangle \langle \lambda' | A | \lambda \rangle}{E_{\lambda'} - E_\lambda - \hbar\omega + i\hbar\epsilon}, \qquad (D.37)$$

where $p_\lambda = \exp(-\beta E_\lambda)/Z$ and $Z$ is the partition function. Similarly, $\chi_{BA}(\omega)$ is given by (D.37) but with $z = \omega - i\epsilon$ replaced by $-z$. Hence, using (D.31) and Dirac's formula (B.20) we obtain the absorptive part of $\chi_{AB}(\omega)$ as

$$\chi_{AB}''(\omega) = \pi \sum_{\lambda,\lambda'} (p_\lambda - p_{\lambda'}) \langle \lambda | B | \lambda' \rangle \langle \lambda' | A | \lambda \rangle \, \delta(E_{\lambda'} - E_\lambda - \hbar\omega), \quad (D.38)$$

and the reactive part as,

$$\chi_{AB}'(\omega) = \sum_{\lambda,\lambda'}^{E_{\lambda'} \neq E_\lambda} (p_\lambda - p_{\lambda'}) \frac{\langle \lambda | B | \lambda' \rangle \langle \lambda' | A | \lambda \rangle}{E_{\lambda'} - E_\lambda - \hbar\omega}$$

$$+ \frac{\delta_{\omega 0}}{k_{\mathrm{B}}T} \sum_{\lambda,\lambda'}^{E_{\lambda'} = E_\lambda} p_\lambda \langle \lambda | B - \langle B \rangle | \lambda' \rangle \langle \lambda' | A - \langle A \rangle | \lambda \rangle. \qquad (D.39)$$

The appearance of the static averages $\langle A \rangle$ and $\langle B \rangle$ in the second term of (D.39), which only contributes at $\omega = 0$, is not obtained directly from (D.37) but is required to ensure that $\chi_{AB}(\omega)$ satisfies the fluctuation–dissipation theorem (D.36).[14]

[14]See Jensen and Mackintosh (1991), Section 3.3.

If the transitions have finite lifetimes $\Gamma_{\lambda\lambda'}^{-1}$, then each $\delta$-function in (D.38) can be replaced by a normalized Lorentzian, eqn (B.5), with full width at half maximum $2\Gamma_{\lambda\lambda'}$ (or some other appropriate line-shape function with unit area) and the absorptive part of the susceptibility may be written

$$\chi_{AB}''(\omega) \simeq \sum_{\lambda,\lambda'}^{E_{\lambda'} \neq E_\lambda} (p_\lambda - p_{\lambda'}) \langle \lambda | B | \lambda' \rangle \langle \lambda' | A | \lambda \rangle \frac{\Gamma_{\lambda\lambda'}}{(E_{\lambda'} - E_\lambda - \hbar\omega)^2 + \Gamma_{\lambda\lambda'}^2}$$

$$+ \sum_{\lambda,\lambda'}^{E_{\lambda'} = E_\lambda} \frac{p_\lambda}{k_{\mathrm{B}}T} \langle \lambda | B - \langle B \rangle | \lambda' \rangle \langle \lambda' | A - \langle A \rangle | \lambda \rangle \frac{\hbar\omega\Gamma_0}{(\hbar\omega)^2 + \Gamma_0^2}.$$

$$(D.40)$$

The scattering described by this expression for $\chi_{AB}''(\omega)$ takes the form of a sum of Lorentzian peaks centred on the transition energies. The term on the second line describes a quasielastic peak with width $2\Gamma_0$.

## D.5   Kramers–Kronig transformation

The real and imaginary parts of the generalized susceptibility are related to one another by what is known as a *dispersion relation*, a property first derived by H. A. Kramers and R. de L. Kronig. The result can be expressed as

$$\chi'_{AB}(\omega) = -\frac{1}{\pi}\mathcal{P}\int_{-\infty}^{\infty}\frac{\chi''_{AB}(\omega')}{\omega-\omega'}\,d\omega', \qquad (D.41)$$

where $\mathcal{P}$ denotes the Cauchy principal value.[15] The relation applies when $\chi'$ and $\chi''$ are real, e.g. when $\hat{B}=\hat{A}^{\dagger}$. The zero-frequency form,

$$\chi'_{AB}(0) = \frac{1}{\pi}\int_{-\infty}^{\infty}\frac{\chi''_{AB}(\omega')}{\omega'}\,d\omega', \qquad (D.42)$$

is proportional to the uniform or bulk static susceptibility, which is measurable by other experimental methods. By combining (D.42) with (D.36) we can write the static susceptibility as

$$\chi'_{AB}(0) = \int_{-\infty}^{\infty}\frac{\widetilde{S}_{BA}(\omega')}{\omega'}\{1-\exp(-\beta\hbar\omega')\}\,d\omega', \qquad (D.43)$$

giving a direct relation between the neutron inelastic scattering spectrum and data from other techniques.

# D.6   Relaxation and spectral-weight functions

In this section we show how to express $\chi''(\omega)$ in a form that is very convenient for the analysis of neutron inelastic scattering spectra. We first define the *relaxation function* $R(t)$ by the equation

$$R_{AB}(t) = \int_{t}^{\infty}\phi_{AB}(t')\,dt'. \qquad (D.44)$$

The relaxation function corresponds to the behaviour of $\langle\Delta A(t)\rangle$ for the case of a constant perturbation that is switched off at $t=0$. This can be seen from (D.22) for the case when $f(t)$ is a constant[16] for $t\le 0$ and zero for $t>0$ — put the upper limit of integration in (D.22) to 0, extend the lower limit to $-\infty$, and with a simple substitution for $t-t'$ one finds that $\langle\Delta A(t)\rangle$ is proportional to $R_{AB}(t)$ as defined in (D.44).

From (D.44) we find

$$R_{AB}(\infty) = 0, \qquad (D.45)$$

and by differentiation,

$$\frac{\partial R_{AB}}{\partial t} = -\phi_{AB}(t). \qquad (D.46)$$

In addition, from (D.25) and (D.46) we see that $R_{AB}(t)$ has the symmetry

$$R_{AB}(-t) = R_{BA}(t). \qquad (D.47)$$

We now return to the generalized susceptibility, and seek to express the absorptive part in terms of the relaxation function. Considering the

[15]The Cauchy principal value is a way of giving a value to certain types of integral which would otherwise not have a definite value. Consider an integral of the form $\int_{a}^{b}f(x)\,dx$, with $x$ real, and suppose that $f(x)$ is undefined at $x=c$, where $a\le c\le b$. The Cauchy principal value of the integral is

$$\mathcal{P}\int_{a}^{b}f(x)\,dx = \lim_{\epsilon\to 0^{+}}\int_{a}^{c-\epsilon}f(x)\,dx + \int_{c+\epsilon}^{b}f(x)\,dx.$$

Similarly, for a rational function $g(x)$,

$$\mathcal{P}\int_{-\infty}^{\infty}g(x)\,dx = \lim_{R\to\infty}\int_{-R}^{R}g(x)\,dx.$$

In both cases the limiting process is symmetric: in the first integral $c$ is approached symmetrically on both sides, and in the second integral the limits approach $\pm\infty$ symmetrically. This is what distinguishes the Cauchy principal value from other limiting processes.

[16]Strictly speaking, for $t<0$, $f(t)$ must slowly increase with $t$ from zero to a constant.

[17]The exp$(-\epsilon t)$ factor and requirement to take the limit as $\epsilon \to 0$ are implicit in what follows.

form given in (D.27), we perform the integration by parts using (D.45) and (D.46), and hence write[17]

$$\chi_{AB}(\omega) = R_{AB}(0) - i\omega \int_0^\infty R_{AB}(t) \exp(-i\omega t)\, \mathrm{d}t. \tag{D.48}$$

It is instructive to examine the zero-frequency limit of this expression. When $\omega = 0$, eqn (D.48) reduces to $\chi(0) = R(0)$, and since the absorptive part of $\chi(\omega)$ vanishes at $\omega = 0$ (because there is no dissipation of energy in a system subjected to a static perturbation) it therefore follows that both $R(0)$ and $\chi(0)$ are real and equal to $\chi'(0)$.

From (D.48),

$$\chi_{BA}(-\omega) = R_{BA}(0) + i\omega \int_0^\infty R_{BA}(t) \exp(i\omega t)\, \mathrm{d}t$$

$$= R_{AB}(0) + i\omega \int_{-\infty}^0 R_{AB}(t) \exp(-i\omega t)\, \mathrm{d}t. \tag{D.49}$$

To arrive at the second line of (D.49) we used the symmetry property (D.47) and changed the sign of $t$ in the integral. Hence, from (D.49) and (D.31), the absorptive part of the generalized susceptibility is

$$\chi''_{AB}(\omega) = \frac{1}{2i}\{\chi_{BA}(-\omega) - \chi_{AB}(\omega)\}$$

$$= \frac{\omega}{2} \int_{-\infty}^\infty R_{AB}(t) \exp(-i\omega t)\, \mathrm{d}t$$

$$= \pi\omega R_{AB}(\omega), \tag{D.50}$$

where $R(\omega)$ is the time Fourier transform of $R(t)$.

The last step to convert (D.50) into a useful form is to examine the normalization of $R(\omega)$. Integrating over $\omega$, and using the integral form of the delta function, eqn (B.14), we have

$$\int_{-\infty}^\infty R_{AB}(\omega)\, \mathrm{d}\omega = \int_{-\infty}^\infty R_{AB}(t) \left\{ \frac{1}{2\pi} \int_{-\infty}^\infty \exp(-i\omega t)\, \mathrm{d}\omega \right\} \mathrm{d}t$$

$$= \int_{-\infty}^\infty R_{AB}(t)\, \delta(t)\, \mathrm{d}t$$

$$= R_{AB}(0)$$

$$= \chi_{AB}(0). \tag{D.51}$$

Therefore, in terms of the function

$$F_{AB}(\omega) = \frac{R_{AB}(\omega)}{\chi_{AB}(0)}, \tag{D.52}$$

which has normalization

$$\int_{-\infty}^\infty F(\omega)\, \mathrm{d}\omega = 1, \tag{D.53}$$

the absorptive part of the generalized susceptibility is given by

$$\chi''_{AB}(\omega) = \pi\omega\chi_{AB}(0)F_{AB}(\omega). \qquad (D.54)$$

If we write in the $\mathbf{Q}$ dependence explicitly then (D.54) reads

$$\chi''_{AB}(\mathbf{Q}, \omega) = \pi\omega\chi_{AB}(\mathbf{Q})F_{AB}(\mathbf{Q}, \omega), \qquad (D.55)$$

where $\chi(\mathbf{Q})$ is shorthand for $\chi(\mathbf{Q}, 0) \equiv \chi'(\mathbf{Q}, 0)$.

Expression (D.55) is seen to be consistent with the dispersion relation (D.42). The function $F(\mathbf{Q}, \omega)$ is known as the *spectral-weight function*.

# D.7 Partial linear response functions

The neutron interaction potential $V(\mathbf{r})$ invariably contains a sum of contributions,

$$V(\mathbf{r}) = \sum_j V_j(\mathbf{r}), \qquad (D.56)$$

due, for example, to scattering from different species, spin components, etc. It is useful, therefore, to consider the *partial* response function $S_{kj}(\mathbf{Q}, \omega)$, defined from (D.32) by

$$S_{kj}(\mathbf{Q}, \omega) = \frac{1}{2\pi\hbar} \int_{-\infty}^{\infty} \langle \hat{A}_k^\dagger(\mathbf{Q})\hat{A}_j(\mathbf{Q}, t)\rangle \, \exp(-\mathrm{i}\omega t) \, \mathrm{d}t, \qquad (D.57)$$

where

$$\hat{A}_j(\mathbf{Q}, t) = \left(\frac{m_\mathrm{n}}{2\pi\hbar^2}\right) \langle\sigma_\mathrm{f}| \, \hat{V}_j(\mathbf{Q}, t) \, |\sigma_\mathrm{i}\rangle \qquad (D.58)$$

and $\hat{V}_j(\mathbf{Q}, t)$ the spatial Fourier transform of $\hat{V}_j(\mathbf{r}, t)$, the Heisenberg operator for the $j$th term in the interation potential (D.56). The total scattering is a weighted sum of the partial response functions.

Following the same procedure as from (D.33)–(D.36), we find that the dynamical part of $S_{kj}(\mathbf{Q}, \omega)$ satisfies the Fluctuation–Dissipation theorem (we omit the $\mathbf{Q}$ dependence[18])

$$\widetilde{S}_{kj}(\omega) = \{1 + n(\omega)\} \frac{1}{\pi}\chi''_{jk}(\omega), \qquad (D.59)$$

where

$$\chi_{jk}(\omega) = \lim_{\epsilon \to 0^+} \int_0^{\infty} \phi_{jk}(t) \exp(-\mathrm{i}zt) \, \mathrm{d}t, \qquad (z = \omega - \mathrm{i}\epsilon), \qquad (D.60)$$

and[19]

$$\phi_{jk}(t) = \frac{\mathrm{i}}{\hbar}\langle[\hat{A}_j(t), \hat{A}_k^\dagger]\rangle; \quad \phi_{kj}(t) = \frac{\mathrm{i}}{\hbar}\langle[\hat{A}_k(t), \hat{A}_j^\dagger]\rangle. \qquad (D.61)$$

If $j \neq k$ then $\chi''_{jk}(\omega)$ is not the imaginary part of $\chi_{jk}(\omega)$ and $\widetilde{S}_{kj}(\omega)$ is in general a complex number.[20] However, from (D.29) and (D.31) it follows that

$$\chi''_{jk}(\omega) = \frac{1}{2\mathrm{i}}\{\chi^*_{kj}(\omega) - \chi_{jk}(\omega)\}, \qquad (D.62)$$

[18] All the functions in this section have argument $+\mathbf{Q}$.

[19] Note that here, interchanging the suffices only exchanges the labels on the operators, not the operators as well.

[20] In the next section we show that $\widetilde{S}_{jk}(\omega) = \widetilde{S}^*_{kj}(\omega)$.

so that the symmetric combination

$$\chi''_{jk}(\omega) + \chi''_{kj}(\omega) = \frac{1}{2i}\left\{\chi^*_{kj}(\omega) - \chi_{kj}(\omega) + \chi^*_{jk}(\omega) - \chi_{jk}(\omega)\right\} \quad \text{(D.63)}$$

[21]Recall that the real and imaginary parts of $\chi$ are defined with phase

$$\chi = \text{Re}\{\chi\} - i\,\text{Im}\{\chi\}.$$

is real and equal to $\text{Im}\{\chi_{kj}(\omega)\} + \text{Im}\{\chi_{jk}(\omega)\}$.[21] Hence,

$$\widetilde{S}_{jk}(\omega) + \widetilde{S}_{kj}(\omega) = \{1 + n(\omega)\}\frac{1}{\pi}\{\chi''_{kj}(\omega) + \chi''_{jk}(\omega)\}$$

$$= \{1 + n(\omega)\}\frac{1}{\pi}\text{Im}\left\{\chi_{kj}(\omega) + \chi_{jk}(\omega)\right\}. \quad \text{(D.64)}$$

Similarly, the antisymmetric combination

$$\widetilde{S}_{jk}(\omega) - \widetilde{S}_{kj}(\omega) = \{1 + n(\omega)\}\frac{1}{\pi}\{\chi''_{kj}(\omega) - \chi''_{jk}(\omega)\}$$

$$= i\{1 + n(\omega)\}\frac{1}{\pi}\text{Re}\left\{\chi_{kj}(\omega) - \chi_{jk}(\omega)\right\}. \quad \text{(D.65)}$$

[22]Only the symmetric combination appears in unpolarized neutron scattering, but the antisymmetric combination is accessible with neutron polarization analysis — see Section 4.4, e.g. eqn (4.75).

In neutron scattering, the functions $\widetilde{S}_{jk}(\omega)$ and $\widetilde{S}_{kj}(\omega)$ with $j \neq k$ only ever occur in the symmetric or antisymmetric combinations.[22]

Partial relaxation and spectral-weight functions can also be defined, and are related to $\widetilde{S}(\omega)$ by

$$\widetilde{S}_{kj}(\omega) = \omega\{1 + n(\omega)\}R_{jk}(\omega)$$

$$= \omega\{1 + n(\omega)\}\chi_{jk}(0)F_{jk}(\omega). \quad \text{(D.66)}$$

# D.8   Symmetry of scattering response functions

Here we establish the behaviour of the various neutron scattering correlation and linear response functions under sign reversals of the arguments and interchanges of the indices. We have already come across some of these, but in this section we give a more comprehensive summary of the formulae valid for partial linear response functions. We also establish the symmetries under sign reversal of $\mathbf{Q}$, and for this reason we display the $\mathbf{Q}$ dependence of all the functions in this section.

Because the neutron–matter interaction potential is a physical observable the corresponding operator is Hermitian, i.e. $V_j(\mathbf{r}) = V_j^\dagger(\mathbf{r})$, and so

[23]This follows from the definition of the Heisenberg operator, eqn (C.47), and the definition of the Hermitian conjugate, eqn (C.23).

$V_j(-\mathbf{Q}) = V_j^\dagger(\mathbf{Q})$. Similarly, the Heisenberg operator satisfies[23]

$$\hat{V}_j(-\mathbf{Q}, t) = \hat{V}_j^\dagger(\mathbf{Q}, t). \quad \text{(D.67)}$$

From (D.58) and (D.67), the corresponding symmetry for $\hat{A}_j(\mathbf{Q}, t)$ is

$$\hat{A}_j(-\mathbf{Q}, t)_{\sigma_f \to \sigma_i} = \hat{A}_j^\dagger(\mathbf{Q}, t)_{\sigma_i \to \sigma_f}. \quad \text{(D.68)}$$

As indicated, there is a reversal in the initial and final spin states of the neutron in this relation. If unpolarized neutron scattering is employed, or if the interaction potential is spin-independent, then relation (D.68) becomes independent of $\sigma_i$ and $\sigma_f$. To cover the general case, however, we display the spin states explicitly from now on.

From (D.68), we see that the correlation function satisfies

$$\langle \hat{A}_j^\dagger(\mathbf{Q})\hat{A}_k(\mathbf{Q},t)\rangle_{\sigma_i\to\sigma_f} = \langle \hat{A}_j(-\mathbf{Q})\hat{A}_k^\dagger(-\mathbf{Q},t)\rangle_{\sigma_f\to\sigma_i}, \qquad (\text{D.69})$$

and identities (C.54) and (C.57) mean its complex conjugate can be written

$$\langle \hat{A}_j^\dagger(\mathbf{Q})\hat{A}_k(\mathbf{Q},t)\rangle^*_{\sigma_i\to\sigma_f} = \langle \hat{A}_k^\dagger(\mathbf{Q})\hat{A}_j(\mathbf{Q},-t)\rangle_{\sigma_i\to\sigma_f}. \qquad (\text{D.70})$$

Hence, from (D.57),

$$S_{jk}^*(\mathbf{Q},\omega)_{\sigma_i\to\sigma_f} = S_{kj}(\mathbf{Q},\omega)_{\sigma_i\to\sigma_f}. \qquad (\text{D.71})$$

From (D.57), (D.69) and (D.70) it follows that the function $\phi_{jk}(\mathbf{Q},t)$ defined in (D.61) satisfies

$$\phi_{jk}^*(\mathbf{Q},t)_{\sigma_i\to\sigma_f} = \phi_{jk}(-\mathbf{Q},t)_{\sigma_f\to\sigma_i} \qquad (\text{D.72})$$

$$= -\phi_{kj}(\mathbf{Q},-t)_{\sigma_i\to\sigma_f}, \qquad (\text{D.73})$$

and similarly, from (D.60) and (D.72), that

$$\chi_{jk}(-\mathbf{Q},-\omega)_{\sigma_f\to\sigma_i} = \chi_{jk}^*(\mathbf{Q},\omega)_{\sigma_i\to\sigma_f}. \qquad (\text{D.74})$$

The reactive and absorptive parts of (D.74) give the separate relations

$$\chi_{jk}'(-\mathbf{Q},-\omega)_{\sigma_f\to\sigma_i} = \chi_{jk}'(\mathbf{Q},\omega)_{\sigma_i\to\sigma_f} \qquad (\text{D.75})$$

and

$$\chi_{jk}''(-\mathbf{Q},-\omega)_{\sigma_f\to\sigma_i} = -\chi_{jk}''(\mathbf{Q},\omega)_{\sigma_i\to\sigma_f}. \qquad (\text{D.76})$$

Repeating the steps leading to (3.48) we find for the partial response function

$$S_{jk}(-\mathbf{Q},-\omega)_{\sigma_f\to\sigma_i} = \exp(-\beta\hbar\omega)\, S_{kj}(\mathbf{Q},\omega)_{\sigma_i\to\sigma_f}. \qquad (\text{D.77})$$

This is the Principle of Detailed Balance in its most general form.

Finally, from (D.46), (D.72), and (D.73), the partial relaxation function satisfies

$$R_{jk}^*(\mathbf{Q},\omega)_{\sigma_i\to\sigma_f} = R_{jk}(-\mathbf{Q},-\omega)_{\sigma_f\to\sigma_i} \qquad (\text{D.78})$$

$$= R_{kj}(\mathbf{Q},\omega)_{\sigma_i\to\sigma_f}, \qquad (\text{D.79})$$

and the partial spectral-weight function satisfies precisely the same relations since, from (D.52), $R_{jk}(\mathbf{Q},\omega)_{\sigma_i\to\sigma_f} = \chi_{jk}(\mathbf{Q})F_{jk}(\mathbf{Q},\omega)_{\sigma_i\to\sigma_f}$. Thus,

$$F_{jk}^*(\mathbf{Q},\omega)_{\sigma_i\to\sigma_f} = F_{jk}(-\mathbf{Q},-\omega)_{\sigma_f\to\sigma_i} \qquad (\text{D.80})$$

$$= F_{kj}(\mathbf{Q},\omega)_{\sigma_i\to\sigma_f}. \qquad (\text{D.81})$$

## D.9   Magnetic scattering

**Magnetic linear response functions**

The magnetic neutron scattering cross-section contains the partial response functions

$$S_{\alpha\beta}(\mathbf{Q},\omega) = \frac{1}{2\pi\hbar} \int_{-\infty}^{\infty} \langle \hat{M}_{\alpha}^{\dagger}(\mathbf{Q})\hat{M}_{\beta}(\mathbf{Q},t)\rangle \, \exp(-\mathrm{i}\omega t)\,\mathrm{d}t, \qquad \text{(D.82)}$$

whose dynamic part may be written [see (D.34)]

$$\widetilde{S}_{\alpha\beta}(\mathbf{Q},\omega) = \{1+n(\omega)\}\frac{\mathrm{i}}{2\pi} \int_{-\infty}^{\infty} \phi_{\beta\alpha}(\mathbf{Q},t)\exp(-\mathrm{i}\omega t)\,\mathrm{d}t, \qquad \text{(D.83)}$$

where [see (D.61)],

$$\phi_{\alpha\beta}(\mathbf{Q},t) = \frac{\mathrm{i}}{\hbar}\langle [\hat{M}_{\alpha}(\mathbf{Q},t),\hat{M}_{\beta}^{\dagger}(\mathbf{Q})]\rangle. \qquad \text{(D.84)}$$

The generalized susceptibility $\chi_{\alpha\beta}(\mathbf{Q},\omega)$ is given by [see (D.60)]

$$\chi_{\alpha\beta}(\mathbf{Q},\omega) = \lim_{\epsilon\to 0^{+}} \int_{0}^{\infty} \phi_{\alpha\beta}(\mathbf{Q},t)\exp(-\mathrm{i}zt)\,\mathrm{d}t, \qquad (z=\omega-\mathrm{i}\epsilon). \quad \text{(D.85)}$$

The dynamical part of $S(\mathbf{Q},\omega)$ is related to the absorptive part of the generalized susceptibility by the Fluctuation–Dissipation theorem (D.59),

$$\widetilde{S}_{\alpha\beta}(\mathbf{Q},\omega) = \{1+n(\omega)\}\frac{1}{\pi}\chi''_{\beta\alpha}(\mathbf{Q},\omega). \qquad \text{(D.86)}$$

For $\alpha \neq \beta$, only the symmetric and antisymmetric combinations of $\widetilde{S}_{\alpha\beta}(\mathbf{Q},\omega)$, eqns (D.64)–(D.65), are accessible to neutron scattering.

**Relation between generalized and magnetic susceptibilities**

The generalized susceptibility tensor $\chi$ contained in the preceding expressions is proportional to the magnetic susceptibility tensor $\chi_{\mathrm{M}}$ in electromagnetism. The latter is defined in SI units by[24]

$$\mathbf{M} = \chi_{\mathrm{M}}\mathbf{H}, \qquad \text{(D.87)}$$

where $\mathbf{H}$ is an infinitesimal magnetic field. The relation between $\chi$ and $\chi_{\mathrm{M}}$ is

$$\chi = \frac{V}{\mu_0}\chi_{\mathrm{M}}, \qquad \text{(D.88)}$$

where $V$ is the volume of the system. We will now establish this result by making use of the correspondence between magnetization components and the generalized force and displacement defined in Sections D.2 and D.3.

The magnetic field acts on the system via the Zeeman interaction. The perturbation on the whole system is described by the Hamiltonian

$$\mathcal{H}_1(t) = -\mu_0 \int \mathbf{M}(\mathbf{r})\cdot\mathbf{H}(\mathbf{r},t)\,\mathrm{d}^3\mathbf{r}. \qquad \text{(D.89)}$$

[24]$\chi_{\mathrm{M}}$ is dimensionless in SI units.

We consider a field of strength $H_\beta(\mathbf{r},t)$ directed along the $\beta$ axis consisting of a single Fourier component, as in Section D.2. The resulting change in magnetization in the $\alpha$ direction is $\langle \Delta M_\alpha(\mathbf{r},t)\rangle$:

$$H_\beta(\mathbf{r},t) = h_\beta(\mathbf{Q},\omega)\exp\{i(\omega t - \mathbf{Q}\cdot\mathbf{r})\} \tag{D.90}$$

$$\langle \Delta M_\alpha(\mathbf{r},t)\rangle = m_\alpha(\mathbf{Q},\omega)\exp\{i(\omega t - \mathbf{Q}\cdot\mathbf{r})\}. \tag{D.91}$$

From (D.87), the associated magnetic susceptibility coefficient is given by

$$m_\alpha = (\chi_{\mathrm{M}})_{\alpha\beta} h_\beta. \tag{D.92}$$

Now, by substituting (D.90) in (D.89) we obtain

$$\mathcal{H}_1(t) = -\mu_0 \int M_\beta(\mathbf{r}) h_\beta(\mathbf{Q},\omega)\exp\{i(\omega t - \mathbf{Q}\cdot\mathbf{r})\}\,d^3\mathbf{r}$$
$$= -\mu_0 M_\beta(-\mathbf{Q}) h_\beta(\mathbf{Q},\omega)\exp(i\omega t), \tag{D.93}$$

and comparing (D.84) with (D.24) we see that $A \equiv M_\alpha(\mathbf{Q})$, and $B \equiv M_\beta^\dagger(\mathbf{Q}) = M_\beta(-\mathbf{Q})$. Therefore, (D.93) is in the same form as (D.14) with generalized force

$$f(t) = \mu_0 h_\beta(\mathbf{Q},\omega)\exp(i\omega t), \tag{D.94}$$

and the equivalent of the generalized displacement $\langle \Delta A(t)\rangle$ may be identified from the spatial Fourier transform of (D.91),

$$\langle \Delta A(t)\rangle = \int \langle \Delta M_\alpha(\mathbf{r},t)\rangle \exp(i\mathbf{Q}\cdot\mathbf{r})\,d^3\mathbf{r}$$
$$= \int m_\alpha(\mathbf{Q},\omega)\exp\{i(\omega t - \mathbf{Q}\cdot\mathbf{r})\}\exp(i\mathbf{Q}\cdot\mathbf{r})\,d^3\mathbf{r}$$
$$= V m_\alpha(\mathbf{Q},\omega)\exp(i\omega t). \tag{D.95}$$

The generalized susceptibility is the ratio of $\langle \Delta A(t)\rangle$ and $f(t)$, and so

$$V m_\alpha = \chi_{\alpha\beta}\mu_0 h_\beta. \tag{D.96}$$

The desired relationship (D.88) follows directly from (D.92) and (D.96).

# References

Abel, C. *et al.* (2020). Measurement of the permanent electric dipole moment of the neutron. *Phys. Rev. Lett.* **124**, 081803.

Aczel, A. A., Granroth, G. E., MacDougall, G. J., Buyers, W. J. L., Abernathy, D. L., Samolyuk, G. D., Stocks, G. M., and Nagler, S. E. (2012). Quantum oscillations of nitrogen atoms in uranium nitride. *Nature Commun.* **3**, 1124.

Alexander, S. (1962). Symmetry considerations in the determination of magnetic structures. *Phys. Rev.* **127**, 420–432.

Altarelli, M. (2006). Resonant X-ray scattering: A theoretical introduction, in *Magnetism: A Synchrotron Radiation Approach*, Beaurepaire, E., Bulou, H., Scheurer, F. and Kappler, J. P. (eds.), Lecture Notes in Physics Vol. 697, pp201–242. Springer, Berlin, Heidelberg.

Andersen, K. H., Cubitt, R., Humblot, H., Jullien, D., Petoukhov, A., Tasset, F., Schanzer, C., Shah, V. R., and Wildes, A. R. (2006). The $^3$He polarizing filter on the neutron reflectometer D17. *Physica B* **385–386**, 1134–1137.

Anderson, I. S., Andreani, C., Carpenter, J. M., Festa, G., Gorini, G., Loong, C.-K., and Senesi, R. (2016). Research opportunities with compact accelerator-driven neutron sources. *Phys. Rep.* **654**, 1–58.

Anderson, I. S., McGreevy, R., and Bilheux, H. Z. (eds.) (2009). *Neutron Imaging and Applications*. Springer.

Andreani, C., Colognesi, D., Mayers, J., Reiter, G. F., and R., Senesi (2005). Measurement of momentum distribution of light atoms and molecules in condensed matter. *Adv. Phys.* **54**, 377–469.

Andreani, C., Krzystyniak, M., Romanelli, G., R., Senesi, and Fernandez-Alonso, F. (2017). Electron-volt neutron spectroscopy: beyond fundamental systems. *Adv. Phys.* **66**, 1–73.

Babkevich, P., Freeman, P. G., Enderle, M., Prabhakaran, D., and Boothroyd, A. T. (2016). Direct evidence for charge stripes in a layered cobalt oxide. *Nature Commun.* **7**, 11632.

Babkevich, P., Poole, A., Johnson, R. D., Roessli, B., Prabhakaran, D., and Boothroyd, A. T. (2012). Electric field control of chiral magnetic domains in the high-temperature multiferroic CuO. *Phys. Rev. B* **85**, 134428.

Babkevich, P., Prabhakaran, D., Frost, C. D., and Boothroyd, A. T. (2010). Magnetic spectrum of the two-dimensional antiferromagnet $La_2CoO_4$ studied by inelastic neutron scattering. *Phys. Rev. B* **82**, 184425.

Baker, M. L., Guidi, T., Carretta, S., Ollivier, J., Mutka, H., Güdel, H. U., Timco, G. A., McInnes, E. J. L., Amoretti, G., Winpenny, R. E. P., and Santini, P. (2012). Spin dynamics of molecular nanomagnets unravelled at atomic scale by four-dimensional inelastic neutron scattering. *Nature Phys.* **8**, 906–911.

Balcar, E. and Lovesey, S. W. (1989). *Theory of Magnetic Neutron and Photon Scattering.* Clarendon Press, Oxford.

Balcar, E. and Lovesey, S. W. (1991). Theory of neutron scattering by atomic electrons: $jj$-coupling scheme. *J. Phys.: Condens. Matter* **3**, 7095–7115.

Becker, K. W., Fulde, P., and Keller, J. (1977). Line width of crystal-field excitations in metallic rare-earth systems. *Z. Physik B* **28**, 9–18.

Becker, P. J. and Coppens, P. (1974). Extinction within the limit of validity of the Darwin transfer equations. I. General formalisms for primary and secondary extinction and their application to spherical crystals. *Acta Cryst.* **A30**, 129–147.

Bée, M. (1988). *Quasielastic Neutron Scattering.* Adam Hilger, Bristol.

Benoit, A., Bossy, J., Flouquet, J., and Schweizer, J. (1985). Magnetic diffraction in solid He-3. *J. Physique Lett.* **46**, L923–L927.

Bertaut, E. F. (1968). Representation analysis of magnetic structures. *Acta Cryst.* **A24**, 217–231.

Biggin, S. and Enderby, J. E. (1981). The structure of molten zinc chloride. *J. Phys. C* **14**, 3129–3136.

Blech, I. A. and Averbach, B. L. (1965). Multiple scattering of neutrons in vanadium and copper. *Phys. Rev.* **137**, A1113–A1116.

Bloch, F. (1936). On the magnetic scattering of neutrons. *Phys. Rev.* **50**, 259–260.

Blume, M. (1963). Polarization effects in the magnetic elastic scattering of slow neutrons. *Phys. Rev.* **130**, 1670–1676.

Blundell, S. J. and Bland, J. A. C. (1992). Polarized neutron reflection as a probe of magnetic films and multilayers. *Phys. Rev. B* **46**, 3391–3400.

Blundell, S. J. and Bland, J. A. C. (1993). Polarised neutron reflection as a probe of in-plane magnetisation vector rotation in magnetic multilayers. *J. Magn. Magn. Mater.* **121**, 185–188.

Boothroyd, A. T., Barratt, J. P., Bonville, P., Canfield, P. C., Murani, A., Wildes, A. R., and Bewley, R. I. (2003). The magnetic state of Yb in kondo-lattice $YbNi_2B_2C$. *Phys. Rev. B* **67**, 104407.

Boothroyd, A. T., Gardiner, C. H., Lister, S. J. S., Santini, P., Rainford, B. D., Noailles, L. D., Currie, D. B., Eccleston, R. S., and Bewley, R. I. (2001). Localized $4f$ states and dynamic Jahn–Teller effect in $PrO_2$. *Phys. Rev. Lett.* **86**, 2082–2085.

Boothroyd, A. T., Mukherjee, A., Fulton, S., Perring, T. G., Eccleston, R. S., Mook, H. A., and Wanklyn, B. M. (1997). High-energy magnetic excitations in CuO. *Physica B* **234–236**, 731–733.

Boothroyd, A. T., Squires, G. L., Fetters, L. J., Rennie, A. R., Horton, J. C., and de Vallêra, A. M. B. G. (1989). Small-angle neutron scattering from star-branched polymers in dilute solution. *Macro-molecules* **22**, 3130–3137.

Bramwell, S. T. and Holdsworth, P. C. W. (1993). Magnetization and universal sub-critical behaviour in two-dimensional $XY$ magnets. *J. Phys.: Condens. Matter* **5**, L53–L59.

Brockhouse, B. N. (1953). Resonant scattering of slow neutrons. *Can. J. Phys.* **31**, 432–452.

Brockhouse, B. N. (1955). Energy distribution of neutrons scattered by paramagnetic substances. *Phys. Rev.* **99**, 601–603.

Brockhouse, B. N. (1983). Slow neutron spectroscopy: an historical account over the years 1950–1977, in *The neutron and its applications*, Schofield, P. (ed.). *Inst. Phys. Conf. Ser., Vol. 64*, 193–198.

Brown, P. J. (2004). Magnetic form factors, in *International Tables for Crystallography*, 3rd ed., Prince, E. (ed.), Vol. C, pp454–461. Dordrecht: Kluwer Academic Publishers. (See also http://www.ill.eu/sites/ccsl/html/ccsldoc.html).

Brown, P. J. and Chatterji, T. (2006). Neutron diffraction and polarimetric study of the magnetic and crystal structures of $HoMnO_3$ and $YMnO_3$. *J. Phys.: Condens. Matter* **18**, 10085–10096.

Burns, G. and Glazer, A. M. (2013). *Space Groups for Solid State Scientists*. 3rd ed., Academic Press, Waltham, MA.

Cabrera, I., Kenzelmann, M., Lawes, G., Chen, Y., Chen, W. C., Erwin, R., Gentile, T. R., Leão, J. B., Lynn, J. W., Rogado, N., Cava, R. J., and Broholm, C. (2009). Coupled magnetic and ferroelectric domains in multiferroic $Ni_3V_2O_8$. *Phys. Rev. Lett.* **103**, 087201.

Caciuffo, R., Amoretti, G., Murani, A., Sessoli, R., Caneschi, A., and Gatteschi, D. (1998). Neutron spectroscopy for the magnetic anisotropy of molecular clusters. *Phys. Rev. Lett.* **81**, 4744–4747.

Carpenter, J. M. and Loong, C. K. (2015). *Elements of Slow-Neutron Scattering: Basics, Techniques, and Applications*. Cambridge University Press, Cambridge.

Caux, J.-S. and Hagemans, R. (2006). The four-spinon dynamical structure factor of the Heisenberg chain. *J. Stat. Mech.*, P12013.

Chadwick, J. (1932). Possible existence of a neutron. *Nature* **129**, 312.

Chatterji, T. (ed.) (2006). *Neutron Scattering from Magnetic Materials*. Elsevier, Amsterdam.

Chudley, C. T. and Elliott, R. J. (1961). Neutron scattering from a liquid on a jump diffusion model. *Proc. Phys. Soc.* **77**, 353–361.

Clementi, E. and Roetti, C. (1974). Roothaan–Hartree–Fock atomic wavefunctions. *Atomic Data and Nuclear Data Tables* **14**, 177–478.

Collins, M. F. (1989). *Magnetic Critical Scattering*. Oxford University Press, Oxford.

Condon, E. U. and Odabaşi, H. (1980). *Atomic Structure*. Cambridge University Press, Cambridge.

Condon, E. U. and Shortley, G. H. (1935). *The Theory of Atomic Spectra*. Cambridge University Press, Cambridge.

Cowan, R. D. (1981). *The Theory of Atomic Structure and Spectra.* University of California Press, Berkeley.

Cowley, R. A., Shirane, G., Birgeneau, R. J., and Guggenheim, H. J. (1977). Spin fluctuations in random magnetic-nonmagnetic two-dimensional antiferromagnets. I. Dynamics. *Phys. Rev. B* **15**, 4292–4302.

Cubitt, R., Saerbeck, T., Campbell, R. A., Barker, R., and Gutfreund, P. (2015). An improved algorithm for reducing reflectometry data involving divergent beams or non-flat samples. *J. Appl. Cryst.* **48**, 2006–2011.

Dahlborg, U., Kramer, M. J., Besser, M., Morris, J. R., and Calvo–Dahlborg, M. (2013). Structure of molten Al and eutectic Al–Si alloy studied by neutron diffraction. *J. Non-Cryst. Solids* **361**, 63–69.

de Gennes, P. G. (1959). Liquid dynamics and inelastic scattering of neutrons. *Physica* **25**, 825–839.

Debye, P. (1947). Molecular-weight determination by light scattering. *J. Phys. Colloid Chem.* **51**, 18–32.

Delaire, O., Kresch, M., Muñoz, J. A., Lucas, M. S., Lin, J. Y. Y., and Fultz, B. (2008). Electron-phonon interactions and high-temperature thermodynamics of vanadium and its alloys. *Phys. Rev. B* **77**, 214112.

Demmel, F., Szubrin, D., Pilgrim, W.-C., and Morkel, C. (2011). Diffusion in liquid aluminium probed by quasielastic neutron scattering. *Phys. Rev. B* **84**, 014307.

des Cloizeaux, J. and Pearson, J. J. (1962). Spin-wave spectrum of the antiferromagnetic linear chain. *Phys. Rev.* **128**, 2131–2135.

Desclaux, J. P. and Freeman, A. J. (1978). Dirac–Fock studies of some electronic properties of actinide ions. *J. Magn. Magn. Mater.* **8**, 119–129.

Diallo, S. O., Azuah, R. T., Abernathy, D. L., Rota, R., Boronat, J., and Glyde, H. R. (2012). Bose–Einstein condensation in liquid $^4$He near the liquid–solid transition line. *Phys. Rev. B* **85**, 140505(R).

Dirac, P. A. M. (1927). The quantum theory of the emission and absorption of radiation. *Proc. Roy. Soc. A* **114**, 243–265.

Doubble, R., Hayden, S. M., Dai, P., Mook, H. A., Thompson, J. R., and Frost, C. D. (2010). Direct observation of paramagnons in palladium. *Phys. Rev. Lett.* **105**, 027207.

Dove, M. T. (2011). Introduction to the theory of lattice dynamics. *Collection SFN* **12**, 123–159.

Egelstaff, P. A. (1992). *An Introduction to the Liquid State (2nd ed.).* Oxford University Press, Oxford.

Ehlers, G., Stewart, J. R., Wildes, A. R., Deen, P. P., and Andersen, K. H. (2013). Generalization of the classical xyz-polarization analysis technique to out-of-plane and inelastic scattering. *Rev. Sci. Inst.* **84**, 093901.

Enderle, M. (2014). Neutrons and magnetism. *Collection SFN* **13**, 01002.

Ewings, R. A., Buts, A., Le, M. D., van Duijn, J., Bustinduy, I., and Perring, T. G. (2016). HORACE: Software for the analysis of data from single crystal spectroscopy experiments at time-of-flight neutron instruments. *Nucl. Instrum. Methods Phys. Res. Sect. A* **834**, 132–142.

Ewings, R. A., Freeman, P. G., Enderle, M., Kulda, J., Prabhakaran, D., and Boothroyd, A. T. (2010). Ferromagnetic excitations in $La_{0.82}Sr_{0.18}CoO_3$ observed using neutron inelastic scattering. *Phys. Rev. B* **82**, 144401.

Fazekas, P. (1999). *Lecture Notes on Electron Correlation and Magnetism.* World Scientific, Singapore.

Fennell, T., Deen, P. P., Wildes, A. R., Schmalzl, K., Prabhakaran, D., Boothroyd, A. T., Aldus, R. J., McMorrow, D. F., and Bramwell, S. T. (2009). Magnetic Coulomb phase in the spin ice $Ho_2Ti_2O_7$. *Science* **326**, 415–417.

Fischer, H. E., Barnes, A. C., and Salmon, P. S. (2006). Neutron and x-ray diffraction studies of liquids and glasses. *Rep. Prog. Phys.* **69**, 233–299.

Fischer, H. E., Neuefeind, J., Simonson, J. M., Loidl, R., and Rauch, H. (2008). New measurements of the coherent and incoherent neutron scattering lengths of $^{13}C$. *J. Phys.: Condens. Matter* **20**, 045221.

Fischer, H. E., Simonson, J. M., Neuefeind, J. C., Lemmel, H., Rauch, H., Zeidler, A., and Salmon, P. S. (2012). The bound coherent neutron scattering lengths of the oxygen isotopes. *J. Phys.: Condens. Matter* **24**, 505105.

Foldy, L. L. (1958). Neutron–electron interaction. *Rev. Mod. Phys.* **30**, 471–481.

Freeman, A. J. and Desclaux, J. P. (1979). Dirac–Fock studies of some electronic properties of rare-earth ions. *J. Magn. Magn. Mater.* **12**, 11–21.

Freeman, A. J. and Watson, R. E. (1962). Theoretical investigation of some magnetic and spectroscopic properties of rare-earth ions. *Phys. Rev.* **127**, 2058–2075.

Funahashi, S., Moussa, F., and Steiner, M. (1976). Experimental determination of the spin-wave spectrum of the two-dimensional ferromagnet $K_2CuF_4$. *Solid State Commun.* **18**, 433–435.

Furrer, A., Mesot, J., and Strässle, T. (2009). *Neutron Scattering in Condensed Matter Physics.* World Scientific, Singapore.

Furrer, A. and Rüegg, Ch. (2006). Bose–Einstein condensation in magnetic materials. *Physica B* **385–386**, 295–300.

Furrer, A. and Waldmann, O. (2013). Magnetic cluster excitations. *Rev. Mod. Phys.* **85**, 367–420.

García Sakai, V., Alba-Simionesco, C., and Chen, S.-H. (eds.) (2012). *Dynamics of Soft Matter: Neutron Applications.* Springer, New York.

Gardiner, C. H., Boothroyd, A. T., Pattison, P., McKelvy, M. J., McIntyre, G. J., and Lister, S. J. S. (2004). Cooperative Jahn–Teller distortion in $PrO_2$. *Phys. Rev. B* **70**, 024415.

Garlea, V. O. and Chakoumakos, B. C. (2015). Magnetic structures. *Experimental Methods in the Physical Sciences* **48**, 203–290.

Gaspar, A. M., Busch, S., Appavou, M-S., Haeussler, W., Georgii, R., Su, Y., and Doster, W. (2010). Using polarization analysis to separate the coherent and incoherent scattering from protein samples. *Biochim. Biophys. Acta* **1804**, 76–82.

Gatteschi, D., Sessoli, R., and Villain, J. (2006). *Molecular Nanomagnets.* Oxford University Press, Oxford.

Glättli, H. and Goldman, M. (1987). Nuclear magnetism. *Methods in Experimental Physics* **23C**, 241–286.

Goodenough, J. B. (1963). *Magnetism and the Chemical Bond.* Interscience Publishers (Wiley), New York.

Goremychkin, E. A. and Osborn, R. (1993). Crystal-field excitations in $CeCu_2Si_2$. *Phys. Rev. B* **47**, 14280–14290.

Griffiths, D. J. (2017). *Introduction to Electrodynamics, 4th ed.* Cambridge University Press, Cambridge.

Groitl, F., Rantsiou, E., Bartkowiak, M., Filges, U., Graf, D., Niedermayer, C., Rüegg, C., and Rønnow, H. M. (2016). A combined radial collimator and cooled beryllium filter for neutron scattering. *Nucl. Instrum. Methods Phys. Res. Sect. A* **819**, 99–103.

Habicht, K., Golub, R., Mezei, F., Keimer, B., and Keller, T. (2004). Temperature-dependent phonon lifetimes in lead investigated with neutron-resonance spin-echo spectroscopy. *Phys. Rev. B* **69**, 104301.

Haldane, F. D. M. (1983*a*). Continuum dynamics of the 1-D Heisenberg antiferromagnet: Identification with the O(3) nonlinear sigma model. *Phys. Lett.* **93A**, 464–468.

Haldane, F. D. M. (1983*b*). Nonlinear field theory of large-spin Heisenberg antiferromagnets: Semiclassically quantized solitons of the one-dimensional easy-axis Néel state. *Phys. Rev. Lett.* **50**, 1153–1156.

Halpern, O. and Johnson, M. H. (1939). On the magnetic scattering of neutrons. *Phys. Rev.* **55**, 898–923.

Hannon, A. C., Gibbs, A. S., and Takagi, H. (2018). Neutron scattering length determination by means of total scattering. *J. Appl. Cryst.* **51**, 854–866.

Hansen, J-P. and McDonald, I. R. (2013). *Theory of Simple Liquids (4th ed.).* Academic Press, London.

Harders, T. M., Hicks, T. J., and Wells, P. (1985). Multiple scattering in neutron polarization analysis experiments. *J. Appl. Cryst.* **18**, 131–134.

Heinloth, K. (1961). Streuung subthermischer neutronen an $H_2O$, $CH_2O_2$ und $C_6H_6$. *Z. für Physik* **163**, 218–229.

Helme, L. M., Boothroyd, A. T., Coldea, R., Prabhakaran, D., Frost, C. D., Keen, D. A., Regnault, L. P., Freeman, P. G., Enderle, M., and Kulda, J. (2009). Magnetic order and dynamics of the charge-ordered antiferromagnet $La_{1.5}Sr_{0.5}CoO_4$. *Phys. Rev. B* **80**, 134414.

Hempelmann, R. (2000). *Quasielastic Neutron Scattering and Solid State Diffusion.* Oxford University Press, Oxford.

Hicks, T. J. (1996). Experiments with neutron polarization analysis. *Adv. Phys.* **45**, 243–298.

Higgins, J. S. and Benoit, H. C. (1994). *Polymers and Neutron Scattering.* Oxford University Press, Oxford.

Hirakawa, K., Yoshizawa, H., Axe, J. D., and Shirane, G. (1983). Neutron scattering study of spin dynamics at the magnetic phase transition in two-dimensional planar ferromagnet $K_2CuF_4$. *J. Phys. Soc. Jpn.* **52**, 4220–4230.

Hirst, L. L. (1997). The microscopic magnetization: concept and application. *Rev. Mod. Phys.* **69**, 607–627.

Holland-Moritz, E., Wohlleben, D., and Loewenhaupt, M. (1982). Anomalous paramagnetic neutron spectra of some intermediate-valence compounds. *Phys. Rev. B* **25**, 7482–7503.

Holstein, T. and Primakoff, H. (1940). Field dependence of the intrinsic domain magnetization of a ferromagnet. *Phys. Rev.* **58**, 1098–1113.

Hotta, T. (2006). Orbital ordering phenomena in $d$- and $f$-electron systems. *Rep. Prog. Phys.* **69**, 2061–2155.

Houmann, J. G., Chapellier, M., Mackintosh, A. R., Bak, P., McMasters, O. D., and Gschneidner, Jr, K. A. (1975). Magnetic excitations and magnetic ordering in praseodymium. *Phys. Rev. Lett.* **34**, 587–590.

Houmann, J. G., Rainford, B. D., Jensen, J., and Mackintosh, A. R. (1979). Magnetic excitations in praseodymium. *Phys. Rev. B* **20**, 1105–1118.

Howe, M. A., McGreevy, R. L., and Howells, W. S. (1989). The analysis of liquid structure data from time-of-flight neutron diffractometry. *J. Phys.: Condens. Matter* **1**, 3433–3451.

Hulthén, L. (1938). Über das austauschproblem eines kristalles. *Arkiv för Matematik, Astronomi och Fysik* **26A**, 1–106.

Hutchings, M. T. (1964). Point-charge calculations of energy levels of magnetic ions in crystalline electric fields. *Solid State Physics* **16**, 227–273.

Ikeda, H. and Hirakawa, K. (1974). Neutron scattering study of two-dimensional Ising nature of $K_2CoF_4$. *Solid State Commun.* **14**, 529–532.

Izyumov, Y. A., Naish, V. E., and Ozerov, R. P. (1991). *Neutron Diffraction of Magnetic Materials.* Springer, New York.

Izyumov, Y. A. and Ozerov, R. P. (1970). *Magnetic Neutron Diffraction.* Plenum Press, New York.

Jalarvo, N., Gourdon, O., Ehlers, G., Tyagi, M., Kumar, S. K., Dobbs, K. D., Smalley, R. J., Guise, W. E., A., Ramirez-Cuesta, Wildgruber, C., and Crawford, M. K. (2014). Structure and dynamics of octamethyl-poss nanoparticles. *J. Phys. Chem. C* **118**, 5579–5592.

Jamer, M. E., Rementer, C. R., Barra, A., Grutter, A. J., Fitzell, K., Gopman, D. B., Borchers, J. A., Carman, G. P., Kirby, B. J., and Chang, J. P. (2018). Long-range electric field control of permalloy layers in strain-coupled composite multiferroics. *Phys. Rev. Applied* **10**, 044045.

Janoschek, M., Klimko, S., Gähler, R., Roessli, B., and Böni, P. (2007). Spherical neutron polarimetry with MuPAD. *Physica B* **397**, 125–130.

Jensen, J. and Mackintosh, A. R. (1991). *Rare Earth Magnetism*. Oxford University Press, Oxford.

Johnson, R. D., Khalyavin, D. D., Manuel, P., Zhang, L., Yamaura, K., and Belik, A. A. (2018). Magnetic structures of the rare-earth quadruple perovskite manganites $R$Mn$_7$O$_{12}$. *Phys. Rev. B* **98**, 104423.

Johnston, D. F. (1966). On the theory of the electron orbital contribution to the scattering of neutrons by magnetic ions in crystals. *Proc. Phys. Soc.* **88**, 37–52.

Johnston, D. F. and Rimmer, D. E. (1969). On the theory of the scattering of neutrons by magnetic ions in crystals. *J. Phys. C* **2**, 1151–1167.

Jyrkkiö, T. A., Huiku, M. T., Lounasmaa, O. V., Siemensmeyer, K., Kakurai, K., Steiner, M., Clausen, K. N., and Kjems, J. K. (1988). Observation of nuclear antiferromagnetic order in copper by neutron diffraction at nanokelvin temperatures. *Phys. Rev. Lett.* **60**, 2418–2421.

Keen, D. A. (2001). A comparison of various commonly used correlation functions for describing total scattering. *J. Appl. Cryst.* **34**, 172–177.

Keller, T., Aynajian, P., Habicht, K., Boeri, L., Bose, S. K., and Keimer, B. (2006). Momentum-resolved electron-phonon interaction in lead determined by neutron resonance spin-echo spectroscopy. *Phys. Rev. Lett.* **96**, 225501.

Kennedy, B. J. and Avdeev, M. (2011). The structure of C-type Gd$_2$O$_3$. A powder neutron diffraction study using enriched $^{160}$Gd. *Aust. J. Chem.* **64**, 119–121.

Kenzelmann, M., Cowley, R. A., Buyers, W. J. L., Tun, Z., Coldea, R., and Enderle, M. (2002). Properties of Haldane excitations and multiparticle states in the antiferromagnetic spin-1 chain compound CsNiCl$_3$. *Phys. Rev. B* **66**, 024407.

Kim, B. J., Jin, H., Moon, S. J., Kim, J.-Y., Park, B.-G., Leem, C. S., Yu, J., Noh, T. W., Kim, C., Oh, S.-J., Park, J.-H., Durairaj, V., Cao, G., and Rotenberg, E. (2008). Novel $J_{\text{eff}} = 1/2$ Mott state induced by relativistic spin–orbit coupling in Sr$_2$IrO$_4$. *Phys. Rev. Lett.* **101**, 076402.

Kisi, E. H. and Howard, C. J. (2012). *Applications of Neutron Powder Diffraction*. Oxford University Press, Oxford.

Kobayashi, K., Nagao, T., and Ito, M. (2011). Radial integrals for the magnetic form factor of $5d$ transition elements. *Acta Cryst.* **A67**, 473–480.

Koester, L., Rauch, H., and Seymann, E. (1991). Neutron scattering lengths: A survey of experimental data and methods. *Atomic Data and Nuclear Data Tables* **49**, 65–120.

Kohlmann, H., Hein, C., Kautenburger, R., Hansen, T. C., Ritter, C., and Doyle, S. (2016). Crystal structure of monoclinic samarium and cubic europium sesquioxides and bound coherent neutron scattering lengths of the isotopes $^{154}$Sm and $^{153}$Eu. *Z. Kristallogr.* **231**, 517–523.

Koster, G. F., Dimmock, J. O., Wheeler, R. G., and Statz, H. (1963). *Properties of the Thirty-Two Point Groups*. MIT Press, Cambridge, Massachusetts.

Kresch, M. G. (2009). *Temperature dependence of phonons in elemental cubic metals studied by inelastic scattering of neutrons and x-rays.* PhD thesis, California Institute of Technology.

Kronmüller, H. and Parkin, S. S. P. (eds.) (2007). *Handbook of Magnetism and Advanced Magnetic Materials, Vol. 3.* John Wiley & Sons, New York.

Kubo, R. (1957). Statistical-mechanical theory of irreversible processes. I. General theory and simple applications to magnetic and conduction problems. *J. Phys. Soc. Japan* **12**, 570–586.

Kubo, R. (1966). The fluctuation–dissipation theorem. *Rep. Prog. Phys.* **29**, 255–284.

Lacroix, C., Mendels, P., and Mila, F. (eds.) (2011). *Introduction to Frustrated Magnetism.* Spinger Series in Solid State Sciences Vol. 164, Springer-Verlag, Berlin, Heidelberg.

Lake, B., Tennant, D. A., Caux, J.-S., Barthel, T., Schollwöck, U., Nagler, S. E., and Frost, C. D. (2013). Multispinon continua at zero and finite temperature in a near-ideal Heisenberg chain. *Phys. Rev. Lett.* **111**, 137205.

Lake, B., Tennant, D. A., and Nagler, S. E. (2000). Novel longitudinal mode in the coupled quantum chain compound KCuF$_3$. *Phys. Rev. Lett.* **85**, 832–835.

Lander, G. H. and Brun, T. O. (1970). Calculation of neutron magnetic form factors for rare-earth ions. *J. Chem. Phys.* **53**, 1387–1391.

Lin, J. Y. Y., Smith, H. L., Granroth, G. E., Abernathy, D. L., Lumsden, M. D., Winn, B., Aczel, A. A., Aivazis, M., and Fultz, B. (2016). MCViNE – an object oriented Monte Carlo neutron ray tracing simulation package. *Nucl. Instrum. Methods A* **810**, 86–99.

Lindgård, P. A., Kowalska, A., and Laut, P. (1967). Investigation of magnon dispersion relations and neutron scattering cross sections with special attention to anisotropy effects. *J. Phys. Chem. Solids* **28**, 1357–1370.

Lisher, E. J. and Forsyth, J. B. (1971). Analytic approximations to form factors. *Acta Cryst.* **A27**, 545–549.

Lorenzana, J., Seibold, G., and Coldea, R. (2005). Sum rules and missing spectral weight in magnetic neutron scattering in the cuprates. *Phys. Rev. B* **72**, 224511.

Lovesey, S. W. (1969). Some aspects of the theory of the scattering of neutrons by magnetic ions. *J. Phys. C* **2**, 470–475.

Lovesey, S. W. (1984*a*). *Theory of Neutron Scattering from Condensed Matter,* Vol. I. Oxford University Press, Oxford.

Lovesey, S. W. (1984*b*). *Theory of Neutron Scattering from Condensed Matter,* Vol. II. Oxford University Press, Oxford.

Lovesey, S. W. (2015). Theory of neutron scattering by electrons in magnetic materials. *Phys. Scr.* **90**, 108011.

Lovesey, S. W. and Rimmer, D. E. (1969). The theory of elastic scattering of neutrons by magnetic salts. *Rep. Prog. Phys.* **32**, 333–394.

Maleev, S. V., Bar'yakhtar, V. G., and Suris, R. A. (1963). The scattering of slow neutrons by complex magnetic structures. *Sov. Phys. Solid State* **4**, 2533–2539 [originally published as Fiz. Tverd. Tela **4**, 3461 (1962)].

Markvardsen, A. J. (2000). *Polarised neutron diffraction measurements of $PrBa_2Cu_3O_{6+x}$ and the Bayesian statistical analysis of such data.* DPhil thesis, University of Oxford.

Marshall, W. and Lovesey, S. W. (1971). *Theory of Thermal Neutron Scattering.* Oxford University Press, Oxford.

McIntyre, G. J. and Stansfield, R. F. D. (1988). A general Lorentz correction for single-crystal diffractometers. *Acta Cryst.* **A44**, 257–262.

Melkonian, E. (1949). Slow neutron velocity spectrometer studies of $O_2$, $N_2$, A, $H_2$, $H_2O$, and seven hydrocarbons. *Phys. Rev.* **76**, 1750–1759.

Meyer, A., Horbach, J., Heinen, O., Holland-Moritz, D., and Unruh, T. (2009). Self diffusion in liquid titanium: Quasielastic neutron scattering and molecular dynamics simulation. *Defect Diffus. Forum* **289–292**, 609–614.

Mezei, F. (1972). Neutron spin echo: A new concept in polarized thermal neutron techniques. *Z. Physik* **255**, 146–160.

Mezei, F., Pappas, C., and Gutberlet, T. (eds.) (2003). *Neutron Spin Echo Spectroscopy.* Lecture Notes in Physics, Vol 601, Springer-Verlag, Berlin.

Mezei, F. (ed.) (1980). *Neutron Spin Echo.* Lecture Notes in Physics, Vol 128, Springer-Verlag, Berlin.

Mitchell, D. P. and Powers, P. N. (1936). Bragg reflection of slow neutrons. *Phys. Rev.* **50**, 486–487.

Mitchell, P. C. H., Parker, S. F., Ramirez-Cuesta, A. J., and Tomkinson, J. (2005). *Vibrational Spectroscopy with Neutrons.* World Scientific, Singapore.

Moon, R. M., Riste, T., and Koehler, W. C. (1969). Polarization analysis of thermal-neutron scattering. *Phys. Rev.* **181**, 920–931.

Moriya, T. (1985). *Spin Fluctuations in Itinerant Electron Magnetism.* Springer-Verlag, Berlin.

Müller, G., Beck, H., and Bonner, J. C. (1979). Zero-temperature dynamics of the $S = 1/2$ linear Heisenberg antiferromagnet. *Phys. Rev. Lett.* **43**, 75–78.

Müller, G., Thomas, H., Beck, H., and Bonner, J. C. (1981). Quantum spin dynamics of the antiferromagnetic linear chain in zero and nonzero magnetic field. *Phys. Rev. B* **24**, 1429–1467.

Nathans, R., Shull, C. G., Shirane, G., and Andresen, A. (1959). The use of polarized neutrons in determining the magnetic scattering by iron and nickel. *J. Phys. Chem. Solids* **10**, 138–146.

Névot, L. and Croce, P. (1980). Charactérisation des surfaces par réflexion rasante de rayons X. Application à l'étude du polissage de quelques verres silicates. *Revue. Phys. Appl.* **15**, 761–779.

Nicholson, L. K. (1981). The neutron spin-echo spectrometer: A new high resolution technique in neutron scattering. *Contemp. Phys.* **22**, 451–475.

Nield, V. M. and Keen, D. A. (2000). *Diffuse Neutron Scattering from Crystalline Materials*. Oxford University Press, Oxford.

Nielson, C. W. and Koster, G. F. (1963). *Spectroscopic coefficients for the $p^n$, $d^n$ and $f^n$ configurations*. MIT Press, Cambridge, Massachusetts.

Osborn, R., Lovesey, S. W., Taylor, A. D., and Balcar, E. (1991). Intermultiplet transitions using neutron spectroscopy, in *Handbook on the Physics and Chemistry of Rare Earths*, Gschneidner, jr., T. A. and Eyring, L. (eds.), Vol. 14, Chap. 93, pp1–61. Elsevier, Amsterdam.

Ott, F. (2014). Neutron surface scattering. Application to magnetic thin films. *Collection SFN* **13**, 02004.

Paddison, J. A. M., Jacobsen, H., Petrenko, O. A., Fernández-Díaz, M. T., Deen, P. P., and Goodwin, A. L. (2015). Hidden order in spin-liquid $Gd_3Ga_5O_{12}$. *Science* **350**, 179–181.

Page, D. I. and Powles, J. G. (1975). The internuclear distance in $N_2$, gas and liquid, by neutron diffraction. *Molec. Phys.* **29**, 1287–1291.

Papoular, R. J. and Gillon, B. (1990). Maximum entropy reconstruction of spin density maps in crystals from polarized neutron diffraction datan. *Europhys. Lett.* **13**, 429–434.

Pappas, C., G., Ehlers, and Mezei, F. (2006). Neutron-spin-echo spectroscopy and magnetism, in *Neutron Scattering from Magnetic Materials*, Chatterji, T. (ed.), Chap. 11, pp521–542. Elsevier, Amsterdam.

Parker, S. F. (2006). Vibrational spectroscopy of $N$-phenylmaleimide. *Spectrochim. Acta A* **63**, 544–549.

Patrignani, C. *et al.* (Particle Data Group) (2016). Review of Particle Physics. *Chin. Phys. C* **40**, 100001. (See http://pdg.lbl.gov).

Pelissetto, C. and Vicari, E. (2002). Critical phenomena and renormalization-group theory. *Phys. Rep.* **368**, 549–727.

Penfold, J. and Thomas, R. K. (1990). The application of the specular reflection of neutrons to the study of surfaces and interfaces. *J. Phys.: Condens. Matter* **2**, 1369–1412.

Petrenko, O. A., Ritter, C., Yethiraj, M., and McK. Paul, D. (1998). Investigation of the low-temperature spin-liquid behavior of the frustrated magnet gadolinium gallium garnet. *Phys. Rev. Lett.* **80**, 4570–4573.

Phelan, W. A., Koohpayeh, S. M., Cottingham, P., Tutmaher, J. A., Leiner, J. C., Lumsden, M. D., Lavelle, C. M., Wang, X. P., Hoffmann, C., Siegler, M. A., Haldolaarachchige, N., Young, D. P., and McQueen, T. M. (2016). On the chemistry and physical properties of flux and floating zone grown $SmB_6$ single crystals. *Sci. Rep.* **6**, 20860.

Placzek, G. (1952). The scattering of neutrons by systems of heavy nuclei. *Phys. Rev.* **86**, 377–388.

Popovici, M. (1975). On the resolution of slow-neutron spectrometers. IV. The triple-axis spectrometer resolution function, spatial effects included. *Acta Cryst.* **A31**, 507–513.

Porter, N. A., Spencer, C. S., Temple, R. C., Kinane, C. J, Charlton, T. R., Langridge, S., and Marrows, C. H. (2015). Manipulation of the spin helix in FeGe thin films and FeGe/Fe multilayers. *Phys. Rev. B* **92**, 144402.

Racah, G. (1943). Theory of complex spectra. III. *Phys. Rev.* **63**, 367–382.

Rainwater, L. J., Havens, Jr., W. W., Wu, C. S., and Dunning, J. R. (1947). Slow neutron velocity spectrometer studies I. Cd, Ag, Sb, Ir, Mn. *Phys. Rev.* **71**, 65–79.

Rauch, H. and Waschkowski, W. (2000). *Landolt–Börnstein, New Series I/16A, Chapter 6.* Springer-Verlag, Berlin.

Raymond, S. (2014). Magnetic excitations. *Collection SFN* **13**, 02003.

Rennie, A. R., Hellsing, M. S., Wood, K., Gilbert, E. P., Porcar, L., Schweins, R., Dewhurst, C. D., Lindner, P., Heenan, R. K., Rogers, S. E., Butler, P. D., Krzywon, J. R., Ghosh, R. E., Jackson, A. J., and Malfois, M. (2013). Learning about SANS instruments and data reduction from round robin measurements on samples of polystyrene latex. *J. Appl. Cryst.* **46**, 1289–97.

Ressouche, E. (2014a). Polarized neutron diffraction. *Collection SFN* **13**, 02002.

Ressouche, E. (2014b). Reminder: Magnetic structures description and determination by neutron diffraction. *Collection SFN* **13**, 02001.

Ressouche, E., Ballou, R., Bourdarot, F., Aoki, D., Simonet, V., Fernandez-Diaz, M. T., Stunault, A., and Flouquet, J. (2012). Hidden order in $URu_2Si_2$ unveiled. *Phys. Rev. Lett.* **109**, 067202.

Richter, D., Monkenbusch, M., Arbe, A., and Colmenero, J. (eds.) (2005). *Neutron Spin Echo in Polymer Systems.* Advances in Polymer Science, Vol. 174, Springer-Verlag, Berlin.

Rietveld, H. M. (1969). A profile refinement method for nuclear and magnetic structures. *J. Appl. Cryst.* **2**, 65–71.

Roinel, Y., Bouffard, V., Bacchella, G. L., Pinot, M., Mériel, P., Roubeau, P., Avenel, O., Goldman, M., and Abragam, A. (1978). First study of nuclear antiferromagnetism by neutron diffraction. *Phys. Rev. Lett.* **41**, 1572–1574.

Rok, M., Bator, G., Sawka-Dobrowolska, W., Durlak, P., Moskwa, M., Medycki, W., Sobczyka, L., and Zamponi, M. (2018). Crystal structural analysis of methyl-substituted pyrazines with anilic acids: a combined diffraction, inelastic neutron scattering, $^{1}$H-NMR study and theoretical approach. *CrystEngComm* **20**, 2016–2028.

Rossat-Mignod, J. (1987). Magnetic structures. *Methods in Experimental Physics* **23C**, 69–157.

Rotter, M. and Boothroyd, A. T. (2009). Going beyond the dipole approximation to improve the refinement of magnetic structures by neutron diffraction. *Phys. Rev. B* **79**, 140405(R).

Rotter, M., Le, M. D., Boothroyd, A. T., and Blanco, J. A. (2012). Dynamical matrix diagonalization for the calculation of dispersive excitations. *J. Phys.: Condens. Matter* **24**, 213201.

Rouse, K. D., Cooper, M. J., York, E. J., and Chakera, A. (1970). Absorption corrections for neutron diffraction. *Acta Cryst.* **A26**, 682–691.

Rowe, J. M., Rush, J. J., de Graaf, L. A., and Ferguson, G. A. (1972). Neutron quasielastic scattering study of hydrogen diffusion in a single crystal of palladium. *Phys. Rev. Lett.* **29**, 1250–1253.

Rüegg, Ch., Cavadini, N., Furrer, A., Güdel, H.-U., Krämer, K., Mutka, H., Wildes, A., Habicht, K., and Vorderwisch, P. (2003). Bose–Einstein condensation of the triplet states in the magnetic insulator TlCuCl$_3$. *Nature* **423**, 62–65.

Santini, P., Carretta, S., Amoretti, G., Caciuffo, R., Magnani, N., and Lander, G. H. (2009). Multipolar interactions in $f$-electron systems: The paradigm of actinide dioxides. *Rev. Mod. Phys.* **81**, 807–863.

Schärpf, O. and Capellmann, H. (1993). The XYZ-difference method with polarized neutrons and the separation of coherent, spin incoherent, and magnetic scattering cross sections in a multidetector. *Phys. Stat. Sol. (a)* **135**, 359–379.

Schmitt, D. and Ouladdiaf, B. (1998). Absorption correction for annular cylindrical samples in powder neutron diffraction. *J. Appl. Cryst.* **31**, 620–624.

Schofield, P. (1960). Space-time correlation function formalism for slow neutron scattering. *Phys. Rev. Lett.* **4**, 239–240.

Schweika, W. (2003). Time-of-flight and vector polarization analysis for diffuse neutron scattering. *Physica B* **335**, 157–163.

Schwinger, J. S. (1937). On the magnetic scattering of neutrons. *Phys. Rev.* **51**, 544–552.

Sears, V. F. (1975). Slow-neutron multiple scattering. *Adv. Phys.* **24**, 1–45.

Sears, V. F. (1986).   Electromagnetic neutron–atom interactions. *Phys. Rep.* **141**, 281–317.

Sears, V. F. (1989). *Neutron Optics.* Oxford University Press, Oxford.

Sears, V. F. (1992).   Neutron scattering lengths and cross sections. *Neutron News* **3:3**, 26–37.

Shamoto, S., Sato, M., Tranquada, J. M., Sternlieb, B. J., and Shirane, G. (1993). Neutron-scattering study of antiferromagnetism in $YBa_2Cu_3O_{6.15}$. *Phys. Rev. B* **48**, 13817–13825.

Shirane, G., Cowley, R., Majkrzak, C., Sokoloff, J. B., Pagonis, B., Perry, C. H., and Ishikawa, Y. (1983). Spin magnetic correlation in cubic MnSi. *Phys. Rev. B* **28**, 6251–6255.

Shirane, G., Shapiro, S. M., and Tranquada, J. M. (2002). *Neutron scattering with a triple-axis spectrometer.* Cambridge University Press, Cambridge.

Shull, C. G. (1968).  Observation of pendellösung fringe structure in neutron diffraction. *Phys. Rev. Lett.* **21**, 1585–1589.

Shull, C. G. and Smart, J. S. (1949). Detection of antiferromagnetism by neutron diffraction. *Phys. Rev.* **76**, 1256–1257.

Shull, C. G., Wollan, E. O., and Koehler, W. C. (1951). Neutron scattering and polarization by ferromagnetic materials. *Phys. Rev.* **84**, 912–921.

Singh, B., Gupta, M. K., Mishra, S. K., Mittal, R., Sastry, P. U., Rols, S., and Chaplot, S. L. (2017). Anomalous lattice behavior of vanadium pentaoxide ($V_2O_5$): X-ray diffraction, inelastic neutron scattering and *ab initio* lattice dynamics. *Phys. Chem. Chem. Phys.* **19**, 17967–17984.

Singleton, J. (2001). *Band Theory and Electronic Properties of Solids.* Oxford University Press, Oxford.

Sinha, S. K., Sirota, E. B., Garoff, S., and Stanley, H. B. (1988). X-ray and neutron scattering from rough surfaces. *Phys. Rev. B* **38**, 2297–2311.

Sippel, D., Kleinstück, K., and Schulze, G. E. R. (1965). Pendellösungs-interferenzen mit thermischen neutronen an Si-einkristallen. *Phys. Lett.* **14**, 174–175.

Sivia, D. S. (2011). *Elementary Scattering Theory.* Oxford University Press, Oxford.

Sivia, D. S. and Skilling, J. (2006). *Data Analysis: A Bayesian Tutorial.* Oxford University Press, Oxford.

Sköld, K., Rowe, J. M., Ostrowski, G., and Randolph, P. D. (1972). Coherent- and incoherent-scattering laws of liquid argon. *Phys. Rev. A* **6**, 1107–1131.

Soper, A. K. (2009).   Inelasticity corrections for time-of-flight and fixed wavelength neutron diffraction experiments. *Molecular Physics* **107**, 1667–1684.

Squires, G. L. (1977). *Introduction to the Theory of Thermal Neutron Scattering.* Cambridge University Press, Cambridge.

Stassis, C. and Deckman, H. W. (1975). Magnetic scattering of neutrons. *Phys. Rev. B* **12**, 1885–1898.

Steiner, M. and Siemensmeyer, K. (2015). Nuclear magnetism and neutrons. *Experimental Methods in the Physical Sciences* **48**, 435–488.

Steinsvoll, O., Shirane, G., Nathans, R., Blume, M., Alperin, H. A., and Pickart, S. J. (1967). Magnetic form factor of terbium. *Phys. Rev.* **161**, 499–506.

Stewart, J. R., Andersen, K. H., Cywinski, R., and Murani, A. P. (2000). Magnetic diffuse scattering in disordered systems studied by neutron polarization analysis. *J. Appl. Phys.* **87**, 5425–5430.

Stewart, J. R., Deen, P. P., Andersen, K. H., Schober, H., Barthélémy, J.-F., Hillier, J. M., Murani, A. P., Hayes, T., and Lindenau, B. (2009). Disordered materials studied using neutron polarization analysis on the multi-detector spectrometer, D7. *J. Appl. Cryst.* **42**, 69–84.

Svensson, E. C., Brockhouse, B. N., and Rowe, J. M. (1967). Crystal dynamics of copper. *Phys. Rev.* **155**, 619–632.

Tasset, F. (1989). Zero field neutron polarimetry. *Physica B* **156–157**, 627–630.

Tasset, F., Brown, P. J., Lelièvre-Berna, E., Roberts, T., Pujol, S., Allibon, J., and Bourgeat-Lami, E. (1999). Spherical neutron polarimetry with Cryopad-II. *Physica B* **267–268**, 69–74.

Taylor, A., Dunne, M., Bennington, S., Ansell, S., Gardner, I., Norreys, P., Broome, T., Findlay, D., and Nelmes, R. (2007). A route to the brightest possible neutron source? *Science* **315**, 1092–1095.

Taylor, A. D., Osborn, R., McEwen, K. A., Stirling, W. G., Bowden, Z. A., Williams, W. G., Balcar, E., and Lovesey, S. W. (1988). Intermultiplet transitions in praseodymium using neutron spectroscopy. *Phys. Rev. Lett.* **61**, 1309–1312.

Taylor, A. E., Pitcher, M. J., Ewings, R. A., Perring, T. G., Clarke, S. J., and Boothroyd, A. T. (2011). Antiferromagnetic spin fluctuations in LiFeAs observed by neutron scattering. *Phys. Rev. B* **83**, 220514(R).

Teixeira, J. (1988). Small-angle scattering by fractal systems. *J. Appl. Cryst.* **21**, 781–785.

Toperverg, B. P. (2015). Polarized neutron reflectometry of magnetic nanostructures. *The Physics of Metals and Metallography* **116**, 1337–1375.

Trammell, G. T. (1953). Magnetic scattering of neutrons from rare earth ions. *Phys. Rev.* **92**, 1387–1393.

Trueblood, K. N., Bürgi, H.-B., Burzlaff, H., Dunitz, J. D., Gramaccioli, C. M., Schulz, H. H., Shmueli, U., and Abrahams, S. C. (1996). Atomic displacement parameter nomenclature. *Acta Cryst.* **A52**, 770–781.

Van Dijk, N. H., Fåk, B., Charvolin, T., Lejay, P., and Mignot, J. M. (2000). Magnetic excitations in heavy fermion $CePd_2Si_2$. *Phys. Rev. B* **61**, 8922–8931.

Van Hove, L. (1954). Correlations in space and time and Born approximation scattering in systems of interacting particles. *Phys. Rev.* **95**, 249–262.

Vineyard, G. H. (1954). Multiple scattering of neutrons. *Phys. Rev.* **96**, 93–98.

von Halban jr., H. and Preiswerk, P. (1936). Preuve expérimentale de la diffraction des neutrons. *C. R. Acad. Sci.* **203**, 73–75.

Wacher, R., Woignier, T., Pelous, J., and Courtens, E. (1988). Structure and self-similarity of silica aerogels. *Phys. Rev. B* **37**, 6500–6503.

Waldmann, O., Guidi, T., Carretta, S., Mondelli, C., and Dearden, A. L. (2003). Elementary excitations in the cyclic molecular nanomagnet $Cr_8$. *Phys. Rev. Lett.* **91**, 237202.

Watson, R. E. and Freeman, A. J. (1961). Hartree–Fock atomic scattering factors for the iron transition series. *Acta Cryst.* **14**, 27–37.

Weiss, R. J. and Freeman, A. J. (1959). X-ray and neutron scattering from electrons in a crystalline field and the determination of outer electron configurations in iron and nickel. *J. Phys. Chem. Solids* **10**, 147–161.

White, R. M. (2007). *Quantum Theory of Magnetism.* 3rd ed., Springer-Verlag, Berlin,.

White, R. M., Sparks, M., and Ortenburger, I. (1965).  Diagonalization of the antiferromagnetic magnon–phonon interaction. *Phys. Rev.* **139**, A450–454.

Wildes, A. R. (2006). Neutron polarization analysis corrections made easy. *Neutron News* **17:2**, 17–25; *ibid.* **18**, 31.

Willendrup, P., Farhi, E., Knudsen, E., Filges, U., and Lefmann, K. (2014). McStas: Past, present and future. *J. Neutron Research* **17**, 35–43.

Williams, W. G. (1988). *Polarized Neutrons.* Oxford University Press, Oxford.

Willis, B. T. M. and Carlile, C. J. (2009). *Experimental Neutron Scattering.* Oxford University Press, Oxford.

Willis, B. T. M., Carlile, C. J., Ward, R. C., David, W. I. F., and Johnson, M. W. (1986). Measurement of the velocity of sound in crystals by pulsed neutron diffraction. *Europhys. Lett.* **2**, 767–774.

Willis, B. T. M. and Pryor, A. W. (1975). *Thermal Vibrations in Crystallography.* Cambridge University Press, Cambridge.

Windsor, C. G. (1981). *Pulsed Neutron Scattering.* Taylor & Francis Ltd., London.

Yamada, T. (1969).  Fermi-liquid theory of linear antiferromagnetic chains. *Prog. Theor. Phys.* **41**, 880–890.

Yarnell, J. L., Katz, M. J., Wenzel, R. G., and Koenig, S. H. (1973). Structure factor and radial distribution function for liquid argon at 85°K. *Phys. Rev. A* **7**, 2130–2144.

Zemach, A. C. and Glauber, R. J. (1956). Dynamics of neutron scattering by molecules. *Phys. Rev.* **101**, 118–129.

Zubáč, J., Javorský, P., and Fåk, B. (2018). Crystal field in $NdPd_5Al_2$ investigated by inelastic neutron scattering. *J. Phys.: Condens. Matter* **30**, 255801.

# Index